REPORT
of the
TRIBUNAL OF INQUIRY
into
THE BEEF PROCESSING INDUSTRY

SOLE MEMBER

The Honourable Mr. Justice Liam Hamilton,
President of the High Court.

DUBLIN:
PUBLISHED BY THE STATIONERY OFFICE.

To be purchased through any Bookseller, or directly from the
GOVERNMENT PUBLICATIONS SALE OFFICE,
SUN ALLIANCE HOUSE, MOLESWORTH STREET, DUBLIN 2.

(Pn. 1007) **Price £20**

ISBN 0-7076-0426-5

TRIBUNAL OF INQUIRY
Appointed by instrument of the Minister for Agriculture and Food dated the 31st day of May 1991.

SOLE MEMBER:
The Honourable Mr. Justice Liam Hamilton, President of the High Court.

REGISTRAR: T. T. McCarthy

TRIBUNAL OFFICE
UPPER YARD
DUBLIN CASTLE
DUBLIN 2.

Tel. No. 6799877

Fax No. 6792082

29th July 1994

To: The Minister for Agriculture, Food & Forestry,
Dublin.

The Report which follows is the Report of the Tribunal established pursuant to a Resolution passed on the 24th day of May, 1991 by Dáil Eireann and on the 29th day of May, 1991 by Seanad Eireann,

1. to inquire into the following definite matters of urgent public importance:

   (i) allegations regarding illegal activities, fraud and malpractice in and in connection with the beef processing industry made or referred to,

      (a) in Dáil Eireann, and
      (b) on a television programme transmitted by ITV on May 13th, 1991.

   (ii) any matters connected with or relevant to the matters aforesaid which the tribunal considers it necessary to investigate in connection with its inquiries into the matters mentioned at (i) above; and

2. Making such recommendations (if any) as the tribunal, having regard to its findings, thinks proper.

Liam Hamilton
The Tribunal

Report of the Tribunal established by the Minister for Agriculture and Food on the 31st day of May, 1991, pursuant to a Resolution passed on the 24th day of May, 1991 by Dáil Eireann and on the 29th day of May, 1991 by Seanad Eireann.

To/ The Minister for Agriculture, Food and Forestry

Table of Contents

CHAPTER ONE

Introduction

(1) A resolution in the following terms was passed by Dáil Eireann on the 24th day of May, 1991, and by Seanad Eireann on the 29th day of May, 1991:

"That it is expedient that a Tribunal be established for:

1. inquiring into the following definite matters of urgent public importance:

 (i) allegations regarding illegal activities, fraud and malpractice in and in connection with the beef processing industry made or referred to:—

 (*a*) in Dáil Eireann, and

 (*b*) on a television programme transmitted by ITV on 13 May, 1991;

 (ii) any matters connected with or relevant to the matters aforesaid which the Tribunal considers it necessary to investigate in connection with its inquiries into the matters mentioned at (i) above;

and

2. making such recommendations (if any) as the Tribunal, having regard to its findings, thinks proper."

(2) The Order appointing the Tribunal was made on the 31st day of May 1991, by the Minister for Agriculture & Food.

After reciting the terms of the resolutions passed by the two Houses of the Oireachtas, the Order provided as follows:—

"1. A Tribunal is hereby appointed—

 (1) to inquire into and report to the Minister for Agriculture and Food on the following definite matters of urgent public importance:

 (i) allegations regarding illegal activities, fraud and malpractice in and in connection with the beef processing industry made or referred to -

 (*a*) in Dáil Eireann, and

 (*b*) on a television programme transmitted by ITV on 13 May, 1991;

(ii) any matters connected with or relevant to the matters aforesaid which the Tribunal considers it necessary to investigate in connection with its inquiries into the matters mentioned at (i) above;

and

(2) to make such recommendations (if any) as the Tribunal, having regard to its findings, thinks proper.

2. The Honourable Mr Justice Liam Hamilton, President of the High Court, is hereby nominated to be the sole member of the Tribunal.

3. The Tribunals of Inquiry (Evidence) Act, 1921, as adapted by or under subsequent enactments, and the Tribunals of Inquiry (Evidence) (Amendment) Act, 1979, shall apply to the Tribunal."

(3) On the 7th day of June 1991 and again on the 16th day of June 1991, the Tribunal caused to be published in the national press an advertisement in the following terms:—

Tribunal of Inquiry

Established in pursuance of a Resolution passed by Dáil Eireann on the 24th day of May 1991 and by Seanad Eireann on the 29th day of May 1991.

A preliminary public sitting of the Tribunal of Inquiry, appointed by the Minister for Agriculture and Food on the 31st day of May 1991, in pursuance of the above Resolution will be held in Dublin Castle (Upper Yard), Dublin 2 on Friday, the 21st day June 1991 at 9.30 a.m.

An announcement will be made later as to the date when the first public Sitting to take evidence will be held.

The Terms of Reference of the Tribunal are as follows:

1. to inquire into the following definite matters of urgent public importance:

(i) allegations regarding illegal activities, fraud and malpractice in and in connection with the beef processing industry made or referred to,

(a) in Dáil Eireann, and

(b) on a television programme transmitted by ITV on May 13th, 1991.

(ii) any matters connected with or relevant to the matters aforesaid which the Tribunal considers it necessary to investigate in connection with its inquiries into the matters mentioned at (i) above; and

2. Making such recommendations (if any) as the Tribunal, having regard to its findings, thinks proper.

Any person interested in the Inquiry, should attend in person or by Counsel or Solicitor.

Any person who desires to give evidence before the Tribunal relevant to the foregoing Terms of Reference should forward his name and address as soon as possible to T.T. McCarthy, Registrar to the Tribunal, at Tribunal Office, Upper Yard, Dublin Castle,

Dublin 2 and should indicate the matters upon which he desires to give evidence and the nature of his evidence.

Any interested person who requires a subpoena to secure the attendance of witnesses at the Inquiry should forward the names and addresses of such witnesses to the Registrar.

Consultation rooms will be available for interested parties on request to Conference Office, Dublin Castle.

BY ORDER OF THE Tribunal

THE 6TH DAY OF JUNE, 1991

(3) The terms of the Resolution were unusually broad in their drafting and scope and, as stated by the Tribunal during the course of its opening remarks, they were so drafted as to enable and indeed to oblige the Tribunal to carry out a wide-ranging and full investigation into the matters entrusted to it.

(4) Miss Christina Loughlin of the Chief State Solicitor's Office was appointed Solicitor to the Tribunal, and Mr Eoin McGonigal SC, Mr David Byrne SC and Mr Raymond Fullam BL assisted by Miss Angela O'Reilly BL were appointed Counsel to the Tribunal. The appointment of such Solicitor and Counsel to the Tribunal was necessary in order to enable the Tribunal to undertake investigations, to have investigations carried out on its behalf, to obtain statements from witnesses, to arrange the attendance of witnesses in due order, to prepare and serve Book of Documents and statements of witnesses on all "interested parties", to present the evidence and examine the witnesses.

(5) Between the date of its establishment and the first preliminary sitting, the Tribunal considered the transcript of the "World in Action" programme transmitted by ITV on the 13th day of May 1991 and the official copies of the Dáil Report of the proceedings therein from February 1988 to May 1991 for the purpose of ascertaining and extracting therefrom the allegations concerning illegal activities, fraud and malpractice in and in connection with the beef processing industry made or referred to therein. This consideration of the official Dáil Reports for the said purpose was made unnecessarily difficult for the Tribunal because of the failure to indicate in the Resolutions passed by the Houses of the Oireachtas the dates upon which the allegations which the respective Houses considered of urgent public importance were made or even indicate the period in which they were made or in any way to particularise the said allegations. As a consequence of such failure the Tribunal was obliged to inquire into all the allegations which it could discover in the official Dáil Reports.

(6) At the first preliminary sitting of the Tribunal held on the 21st day of June, 1991 Counsel read out a Statement of the Allegations made or referred to in Dáil Eireann and which had been extracted by the Tribunal from the Transcript of the "World in Action" programme and the official reports of proceedings in Dáil Eireann.

(7) At the conclusion of the reading of the Statement of Allegations, the Tribunal invited and received applications by interested parties for representation at the Inquiry.

(8) The Tribunal accepted that the following were "interested persons" within the meaning of the Tribunals of Inquiry (Evidence) Act 1921 and allowed them to be represented at the Inquiry:

(*a*) The Attorney General and all State Authorities who were represented by Mr H Whelehan SC, Mr Henry Hickey, Mr Gerry Danaher BL and Mr Colm Ó hOisin BL instructed by Mr John Corcoran of the Chief State Solicitor's Office.

Subsequently Mr Whelehan was appointed as Attorney General and was replaced by Mr Conor Maguire SC.

(*b*) Goodman International and its subsidiary companies involved in the beef processing industry who were represented by Mr Dermot Gleeson SC, Mr Peter Kelly, SC, Mr Michael Collins, BL, and Mr Ian Finlay BL who were instructed by Messrs A & L Goodbody, Solicitors.

(*c*) Mr Larry Goodman, who was represented by Mr Seamus McKenna SC and Mr Donal O'Donnell BL instructed by Messrs A & L Goodbody.

(9) The allegations which the Tribunal was established to inquire into were serious and wide-ranging and were made against Mr Larry Goodman, Goodman International and its subsidiary and associated companies, members of the Government including the then Taoiseach, Mr Charles J Haughey and the then Minister for Industry and Commerce, Mr Albert Reynolds, officials of various State authorities, such as the Department of Agriculture, the Department of Industry and Commerce, the Revenue Commissioners, the Customs and Excise and many others.

(10) Because of the seriousness of the allegations and because of the potential damage to the reputations and good names of the persons against whom the allegations were made, and the risk of personal hurt and injustice to any person involved in the inquiries, the Tribunal was, from the outset, concerned and indeed obliged to have regard to the principles of natural justice in the conduct of its inquiries and to ensure that fair procedures were adopted by it.

(11) In the course of the Report of the Royal Commission on Tribunals of Inquiry under the Chairmanship of The Right Honourable Lord Justice Salmon (1966) the Commission stated and recommended that the following cardinal principles should be observed to minimise the risk of personal hurt and injustice to any person involved in the inquiries:—

(i) before any person becomes involved in an inquiry, the Tribunal must be satisfied that there are circumstances which affect him and which the Tribunal proposes to investigate.

(ii) before any person who is involved in an inquiry is called as a witness, he should be informed in advance of allegations against him and the substance of the evidence in support of them.

(iii) (a) he should have adequate opportunity of preparing his case and of being assisted by legal advisers.

(b) his legal expenses should normally be met out of public funds.

(iv) he should have the opportunity of being examined by his own solicitor or counsel and of stating his case in public at the inquiry.

(v) any material witnesses he wishes called at the inquiry should, if reasonably practicable, be heard.

(vi) he should have the opportunity of testing by cross-examination conducted by his own solicitor or counsel any evidence which may affect him.

(12) In re Haughey 1971 IR p 217, stated that the minimum protection which should be offered to a person against whom allegations are made is that:-

(*a*) he should be furnished with a copy of the evidence which reflected on his good name;

(*b*) he should be allowed to cross-examine, by Counsel, his accuser or accusers;

(*c*) he should be allowed to give rebutting evidence; and

(*d*) he should be permitted to address, again by Counsel, the Committee in his own defence.

(13) In order to comply with these requirements the Tribunal and its staff were obliged to carry out extensive and detailed inquiries, examine critically the submissions made by interested parties, interview potential witnesses and obtain statements from them, determine what evidence was relevant and admissible, consider and evaluate a massive amount of documentation received from interested parties and in particular from various Government Departments. The nature and extent of the inquiries carried out by the Tribunal are illustrated in some detail elsewhere in this Report.

(14) Arising from such inquiries, statements had to be prepared, incorporated into book form and including numerous exhibits and served on interested parties, particularly those against whom allegations were made.

(15) All of this was time consuming and while this work was in progress a further number of preliminary public sittings were held on the 26th day of July 1991, the 26th day of August, 1991 and the 30th day of September, 1991.

(16) These latter preliminary sittings dealt mainly with the applications for representation at the Inquiry and further submissions in regard thereto.

(17) Submissions with regard to representation were made on behalf of Granada Television by Mr Niall Fennelly SC instructed by McCann Fitzgerald & Co Solicitors, Mr Brian McCracken SC instructed by Spring, Murray & Co Solicitors on behalf of Deputy Dick Spring, Mr Frank Clarke SC instructed by Murray, Sweeney & Co on behalf of Zachariah Al Taher, Mr David Hamilton SC instructed by Donal J. Hamilton, Solicitors on behalf of the Amalgamated Transport and General Workers Union, Mr Richard Kean, BL, on behalf of SIPTU, instructed by Bowler, Geraghty and Company Solicitors, Mr Michael Gray BL instructed by Henry P Kelly & Co Solicitors on behalf of the United Farmers Association, and Mr Michael White Solicitor on behalf of Deputy Tomás MacGiolla and Deputy Patrick Rabbitte.

(18) On the 26th day of July 1991 limited representation was granted to:-

(*a*) SIPTU, for whom Mr Ercus Stewart SC, and Mr Richard Kean BL (instructed by Bowler Geraghty and Co) appeared on the basis that they represented a considerable number of members employed in the AIBP plants in Bagenalstown,

Nenagh, Cahir, Waterford, Rathkeale and Donegal and allegedly involved in illegal and irregular practices.

(*b*) Amalgamated Transport and General Workers Union, for whom Mr David Hamilton SC and Mr Barry Hickson BL (instructed by Donal J. Hamilton and Co Solicitors) appeared on the same basis as granted to SIPTU.

(*c*) Mr Zachariah Al Taher, beneficial owner of the Taher Group of companies, for whom Mr Frank Clarke SC and Mr Bill Shipsey BL (instructed by Murray Sweeney & Co Solicitors) appeared.

(*d*) United Farmers Association for whom Mr Paul Callan SC and Mr Denis Vaughan Buckley SC and Mr Michael Gray BL (instructed by Henry P Kelly & Co, Solicitors) appeared. They had sought full representation as interested parties before the Inquiry but representation was only granted on a limited basis.

(*e*) Mr Liam Marks, for whom Mr Eamonn Coffey BL (instructed by Johnston Lavery and McGahon, Solicitors) appeared on the basis that he was allegedly involved in the commission of certain irregular and illegal activities to be inquired into by the Tribunal.

(19) On this date, Mr Tomás MacGiolla and Mr Patrick Rabbitte TD, for whom Mr Adrian Hardiman SC and Mr Thomas O'Connell BL appeared, instructed by Michael D. White, Solicitor, merely sought representation whilst giving evidence before the Tribunal and the application was granted. Such right of representation was subsequently extended to cover the periods during which evidence was being given in relation to the specific allegations made by either of them.

(20) On this date, Mr McCracken, SC who appeared for Mr Dick Spring, TD, reserved his position with regard to the nature and extent of the representations which he would seek.

(21) The Tribunal had hoped to start its public hearings on the 26th day of August 1991 but was obliged to adjourn such hearings in order to enable the Tribunal to continue the preparation of documents and the taking of statements from witnesses which had to be served on the interested parties before the public hearings could begin and in ample time to enable them to deal with the allegations contained therein.

(22) Despite its best endeavours the Tribunal was unable to serve the necessary documentation on the Chief State Solicitor, on behalf of the Attorney General and all the State Authorities, and on Messrs A & L Goodbody, on behalf of Goodman International and its subsidiary and associated companies and Mr Larry Goodman, until the afternoon of Friday the 27th September, 1991 and when the Tribunal sat on the 30th day of September 1991, it adjourned its proceedings until the 9th day of October, 1991 in order to enable the parties to consider the huge amount of documentation served on them.

(23) At the hearing on 30th September 1991, Mr McCracken on behalf of Mr Spring TD sought representation before the Inquiry when the Tribunal was dealing with the allegations made by his client and this application was granted. In addition Mr Jim Fairbairn, for whom Mr Paul Fogarty BL (instructed by Messrs Moran and Ryan, Solicitors)

appeared, was granted limited representation when matters in which he was alleged to have been involved were being dealt with by the Tribunal.

(24) On the 9th day of October 1991 Counsel to the Tribunal opened proceedings by setting forth in detail the allegations to be inquired into by the Tribunal and which had been served on the interested parties. A copy of this Book of Allegations is contained in Appendix 1.

(25) Counsel on behalf of Goodman International and Counsel on behalf of Mr Laurence Goodman made certain submissions to and sought certain rulings from the Tribunal.

In connection therewith, the Tribunal heard submissions from Counsel for the Attorney General and State authorities, Counsel for Mr Spring TD, Counsel for Messrs Rabbitte TD and MacGiolla TD and Counsel to the Tribunal.

The Tribunal ruled on the matters raised in these submissions on the 10th day of October, 1991.

(26) The Tribunal then adjourned its hearing in regard to these matters because it was indicated to it by Counsel to Goodman International that it was proposed to challenge its rulings in the High Court.

(27) On the 11th day of October 1991 Counsel on behalf of Goodman International and Mr Laurence Goodman sought and obtained from the High Court leave to apply for the following reliefs:-

(i) A Declaration that the first named Rcspondent cannot lawfully enquire into and determine matters which are the subject of Civil Litigation;

(ii) A Declaration, that the first named Respondent cannot lawfully enquire into or consider, matters which are or may be the subject of purely private disputes;

(iii) A Declaration that the first named Respondent cannot lawfully enquire into, consider or determine matters which have already been the subject of criminal prosecution;

(iv) A Declaration that the first named Respondent cannot lawfully enquire into and/or determine the truth or otherwise of allegations of criminal conduct;

(v) A Declaration that the first named Respondent, in conducting his enquiry, cannot lawfully consider for the purposes of determination, evidence which is not admissible in Courts established under the Constitution;

(vi) A Declaration that no evidence should be led which potentially adversely affects the good name or property rights of a party unless it is evidence which would be admissible in Courts established under the Constitution;

(vii) A Declaration that where doubtful or questionable evidence which has the potential for adversely affecting the good name or property rights of a party before the Tribunal is sought to be led, that such evidence ought first be heard and ruled upon in a private sitting of the Tribunal.

(viii) A Declaration that the first named Respondent is obliged to furnish and identify to the Applicants statements of the evidence which it is proposed to adduce and to afford to the Applicants a reasonable time to consider such statements.

(ix) An Order of Prohibition preventing the first named Respondent from proceeding with the enquiry other than in accordance with the Declarations aforesaid.

(x) Such further and other relief as to this Honourable Court shall seem fit or necessary.

The proceedings were heard by the High Court on the 15th and 16th days of October 1991 and judgment thereon was delivered on the 21st day of October 1991 and the application made on behalf of Goodman International and Mr Laurence Goodman was dismissed and the relief sought was refused.

The applicants appealed to the Supreme Court against the Order of the High Court and their appeal was heard by the Supreme Court on the 24th day of October 1991 and successive days concluding on the 30th day of October 1991.

Judgment by the Supreme Court was delivered on the 1st day of November, 1991, dismissing the said appeal and affirming the judgment and Order of the High Court.

Because the applicants in this case had raised issues affecting the jurisdiction of the Tribunal and the procedures adopted by it, the Tribunal had adjourned its public hearings pending the final determination of these issues.

(28) The Tribunal resumed its hearings on the 7th day of November, 1991.

(29) The Tribunal heard evidence on 226 days and oral submissions at the conclusion of the oral evidence for 5 days.

(30) All the oral testimony, and submissions, was taken down by stenographers, transcribed over-night and is delivered with this Report.

(31) 475 witnesses gave oral testimony before the Tribunal.

A chronological list of witnesses was prepared by the Tribunal Registrar and this together with transcript references in which the witness's oral testimony is to be found is contained in Appendix 2.

(32) With regard to the manner in which oral testimony was adduced before the Tribunal, the Tribunal adopted the procedure which had been followed in recent Inquiries of a similar nature, such as the Whiddy Island, Stardust and Kerry Babies Inquiries, which was that:—

(i) All witnesses were called by the Tribunal's counsel and first examined by him;

(ii) they were then available for cross-examination by Counsel or Solicitor for the parties to whom the right of representation had been granted and who had a legitimate interest in the evidence of the witness in the appropriate order;

(iii) If necessary, they were then cross-examined by Counsel to the Tribunal.

In a number of isolated instances this procedure was varied by permitting Counsel or Solicitor appearing for a particular witness to lead the evidence.

Prior to their giving evidence, a written statement of the witness's evidence and all relevant documentation was circulated to all the parties interested by the Solicitor to the Tribunal.

(33) In approaching the consideration of the evidence adduced or sought to be adduced before the Tribunal, the Tribunal at all times had regard to its function which, as described by the Chief Justice in the case of Goodman International and Laurence Goodman -v- The Sole Member of the Tribunal of Inquiry, was to carry out a "simple fact-finding operation" and then to report thereon to the Legislature through the Minister for Agriculture and Food, and in so doing had regard only to properly admitted evidence which had been, where necessary, subjected to cross-examination by Counsel or Solicitor appearing for any party likely to be affected thereby.

(34) Many of the written submissions or statements made to the Tribunal contained material which was based on rumour or hearsay. In connection with this material the Tribunal adopted the same approach as did the Tribunal of Inquiry into dealings in Great Southern Railway Stock referred to in the case of the previous paragraph of this Report viz it has sifted through rumour and hearsay but relies only on evidence, properly admitted, for its findings.

(35) In the production to and hearing of evidence before the Tribunal, Counsel to the Tribunal and the Tribunal were at all times conscious of the obligation to ensure that all the requirements of fair procedures had been complied with. The relevant 'interested parties' likely to be affected by such evidence were furnished with a copy of the evidence intended to be given and all relevant documentation. The witnesses were subject to cross-examination. Any party affected by such evidence was afforded the opportunity, if he so wished, of calling rebutting evidence and of addressing the Tribunal.

(36) During the course of the proceedings before the Tribunal, a number of applications for representation on a limited basis were made to the Tribunal, by or on behalf of witnesses involved in the inquiry and whose reputations could be affected by the findings of the Tribunal, and in the interest of fair procedures such applications were acceded to. A full list of the parties who were granted such representation is set forth in Appendix 3.

(37) Included in this list is the representation granted to Deputy Desmond O'Malley. In his then capacity as Minister for Industry and Commerce he was represented by Counsel for the State Authorities but a number of the allegations with regard to the administration of the Export Credit Insurance Scheme being inquired into by the Tribunal had been made by him in his capacity as a Dáil Deputy in Dáil Eireann and in that capacity, he was granted separate representation while these allegations were being inquired into.

(38) The granting of representation at the Tribunal to so many parties was necessitated by the far reaching inquiries which the Tribunal was obliged to undertake in pursuance of the resolution passed by both Houses of the Oireachtas and by the obligation placed on the Tribunal to follow fair procedures as outlined herein. Every witness who was likely to be affected by the findings of the Tribunal was entitled to be legally represented before the Tribunal.

(39) Section 6 of the Tribunals of Inquiry (Evidence) (Amendment) Act, 1979 provides that:—

> "(1) Where a Tribunal, or, if the Tribunal consists of more that one member, the chairman of the Tribunal, is of opinion that, having regard to the findings of the Tribunal and all other relevant matters, there are sufficient reasons rendering it equitable to do so, the Tribunal or the chairman, as the case may be, may by order direct that the whole or part of the costs of any person appearing before the Tribunal by counsel or solicitor, as taxed by a Taxing Master of the High Court, shall be paid to the person by any other person named in the order.
>
> (2) Any sum payable pursuant to an order under this section shall be recoverable as a simple contract debt in any Court of competent jurisdiction.
>
> (3) Any sum payable by the Minister for Finance pursuant to an order under this section shall be paid out of moneys provided by the Oireachtas.

(40) In the course of his judgment in the case of Goodman International and Laurence Goodman -v- The Sole Member of the Tribunal of Inquiry, the late Mr Justice McCarthy in dealing with the provisions of Section 6 stated that:—

> "(c)S.6 : The liability to pay costs cannot depend upon the findings of the Tribunal as to the subject matter of the Inquiry.
>
> When the inquiry is in respect of a single disaster, then, ordinarily, any party permitted to be represented at the inquiry should have their costs paid out of public funds. The whole or part of those costs may be disallowed by the Tribunal because of the conduct of or on behalf of that party at, during or in connection with the inquiry. The expression 'findings of the Tribunal' should be read as the findings as to the conduct of the parties at the Tribunal. In all other cases the allowance of costs at public expense lies within the discretion of the Tribunal, or, where appropriate, its Chairman."

The Chief Justice stated that:—

> "with regard to the other questions raised concerning the construction of Ss 4, 5, and 6 of the Act of 1979 I have had the opportunity of reading the judgment which is about to be delivered by McCarthy J, and I agree with it".

(41) It is quite clear from this judgment that in the exercise of its discretion to direct the payment of the whole or part of the costs of any person appearing before the Tribunal by Counsel or Solicitor, the Tribunal cannot have regard to any of the findings of the Tribunal on the matters being inquired into by it but is only entitled to consider

> "the conduct of or on behalf of a party at, during or in connection with the inquiry"

and that unless such conduct so warrants, a party permitted to be represented at the inquiry should have their costs paid out of public funds.

(42) During the course of the hearings before it, the Tribunal was obliged to make rulings on such matters as the right to representation before the Tribunal, the alleged right of certain journalists and reporters to claim privilege in respect of the sources of information published by them, the common law and constitutional rights of members of the Houses

of the Oireachtas to refuse to disclose the sources of information on the basis of which statements were made therein, and the absolute confidentiality of discussions, as distinct from decisions, of the Cabinet.

(43) A number of these rulings were challenged in the High Court and Supreme Court and the Rulings of the Tribunal were upheld in the cases of:—

(i) Boyhan and Others -v- Tribunal of Inquiry into the Beef Processing Industry (reported at 1992 ILRM 545) which dealt with the right of representation.

(ii) Goodman International and Laurence Goodman -v- The Sole Member of the Tribunal of Inquiry into the Beef Processing Industry, Ireland and The Attorney General (reported at 1992 21R 542), which involved a challenge to the constitutionality of the Tribunal and procedures followed by it.

(iii) Kiberd and Carey -v- The Tribunal (reported at 1992 2IR 257), which dealt with the question of the privilege claimed by journalists to refuse to disclose sources of information.

(iv) Attorney General -v- The Tribunal, Goodman International and Laurence Goodman and Dick Spring TD, Pat Rabbitte TD and Tomás MacGiolla (Notice Parties) which dealt with the constitutional privileges of members of the Oireachtas.

The challenge by the Attorney General to the ruling made by the Tribunal with regard to its right to inquire into discussions at the Cabinet was upheld by the Supreme Court in the case of the Attorney General -v- The Sole Member of the Tribunal of Inquiry into the Beef Processing Industry, The Honourable Liam Hamilton (Reported at 1993 ILRM 81).

(44) By virtue of the ruling of the Supreme Court on the question of absolute privilege from disclosure of discussions which took place at meetings of the Cabinet, the Tribunal was prohibited from inquiring as to such discussions and members of the Cabinet were equally precluded from giving evidence with regard thereto.

(45) Ms Susan O'Keeffe, who was the main researcher for the programme transmitted by ITV and which contained many of the allegations inquired into by the Tribunal, gave evidence before the Tribunal and when requested so to do refused to give the names and addresses of all persons interviewed by her in the course of her research into the programme and Granada Television refused to make available to the Tribunal all film and rushes in their possession which were unused in the transmission of the programme and the notes taken and memoranda prepared by Ms O'Keeffe.

(46) The Tribunal had made rulings with regard to the nature and extent of the privilege enjoyed by members of each House of the Oireachtas by reason of the provisions of the Constitution which included the privilege of not disclosing the sources of the information available to them and relevant to the matters being inquired into by the Tribunal.

(47) Because of the failure to have the information referred to at 45 and 46 above, Counsel for Goodman International and associated companies and Counsel for State authorities submitted to the Tribunal that their constitutional rights were not being vindicated and

that irrespective of the evidence given or to be given the Tribunal was not entitled to proceed with its inquiries.

(48) In rejecting this submission, the Tribunal again emphasised and does so again that its findings would not be based on hearsay evidence but on the basis of direct evidence given on oath and which had been subject where necessary to cross-examination.

(49) This ruling was challenged by Goodman International and Laurence Goodman in an application for Judicial Review in the High Court to which application the Attorney General and Deputies Spring, Rabbitte and MacGiolla were Notice Parties.

(50) It appears from the judgment of Mr Justice Geoghegan delivered on the 18th day of February 1993 that it was submitted on behalf of the applicants that the Tribunal was in breach of its constitutional obligations to vindicate the good name of each of the applicants, viz Goodman International and Laurence Goodman, by not expressly terminating all further inquiry into any of the allegations made in the Dáil or in the World in Action programme and not giving an immediate express vindication of the applicant's good name.

(51) This application was refused in the High Court but an appeal from such Order is still pending in the Supreme Court.

(52) The Tribunal is however satisfied that it has fulfilled its responsibilities to all parties to ensure that the proceedings before it were conducted in accordance with the principle of constitutional justice and fair procedures and has had due regard to the rights, constitutional and otherwise, of all parties and witnesses appearing before it.

(53) Again it must be emphasised that the function of a Tribunal appointed under the Act is to conduct an inquiry to establish the truth or otherwise of the matters which have been referred to it by the Oireachtas. In determining these matters, the Tribunal has relied only on evidence given before the Tribunal, which was legally admissible and accepted by the Tribunal.

(54) In view of the nature of some of the allegations the Tribunal sought particulars of contributions made, by companies or persons engaged in the food processing industry, to Political Parties, Ministers and a number of individual members of Dáil Eireann, from the parties concerned and the companies making the contribution.

The Tribunal received full co-operation from all parties concerned and the relevant details were supplied as requested by the Tribunal. The Tribunal does not intend to refer further to this matter or report thereon as the Tribunal is satisfied that such contributions were normal contributions made to Political Parties and did not in any way affect or relate to the matters being inquired into by the Tribunal.

(55) As stated 475 witnesses gave evidence before the Tribunal on different matters relevant to the matters being inquired into by the Tribunal which evidence is contained in approximately 452 books of transcripts which are delivered with the Report. It would be impossible for the Tribunal in the course of its Report to refer to or deal with the entire of such evidence. It has however in the course of this Report dealt with the evidence, oral

and documentary, upon which it has relied to establish the facts upon which the Report is based. All the other evidence received careful consideration from the Tribunal.

(56) The function of the Tribunal was to 'carry out a simple fact finding operation' into the truth or otherwise of the allegations regarding illegal activities, fraud and malpractice, and in connection with the beef processing industry made or referred to in Dáil Eireann and on a television programme transmitted by ITV on the 13th May, 1991 and in presenting this Report has sought to confine its role to that function.

CHAPTER TWO

Inquiries made by Tribunal

This Chapter is intended to give a general picture of the nature of the inquiries made by the Tribunal and the steps taken by it to obtain the information necessary to enable it to fulfil its function. It is not intended to be all embracing.

Subsequent to its establishment on 31st May, 1991, the Tribunal, in pursuance of its inquiries and in addition to publishing the advertisement in the National Press set forth in the previous Chapter wrote to a number of persons and bodies seeking information and evidence in relation to the matters into which it was obliged to enquire. A list of the persons/bodies from whom the Tribunal sought information or assistance is attached at Appendix 4. Many of these written to submitted statements and or documentation to the Tribunal which formed the basis of the Books of Documents ultimately served on persons represented before the Tribunal.

In June, 1991, the Tribunal wrote to Granada Television which had broadcast the World in Action programme referred to in the Resolution passed by both Houses of the Oireachtas. In July, 1991, the Tribunal also wrote to Ms Susan O'Keeffe, programme researcher. The following participants in the programme of 13/5/'91 were written to by the Tribunal:—

Patrick McGuinness
John Tomlinson
Barry Desmond
Thomas Ruddy
Ray Mac Sharry
Joe Carey — EC Court of Auditors
Brendan Solan (Programme of 22/7/1991).

Replies were received from all those contacted. In addition submissions were received from Granada Television, Patrick McGuinness and Barry Desmond. The EC Court of Auditors forwarded copies of a number of its reports for the Tribunal's attention.

Between June — August, 1991, The Tribunal wrote to all the members of the Oireachtas who had made or referred to allegations in Dáil Eireann and Seanad Eireann, inviting them to submit to the Tribunal all relevant information in their possession.

The Tribunal also wrote to all persons and organisations referred to in the allegations and invited them to submit, to the Tribunal, all relevant information in their possession:

Members and former members of government were requested to make statements on any allegations relating to their area of responsibility and other matters arising during the course of the Tribunal and did so on receipt of such request. All such members required by the Tribunal to give evidence before it did so.

Information was sought, at various stages, by the Tribunal from the following Departments of State or State bodies.

Department of Agriculture & Food
Department of Industry & Commerce
Department of Finance
Department of Social Welfare
Department of Foreign Affairs
Department of Labour
Revenue Commissioners
Customs & Excise
Comptroller & Auditor General
Department of the Taoiseach
Garda Siochana
European Court of Auditors
Secretary to the Government
Central Statistics Office
Chief State Solicitors Office
Industrial Development Authority
CBF
Industrial Credit Corporation
RTE.
Director of Public Prosecutions

and all relevant information sought by the Tribunal was supplied in response to such requests.

The Tribunal also wrote to the EC Commission and again all necessary information was made available to the Tribunal.

Files dealing with relevant matters were submitted to the Tribunal by the following bodies and organisations;

Department of Agriculture & Food
Department of Industry & Commerce
Peter Fitzpatrick, Examiner to G.I. (Reports)
EC Court of Auditors (Reports)
Goodman International
Comptroller and Auditor General
Department of Foreign Affairs
Garda Siochana
Industrial Development Authority
Revenue Commissioners & Customs and Excise.

In addition documentation and file extracts were received from the Department of Finance and the Department of the Taoiseach.

A number of persons and organisations forwarded submissions to the Tribunal. These gave background information on their organisation's involvement with the beef industry and/or provided information on the various matters coming within the Tribunal's terms of reference.

The Veterinary Inspectors and Agricultural Officer grades employed in all plants engaged in beef processing were asked to notify the Tribunal of any instances of fraud, malpractice or irregularities in plants under their control. A letter in these terms was forwarded to Veterinary Inspectors employed directly by the Department of Agriculture (73), Temporary Veterinary Inspectors (340), Agricultural Officers (141), Classification Officers (54), Higher Agricultural Officers (90), Supervisory Officers (31) and Senior Supervisory officers (1).

Replies were received from more than 60% of permanent Veterinary Inspectors. Only 25% of those in Agricultural officer grades responded to the Tribunal's letter. Of the 340 Temporary Veterinary Inspectors contacted, 25 replies were received.

In addition the Tribunal wrote directly to as many employees and former employees in the beef industry as could be identified. They were asked to make available any evidence in relation to the following general matters:

(*a*) Irregularities in the meat processing business

(*b*) Method of payment of employees

(*c*) Non disclosure of payment of employees

(*d*) Whether contracts of service exist between the meat company and the employees.

In all over 18,000 employees and former employees were written to during the course of the Tribunal.

The response rate from current and former employees to the Tribunal's general letter was poor with written replies received from less than 4% of those contacted.

The Tribunal sought documentation or other relevant material from reporters in the following newspapers / magazines who had written articles concerning the beef processing industry:

The Irish Times
The Irish Independent
The Sunday Independent
The Sunday Press
Business and Finance
The Sunday Tribune
The Sunday Business Post.

A letter also issued to reporters in RTE who had been involved in the making of a programme concerning the beef industry.

The Tribunal received a submission from Mr Jerry O'Callaghan, RTE who had investigated alleged irregularities in the beef industry.

The Tribunal asked for searches to be carried out at the Companies Registration Office in respect of approximately 85 companies. A small number of searches were also conducted at the Isle of Man, Northern Ireland, U.K. and Channel Island Companies Registration Offices.

The Tribunal heard evidence from 475 witnesses. These included public servants, persons working in the beef industry and those who had made allegations as outlined in the terms of reference. A list of witnesses prepared by the Tribunal's Registrar is contained in Appendix 2. A list of exhibits is also given at Appendix 5.

From the files, submissions and statements made available to it, the Tribunal compiled books of documents under various subject headings. In all the Tribunal compiled 52 volumes of documentation consisting of over 150 books, together with additional material not presented in book form. These documents were served on parties who had been granted legal representation before the Tribunal. Those who were granted limited representation were served only with documents relating to areas for which they had such representation.

In addition the Tribunal made available to the Department of Agriculture and Food the information and documents which it had received in relation to alleged illegal activities at the AIBP Plant in Rathkeale and made available to the Revenue Commissioners the information and documents which it had received in relation to instances of alleged tax evasion.

CHAPTER 3

The World in Action Programme

1. The television programme referred to in the Resolution passed by the Houses of the Oireachtas had been devised and filmed by a "World in Action" team of broadcasters and was transmitted by Granada Television on the evening of the 13th day of May, 1991.

2. This programme, which received widespread publicity, was stated to be an investigation of Europe's "Mr Meat" viz Larry Goodman and contained many serious allegations against Larry Goodman and the companies which he controlled, including Anglo Irish Beef Processors plc and by implication members of the Government and the Fianna Fáil Party.

3. These allegations can be briefly summarised as follows:—

(i) Abuses of the system under which subsidies are paid by the European Economic Community to those engaged in the beef processing industry and in particular Anglo Irish Beef Processors plc by:

(*a*) Falsification of documents which provide the basis for the payment of such subsidies.

(*b*) Use of bogus stamps to alter the classification of animals being processed;

(*c*) Switching of meat taken into intervention the property of the Intervention Authority and the substitution therefor of inferior product;

(*d*) Falsification of weights shown on cartons of beef;

(ii) Abuses of the Export Refund subsidy system by:

(*a*) failing to comply with the contractual requirements of Middle-East customers with regard to Halal slaughtering of beef exported to such countries and the unauthorised use of Islamic stamps, in the possession of the Company, to show compliance with this requirement;

(*b*) Re-boxing of meat purchased from the Intervention Agency for the purpose of misleading customers.

(iii) Abuses of the Aids to Private Storage Scheme at the AIBP factory in Waterford, by:

(*a*) Falsification of weights;

(*b*) Addition of poor quality meat;

(*c*) Attempting to conceal the extent thereof by:

(i) altering case weights at the Cold Store,

(ii) preparing a plan at Senior Management level within the Goodman Group to limit the extent of the damage to the Goodman Group which proved abortive.

(iv) Allegations of Political Influence.

(i) That Larry Goodman and his companies had "the right connections at the right places that could basically control any investigation that would be put in place."

(ii) That, though the National Governments of each individual country are responsible for tackling fraud on the European taxpayer, Larry Goodman had, in the Irish Government, some of his strongest supporters.

(iii) That the links between the then Taoiseach, Charles J Haughey TD and Larry Goodman went "back a long way": that Larry Goodman gave money to the Fianna Fáil Party and the then Taoiseach publicly promoted Goodman at the very time that Customs investigators were warning that Goodman's operations were strongly suspected of involvement in fraud.

(iv) That a major European investigation into the operation of Goodman companies was prevented by assurances from the Irish authorities that they themselves had a wide ranging investigation of Goodman in hand and that there is no evidence of any such investigation.

(v) That the Customs report on the Waterford investigation was withheld from the Gárda Fraud Squad for a period of eighteen months even though the Customs authorities had recommended the instigation of criminal proceedings.

(vi) The Commissioner Ray Mac Sharry had sought the assistance of the Dutch Agriculture Minister, Herik Braks to approach a Dutch Bank, Amro, to withhold proceedings against Goodman.

(vii) Abuses of the Tax system, by

(i) having a company wide scheme of under the table payments to employees;

(ii) making out cheques against bogus invoices, having same endorsed by Goodman employees, cashed at local branches of the Allied Irish Bank and the cash received distributed to employees, the amount involved being approximately £3 million per annum.

4. The programme contained interviews with:

(*a*) Patrick McGuinness, a former employee of the Goodman Group of Companies.

(*b*) Thomas Ruddy, a former employee of the Goodman Group of Companies.

(*c*) Barry Desmond M.E.P.

(*d*) John Tomlinson M.E.P.

(*e*) Joe Carey, a member of the European Court of Auditors;

and the statements made by them in the course of such interviews must be considered in the context of the programme and the statements made by the presenter of the programme.

5. At the outset of the programme the presenter stated that the programme was investigating the operations of Europe's "Mr Meat", the man who makes his money putting the beef into Britain's supermarkets while picking up millions from taxpayers who have been handing out money for nothing.

At the outset of the programme the presenter then stated that Europe's beef business is built on taxpayer's subsidies, that Europe's taxpayers in that year alone would have "to cough up some £3 billion". With such huge amounts of money on offer, "it is little wonder that the subsidy system has been wide open to abuse". Mr Carey concurred with such statement, stating that "there is no doubt that it is seriously open to fraud".

6. The Presenter then proceeded to state that "European beef subsidies are not paid to the farmers who breed the cattle, instead they go to the people who buy them, the beef processors. A subsidy system which was designed to support those working on the land has turned into a Welfare State for industrialists" and identified Larry Goodman as one of the industrialists to whom the presenter referred by stating that Larry Goodman's company, Anglo Irish Beef Processors has made him Europe's biggest beef baron.

7. The presenter then stated that Larry Goodman publicly maintains that his business has always been run to the highest of standards and proceeded to state that

> "Within Goodman's factory, it has been an open secret that this is not the case but until tonight no Goodman employee has dared to speak openly."

8. This served as an introduction to Patrick McGuinness, an accountant who had worked at a number of Goodman factories and at the Goodman Head Office who in the course of a number of contributions to the said programme stated, inter alia that:-

(i) "The philosophy of the Company is basically profit maximisation. You can only make so much money by doing it right but it is so easy to make much more by abusing the system, all the factories did it..."

(ii) "Mr Goodman set the tone because he controlled the Company very tightly".

(iii) "The intervention system was vital to the Goodman companies because at the end of the day that is where the profits came from.

(iv) "....abuses within the system that were ..... institutionalised within the factories."

(v) "The whole basis of this import system is documentation and if the documentation shows either more product or higher quality product than it is, they are going to get paid more. All intervention product has to be weighed in and the weights have to be recorded on a document called the IB4. One of the ways of changing the weights was basically to reproduce an IB4, this would be a duplicate copy and what would happen is that the same details would be written down except that the weights would be increased by a certain number of kilos. If there was any special notations such as signatures or other notation or even blood on the original document, this was put on to the duplicate."

(vi) "It was very easy to change the grades, with a knife you cut off the grade that is marked on the animal and you can then put any other grade you like on it. You would have your own stamps at the factory".

(vii) ".....all grading stamps were supposed to be tightly controlled by the Department of Agriculture".

(viii) "The best way to get rid of a bad product was getting it into intervention, by switching the product...."

9. The presenter then stated that "one of the reasons why no Goodman employee has spoken out until now, was a company wide scheme of under the table payments. Cheques were made out against bogus invoices, endorsed by Goodman's employees and cashed at local branches of the Allied Irish Bank"

and Mr McGuinness stated that:—

"The payments were made quarterly... and were paid to everybody in the Company basically" and amounted to a sum of £3 million per year.

10. Dealing with Export Refunds the presenter stated that:—

"Goodman took great advantage of export refund subsidy. With the backing of Irish politicians including the Prime Minister, Charles Haughey, he went after huge contracts in Libya, Egypt, Iran and Iraq".

"Most Middle East customers want their animals killed by Islamic ritual slaughter known as Halal. But this method needs more effort than the usual technique so, at Goodman plants that didn't always happen".

Subsequent to this statement, Mr McGuinness stated that:—

"You can just say that you slaughtered Halal, and then you would have various stamps and forms to mark the livestock that it was slaughtered Halal. These were Islamic stamps with Islamic writing on them to show that it was Halal slaughtered".... we had our own stamps."

11. Coming to the contracts with Iraq the presenter stated that:

".....Goodman's biggest subsidised coup came when he linked up with Saddam Hussein. The butcher of Baghdad needed millions of tonnes of beef to feed his troops and Europe's "Mr Meat" was just the man to sell it to him."

And in this connection Mr McGuinness stated:—

> "There was many different sources for the meat, there was fresh Irish meat, there would have been intervention meat, there would have been frozen meat, it may have been Halal slaughtered, it may not have been Halal slaughtered, it could have been cow, it could have been bull, it could have been anything."

12. The presenter then stated:

> "Boxes of old frozen meat from European intervention stores were brought by the truck load to the Goodman owned Ulster Cold Stores Craigavon, Northern Ireland. Here a transformation took place. For a solid eighteen months, old frozen meat was turned into new."

Thomas Ruddy stated:—

> "All of it was reboxed as killed within the last week or two, slaughtered within the last week or two....."

Patrick McGuinness stated that:—

> "The whole system was that you switch product to show what the customer wanted. If that meant reboxing, you reboxed the meat to show what the company thought he was getting."

13. The programme also referred to the Customs investigation at the AIBP factory in Waterford.

The presenter stated:

> "In all his years of business, Goodman's companies have rarely been investigated. But in 1986 they came under Customs scrutiny at the AIBP factory in Waterford. At the time, this investigation was kept secret. The Customs men quickly found that weights had been falsified and poor quality meat added to some 70,000 boxes of frozen beef bound for the Middle East."
>
> Goodman has always maintained that this was the work of a sub-contractor, that he was innocent of any wrong doing but World in Action has discovered that on at least two occasions Goodman's own managers tried to obstruct the Customs investigation."

In this connection Mr McGuinness stated:—

> "There was a massive panic within the company and a plan was put forward as to how the damage could be limited. The plan was basically agreed between our people and the Customs people at their Head Office, that a certain sample of good product would be selected for thawing out and for investigation. This was a deliberate scheme to contain the damage because of the explosive nature of the investigation ......was agreed at senior management level".

He further stated that

> "That particular scheme didn't work out correctly because of the local customs agents basically kicked up a fuss."

14. The presenter further stated that:—

MALE PRESENTER VOICE OVER

"In the customs case summary obtained by World in Action the investigators highlight a second attempt by Goodman employees to undermine the investigation.

"Goodman employees attempted to disguise the extent of the fraud by altering case weights at the cold store."

"The customs men concluded that while "... it has not proved possible to establish conclusively whether or not AIBP Ltd. were knowingly involved in this fraud, they are not an incompetent or inefficient organisation and it is strongly suspected that they were aware."

"Customs recommended the instigation of criminal proceedings, yet the Fraud Squad were unable to get their hands on the customs report until eighteen months later.

FEMALE PRESENTER VOICE OVER

"Was there a fear among Goodman employees or within the company that you could get caught doing all these things."

PATRICK MCGUINNESS

"There was always a fear of being caught obviously, doing something that is basically illegal. But there was also a feeling that we were invincible, we had the right connections at the right places, that could basically control any investigation that would be put in place. If the investigation had revealed the extent of the problem, it would have led to loss of the guarantees that had already been placed which could result in tens of millions of pounds".

CHAPTER FOUR

Allegations Referred to in Dáil Eireann

1. Many of the allegations made on the ITV programme had previously been made or were subsequently referred to in Dáil Eireann.

2. The Tribunal was obliged to inquire into allegations regarding illegal activities, fraud and malpractice in and in connection with the beef processing industry made or referred to in Dáil Eireann.

The terms of the Resolution passed by both Houses of the Oireachtas were not helpful to the Tribunal as they did not specify the dates on or periods within which the allegations regarded as of urgent public importance were made and the Tribunal was obliged to read through Dáil Reports for the purpose of endeavouring to ascertain what allegations were the subject of the said motions.

It would have been more desirable and indeed more correct and in accordance with the provisions of the Tribunals of Inquiry (Evidence) Act, 1921 if the Resolution passed by both Houses of the Oireachtas had specified in detail the definite matters of urgent public importance which they had resolved should be inquired into.

3. The allegations made in Dáil Eireann by various members of the House are set out in the Book of Allegations contained in Appendix 1.

4. The Book of Allegations contained certain allegations which are set forth under the following headings viz.

"Goodman and the Banks"	Allegations	2-7 inclusive
"The Cyprus Loan"	Allegations	1-5 inclusive
"Goodman and Classic Meats"	Allegations	1-6 inclusive

While these allegations were originally included in the Book of Allegations to be inquired into, the Tribunal subsequently decided that they did not come within the terms of the Resolution passed by the Houses of the Oireachtas.

In the course of his judgment in Goodman International and Laurence Goodman -v- Mr Justice Liam Hamilton [1992] 2IR 591, the Chief Justice stated:—

> "I would accept... that there is no indication of an intention on the part of the Tribunal, and no obligation, having regard to the Terms of the Resolutions in pursuance of which it is acting, for the Tribunal to inquire into any private matter or dispute which has not also got a public effect .... The question whether in respect of any individual item of allegation, it has got a public connotation is a matter to be dealt with and determined as the proceedings of the Tribunal continue."

In the course of his judgment in the same case, the late Mr Justice McCarthy at [1992] 2IR 604 stated:—

> "It may be that such a Tribunal could be restrained from inquiring into civil disputes that only involve private parties and where there is no public element".

The Tribunal satisfied itself that there was no public element or public connotations involved in the allegations with regard to the Banks, the Cyprus Loan and the ownership of Classic Meats, which had been dealt with by the Fair Trade Commission and being so satisfied did not follow further inquiries into them.

5. So far as the Tribunal has been able to ascertain from a consideration of the Dáil Reports, apart from a reference to a major investigation into the Charleville plant of the Halal associated United Meat Packers Exports Company in relation to export refunds, all the allegations involving illegal activities, fraud and malpractice made in Dáil Eireann relate to Goodman International, its associated companies and employees and the alleged co-operation of and condonation or cover up by the State authorities allegedly due to the influence of Larry Goodman and his political connections with the then Fianna Fáil Government.

6. The allegations made in Dáil Eireann cover many of the allegations made in the television programme and may be summarised as follows and under the following headings:—

(i) Abuses of the system under which subsidies are paid by the European Economic Community to those engaged in the beef processing industry.

(*a*) Maintenance of an entire production line in Nenagh designed for taking stamps from frozen carcases and re-stamping and re-packing them, — made by Deputy Pat Rabbitte on the 15th May 1991;

(*b*) Change of labels on meat in different parts of the country by a team moving about to do this job on behalf of Goodman companies, — made by Deputy Tomás MacGiolla on the 9th day of March, 1989;

(*c*) The removal and changing of stamps, dressings and labels on beef carcases in a plant on the 12/13 January 1989 seen by a journalist and notified to Department of Agriculture and Food, -made by Deputy Barry Desmond on the 12th day of April 1989;

(*d*) Attempted use of South African Customs stamps to defraud the Department of Agriculture which resulted in the conviction of a close aide of Larry Goodman who was found in possession thereof, made by Deputy Dick Spring on the 15th day of May, 1991 and by Deputy Pat Rabbitte on the 28th August 1990.

(*e*) Illegal labelling of meat carcases in the Eirfreeze factory in the North Wall by changing labels and dates of slaughter on meat which resulted in the shut down of the plant by Inspectors from the Department of Agriculture and Food, made by Deputy Tomás MacGiolla on the 9th March 1989 and on the 15th day of May 1991;

(*f*) Carrying out grotty repackaging and restamping operations in Goodman plants in operations heavily subsidised by the Irish taxpayer, thereby putting Ireland's reputation for quality at risk, — made by Deputy Pat Rabbitte on the 24th May 1991.

(*g*) Engaging in a Carousel operation, — made by Deputy Pat Rabbitte on the 24th May 1991.

(*h*) Engaging in serious irregularities in connection with the operation of the 1986/87 Aids to Private Storage Scheme at plants in Waterford and Ballymun, made by Deputy Barry Desmond on the 19th day of March 1989 and by Deputy Dick Spring on the 15th day of May, 1991.

(ii) *Failure of regulatory authorities and allegations of political influence in relation to alleged abuses of the system.*

(*a*) The Department gave advance notice of inspections at meat plants and in particular at Foynes on the 15th and 16th day of April, 1989, made by Deputy Dick Spring on the 28th April, 1989.

(*b*) Almost all of the samples taken in Foynes had trimmings in them or were otherwise suspicious, made by Deputy Dick Spring on 28th April 1989.

(*c*) The regulatory authorities turned a blind eye on (Goodman's) dubious business practices — the false labelling and accounting, the commercial arrangements involved in the disposal of offal and so on, made by Deputy Dick Spring on the 28th day of August 1990.

(*d*) The Department of Agriculture did not diligently assist the Garda Fraud Squad in relation to the Waterford and Ballymun investigations and ignored the request made for the release to the Fraud Squad of the Department's file in relation to the investigation, made by Deputy Dick Spring on the 15th day of May, 1991.

(*e*) Notwithstanding their knowledge of the irregularities at Waterford and Ballymun and the prosecution of Mr Nobby Quinn in relation to the bogus South African stamps, the Department (and the Minister) was prepared to release bank guarantees of up to £20m (frozen because of the irregularities at Waterford) as part of the overall deal in the Examinership, made by Deputy Dick Spring on the 15th day of May, 1991.

(*f*) The Department of Agriculture and Food and prosecuting Counsel seemed very reluctant to pursue the charges against Eirfreeze and AIBP with any vigour, on the 30th July 1990 and in particular the issue of fraud and forgery about which the Garda were not informed, made by Deputy Tomás Mac Giolla on the 15th day of May 1991.

(*g*) The regulatory and control procedures for the Irish Beef Industry are not satisfactory and in particular the Government have failed in their responsibility of rooting out those people who have turned the beef industry into an object of scandal and disgrace. The Government have covered up the illegal and improper activities in the beef industry since 1987, made by Deputy Spring on the 15th day of May 1991.

(*h*) There was official indifference to the climate of fraudulent practices that characterised the Goodman group. According to one public official, the whole ethos was "do not interfere, do not make trouble, this man is doing a great job." If you hoped to be promoted the last thing you wanted to do was start shouting foul at Larry Goodman — made by Deputy Rabbitte on the 15th day of May 1991.

(*i*) The Department failed to make proper arrangements to give Customs officials sufficient notice of export consignments to allow them to carry out detailed examinations made by Deputy Eamonn Gilmore on the 21st day of June 1990.

(iii) *Tax evasion and Political influence in regard thereto.*

(*a*) A great many Goodman workers were on the dole and were being paid under the counter, made by Deputy Pat Rabbitte on the 15th May 1991.

(*b*) Because of Goodman's political connections, the Revenue Commissioners turned a blind eye to the type of "remuneration packages" enjoyed by senior executives and a non-return of PAYE and PRSI to the Exchequer for many workers because of the operation of the contract system for a large proportion of the Goodman workforce — made on the 28th August 1990 and repeated on the 15th May 1991 by Deputy Pat Rabbitte.

(*c*) In the Finance Act, the Government made a special arrangement to enable Mr Goodman to avail of high coupon finance (in respect of Section 84 loans) to fund speculative ventures abroad, made by Deputy Pat Rabbitte on the 15th May 1991 and because of its use outside the State to fund speculative ventures, it amounted to tax evasion warranting prosecution.

(*d*) Mr Goodman got special concessions in regard to tax from the Government. He got a concession of £4 million from the Revenue Commissioners, which was 50% of the tax bill he owed and which did not include interest, made by Senator Thomas Raftery in Seanad Eireann on the 29th day of May 1991.

(*e*) In return for the Revenue Commissioners agreeing not to take proceedings against Mr Goodman or his company in respect of large scale tax evasion practices going back over many years. Goodman International paid the Revenue Commissioners £4 million in respect of all outstanding liabilities and penalties, a settlement which was by far the largest of its kind in the history of the State, made by Deputy Dick Spring, on the 15th day of May 1991.

(*f*) The Government's support for Goodman included changes in the tax laws to enable a substantial amount of Mr Goodman's income from beef processing to be taxed at 10% manufacturing rate, made by Deputy Dick Spring on the 18th August 1990.

(iv) *Goodman, the Industrial Development Authority and political influence.*

(*a*) The Goodman organisation was chosen as the hub around which Fianna Fáil had built its development policy for the food industry, including beef, dairying and sugar. Government funding commitments to him of between £200 and £250 million in 1987 had given him "tremendous credit" in raising finance wherever he wished to go and he had also received IDA grants of up to £25 million. The Taoiseach himself directly intervened with the IDA to drop the performance clause in the case of grants to the Goodman Company, made by Deputy MacGiolla on the 9th day of March 1989.

(*b*) In June of 1987 the Government decided against the wishes of the IDA to give £25 million to Laurence Goodman, made by Deputy Barry Desmond on the 9th March 1989.

(*c*) When Goodman applied for assistance for a Five Year Plan for the Beef Industry, the grant package was rushed through by the IDA under political pressure and was rushed through the Department of Finance under similar political pressure with the Taoiseach's own personal and improper interference, made by Deputy John Bruton on the 24th May 1991.

(*d*) Enormous political pressure from the highest possible level was brought to bear on the Goodman Group and the IDA to announce the expansion programme of 1987 before details had been worked out, solely as a PR exercise for the Taoiseach and his Government, made by Deputy Sean Barrett on the 28th August 1990.

(*e*) The decision on the part of the Government to rely solely on Goodman to develop the beef industry was downright irresponsible and was made at considerable expense to the taxpayer, made by Deputy Sean Barrett on the 18th day of December, 1990.

(*f*) The entire board of the IDA at one stage threatened to resign over this grant to expand an industry that already had a surplus processing capacity, made by Senator Thomas Raftery on the 29th May 1991.

(v) *Abuse of Export Credit Insurance Scheme*

The allegations in regard to this aspect of the Inquiry were made by Deputy Desmond O'Malley, Deputy Pat Rabbitte and Deputy Dick Spring and may be summarised as follows:—

(*a*) The provisions by the State of Export Credit Insurance cover on the sale of beef to Iraq in 1987 and 1988 of an amount in excess of the amount actually exported was:

(i) in breach of the terms of the Export Credit Insurance Scheme;

and

(ii) constituted a substantial abuse amounting to a fraud on the taxpayer, the scale of the abuse and of the potential liability of the State being unprecedented, made by Deputy Desmond O'Malley on the 10th May 1989.

(*b*) The provision in 1987 and 1988 of between one fifth and one third of all Export Credit Insurance cover available with over 80% going to Goodman:

(i) amounted to abuse of the scheme, and

(ii) excluded fair competition from within the State which aggravated the scandal, made by Deputy Desmond O'Malley on the 10th May 1989.

(*c*) Allowing just two companies, of which by far the larger and more substantial was Goodman, cover, under the Export Credit Insurance scheme for beef exports to Iraq, so considerably in excess of their actual exports to that country, was an act of blatant favouritism and had the effect of strengthening further the already strong position of Goodman (to whom members of the Government were extremely personally close) as the dominant group within the beef processing and allied trades, contrary to the interests of farmers and employees and of exporters in other business sectors, made by Deputy Desmond O'Malley on the 10th May 1989.

(*d*) The decision taken in 1987 by the Fianna Fáil Government to reinstate Export Credit Insurance was taken against the best professional advice available to the Government, made by Deputy Dick Spring on the 28/8/1990 and reported on by him on the 15th May 1991 and by Deputy Pat Rabbitte on the 24th May 1991.

(*e*) In respect of Goodman's Export Credit Insurance Policy declarations were made that only beef with its origins in the Republic of Ireland would be covered, nevertheless very large quantities of non-Irish beef were included in shipments purporting to be covered by that policy, made by Deputy Desmond O'Malley on the 28th August 1990.

(*f*) Conscious decisions were taken to give one conglomerate (Goodman) more than 80% of the available cover in that market, disadvantaging rivals and exporters in other products, made by Deputy Rabbitte on the 24th May 1991.

(*g*) The granting of Export Credit Insurance was a political decision and depended on whether "you were a member of the club" and Mr Goodman, when he heard that Halal had been granted a slice of the Export Credit Insurance, intervened with the Taoiseach who caused the Minister for Industry and Commerce, Albert Reynolds TD, to cancel the allocation of such insurance and to inform the Chief Executive of Halal, made by Deputy Rabbitte on 15th May 1991.

(vi) *Allegations of Political Influence*

In addition to those set forth herein, further allegations were made in Dáil Eireann as follows:—

(*a*) The extraordinary recall of the Dáil and Seanad in August 1990 had as much to do with the integral link between Fianna Fáil and the Goodman organisation as it has with protecting a key Irish industry: made by Deputy Rabbitte on the 28th day of August 1990.

(*b*) The Companies (Amendment) Bill, 1990 represented only Goodman's third choice proposal, arising from meetings held with the Taoiseach, the first being a £300 million rescue package which Mr Goodman demanded the Government should underwrite, the second involving an approach by Mr Goodman's friends in Cabinet to the EC Commissioner, Mr Mac Sharry in an attempt to persuade him to bring forward an EC plan that would be of similar assistance to Mr Goodman but which would be cosmetically packaged as being in the interest of the total industry: made by Deputy Rabbitte on the 28th August 1990 and repeated on the 15th May 1991.

(*c*) Goodman successfully intervened with the Taoiseach to cause the Government to reverse a decision to increase the budget to be given to CBF the meat marketing board, in 1988, in order to shut out the prospect of markets being expanded for his competitors, made by Deputy Rabbitte on the 25th October 1990.

(*d*) Charles Haughey publicly promoted Goodman. At the very time the Customs investigations were warning that Goodman's operations were strongly suspected of involvement in fraud, the Irish Prime Minister was endorsing Goodman for millions in Irish and European grants: made by Deputy Spring on the 15th May 1991.

(*e*) There was political interference in the work of Agricultural Officers and Customs men in attempting to investigate suspected breaches of EC regulations: made by Deputy Pat McCartan on the 24th February 1988 and by Deputy Tomás MacGiolla on the 9th March 1989.

(*f*) It has been suggested that Goodman was subjected to a lesser degree of Customs inspection than other commercial operations (especially in regard to container loads going North) and that he was able to virtually close off the port of Greenore to other people when he was exporting meat: made by Deputy Rabbitte on the 15th day of May 1991.

(*g*) Goodman had been allowed to "cherry pick" the best of the ICC property portfolio, because he was on the "inside political track" before any other party became aware of these properties: made by Deputy Pat Rabbitte on the 28th August 1990.

(*h*) Knowing the inside political track had enabled him to get access to exceptional lines of credit and to benefit from risky but profitable Middle East contracts, confident in the knowledge that he was guaranteed by the Government so long as Fianna Fáil remained in power: made by Deputy Pat Rabbitte on the 28th August 1990.

(*i*) Fine Gael's attitude to Goodman was uncommonly acquiescent, a consideration affecting their attitude being the receipt of a donation of £60,000 from Goodman in 1988, made by Deputy Pat Rabbitte on the 15th October 1990.

CHAPTER FIVE

Beef Industry and Common Agriculture Policy Support Systems

As a result of these allegations, the Resolutions referred to were passed by both Houses of the Oireachtas and this Tribunal was established to inquire into allegations of illegal activities, fraud and malpractice in the beef processing industry and many of the allegations related to the alleged activities of Mr Laurence Goodman and his various companies engaged in the food processing industry. During the course of this Report, these companies will be referred to at different stages as Goodman International, AIBPI or AIBP though the Group is comprised of approximately thirty-five separate units.

Insofar as the beef processing industry is concerned, the allegations mainly refer to alleged abuses of the systems under which subsidies are paid by the European Economic Community to those engaged in the beef processing industry.

The importance of this industry in the economic life of this State cannot be overstated and the role of Laurence Goodman and the companies controlled by him in its development has been considerable.

The beef processing industry is part of the agricultural industry which is of central economic and social importance in Ireland and accounts for about 10.5% of Gross Domestic Product. There are 165,000 people directly engaged in agriculture with a further 40,000 people employed in the food and drink processing sector. Agricultural exports represent about 16% of total exports. (When exports of food and beverages are included, this figure increases to 25%). Inputs used in agriculture are valued at some £1.3 billion annually. These include materials and services, the vast bulk of which originate in Ireland.

Production of beef cattle takes place mostly on family farms and beef herds are quite small on average. There are about 100,000 farmers involved on a full-time or part-time basis in cattle production. About 1.4 million head of beef cattle are slaughtered each year (— in approximate percentages, steers 57%, heifers 25% and cows 18%). Steer beef is mostly for export outside the European Community, while heifer and cow beef is mainly consumed in Ireland and in other EC member States. Because Irish cattle production is predominantly based on grass, the highest proportion of cattle is fattened in the summer and slaughtered in the autumn and winter. Unlike other member States of the EC, where patterns of production are very different to those in Ireland, this country is over 600% self-sufficient in beef.

Slaughtering and processing of beef cattle is a significant element of the country's industrial sector. The export beef industry employs some 4,500 people on a permanent basis. This figure increases during peak production times to around 6,000. Beef and cattle exports are valued at approximately £700m a year excluding export refunds, which represents some 5% of total exports.

There has been a trend toward the development of value-added product and away from live cattle export. In recent years, export of boneless beef and vacuum packed product has increased. The vacuum packed share of total output rose from 3% in 1980 to 20% in 1990.

Vacuum packed sales of Irish beef have increased dramatically in recent years:—

1984 — 25,606 tonnes (carcase weight equivalent) Value £61m.
1992 — 113,300 tonnes (C.W.E.) Value £265m.

This more than four-fold increase has been achieved against a background of falling EC consumption, increased pressure from competing meals, alternative protein sources and the growth in the convenience food sector. Such a striking commercial performance could not have taken place if the quality of the product were in doubt and concerns which have been expressed about the quality of Irish meat exports are, in the view of the Tribunal unfounded: it is a product of the highest quality, justifiably commanding a premium price on international markets both within the EC and worldwide, and remains the country's most successful foreign export commodity.

The underlying strength of the Irish cattle industry, deriving from the economics of raising cattle on grass, had long been recognised but what is distinctive about the last 25 years is that Ireland, long a successful producer of live cattle, has become a successful processor and international trader as well. The benefits to the Irish economy, by way of value added and job creation, deriving from the export of processed beef rather than the export of live cattle are self evident.

The Goodman Group has played a significant role in this development and the technical standards of the plants, owned and operated by them, compare more than favourably with meat plants throughout the community:

—The Group slaughtered 2.3 million cattle in Ireland between 1986 and 1991, representing 24.8% of the national kill and 29.5% of the export kill.

—The Group killed its largest share of Irish cattle in 1988 (31%), the year in which cattle prices were at their highest in the period 1986 to 1991.

—Up to August, 1990 AIBP had exported Irish beef to 77 different countries.

—Within the European Community, AIBP supplied virtually every major supermarket group; in the United Kingdom it is by far the largest supplier of beef to all the major supermarket groups.

—AIBP's boneless vacuum packed exports grew by 300% between 1987 and 1991.

—Commercial beef boned by AIBP increased by a factor of 18 times between 1981 and 1991. In any one year, the Group produces 48 million consumer packs for supermarkets worldwide.

—Commercial beef, boned by the meat division, can be purchased in 14,000 retail outlets throughout Europe.

It further appears from the evidence of Mr Gerry Thornton, Deputy Chief Executive of the Meat Division of the Goodman Group, that:—

—At the beginning of 1986 AIBP operated 7 meat plants, namely the plants at Dundalk, Dublin, Bagenalstown, Cahir, Nenagh, Newry and Enniskillen.

—Meat Division now consists of Head Office at Ravensdale and thirty-five different operating units namely:—

11	Beef Slaughtering Units.
11	Beef Deboning Operations
4	Lamb Slaughtering Units
2	Lamb Deboning Units
3	Cold Stores
1	Pork Slaughtering Unit
1	Pork Deboning Unit
1	Cannery
1	Feed Lot

—From a turnover of £250 million in 1985, Meat Division had a turnover in excess of £520 million in 1992;

—From employing 832 people in 1985, Meat Division now employs 2,120 people;

—From hiring 532 independent contractors in 1985, Meat Division now uses the services of 906 independent contractors;

—In 1985 Meat Division deboned 126,000 cattle. In 1992 Meat Division deboned in excess of 336,000 cattle;

—In 1985 Meat Division produced in excess of 4 million primal cuts by the deboning process. In 1992 Meat Division produced almost 16 million primal cuts.

This record establishes a significant contribution to the development of the food processing industry job creation and Irish exports.

Having regard to the nature of the allegations made against the Goodman Group of Companies, and the widespread publicity given to such allegations and to the evidence given at this Tribunal, it is significant that there were no complaints from consumers or purchasers of either commercial, intervention or Third Country beef sold and exported by the Goodman Group of Companies.

The development of the food processing industry was assisted to a very considerable extent by the various market support schemes introduced by the European Economic Community which will be described at a later stage of this Report, particularly those in regard to Intervention, Aids to Private Storage and the Export Refund Subsidy Scheme.

The Goodman Group of companies engaged in the food processing industry availed of the supports available under these schemes but the evidence adduced before this Tribunal clearly established that it relied less on the Intervention Scheme than any other company engaged in the food processing industry.

As stated by Mr Laurence Goodman in the course of his evidence, "the commercial sale of beef was the bedrock of his companies' business" and they were less reliant on Intervention than any other Irish company.

In the period 1987-1989, intervention sales by the Goodman Group represented 11% of the Republic of Ireland turnover of the companies and 8% of the British Isles.

Turnover in 1988 intervention sales by the companies represented 6% of turnover in the Republic of Ireland and 4% of British Isles turnover.

Such figures do not support the statement made by Mr Patrick McGuinness on the ITV programme that

> "The Intervention system was vital to the Goodman companies because at the end of the day that is where the profits come from."

The benefits to the Irish economy arising from the application of such schemes is set forth in the evidence given to the Tribunal by Mr Michael Dowling, Secretary to the Department of Agriculture and Food.

He states that:—

> "Between 1973 and 1992, 2,116,508 tonnes of beef were purchased by the Minister for Agriculture in his role as Intervention Agent. This was equivalent to over 12 million sides of beef or 6.3 million cattle. The Department made payments totalling £4.097 billion in respect of the purchase of this beef. Average annual payments for the purchase of beef amounted to about £200m. In 1991 payments of £591m were made in respect of 262,000 tonnes of beef, accounting for the disposal of almost three quarters of a million cattle. It was the highest ever annual intake into intervention."

The quantities purchased and payments made were set forth in the following table produced by him.

Intervention Beef Purchases and Quantities Deboned
1973 — 1992

Year	Total Purch. (t)	Value of Purch. (£)	Qty. Deboned (t)	Qty Prod. (t)	Yield (%)
1973	2,383	1,588,662	Nil	Nil	N/A
1974	121,682	82,719,214	67,864	43,149	63.6
1975	136,635	111,196,217	68,910	44,157	64.1
1976	70,916	69,717,306	41,118	26,809	65.2
1977	90,897	112,362,713	56,026	36,696	65.5
1978	86,148	120,127,713	58,582	38,274	65.3
1979	89,458	132,349,382	59,934	39,157	65.3
1980	102,188	160,993,459	62,440	40,761	65.3
1981	47,476	81,942,312	35,058	23,090	65.9
1982	60,510	117,125,986	40,323	26,641	66.1
1983	60,506	142,257,776	43,511	28,849	66.3
1984	66,562	158,257,845	44,998	30,399	67.6
1985	73,568	181,166,942	55,019	37,488	68.1
1986	129,759	372,146,094	87,653	59,841	68.3
1987	100,530	268,145,044	72,132	49,278	68.3
1988	59,846	114,954,362	48,209	32,960	68.4
1989	77,515	182,437,846	59,415	40,637	68.4
1990	230,638	545,494,072	214,339	146,630	68.4
1991	262,094	591,056,818	256,710	175,928	68.5
1992p	247,197	550,812,915	241,839	165,871	68.6
TOTAL	2,116,508	£4,096,852,678			

p=Provisional

In the same period intervention beef, either in bone-in, boneless or canned form was sold from intervention stocks by the Department, at a total sales value of £1.657 billion. On average, beef to the value of £82,000,000 was sold each year. In 1992 sales of over 170,000 tonnes were effected at a sales value of over £200m.

Details of such annual sales are set out in following table produced by Mr Dowling.

Intervention Beef Sales 1973 — 1992

Year	Bone-in Sales		Boneless Sales		Canned Beef Sales	
	Tonnes	£	Tonnes	£	Tonnes	£
1973	Nil	Nil	Nil	Nil		
1974	28,159	14,161,789	861	634,880		
1975	50,447	26,795,519	53,829	38,165,481		
1976	46,876	32,255,196	38,447	37,729,467	750	716,284
1977	15,492	13,078,417	17,226	19,441,785	1,710	2,191,012
1978	40,982	36,991,562	40,347	64,978,710	301	250,681
1979	20,824	22,742,729	29,121	42,334,468	2,281	1,490,029
1980	40,890	52,874,624	44,888	66,231,542		
1981	29,592	43,020,905	40,646	76,391,605		
1982	11,912	17,467,078	25,473	42,240,570		
1983	20,030	28,768,147	17,504	31,522,650		
1984	2,285	3,308,527	14,049	27,949,617		
1985	7,937	8,373,141	22,328	38,222,055		
1986	54,255	55,068,581	58,253	62,427,684		
1987	10,185	13,660391	54,574	98,766,992		
1988	30,855	40,063,586	53,195	94,271,733		
1989	27,815	43,595,194	48,582	51,895,613		
1990	19,087	19,579,601	43,161	70,332,684		
1991	496	359,674	90,267	111,307,803		
1992p	12,639	11,513,976	158,067	189,194,990		

p=provisional
Total Sales Value = £1,657,366,981

In addition the Department paid aid towards the private storage of 569,152 tonnes of beef. The total aid payment amounted to £195,588,140. Details of the annual aid payments are set forth in the following table.

Aid for the private storage of beef
Quantities and Amounts Paid

Year	Quantity (tonnes)	Amount Paid £
1973	Nil	Nil
1974	150	7,300.00
1975	1,445	323,575.05
1976	11,964.47	3,144,854.96
1977	22,639.472	6,493,729.43
1978	11,626.162	3,183,625.38
1979	10,619.207	3,601,865.37
1980	4,589.047	1,717,529.37
1981	417.828	193,870.63
1982	521.717	250,567.48
1983	104.593	52,867.87
1984	854.927	339,389.98
1985	71,939.331	27,525,665.69
1986	82,570.27	28,756,469.10
1987	59,011.2272	15,984,199.22
1988	77,103.1901	24,084,521.98
1989	125,531.1608	42,198,767.07
1990	87,751.4755	37,405,163.50
1991	312.9018	324,178.73
Total	569,151.9794	195,588,140.81

Payments of export refunds and monetary compensatory amounts in the period 1973 to 1992 totalled £2.464 billion. The annual details are provided in the table hereunder.

	Export Refunds	MCAs
Year	£	£
1973		211,747
1974	3,303,841	-899,004
1975	10,104,829	−10,828,275
1976	14,046,657	−15,932,356
1977	13,575,235	34,720,745
1978	14,922,239	72,689,205
		−14,032,992
1979	26,871,041	33,440,683
		-3,612,895
1980	82,039,738	7,368,814
		-73,519
1981	114,987,696	171,186
		-69,577
1982	131,183,340	665,920
		−10,155
1983	169,096,394	1,227,970
		-907,504
1984	196,970,605	192,586
1985	232,351,138	1,874,599
		-517
1986	183,509,615	24,135,169
		-7,338,592
1987	177,490,060	72,737,475
1988	196,710,838	24,906,667
1989	246,095,585	4,873,336
1990	142,226,853	7,078,696
1991	113,082,465	113,974
1992	139,997,612	34,201
Total	£2,208,568,521	Paid £286,442,973
		Received £53,705,386

In addition variable premium payments amounted to £92,671,534 for the years 1975 to 1990.

The total financial responsibility of the Department in respect of the above measures from 1973 to 1992 amounted to £8.506 billion.

As stated by Mr Dowling "the scale of the operation of the intervention measures in the beef sector in Ireland was enormous."

It is against this general background that the Tribunal inquired into the allegations of abuse of the Community Common Agricultural Policy Support Funds.

For the proper understanding of same the Tribunal was obliged to consider all the relevant regulations and practices in relation thereto and considers it desirable to give a comparatively short summary of the relevant schemes and their mode of operation and will do so prior to dealing with the specific allegations in regard thereto.

Before doing so however, the Tribunal proposes to report on the allegations made with regard to the re-introduction of the Export Credit Insurance Scheme and the allocation of cover in pursuance thereof by the Minister for Industry and Commerce, the allegations made with regard to tax evasion and avoidance, the allegations made with regard to the IDA and the Five Year Development Plan with its subsidiary issue in relation to Section 84 borrowings and a number of other allegations none of which relate to alleged abuses of the EEC system of supports.

CHAPTER SIX

Export Credit Insurance

Many of the allegations made in Dáil Eireann and set forth in Chapter hereof relate to the operation in Ireland of the Export Credit Insurance Scheme. To appreciate the nature and affect of such allegations it is necessary to understand the operation of such a system.

The Insurance Acts 1953 to 1988 gives to the Minister for Industry and Commerce power, for the purpose of encouraging exports, to give guarantees with respect to insurance risks in connection with external trade.

This section has been amended on a number of occasions by subsequent Insurance Acts the most recent of which was the Insurance Act, 1983.

Section 1 of the Insurance Act, 1983 states:—

> "The following subsection is hereby substituted for subsection (1) of Section 2 of the Insurance Act, 1953:
>
> (1)(*a*) For the purposes of encouraging the exportation of goods and the provision of such services as are specified from time to time from by order made by the Minister, the Minister, with the consent of the Minister for Finance, may make arrangements for giving to, or for the benefit of, persons carrying on a business or profession in the State guarantees in connection with the export, manufacture, treatment or distribution of goods, the provision of services or any other matter which appears to the Minister conducive to that purpose"

Section 2 (3) of the Insurance Act, 1953 stated that:—

> "The aggregate amount of the liability at any time of the Minister for principal moneys in respect of arrangements under this Section shall not exceed two million pounds"

There were a series of acts in the period 1953 to 1988 the sole purpose of which was to increase the statutory limit of liability for Export Credit Insurance and guarantees.

At all time material hereto the statutory limit of liability for Export Credit Insurance and guarantees was £300 million until that statutory limit was increased to £500 million by the Insurance (Export Guarantees) Act of 1988.

Export Credit Insurance is one of the approved classes of insurance under EEC Council Directive 73/239/EEC (Class 14). EEC Council Directive 73/239/EEC was adopted into Irish Law by Statutory Instrument 115 of 1976 entitled "European Communities (Non Life Insurance) Regulations 1976."

In the period 1953 to 1971, the Export Credit Insurance Scheme provided for by the aforesaid insurance acts was operated on behalf of the Minister for Industry and Commerce by a number of Irish insurance companies.

In 1971 the Insurance Corporation of Ireland plc was given the sole agency in respect thereof by the Minister for Industry and Commerce.

That sole agency remained in place until the 4th day of November 1987 when it was replaced by a new Agency Agreement made between the Minister for Industry and Commerce and the Insurance Corporation of Ireland plc, which said agreement continued to grant a sole agency in respect of the operation thereon to the Insurance Corporation of Ireland plc.

The main points from the Agency agreements are as follows:

"(i) The Insurance Corporation of Ireland plc ("ICI") agreed not to sell Export Credit Insurance for its own account as long as it acted as agent for the Minister:

(ii) The Export Credit Insurance policies could be issued to persons exporting goods and/or services manufactured or produced in Ireland:

(iii) ICI undertook to use the resources necessary to ensure that all reasonable care, skill and judgement was exercised in the operation of the Scheme:

(iv) ICI undertook to endeavour to ensure that, taking one year with another, the scheme would involve no net loss to State funds:

(v) It was ICI's function to ensure that the maximum potential liability of the Minister in respect of Export Credit Insurance policies should not at any time exceed the sum provided by the Insurance Acts:

(vi) ICI were to keep two separate sets of accounts — a "Number 1 Account" and a "Number 2 Account". The Number 1 Account related to normal insurance business, exclusive of the insurance included in the Number 2 Account. The Number 2 Account related solely to business done on the basis of special Ministerial or Government decisions. The ICI were to provide insurance for the Number 2 Account only when specified by the Minister.

Section 15 of the Agency agreement provides:—

> "**Underwriting practices**
>
> "The Company will keep two separate sets of accounts Nos. 1 and 2 respectively, which shall be shown separately on all returns to the Minister subject to the terms and conditions hereunder:
>
> "The No. 1 account shall be related to normal insured business exclusive of the insurance included in the No. 2 account.
>
> "The No. 2 account shall be related to business done on the basis of special Ministerial or Government decision.
>
> "The Company shall provide insurance for the No. 2 account only where specified in writing by the Minister"

In addition to the provision of insurance ICI was authorised to issue guarantees on behalf of the Minister to Associated Banks, the Non-Associated Banks and the Industrial Credit Corporation. Depending on whether the transaction was a short-term or a medium-term transaction, the Schemes under which ICI issued these guarantees were known as:—

> The Short-Term Finance Scheme ("STFS")
> The Medium-Term Finance Scheme ("MFTS")

With regard to the Short-Term Finance Scheme section 21 of the agreement provided that:—

> "The Company may issue guarantees on behalf of the Minister under the Short-Term Finance Scheme to the Associated Banks, the Non-Associated Banks, the Industrial Credit Corporation and such other Financial Institutions as may be agreed from time to time, subject to the conditions specified hereunder:—
>
> (*a*) the person or company requesting such a guarantee shall have a current and valid Export Credit Insurance policy.
>
> (*b*) payment shall be provided for in each case by an eligible instrument which shall be either a Bill of Exchange drawn by the exporter on a buyer, or a Promissory Note issued by the buyer.
>
> (*c*) a premium in addition to that charged for the Export Credit Insurance shall be charged to the policy holder for the guarantee. The premium level shall be as defined in the Guidelines."

With regard to the Medium-Term Finance Scheme Section 22 of the agreement provides that:—

> "The Company may issue guarantees on behalf of the Minister to the Associated Banks, the Non-Associated Banks and the Industrial Credit Corporation and such other Financial Institutions as may be agreed from time to time, in respect of the export of capital goods, under the Medium-Term Finance Scheme subject to the following;
>
> (*a*) the person or company requesting such a guarantee shall have a current and valid Export Credit Insurance policy.

(*b*) payment shall be provided for in each case by an eligible instrument which shall be either a Bill of Exchange drawn by the exporter and accepted by the buyer or a Promissory Note issued by the buyer.

(*c*) a premium in addition to that for the Export Credit Insurance shall be charged to the policy holder for the guarantee. The premium level shall be as defined in the Guidelines

(*d*) before a guarantee is given by the company under the Medium-Term Finance Scheme which normally involves an element of State subsidy, the approval of the Minister must be obtained as laid down in the Guidelines."

Short-Term Finance Scheme (STFS)

The Short-Term Finance Scheme is available by virtue of the terms of this agreement for transactions where the exporter in addition to holding Export Credit Insurance policy is dealing by way of Bill of Exchange or Promissory Note with a buyer in good standing.

The first step for ICI is to issue to the bank an unconditional guarantee of the exporter's debts. The exporter can then obtain, from a participating bank, up to 90% of the face value of the bills or notes subject to a revolving limit which will have been agreed between the discounting bank and ICI. Thus, finance can be made available to the exporter from the date the goods are shipped and his working capital requirements will be eased accordingly.

On the maturity date of the bills or notes when payment is received, the participating bank will retain the amount advanced to the importer plus interest on the amount from the date of his original advance. Any balance remaining will be paid by the bank to the exporter.

If the foreign buyer defaults, the bank will nonetheless be paid in full by ICI under the unconditional guarantee 6 months after the due date of the bills or notes. In certain circumstances, ICI will be entitled to a repayment of monies paid by it to the participating bank. This would arise, for example, where the exporter had failed to comply with the terms of the Export Credit Insurance policy and, accordingly, a claim would not arise under the policy. However, by virtue of the stand alone bank guarantee, ICI would have to pay 90% of the face value of the bills or notes to the participating bank and would then seek recourse from the Irish exporter.

Medium-Term Finance Scheme (MTFS)

Where large once off type contracts are insured under a "specific contract" Export Credit Insurance policy finance is also available through provision of a medium-term guarantee to a participating bank covering the outstanding amount. The officially supported terms which may be offered for the export of goods under such contracts are governed by the International Arrangement on Guidelines for Officially Supported Export Credits (hereinafter called "the Consensus") which is administered by the OECD. The usual credit terms are 2 — 5 years. A minimum 15% down payment is required and cover is provided for 90% of the outstanding amount. Payment on the outstanding balance must be made on regular and equal instalments not less frequently than every 6 months. The

Consensus also set minimum interest rates which may be offered to the buyer of an officially supported contract.

Export Credit Insurance

The availability of an Export Credit Insurance Policy was a necessary prerequisite to obtaining the benefits conferred by the Short-Term Finance Scheme and the Medium-Term Finance Scheme.

Consequently the benefits deriving from the Export Credit Insurance Policy are manifest, namely, indemnity against the agreed percentage of loss due to non payment and payment of up to 90% of the face value of Bills of Exchange or Promissory Notes from the dates from which the goods are shipped, thereby lessening demands on working capital.

In addition, and quite independent of the availability of Export Credit Insurance in the case of exports to countries such as Iraq to which the export refund scheme applied subsidies by way of export refund were payable when the product exported came under customs control.

The allegations made in Dáil Eireann with regard to the operation of the scheme relate to the alleged reintroduction of the scheme in 1987 and to the allocation of Export Credit Insurance for beef exports to Iraq during the years 1987 and 1988, which allocations were made, as provided for in the agreement, on the basis of "Special Ministerial or Government decisions" and were specified in writing by the Minister and related to number 2 account business to which normal commercial considerations did not apply and which was operated by the Minister for Industry and Commerce in "the national interest".

The creation and operation of this account did not begin in 1987 and it is clear from the documents made available to the Tribunal that, certainly from April 1983, the provision of Export Credit Insurance in respect of exports to Iraq could not be justified and was not sought to be justified by the ICI on purely commercial grounds but was at all times regarded as "a national interest" case, the provision for such insurance to be a matter for consideration and an authorisation by the Minister for Industry and Commerce.

In order to understand the events of 1987 and 1988 it is necessary to outline briefly the situation which existed prior thereto with regard to trade with Iraq and the allocations of Export Credit Insurance in regard thereto.

It appears from a detailed Memorandum for Government dated the 16th day of December 1982 that;

"(1) Irish exports to Iraq had expanded dramatically both in range and value over the previous two years;

(2) That whereas the total exports from Ireland to Iraq in the year ended December 1981 reached £21.4 million, it was estimated that total exports reached £50.6 million from the period January — October 1982;

(3) In October, 1981 an Ireland/Iraq Co-operation Agreement had been signed in Baghdad which had committed both countries to joint efforts to increase and strengthen their relations in the economic, commercial, scientific and technological fields. Export Credit Insurance plays an important role in export trading particularly in the conditions of financial insecurity and political instability that are characteristic of certain international markets;

(4) An application for Export Credit Insurance had been received from Anglo-Irish Meat Company Limited in respect of a $27.210 million contract (IR£20 million) with the State Enterprise for Agricultural Products Trading, Baghdad, Iraq;

(5) The purpose of the request for Export Credit Insurance was to insure the Irish exporter against non-payment by the Iraqi State Company due to either commercial or political default. The policy also enabled the exporter to obtain finance on shipment from their Bank by the assignment of their Export Credit Insurance Policy to the Bank. The Anglo-Irish Meat Company required Export Credit Insurance during that year as their Banker's, Allied Irish Banks Limited were unable because of the unstable situation in Iraq to obtain confirmation on an Iraqi letter of credit. During the previous year Export Credit Insurance was not required by Anglo-Irish Meat Company Limited because the Company's payment was made in cash on shipment;

(6) Anglo-Irish Meats Limited had established a long trading relationship with buyers in Iraq. In 1981 they fulfilled a contract valued at $20 million with the State Enterprise for Agricultural Products Trading.

(7) The request for Export Credit Insurance from Anglo-Irish Meats Limited was a direct consequence of extended credit terms being offered by their competitor's to Iraqi buyers and in order to remain competitive in the Iraqi market Anglo-Irish Meats Limited must be prepared to offer competitive extended terms;

(8) The support offered to Irish exporters by their Export Credit Insurance Policy was of crucial importance in reducing the risk element as well as facilitating the very necessary export financing."

This Memorandum for Government had been prepared because a request for Export Credit Insurance had been made to the said Department in November-December 1982 in respect of a hoped for contract for the supply of beef for $27,210,000 to the State Enterprise for Agriculture Products Trading.

On the 21st day of December 1982 the Government agreed that an Export Credit Standard Shipments Policy should be granted to the Company on the conditions outlined in paragraphs 14 (i) and 14(iii) and paragraph 15 of the said Memorandum and such decision was communicated to the Department by the Secretary to the Government.

On the 3rd day of February 1983 Anglo-Irish Meats informed the Department of Trade, Commerce and Tourism that the contract that they had hoped to obtain had been placed with South American competitors and no cover was taken up.

The question of the provision of Export Credit Insurance for exports to Iraq was kept under constant review by the Government and the Minister for Trade, Commerce and Tourism.

On the 17th day of June 1983 the Government decided

"(1) that Export Credit Insurance should be provided to Irish Exporters dealing with Iraq on short-term business on the conditions stipulated in paragraph 8.3 of the Memorandum, subject to the consent of the Minister for Finance being required for the provision of Export Credit Insurance for any proposal which would cause the total value of business with Iraq covered by Export Credit Insurance to exceed £12,000,000;"

By letter dated the 15th day of December 1983 the Minister for Finance conveyed sanction for the limit of the total value of the short term business with Iraq covered by Export Credit Insurance to be increased from £12 million to £15 million, such insurance to be provided in accordance with the terms of the Government decision of the 17th June 1983 (excluding the limit of £12 million set in that decision).

On the 10th day of September 1984, the Minister for Finance sanctioned an increase from £15 million to £25 million in the limit for short term business with Iraq covered by Export Credit Insurance on the strict conditions that the limit will be introduced on a phased basis over the next six months and that, in the event of any default by the Iraqis during that period, the limit at the time would be frozen.

On the 27th day of May 1985 the Minister for Finance sanctioned the increase in the cover for Export Credit Insurance provided for the Iraqi market from £25 million to £35 million.

On the 6th day of February 1986 the Government:—

"(1) approved special Export Credit Insurance arrangements under strict and specific provisions, to give cover for exports which may not be acceptable under the normal scheme but which nevertheless have an assumable risk:

(2) agreed, in relation to any claims arising under the Special Scheme:

(*a*) that they will be borne by the Exchequer i.e. other than on the normal Export Credit Insurance Scheme or out of the existing resources of Departmental votes, and

(*b*) that the Minister for Finance will decide, in consultation with the Minister concerned and the Minister for Industry, Trade, Commerce and Tourism how they are ultimately met;

(3) agreed that there will be separate accounting arrangements for the normal scheme and the special scheme — those arrangements for the special scheme will be generally on the lines of that for the normal scheme with particular reference to the need to avoid over-dependence by any company on a single market;

(4) agreed to increase the limit to £70,000,000 of cover offered in special arrangements in respect of Iraq".

At the time of the aforesaid increase in the amount of cover available in respect of exports to Iraq the total exposure was as follows:—

£2.5 million when the amount was increased to £12 million.
£4.700,297 million when the amount was increased to £15 million.
£14.099 million when the amount was increased to £25 million.
Almost £25 million when the amount was increased to £35 million.
£33.197 million when the amount was increased to £70m.

While the total Export Credit Insurance Exposure in Iraq on the 7th February, 1986 was £33,197,725, only £10,480,487 thereof related to beef exports, leaving an exposure of £22,717,238 in respect of other manufactured goods and services. The exposure in relation to beef exports was as follows:—

Dantean International (Hibernia) £7,959,479 Nenagh Chilled Meats (a Goodman Company) £2,521,008.

At that time the exposure in respect of beef exports was less than one third of the total exposure in respect of Export Credit Insurance allocated in respect of exports to Iraq and was approximately 3.5% of the statutory limit of £300m.

Between March 1983 and March 1987 the majority of applications for Export Credit Insurance in respect of exports to Iraq related to non-beef produce.

However a number of enquiries concerning the availability of Export Credit Insurance of beef exports and applications therefor were made during this period.

These were made by or on behalf of Anglo-Irish Meats and Hibernia Meats (Dantean International) who appeared to be the only Irish Companies exporting beef to Iraq during this period. These consisted of:

"(1) An enquiry made on behalf of Anglo-Irish Meats in March 1983 with regard to a tender for a contract worth $30 million with the State Enterprise for Agriculture Products Trading, Baghdad.

(2) An enquiry made on behalf of Hibernia Meats in June 1984 in regard to the availability of Export Credit Insurance in respect of a potential contract worth $6.7 million.

(3) An enquiry made on behalf of Anglo-Irish Meats with regard to the availability of Export Credit Insurance in respect of contracts worth between £7.5 million and £10 million which they were hopeful of obtaining.

(4) A request made on behalf of Hibernia Meats in November 1984 for Export Credit Insurance cover in respect of a contract of £5 million.

(5) A request made on behalf of Hibernia Meats in May 1985 for Export Credit Insurance in respect of a contract for £6.6 million.

(6) An enquiry on behalf of Hibernia Meats Limited on the 5th day of June 1985 with regard to the availability of Export Credit Insurance with regard to a contract for about $11 million which they were in the position to obtain if insurance cover was available.

(7) An enquiry on behalf of Nenagh Chilled Meats (a company in the Goodman Group) in June 1985 with regard to the possibility of obtaining Export Credit Insurance for a $18 million contract in Iraq.

(8) An enquiry made on the 3rd September 1985 on behalf of Anglo-Irish Meats with regard to the availability of Export Credit Insurance in respect of an $18 million contract with Iraq.

(9) An enquiry made on behalf of Hibernia Meats on the 5th day of February 1986 with regard to the availability of Export Credit Insurance in respect of the $15 million contract that the company was confident of negotiating.

(10) A request for cover from AIBP of contract value of $29 million which the company was in a immediate position to conclude subject to the availability of Export Credit Insurance.

(11) An application on behalf of Hibernia Meats on the 6th May 1986 for £2 million cover and subsequently on the 12th of May 1986 an application for Export Credit Insurance in respect of a £15 million contract he was tendering for at that time.

(12) An application made on the 30th May 1986 on behalf of AIM requesting Export Credit Insurance cover in respect of $32 million contract that had been signed.

(13) A request made on the 28th July 1986 on behalf of AIM for cover for a IR£34 million contract in Iraq which was expected to be signed within a few weeks and which contract was subsequently signed

(14) A request made on the 24th October 1986 on behalf of AIBP for Export Credit Insurance in respect of a contract for the supply of beef valued $28 million that AIBP had been asked to supply to Iraq."

It was not possible to grant Export Credit Insurance in the amount sought by the applicants therefor.

The Insurance Corporation of Ireland was, by virtue of its agreement with the Minister, obliged to endeavour to ensure that, taking one year with the other the operation of the Export Credit Insurance Scheme would involve no net loss to State funds and to ensure that the maximum potential liability of the Minister in respect of Export Credit Insurance Policies should not at anytime exceed the sum provided by the Insurance Acts.

Prior to April 1983 the Scheme was operated by the Insurance Corporation of Ireland on a strictly commercial basis and in accordance with these obligations and effectively without reference to the Minister of the Department.

By letter dated the 12th day of April 1983, the Department informed the Insurance Corporation of Ireland that:

> "Because of the deteriorating economic situation in Iraq I feel it would now be advisable that you consult the Department on all limit applications for this market in excess of £250,000 prior to the issue of a limit approval"

In the course of a telex dated the 27th day of April 1983 forwarded to the Department the Insurance Corporation of Ireland reviewed the situation in regard to Iraq and recommended that "we completely suspend all cover for Iraq immediately except where there is a guarantee of payment issued outside that country in the form of a bank guarantee or a confirmed irrevocable letter of credit.

The Minister then submitted to the Government a Memorandum for Government dealing with "Conditions Applying to the Provision of Export Credit Insurance in respect of Trade with Iraq" and dated the 15th day of June 1983.

In this Memorandum the Minister sought a decision from the Government in favour of providing Export Credit Insurance to Irish exporters dealing with Iraq, in spite of the recommendation contained in the Telex dated the 27th April 1983 from the ICI.

The Memorandum set forth in detail the reasons for such recommendation and referred, inter alia to:—

(*a*) in the official policy of encouraging Irish export to diversify away from traditional (mainly European) markets into new markets and stated;

> "many of these new markets are, however, volatile and very often there must be a long lead time (more simply regarded as a medium to long term investment period) before the credibility of exporters is established and real returns, in the form of substantial export orders, are received. Indeed, it is often an inevitable event that, in the course of such market development, some initial losses are incurred"

(*b*) the increasingly important role of Export Credit Insurance in export trading, particularly in financial insecurity and political instability that currently characterise certain international markets;

(*c*) the fact that Ireland's expanding trade with Iraq could be irreparably damaged and its credibility as a friendly and serious trading partner considerably tarnished if, having sought at every opportunity to maximise Iraqi goodwill towards Irish exporters with a view to obtaining optimum export orders, it would now seem to prevaricate in the light of Iraq's current economic difficulties, and to the fact that "there would also be significant scope for the charge that, having encouraged Irish exporters to deal in the market, the Government was now breaking good faith with the same exporters by withdrawing what is, in effect, a vital support in realising export orders."

This Memorandum incorporating not only the views of the Department of Trade, Commerce and Tourism but also of the Department of Foreign Affairs and the Department of Finance led to the Government decision of the 17th day of June 1983 already referred to and led to a more flexible approach in the allocation of Export Credit Insurance in respect of exports to Iraq which would no longer be made by the Insurance Corporation of Ireland on a purely commercial basis but in the case of all cover in excess of £250,000 would be made as result of a decision by the Minister for Trade, Commerce and Tourism in the national interest.

The Minister was obliged to ensure that, save with the consent of the Minister for Finance, no proposal which would cause the total value of business with Iraq concerned by Export Credit Insurance to exceed the sum of £12,000,000 which sum was increased from time to time as set forth above and to £70,000,000 on the 6th day of February, 1986.

Having regard to the extent of cover sought in respect of beef exports to Iraq this meant that all decisions with regard to the allocation of Export Credit Insurance in respect of beef exports to Iraq were made by the Minister who was at all times obliged to ensure that the total value of business with Iraq covered by Export Credit Insurance did not exceed the agreed amount, which agreed amount covered not only beef exports but all exports to Iraq.

Because of such constraints the Minister was unable to accede to or to accede in toto to the requests hereinbefore referred to during this period viz March 1983 to March 1987. The only Export Credit Insurance granted in respect beef exports to Iraq were:—

Dantean International (Hibernia) £7,959,479
Nenagh Chilled Meats (a Goodman company) £2,521,008

The availability of such insurance was regarded by these Companies as of considerable importance and both companies made regular representations to the Minister and other members of the Government for an increase in the amount available for such insurance, increases in the level of cover and the period of cover.

These representations were made by letters to and meetings with the Taoiseach, Garret Fitzgerald TD and individual members of the Government, including the Minister for Finance Alan Dukes TD, the Minister for Industry, Trade, Commerce and Tourism John Bruton TD and the Minister for Industry and Commerce, Michael Noonan TD

These contacts between beef exporters and members of the Government arose because of the realisation by beef exporters that as a result of the Government decision of the 17th day of June 1983 all decisions with regard to the allocation of Export Credit Insurance in respect of the export of beef to Iraq were to be made by the Minister for Industry, Trade, Commerce and Tourism (subsequently the Minister for Industry and Commerce) and not by the Insurance Corporation of Ireland, who were only concerned with the administrative details of writing the policy, and the decision to allocate and the terms of the allocation were made by the Minister.

While the Government and the relevant Minister were anxious to encourage the development of trade, including the export of beef to Iraq and had, as hereinbefore set forth, increased the amount available for such cover from £12m to £70m, and up to February 1986 had approved cover which led to an exposure in respect thereof of £33,197,725, by May 1986 concern was being expressed in the Department of Industry and Commerce concerning Iraq's capacity to pay the debts.

At that time applications for Export Credit Insurance had been received from Anglo Irish Meats and Hibernia Meat Packers and the Minister for Industry and Commerce contemplated granting cover in the sum of £15m, being as to £10m to Anglo Irish Meats and as to £5m to Hibernia Meat Packers.

A Departmental Conference was held on the 15th May 1986 for which Mr Fahy of the Export Credit Section prepared a note in the course of which he recommended that the ceiling then existing be frozen and no further requests for cover be entertained until the position regarding the re-scheduling of Iraqi debts and the future Iraqi repayment record became clearer.

It appears from the Departmental note of this meeting that the Minister agreed that

"(*a*) in regard to the two beef contracts we should neither approve nor reject these applications but stay our hand;

(*b*) we should effectively close the account on Iraq except in cases of extreme hardship;

(*c*) as regards the £5m insurance committed but not yet taken up by exporters we should honour this business;"

On the 27th day of May 1986, Brian Britton, the Financial Director, Anglo Irish Meats, rang the Department of Industry and Commerce and informed them that his company had secured the contract for £29m and inquired about Export Credit Insurance. He was informed that no decision was being made as the situation in Iraq was under review.

Laurence Goodman then wrote to the Taoiseach on the 28th May 1986 and on the 30th day of May 1986 to the Minister for Industry and Commerce, enclosing a copy of his letter to the Taoiseach.

An Aide Memoire was then prepared for and submitted to the Government but it was decided that the Government would consider the position, particularly in regard to Iraq on the basis of a factual Memorandum from the Department of Industry and Commerce.

This Memorandum contained the statement that:—

"12. Two large Irish Beef exporters to Iraq, Anglo Irish Meats and Hibernia Meats are pressing for further cover in respect of new contracts. The Minister for Industry and Commerce has grave reservations about giving cover..."

In the introduction to the said Memorandum the Minister stated:

> "on the basis of his view of the Iraqi market, he is not prepared to recommend any further export credit cover to Iraq unless or until the Iraqi debt repayment position has been clarified to his satisfaction".

The contents of the said Memorandum were noted by the government on the 16th July 1986.

On the 28th July 1986, the Minister for Industry and Commerce Michael Noonan TD, the Secretary to the Department of Industry and Commerce and Mr Fergus Walsh of the Export Credit Section of the Department met Mr Lawrence Goodman and Mr Brian Britton of Anglo Irish Meats.

At this meeting the question of Export Credit Insurance was discussed and Mr Goodman sought Export Credit Insurance in respect of a contract for the supply of beef to Iraq worth £34 million which he hoped to secure but the Minister was not prepared to offer any further Export Credit Insurance for Iraq until such time as the situation was seen to be sufficiently improved to enable such cover to be put in place.

Despite subsequent and further representations by Mr Britton no cover was granted.

On the 24th day of October 1986, Mr Goodman wrote to the Minister for Industry and Commerce in connection with Export Credit Insurance generally and in particular with regard to a new contract being offered to his Company for the supply of 13,670 tons of boneless beef and requesting a meeting with the Minister, which meeting took place on the 4th November, 1986.

On the 6th day of November a meeting was held between the Secretary and Mr Fergus Walsh of the Department and Mr Brian Britton of Anglo Irish Meats with regard to the provision of Export Credit Insurance in respect of this contract for £34m.

Subsequent to this meeting consideration was given by the officials of the Department of Industry and Commerce to the question of granting a small amount of Export Credit Insurance in respect of exports to Iraq and a Memorandum for Government was drafted.

However on the 12th day of February 1987, the Minister for Industry and Commerce, acting on the advice of Mr Liam Kilroy decided that no further cover should be granted in respect of exports to Iraq and this advice appears to have been based on the general deteriorating financial and military situation in Iraq and that there were a number of overdue payments from Iraq and the submission of the Memorandum to Government was not proceeded with.

Mr Britton continued to make contacts with the Departments and on the 25th February 1987 wrote to the Department in the following terms:—

> "Re: *Trade with Iraq*
>
> Dear Mr Donlon
>
> Further to meetings with the Minister for Industry and Commerce by Larry Goodman during 1986 and subsequent correspondence, our Group were asked to keep

you fully informed on progress relating to contracts of boneless beef which we are supplying to Iraq. You may recall that we expressed grave concern about the problems being experienced by our Group in obtaining Export Credit Insurance for trade with Iraq.

I now understand from members of your Department that a fresh look is being taken on Export Credit Insurance for Iraq. If that is the situation, could I take this opportunity to remind you that the Minister indicated that he would look at our Group's application favourably if the department were to restart allocating Iraqi Export Credit Cover.

Notwithstanding the lack of support from your department and various financial institutions, our Group are currently in the process of supplying our 1986 contract for 13,670 tonnes of boneless beef. We have committed ourselves to this contract in order to maintain a market opened up by us in the early 1980's which market has established a premium for Irish beef. All of the product being supplied is being processed and prepared in our own Group factories, giving substantial employment in this country.

Our excellent relationship with the Iraqis has been proved by the receipt this week of funds clearing our 1985 contract up to date. This has enabled the release of the $3m allocation which your department gave to that $18m contract.

Within our own industry, not alone are we the only Irish meat processor shipping to Iraq, but we are also the only Irish meat processor with a signed contract for that country. We believe that recognition deserves strong support from your department.

In regard to our current contract for $28m, we are more than half way through our shipment schedule, having shipped 7,060 tonnes to date out of 13,670 tonnes. We believe that we have shown the commitment requested by both the Minister for Industry and Commerce and the Minister for Agriculture last autumn, which has been of valuable assistance to the Irish economy. In return, we are now formally requesting your Department for the substantial support on Iraqi risk, promised should the Department's position on Iraq change.

I am meeting Liam Kilroy and Eugene Forde from your department on Tuesday next, 3 March and look forward to a favourable response from them to this request."

It is important to note the statement made in this letter that:—

"All of the product being supplied is being processed and prepared in our own Group factories, giving substantial employment in this country."

Following receipt of this letter a meeting was held in the Department between Mr Ted O'Reilly, Assistant Secretary, Mr Liam Kilroy and Mr Eugene Forde of the Department and Mr Brian Britton of Goodman International, during the course of which the Department's then policy was repeated viz that while the question of Export Credit Insurance was being kept under continuous review, the Minister had decided that no further cover would be given in the then existing circumstances.

This position was confirmed in writing by the Department's reply dated the 5th March 1987 to the letter dated the 25th day of February 1987 which was in the following terms:—

"Dear Mr Britton

"Thank you for your letter (Ref. BB/GB) of 25 February, 1987 to the Secretary about the question of export credit insurance cover for beef supplies to Iraq. As noted in your letter and as confirmed to you here on Tuesday 3rd March the question of further credit to Iraq is the subject of continuous review in this Department. I also confirm what I told you to the effect that arising from a very recent review the Minister has decided that no further cover be given to any exports to Iraq.

"In taking this decision the Minister, of course, recognises the problems for Goodman International Ltd and, indeed, various companies in other sectors which have sought cover.

"Notwithstanding some repayments, such as the one to which you refer, the Minister is satisfied that the totality of difficulties facing the Iraqis, reflected in the non-payment of significant amounts currently due to Irish exporters, are such as to preclude him from extending further cover. The position will be kept under review and reconsidered at appropriate intervals. Such reconsideration will, of course, take due account of your interests and those of other exporters.

"I would have to comment that there are parts of your letter which I find distasteful and consider to be rather less than helpful. In the interests of accuracy I should point out that the Department's records of the Minister's comments to your request for cover during 1986 do not refer to his indicating a promise of favourable treatment for your application such as is suggested in your letter. Rather was the Minister indicting his desire to assist Goodman International, and other interested exporters as much as possible with their export efforts; always referring to the extreme difficulties which lay within Iraq itself. As you know these difficulties have been exacerbated in the last few months.

"I regret very much that you find it necessary to commit to paper a charge of "lack of support" from this Department in view of our efforts to support all exporters in what is recognised to be a financially squeezed market. I can only hope that you found Tuesday's meeting of some assistance to you and, if it was otherwise, I invite you to come in again for further discussions.

"For the record the application from Goodman International Ltd is not the only one we have received from Irish meat exporters for cover in Iraq."

As of this point of time viz 5/3/1987, no further cover was being granted in respect of exports to Iraq because, as stated in the said letter, the Minister was satisfied "that the totality of difficulties facing the Iraqis, reflected in the non-payment of significant amounts currently due to Irish exporters are such as to preclude him from extending further cover."

At that stage, the outstanding liability of the Minister in respect of exports to Iraq was £30m.

In an internal note made on the 8th day of March 1987, Mr Britton noted the receipt of this letter and stated that he intended waiting until the new Minister was appointed before taking further action.

On receipt of this note Mr Larry Goodman directed Mr Britton:

> "to keep in touch, keep Goodman International ahead of other exporters, find a way to achieve this."

At this point of time Goodman International were in the process of fulfilling the contract referred to in Mr Britton's letter to the Department and which they had undertaken without the benefit of Export Credit Insurance.

However the availability of Export Credit Insurance for this market viz Iraq was an important consideration not only to the Goodman Group but to all potential exporters of products to that country.

There is no doubt but that Goodman International had over the years expended considerable effort in establishing a foothold in and developing a market for the export of beef to the Iraqi market and had been successful therein and in developing relationships with the Iraqis.

On the 7th April 1987 AIBP received a telex from The Iraqi Company for Agricultural Products Marketing (ICAPM) one of the State bodies in Iraq authorised to purchase beef, inviting them to participate in their tender 2/87 for the purchase of beef which had a closing date on the 9/4/1987.

The tender described Mr Goodman as the "famous friend of ICAPM"

Anglo Irish Beef Packers (INT) Ltd replied with a tender offer for 2/1987 and as required placed a bid bond in the sum of 1,200,000 Deutschmarks with the Rafidain Bank.

By telex dated the 14th April 1987, ICAPM requested AIBP to send their representatives to Baghdad for negotiation. AIBP confirmed that their representatives would be in Baghdad by the 25th April 1987 and the offer to tender was then extended until the 7th day of May 1987.

Consequently there was at that time a real probability of the successful negotiation of substantial contracts for 1987.

On the 11th day of March 1987, there had been a change of Government and Albert Reynolds TD was appointed Minister for Industry and Commerce by the new Taoiseach, Charles J. Haughey TD.

In the course of the chapter of this Report dealing with the IDA, the Tribunal will deal with a meeting between the Taoiseach and representatives of the Government, including the Minister for Industry and Commerce, and Messrs Goodman and Britton held on the 9th day of April 1987 with regard to a proposed development plan for the Beef Industry.

The question of Export Credit Insurance for beef exports to Iraq did not form part of that discussion but when the business of that meeting was concluded, Mr Goodman availed of the opportunity to discuss it with the Minister for Industry and Commerce.

This was a short, general discussion and the Minister agreed to look into the matter but did not give any commitment to re-introduce it or to make it available to Goodman International.

On the 13th of April 1987 a *"Note for Minister's Information on Export Credit Insurance for Iraq"* was prepared and a copy given to the Minister, Mr Reynolds TD.

This Note for Minister's Information referred to the outstanding liability of IR£30m then existing at the time of the previous Minister's decision made in February, 1987 and to the payment of $1,994,685 to Dantean International Ltd (Hibernia Meats) and US$4m to Nenagh Chilled Meats Ltd (Anglo Irish) subsequent to the making of that decision.

It pointed out that applications for Export Credit Insurance then current amounted to approximately IR£30m and included an application from Anglo Irish Meats for IR£20m and from Dantean Meats for $2m, which applications had been lodged with the Department in 1986.

The note concluded with the following recommendations:—

> "Despite the recent payments by the Iraqis and the reduction in the States exposure to IR£24m, the amounts overdue i.e. IR£5m and the amounts due to arise for payment before end June, 1987 i.e. IR£2.5m are considerable. It is recommended therefore that the current policy on Iraq i.e. no further cover, should be continued for the present. Developments on payments from Iraq will be monitored closely. If the payment record improves substantially the question of further cover can be reconsidered again".

This was, in effect, a reiteration of the views expressed to and recommendations made to the previous Minister Michael Noonan TD.

The Minister for Industry and Commerce directed his officials to meet with representatives of Goodman International to discuss with them the facilities which he was contemplating making available to them under the Export Credit Insurance Scheme for the export of beef to Iraq.

His approach to the re-introduction of the Scheme was extremely cautious and he contemplated the giving of cover in respect of $\frac{1}{3}$ of the amount of the contract with the company carrying $\frac{1}{3}$ and its bankers $\frac{1}{3}$.

On the 16th of April, 1987 a meeting was held between Mr T. J. O'Reilly and Mr Fergus Walsh of the Department of Industry and Commerce and Mr Brian Britton representing Anglo Irish Meats.

During the meeting Ted O'Reilly informed Brian Britton:—

> "*a*) that the Minister had decided that Anglo-Irish Meats would be given some facilities in the Export Credit Insurance Scheme for trade in beef with Iraq.
>
> *b*) that the Minister had directed a meeting would take place to explore the present situation.

c) that the Minister was prepared to consider facilitating Anglo-Irish for one third of their contract value.

d) that the Minister thought that the Department should be prepared to speak to Anglo-Irish bankers to assume a further one third on the exposure of Anglo-Irish's contract.

e) that following his Minister's direction he was anxious to see to what extent the Department could be of help.

f) requested whether Anglo-Irish exported live cattle to any country abroad.

g) requested whether Anglo-Irish Meats imported cattle or meat with a view to subsequently exporting such produce."

Mr Britton's response to this was:

"*a*) Anglo-Irish did not export live cattle to any destination;

b) their export business was solely meat;"

That the only circumstance in which Anglo-Irish imported beef with a view to its subsequent export from Ireland were:—

"i) where the import of already prepared and dressed cuts of meat to make up an order for export which was incomplete; and

ii) the import of unprepared meat which was subsequently prepared and dressed for export;"

But he added that this formed a very small percentage of Anglo-Irish's total meat exports.

Brian Britton went on to outline the history of the company's trade with Iraq and in particular confirmed that apart from one payment of £1.324 million which was due on the 30th of April 1987 that all other sums due on the $18 million dollar contract shipped between September 1985 and April 1986 had been remitted to the company. He explained that the Export Credit Insurance Scheme had covered $4m of that contract and this exposure had been cleared. Mr O'Reilly accepted that this demonstrated a good relationship between the Goodman company and the Iraqis.

Mr Britton then indicated that the 1986/1987 contract was for US$28 million but that no Export Credit Insurance had been given by the Minister in respect of that contract and that Anglo-Irish's banks had been unwilling to carry any portion of the risk either. Despite this the company had gone ahead with the contract and carried the full risk themselves with shipments commencing in September 1986.

As of 16th April 1987 $22 million of the $28 million contract had been shipped, and it was anticipated that the shipping of the product due on foot of the contract would be completed within 4 weeks of the 16th of April 1987. He explained that payment was on one year irrevocable letter of credit and the payments therefore would become due in September 1987. The company was not seeking retrospective cover on the $22 million that was already shipped nor on the balance of the $6 million but he was seeking State support in respect of a new contract for which the company was now tendering.

This contract was for a value of between US$32 million and US$35 million with shipments commencing in September 1987. The company was not prepared to carry the full risk. He argued that full Export Credit Insurance should be allowed emphasising the following:-

"*a*) the company had an excellent payment record to date in Iraq with no overdue payments;

b) the company had shown its own faith in the market by carrying the full $28m exposure on the previous contract;

c) the company employed 1,500 people and that the new contract is important to maintain employment at that level;

d) the companies exports were of the order of 4% of Ireland's total exports and as such the company should be encouraged whenever possible;

e) the company's exports had helped the cattle industry in Ireland;

f) the company would be obliged to put up a bid-bond of 5% of the contract value when tendering and if the contract is secured a 10% performance bond would then have to be put in place by the company; the fact that the company were prepared to do this without seeking cover for either bond was a further indication of their confidence in the market as it constituted a further exposure on the company;

g) that in the country's interests the market should be maintained by the export of beef to Iraq. The French, who are also bidding and who had a considerable share in the market had the full support of the French Government."

In reply to these arguments Ted O'Reilly proposed;

"(i) that cover be made available on one-third of the new contract value estimated at $11 million US approximately;

(ii) that the balance of the cover be carried one-third by the company and one-third by Anglo-Irish Banks;

(iii) that the Department would exhort the companies' banks to assume one-third of the risk;

(iv) that the Department would not give any letter of guarantees, or letter of comfort to the banks in respect of any exposure that they took on board;"

Mr O'Reilly noted that the cover now being sought by the company would not be required until September of 1987 when shipments would commence.

This offer was tentatively welcomed by the company and this offer was to become a matter of further negotiations when an overall package was being put together.

It was made clear at this time by the representatives of the Minister that Goodman International should not give the impression to other exporters that the Department was open for Export Credit Insurance to Iraq.

The meeting concluded with the Department indicating that it would confirm its offer in writing. However, there were no further negotiations between the Department officials and representatives of Goodman International with regard to this proposal.

Following the meeting on the 16th of April 1987, the Minister discussed with his officials the question of providing Export Credit Insurance Cover for Iraq. Arising from the discussion and having assessed the information available to him the Minister directed that Export Credit Insurance Scheme was to be made available to all exporters for all exports to Iraq subject to the following conditions:—

a) that cover was to be made available up to the ceiling of £45 million;

b) that a pragmatic approach was to be adopted in the allocation of the cover and that companies which already had payments overdue from Iraq should be given no further cover until amounts overdue had been cleared;

c) that in the specific case of Anglo-Irish Meats we should offer to cover one-third of the proposed contract, with the banks and Anglo-Irish carrying one-third each also;

d) that proposals for providing cover to Iraq be devised on the above basis;

The ceiling of £45m included a sum of £10m which had been allocated to the PARC Hospital project.

At this time the Departments' exposure to Iraq was £14 million which meant that there was a further £21 million available. The proposed contract for Anglo-Irish was somewhere in the region of US$33 to US$35 million and by the Department's calculation it left approximately £11 million which could be made available to other applicants. (i.e., allowing £10 million for the Anglo-Irish contract).

From a Department minute bearing date the 30th of April 1987, prepared by Mr Kilroy at that time excluding Anglo-Irish Meats, and PARC (a hospital concern), there were applicants in respect of alleged contracts of £11.5 million approximately and allowing 75% cover meant a commitment of IR£8.5 million. These applicants were mainly non-beef exporters with one beef exporter Dantean Meats (part of the Hibernia Meats Group) seeking cover on a £2.4m. contract.

One of the directives of the Minister at this time was that cover was not to be given to exporters where amounts were currently overdue to the company. This particularly affected two companies, one of which was in the beef sector, Dantean International Ltd (part of the Hibernia Group) were overdue on £5.2 million.

Mr Kilroy, in his minute, recommended that the Department proceed to offer cover to the five "acceptable" applicants to a total sum of £3.4m. These five were not beef exporters and the other non-beef companies owed money. Mr Kilroy further recommended that Dantean be advised of the new situation and that they be encouraged to pursue recovery of overdue payments. He further recommended that a sum of £5m. cover be retained for a period of three months, so that it would be available to the two companies should the repayments happen. In all cases Mr Kilroy was proposing a maximum credit period of 12 months.

The Department in implementing the directive of the Minister:—

(*a*) undertook to provide Anglo Irish Meats with £10m. approximately in cover i.e., $\frac{1}{3}$ of the latest Anglo Irish contract for US$33/US$35m.

(*b*) make available to the five "approved" non-beef applicants £3.4m. in Export Credit Insurance and provide a sum of £2.5m. for other applicants as received under the Export Credit Insurance Cover Scheme; and;

(*c*) rather than shut the door on the two companies that had sought facilities under the Export Credit Insurance Scheme and for whom money was overdue decided that they would allow the facility to be taken up if the monies outstanding were repaid within three months.

The reasons given for taking up this position were:—

a) that Dantean Meats might otherwise claim that Anglo-Irish were getting unjustified favourable treatment; and;

b) that (non-beef company) has a long history of doing business with Iraq and the Department was anxious that this market share should not be lost.

In evidence Mr Reynolds said:—

> "The reintroduction of Export Credit is a decision within the statutory authority of the Minister for Industry and Commerce, and, I didn't discuss it with any of them before I reintroduced it. I carried out my own evaluation and I made my decision. I did not make Members of the Government aware that I had reintroduced it at that particular time. So far as informing other traders was concerned I view my position as Minister for Industry and Commerce as the person who makes a decision, and makes the policies, and the Department Officials carry out the policies and the decisions."

The Minister elaborated on his reasons in evidence as follows;

> "My reasoning for offering the restricted cover was that I was taking into account the advice given to me by the Department and indeed it was a question of trying to strike the right balance and in any situation that you are going into, new or uncharted waters, you put your toe in the water first and find out what it is like and you go on from there. There was a two-pronged approach to the Iraq market. One was for beef, 6 million and the other was for the smaller companies, clothing and food and pharmaceutical companies, because they needed support as well. It was restricted in that manner and lets see how they performed. You let the policy develop and change it along the way if changes were needed, that is always my view. So far as Hibernia was concerned, unlike AIBP, who had an outstanding track record, that never had a default in payments, they never cost money for the Department. Hibernia had money outstanding and I said, cover shouldn't be extended until a payments record improved in that area. It was one of the conditions I laid down when I reintroduced it and that was the position with Hibernia Meats."

In May 1987, Mr Oliver Murphy, Managing Director of Hibernia Meats Ltd met with Mr Ray Mac Sharry, the then Minister for Finance, and made representations to him with regard to Export Credit Insurance and this meeting was followed by a letter dated 19th May 1987, which was as follows:

> "**Re: Export Credit Insurance.**
>
> "Dear Minister,
>
> "Thank you for meeting me last Thursday at such short notice — I found our meeting very worthwhile.
>
> "As I mentioned I consider that a more comprehensive Export Credit Insurance Scheme would be attractive to meat exporters — and, indeed, to all Irish exporters. As you are aware the Scheme is currently operated by the Insurance Corporation of Ireland (ICI) on behalf of the Department of Industry and Commerce.
>
> "We are currently in negotiations with the Iraqi State purchasing organisation for an order worth 20 million US Dollars. The Iraqi authorities are currently seeking credit facilities for two years but I believe we can negotiate this to a single year's credit facility. I have been in virtual constant communication with the ICI and am informed that they are not in a position to offer facilities over and above their present exposure to Iraq. I understand any change in this position would require Government approval.
>
> "You are fully aware of recent reforms in the EEC Beef Regime and the likelihood of further undermining of the price support mechanisms. This contract which we are in a position to finalise with the Iraqi authorities would be of great benefit in supporting Irish producer prices later in the year and would considerably compensate for reduced supports from Brussels. The proposed contract with Iraq is for boneless beef so it also has the greatest added value and employment factors.
>
> "The current Export Insurance Scheme covers only 75% of the risk and this should be increased to cover 95% of the risk as it is not possible to get banking confirmation of a letter of credit even for 25% I think you will agree — considering the margin the meat industry operates on — that it is not reasonable or possible for a company to carry 25% of the exposure.
>
> "As the Iraqi authorities will be finalising their purchasing requirements in the course of the next four weeks it is very important that the position be reviewed without delay. Subject to the availability of credit insurance facilities I am confident we can successfully conclude this contract.
>
> "I can assure you that we have exhausted all commercial banking possibilities in order to execute this contract — however they are not prepared to co-operate unless they are covered by the scheme.
>
> "I would appreciate if you could arrange to have the current position relating to Iraq reviewed as a matter of urgency — also the question of cover for 95% of the risk. In the medium term the question of a more comprehensive scheme might be looked at

as I am satisfied that it would be of major assistance in increasing the volume of Irish exports."

A copy of this letter was forwarded to the Minister for Industry and Commerce on the 25th May 1987.

On the 10th day of June 1987 the Department of Industry and Commerce informed ICI as follows:—

"COVER FOR EXPORTS TO IRAQ

We have now received ministerial approval for provision of further Export Credit Insurance for Iraq. The following are the details of cover to be provided:—

1. Conditions to apply generally:—

   (A) All insurance for Iraq remains categorised as No. 2 a/c business.

   (B) No commission shall be payable to intermediaries for this business.

   (C) Full premium is payable before provision of Insurance/Finance.

   (D) All shipments to be secured by IL/C issued by Central Bank of Iraq.

   (E) Insurance to be restricted to political risk only.

   (F) Insurance and guarantees may not be revolved i.e. cover must relate to named shipments, total value not exceeding contract values specified below.

   (G) ICI may take 15% admin fee and 5% profit margin on this business based on one quarter of the premium rate charged in each case.

2. Cover may be offered to the following companies on the following terms:—

   (A) 70% Cover

   (B) Maximum credit period of 1 year

   (C) Claims waiting period of 1 year

   (D) Premium of 4%

Company Name	Contract Value	IR£ Exch Rate (Approx)	IR£ Value	70% Cover
Wyeth Ltd	USD 4M	1.48	IR£2.7M	IR£1.89M
Glenabbey Ltd	Stg£67,000	0.90	IR£74,500	IR£52,150
Johnson Wellscreens	Stg£29,000	0.90	IR£32,500	IR£22,750
Medisco Ltd	Stg£18,000	0.90	IR£20,000	IR£14,000
Manford Clothing	USD1.2M	0.90	IR£0.81M	IR£0.567M
Wilson & McBroinn	Stg2M	0.90	IR£2.23M	IR£1.561M
Lastra	IR£1M		IR£1M	IR£0.7M
CS Laird	Stg£12,000	0.90	IR£13,500	IR£9,450
Antigen	Stg£350,000	0.90	IR£389,000	IR£272,300

3. Anglo Irish Meats have received a commitment to cover up to a maximum of IR£10 million. This will not be taken up until September 1987. Terms of cover have yet to be agreed. ICI take no action for present."

On the 17th day of June 1987, and by letter of that date, Hibernia Meats Ltd sought from the Insurance Corporation of Ireland cover for a two year period of contracts valued at $46m which the Company was negotiating with the Iraqi purchasing agencies.

By letter dated the 9th July 1987 the Minister for Industry and Commerce replied to the Minister for Finance as follows:—

"I refer again to your representations of 25 May, 1987 on behalf of Mr Oliver Murphy, M.D., Hibernia Meats Ltd about export credit insurance for Iraq and, in particular, the question of increasing the cover from 75% to 95%.

"The position is that Government Decision S.15005B of 17 June, 1983 specified cover at 75%. In a re-assessment of the situation which I did in May this year I decided that the risk was such that cover should be reduced from 75% to 70%. Moreover I also decided that cover was not to be given to any exporter who was not getting paid debts due from Iraq.

"An entity, Dantean International Ltd, which is part of the Hibernia Meats Group, have outstanding insured debts of US $6.8m due from Iraq. If these debts remain unpaid the Exchequer will be faced with claims of IR £3.5m. in the period August — December, 1987.

Mr Murphy's letter to you of 19 May, 1987 refers to an order of US $20m. but a later letter of 17 June, 1987 to ICI from Mr Murphy mentions an order of US $46m. While it is most probable that I could not consider cover for an order in that sum the best that I can do is to say that if Dantean were fully paid I would be prepared to look again at the matter. However, it must be clearly understood by all concerned that in the light of the decisions mentioned above there are no circumstance in which cover of 95% — the basic request of Mr Murphy to you — could be given.

I enclose for your personal information a note about the situation on Iraq."

The note referred to in the said letter was as follows:—

"The background to this matter is that in the years to the beginning of 1986, the overall limit on the amount of export credit which was provided for the Iraqi market was periodically increased. In February, 1987 the limit stood at IR £70m. However, in May 1986 the then Minister decided to suspend further provision of cover when it became clear that the fall in oil prices was severely affecting Iraq's ability to pay for its imports. Iraq was defaulting on debts owed, it was seeking to reschedule major debts which previously had already been rescheduled and payment delays were beginning to arise for Irish exporters. At that time, our level of exposure stood at £30m.

"Since then it was generally accepted that exposure for Iraq should be gradually reduced. The matter was reviewed in May this year when our total exposure stood at £12m. It was decided to re-open insurance cover for Iraq up to a maximum limit of £35m. Moreover, cover was to be extended only on very stringent terms — (i) premium considerably in excess of the norm, (ii) cover for a maximum of 70% of contract, (iii) credit period maximum of one year and (iv) a claims waiting period of 1 year. In addition, cover would only be provided to exporters who have a current record of getting paid in Iraq.

Existing liabilities and the new offers of cover now puts our exposure at £30m."

and represented the view of the Minister for Industry and Commerce with regard to the provision of Export Credit Insurance in respect of exports to Iraq at that time, which was that:—

"(i) Export Credit Insurance in respect of exports to Iraq should be subject to a maximum limit of £35m:

(ii) such cover was to be extended only on very stringent terms viz

(*a*) the premium should be considerably in excess of the norm

(*b*) the cover would be for a maximum of one year

(*c*) there should be a claims waiting period of 1 year, and

(*d*) cover would only be provided to exporters who had a current record of getting paid in Iraq."

The decisions in this regard had been taken by the Minister for Industry and Commerce in spite of the recommendation of the Department contained in the Memorandum dated the 13th day of April 1987 that the then "current policy on Iraq i.e. no further cover, should be continued for the present" but in view of the stringent conditions imposed by him, the Minister for Industry and Commerce obviously had regard to the recommendation hereinbefore referred to and considered that Export Credit Insurance in respect of beef exports to Iraq should be granted in a very limited and restricted basis.

While the discussions between Mr Ted O'Reilly and Mr Brian Britton held on the 16th April 1987 had related to a proposed contract for approximately US$32m./ US$35m. the situation, in Baghdad, was rapidly changing and as a result, the negotiations envisaged never took place.

At the invitation of the Iraqi buyers, Mr Laurence Goodman had returned to Baghdad to resume negotiations in connection with the contract 2/1987.

These resumed negotiations were successful and on the 6th day of May 1987 he telexed his Secretary with the following message:—

> "Pls. contact Brian with following message:—
>
> "He is to phone Albert Reynolds, at home if necessary, and before next cabinet meeting as we understand Gov Decision will be given on his proposal to open up and increase facility for Iraq. Advise A.R. that we will require a very substantial amount for here, i.e., if they are to give 50%. We will require 50m or if it is 33% we will require 33m advise A.R. that I will contact him immediately on my return. Suggest Brian diplomatically remind A. R. of discussion with me on restricting cover to us as the only continuous supplier to this market.
>
> "It is of critical importance that Brian contact A. R. today (Wed) re. above. A.R. telephone nos. are as follows:—
>
> (i) Dublin (Flat) ....
>
> (ii) Office (Longford) ....
>
> (iii) Home ...."

It is clear from the terms of this telex that Mr Goodman believed that the Minister for Industry & Commerce proposed to open up and increase Export Credit Insurance in respect of beef exports to Iraq and that he should, at the earliest possible opportunity, be advised that Goodman International would "require a very substantial amount" for Iraq.

It is clear also that he was at that time confident that the value of the contract which he had negotiated would be in the region of £100m because he stated that if the Government were to cover 50% of the contract, they would require £50m. or if they were to cover 33%, they would require £33m.

It is also clear that he had previously discussed with the Minister, the possibility of restricting cover to Goodman International "as the only continuous supplier to the market".

On receipt of this message Mr Britton endeavoured to contact the Minister for Industry and Commerce but due to his unavailability was unable to do so.

Mr Britton prepared and left a memo for Mr Goodman, on his return from Iraq, advising him of his failure to contact the Minister and stated:—

> "As he has been out of the country until now I obviously couldn't speak to him. Now that you are back in the country and because of the delicacy of the matter discussed

by you with him, regarding restricting Iraqi cover to us as the only supplier, I have left a message for Albert Reynolds to ring you at home over the weekend. When his Private Secretary asked me for your number I declined to give it to him because of confidentiality but intimated to him that I believed that Albert Reynolds already had it. If he cannot get you I have taken the precaution of also leaving my home number with his Secretary.

Contact me at home, if you would like to discuss this."

In spite of the urgency expressed in this exchange, neither the Minister for Industry & Commerce nor Mr Goodman have any recollection of any contact made or meeting held on the weekend subsequent to that message.

On the 13th day of May 1987, the Minister for Industry and Commerce, met with Mr Britton and there was a discussion with regard to various aspects of the I.D.A. Development Plan and further meetings were held with Mr Goodman and Mr Britton at which the same project was discussed on the 19th day of May 1987 and on the 21st day of May 1987.

While the main purpose of these meetings related to various aspects of the negotiations between the I.D.A. and Goodman International and difficulties arising out of the negotiations in progress, the Minister, while having no specific recollection of the details, stated that:—

"I have little doubt that any opportunity they got they would be bringing me up to date on how their negotiations were going on in Iraq because they are constantly over and back and you know, had a strong presence there and I have no doubt at any meeting they would be, I expect to be appraising me of the progress they were making in that particular area." (T.133A — Q. 99.)

It is most unlikely that having been successful in negotiating the largest contract for the export of beef from this State with a potential value of £100m that Mr Goodman would not have so informed the Minister and given particulars of the Export Credit Insurance required which was considerably in excess of that previously indicated.

When Mr Goodman had left Baghdad in May 1987 he believed that he had concluded a contract with ICAPM for the supply of 30,000 tonnes of boneless beef together with the possibility of a further 10,000 tonnes.

However before the contract was actually signed, a change of circumstances had occurred in Baghdad. A substantial part of the responsibilities of ICAPM, with whom Mr Goodman had been engaged in negotiations, had been transferred to another Iraqi State Company entitled The State Organisation for Grain and Foodstuff Trading (SOGFT).

As recorded by Ambassador McCabe.

"In mid June when Aidan Connor arrived to conduct the final stage of negotiations with ICAPM it became apparent that a substantial part of that had been transferred to the State Organisation for Grain and Foodstuff Trading, and that he was going to have to negotiate with both enterprises. The Embassy assisted in briefing Mr Connor on the implications of this changed situation and indeed while the final negotiations were underway the Embassy was in touch with the two enterprises to make it clear that Goodman International had the support of the Embassy. For at all stages Goodman International were advised of the problematic payment situation in Iraq and the importance of adequate Export Insurance Cover. The Ambassador also makes it clear that since the Embassy opened that they had given such logistical help as could be given to the company during visits by its representatives and in particular in discussions with the Minister for Agriculture and Trade here earlier, the Ambassador made it clear our support for the company activities in Iraq."

On the 15th day of June 1987 Mr Goodman met the Taoiseach Charles J Haughey TD in connection with the I.D.A. Development Plan and both say that they did not discuss the question of Export Credit Insurance or the state of the contract being negotiated in Baghdad.

On the 26th day of June 1987 the powers of the Minister for Industry and Commerce under the Insurance Acts were by order delegated to the Minister for Trade & Marketing Seamus Brennan TD but by virtue of the terms of Section 2 (1) of the Ministers and Secretaries (Amendment No. 2) Act 1977, the powers which had been delegated also remained vested in the Minister for Industry and Commerce who, in fact, continued to exercise these powers in relation to the allocation of Export Credit Insurance for exports to Iraq.

On the 2nd day of July 1987 Anglo Irish Beef Processors International Limited (hereinafter referred to as AIBP) signed a contract for the export of beef to Iraq with a total value of $134,500,000. This contract was the largest ever negotiated by the Goodman Group and as stated by Brian Britton, the Financial Controller of the Group, was of such a scale that it was imperative "that the Group continue its prudent policy of arranging Export Credit cover for all such foreign receipts."

On the 8th day of July 1987 Lawrence Goodman met the Minister for Industry and Commerce. According to the Minister's evidence, it was for the purpose of "updating him in relation to Export Credit Insurance" but he was also keen on discussing other matters such as the re-designation of County Louth and the obtaining of F.E.O.G.A grants in relation to the Development Plan.

The Minister however has no clear recollection as to whether Mr Goodman informed him of the nature and extent of the contract.

On the 14th day of July 1987 Mr Goodman met with the Taoiseach Charles J Haughey TD.

Mr Haughey stated that he had no recollection of discussing Export Credit Insurance with Mr Goodman and in particular and more specifically, had no recollection whatsoever of having been informed by Mr Goodman about the recent signing of the $134.5m contract.

Mr Goodman stated that he would have discussed all matters that were of importance to him whenever he got the opportunity but had no clear recollection of discussing Export

Credit Insurance with the Taoiseach on any specific occasion but emphasised that he would have discussed it whenever he got the opportunity.

On the 15th day of July 1987 Mr Joe Timbs of the Export Credit Section of the Department of Industry and Commerce requested his officials to prepare a general note of the position with regard to Iraq for the benefit of the Minister for Industry and Commerce.

This resulted in the "*Note for the Minister's Information*" entitled "*Export Credit Insurance for Iraq — Current Position*" dated the 30 July 1987 and signed by Mr O'Mahony of the Export Credit Section.

This note which set out in full detail all particulars with regard to the then existing position which showed that if all applications were granted the £35m ceiling would have been breached and concluded with the recommendation that:—

"The Export Credit Section's view of Iraq is that no further cover should be provided for Iraqi business."

The Minister for Industry & Commerce was in hospital from the 23rd day of August to the 29th day of August 1987. During the course of this stay he was, when possible, dealing with the business of his Department.

The relevant files which needed attention were brought to him by Mr Dominick McBride, an official of the Minister's Administrative Staff, who was acting for the Minister's Private Secretary, who was on holidays at the time.

Among the matters which he brought to the attention of the Minister on the 28th day of August 1987 was a Summary of the then existing position with regard to the allocation of Export Credit Insurance to Iraq which showed a potential State exposure in respect, thereof, of £34.74m and as the approved ceiling for such exposure was £35m. and contained a recommendation from the Export Credit Section of the Department that no further cover should be provided by the Department.

Mr McBride stated that he showed the Summary to the Minister and, that he, the Minister, agreed with the recommendations. On the 31st day of August 1987 Mr McBride made a note on this document which he signed and dated.

In the course of his evidence, the Minister for Industry and Commerce stated that he saw Mr McBride on the 28th day of August 1987 but has no recollection of seeing the said Summary or agreeing with the said recommendation. He stated that:—

"I have no recollection whatsoever. There was something like a couple of hundred files there. As I told you, I wasn't concerned with dealing with anything except something that required urgent decision and there was no urgency about that decision. It was a note sent up to me for my perusal along with a lot of others and I am sure that I just looked at it and left it aside."

In view of the large number of files, considered to require the Minister's attention during his period in hospital, a couple of hundred according to Mr McBride, the Minister's attitude to the note brought to his attention by Mr McBride is perfectly understandable as the matter raised therein did not, at that stage on the 28th August 1987, require any decision.

However, the situation changed between the 28th day of August 1987 and the 31st day of August 1987 because on that day Mr McBride noted:-

> "I understand from the Minister that Anglo Irish Meats will today be making a submission for consideration by the Dept. The Minister will make a decision on Iraq following on examination of the Anglo Irish Meats proposals. The Minister has indicated that the Anglo Irish proposal will require Government approval and he wants a memo for Wednesday's Government meeting. The Minister is unlikely to attend the Government meeting and the Taoiseach will be taking this item at Government."

This note would appear to contradict the earlier note of that date signed by Mr McBride who stated that, to the best of his recollection, when the contradiction was brought to his attention by Mr Ted O'Reilly, the Assistant Secretary, he drew a line through the first note and signed it.

There is no doubt whatsoever, but that the Minister for Industry & Commerce, on the 31st day of August, 1987, informed Mr McBride that an application would be made by AIBP and of the steps to be taken in regard thereto.

In his evidence, Mr Reynolds stated that on the 31st day of August 1987 he was aware that the application would be made by AIBP and that he ordered an Aide Memoir be prepared by his Department.

He stated that:—

> "To the best of my recollection, I don't know who told me about it but the fact is that the recollection that sticks in my mind is the one of the letter, the details of the letter, the size of the contract and I am not here to speculate on who might or might not (have told me). The fact was I was told. I acted upon it, I sent the matter to Government. The Government decided and I carried out that decision".

At the time that the Minister for Industry & Commerce gave his instructions to Mr McBride on the 31st day of August 1987 he must have been aware of the value of the contract and the amount of Export Credit Insurance that was required because he stated that Government approval would be required. Government approval would only be required if it was necessary to increase the ceiling above £70m. which had been the ceiling fixed by the previous Government.

Having regard to the size of the contract, the largest ever negotiated by the Goodman Group, it is most unlikely that the information with regard thereto would have come from anybody other than Mr Larry Goodman or Mr Brian Britton either directly or through the then Taoiseach, Mr Charles J. Haughey TD.

In the course of his evidence, Mr Brian Britton stated that:

> "I didn't know how Mr Reynolds knew it."

and

"All I can say categorically is that I did not speak with Mr Reynolds in the weeks prior to the 31st August 1987"

The relevant matter with regard to this aspect of the Inquiry is contained in the cross-examination of Mr Goodman by Mr Durkan B.L. and is as follows:-

"1. Q. Good morning, Mr Goodman. If I can take you back to where we were yesterday; I was dealing with the time when you had signed the contracts in Iraq in July 1987.

A. Yes.

2. Q. The letter applying for cover in regard to those contracts is dated the 31st of August 1987, and was sent in by Mr Britton.

A. Yes.

3. Q. In the weeks leading up to the sending in of that letter, did you lobby politicians, did you lobby members of government, to make sure that your application would be successful?

A. I would have, yes.

4. Q. Who did you lobby?

A, I would have been in touch, I would have thought that it would be highly likely I would be in touch and lobbying before going in relation to a major contract, with Minister Reynolds. We are talking about July 1987 now.

5. Q. Yes, and what I'm asking you is the letter of the 31st of August, in the month of August, can you recall were you in touch with politicians, with members of the Government?

A. I can't recall specifically the day, but I would have been. It would have been normal for me to lobby, if I were going on a major trip. I would try and make contact to see what the up to date situation was if I were going on a major trip.

6. Q. What I'm trying to get at is this, in regard to the actual application for the 134.5 million cover, which was going on the 31st of August, did you contact anyone and say "This application is going in. I would like it to be supported".

A. I can't recall that. As I'm saying, it would be highly likely, in relation to a contract of that size that early sometime in May or coming back from that trip, that I would have made contact, or tried to make contact with Minister Reynolds.

7. Q. Have you any recollection at the time, or about the time of the application of 31st of August, of going in, or being in touch with Mr Reynolds?

A. No. Mr Britton would have been handling the administration side of things. Once I would get the contract I would hand it over to him, or to someone in the International Division, in those days.

8. Q. Were you in touch with Mr Haughey, by any chance, about this?

A. Not unless I was meeting with him on something else. If I was meeting with him I would use the opportunity to say that we had applied and what can be done about the insurance.

9. Q. Can you recall whether you were meeting with him about anything else?

A. There were some meetings about the five year plan. I am not sure of the dates, but I have the dates here somewhere.

10. Q. The reason I am asking you this is can you see, from whatever records you have there, whether you in fact met Mr Haughey?

A. I have just notes in relation to dates of meetings here.

11. Q. Well, it's clear from the evidence that Minister Reynolds knew that the Goodman application for 134.5 million was going to come in to the Department, in or about the 1st of September, because he told his officials to expect it. Do you know how he knew that it was coming in?

A. No, I wouldn't be aware of that. I may have said it to him. I don't know.

12. Q. What I want to know is, if you did say it to him, when did you say it to him?

A. Probably when I would have come back from Iraq, I would have made contact and said that we had managed to conclude the business "what's happening on the Export Credit Insurance, when are we going to be awarded it?" It would have been that type of conversation.

13. Q. But you have no specific recollection of being in touch with Mr Reynolds, or Mr Haughey?

A. No, but it's likely I would have been.

14. Q. I see, in or about the 31st of August, when the application actually went in, did you not feel at that stage, when the application went in and there was actually a concrete application on the table, if I can put it that way did you not feel a need to contact either Mr Reynolds or Mr Haughey, in support of that application?

A. I would have been more anxious to do that after I got the contract, which would have been in or about May and if it was finalised, in July. I would be trying to edge things forward at that point in time, but more so when Mr Britton would be dealing with the administration end of it, and I tended to do the deals, even acquisitions and once they were done I would pass on the administration to somebody else and hope that they would move it from there."

From this it appears that Mr Goodman has no clear recollection of when or how the Minister for Industry and Commerce became aware of the fact that AIBP were on the 31st day of August 1987 going to apply for Export Credit Insurance.

Subsequent to the signing of the contract on the 2nd July, 1987 Mr Goodman met the then Taoiseach Charles J Haughey on the 14th July 1987.

In his evidence Mr Haughey stated that that meeting was in relation to the IDA plan and particularly in relation to the building of a new meat processing plant in Tuam which was of very great interest to him and that he was pressing Mr Goodman to get on with it and has no recollection of any discussion with regard to the Iraqi contract on Export Credit Insurance.

In Mr Haughey's diary there was an entry in respect of a meeting with Mr Goodman for 9 am on the 28th August 1987 but a line is drawn through the name Goodman.

Mr Haughey has no recollection of this meeting and stated that the fact that there was a stroke through it indicated that it did not take place and had earlier stated that he was dubious about that meeting

> "because in my diary there is a stroke through that — it was the 28th August and other things being equal.... I would be on Inisvickalaune".

However he stated that if did happen

> "it was simply again me pressing Goodman to get on with it".

It is difficult to understand how nobody recollects the circumstances under which notice was given to the Minister for Industry and Commerce that the Goodman Group would be making an application for Export Credit Insurance in respect of the largest contract for the export of beef ever negotiated by an exporter within the State.

It appears from the evidence of Mr Donlon, Secretary to the Department of Industry and Commerce that the first indication that there was going to be an application came directly from the Minister for Industry and Commerce through Mr McBride.

On the 31st day of August 1987, Brian Britton Deputy Chief Executive of Goodman International wrote to the Minister for Industry and Commerce in the following terms:—

> "We wish to apply for 100% Export Credit Insurance for the supply of 40,000 +/– 5% tons of Boneless Beef to Iraq with a total value of USD 134,500,000 under contracts recently negotiated by our subsidiary company Anglo Irish Beef Processors International Limited.
>
> Appendix One sets out the salient features of the contracts which may be summarised as follows:

1. Total Sales Price Receivable	134,500,000 USD
2. Total E.E.C. Export Refunds Receivable	111,500,000 USD
	246,000,000 USD

3. Sales proceeds payable 1 year after shipment
4. Shipment will be in equal instalments between September 1987 and June 1988.

This contract is of major importance to both our company and to the Irish economy generally for the following reasons:

(*a*) The contract will generate substantial foreign earnings for both company and country with consequent benefits for both.

(*b*) The Iraqi market is a major outlet for Irish beef. Our Group opened up the market some years ago and we have consistently increased our tonnages to that country in spite of major competition from Europe, particularly France.

Based on our past experience in dealing with them, we are confident that payment for the contract will be received from Iraq. The economic situation has improved recently and the recent opening of a second oil pipeline will significantly increase that country's foreign currency earning capacity. Iraq consistently gives priority to payment for imported foodstuffs as has been proved by this company's record in the past and is evidenced by the attached letter to AIBP International Ltd (Appendix Two) from the Iraq Ministry of Agriculture and Agrarian Reform, prepared at the request of the Deputy Prime Minister, confirming that payments on our recent contracts will be made promptly on the due dates.

This contract, the largest ever negotiated by the Group, is of such a scale that it is imperative that the Group continues its prudent policy of arranging Export Credit Cover for all such foreign receipts.

We trust this application will get a most favourable response from your Department. Because of the sensitivity of dealing with a Country such as Iraq, we would ask that no publicity whatever be given to our request."

This letter was received in the Department on the 1st day of September 1987 but the Minister's office was aware on the afternoon of the 31st August 1987 that the application would be made.

On the morning of the 1st of September the Minister for Industry and Commerce spoke to Mr Timbs of the Department of Industry and Commerce by telephone and directed that an Aide Memoire be prepared and be given to the Taoiseach as soon as possible. The Aide Memoire was prepared by Mr Timbs and approved by Mr O'Reilly the Assistant Secretary and given to the Runai Aire on the 1st of September 1987 for transmission to the Taoiseach's office.

The Aide Memoire was given to the Taoiseach on the morning of the 2nd of September 1987.

On the same day a copy of the Aide Memoire was given to the Private Secretary to Minister of State Seamus Brennan together with a minute prepared by Mr O'Reilly in the following terms:

"In completing the memo I was very conscious that no proposal was being made but Mr Timbs told me that this was Minister Reynolds' wish. There are obvious proposals which could be made".

The Aide Memoire and minute of Mr O'Reilly were seen by Minister of State Seamus Brennan on the 2nd of September 1987.

No decision, formal or informal is recorded for the Cabinet meeting held on the 2nd September 1987 concerning the question of the Export Credit Insurance Scheme and it would appear to have been referred back to the Department of Industry and Commerce because it did not seek any decision from the Government.

On the 4th of September 1987 the Secretary had a discussion with the Minister for Industry and Commerce following which the Secretary asked his officials to have the Aide Memoire redrafted along the lines indicated by the Minister and put in the form of a Memorandum for Government. The Memorandum for Government was finalised and cleared by the Secretary on the 7th September 1987 during Mr O'Reilly's absence abroad with Minister of State Brennan, The Memorandum was transmitted to the Runai Aire for circulation.

The Memorandum for Government was dated the 7th of September 1987, entitled "*Export Credit Insurance for Iraq*" and was in the following terms:—

"1. *Decision Sought*:

The Minister for Industry and Commerce seeks the approval of the Government to:

(*a*) raise the ceiling for insured exports to Iraq from the previous Government approved ceiling of £70m. to £150m. — the conditions of cover to be those as set out in Paragraph 6.

(*b*) draft legislation to increase the ceiling for Export Credit Insurance which may be given for all exports to all markets from the present ceiling of £300 to £500m.

2. The governing legislation provides that the ceiling for Export Credit Insurance which may be given for all exports to all markets is £300m. As at 30 June, 1987 the Minister's total liability was approximately £194m.

5. For business insured on those conditions prior to July 1986 Iraq has been paying erratically — some payments being received on time, others after delays of 6 months or more. The first 3 claims in respect of Iraqi business paid between December 1986 and August 1987 amounted to £1.4m. but these amounts were received from Iraq within a short subsequent period. Under normal circumstances, claims are paid 6 months after default by the buyers. Delays are still continuing and if no further payments are received claims totalling £3.54m. will have to be paid between now and December 1987. What seems to be happening is that while one year credit was given the Iraqis are actually taking about 18 months credit.

6. In May, 1987 when total exposure in Iraq stood at £22m. the Minister decided that limited cover was to be made available for Iraq subject to even more stringent conditions and to an overall ceiling of £45m. The conditions were;

   (*a*) 70% cover maximum on any contract;

   (*b*) a maximum credit period of one year;

   (*c*) a claims waiting period of 12 months as opposed to the normal 6;

   (*d*) a minimum premium rate of 4% of full contract value as opposed to the usual .04% for good risk countries generally; and

   (*e*) cover to be provided only to exporters in respect of whom claims had not arisen in Iraq or who had subsequently secured payment.

   Offers of cover under those conditions were made to exporters within the £45m. ceiling. The take up was very slow, partly due to the stringent conditions and partly due to not finalising contracts. Taking into account commitments already made to give cover, present exposure in Iraq amounts to £25.14m.

7. A number of companies have recently sought new cover in Iraq. Based on cover of 70% the provision of insurance for these new applicants would increase exposure by a further £96.51m. — details as follows.

Name of Company	Contract Value	70%
	IR£	IR£
Non-beef exporter	03.9m	02.75m
D Heyer Meats (Beef)	11.38m	07.97
Anglo Irish Beef Processors International Ltd (Beef)	91.53m	64.07m
Hibernia Meats (Beef)	30.14m .90m	21.09m 00.63m
TOTAL	137.85m	96.51m

   To agree to the above new requests would bring the Minister's total liability in Iraq to £121.65m, representing 40% of worldwide exposure, and under the Scheme in the aggregate to £295m, leaving only £5m. for new insurance worldwide under the present ceiling.

8. When the Minister re-opened cover in May 1987, many of the applicants sought less stringent terms. In one case (........) a credit period of two years was refused and in another (Hibernia Group) cover was refused on the grounds that a subsidiary company had monies overdue from Iraq. In the case of the present and largest application from Anglo Irish Beef Processors, 100% cover has been sought on 1

year credit. In 1986 Anglo Irish signed a £19.3m. beef contract for Iraq with payments due to begin in September 1987. While a request for Export Credit Insurance for this contract was refused, the exporter fulfilled it. No claims have arisen in Anglo Irish businesses covered by the scheme, in any market.

9. The question of increasing the ceiling for cover in Iraq must be viewed against present uncertainties in the Gulf region. The arguments could be summarised as follows:

For

(*a*) While 3 claims have been paid in respect of default by Iraq the monies were subsequently received within a few months.
At present no claims have been paid where funds have not subsequently been recovered — but see Paragraph 5 above about the position between now and end year.

(*b*) A leading beef exporter has pointed out that the Iraqi market is a major and increasing market for Irish beef, despite major competition from Europe, and that the beef contracts generate substantial foreign earnings by both Irish companies and the country.

(*c*) Premium income at 4% of contract values approximates £5.5m. representing over 70% of the cumulative deficit under the scheme as at 31/12/86.

Against

(*a*) The recent escalation of hostilities in the Gulf must further drain already strained Iraqi resources.

(*b*) Present applications plus existing commitments would mean that insurance for Iraq would constitute 40% of worldwide exposure.

(*c*) To provide cover under the conditions in paragraph 6 would invariably involve additional requests from companies who have not approached us on the basis that they know we are effectively off cover.

(*d*) While it is difficult to ascertain precisely what other export credit agencies are doing credit lines are being reopened for the Iraqi market by many OECD countries on the basis of agreements with the Iraqis to reschedule existing debts e.g. the UK agreed credit lines of Stg £575m up to y.e. 1987 are being renegotiated at present. In any event most OECD credit agencies operate an extremely restrictive cover policy or in some cases are totally off cover.

(*e*) If we were to substantially increase our credit line to Iraq and if their financial situation deteriorated further, we could be asked by the Iraqis to reschedule involving payment of claims to exporters, a moratorium on rescheduling repayments and payments spread over a number of years.

10. The Minister is of the view that substantial export markets in certain sectors exist in Iraq. While delays in payments have occurred, there are no claims outstanding at present. In view of the above and the fact that some OECD countries are considering re-opening lines of credit on the basis of rescheduling the Minister feels

that exporters to Iraq could be put at a serious disadvantage were credit facilities not available. He would point out that the terms under which he is proposing that cover be offered are very restrictive and expensive.

11. The Tanaiste and the Ministers for Finance and Agriculture have been provided with a copy of this Memorandum."

This Memorandum was dated the 7th of September of 1987.
This Memorandum was received by the Minister of State Seamus Brennan and in a minute dated the 7th of September his Private Secretary notes:—

"In relation to the Memorandum for Government on Export Credit Insurance for Iraq the Minister requested that I inform you that he discussed the question of the Export Credit Scheme with the Taoiseach recently and informed him that he was concerned at the losses incurred and was reviewing the operation of the scheme."

This minute was for the private secretary to the Minister for Industry and Commerce.

On receipt of the minute of the same date, Finbar Kelly, Secretary to the Minister for Industry and Commerce notes:—

"I brought the full text of the above minute to the Minster's attention this evening as you may be aware the memo for the Government on Export Credit Insurance for Iraq is listed for consideration at the Government meeting on the 8th of September 1987."

The Memorandum was sent to the Government Secretariat with an *urgency certificate* on the 7th September for consideration on the 8th September. At the same time, it was circulated to the

(i) the Tanaiste;

(ii) the Minister for Finance;

(iii) the Minister for Agriculture;

This short notification did not allow the respective departments of those Ministers to respond in writing to the sponsoring Department as would normally occur.

The officials in the Department of Finance prepared a response dated the 7th of September which would have been available to the Minister of Finance. This response would not have been available to any other members of the Government.

The Department of Finance received the Memorandum on the afternoon of the 7th of September and their response was as follows:—

"The officials recommend strongly that you oppose the Minister for Industry and Commerce's proposals for the following reasons;

3. The present "official" ceiling for cover for Iraq is £70m. This limit was decided upon by the Government in February of 1986 and represented a major increase from the previous limit of £35m. However the extent of cover has never remotely reached the £70m "official" limit, as in the spring of 1986, the Iraqis started to default in their payments. Effective cover even now amounts to only £25m. In effect, what the Minister for Industry and Commerce is seeking is an increase in cover from *£25m* to *£150m.*

4. Even in normal circumstances such an increase would be fraught with risk as it would greatly increase the exposure of the Scheme and hence the Exchequer. In the present case the risk is even greater. A very high proportion of the increased cover would relate to Iraq. As recent events in the Persian Gulf have illustrated, the Iraqi situation is extremely volatile. The Iraqis have, to date, been erratic in fulfilling their obligations. A deterioration in the country's military and economic position could lead to its defaulting on its foreign debts. If this were to happen (and if the Minister for Industry and Commerce's proposal to increase the ceiling of cover to £150m had already been accepted), *the Exchequer would be at a loss for a considerable sum possibly of the order of £120m.*

5. *In essence, the Minister for Industry and Commerce's proposals are too much of a gamble with the Exchequer's resources.* You should seek to have the effective limit of cover for Iraq confined to £45m, (£20m above the present level of exposure) under the conditions as set out in paragraph 6 of the Memorandum. This roughly represents the limit beyond which the Exchequer should not go.

6. The second of the Minster for Industry and Commerce proposals — to increase the ceiling for all markets from £300m to £500m follows from the first. If the first proposal is not accepted by the Government, there will be no need for the second one."

The above document was a document prepared by Mr Quigley of the Department of Finance for the advice of his own Minister and was not included in the Memorandum for Government.

The Memorandum for Government did not disclose that the most recent advice available to the Minister for Industry and Commerce was that no further cover should be provided in respect of exports to Iraq.

What was being sought from the Government was approval to raise the ceiling for uninsured exports to Iraq from the previous Government's approved ceiling of £70m to £150m.

Having regard to the decision made by the Minister for Industry in May 1987 to limit the availability of Export Credit Insurance cover in respect of exports to Iraq to £45m (inclusive of £10m to the PARC Hospital), this represented a massive increase and was wholly prompted by this application from AIBPI.

On the 8th day of September 1987 the Government decided:—

"1. that the ceiling for insured exports to Iraq should be raised from £70,000,000 to £150,000,000; and

2. that the question of increasing the ceiling for Export Credit Insurance generally might be considered further at a later date, as and when the need arises;"

As appears therefrom, the Government made no decision on the question as to whether the statutory limit should be increased.

On the 8th of September 1987 at 4 p.m. the Minister for Industry and Commerce, Mr Reynolds, met with Mr Oliver Murphy of Hibernia Meats Limited and Mr Pascal Phelan of Master Meats Ltd This meeting had been arranged by Mr Phelan. The purpose of the meeting was to discuss the question of both companies being granted Export Credit Insurance cover for their exports to Iraq and Iran.

Prior to this meeting the position was that Hibernia Meats Ltd, had written on the 19th of May 1987 to Minister Mac Sharry indicating that they were negotiating a contract for 20 million US dollars and seeking his support for Export Credit Cover and on 17th of June 1987 had written to ICI requesting cover in respect of a contract for 46 million US dollars.

The view of the Department in August of 1987 was that no Export Credit Insurance should be given to Hibernia Meats Ltd, until all sums outstanding on previous export claims had been paid by Iraq.

In fact arising from the summary prepared by the Department dated the 24th August, 1987 the Minister had instructed his Department officials to inform Hibernia International Limited that no further Export Credit Insurance would be considered for the company whether to Iraq or Iran until such time as their outstanding payments in Iraq had been corrected.

On the 8th of September Mr Murphy advised the Minister that the company was negotiating a contract for US$46 million.

During the course of the meeting, the Minister indicated that he would allocate a facility of £10m. to Hibernia Meats Ltd, with a credit period of 18 months and also the sum of £10m to Master Meats on the same terms. However, at the same time, the Minister emphasised the necessity and importance of having the outstanding payments cleared. He informed Mr Murphy that the details of the facility would be communicated to his company by the Department.

On the 9th of September following the meeting on the 8th, Mr Oliver Murphy wrote to the Minister for Industry and Commerce in the following terms:—

"I wish to thank you for meeting Pascal and I yesterday, Tuesday 8th September. I would also like to take this opportunity of wishing you a speedy and full recovery.

"Further to our discussions regarding Iran/Iraq I wish to clarify the following:—

RE: *IRAN*

Hibernia has signed with Iran for production this year from commercial stocks a total contract of 19000 metric tonnes. Master Meat Packers are taking 5000 metric tons approximately of this quantity.

Due to the situation in the Gulf bankers are becoming nervous in arranging facilities to enable us to produce the quantity contracted. Iran as you know has not to my knowledge defaulted at any stage and as I informed you we already have to hand an irrevocable Letter of Credit opened by Bank Melli, Tehran to Bank Melli, London for an amount of 25,940,250 US Dollars which covers the 12500 metric tons.

The contract for 5000 metric tonnes together incidentally with a contract for 1500 metric tonnes boneless forequarters were signed last Saturday, 5th September 1987. Having put in place the Performance Bond I would expect the Letter of Credit to be opened in 4/5 weeks from receipt of Bank Melli in Tehran of the Performance Bond.

Essentially I am requesting a roll over facility of Credit Insurance from ICI of US Dollars 10 million in respect of a total contract with an invoice value in the case of Hibernia for:—

(*a*) Bone-in Contract 12,500 metric tons of US Dollars 25,940,250.00

(*b*) Boneless Forequarters 1,500 metric tons of US Dollars 3,418,500.00

Master Meat Packers, in respect of the 5000 metric tonnes Bone-in back to back contract with a Letter of Credit value *9,900,000* US Dollars, are looking for credit cover of 5 million US Dollars rolled over.

I think we explained satisfactorily to you the clause re radiation and it is better to have it included than not.

For your guidance Coface in France are prepared to cover Iranian exposure at 0.03%. On checking with the ICI they have no indication rate for Iran but the above should act as a guideline. Primarily the request for credit insurance is to enable both parties involved make the necessary banking arrangements in respect of the contract.

RE: *IRAQ*

For your information the first Irish Company to export to Iraq was Hibernia Meats Ltd We have been constantly seeking Export Credit Insurance for Iraq and I enclose for your information correspondence with the Insurance Corporation of Ireland dated 17/6/87 and a letter to the Minister for Finance of 19/5/87. If your officials check with the Insurance Corporation of Ireland (contact Mr John Barton) it will be confirmed to you that for the past 12 months we have been in correspondence with them re this matter.

I wish to thank you for your decision yesterday in relation to a facility for Iraq of 10 million which we will try to limit to 18 months but in reality the Iraqi Purchasing Authorities are looking for 2 years.

You will note from my correspondence with the ICI that as of last June we sought credit insurance cover for a total of US Dollars 46 million for a two year period.

If this matter is being actively considered I would like you to keep our request in mind.

In the meat industry, where margins vary from one per cent to three per cent nett of turnover, I do not think that 70% cover is realistic. It may well be for the pharmaceutical industry where margins are acknowledged as being significantly larger. In my opinion credit insurance in respect of meat exports should cover at least 90% as it is difficult for the industry to operate otherwise.

My apologies for writing at length, however, I do think in order to see matters in perspective that it may be helpful to you in arriving at a conclusion.

I do wish to assure you that we are using our best endeavours to have the last credit facility fully cleared as quickly as possible and you will have noticed some progress on that in recent weeks.

If you wish to meet me at any time to give you a commercial view and appraisal of either Iran/Iraq in respect of Export Credit Insurance I will be pleased to do so.

Again my thanks for meeting Pascal and I, yesterday, and we very much appreciate your response."

On the 11th of September 1987 Mr Ted O'Reilly, Mr Joe Timbs and Mr Dermot O'Mahony of the International Trade Division met with Mr Brian Britton and Mr Aidan Connor of Goodman International to discuss "the terms for Insurance for their contracts," following the Government decision on the 8th of September 1987 that the ceiling for insured exports to Iraq should be raised from £70 million to £150 million.

It was indicated at the beginning of the meeting by Mr Britton that the terms of insurance should be favourable to AIBP because the company had already demonstrated its good relations with Iraq by successfully completing a contract at its own risk the previous year. Mr Britton also mentioned that the £10 million cover requested by the company in March 1987 was now withdrawn and that cover was now required for a contract of $134.5 million US (91 million Irish) excluding export refunds.

Mr O'Reilly stated that the terms of cover which were recently offered were:—

1. All payments to be secured by 360 day irrevocable letter of credit;

2. Insurance cover to be limited to 70% of the contract value;

3. A 12 months claims waiting period;

4. Premium rate to be 4% of gross contract value and that these terms were now on offer to AIBP;

These were the conditions set forth at Paragraph 6 of the Memorandum for Government dated the 7th day of September 1987 and Mr O'Reilly in his evidence stated that his "starting point was the terms identified in the Memorandum for Government". These terms were not however acceptable to Mr Britton.

Mr O'Reilly then decided to "go away from the conditions imposed by the Government and seek to get the best conditions he could."

Obviously in adopting this course Mr O'Reilly did not consider himself or the Minister for Industry and Commerce, on whose behalf he was negotiating to be bound by the conditions set forth in the Memorandum for Government.

After lengthy discussions the parties agreed;

1. claims waiting period of 6 months;
2. 80% cover;
3. *a*) premium of 1% gross contract value at commencement plus 2% of gross contract value in the event of a claim;

   b) the 2% to be payable by way of a deduction from the amount paid in claims;
4. all payments to be secured by 360 day irrevocable letter of credit;
5. the package to be subject to the strictest confidentiality by both parties;

Mr O'Reilly informed Mr Britton that he would be recommending this package to the Minister who would a make the final decision and it was said by Mr Britton that he would be speaking to Mr Goodman that night and he did not see any problem in obtaining his approval. It was further agreed that Anglo-Irish Beef Processors would be informed of the Minister's decision as soon as possible and that full terms would be detailed in a written communication to ICI from the Department.

Mr O'Reilly rang the Minister for Industry and Commerce on the Friday night at the conclusion of the meeting and informed him of the outcome.

On Monday the 14th of September 1987 Mr O'Reilly telephoned the Minister for Industry and Commerce telling to him that he felt that he was "beaten" on the claims waiting period and that if the Minister wished he would go back and renegotiate it. The Minister told him not to.

Mr O'Reilly made this telephone call to the Minister because, as he stated in evidence,

> "over the weekend I was conscious of the fact that I had departed to a considerable degree from the conditions in the Government Memorandum and on which the Government had given its decision and, in particular I felt very unhappy about the claims waiting period."

As the result of the agreement reached at that meeting Mr O'Reilly wrote to Mr Britton on the 8th day of October 1987 in the following terms:

"Dear Brian

I am directed by the Minister for Industry and Commerce to refer to your letter of 31 August 1987 applying for Export Credit Insurance in respect of beef exports to Iraq valued at US $134.5 million and to confirm that he has agreed that Export Credit Insurance will be provided on the following terms:

1) cover to be provided for 80% of contract value exclusive of export refunds; the Minister's maximum liability will be the Irish pound equivalent of 80% of US $134.5 million,

2) claims waiting period of 6 months,

3) payment by 360 day irrevocable letter of credit issued by Rafidain bank,

4) (*a*) premium of 1% contract value payable on writing policy (i.e. 1% of US $134.5m).

   (*b*) a further premium of 2% of contract value is payable in the event of a claim which will be non-refundable irrespective of the size of the claim or whether there is a subsequent recovery, this further premium to be deducted from the claim payment.

If you would like to apply for this insurance you should contact Mr Pat Leamy or Mr John Barton at the Insurance Corporation of Ireland plc who will be able to deal with your application.

If there is anything that you would like clarified don't hesitate to contact me.

Yours sincerely
TED O'REILLY
Assistant Secretary"

✗ These conditions were substantially better than those set forth in the Memorandum for Government.

The Memorandum for Government had sought the approval of the Government to

(*a*) raise the ceiling for insured exports to Iraq from the previous Government approved ceiling of £70m to £150m — the conditions of cover to be those as set out in Paragraph 6 (of the Memorandum).

Paragraph 6 of the said Memorandum stated that:—

"In May, 1987 when total exposure in Iraq stood at £22m, the Minister decided that limited cover was to be made available for Iraq subject to even more stringent conditions and to an overall ceiling of £45m. The conditions were:—

(*a*) 70% cover maximum on any contract;

(*b*) a maximum credit period of one year;:

(*c*) a claims waiting period of 12 months as opposed to the normal 6;

(*d*) a minimum premium rate of 4% of full contract value as opposed to the usual .04% for good risk countries generally; and

(*e*) cover to be provided only to exporters in respect of whom claims had not arisen in Iraq or who had subsequently secured payment."

It was on the basis that cover would be granted on those terms that Minister sought the approval of the Government to raise the ceiling for insured exports to Iraq. This fact is emphasised by

(1) the reference at 9(*c*) of the Memorandum which deals with the arguments for the granting of the approval to the fact that

"Premium income at 4% of contract values approximates £5.5m representing over 70% of the cumulative deficit under the Scheme as at 31/12/1986, and

(2) the Minister pointing out at the end of Paragraph 10 of the said Memorandum "that the terms under which he is proposing that cover be offered are very restrictive and expensive."

The recorded decision of the Government as appears from the letter dated the 8th September 1987 was

"that the ceiling for insured exports to Iraq should be raised from £70m to £150m"

This decision does not deal with the conditions set forth in the Memorandum for Government and which provided the basis for the approval sought.

As a result of the decision of the Supreme Court in the case of the Attorney General -v- The Sole Member of the Tribunal of Inquiry into the Beef Processing Industry, the Honourable Liam Hamilton, on the issue of "Cabinet Confidentiality" the Tribunal was prevented from inquiring into the question whether or not the approval granted by the Government was based on or subject to the imposition of the terms upon which such Export Credit Insurance would be granted.

In the course of his judgement, the Chief Justice stated:—

"I would, therefore, conclude that the claim for confidentiality of the contents and details of discussions at meetings of the Government, made by the Attorney General in relation to the inquiry of this Tribunal is a valid claim. It extends to discussions and to their contents, but it does not, of course, extend to the decisions made and the documentary evidence of them, whether they are classified as formal or informal decisions. It is a constitutional right which, in my view, goes to the fundamental machinery of government, and is, therefore, not capable of being waived by any individual member of a government, nor in my view, are the details and contents of discussions at meetings of the Government capable of being made public, for the purpose of this Inquiry, by a decision of any succeeding Government."

The terms of the cover granted to AIBPI as set forth in Mr O'Reilly's letter dated the 8th day of October 1987 to Mr Brian Britton were substantially better than those set forth in the Memorandum for Government and which had previously applied.

Cover was granted in respect of 80% of the value of the contract instead of the 70% maximum contract referred to in the Memorandum for Government: a claims waiting period of 6 months was agreed instead of the 12 months period referred to in the Memorandum for Government and premium rate of 1% in lieu of the minimum premium rate of 4% of full contract value referred to in the Memorandum for Government.

The further premium of 2% of contract value was only to become payable in the event of a claim and would be charged in the form of a deduction from the claim paid irrespective of the size of the claim or whether a subsequent recovery is made.

The main effect of this agreement, apart from the increase in the amount of cover and the limitation of the claims waiting period to six months, is that the premium payable on the issue of the policy was $1.345m rather than $5.380m which would have been payable if the conditions set forth in the Memorandum for Government had been adhered to.

Similar terms were granted to Hibernia Meats subject only to the variation that the cover was limited to 70% of the full contract value because they were obliged to grant 18 months credit in respect of a contract value $46m which meant that the premium payable as a result of such offer was $.46m rather than a premium of $1.84m which would have been payable if the conditions set forth in the Memorandum for Government had been adhered to.

On the same day Mr Timbs wrote to Mr Colligan of the Insurance Corporation of Ireland in the following terms:—

> "Dear Paul
>
> I am writing to you to set out the terms on which Export Credit Insurance has been agreed for the Anglo Irish Beef Processor's (AIBP) contract in Iraq. A copy of letter dated 30 August, 1987 from Goodman International is attached for information. The contract value, exclusive of EEC refunds, amounts to USD 134.5 million and the terms approved by the Minister are:—
>
> (1) Cover to be provided for 80% of contract value net of export refunds; our maximum liability will be the Irish pound equivalent of 80% of USD 134.5 million.
>
> (2) A claims waiting period of 6 months to apply.
>
> (3) Premium payable up front of 1% of insured contract value (i.e. 1% of USD 134.5 m). A further non-refundable premium of 2% of insured contract value is payable in the event of any claim. This premium will be charged in the form of a deduction from the claim paid irrespective of the size of the claim or whether a subsequent recovery is made.
>
> (4) Payment to be secured by 360 day Irrevocable Letter of Credit issued by Rafidain Bank, Iraq.
>
> (5) No commission will be payable to intermediaries.
>
> (6) Management fee and profit to be retained by ICI will be 20% of one quarter of the 1% premium amount.

(7) Cover will relate to named shipments; in addition the exporter will be obliged as a condition of his policy to inform ICI, within 7 days, of dates and values/tonnages of shipments and payments.

(8) A short-terms finance guarantee to be provided in accordance with (1) above if sought.

(9) As a condition of the policy, the exporter to demonstrate to the satisfaction of ICI/Department that the beef is 100% sourced and processed in the Republic of Ireland.

(10) The existing causes of loss covered in Iraq will apply.

Yours sincerely
Joe Timbs
Principal Officer"

As will be seen from Mr Timbs' letter to Mr Colligan, it was expressly provided at (9) thereof that:—

"As a condition of the policy, the exporter to demonstrate to the satisfaction of ICI/Department that the beef is 100% sourced and processed in the Republic of Ireland".

At this stage, agreement had been reached between the Department of Industry and Commerce and Goodman International on the terms to be applied to the allocation of Export Credit Insurance in respect of this contract for $134.5m viz 80% on the basis of 12 months credit being the Irish pound equivalent of $107.60m which at that time was £69.42m.

On the 8th day of September 1987 as already stated the Minister for Industry and Commerce had verbally indicated that he would allocate to Hibernia Meats Ltd and Master Meats Ltd Export Credit Insurance to each of these companies in the sum of £10m each with a credit period of 18 months but cover would be restricted to 70% of the contract value.

Subsequent to that meeting on the 8th September 1987 Mr Murphy of Hibernia Meats wrote to the Minister adverting to the fact that they had in June 1987 sought cover from the ICI for a total of $46m with a two-year credit period and stating

"If this matter is being actively considered I would like you to keep our request in mind"

On the 18th September 1987, Mr Timbs and Mr O'Mahony of the Department of Industry and Commerce met with Mr Oliver Murphy of Hibernia Meats Ltd to discuss the terms of the allocation of £10m which the Minister had agreed to allocate and which Mr Timbs stated were:—

(i) 70% cover ;

(ii) Premium of 1% payable up-front plus a further 2% payable in the event of a claim;

(iii) All payments to be secured by 18 months irrevocable letter of credit.

(iv) A claims waiting period of 6 months to apply.

At this meeting Mr Murphy stated that he had heard from reliable sources that a large beef exporter had obtained very favourable terms from the Department in respect of the allocation of Export Credit Insurance in respect of beef exports to Iraq and had obtained cover for exceeding the £10m which his company had obtained.

He stated that he wished to record his deep dissatisfaction with the situation where one beef exporter appeared to have taken the lions share of Export Credit Insurance Cover for Iraq.

He sought additional cover but Mr Timbs stated that he did not have authority to agree additional cover but would inform the Minister of the request.

Following the meeting between the Department of Industry and Commerce and Mr Oliver Murphy of Hibernia Meats Limited on the 18th September 1987, Mr O'Mahony of the Department prepared a memo raising a number of issues:—

"(1) If we are to provide Export Credit Insurance in Iraq on an equitable basis for all beef exporters then the terms which Hibernia Meats Ltd would be entitled to on a 360 days ILC or

(i) 80% cover

(ii) premium of 1% plus 2% (in event of claim),

(iii) claims waiting period of 6 months.

These are the terms offered to Anglo Irish Meats on their full contract value. Hibernia Meats Limited have a contract value at 46 million US dollars or exposure on the above terms would be $46 million at 1.46 exchange multiplied by 80% equals £25.2 million Irish. Premium would be £315,000 Irish plus £630,000 Irish.

(2) I do not think that there are any grounds for refusing the terms detailed above to Hibernia Meats. (Except where the credit period is 18 months in which case they get 70% cover).

The point at issue therefore is whether we should limit our liability (to £10 million or £20 million) on this contract when there is no such limit in place on the Anglo-Irish deal. If the existing limit (£10 million Irish) which was agreed by the Minister, is to be altered, then the Minister's approval will once again be required."

This minute lead to a discussion within the Department and the view was expressed that having regard to the effect that there was money at that time outstanding and that there were sums due between September and December of approximately £5.2 million that the level of cover already offered should not be increased.

On the 30th of September 1987 Mr Fanning in a minute to Mr Walsh in the Department of Industry and Commerce noted

"As regards Hibernia's request for cover on a US$46 million contract in Iraq, the company have been offered cover subject to a IR£10 million ceiling on total exposure to be borne by the Minister. The company are however unhappy with this offer and would like the ceiling to be increased to IR£20 million. The company have cited the favourable terms quoted to another major beef processor and would like to be treated equitably by the Department.

The point is however that we have incurred losses of approximately IR£1.3 million on claims in respect of defaults on the Dantean contract in Iraq (Dantean is a subsidiary of Hibernia) and my view is that:—

(i) We should withhold offering insurance cover to Hibernia until all overdue Dantean debts are paid; and;

(ii) Tell Hibernia that we will be prepared to consider their request for cover in Iraq up to a IR£20 million ceiling if they are successful in recovering overdue debts."

On the 6th of October Mr Murphy informed the Department that he anticipated that money in the amount of £2.38 million due from Iraq would be paid within a short period of time.

On the 7th of October 1987 Mr Timbs wrote to Mr Murphy of Hibernia Meats Ltd offering him cover Export Credit Insurance on a contract in Iraq for Irish beef.

The letter was in the following terms:—

"Mr Oliver Murphy,
Director,
Hibernia Meats Ltd,
Sallins,
Co. Kildare.

Dear Mr Murphy,

I am writing to you following your meeting with Minister Reynolds and our subsequent discussions about Export Credit Insurance for Iraq.

I can now confirm that the following terms of insurance are available for your contract in Iraq:

(i) 70% cover up to a maximum State liability of IR£10m.

(ii) (*a*) Premium of 1% of contract value covered (i.e. 1% of IR£14.28m.) payable upfront.

(*b*) A further premium of 2% of contract value is payable in the event of a claim which will be non-refundable irrespective of the size of the claim or whether or not there is a subsequent recovery and which will be deducted from the claim payment.

(iii) All payments to be secured by 18 months irrevocable letter of credit.

(iv) A claims waiting period of 6 months to apply.

If you obtain 360 day irrevocable letters of credit, the Minister may consider amending the above terms of insurance."

However on the 15th day of October 1987 the Minister for Industry and Commerce decided that the additional cover sought by Hibernia Meats Ltd would be allocated and that they should be told that they were expected to recover all the money due in respect of previous cover before the end of December 1987.

In accordance with the Minister's direction Mr Timbs wrote to Hibernia Meats Ltd on 23rd October 1987 as follows:—

The letter was in the following terms:—

"I am directed by the Minister for Industry and Commerce to refer to your representations about Export Credit Insurance for Iraq. The Minister has decided that the Export Credit Insurance cover will be available for the full value of your contract in Iraq.

I can therefore confirm that the following terms of insurance are available for your contract. These terms replace those given in my letter to you of 7th October, 1987.

(1) 70% cover on a contract value of US$46 million.

(2) (*a*) Premium of 1% of contract value covered payable up front.

(*b*) A further premium of 2% of contract value is payable in the event of a claim which will be non-refundable irrespective of the size of the claim or whether or not there is a subsequent recovery and which will be deducted from the claim payment.

(3) All payments to be secured by 18 months irrevocable letter of credit issued by an approved bank.

(4) A claims waiting period of 6 months to apply.

If you obtain 360 day irrevocable letters of credit, the Minister may consider amending the above terms of insurance.

The Minister has also asked me to say that he expects that you will continue to make every effort to recover the remaining monies due from Iraq on the Dantean contract."

On the 22nd of September 1987 Mr Pascal Phelan of Master Meat Packers Group wrote to the Department of Industry and Commerce

"Further to our recent meetings, I confirm that we have our Export Credit Insurance requirement for export of frozen boneless beef to Iraq for £10,000,000. This represents a sale of 4,000 tonnes."

I should appreciate it if you would let me know availability rate and other details at your earliest convenience".

In reply Mr Timbs wrote on the 22nd October, 1987 to Mr Phelan in the following terms:—

"I am directed by the Minister for Industry and Commerce to refer to your letter of 22 September, 1987, applying for Export Credit Insurance in respect of beef exports to Iraq valued at IR£10m. and to confirm that he has agreed that Export Credit Insurance will be provided on the following terms:

(i) cover to be provided for 70% of contract value exclusive of export refunds; the Minister's maximum liability will be the IR Pound equivalent of 70% of IR£10m;

(ii) a claims waiting period of 6 months;

(iii) payment by 18 months Irrevocable Letter of Credit issued by an approved bank;

(iv) (*a*) premium of 1% of contract value payable on writing policy (i.e.1% of IR£10m.);

(*b*) a further premium of 2% of contract value is payable in the event of a claim which will be non-refundable irrespective of the size of the claim or whether there is a subsequent recovery; this further premium to be deducted from the claim payment.

If you obtain 360 day irrevocable letters of credit, the Minister may consider amending the above terms.

If you would like to apply for this insurance you should contact Mr Pat Leamy or Mr John Barton at the Insurance Corporation of Ireland plc, who will be able to deal with your application."

As a result of such letters to AIBP, Hibernia Meats Ltd and Pascal Phelan of Master Meats, the exposure in respect of Export Credit Insurance had been increased by approximately £100.83m by the 23rd day of October 1987.

On the 15th September 1987 Messrs John Egan and Vincent Burke of Agra Trading Ltd met with Messrs John Fanning and Dermot O'Mahony of the Department of Industry and Commerce and requested Export Credit Insurance cover in respect of a proposed contract for the supply of 5000 tonnes of boneless beef valued at $17m with a credit period of two years.

They were informed that their application would be facilitated in respect of a credit period of 18 months with 70% cover.

By letter dated the 28th day of October 1987, Agra Trading Ltd wrote to the ICI referring to this meeting and stated that

"We were informed that Agra Trading Limited's application for Export Credit Insurance covering 70% of a contract valued at USD$17m would be available with a credit period of up to 18 months.

Our representatives, Messrs O'Halloran and V. Burke are travelling to Baghdad next week with a view to finalising the contract and I will be in contact with you on their return with final details."

On the 21st October 1987 Halal Meat Packers telephoned ICI seeking to be included in Export Credit Insurance Cover Scheme which they had become aware had been re-opened and was being allocated to processors with contracts in Middle Eastern countries.

On the 22nd of October 1987 ICI received a telephone query from the Minister for Finance's office on behalf of Halal Meat Packers. Halal apparently had been seeking the Minister's assistance in helping their application to ICI for inclusion on the Export Guarantee Scheme being provided to various meat exporters signing large contracts at that time in the Middle Eastern countries. Halal Meat Packers were informed that they should prepare their application documents as quickly as possible and apply in the same way as other applicants had done.

ICI notified the Department of Industry & Commerce of the pending application on the same day.

On the 26th of October 1987 Halal Meat Packers applied in writing to the Insurance Corporation of Ireland for Export Credit Insurance in respect of a contract value of $25 million for beef exports to Iraq. The company said they were presently tendering for a contract of that value but had not yet signed.

On the 4th of November 1987 the Minister for Industry & Commerce wrote to the Minister for Finance replying to his representation on behalf of Halal Meat Packers in the following terms:—

> "I wish to refer again to your representations on behalf of Mr P. J. Clarke of Halal Meat Packers, Ballyhaunis, Co. Mayo, regarding Export Credit Insurance for Iraq.
>
> "The Government, as you know, in recognition of the importance of the Iraqi market to Irish exporters, have decided to raise the ceiling on insured exports to Iraq. You will appreciate nonetheless that the provision of insurance cover for Iraq must continue to be subject to stringent terms and conditions which reflect the risks involved in underwriting exports to that country. Thus, for example given the volatile situation in Iraq and the ongoing tensions in the Gulf, the maximum credit period which I have generally been prepared to underwrite is 1 year.
>
> The recent proposal from Halal Meat Packers involves a credit period of 2 years and I am not prepared to underwrite contracts including this credit term. The Export Credit Insurance Scheme has become an increasing burden on the Exchequer in recent years due to a significant increase in claims. Accordingly I am anxious to ensure that the State is not over exposed in the operation of the scheme.
>
> What I am prepared to offer Halal Meat Packers, however, is insurance cover for Iraq on credit terms of 18 months. I am conscious that exporters such as Halal are having difficulties is obtaining orders in Iraq based on 1 years credit and I am now willing to consider 18 months credit terms in certain circumstances. If the company wishes to avail of this offer they should submit a fresh application to the Insurance Corporation of Ireland plc. who operate the Export Credit Insurance Scheme on my behalf".

On the 26th day of October 1987 Halal Meat Packers (Ballyhaunis) Ltd applied to the Insurance Corporation of Ireland for Export Credit Insurance cover in respect of a contract worth $25 million in Iraq. In the course of their said letter they sought cover for a two year period from date of shipment.

This company had sought the assistance of the then Minister for Finance, Ray Mac Sharry TD in obtaining such Export Credit Insurance and representations were made by him to the Minister for Industry & Commerce.

By letter dated 4th November 1987 the Minister for Industry & Commerce informed the Minister for Finance that he was not prepared to underwrite contracts including a credit term of two years. He informed him that he was prepared to offer Halal Meat Packers insurance cover for Iraq on credit terms of 18 months and stated that:—

> "If the company wishes to avail of this offer they should submit a fresh application to the Insurance Corporation of Ireland plc who operate the Export Credit Insurance scheme on my behalf."

By telex dated the 4th day of November 1987, Halal confirmed to the Department of Industry & Commerce that they would be prepared to accept an 18 month credit term.

By letter dated the 5th day of November 1987, Halal confirmed to the Insurance Corporation of Ireland their application for cover as contained in their letter dated the 26th day of October 1987 subject to the variation that the request was for credit for a period of 18 months rather than the 24 months originally sought.

By telex dated the 5th day of November 1987, Halal informed the Insurance Corporation of Ireland that they were seeking cover of 90% of the value of the contract and requested them to seek such increase in the level of cover.

By telex dated the 5th day of November 1987, the Department of Industry & Commerce informed the Insurance Corporation of Ireland that:—

> "This company may be given cover on the following basis subject to your usual underwriting requirements for Iraq:—
>
> (1) Contract value US$25m (exclusive of EEC refunds).
>
> (2) Level of indemnity — 70%.
>
> (3) Payment by 18 months IL/C.
>
> (4) Six months claims waiting period."

On the 5th November 1987 John Stanley of Halal contacted Mr O'Mahony of the Export Credit Section of the Department of Industry & Commerce seeking cover for 90% of the value of the contract but was informed that nobody had succeeded in getting such terms since 1983.

However they informed him that if they succeeded in limiting the credit period to 360 days that consideration would be given to increasing the percentage of cover.

On the 16th day of November 1987 Halal by communication to the Insurance Corporation of Ireland sought revision of their application for Export Credit cover relative to Iraq as follows:—

(*a*) $US18.6m contract with credit terms of 12 months and a level of indemnity of 80%;

(*b*) $US6.4m contract with credit terms of 18 months and a level of indemnity of 70%.

On the 17th of November Halal informed the Insurance Corporation of Ireland, by telex, that they were discussing an increased contract and as a consequence they sought an increase in the overall level of cover from US$25m which they had been offered to US$37.2m. The said telex was in the following terms:—

> "As you are aware we are currently discussing our contract with the Iraqis. Our source in Iraq has presently come back to us with a view to topping up our original contract by 4,000 tons giving a total of 12,000 tons. We therefore, require an increase in the overall level of cover from US$25 million to US$ 37.2 million.
>
> "We are negotiating a reduction in credit terms from 18 to 12 months in respect of this US$37.2 and we require 80% level of indemnity. Shipments will take place during the first quarter of 1988."

On the 13th day of November, 1987 the Minister for Industry and Commerce met Mr L Goodman, who had requested the meeting to discuss the question of the availability of Export Credit Insurance.

In anticipation of such meeting the Export Credit Section of the Department of Industry and Commerce prepared a briefing Memorandum for the Minister.

This Memorandum was as follows:—

> "1. On the 8th of October 1987 Goodman International were offered Export Credit Insurance in respect of beef exports to Iraq valued at US$134.5 million on the following terms:—
>
> (i) cover to be provided for 80% of contract value exclusive of export refunds. The Minister's maximum liability will be the IR£ equivalent of 80% of $134.5 million,
>
> (ii) claims waiting period of 6 months;
>
> (iii) payment by 360 day irrevocable letter of credit issued by Rafidain Bank;
>
> (iv) (*a*) premium of 1% contract value payable on writing policy (i.e., 1% of US$134.5 million)
>
> (iv) (*b*) a further premium of 2% of contract value is payable in the event of a claim which will be non-refundable irrespective of the size of the claim or whether there is a subsequent recovery; this further premium to be deducted from the claim payment."
>
> Since that offer (had been made by letter of the 8th of October), Aidan Connor of Goodman International had arranged a meeting with ICI for the 4th of November 1987 to discuss the provisions of an Insurance Policy but subsequently cancelled the meeting. ICI have not had any further communication with the group.

2. Export Credit Insurance has also been formally offered for the following beef contracts in Iraq:—

Exporter	Contract Value	% Cover	Credit	Date of Offer	State Exposure
Hibernia Meats	$11.5m	80%	12 mths	11/11/87	£06.13m
	$34.5m	70%	18 mths	23/10/87	£16.10m
Master Meat Packers	IR£10m	70%	18 mths	22/10/87	£07.00m
Halal Meats	$25m	70%	18 mths	05/11/87	£11.70m

$1.50 = IR£1.00

3. No formal commitment has been given to the following company. The company were asked to make formal application when they got their contract and were told they would receive equal treatment.

Exporter	Contract Value	% Cover	Credit	Date of Offer	State Exposure
Agra Trading	$17m.	70%	18 mths	15/9/87	—

4. It is understood that Taher Meats will be looking for cover for Iraq. No details available as yet.
5. Of the ceiling of IR£150 million for Iraq approximately IR£140 million has been allocated."

Prior to his meeting on the 13th with Mr Goodman, the Minister enquired of Mr Timbs what the present position was in relation to take up of cover for Iraq and how this effected the overall ceiling in legislation. Mr Timbs informed the Minister that:-

"on the basis of a dollar exchange rate of 1.6 and a sterling exchange rate of 0.89 the potential State exposure on contracts with Iraq amounted to approximately £132.4m. This figure varied with exchange rate movement. Furthermore of the £132.4m. only £17.2m was actual exposure in that none of the beef exporters which represent the bulk of the business had actually signed up and paid premiums as yet."

It was explained to the Minister that applications from Agra and another company would increase the potential exposure by about £8 million.

The Minister was informed that ICI were calculating the up to date exposure in relation to the overall legislative ceiling but that it would be at that time in the region of £200 million and that all the Iraqi business would obviously push the exposure above that £300 million ceiling: that all the potential exposure would not happen at the same time but would build up over the next few months as shipments were made, and that the Department would be making a detailed submission which would involve a draft Memorandum for Government seeking to have the ceiling increased.

The Minister of Industry & Commerce met Laurence Goodman on the 13th November 1987. At that point of time, the Minister for Industry and Commerce had, as appeared from the Memorandum prepared for him, authorised that allocation of Export Credit Insurance to Goodman International (AIBP), Hibernia Meats Ltd, Master Meat Packers Limited and while no formal commitment had been made, Agra Trading Ltd had been informed by the Department of Industry and Commerce officials that, if they were successful in obtaining the contract, which they had stated they were tendering for, viz the sale of beef worth $17m they would receive Export Credit Insurance cover on the same terms viz 70% cover in respect of an 18 month credit period.

This meeting between the Minister for Industry and Commerce and Mr Goodman is of importance and it is desirable that the Minister's account of what transpired at that meeting should be stated in full;

> "146 A. Yes, on the 13th of November I had a meeting with Mr Goodman. It's one he had phoned in in advance looking for and he wanted to see me about seeking additional cover. Before that meeting I spoke to Mr Timbs and asked him what the present position was in relation to the take up of cover for Iraq and how we stood as regards the overall statutory ceiling. I have seen Mr Timbs' note of this meeting that was given here in evidence in which he says that he informed me that the potential state exposure in contracts with Iraq amounted to approximately 132.4 million but that only 17.2 million was as yet actual exposure. According to the note, Mr Timbs also told me that another beef exporter namely, Agra had been in contact with the Department. It was hoping to obtain a contract with the Iraqis in respect of, it would be seeking Export Credit Insurance and which would, if granted increase our potential exposure about a further 8 million approximately.
>
> As regards the statutory ceiling, Mr Timbs note states that he informed me that he understood our total exposure to be in the region of 200 million. In relation, Mr Timbs' note also records him as telling me all that the Iraqi business would push us over the 300 million statutory ceiling but that all the potential exposure in Iraq would not happen or be written at the same time.
>
> In any event, it was always intended to go to Government to have the statutory ceiling increased when required. You recall that we had applied to the Government on the 8th of September and the Government had not agreed to it at the 8th of September. It was always the intention to go back to the Government for that. Now, what I accept the main trust of Mr Timbs note. My recollection is that before the meeting, I asked Mr Timbs what spare capacity there was in regard to Iraq. And that would be a logical question for me to ask going into a meeting to know what the position was before I go into a meeting especially when I knew what the purpose of the meeting was. He told me there was cover for 30 million dollars but that this amount included Agra who informed me had been told that they would have an application considered if they produced a contract. Now, in relation to Agra I might as well just deal with it and get it out of the way.
>
> The Agra officials or Agra representatives had a meeting with the Department on the 15th of September and they had gone off to Iraq with a clear understanding and rightly so that if they came back with a contract they would get the same consideration as everybody else. But I have to say they never formally applied for cover and they never reverted back

to the Department after their visit to Iraq. Just to clear the position with them. So,that's a decision. So, I went into this meeting with Mr Goodman, 30 million dollars available. No more. And that was it and it was on that basis that I held that meeting and at the meeting Mr Goodman confirmed that he was seeking further cover in respect of what I understood to be a sizable extension of his existing contract. I can not remember what Mr Goodman said was the amount of this extra business except that it was more than the cover for 30 million dollars for which Mr Timbs had told me was actually still available within the ceiling.

I told Mr Goodman that there was only limited cover available and he indicated that the best that could be done for him was to provide cover in respect of an additional 30 million dollars. When I so informed Mr Timbs he said that this would not leave room for Agra if they had obtained their contract. My view was that it made more sense to allocate the cover to the man who had the business and indeed if Agra whose involvement with Iraq seemed to me to be at a very early stage, if they did succeed in obtaining a contract I was willing to go back to the Government to facilitate them.

Now, the decision was talking to somebody who had the business, or waiting for somebody who might get the business. I made the decision and told Mr Goodman that the 30 million dollars was available and no more was available despite the fact that he said what he would be looking for was significantly in excess of 30 million dollars.

147 Q. And the reason you limited it to the 30 million dollars was because of the information which Mr Timbs had given you in relation to the ceiling?

A. Precisely. I see that Mr Timbs noted that Mr Goodman had complained during the course of this meeting that both Halal and Agra were causing him difficulties in Iraq by cutting prices. I take the view that such competition between Irish exporters can only be of benefit to foreign consumers, foreign consumers. It is against, in my view, the national interest and the national economic interest to allow foreign consumers the benefits of lower prices. I fail to see how Mr O'Malley's view to the contrary can be justified.

Furthermore, in the Iraqi market, there is no question of normal competition obtained because there is only one purchaser — the State. So the State has the obvious advantage of numerous people coming into sell into a market where there is only the one buyer. They can put one against the other. We learned this dear lesson in Egypt, I think in 1985. We destroyed the market ourselves. That's my view.

Now, you can understand and indeed, every time Mr Goodman had a meeting with me he always made the point that he believed he was the only person who was entitled to all the cover that was available in Iraq and understandably so. That's his argument. I have to say that I was sceptical and would always be sceptical of anybody complaining to me about competition, you know, interfering with their line of business. This is one commercial operator complaining about another. I would be sceptical about that and no evidence was produced to show it but never the less, I would always take it on board and keep it there but I would have to see more evidence of it before I would pay any great attention. But, that's the position in relation to that. Nothing unusual in my view in Goodman making such a complaint because as I said, he believed he pioneered the market, he spent his money, he invested his money, he invested management time over the years and consequently felt the market should be his. It's important which of them went into the market first them or Hibernia. Certainly the two of them were and the two of them were being supported. I believe that's the right thing to do and however, that's the position and I want to reject categorically, reject categorically that that meeting as suggested by Mr O'Malley to this Tribunal that it was the genesis, as he called it, the genesis of a particular policy to be pursued by me. How can anybody seriously assert that what happened at a meeting that you weren't present at or whose expected to believe that. I was at the meeting. I am saying what the policy was and I know what the policy was and I am saying there was no question of a genesis in relation to it but I would be sceptical of anybody making complaints like that. The situation was a development one as far as I was concerned in Iraq. Halal and Agra, they were not in the market when I was appointed. They weren't in the market, I believe, in my predecessor's time and the scale of competition between Irish exporters was far less at that stage, was much less than it later became in 1988 but at this point, we are dealing with the latter part of 1987.

148 Q. The policy you speak about is what has become known as the managed policy?

A. If somebody wants to define managed policy to me, I want to hear the definition because I have only come across that definition when I started to read about that Tribunal. If somebody wants to define what they mean by managed policy I want to hear them. Maybe it'll come up later in the Tribunal but —.

149 Q. Clearly, were you concerned there was no special policy in relation or was there any special policy in relation to the way in which you allocated the Export Credit Insurance during this period?

A. Well, let's just look at what I did during this period. First of all, I dealt with an application from AIBP. Early on April, May. That a 20 million cover, 6 million cover, a third, a third, a third. It didn't take it up, wasn't used for whatever reasons. That's their business. I came into September with an application in front of me for 134 million odd dollars for cover. That could not be accommodated within the ceiling that was available to me as Minister for Industry and Commerce. I went to the Government.

The Government took the view that they would increase the ceiling from 70 to 150 million to deal with the applications at hand and they were the applications. The application on hand at that time was AIBP, and talk from Paschal Phelan about Hibernia but Hibernia always had one, they are, they are always looking for an extension of what they had. So the Government meeting decides to increase the ceiling to 150 million and the details were given in the Memorandum that went to Government. Negotiations took place. The AIBP group were given cover for, 80 percent cover on a 12 month period with the conditions laid out. Hibernia Meats and Master meats were given cover on different terms and in fact, could be legitimately argued on better terms than the Goodman Organisation as far as Iraq was concerned. 70 percent cover but an 18 month period which was more attractive to the Iraqis.

Now, at this point, I had, I was, I had already an offer out to Halal for 25 million dollars. So, instead of what has been generally put out from this Tribunal, is that only two companies in 1987 were offered cover. That is not correct. AIBP was offered cover. Hibernia Meats was offered cover. Master Meats was offered cover. Halal Meats were offered cover. Four, not two. And the position was clearly that was the position. There was no confinement and that date to suggest to me was a genesis of a new policy. That was the policy I was following and I hope everybody knows that."

Mr Goodman's account of that meeting is as follows:—

"326 Q. And what happened at that meeting?

A. At that stage we were in negotiations with SCOFT, which is the Company that had just taken up the 30,000 tonnes, and we were negotiating against a background of the Irish Joint Commission that had taken place in Dublin, that situation, and SCOFT had gone out to tender for 5,000 tonnes but we knew, from our close association with the various companies and organizations there, that they were, in fact, going to buy substantially larger quantities.

327 Q. Did you discuss with Minister Reynolds, at that time, the question of a contract?

A. I would have, I'm sure, yes.

328 Q. Can you recollect the nature of that discussion?

A. Yes. Well, we would have said that we were in discussions in relation to repeat business with SCOFT.

329 Q. I think you are aware that Mr Timbs, a previous witness a suggested that you told the Minister that you had obtained a contract for 30 million US?

A. No, there wasn't any quantity for 30 million. As I have laid out here in 184, "I returned to Ireland on Friday the 13th of November and I met with Minister Reynolds and I have been shown a note of Mr Timbs' discussion with Mr Reynolds on the 13th of November which suggests that I told the Minister that I had obtained a further contract for 30 million dollars. There was never any contract for this amount and I am quite sure that I did not say to the Minister that there was any such contract in existence. We were anticipating a further contract with SCOFT for 30,000 tonnes, $ 105 million, and a repeat contract with ICAPM and I am satisfied that I made Mr Reynolds aware of the potential level of the future business and that he told me, at that moment, there was only $30 million available and that the balance would be allocated to AIBP if the contracts materialized and subject to the approval by Government and the Dail to the increase in the overall limit of Export Credit Insurance". So, we are discussing in Baghdad, prior to my coming home, a repeat of the business with SCOFT, a repeat of the business with ICAPM based on our various discussions with the various clients during the Baghdad Fair. Now, it's against that background that I would have been saying to the Minister that we were expecting to do business of between a 150 and 200 million dollars for the following year with the two clients."

As appears from both the Minister's and Mr Goodman's account of this meeting there was no question of Mr Goodman seeking Export Credit Insurance in respect of a $30m contract; in the words of the Minister he was seeking cover "in respect of a sizeable extension of his existing contract" and according to Mr Goodman he informed the Minister that they "were expecting to do business of between a $150m and $200m for the following year with the two clients" viz SCOFT and ICAPM, the two Iraqi Trading Companies.

In the circumstances outlined in his evidence, the Minister agreed to allocate Export Credit Insurance to Goodman International in the sum of $30m which was the entire amount available under the existing ceiling of £150m.

Mr Goodman stated in evidence that the Minister stated that:—

> "the balance would be allocated to AIBP if the contracts materialised and subject to the approval by Government and the Dail to the increase in the overall limit of Export Credit Insurance".

The question of whether or not such promise was made is relevant to the High Court proceedings instituted and hereinafter referred to and will, no doubt, be determined in such proceedings.

On the 13th November 1987 the Minister informed Mr Timbs that this additional Goodman contract should be covered on the usual terms on the following Monday 16th of November 1987.

This brought the level of exposure up against the ceiling of £150m. and meant that the Agra Trading Co. Ltd application could not be accommodated.

When this was pointed out to the Minister, he stated that if Agra Trading Ltd obtained a contract the question of the ceiling could then be reconsidered.

It appears from Mr Timbs note dated the 16/11/1987 that the Minister considered or had been informed that Agra Trading and possibly Halal were causing difficulties in Iraq by cutting prices and this is consistent with the Minister's evidence.

Mr Colm O'Halloran of Agra Trading Ltd, and Mr Mohammed Khalid of Halal Meat Packers Ltd, Ballyhaunis had been in Baghdad during the course of Mr Goodman's visit to Baghdad seeking to obtain contracts for the supply of beef to ICAPM and SCOFT.

Mr Goodman had expressed his unhappiness at their presence in Baghdad to the Irish Ambassador to Iraq, Mr McCabe and Mr McCabe sent a telex in the following terms to the Assistant Secretary Swift of the Department of Foreign Affairs:—

"*FOLLOW-UP TO IRELAND — IRAQ J.C.*

1 Further to our recent messages on this subject, Mr L. Goodman, Chairman of Goodman International, dined with me last night. He and a team from AIBP have been in Baghdad since Sunday last for discussions with AIBP's two Iraqi clients, The State Company for Foodstuff Trading and the Iraqi Company for Agricultural Products Marketing. Mr Colm O'Halloran of Agra Trading Ltd, Dublin, and Mr Mohammed Khalid of Halal Meat Packers Ltd Ballyhaunis are also in Baghdad for the last week, seeking business with the two clients of AIBP. Mr Goodman is very unhappy about this. He fears (*a*) Price cutting, (*b*) that AIBP, which is doing business on a contractual 12 months deferred basis with its two Iraqi clients, will be disadvantaged if its Irish competitors secure business with one or both of AIBP's clients on the basis of Irish insurance cover for 18 months deferred payments.

2. Mr Goodman met with the Iraqi Minister for Transport and Communications, Mr M Hamza on 10th November. (I understand that this meeting was arranged as a follow-on to a meeting which Mr Goodman had with Minister Hamza en marge of the September J.C. and, I am told, on the basis of a telex from Minister Brennan to Minister Hamza). According to Mr Goodman, Minister Hamza at this 10 November meeting referred to the agreed minutes of the J.C. and told Mr Goodman that the Iraqi authorities expected AIBP to supply beef now on an 18 months deferred payment basis. (This despite the fact that Para. 5 of the J.C. agreed minutes, where it speaks of increasing credit terms from the existing period of 12 months, refers clearly to "new business". I had advised Mr Goodman before the meeting that Minister Hamza might take this line — see Paras. 3 and 6 of our C181 of 6 November). Mr Goodman told me that he pointed out firmly to Minister Hamza that AIBP's contracts with their two Iraqi customers contain contractual provisions for 12 months deferred payment, and that AIBP could not be expected to change these provisions post-facto. Minister Hamza also told Mr Goodman (as he told me on 5 November — Para 5 of our C181) that he is awaiting a letter from Minister Brennan arising out of Para 5. of the agreed minutes of the J.C.

You may wish to copy this message to Minister Brennan's office."

It appears from Ambassador McCabe's evidence that Mr Goodman was concerned about two aspects of the presence of representatives from Agra Trading Ltd, and Halal Meat Packers Ltd,

(i) that it would lead to price cutting; and;

(ii) that AIBPI would be disadvantaged if Agra and Halal were in a position to offer a credit term of 18 months, whereas AIBP were doing business on the basis of 12 month credit.

This was not the first time that Mr Goodman has expressed his reservations about and objection to the presence of his competitors in Baghdad.

During the months of June and July of 1987 Agra Trading Limited a Beef Processor had been in touch with the Trade Section of the Department of Foreign Affairs.

The company wrote on the 9th of June 1987 to Tom Bolster, an official in the Department of Foreign Affairs, indicating that they would like to visit Iraq to discuss the export of beef with two companies (i) the Iraq Company for Agricultural Products Marketing and (ii) the State Organisation for Grains and Foodstuff Trading and requested assistance in obtaining a visa from the Iraqi Government.

On receipt of the letter Mr Bolster contacted the Embassy in Iraq by telex of the 9th:—

> "Mr Colm O'Halloran of Agra Trading would like to visit
>
> (i) Iraq Company for Agricultural Marketing Baghdad; and;
>
> (ii) State Organisation for Grains and Foodstuff Marketing.
>
> In order to get a visa he would need an invitation from one or both of the above. Grateful if you would make approach in support of Mr O'Halloran's request and inform me of outcome. Ideally visa would be applied for in London".

The Embassy of Ireland in Baghdad replied to this telex requesting background information on Mr O'Halloran and Agra. This information was relayed by telex of June 16th 1987.

> "Agra was established by Friedhelm Danz in 1973 and since then has been trading successfully and extensively both within the community and with considerable number of third countries particularly USSR and Middle East. They are well known to the Department. O'Halloran believes he met you in your former posting. In recent past Government Ministers had made representations on behalf of the company to Eastern Bloc Governments".

A further telex was sent to the Embassy in Baghdad on the 30th of June by Mr Bolster of the Department of Foreign Affairs.

> "Colm O'Halloran of Agra has been enquiring about progress with his request for introduction to companies and invitation for visa purposes. What is latest position please?".

This telex was immediately replied to by the Irish Ambassador to Iraq in Baghdad and was received by Mr Bolster on the 2nd of July 1987:—

> "Your number H20 Re. Agra bears on most delicate matter which is currently active and which you may not be aware of. I am anxious to avoid any possibility of the Department being placed in a difficult position at home. For background, and to facilitate further consideration, I strongly recommend that you telephone me immediately at office or up to 12.00 noon Irish time or afterwards at home".

Mr Bolster telephoned the Ambassador as requested and was informed by the Ambassador:—

"1). that everything he had to say was totally confidential;

2). that the Embassy had not acted on the request vis-a-vis Agra because representatives of Goodman International had been in Baghdad for the past two weeks in major negotiations with the same two Iraqi bodies on the same product;

3). That Mr Goodman himself, had been in Iraq for the past few days and would be signing an extremely large contract on the 2nd of July after very difficult negotiations;

4). On the 1st of July Goodman told the Ambassador some details of the contract including quantity and value but added that he did not wish the information published because he did not want to create "farmer euphoria";

5) Ambassador McCabe did not communicate the details of the contract to Mr Bolster as they were not material to the issue;

6) Goodman further told the Ambassador that he had an agreement with CBF (Tony O'Sullivan previous Chief Executive and Paddy Moore current Chief Executive) that CBF would not introduce other Irish exporters to markets that he had prospected thereby avoiding the situation where others could undermine or undercut his position he did not want the situation to happen in Iraq which had happened in Egypt and Iran;

7) Goodman informed the Ambassador that he did not inform the Department of Agriculture of the details of his business and stressed his concern that nothing be done which would compromise the confidential nature of his conversation with me."

As a result of the matters disclosed by Ambassador McCabe, Mr Bolster made certain inquiries and reported as follows to Assistant Secretary Swift.

"1. Please see attached copies of telexes between myself and Embassy Baghdad. In summary they request Embassy to get invitations from two Iraqi purchasing boards for representatives of Agra Trading. In response to Ambassador McCabe's telex I telephoned him as requested.

2. The issue is whether we should now persist through Embassy Baghdad in trying to get an invitation for Agra, or, if not, what response/explanation we should give to Agra.

3. Ambassador McCabe told me, stressing that everything he had to say was totally confidential, that the Embassy had not acted on my request vis-a-vis Agra because

representatives of another Irish company (Goodman) had been in Baghdad for the past two weeks in major negotiations with the same two Iraqi bodies on the same commodity (meat) and that in fact Mr Goodman himself who had been in Iraq for the past few days would be signing an extremely large contract on 2 July after very difficult negotiations.

4. Ambassador McCabe said that he had had Mr Goodman to dinner on 1 July and the latter had informed him of some details of the contract including quantity and value, Mr Goodman said that he did not want this information publicized because he did not want to create "farmer euphoria" etc. Ambassador McCabe did not communicate the details to me and they are not in fact not material to the present issue.

5. The issue is precisely that Mr Goodman told Ambassador McCabe that he (Goodman) had an agreement with CBF (Tony O'Sullivan previous Chief Executive, and Paddy Moore, current chief executive) to the effect that CBF would *not* introduce other Irish exporters to markets that he (Goodman) had prospected and thereby avoid the situation where others would come in on his coat-tails and undermine or undercut his position. Mr Goodman referred to occasions in the past where precisely this had happened (Egypt, Iran etc). He was particularly anxious now that it not happen in Iraq. Ambassador McCabe does not appear to have questioned these statements critically, and his position may be understandable to the extent that he did not want to take any action which might have been counter-productive at a very delicate stage.

6. However the meat business is extremely competitive and even cut throat and Mr Goodman's statements are entirely *ex parte*. We are not aware of any such agreement with CBF and our policy and practice is to deal with all exporters in an even-handed way. The owner of Agra-Trading, Mr Dantz, one of Mr Goodman's rivals, has openly recognised the even-handedness of civil servants in this regard but expressed extreme concern sometime ago to Mr McDaid that the "political interference" that had worked against him in other middle Eastern Markets, would also be expressed in respect of the Soviet Union a market which he had opened up!

7. I drew Ambassador McCabe's attention to the fact that he had only heard Mr Goodman's half of the story, that we had no awareness of the agreement referred to, that there was a long and complicated history of rivalry involving both companies and several others in several export markets and that while the delicacy of his (the Ambassador's) position was appreciated at this delicate juncture nonetheless the matter might have to be examined by us in Dublin in conjunction with D/Agriculture and CBF and that we might have to revert to him.

8. Ambassador McCabe said that Mr Goodman told him that he did *not* inform D/Agriculture of the details of his business and again went on to stress his concern that nothing be done which would compromise the confidential nature of his conversation with me. I rejoined by saying that in so far as I was concerned the actual details could wait until they would appear overtime in trade statistics and that any action being contemplated would endeavour not to make any difficulties for the business. Nonetheless we would have to examine and reconfirm to him as necessary our policy of dealing with all exporters in an even-handled way.

9. I was not aware that Goodman representatives were in Baghdad over recent weeks and it is perhaps unfortunate that we were not informed of this aspect earlier by the Embassy. However, if Mr Goodman decides to publicise his contract over coming days or weeks, Agra could interpret lack of action by the Embassy in their regard as favouritism or whatever towards Goodman. It is appreciated that negotiations with the Iraqis by several Irish parties simultaneously might have put the Irish side over all into a "dutch auction" situation but that is not an aspect of direct concern to us, and the current situation where one exporter gets treatment to the exclusion of others is from our point of view much more serious.

10. I am not sure if anything of value can be gained by checking out the "agreement", with CBF but, have started enquiries with Department of Agriculture. I recommend that confirmatory instruction be issued to Embassy Baghdad as soon as possible to introduce Agra as originally requested and to seek visas for them. This may be essentially *pro forma* in the sense that Iraqi business is now spoken for by another company at least for this year, but it is the minimum required to maintain our standing and credibility vis-a-vis of exporters in general.

T. Bolster
Trade Section
3 July 1987

PS My interlocutor in Department of Agriculture was not able to check "agreement" matter with CBF directly, because of absences of personnel on leave. However, he discussed it with Assistant Secretary Hoey who said that such an agreement should not exist and if it did Department of Agriculture could not stand over it. Assistant Secretary Hoey qualified the remarks by saying that CBF might have a nuanced role when the introduction of several Irish exporters to a single market might only lead to fruitless infighting, but that CBF should apply their goal of "expanding and developing" markets where such potential existed and that they should not and could not be seen as building up Irish monopolies. Addressing this point I put it to my Department of Agriculture interlocutor that Iraq, being a centralised State Trading country, there was a monopoly-type situation there. His reaction was that while that might be the case it would be desirable to assist any interested exporter with introductions to that market."

On receipt of this Memorandum, Mr, now Ambassador, Swift took certain steps which he described in evidence as follows:

407 A. "This matter came directly to me. I spoke to the Department of Agriculture as the lead Department on matters relating to the meat industry. I asked for views, specifically on the question of the so-called understanding or arrangement with C.B.F. My contact, I think it was Deputy Secretary Mockler in Agriculture spoke to a number of his colleagues including eventually the Secretary of that Department. I spoke to my Secretary and at a later stage, not very much later, but slightly later, directly to my Minister. Eventually, on the 15th of July, I issued the instruction as I had been asked for, saying that I had received a full report from Mr Bolster

on the various points in issue, that I had discussed the matter widely within our own Department and with the Department of Agriculture and I confirmed that the Ambassador should approach the local authorities and support the request of Mr O'Halloran for an invitation and for the visas.

408 Q. I think at paragraph 9 at page 3 of the first document, it sets out really....- Does that set out the Department of Foreign Affairs view so far as the Embassy was concerned?

A. Yes. I can say that that is a fair summary of the Department of Foreign Affairs views. I have to say that I would like to add something to it if you thought it would be useful for the Tribunal in the sense that I think there is a very real problem here. This is an example of a problem, and I would like to describe, if you like, how I saw then the problem and, I think, continue still to see it. I felt at the time — and I used this at my discretion with others in my Department — that there were two basic principles that had to be followed in deciding how to deal fairly with Irish competitors' interest in the same market. The first principle is the straightforward one of non-discrimination and equality of treatment, that we couldn't refuse to help any reputable firm who came to us looking for assistance and that we should make every effort to treat every firm on an equal footing, that it wasn't part of our business to help certain companies monopolise in certain foreign markets. However, I felt then, and I still feel, that that principle has to be taken together with another principle which is that the help given by the State, whether it is the State acting through the agency of, say, a body like Coras Trachtala or through an Embassy abroad or the Department of Foreign Affairs or the Department of Industry & Commerce at home, that the help given by the State must be as concrete and as effective as possible and, in order to be effective and concrete, that help must be geared to judgments about the particular situation in question and the particular chance of success of the individual firms. In other words, that while equality of treatment is the first and overriding principle, that equality or fairness doesn't necessarily imply in all circumstances and in all conditions that the treatment must be precisely similar to different companies, and I would refer back to what I said previously in my report regarding the fifth joint commission. In the fifth joint commission you had 21 different firms employed in Baghdad at the time. the leader of the Irish delegation — with, I have to say, full support from myself at least and I think from all of the civil servants involved — had taken a decision that within those, the various firms who were seeking to do business, it was reasonable to pick out those firms who we could help in a precise and concrete way during the two or three days we were there. Applying that sort of general principle to these conditions the answer was clear, that we should help Agra if they asked for what was, after all, a minimal type of help, that is to say help in relation to the issue of invitations and visas. And that we should do this even if there were views expressed by the competitor of Agra that this was, shall we say, helping a competitor to exploit certain possibilities that had been created in the first place by the first firm in the market."

On the 9th July 1987, Mr Bolster contacted the Ambassador and reported to Assistant Secretary Swift as follows:—

" *Strictly confidential*

Assistant Secretary,

Following our discussion on various points of trade with Iraq, I telephoned Ambassador McCabe in Baghdad on 8/7 and asked for details of the recently signed meat contract explaining in outline the context in which I was making the enquiry (D/Finance Memorandum) and notwithstanding the bond of confidentiality under which Mr Goodman has placed the Ambassador when he had spoken to him last week.

Ambassador McCabe replied to my question by referring to the "structure of confidentiality and trust" which Mr Goodman had placed on him and the latter's concern at details not being disclosed in Ireland to the media (to avoid farmer euphoria) or even to the Department of Agriculture in whom he was not wont to confide details of his business. For these reasons, and fearing that any disclosure could be sourced to him, Ambassador McCabe was most concerned that what he was about to say not be placed in any Memorandum which would be circulated outside the Department. I explained that it was not envisaged that the details he would supply would actually be included in the Memorandum but that they could be presented in some suitable form to the Tanaiste who could draw on "this argument" if necessary at the Cabinet table.

Having repeated his concerns, Ambassador McCabe said that according to his information, two contracts had been signed: one for 10,000 tons with one purchasing agency and another for 30,000 with a second agency. The contract is to be filled over the next twelve months and the value of the contract according to Mr Goodman was between £120m and £130m (sic:). I have drafted "format" or "version" of this information on the attached sheet. Ambassador McCabe said that his position could be eased on this matter should it be possible for somebody in Dublin to contact Mr Goodman's company and establish an outline independently.

It may not be possible to do this directly or in full but one possible avenue could be to talk with the Export Credit Section of D/Industry and Commerce and probe whether they had received any approach form Mr Goodman for insurance cover. I suspect that they would only be willing to cover a fraction of the amount stated above (£10m or £20m) but a positive response from them even on this basis, and one might glean more, would be most useful in terms of responding to D/Finance assertions of minimal interest etc in the Iraqi market.

I also availed of the opportunity to speak with Ambassador McCabe on the matter of assisting Agra Trading (a rival meat company) with introductions and visas for

Iraq. Ambassador McCabe referred to the comments which he had made to me last week and on which I reported to you at that time. I pointed out that the question of an "agreement" with CBF had been checked with D/Agriculture and they had informed us that they were not aware of an agreement nor would they stand over one if such existed. Ambassador McCabe reminded me about Mr Goodman's remarks to the effect that other companies had come in after him to the Egyptian and Iranian markets and spoiled his position by undercutting him and then not being able to deliver. I pointed out that Mr Goodman had not referred to the Soviet Market where he and Agra Trading were in quasi reverse positions. I also mentioned that you and I had discussed the matter and that I was conveying your instructions.

Ambassador McCabe then said that he would like these instructions in writing. He repeatedly emphasised to me that he had no hesitation about carrying out the Department's instructions but wished us to be aware of the risk which could be involved for the Department if Mr Goodman learnt of the assistance we offered to other companies and was angered at the risk of possible upset for his business. Ambassador McCabe said to me twice that Goodman had said to him that he (Goodman) "would put it about in the Dublin media that the Department of Foreign Affairs had queered his pitch in Iraq should we introduce other parties to the market".

With reference to these remarks may I recall the remarks made by Mr Dantz of Agra Trading to Mr McDaid and Mr Cassidy (D/Ag) in Moscow viz that he was satisfied with the even-handed treatment that he received from officials but felt that the political cards were stacked against him. These remarks were made during the term of the previous Government.

I said to Ambassador McCabe that I would report to you on this conversation and take particular note of his request for instructions in writing on this matter. I also pointed out again that Mr O'Halloran of Agra had indicated to me that he had received an indication from the Iraqi Agency or Board that they were prepared to meet him and that the Embassy's assistance and support was now required to expedite the invitation and visa process (i.e. have the Iraqi Foreign Ministry or the Iraqi Agency inform Iraqi Embassy London and give them the necessary visa authorisation).

I concluded this part by saying that we would revert to Ambassador McCabe as quickly as possible. Before finishing Ambassador McCabe, bearing the general role and concerns of his Embassy in mind, expressed the hope that the Department was fully aware of the PARC Hospital contract in Baghdad and repeated the following details.

PARC are currently carrying out a contract for the two year period 1987-88 which is worth approximately US dollars 40m per annum. The contract will expire at the end of 1988. The Iraqis have indicated to PARC that a decision will be made this Autumn as to whether the contract for the period after 1988 will be put out to tender but they have also indicated that they could decide to negotiate separately and bilaterally with PARC for the next extension. PARC naturally hope that the Iraqis will opt for the latter course. This contract of course falls within the services area and therefore these monies will not appear in the visible trade statistics. Even though Ambassador

McCabe did not say so explicitly, apart from wishing to have the Department fully aware of the value to the country of the current PARC contract, I believe it can be taken as understood that any negative decision in regard to our diplomatic presence in Baghdad could have a detrimental effect on PARC's negotiating position for the next round.

T. Bolster
9 July 1987

P.S. PARC confirmed today on the telephone the monetary amount of the contract. There are approximately 250 Irish citizens in Baghdad working on it . D/Industry and Commerce (Export Credit Section) have only had tentative enquire so far from Mr Goodman about £20m cover. They expect to hear from him again in September."

On the 17th day of November 1987, the position with regard to Export Credit Insurance for contracts in Iraq was:

"(i) Existing Exposure (i.e. on contracts already insured)	£17.18m
(ii) Commitments (i.e. offers of Insurance made but not yet taken up)	£133.48m
TOTAL	£150.66m

In respect of beef exports the commitments were expressed to include the following:—

Anglo Irish	$134.5m @ 80%	£69.42m
	$30m 80%	£15.48m
Hibernia	$11.5m @ 80%	£5.94m
	$34.5m @ 70%	£15.58m
Master Meat Packers	$10m @ 70%	£7.00m
Halal	$25m @ 70%	£11.29m
	TOTAL	£124.71m

The balance of £8.77m was in respect of non-beef exports.
Not included in this table are the applications for cover from Agra Trading viz £9.290m and the additional cover sought by Halal."

While the offers made in respect of exports to Iraq had approached but not exceeded the authorised cover, namely £150m, concern was being expressed in the Department of Industry & Commerce that the statutory limit of £300m in respect of all exports was in danger of being breached.

In the course of a minute dated the 25th day of November 1987 from Mr Walsh to Mr Timbs, Mr Walsh stated that:

"If all the offers made are taken up, worldwide exposure will gradually increase to in excess of £350m, £150m of which will be in Iraq (approximately 43%). By no stretch of the imagination could this be considered a balanced portfolio."

In a minute dated the 27th day of November 1987, from Mr Timbs to Mr O'Reilly, Assistant Secretary, Mr Timbs referred to this minute and stated, inter alia, that:—

"As offers made now bring us to the ceiling of £150m set by the Government for Iraq, we have to either increase the ceiling or close the market to further cover and refuse the applications on hand. In view of the amount of Iraqi business relative to the overall size of the scheme, it would be difficult to justify a further increase in the ceiling.

"I accordingly agree with Mr Walsh that we now close the market for export credit business and inform those who have applied that as the ceiling has been reached, cover cannot be made available."

This situation is confirmed by the letter dated 30th day of December 1987 from the Chief Executive of the Insurance Corporation of Ireland to Mr O'Reilly the Assistant Secretary in the Department of Industry & Commerce. In the course of this letter he stated:-

"As you know we are in the course of complying with your Department's instructions to issue policies of credit insurance on your Minister's behalf to exporters to Iraq and Iran. This follows the Government's decision in September to extend its credit limits for the two countries to IR£150million and IR£15million respectively."

"Although no actual policies have yet been issued, we have now been asked to draw up four policies covering meat exports to Iraq with an insured value of up to IR£146 million and two policies for Iran amounting to IR£24 million (which would exceed the country limit if all utilised at once). As the existing policies in force within the Export Credit Insurance Scheme cover approximately IR£200 million of debt and the present statutory limit for the Scheme as a whole is only IR£300 million, I thought it appropriate to write to you, before we are asked to issue the new policies, to draw attention again to the immediate requirement to increase the overall Scheme and Iranian limits to cover the new exposures."

Independently of the allocations of Export Credit Insurance authorised by the Minister for Industry and Commerce in respect of the exports of beef to Iraq, it would appear from their letter from ICI, that the existing policies in force within the Export Credit Insurance Scheme covered approximately £200m. in debt.

If the commitments then existing, but not yet taken up, were to be honoured amounting to £133.48m. then the statutory ceiling of £300 m. would be breached.

In that situation though the Minister for Industry & Commerce was prior to his meeting with Mr Goodman informed that $30m. was available for allocation, it would appear that while it was available under the £150m. ceiling for exports to Iraq, it was not available under the statutory ceiling of £300m.

This appears from an examination of the figures and is confirmed by the terms of the note faxed by Mr Timbs to AIBP International on the 10th February 1988 in the following terms:—

> "Further to our recent telephone conversation I confirm that the Minister has agreed that a further US$30m. contract value covered at either 80% or 70% depending on the credit period has been allocated to AIBP International in respect of contracted beef sales to Iraq, subject to the enactment of amending legislation to increase the aggregate amount of the Minister's liability under the Export Credit Schemes. As I mentioned to you, the US$30m. contract value is available within our ceiling for Iraq but will not be available under our overall legislative ceiling until amending legislation has been enacted."

Indeed the position with regard to exposure under the Export Credit Insurance Scheme in respect of exporters, other than exports to Iraq, was known to the Department of Industry and Commerce prior to the preparation of the Memorandum for Government dated the 7th day of September 1987.

This Memorandum after dealing in paragraph 7 with a number of applications for such insurance which would result in an additional exposure of £96.51m went on to say that:—

> "To agree to the above new requests would bring the Minister's total liability in Iraq to £121.65m, representing 40% of worldwide exposure and under the Scheme in the aggregate to £295m, leaving only £5m for new insurance worldwide under the present ceiling"

The application referred to in the said paragraph involved an increase in the State's exposure of £96.51m and such amount would increase the total exposure to £295m: this meant that the then existing exposure was £198.59m.

The allocations made by the Minister for Industry and Commerce subsequent to the Government decision of the 8th September 1987 to AIBPI (£71.43m), Hibernia Meats Ltd (£22.23m) and Master Meat Packers Ltd (£7m) increased that exposure by £100.66m, making a total of £299.65m, just short of the maximum statutory ceiling of £300m.

On the basis of these figures the allocation of cover to Halal contained in the November 1987 telex from the Department of Industry and Commerce to ICI viz 70% of $25m (£11.7m) would, if granted, have breached the statutory ceiling and there was no availability within that ceiling to make any allocation of cover to either Agra Trading Ltd or any additional cover to AIBPI (Goodman International) 

As the allocations made to these three companies AIBPI, Hibernia Meats Ltd and Master Meat Packers Ltd had increased the State's exposure under the scheme to £299.65m, this meant that there was no Export Credit Insurance available, not only for other beef exporters, but for any other exporters, unless and until the statutory ceiling was increased.

The Memorandum for Government had sought the approval of the Government to draft legislation to increase the ceiling for all markets from £300m to £500m but the Government made no decision on this application and as a consequence the Minister for Industry and Commerce was obliged to operate the Scheme within the limit of £300m worldwide.

However, Mr Ted O'Reilly Assistant Secretary on the 8th day of January 1988 replied to the letter dated the 30/12/1987 from the ICI and stated inter alia, that

> "I can assure you that the matter of amending legislation is now having immediate attention and will be introduced in the Oireachtas as early as possible."

On the 17th day of November 1987, Halal telexed ICI in the following terms:—

> "As you are aware we are currently discussing our contract with the Iraqis. Our source in Iraq has presently come back to us with a view to topping up our original contract by 4,000 tonnes giving a total of 12,000 tonnes. We, therefore, require an increase in the overall level of cover from US Dollars 25m. to US Dollars 37.2m.
>
> "We are negotiating a reduction in credit terms from 18 to 12 months in respect of this US$37.2 million and we require an 80% level of indemnity. Shipments will take place during the first quarter of 1988".

On the 18th November 1987 there was a meeting between officials of the Department of Finance and the Department of Industry and Commerce for the purpose of discussing the operation of the Export Credit Insurance Scheme. In the course of a wide ranging discussion on the Scheme the following matters relevant to the issues before the Tribunal were noted by Mr Quigley of the Department of Finance, being:—

> "4 (iii) In September, 1987, the Government effectively raised the ceiling on cover for Iraq from £25million to £150million. At present, the State's exposure in Iraq amounts to £17 million but commitments have been made to cover £133 million (£85 million of which relates to Goodman International). It is likely that the State exposure up to the limit of £150 million, will rise gradually over the next 6 — 8 months. The Iraqis will not have to pay for the exports until 12 — 18 months have elapsed from receipt of the exports and, if default occurs, the Exchequer does not have to meet claims until a further 6 — 12 months have elapsed. To date, none of Goodman International's exports to Iraq have been defaulted upon. Nevertheless, it is plain that the Exchequer is entering upon substantial contingent liabilities. (Mr Walsh stressed that his Department had been just as much opposed to the increase in the limit as this Department);
>
> (iv) The current statutory limit on the Exchequer liability is £300 million. This will have to be increased in the New Year to at least £450 million. This Department could expect to receive a draft Memorandum on the subject shortly."

On the 24th day of November 1987, Mr Stanley of Halal Meat Packers met Mr Hanney and Mr Dermot O'Mahony of the Export Credit Section of the Department of Industry & Commerce and informed them that his company had negotiated a contract to supply $18 m. of hindquarter beef to Iraq on 360 day Letters of Credit terms and that shipments could start immediately once Export Credit at 80% cover was put in place.

He informed Mr Hanney and Mr O'Mahony that the Letters of Credit were at the Halal premises.

On the 27th day of November 1987, Mr Barton of the ICI informed Mr O'Mahony that he had been speaking to Mr Stanley and that the position was that he did not have the

Letter of Credit as represented to Mr O'Mahony at the meeting on the 24th day of November 1987.

The Minister for Industry & Commerce was dissatisfied with regard to the Halal situation and the uncertainty with regard thereto and sought the advice of the Attorney General as to whether or not there was a binding offer of Export Credit Insurance to Halal.

On receipt of the Attorney General's opinion, the Minister for Industry & Commerce decided to withdraw the offer of Export Credit Insurance as it had not been accepted in the terms in which it was offered and instructed his Private Secretary to inform Halal of his decision.

This decision was made by the Minister for Industry & Commerce despite the note from Mr Timbs that which stated that:—

> "I can hardly disagree with attached advice from a legal standpoint. However, two vital points are:
>
> 1. We have never refused to cover when an offer was made even though the offer often resulted in subsequent negotiations;
> 2. The advice, in the last sentence, says "if the Minister does not accept the revised application before him. The companies at the moment are seeking urgent agreement for 80% contract on $18m contract with 12 months credit which is in accordance with our criteria".

This decision was made by the Minister and in view of the allegations made in regard thereto it is desirable that the Minister's evidence in regard to his reasons for that decision be printed in full:—

> "A. Towards the end of November, yes, I was dissatisfied with the entire Halal situation. In my view, they had been made a reasonable offer and had failed to accept it. I expressed my dissatisfaction to the Secretary of the Department and it must be noted that Mr Donlon, as Department Secretary, was also the accounting officer and my recollection is that either he or Mr Timbs or both of them expressed concern to me that if I withdrew the offer of cover to Halal that the Department could end up with litigation. Now, I just want to explain to the Tribunal that in all the ministerial departments, all the ministerial positions that I held, in every department I was always very conscious and, I have no doubt, every other Minister would, that any concern expressed by a Departmental Secretary, who, after all, is the accounting officer, who, after all, has to appear before the Comptroller and Auditor General to account for every scrap of financial expenditure, I was always very conscious and every Minister would be to insure that they were satisfied with every action being taken because they are really accountable for it afterwards and once that sort of concern would be expressed to me, fine, I would take note of it and insure that whatever level of comfort or whatever advice they needed or whatever I could do to alleviate any concern I would certainly do it. So, in relation to that and following on that, my view was that there was no question that a contract existed between Halal and the

Department as Halal had never accepted the offer of cover made to them. Now, my understanding of contract, and it has always been that you make an offer and it's accepted and if the offer is not accepted you don't have a contract. That's the essence of contract law as I have unfortunately learned the hard way once in my lifetime when I had to pay to find out, but, however, education in the university of life sometimes is useful later on. So, in order to be absolutely sure where the Department stood I took the advice of the Attorney General on this matter and it also transpired that Halal, having earlier told the Insurance Corporation of Ireland that they had a letter of credit at their premises, later had to admit that this was not true. In fact, the Insurance Corporation advised the Department to make Halal produce the contract, that they also said they had a letter of credit which they claimed to have. I informed Mr Donlon and Mr Timbs, on the 27th of November, that Halal did not have a commitment as to Export Credit Insurance cover for Iraq and I instructed by private secretary to communicate the position to Halal. I also instructed him to offer Halal a meeting for the 3rd of December with myself and them to go over all the facts of the situation. I might say to the Tribunal, at this stage, that here was a situation where, first of all, it started off with Halal looking for cover for two years, then, eighteen months after I had written to Minister Mac Sharry about it, and on the same day after getting that they come back looking for 90% cover. That was on the 5th of November. On the 16th of November they come back looking for $18.6 million for a 12 month contract which they said they had. A $6.4 million at 18 months cover for 70% and on the exact same day, on the 16th of November, they come back looking for an increase in the amount to $37.2 million. Could anybody seriously believe that this company knew exactly where they were going with that number of changes and, finally, to try and mislead the Department, the ICI and myself that they had a contract when they didn't have a contract, that they had a letter of credit and stated it which they didn't have when they were asked to produce it. My experience in business is that your word is your bond. I will leave it at that.

183 Q. Were they all matters which you gave consideration to at the time?

A. I sat down in relation to the 27th of November? I was certainly absolutely convinced that there was something seriously wrong in a company that had made so many changes even on the same day, two and three changes on the one day. Now, I have experience of running a business and two or three changes in the one day, figures, terms and everything else, certainly rings a question mark in your mind. The Minister for Industry & Commerce is a busy man in normal terms. He has to deal with all sorts of problems coming up to him and all sorts of files. He cannot be running around day after day listening to various changes, this, that and the other. You expect companies doing business with you and that expect to do business in the international market to know what they want. I'm sorry to say this company did not appear to know what they wanted. I then had a meeting on the 3rd of December at which all the principals of the

Halal company sat down in the Dail with myself and the secretary of the Department, John Donlon and Mr Timbs, and we went through every aspect of this right down the line. I pointed out to them that they had been treated reasonably and no different to anybody else, but what were they at in chopping and changing. The attitude at that meeting, I have to say, was one of more to try and establish that a legal situation existed between the Department and Halal really than talking about business but, having said that, the final position that I remember distinctly being put to me was this, look, Minister, we went out and we bought X thousand head of cattle, we killed X thousand head of cattle, we killed them to the Halal cut, we stored them, we wanted our Export Credit and we wanted our contract and I said it was at that point I said show me the contract for what you have done, show me the contract you want covered. They weren't able to produce it and I pointed out that they said they had it. I then went on and I can tell you it was a fairly heated meeting, to be honest with you, and I took them down the line and it was a strange thing, a strange thing to me for a company that here was a company who did everything by letter or fax but when you came to ask them to produce the document where they said that they had accepted the offer they said, ah well, that was done on the phone. That was not the way to do business for a company that had apparently done everything else the other way. Having said that, I could not, in all conscience, accept the situation that a company would come in to me and say we went out into marts and fairs of Ireland and we bought X amount of heads of cattle, we killed them, we put them in store to a specific cut without having an order, without having a contract, and, as they said themselves, without having Export Credit and I have to say, and I said to the Chief Executive that day, Mr John Stanley, who I understand was here, I said "Mr Stanley, if you were working for me and committed a company to that type of expenditure without knowing where you were going, I'm afraid you wouldn't be working for me the following week".

184 Q. And I think you have indicated, anyway, your final position was your final position that if they came to you with a contract you would be prepared to consider?

A. Yes. They said they could produce it. I said come back and show me when you have got it. Where was the letter of credit that was supposed to be available? It never turned up. Four and a half years later in this Tribunal Mr Khalid was here and he had to admit they hadn't a contract. So, I mean, I will leave it to yourselves.

By telephone on the evening of the 27th November 1987, Mr Finbarr Kelly, the Private Secretary to the Minister informed Halal that the offer of export credit was withdrawn.

As outlined by Mr Reynolds in his evidence, the Secretary, Mr Donlon and Mr Timbs met with the principals of Halal on the 3rd December 1987 and after the discussion described by him, he informed them that if they produced a contract he would consider their application for Export Credit Insurance.

On that evening and subsequent to that meeting, Mr Rafique of Halal Meat Packers (Ballyhaunis) Ltd wrote to the Minister for Industry and Commerce as follows:—

"3rd December 1987

Mr Albert Reynolds TD,

Minister for Industry & Commerce,
Kildare Street
Dublin 2.

Dear Minister

I would like to express my gratitude for meeting us at such short notice. I fully appreciate that your time is valuable and that you must have many other government matters to attend to.

Please let me firstly take this opportunity to introduce you to the Halal Group of companies and the history of the Group.

Brief Group History

The Sher family originally came from Pakistan and are now Irish and British naturalised citizens.

Halal Meat Packers commenced operations in Ireland during 1974 by acquiring an old sheep abattoir. This plant was located at Clare Road, Ballyhaunis, County Mayo, which is the present location of the Group's headquarters.

Since the initial plant was opened, the Group has expanded in Ireland, the United Kingdom and France. This was achieved by both building new abattoirs and acquiring established facilities throughout Ireland and the UK. The Group were assisted by both Member State Governments and the European Agricultural Guidance and Guarantee Fund.

A summary of the Group's meat processing facilities in Ireland follows:—

	Beef per Week	*Sheep per week*
Ballyhaunis	4,000	20,000
Sligo	2,000	—
Ballaghaderreen	2,000	—
Camolin	—	20,000

Banagher: Convenience food processing and cold store facilities.

In Ireland, during the last seven years, Halal Meat Packers have been responsible for processing and exporting in excess of 40% of the sheep throughput and approximately 10% of the beef throughput.

Markets

The Sher family, originally from Pakistan, are Arabian by descent and Muslim by religion. The family are well established and involved in the agricultural, industrial and political life of their home country.

All throughput at Halal plants is ritually slaughtered in accordance with the Muslim rite and throughout the world Muslims may only eat "Halal" meat. Due to the cultural and religious background of the family, the Group enjoys free access to North

African and Middle Eastern markets. The Group's ritual facilities in Ireland are the largest in Western Europe and are fully recognised by all Muslim communities throughout Europe and the Middle East.

It was to meet the demands of this market that the Group expanded in the late 1970s and 1980 and will continue to expand during the coming years.

In addition to these markets, the Group has well established markets in the United Kingdom and other EEC countries, especially France, Germany, Belgium, Italy and Spain.

Export Credit Insurance

Every year, since commencing operations, the Group has taken out an Export Credit Insurance policy. To date no claims have been made on this policy. No doubt the Insurance Corporation of Ireland who operate the scheme on behalf of the Government have made you aware of this. This cover has been provided in respect of the annual company turnover of £120 million approximately.

We are currently selling boneless beef to the Iraqi market and have requested Export Credit Insurance cover relative to same for which you have kindly agreed to provide facilities of $25 million for eighteen months at a 70% level of indemnity.

You should note that the Company participated again this year in the EEC Private Storage Scheme — APS and managed to store 11,000 tonnes of bone-in beef over a ten week period.

It is this beef that we are presently negotiating to sell to the Iraqi market and we already have in hand a contract to sell 6,000 tonnes of beef. The total value of this contract is $18.6 million and will be paid for out of the proceeds of a 360 day letter of credit.

In addition, we have tendered for a second 5,000 tonnes which would have approximately the same value. The Iraqis have requested an eighteen month credit period in respect of this contract. This, I understand, is as a result of discussions that took place between a delegation from Iraq and the Irish Government.

We now require your confirmation that Export Credit Insurance cover will be provided to us in respect of this market.

Yours faithfully
HALAL MEAT PACKERS (BALLYHAUNIS) LIMITED
Chaudry Sher Mohammad Rafique,
CHIEF EXECUTIVE"

On Friday evening the 4th December 1987 the Minister for Industry and Commerce met Mr F Walsh and Mr D O'Mahony Export Credit Section, to discuss the Halal Meat Packers contract in Iraq.

The Minister informed Mr Walsh and Mr O'Mahony that the position is that Halal did not have a commitment on Export Credit Insurance for Iraq and that this had also been conveyed to the company last week by the Minister for Finance. The Minister said that if Halal press the point they should be told firstly, that no commitment on cover exists, and secondly, that the Minister having scanned Mr Stanley's letter, considered that the statements made in the letter are inconsistent, but that he will revert with his observations when he has had time to fully consider the matter.

This minute was prepared and signed by Mr O'Mahony on the 7th December and shown to the Minister by his Private Secretary Mr Kelly on the 9th December 1987.

On the instructions of the Minister, the Private Secretary requested the Department of Foreign Affairs to ascertain from the Iraqis what Irish firms had signed contracts in Baghdad.

On the 9th of December a telex was sent from Foreign Affairs in the following terms:—

> "9/12/87
>
> TO BAGHDAD FROM HQ
> FOR AMBASSADOR FROM T LYONS
>
> PSM in Dept of Ind and Com has requested that Embassy approach Iraqis to ask what Irish firms have signed contracts for supply of beef to Iraq as of last Thursday evening. Information is required by this evening our time. Grateful if Embassy would make necessary approaches and let us know position."

Ambassador McCabe phoned Mr Lyons of the Department of Foreign Affairs on receipt of the telex of the 9th December. The Ambassador referred him to the report on the Baghdad Fair (Telex of 2nd December 87). The Ambassador was concerned that AIBP might be inclined to complain about help given to other exporters not previously active in this market. At present there are representatives of Halal and Hibernia in Baghdad and AIBP had enquired specifically about assistance given to them.

Ambassador McCabe thought there might be an understanding between the three companies that they would not undercut each other. To date no contract has been signed by any of the three in relation to the new (unpublished) tender.

On the same day a telex from J. Rowan, Baghdad was sent to Mr Lyons:—

> "Irish companies which had concluded contracts to supply beef to Iraq are AIBP (June last) and, to the best of our knowledge, Hibernia which we believe supplied French beef to the Market. However we have no specific information on this latter contract and the company has not been in touch with us. State Company for Foodstuffs Trading Director, Mrs Amal Aziz, told us today that nothing has yet been concluded in regard to the most recent beef tender.
>
> Halal and Agra representatives returned to Baghdad last week and, as far as we are aware, are continuing to negotiate on the tender".

On the 11th December Mr Finbar Kelly, Secretary to the Minister for Industry and Commerce sent a further telex number 78

> "Re: contract for supply of Irish beef to Iraq, the Minister for Industry and Commerce has asked if there has been any further developments since your telex of the 9.12.'87. He specifically asked what is the position in regard to the two companies mentioned

in the last para of your telex (Halal and Agra) i.e. have they signed contracts. Urgent reply required and Minister would be grateful if you would keep him posted on developments."

That telex was replied to on the 13th of December 1987

"1. Your number 78 of Dec 11 refers.

2. Only developments since our message of 9 December are that reps of two Irish companies mentioned in para 2 of that message left Iraq 10 December without signing contracts according to what they told us.

3. Representatives of both companies told us that State Company for Foodstuff Trading is seeking quotations at a price per tonne which both companies regard as too low for them to meet. We have some doubts whether the State Company will succeed in securing quotations at the price in question from European suppliers. We would not be surprised, therefore, if the representatives of Irish companies returned to Iraq for further discussions before too long.

4. We will advise of any relevant development we become aware of."

On the 16th of December 1987 the Ambassador replied

"Grateful if you could urgently pass this very confidential message to Minister Reynolds via his Private Secretary Mr Finbar Kelly (telex message number 78 of 11th of December to me from Mr Kelly refers).

Mr Mohammed Khalid of Halal Meat Packers telephoned me today. He said that the Department of Industry and Commerce had turned down his company's request for use of the Irish Government's Export Insurance facility for Iraq, and that in doing so the Department of Industry and Commerce had told him that this was on the basis that I had informed the Department that Halal had not secured any contract in Iraq (you will recall that in my message of 13th December I made clear that the Embassy would not be surprised if representatives of Halal returned to Iraq before too long for further negotiations with the State Company for Foodstuffs Trading).

Mr Khalid spoke of his company being discriminated against at the political level, indicating his view that his company's efforts to obtain Irish Insurance cover were being blocked by an Irish competitor which, he said, had more influence at the political level.

Mr Khalid went on to give me an amount of information most of which he had not previously made available to the Embassy. This information can be summarised as follows:—

(*a*) On the 13th September 1987 an French Company SCOA International, signed a contract with the Iraqi State Company for Foodstuffs Trading for supply of 6,000 tonnes of beef. In signing this contract SCOA International had an agreement with Halal that the 6,000 tonnes of beef would be supplied from Ireland by Halal. Mr Khalid said that Halal is obliged to supply this 6,000 tonnes with or without cover. Although he was not fully explicit on the point he gave me the impression that this meat has either been delivered in Iraq or is on its way now.

(*b*) On the 22nd of September 1987 Halal bought 6,340 tonnes of beef from the Irish Intervention Agency. The purchase was specifically for the Iraqi market according to Mr Khalid a French company CED has a contract to supply 12,000 tonnes of beef to Iraq. Hibernia Meats, which is 50 per cent owned by CED, have Irish Export Insurance cover available according to Mr Khalid. Halal Meats, in association with Hibernia used Hibernia's Insurance cover to fulfil 6,340 of the CED contract for 12,000 tonnes.

(*c*) Halal is now seeking to negotiate a contract to supply directly to the Iraq State Company for Foodstuff Trading 10,000 tonnes of Irish beef. According to Mr Khalid, his company intends to continue negotiations on this possible contract. The price being offered by the potential customer is too low at present. Mr Khalid hopes that the customer will be obliged to raise it because of difficulties in obtaining good quality meat at the price now on offer. Mr Khalid indicated that Halal has 10,000 tonnes in APS stock at present and that the company sees the Iraqi market as the most promising destination for this meat.

Mr Khalid indicated that his company's current applications for Irish Export Insurance cover relate to (A) and (C) above.

When I asked Mr Khalid why he not previously informed the Embassy of (*a*) and (*b*) above he said that he had seen no reason to involve us.

Since this message is confidential I would particularly request that its contents not be quoted in any discussions between the Department of Industry and Commerce or ICI and Halal Meat Packers".

Goodman International signed a contract with the State Company for Foodstuff Trading on the 13th day of December 1987 for $105m (contract No. 6/88) and Mr Aidan Connor spoke with Mr Timbs of the Department on the telephone and Mr Timbs record of such telephone call is:—

> I spoke to Mr Aidan Connor, Goodman International. I informed him that it was imperative that he conclude the Export Credit Insurance arrangements for the US $134.5m. beef contract for Iraq within the next week. He indicated that he was ready to sign the relevant papers and pay premium. He mentioned, however, that he still required cover for a further US $52m contract. I informed him that as of the moment the maximum cover which I could make available, subject to Ministerial approval, was US $30m. at 80% cover. Furthermore, we would need formal application in respect of the additional amount of cover now required. Mr Connor indicated that insofar as he was aware other interests seeking contracts in Iraq had been unsuccessful."

On the 27th day of November 1987 Mr Aidan Connor Deputy Chief Executive AIBPI wrote in the following terms to the Secretary to the Department of Industry and Commerce as follows:—

"*For the attention of Mr Joe Timbs*

The Secretary
Department of Industry & Commerce
Kildare Street
Dublin 2

Dear Sir

Re: *Export Credit Insurance-Iraq*

You will recall that in September of this year you approved insurance cover for AIBP in respect of a beef contract with Iraq valued at USD 134.5 million. AIBP has now secured an extension of our existing contract valued at USD 52 million and seeks insurance cover on this amount.

The terms of the existing policy are as follows:

Period of Cover	12 months
Amount of Cover	80% of contract value
Claims waiting	6 months
Premium	1% of contract value plus a further 2% in the event of any claim

The terms of our new contract will require similar cover to the above except in relation to the credit period which is now set at 18 months following the recent agreement at the Iraqi-Irish joint Commission.

To understand how important this market is, both to AIBP and Ireland as a whole, it may be helpful to outline the recent history of beef sales to Iraq.

1. Iraq is a large market in world terms, importing 120,000 tonnes of beef annually (as a comparison, the Iraqi market is twelve times larger than Libya.
2. AIBP was the first Irish Company to gain entry into the Iraqi market. It took three years and involved a huge deployment of resources, both in human and financial terms to finally achieve the breakthrough.
3. AIBP has now supplied the market for seven years. Some three years ago, the Iraqis sought extended credit from their suppliers. In response to this, our competitors gained government support for their marketing efforts, e.g., COFACE in France, Petrobras in Brazil (an oil barter arrangement), ECGD in the United Kingdom. AIBP continued to supply the market with no Bank support and very little support from the Irish Government. In doing so, we took a calculated risk in order to preserve the Irish identity in the market place.
4. Due to the excellence of both product and service provided by AIBP to Iraq, we have successfully established a premium in the market for Irish beef. In addition, we have increased substantially our market share despite severe competition from Brazil and France.
5. It is absolutely vital that we supply this contract to keep our competitors, principally the Brazilians, from regaining their previously commanding position in the market. Because of the major depreciation of the US dollar against European currencies, our price to Iraq has declined by 10% in real terms over the last two months. At the same time, all South American currencies have depreciated against

the US dollar therefore providing our Brazilian competitors with a two part advantage.

6. Good management dictates that we should minimise any risk exposure and thus we seek export credit cover on this additional quantity.

7. From a marketing perspective, it is imperative that the Iraqis see a united front from the sellers of Irish beef in order to preserve the price premium now clearly established. The Brazilian exporters openly compete with one another in Iraq and this fact has been exploited in full by the Iraqis as is evidenced by the successive reductions in selling price accepted by the Brazilians in recent tenders. For Ireland, a single voice is an essential marketing tool to prevent such an occurrence. Because of our history in the market, AIBP should be that voice and I would therefore request that your Department reject sundry applications for credit from various Irish suppliers in order to prevent a repetition of the Brazilian experience.

From the above, it is evident that AIBP needs full government support in the supply of this contract and I would ask for your most favourable consideration of our request which is summarized as follows:—

Value of the contract	USD 52 million
Period of credit	18 months
Credit cover	80% of contract value
Claims waiting	6 months
Premium	1% of the contract value plus a further 2% in the event of any claim

I look forward to your early response.

Yours faithfully
AIDAN CONNOR
DEPUTY CHIEF EXECUTIVE"

There is considerable uncertainty with regard to the receipt of this letter by the Department of Industry & Commerce. While it is addressed to the Secretary of the Department and expressed to be for "For the attention of Mr Joe Timbs", there is no record of its receipt in the Department by either the Secretary or by Mr Timbs.

There is no doubt that this letter constituted a formal application for cover in respect of a $52m contract.

Mr Timbs does not appear to have been in possession of or have seen this letter prior to the 16th day of December 1987 because on that date he spoke to Mr Connor for the purpose of impressing on him the necessity to conclude the Export Credit Insurance arrangements for the $134.5m beef contract for Iraq.

During the course of this conversation Mr Connor mentioned to Mr Timbs that he still required cover for a further US$52m contract.

Mr Timbs informed Mr Connor that the maximum cover which he could make available, subject to Ministerial approval, was $30m at 80% cover. Furthermore he would need, as he stated, formal application in respect of the additional amount of cover now required.

Mr Connor did not refer to the fact that by letter dated the 27th day of November 1987 he had made formal application for such cover and if Mr Timbs had received the said letter he would have been aware of this application.

Though this is denied by the Minister for Industry & Commerce, Mr Timbs stated in evidence that towards the end of December 1987 he received for the first time a copy of this letter from the Minister for Industry & Commerce and discussed terms thereof with him.

According to Mr Timbs' evidence the letter applied for Export Credit Insurance for $155m.

It is now clear that Mr Timbs must have been mistaken in this regard because the original application was for Export Credit Insurance for $52m and obviously related to the offer of cover in the sum of $30m which the Minister had agreed to make available to Mr Goodman during the course of the meeting on the 13th day of November 1987.

In the course of his evidence, the Taoiseach, Mr Reynolds stated that he had no recollection of receiving the original or a copy of either version of the letter bearing date of the 27th November 1987 or of producing it to Mr Timbs in late December 1987 and discussing the contents thereof with him.

The Tribunal is satisfied that his evidence in this regard is correct and that Mr Timbs account is mistaken, possibly due to faulty recollection.

Mr Timbs had had a meeting with Mr Aidan Connor of AIBP, the writer of the letter, on the 23rd day of November 1987 in the Department of Industry and Commerce, at which the provision of further Export Credit Insurance to AIBP in respect of exports to Iraq was discussed.

Mr Connors note of this meeting is as follows:—

"***Joe Timbs — Department of Industry and Commerce***

23/11/87

Meeting to encourage further credit for Iraq.

Proposal put JT as follows:—

1. Contract Value $52m.
2. Period/January 88 to January 89 (overlapping existing contract).
3. Credit period/18 months.
4. Other Terms and Conditions / As before 80% cover, six months claims waiting).
5. Premium — As before/1% on contract value. Further 2% on event of a claim.

Problem:— JT has only $30m contract available under guidelines at the moment. He will speak to the Minister and revert to me.

> Existing letters of offer expire between mid-December 1987 and early January 1988."

The original letter dated the 27th November 1987, which was expressed to be "for the attention of Mr Joe Timbs", contained a request which was summarised in the said letter as follows:

Value of the contract	USD 52million
Period of credit	18 months
Credit cover	80% of contract value
Claims waiting	6 months
Premium	1% of contract value plus a further 2% in the event of any claim

The letter of the 27th November 1987 written by Mr Connors was obviously a formal written application in relation to matters that had been discussed in detail with Mr Timbs on the 23rd November 1987 and was intended for Mr Timbs and not for the Minister for Industry & Commerce.

It is most improbable that a letter addressed to the Secretary, Department of Industry and Commerce, expressed to be "For the attention of Mr Joe Timbs" and relating to a contract which had been discussed with Mr Timbs some five days earlier, would have been given to the Minister for Industry and Commerce and even more improbable that he would have retained possession of it until late in December 1987.

Mr Timbs took exception to the terms of one paragraph contained in the letter and having discussed the matter with Mr Connor on the telephone, returned the letter to him for the purpose of having the offending paragraph deleted.

This paragraph was as follows:—

> "From a marketing perspective, it is imperative that the Iraqis see a united front from the sellers of Irish beef in order to preserve the price premium now clearly established. The Brazilian exporters openly compete with one another in Iraq and this fact has been exploited in full by the Iraqis as is evidenced by the successive reductions in selling price accepted by the Brazilians in recent tenders. For Ireland, a single voice is an essential marketing tool to prevent such an occurrence. Because of our history in the market, AIBP should be that voice and I would therefore request that your Department reject sundry applications for credit from various Irish suppliers in order to prevent a repetition of the Brazilian experience."

The views expressed in this paragraph undoubtedly represent the views of Mr Goodman and the Goodman organisation, that as they had opened the market for the sale of beef in Iraq, they should be the only Irish suppliers to be afforded assistance in the development of this market and that no other supplier should be facilitated in any way by the State whether by way of Export Credit Insurance or otherwise.

On the 6th day of January 1988, the amended letter was received in the Department of Industry & Commerce and bore the date the 27th day of November 1987, the offending paragraph was removed but in lieu of an application for cover for $52m, the application was for $155m and this application was treated by the Department of Industry & Commerce as having been made on the 27th day of November 1987.

In view of the allocations of Export Credit Insurance cover already made and referred to herein the Export Credit Insurance sought, whether it be $52m or $155m was not available because of the statutory limit of IR£300m.

By letter dated the 13th day of November 1987 the Insurance Corporation of Ireland were informed by Mr Fitzpatrick, Managing Director of Taher Meats Ltd that his company was then engaged in talks with Iraq for the supply of approximately 6,000 to 7,000 tonnes of meat, the total value of which was in excess of IR£14m for which they requested cover or whatever was possible in the circumstances in order that they may be able to compete for this order.

On the 8th day of December 1987 Mr Fitzpatrick met Mr Walsh and Mr O'Mahony of the Export Credit section of the Department of Industry & Commerce and informed him that Taher Meats had passed up the contract valued at £14m referred to in the letter to the ICI as Export Credit Insurance was not available at the time. He stated however that a further contract valued at £10m or £11m had been offered and they sought Export Credit Insurance in respect thereof.

At this point of time no offer of cover could be made to either Agra Trading Ltd or Taher Meats Ltd because the permitted ceiling had been reached and the previous offer made to Halal had been withdrawn.

On the 19th of January, 1988 ICI requested a meeting with the Department of Industry and Commerce to discuss problems which had been identified in relation to the £300 million allowed under the Insurance Act 1953 as amended. The only significant details to emerge from this meeting so far as this Tribunal is concerned is that Mr Timbs at an early stage in the meeting made the comment that Larry Goodman had been in touch with them recently looking for yet more cover for Iraq. The amount involved was approximately the same as the last application ($134.5m.) but that no decision had yet been taken and that matter had been passed to Government for a decision.

The meeting considered at the current exposures and commitments under the scheme.

The following figures were noted:

Existing liabilities under policies...........	£185.0m.
Outstanding offers to Iraq....................	£124.0m.
Outstanding offers to Iran...................	£11.5m.
	£320.5m.

ICI commented that "the limit is blown". It was also pointed out that the £124m outstanding Iraq offers was understated by approximately £10m. as Anglo Irish Meats contract gives an option of cover in DMs or $s. Both sides agreed that the legislation needed to be changed urgently.

An internal Memorandum between Frank Mee and Bob Frewen of the ICI dated the 26th of January reflects the concern of ICI at this period:

> "At our meeting with Ted O'Reilly and Joe Timbs on 19th January 1988, Joe Timbs mentioned that Larry Goodman had been in to the Department recently looking for extra cover for Iraq. The amount involved was roughly the same as the current application ($134.5m). No decision has yet been taken and the matter is before the Government.
>
> At another stage in the meeting Ted O'Reilly mentioned that the Department would very much appreciate a "thesis" prepared by ICI on any country which we do business with.
>
> I feel that this is an ideal opportunity for us to prepare a detailed assessment on Iraq, and possibly Iran. We have been sending over information piecemeal over the past few weeks so most of the information should be to hand. When we are sending the "thesis" we should link it to Joe Timb's comment re Goodman's request for extra cover.
>
> As I mentioned earlier the Administrator's instructions are to take every possible opportunity of expressing our reservations on doing business with Iraq and Iran."

The ICI supplied to the Department of Industry and Commerce a Credit Assessment of Iraq dated the 29th of January of 1988, and stated that:-

> "In our opinion that the perceived success of Irish exporters in securing contracts to Iraq is largely due to the unwillingness of other countries to supply goods to Iraq.
>
> We strongly recommend that no further credit be offered to Iraq under the Export Credit Scheme by the Minister for Industry and Commerce. Indeed it is also our opinion that the current ceiling on insured exports to Iraq of IR£150 million is not justified as it is our fear that claims, and rescheduling, will arise on existing commitments unless there is an immediate end to the Iran/Iraq war and a consequent dramatic improvement in Iraq's financial position."

On the 17th of February 1988 the Insurance Corporation wrote to AIBP indicating that the ceiling of £300 million had been reached under the Export Credit Insurance Scheme. The letter continued:—

> "At the moment, the legislation provides for a maximum liability of IR£300 million for all amounts included under the Scheme. If all applications for Export Credit Insurance, including your own, were covered for the full amounts the £300m. legislative ceiling would be exceeded. We understand that the Department of Industry & Commerce is aware of this problem and is taking the necessary steps to rectify the situation.
>
> Until such time as the Department has regularised the position, we propose issuing future cover limits in respect of our unutilised capacity on a shipment by shipment basis".

In a meeting on the 17th of February 1988 between the Department of Industry and Commerce and ICI numerous matters in relation to the contract between AIBP and ICI were discussed. Mr Barton confirmed that they had insisted on a statement that all goods must be produced in the Republic of Ireland and evidence retained by Anglo-Irish if called on to produce such evidence by ICI or the Department.

On the 18th of February 1988. Joe Baragwanath on behalf of AIBP sent to the ICI letters referring to numbered letters of credit

> "We confirm that beef supplied under the above Letters of Credit will be the produce of the Republic of Ireland and we will retain evidence to prove this."

On the 18th of February 1988, Aidan Connor prepared a memo for Mr Larry Goodman in the following terms:—

> "1. A number of events have recently occurred on the above subject which cause me great concern and have potentially serious implications for our 1988 production / shipping schedule to our Iraqi customers.

To bring you up to date I attach a number of documents which are self explanatory:—

> "*a*) A original letter of offer dated the 18th of December 1987 from ICI in relation to our 1987 contracts.
>
> "*b*) Letter dated the 17th of February 1988 which, in effect, qualifies considerably the support promised in the original letter.
>
> "*c*) Letter of application dated the 27th of November 1987 for cover on the 1988 contracts which we have negotiated.
>
> "*d*) Unsigned reply dated the 10th of February 1988 from Joe Timbs Department of Industry & Commerce to my letter of 27th November 1987.

My concerns are:

1. Just today we eventually managed to get our hands on the policy which is the subject of the offer dated the 18.12.87 (*a*) above) relating to the 1987 contracts.
2. I read their letter of 17th of February 1988 as essentially a replacement of the overall cover by a shipment to shipment cover which is totally unsatisfactory and inappropriate for our contracts which are a fixed commitment for the next six months.
3. My letter to the Department (27th of November 1987) asked for cover of the 1988 contracts which at that time was tentatively agreed. In response to that letter, you told me that, on your information, there was $58 million cover available immediately and the balance would follow in due course.
4. Based on your information, we proceeded to firm up the contracts and indeed were obliged to extend 18 months credit on the basis of the agreement on credit reached at the Iraqi Irish Joint Commission.
5. I was concerned at the lack of formal response to my letter of the 27.11.87 from the Department and eventually, after a number of calls, I managed to get the commitment contained in (*d*) attached (10th February 1988). This in itself is scarcely official since it is unsigned.
6. I am even more concerned that on the basis of this reply it looks as if there will

be no further credit beyond the $30m for 1988 contracts, leaving us with a $125m uncovered.

As a result of the above, we are faced with some stark choices:

1. Cancel some of the contracts to reduce the risk. This will undoubtedly lead to a severe loss of creditability in the eyes of our customers and guide them towards the South Americans and/or French who, of course, have almost an unlimited Government backing.
2. Move our production from Ireland to South America or France and hope to avail of some of the Government cover there. This move would entail investment in plants but would certainly yield a reasonable pay-back.
3. Lobby the Irish Government for more support and press on with our production/ shipping plans from Ireland.

"I would appreciate your comments as a matter of urgency since the first shipment date for 1988 contracts is March (1987 and 1988 deliveries overlap from March through June '88)".

On the 19th of February 1988, ICI wrote to Mr Oliver Murphy of Hibernia Meats Limited and refer to their letter of the 22nd of December of 1987 and Industry and Commerce's letter of the 11th of November 1987 (which letters refer to the offer of cover on US$46 million US business). In the letter of 19th February 1988 ICI said that the offer of cover expires 60 days after signing the Iraqi contract. They sought certain information to enable them to issue the policy of insurance. The letter ended

"If you wish to take up this offer please contact us immediately. If we do not hear from you by Wednesday 24th February we will presume that you are not accepting our offer of cover and accordingly will be withdrawing this offer from that date."

On receipt of this letter, Mr Murphy arranged to meet Mr Timbs and Mr Donnelly of the Department of Industry and Commerce which meeting occurred on the 23rd February 1988.

Mr Donnelly's report dated the 26th February 1988 deals with the discussions at such meeting and the relevant portions thereof are set forth hereunder.

"**REPORT OF MEETING**

Subject: Export Credit Insurance cover for Iran and Iraq

Date: 23 February, 1988

Venue: D/Industry and Commerce

Present: Mr Joe Timbs, Mr Gerry Donnelly } D/Industry and Commerce

Mr Oliver Murphy, Mr Bernard Maguire } Hibernia Meats

1. Mr Murphy referred to a letter dated 19 February, 1988 which he had received from the Insurance Corporation of Ireland. The opening paragraph of this letter stated that the offer of cover on the company's exports to Iraq expired 60 days after the signing of the contract. Mr Murphy said that this could not be correct and requested clarification of the situation. Mr Timbs agreed that this was an error and that the cover expired 60 days after the date of offer. Mr Murphy said that on that basis the offer of cover expired the following day, 24 February, 1988. At Mr Murphy's request Mr Timbs agreed to roll over the offer of cover for a further 60 days, i.e., until 24 April, 1988.

3. Mr Murphy indicated that he was anxious to pay the full premium on the entire $47m, contract value in Iraq i.e., the $28m. worth of business already obtained and the $18m. worth which they were confident would be signed up in the immediate future. Mr Timbs agreed that this would be done but indicated he would prefer if it waited until the remaining issues associated with the policy and the terms of cover had been sorted out hopefully by the end of this week. It was also agreed that premium on the balance of $18m. would be refundable if not utilised.

4. Mr Murphy asked that the terms for the balance of $18m. would be on the basis of payment of claims in dollars also. Mr Timbs said he had no decision on this matter but was continuing to examine it.

5. Mr Timbs agreed to write to Mr Murphy on the outstanding issues as soon as possible.

---

26 February 1988.

c.c. Mr J Fanning,
Mr J Hanney,
File."

"Mr Timbs wrote to Mr Murphy on the 9th March 1988 as follows:—

"9th March 1988

Mr Oliver Murphy
Managing Director
Hibernia Meats Ltd
Sallins
Co Kildare

Dear Oliver

I refer to our recent discussions and confirm agreement to extend the offer of Export Credit Insurance cover in respect of contracts for the export of beef to Iraq valued at US $46m. for a further 60 days. This offer will now expire on 24 April, 1988.

I understand that the initial contract for $28m will be shipped by mid-1988. In relation to the remaining $18m. not yet the subject of firm contract, the position is that cover will only be available for this amount on condition that all shipments are made to Iraq by end-September, 1988.

Your sincerely

____

Joe Timbs"

The Tribunal refers to this meeting to illustrate the fact that though on the 15th day of October 1987 the Minister for Industry and Commerce notified Hibernia Meats Limited that he had decided to make Export Credit Insurance available for their contract in Iraq in the sum of $46m, Hibernia Meats Limited had not by the 23rd February 1988 succeeded in negotiating contracts for the full amount thereof but had merely negotiated a contract for $28m and that in spite of that the cover granted by the Minister for Industry and Commerce was continued.

Though Mr Timbs had on the 22nd day of October 1987 conveyed to Mr Phelan of Master Meat Packers Limited confirmation of the allocation of cover in respect of beef exports to Iraq valued at £10m, he had not by the 19th day of July 1988 availed of such cover.

On the 19th of July 1988 the Insurance Corporation of Ireland wrote to Pascal Phelan indicating that his £10 million Irish was due to expire on the 26th of August of 1988 and indicating the terms of the insurance being offered to him, which was £10 million Irish with an indemnity of 70% exclusive of export refunds and an 18 month irrevocable letter of credit issued by the Rafidain Bank Baghdad. This was eventually replied to on the 20th of July 1988 by Pascal Phelan indicating that the terms in the letter of July 19th were acceptable but indicating that the policy would have to cover shipments up to July of 1989.

On the 22nd of November of 1988 Mr Timbs the Principal Officer met the Minister for Industry and Commerce concerning the position on Export Credit Insurance for Iraq. It was confirmed to him that the Department of Finance approval for increasing the ceiling to £270 million while it had been sought no response had yet been received. The Minister indicated that he intended to speak to the Minister for Finance about the matter. Mr Timbs pointed out to the Minister that Hibernia Meats had been pressing very strongly for approval of a £10 million cover which had originally been allocated to Master Meat Packers. MMP had produced the beef but for marketing reasons it was being supplied through Hibernia Meats. Shipments were being made this week and in early December the Minister had on the 21st of October 1988 agreed that this £10 million cover could be allocated to Hibernia but as part of the overall additional £20 million for 1989 on the basis that there was room within the present ceiling of £150 million for Iraq to accommodate the Hibernia request the Minister indicated that the £10 million cover could now be formally allocated to Hibernia.

On the 25th of February 1988, there was a meeting between John Stanley of Halal Meats and Gerry Donnelly and John Fanning of the Export Credit Section. The purpose of this meeting was to try and obtain Export Credit Insurance for beef contracts in Algeria and

Iraq. So far as Iraq was concerned Mr Stanley stated that he had no clear understanding as to why his Export Credit Insurance for Iraq was not approved by the Minister in Christmas 1987. He was now in a position where he had been invited by the Iraqis to tender for 15,000 tons of boneless fore-quarter beef value approximately $40 million US. He did not intend tendering for the full amount but expects to limit his tender to about US$25 million worth of beef on an 18 month irrevocable letter of credit to be opened by an approved Iraqi bank.

He had received from Baghdad a telex (9/2/1988) in the following terms:—

> "Here is Iraqi Co. for Agricultural Products Marketing. We are pleased to inform you that we have now tender to supply your Co. (15,000) M/T Boneless Beef meat (forequarter). The closing date is 24.3.1988. We like that your company to participate in this tender, and please attach with your offer B.B. (Bid Bond) as 10/0 from total amount of your offer. And for more information, you can take the general conditions and specifications from our head office."

Industry and Commerce indicated that the ceiling on insured exports to Iraq had been reached and no further cover was available.

Mr Stanley said that Halal would be submitting an application through ICI for Export Credit Insurance for Iraq notwithstanding the fact that the existing ceiling had been reached.

He appreciated that the decision to provide insurance cover for the Iraqi market rested with the Minister and he requested that Halal's application be given full consideration by ICI, the Department and the Minister.

On the 26th of February 1988 Halal Meat Packers wrote to ICI:—

> "Once again we have been requested by the Iraqi Company for Agricultural Products Marketing to tender for 15,000 tonnes of forequarter boneless beef. Our invitation to tender is attached. It is our intention to tender for between 8,000 to 10,000 tonnes.
>
> The conditions relative to the sale are:—
>
> (i) Delivery will take place between June 1988 and May 1989;
>
> (ii) Payment will be monthly by Letter of Credit opened by Rafidain Bank;
>
> (iii) The General Conditions of the Contract do not specify the term of credit, they require;
>
> (iv) The contract value will be $25m. approximately.
>
> Please consider this as our application for Export Credit Insurance cover in respect of this sale valued at $25m. approximately.
>
> We expect to hear from you shortly detailing the premium charge and what period of credit you would be in a position to offer us."

Though Taher Meats Limited had been refused Export Credit Insurance, they continued to make representations in regard thereto and on the 5th day of May 1988 their representatives met the Minister for Industry and Commerce.

Taher Meats Ltd had on the 30th March 1988 written to the ICI, with copy to the Department of Industry and Commerce in the following terms:—

> "Taher Meats Limited is primarily engaged in the meat and livestock business. The Chairman of the company Mr Naser Taher, a Jordanian gentleman, is well known in the Irish Meat Trade.
>
> The company is very strong on the marketing side with particular emphasis on North African and Middle Eastern Countries. The company has offices in Cairo, Egypt and Amman in Jordan which gives us a unique and strong market position. We also have an office in London where our sales force sell our canned goods to major supermarkets. On major contracts we allocate personnel to the buyer's country as we find from experience that any problems that arise can be overcome speedily.
>
> We slaughter cattle at Roscrea, formally owned by the Purcell family, and prior to that, owned by the Crowley family. We produce up to 200,000 cans of meat every week for orders in the UK. We are about to embark on a major development programme in Roscrea and within two to three weeks, we will be submitting proposals to the IDA/SFADCO with expenditure of the order of IR£12.5m. over two years. A minimum of IR£5.5 m. will be spent at Roscrea over the next six months in modernising our facilities and increasing our product range in the canned goods.
>
> In the Autumn of 1987 the company was very active in buying beef and storing it under the Aids for Private Storage Scheme. Indeed, we were so active that one well known meat trader, who also has an abattoir, has actually "accused" Taher Meats Limited for being the reason for farmers getting more for their cattle in the back end of 1987. On this one, we will stand "accused". Most of the cattle that we slaughter at Roscrea were de-boned and packed for third country markets.
>
> The company's longer term strategy is to build-up a strong, viable meat producing and processing company. As mentioned, we have an abattoir at Roscrea, and we intend to add to our production facilities by either building new or acquiring old.
>
> In our previous letters to you of 13 November 1987, 7 March 1988 and 16 March 1988, we mentioned that we were actively concentrating on two specific markets, namely Iraq and Angola. No doubt you are aware that winning success in the market place is a long, slow process. Four of our Executives have visited Angola in recent months and we are actively pursuing the deal as mentioned in our letter of 7 March 1988. One of our sales force has spent five of the last six weeks in Iraq and our Chairman. Mr Naser Taher, has been there twice in the last four weeks. The contract currently being negotiated in Iraq is in two parts and part one is worth IR£28m. and part two, IR£12.5m. and both these contracts are additional to the business being done by other Irish companies.
>
> We need ICI cover for this business and we feel very strongly about obtaining such cover. We hasten to add, that we fully appreciate your position, but we also know

that since our original application, a great amount of beef has been shipped to Iraq from Ireland.

1. Mr Naser Taher is a wealthy Jordanian business man who choose to come to Ireland and invest substantially in Irish business and Irish people.
2. Mr Taher asks for, and expects, no more or less favourable terms than competitors.
3. His marketing strengths and worldwide experience gives him the edge over most competitors. In two recent tenders i.e. Egypt and Algeria, contracts had been agreed and in the case of Algeria a contract had been signed with our company, but because other Irish companies undercut our prices by between 150 and 200 dollars per tonne, we did not secure the business.
4. In most third country markets he speaks the language of the buyer and undoubtedly this gives him an added advantage.
5. Mr Taher has purchased an abattoir in the hear of Ireland's cattle producing area — Tipperary. It is his intention to invest substantially in Roscrea, and as mentioned, to expand his interest in fixed assets in the meat trade.
6. Assuming Mr Taher expands his interest in the meat trade, he would become good competition for existing operators and would most assuredly be welcomed by the Irish producers of cattle.
7. I myself, have spent a lifetime in the Irish meat trade and am well known to both producers of cattle and buyers of beef. I jointed Mr Taher last July as Managing Director of his meat interest. I assured him at that time that Ireland needed people like himself and welcomed foreign investment in general and that the Government and State and semi-State Agencies would treat all investors, whether they were Irish or not, as equal.
8. As you will appreciate an answer to our request for cover is urgently required as major decisions must be taken on the outcome."

The Department of Industry & Commerce on receipt of this letter from Taher Meats Ltd, wrote to ICI

"I am writing just to ensure that there are no misunderstandings in relation to Iraq. The position is, as I have told all the meat companies who have contracted me, that we are not open for business for meat contracts for Iraq and do not envisage being so in the future. It would be as well that Taher be left in no doubt on this score."

Present at the meeting of 5th May 1988 was Dr Sean McCarthy a Junior Minister in the said Department who had been making representations on behalf of Taher Meats Limited.

As appears from the notes of the meeting taken by Mr Timbs, the Minister for Industry and Commerce, having heard their representations, informed them that as the aggregate liability under the credit and finance schemes had reached the ceiling set by legislation he could not consider either of the Companies applications, regardless of the risk assessment.

The Minister, at this meeting, made it clear that the aggregate liability under the credit and finance schemes had reached the ceilings set by legislation and he could not consider either of the companies' applications regardless of the risk assessment. He indicated that the matter would be going to Government but that he expected opposition from the Department of Finance and that accordingly it might take sometime for the necessary legislation to be enacted.

The Minister indicated that the ceiling for Iraq had been reached and that, in his view, exposure in that country was now at the maximum level acceptable but that the matter would be brought to Government.

The Minister concluded the meeting by asking the visitors mainly the representatives of Taher Meats Limited to provide him with a detailed case which he might use in discussion with Government.

At this time the following meat processing companies were seeking Export Credit Insurance in respect of the export of beef to Iraq.

(i) AIBPI in respect of contracts with a total value of $155m.

(ii) Agra Trading Ltd in respect of a contract valued at $17m.

(iii) Taher Meats Ltd in respect of a contract valued at £11m.

(iv) Halal Meats in respect of a contract valued at $25m.

The Companies concerned were in the course of negotiations with the relevant Iraqis authorities in respect of such contracts.

If cover were granted as sought and on the basis of 70% cover in respect of an 18 month credit period, the State's liability would be increased by £93.905m.

The granting of Export Credit Insurance on this scale at this time was not possible having regard to the statutory ceiling of £300m.

In view of the then existing situation, Mr Gerard Donnelly of the Export Credit Section of the Department of Industry and Commerce prepared on the 29th February 1988 a Memorandum entitled *"Ceiling on Export Credit Insurance for Iraq"* for the benefit of the Minister.

A summary of this Memorandum was also prepared on the 29th February 1988 and it is desirable that the summary thereof be set forth herein as follows:

"SUMMARY OF MEMORANDUM

Ceiling on Export Credit Insurance for Iraq

1. ***Decision Sought***

   It is necessary to consider whether there is a case for increasing the export ceiling of IR£150m. on insured exports to Iraq in order to accommodate applications for cover currently on hands from Irish exporters.

2. ***Current Position***

The existing ceiling of IR£150m. was set by the Government at their meeting on 8 September 1987. Existing exposure and commitments amount to a contingent liability of IR£141m. This leaves a balance of IR£9m. to cater for new applications.

3. ***New Applications***

Applications currently on hands, would, if approved, increase the contingent liability by IR£59.26m. An increase of £50.26m. in the existing ceiling would be required to accommodate these new applications.

4. ***Application from Anglo-Irish Beef Processors (AIBP)***

The bulk of the additional liability would arise as a result of an application from AIBP for increased cover on exports to Iraq. AIBP have already been provided insurance cover on exports valued at DM 257.1m. (contingent liability IR£76.75m). They have also been offered insurance to cover a further US$30m. sales to Iraq and are now looking for additional insurance to cover additional sales of US$125m. to Iraq. The additional US$125m. sales would, if insured, give rise to an increase of IR£54.7m. in the Minister's liability over and above existing commitments.

5. ***Possibility of Additional Spare Capacity***

It is possible that additional spare capacity may arise within the existing £150m. ceiling.
This would result from,

(*a*) settlement of overdue debts by the Iraqis (maximum IR£4.2m.),

or

(*b*) existing offers of insurance not being taken up by exporters (amount indeterminable).

6. ***Conclusions and Recommendations***

(*a*) A current credit assessment of Iraq conducted by ICI concludes that Iraq is a high risk market and strongly recommends that no further credit be offered to Iraq under the Export Credit Insurance Scheme;

(*b*) Ireland has a liberal attitude with Iraq compared with our EEC partners. Netherlands, Spain, Greece, Italy and Denmark have all suspended cover. Limited cover only is available in the UK, Germany, France and Belgium;

(*c*) The existing statutory limit for all exports under the Scheme is IR£300m. Current exposure in respect of exports worldwide is £185m. In order to accommodate the existing ceiling of IR£150m. on exports to Iraq an increase in the statutory limit will be necessary and amending legislation will be submitted for Government approval shortly (to increase the statutory limit to IR£500m.). A £150m. liability in Iraq represents 40% of total liability worldwide and some 30% of the proposed new statutory limit. This is already considered to be excessive.

(*d*) Spare capacity within the existing ceiling amounts to £9m. £4.56m would be required to meet applications currently on hands from non-beef exporters. It is advisable that the remaining £4.4m be set aside to allow for possible additional liability arising from exchange rate fluctuations and from the 5% variation clause in the existing AIBP contract.

It is therefore Recommended

(1) that we maintain the IR£150m. ceiling as our absolute maximum liability for Iraq;

(2) that current spare capacity within this ceiling be allocated on the basis outlined in (d) above; and

(3) AIBP be informed that no additional insurance cover is available above the US$30m. sum already offered for insurance.

G Donnelly
29th February 1988"

The Appendix 1 to the said Memorandum set forth the commitments existing as of January 1988 with regard to the granting of Export Credit Insurance for beef exports to Iraq as follows:—

3. ***Breakdown of Commitments — January 1988***

Company	Contract Value	Exch Rate	Credit Period (Months)	%	Liability IR£
AIBP	US$30m	1.60	18	70	£13.125m
Hibernia	US$411.5m	1.6	12	80	£5.75m
	US$17.0m	1.6	12	80	£8.50m
	US$17.5m	1.6	18	70	£7.66m
Master Meat Packers	IR£10m.	—	18	70	£7.00m

4. ***New Applications on Hands***

Anglo Irish	US$125m.	—	18 months	70%	£54.7m.u

5. ***Applications which are not being provided with Insurance Cover***

Company	Contract Value	Credit Period	Level of Indemnity	State Liability IR£
Agra Trading (Beef)	US$17m.	18 months	70%	£ 7.44m.u
Taher Meats	IR£11m.	18 months	70%	£ 7.7m.
Halal Meats	US$25m.	18 months	70%	£10.94m.

This Memorandum, together with minutes thereon, written by Mr Timbs and Mr O'Reilly were submitted to the Minister for Industry & Commerce by the Secretary on the 22nd day of March 1988. These minutes are of importance and merit inclusion in this Report.

"Mr Ted O'Reilly
Assistant Secretary

Detailed Memorandum, with summary, on the question of Export Credit Insurance for Iraq is submitted across please.

I support the recommendations in the Memorandum. In addition to the request from AIBP for cover in respect of additional sales of US $125m. I understand that Halal Meats intend applying for cover in respect of US $25m. contract and Hibernia Meats would apply for insurance cover in respect of a further US $50m. sales to Iraq. It is also likely that were it known that we were again open for insurance Agra Trading and Taher Meats would also apply for additional substantial cover.

I spoke to Minister Reynolds, at his request, about Export Credit Insurance for Iraq and advised him of the Memorandum across which I indicated would be submitted to him formally.

Joe Timbs
March, 1988.

Mr O'Reilly
J.T. 9/3

1. Secretary
2. Runai Aire
3. Private Sec. to Minister S Brennan.

As I write the fundamental issue is that the existing statutory limit £300m. has been practically reached. The aggregate amount of the existing exposure, in terms of insurance business written, is £290m.

Of this £114m ..... to Iraq. Therefore the balance within the Iraq limit (£150 — £114 = £36m) cannot now be allocated. It must await the passage of the legislation which it is expected to put before Government in a week or so.

Common sense would suggest that with Iraq at £114m. accounting for 40% of the total £290m. business written the portfolio is dangerously unbalanced.

The situation in the Iran/Iraq war is deteriorating daily. A delicate truce of a few days duration has on Sunday been shattered by Iraqis firing missiles at Tehran in the battle of the cities. This phase would seem to be more dangerous for our purposes than engagements in the provinces or at sea.

As the figures stand at present there can be no question of giving cover to AIBP now. The balance remaining under the statutory limit cannot be deemed to be available for them. Otherwise there would be the situation that the Government is closed for Export Credit Insurance business in every country except Iraq! I have asked that the position today be confirmed or ascertained from ICI and that the figures be clearly identified. Competing demands would have to be looked at closely but in general I would suggest that the balance remaining £10m., must be left available for exports everywhere except Iraq. There will be considerations in some peoples minds about the state of contracts in Iraq but I suggest that we would look silly if we had to refuse cover to an exporter for some country where there are no such political disturbances evident on the grounds that we have to give it to Iraq.

On the question of giving further cover to meat exporters to Iraq there are some basic considerations. It is necessary to try to ensure the best possible price for Irish meat but it is also necessary to be seen exercising equity in the allocations of cover. Within those factors there is the further consideration as to what extent the State should be prepared to go in supporting one individual entity. The outcome of continuing indefinitely is to increase the dominance of that entity with obvious consequences. Incidentally Goodman International have let over 100 people go at Bailieboro.

If it were decided to go that road there should be no difficulty about increasing the £150m. limit which would, as mentioned, only be operative after the legislation. This could be done, as it was before, with the approval of the Minister for Finance.

The critical issue is (i) whether to do so and (ii) if the decision on (i) is YES by what amount. It seems to me that it cannot be for AIBP alone. There are other applicants who say that they have contracts or that they have been invited to tender. On whatever additional amount of cover might be provided for AIBP in the event of extending the £150m. limit it seems to me that , as their increased business is magnifying the State's risk, they would have to accept punishing terms. The entity operates on such a scale that the new business and the risks attaching to it should be borne in three equal segments, (i) by AIBP (they carried all their risk in 1985/86), (ii) their banks and (iii) the State. As regards (iii) we could then negotiate terms that would have to be very stringent and would have to be more demanding than those in the existing bargain.

The same kind of terms would have to be required of other meat firms, who at present enjoy commitments on the same terms as AIBP, if they were to be given extended cover under new limits. On the other hand equity would seem to require that any cover commitment given under extended limits for Iraq should be on the same terms as that originally given in commitments to AIBP and others.

Because it is obviously wrong in terms of a balance in the total exposure I would be opposed to seeking extended cover for Iraq. The real benefits of the business in Iraq are assumed to exist. I haver never seen any analysis of them in precise terms or whether such benefits might be obtained by exports to another country. One development is clear; the more contracts that Irish meat entities get in Iraq the more they will expect Export Credit Insurance cover and the more will the State's exposure in this obviously risky market be increased. Another obvious factor is the consideration whether Irish entities are getting the business because other countries do not provide insurance.

Ted O'Reilly
March, 1988."

"Minister

"Memo on extended cover for Iraq is submitted across for consideration together with proceeding minutes.

"Given the magnitude of business now being undertaken in this market, claims or indeed one claim could have a critical effect on Exchequer finances in any particular year."

JD
22/3"

On the 31st day of March 1988 Mr Timbs wrote to the Insurance Corporation of Ireland stating:—

"I am writing just to ensure that there are no misunderstandings in relation to Iraq. The position is, as I have told all the meat companies who have contacted me, that we are not open for business for meat contracts for Iraq and do not envisage being so in the future."

As of the 31st day of March 1988 the Department of Industry & Commerce was "not open for business for meat contracts for Iraq" and did not "envisage being so in the future."

At that time, decisions had been taken, for various reasons, not to grant the Export Credit Insurance which had been sought by them to Agra Trading Ltd, Taher Meat Packers Ltd, and Halal.

By the letter dated the 27th day of November 1987 (amended and back dated), AIBP had sought insurance cover for a further $155m in respect of contracts for the sale of beef to Iraq. The Minister for Industry & Commerce had informally offered insurance cover in the sum of $30m in respect of these sales in November 1987 to AIBP, so that the net additional cover sought by AIBP for such contracts was $125m.

It was pointed out in the aforesaid Memorandum prepared by the Export Credit section of the Department of Industry & Commerce and submitted to the Minister for Industry & Commerce that:—

"(1) an increase in the ceiling of £150m would be required to facilitate this request;

(2) the provision of such insurance cover for the amount of $125m would involve a net increase of £54.7m in the Minister's liability over and above existing commitments.

(3) it was not possible to facilitate this contract within the £150m ceiling at either 70% or 80% level of indemnity.

(4) Halal Meats, Agra Trading and Taher Meats had sought Export Credit Insurance involving a total liability of some £26m.

(5) Halal Meats were informed by the Minister for Industry & Commerce that insurance cover could not be considered until such time as Halal had provided clear evidence that they had a contract in Iraq and that Agra Trading and Taher Meats were told that no further insurance cover was available within the £150m ceiling.

(6) that if the Government decided to increase the £150m ceiling, it might wish to take account of the position of other exporters such as Halal, Agra Trading and Taher Meats in setting the new ceiling."

As previously stated this Memorandum recommended strict adherence to the £150m ceiling and that AIBP be informed that no further insurance cover is available above the $30m already offered at 70% indemnity by the Minister for Industry & Commerce.

Though Taher Meats Limited had been refused Export Credit Insurance, they continued to make representations in regard thereto and on the 5th day of May 1988 their representatives met the Minister for Industry and Commerce.

On the 7th of July 1988 the Minister through his Private Secretary wrote to Mr Fitzpatrick of Taher explaining why at that time it would not be possible to give him Export Credit Insurance:—

> "The position with regard to Export Credit Insurance for Iraq is, as explained to you by the Minister, that the ceiling on exposure for that country has effectively been reached. The small amount of cover remaining available is being allocated to small contracts, of a size which would be of no interest to meat exporters. The Minister regrets, therefore, that export credit insurance cover cannot be made available for the Taher Meats contract or indeed any other major contract in Iraq for the foreseeable future".

On the 24th day of October 1988, Taher Meats wrote to the ICI in the following terms:—

> "I write to inform you that we can no longer, as a Company, accept the treatment been given to us by the Insurance Corporation of Ireland. I write out of exasperation and anger at the way our best efforts to obtain insurance cover for Iraq have been persistently frustrated and sometimes ignored by your company.
>
> On 13 November, 1987, I wrote to your Company seeking cover for IR£14 million on an Iraqi Letter of Credit opened by Rafidain Bank with a 540 day delay payment term. I spoke with you several times on the telephone and visited you on a number of occasions in your office, in connection with this matter, but all to no avail as no offer was made to us.
>
> On 07 and 16 March, 1988 I made two further requests in writing for cover, but again, whilst not receiving a reply from ICI, we did not obtain cover.
>
> Following our meeting on 29th March, 1988 I wrote to you on 30 March giving details on Taher Meats and the investments which we proposed embarking on at our abattoir at Roscrea, Co. Tipperary. Again, I requested cover for Iraq and again, no success followed from that letter.
>
> On 08 June last, I wrote to you confirming that we had signed a US$4.8 million contract with Iraq and I sought cover for this amount. Unfortunately, no cover was again forthcoming.
>
> On 14 October, 1988 I wrote to you enclosing copy of a Letter of Credit opened by Rafidain Bank and confirmed by Gulf Bank, London. I also enclosed a copy of Gulf Bank's commitment with that letter. The purpose of this letter was for information only, in order to prove to you that, not alone did we have contacts in Iraq, but we were also capable of doing business.
>
> We are now sick and tired of our singularly unfair treatment from ICI in regard to cover for Iraq. I thought that when our Government took over the ICI liabilities etc., and the Company itself from AIB, that it became a Semi-State/State owned company. However, on the Iraqi market, it appears to act, in the main, for just one company. Why is this so? Why can't we have fair play and equality in the use of the Insurance Fund for Export Guarantees? We will not be cutting across any other Irish exporter on the Iraqi market. Our business is new and will be in addition to existing business there. When we made our first application nearly one year ago, we were told that while the cover for Iraq was exhausted, that we could expect to be offered cover when Rafidain L/Cs, with deferred payment terms, reached maturity. What is the position on payments from Iraq and when can we expect an offer?
>
> Unlike other Irish companies seeking business in Iraq, we are currently supplying beef without the aid of ICI cover, albeit the cost to the Company is in excess of IR£1 million. We have people on the ground in Iraq who work in very close liaison with both the Iraq Company for Meat Production and Marketing and also the Rafidain Bank. Any problems that may arise can easily be sorted out within hours.
>
> As mentioned, previously, the Company intends spending about IR£8 million at its abattoir in Roscrea. When this expenditure is completed, our employment level there will be in excess of 300 people. Our Company has submitted to the I.D.A. a Five Year Development Plan. When completed, our investment in the country will be of the order of IR£30 million employing nearly 1,000 people.

Mr Naser Taher, who is the principal shareholder in Taher Meats, is totally confused in regard to this ICI business and cannot, for the life of him, understand how only one or two meat companies are able to secure cover from ICI. He wishes to know what is the secret, what does he have to do, who does he have to see, who does he have to talk to in order that we might get cover for our Iraqi business? He only asks for equal opportunity for competition in this market.

Since our first letter to you nearly one year ago, about 20,000 tonnes of beef have been shipped to Iraq from Ireland, most of it exported by one company. Some of this meat was only contracted for in June of this year. How come cover was available for this beef? In 1986/1987, a total of over 28,000 tonnes was exported from Ireland to Iraq. Are we to take it that the fund has been pre-booked by a chosen few, or are we to take it that we, as a Company, are never going to get ICI Export Credit Insurance for Iraq, for what ever reason? Consequently, should the Company be considering a move to some other country as a better investment prospect where equality of treatment will be guaranteed?

As an Irishman, who once had to emigrate, I would be appalled if foreign investors were effectively told that they were not wanted in Ireland. I would have thought that with the state of the agri-business sector at the moment in Ireland, that we needed more, not less, competition.

Over the next year, we are in a position to conclude IR£30 million of business in meat with Iraq. Should we obtain ICI cover on this, the business will bring about, inter alia:

i) substantial investment in capital expenditure in Roscrea;

ii) increased jobs in year round working in our abattoir at Roscrea; and

iii) more competition for producers of cattle in Ireland.

Yesterday, we put in place with the Rafidain Bank, a Bid Bond to enable us to tender for a contract which requests closing offers by close of business to-day.

We now formally request the facilities for IR£30 million under the Export Credit Insurance Scheme to cover business in Iraq with 540 day delayed payment terms. We know that the Insurance (Export Guarantees) Bill, 1988 which passed all its stages in the Dail and the Senate, increased the fund from IR£300 million to IR£500 million. Has all this money been allocated? Why can't we even have a small percentage of the overall fund? 10 percent, 8 percent, 6 percent?

Your earliest reply to the above matter is of the utmost importance.

Yours sincerely

____

A.C. Fitzpatrick
Managing Director."

On the 2nd day of November, 1988 Mr Fitzpatrick and Mr Taher met the Minister for Trade and Marketing, Mr Brennan, to discuss Export Credit Insurance. Mr Donnelly prepared a report of the meeting which was as follows:—

"REPORT OF MEETING

1. On Wednesday, 2 November, 1988 the Minister for Trade and Marketing, Mr Seamus Brennan TD, met with Mr Gus Fitzpatrick and Mr Naser Taher of Taher Meats Ltd, concerning the provision of Export Credit Insurance for Iraq.
2. The company representatives complained that they had been seeking Export Credit Insurance cover on beef contracts in Iraq from ICI for the past twelve months or so but without any success. They were at present completing an existing contract for the supply of US$5m. worth of beef to Iraq in respect of which they had had to obtain cover in the private market in the UK. This had been extremely expensive and the need to obtain cover in this way in the future would impact on their ability to win new contracts in Iraq. Mr Taher reminded the Minister that when they originally applied for insurance from ICI they had been told that insurance cover could not be provided unless the company had a contract. The company subsequently successfully negotiated a contract in respect of which they were required to put up a bid bond. However, when they returned to ICI they were then informed that cover was no longer available. As a result the company were not in a position to proceed with the contract and the bid bond is now in danger of being confiscated. This would clearly represent a tremendous loss to the company.
3. In response Minister Brennan indicated that Minister Reynolds took a significant and direct interest in the Export Credit Insurance Scheme and he undertook to brief him fully on the case made by Taher Meats. Minister Brennan appreciated fully the company's feelings on the matter. Minister Brennan indicated however that the current limit on exports to Iraq was to all intent and purposes reached and no further cover was available at present. Mr Taher indicated that he had been informed by the Iraqi authorities that a "major meat exporter" had been given additional cover worth IR £80m. by the Irish Export Credit Agency. Minister Brennan said that no such offer could have been made within the existing limit of £150m.
4. Messrs. Fitzpatrick and Taher thanked the Minister for receiving them and looked forward to positive developments in relation to their application for cover in Iraq.

---

3 November 1988.
c.c. Secretary
P.S. to Minister
P.S. to Minister Brennan
Mr T. O'Reilly, Asst. Sec.,
Mr J. Timbs
Mr L. Kilroy
Mr J. Fanning
Mr J. Hanney"

In January 1988, the Department of Industry and Commerce began preparing the Memorandum for Government to seek an increase in the aggregate amount of liabilities which the Minister could assume in connection with the Insurance Acts from £300m to £500m.

The draft Memorandum was circulated to the other Departments concerned and their observations sought and obtained, which were incorporated in the Memorandum for Government dated the 2nd June 1988.

In a Memorandum prepared for the information of the Minister for Agriculture and Food on the Memorandum on the increase from £300 to £500 million the Department strongly supported the Memorandum for the following reasons:—

"1. Third country markets are very important markets for agricultural exports particularly beef. In 1987, 47% of our total beef exports went to Third countries, mainly Egypt, Iran and Iraq. Trade with these countries involves a higher risk than normal for exporters because of the political situation in the Middle East and to a lesser extent in North Africa. Exporters to these destinations require Export Credit Insurance Cover at reasonable cost.

2 The increase in the Export Credit Insurance limit for Iraq has helped to ensure the continuance of our very high level of trade with Iraq. It is essential that the increase in the availability of cover for exports to Iraq does not reduce the cover available to other destinations.

3 The ability of Irish Agricultural exporters to continue to trade with third countries will be greatly influenced by the availability of Export Credit facilities. Indeed the expansion of this trade and the diversification into the value added product area, which is our objective, will increase the demand for insurance cover.

4 Demands by third country's importers of dairy products for credit terms which we could not meet have resulted in the loss of some trade. The availability of additional credit may help in regaining some of this trade.

5 In supporting the proposal we would, however, urge that, in the allocation of the additional resources, a policy of positive discrimination in favour of agricultural products, which are entirely indigenous, should apply".

On the 2nd day of June 1988 the Minister for Industry and Commerce submitted a Memorandum for Government, dealing with *"Increase in the Statutory Limit on Liability under the Exports Credit Insurance and Finance Schemes."*

It appears from this Memorandum that the decision sought was:

"1. The Minister for Industry and Commerce requests the Government to approve:—

(1) An increase in the aggregate amount of liabilities which the Minister may assume in respect of guarantees in connection with exports under the Insurance Act 1953 as amended from IR£300m to IR£500m.

(2) The text of a Bill, which would enable the increase sought to be implemented.

(3) The presentation and circulation of the Bill to the Oireachtas."

This Memorandum did not seek an increase in the ceiling in respect of exports to Iraq but continued as follows:—

"***Reason for Proposal***

2. Under the Insurance Act, 1953 the Minister for Industry and Commerce may, with the consent of the Minister for Finance, make arrangements for the giving of

guarantees for the purposes of encouraging exports. This legislation has led to the development of the Export Credit Insurance and Finance Schemes. The principal aims of the Schemes are to assist and encourage Irish exporters by providing them with protection against non-payment by foreign buyers due to various political and commercial risks, and to help maintain the competitiveness of Irish exports by providing access to export credit finance. Since 1971, the Schemes have been administered by Insurance Corporation of Ireland plc as agent of the Minister.

3. The Insurance Act, 1953 provided that the maximum liability which the Minister could assume at any one time in respect of insured/guaranteed exports amounted to £2m. Since then the maximum liability was increased in August, 1961 to £5m, in April, 1969 to £10m, in June 1971 to £30m, in December 1978 to £100m, and in December 1981 to £300m., respectively.

4. The Government decided (S.15005C) on 8 September 1987

   (1) that the ceiling for insured exports to Iraq should be raised from £70m. to £150m and

   (2) that the question of increasing the ceiling for Export Credit Insurance generally might be considered further at a later date, as and when the need arises.

5. The maximum liability of the Minster under the Export Credit Insurance and Finance Schemes has now reached nearly £298m. Insofar as Iraq is concerned, total exposure plus commitments amount to £145m. As well as the increase in demand for cover for the Iraqi market, an increase in general demand for Export Credit Insurance for markets worldwide is likely to occur over the next few years as exports grow. Total exports reached a record £10,500m. in 1987. Accordingly, the Minister considers it desirable to increase the aggregate amount of liabilities which he may assume in respect of guarantees under the Insurance Act 1953 as amended from £300m to £500m.

6. An amendment to the Insurance Acts will be necessary to give effect to the Minister's proposal. An appropriate Bill has been drafted by the Parliamentary Draftsman and is attached at Appendix 1.

Staffing and Cost Implications:

7. There are no staffing implications associated with the proposal. The raising of the ceiling allows for an increase in the aggregate amount of exports which can be insured. The Exchequer implications of such an increase are (*a*) an increase in premium income and (*b*) a likelihood of an increase in the amount of claims in the future.

Consultation with other Ministers

8. The Tanaiste and Minister for Foreign Affairs and the Minister for the Marine have no observations to offer on the Memorandum.

9. The Minister for Agriculture and Food strongly supports the proposal to increase the liability limit for Export Credit Insurance from £300m to £500m. He states that if agricultural exports are to be maintained and expanded in particularly difficult markets it is essential that adequate export credit facilities be made available. In the allocation of the additional cover the Minister for Agriculture and Food urges that a policy of positive discrimination in favour of value enhanced agricultural

products should be adopted in view of the importance of agricultural exports to the economy and the entirely indigenous nature of these products.

The Minister for Industry and Commerce notes the views of the Minister for Agriculture and Food and recognises the need to ensure that the current significant level of export credit facilities afforded to agricultural exports is continued. He pointed out, however, that this objective must be set against the need for adherence to sound commercial underwriting procedures, the development of a balanced portfolio of risk and the continued provision of export credit facilities to Irish exporters generally.

10. The Minister for Finance has raised no objection to the submission of the Memorandum to the Government subject to the inclusion in the Memorandum of the following observations:

The Minister for Finance is concerned at the deficit which has accumulated under the Schemes in recent years. Any increase in the statutory limit on liability increases the potential for default and thus entails greater risk for the Exchequer. The Minister would stress the need for tight procedures and rigorous assessment of all proposals under these Schemes in order to ensure that further demands on the Exchequer's resources are kept to the absolute minimum".

The Minster for Industry and Commerce would make the following comments on the observations of the Minister for Finance.

Total insured exports under the scheme since 1971 amount to some £4.5 billion. The cumulative deficit figure at the end of 1987 amounted to £12m or 0.26% of turnover covered since 1971. This compares very favourably with experiences elsewhere. Approximately £9m. of the deficit is in respect of sovereign debts arising out of foreign currency shortages in buyer countries. There is a reasonable prospect of recovering a significant proportion of this sum. In addition positive efforts are being made to recover debts which have arisen through commercial default.

Notwithstanding the above the Minister for Industry and Commerce is in full agreement with the Minister for Finance's views in regard to further demands on Exchequer resources being kept to a minimum. He intends to ensure the adherence to strict and proper procedure with a view to attaining this objective."

It appears from this Memorandum that the decision sought was:

"1. The Minister for Industry and Commerce requests the Government to approve:—

(1) An increase in the aggregate amount of liabilities which the Minister may assume in respect of guarantees in connection with exports under the Insurance Act 1953 as amended from £300m to £500.

(2) The texts of the bill which would enable the increase sought to be implemented.

(3) The presentation and circulation of the Bill to the Oireachtas."

This Memorandum did not seek an increase in the ceiling in respect of exports to Iraq.

On the 8th day of June, 1988 the Secretary to the Government wrote to the Private Secretary of the Minister for Industry & Commerce in the following terms:—

> "I am to refer to the Memorandum ref. ECI/RT-21 dated 2 June 1988 submitted by the Minister for Industry & Commerce with the text of the Insurance (Export Guarantees) Bill, 1988 and to inform you that, at a meeting held today, the Government:
>
> (1) approved the text of the Bill, and
>
> (2) authorised the Minister to present the Bill to Dáil Eireann and to have it circulated to Deputies
>
> on the understanding that further demands on the Exchequer under the Export Credit Insurance and Finance Schemes would be kept to an absolute minimum by the use of strict procedures and the rigorous assessment of proposals for guarantees."

The Insurance (Export Guarantees) Bill 1988 had passed all stages in the Dail and Senate when on the 7th day of July 1988 the Private Secretary to the Minister for Industry & Commerce, wrote to Mr Fitzpatrick, Managing Director, Taher Meats Ltd in the following terms:—

> "Dear Mr Fitzpatrick
>
> The Minister for Industry & Commerce, Mr Albert Reynolds, TD, has asked me to refer again to your recent letter concerning Export Credit Insurance cover for the sale of meat to Iraq.
>
> The purpose of the Insurance (Export Guarantees) Bill 1988 which has now passed all stages in the Dáil and the Senate is to increase from £300m to £500m the aggregate amount of liabilities which the Minister for Industry & Commerce may assume at any one time under the Export Credit Insurance Scheme. Where individual country limits apply, as in the case of Iraq, they are not automatically increased as a result of this legislation.
>
> The position with regard to Export Credit Insurance for Iraq is, as explained to you by the Minister, that the ceiling on exposure for that country has effectively been reached. The small amount of cover remaining available is being allocated to small contracts, of a size which would be of no interest to meat exporters. The Minister regrets, therefore, that Export Credit Insurance cover cannot be made available for the Taher Meats contract or indeed any other major contract in Iraq for the foreseeable future."

In spite of the increase in the statutory ceiling from £300m to £500m, the ceiling already fixed in respect of exports to Iraq, namely £150m remained and no allocations of cover were made.

The Fifth Session of the Irish/Iraq Joint Commission was due to take place in Baghdad in the week beginning the 7th of November 1988. The Irish delegation to that Commission was to be led by the Minister for Trade and Marketing, Mr Seamus Brennan TD.

The Baghdad Trade Fair was due to take place in Baghdad from the 1st to 15th November 1988 and a number of Irish companies were to be represented thereat.

In September 1988 the Export Credit Section of the Department of Industry & Commerce was carrying out a detailed review of Export Credit Insurance for Iraq.

On the 16th day of September 1988 Mr Timbs of the International Trade Division of the Department of Industry & Commerce wrote to Ms Kerrigan of the Department of Agriculture seeking the views of that Department as to whether the £150m ceiling should be maintained, increased or gradually reduced.

In particular he sought the views of the Department of Agriculture as to whether the ceiling should be increased to facilitate further beef contracts and, if so, by how much.

He also sought the proposals of the Department on what Irish meat companies should be considered for cover and how any available cover should be allocated.

By letter dated 6th October 1988 the Department of Agriculture informed Mr Timbs that, in their view, all contracts for beef exports to Iraq should be granted insurance cover if they met the regulation and document requirements of the Department of Industry & Commerce.

However, they stated that

> "it would not be possible for this Department to make proposals on which Irish meat companies should be considered for cover nor on how any available cover should be allocated."

By letter dated 27th September 1988 the Insurance Corporation of Ireland wrote to Mr Timbs in the context of the Irish/Iraqi Joint Commission Talks in which they expressed their views with regard to the credit worthiness of Iraq. At the end of a detailed review of the position they stated that:—

> "In summary, our recommendation remains unaltered from that previously advised and although difficult the Minister should strive to reduce our present exposure as soon as possible."

The full context of their Report is as follows:—

> "Mr Joe Timbs
> Principal Officer
> Foreign Trade Section
> Department of Industry and Commerce
> Kildare Street
> Dublin 2
>
> 27th September, 1988
>
> *RE: Irish/Iraqi Joint Commission Talks*
>
> Dear Joe
>
> I note the contents of your letter of the 16th September setting out the background details on the above. Most of the information contained in the earlier country profile report on Iraq submitted to you last year continues to be valid. Although the amount

of debt owing to other countries by Iraq has grown considerably and Iraq has rescheduled, it must be said that there is some more hope for the economy now that there is a truce in the Iran/Iraq War. It could not be a more difficult time to assess Iraq from a credit viewpoint.

Whilst "the War" has ceased, the truce is uneasy. There is skirmishing on the Iranian frontier and, it is believed, heavy fighting in Kurdistan. The situation is so volatile that it is changing on a daily basis. Accordingly, I believe the best approach to take is to examine the factors which will have the biggest effect on the Iraqi political and economic environment.

Economic Circumstances

There are signs that there is now some confidence in the Iraqi economy. Over the past few years, despite fighting a major war, the Iraqis have not neglected their infrastructure and have tried hard to maintain their infrastructural development. The philosophy behind this was to satisfy the populace and keep it happy for the duration of the War. This was typified by the "gun, butter and videos" slogan of President Saddam Hussein. Although there was some hardship the people were prepared to accept it in a time of war. Whether they will continue to do so now that there is a truce, remains to be seen.

This infrastructure will help speed recovery from the ill-effects of the war with Iran. The country is very heavily in debt and in recent years depended on aid, mainly from Saudi Arabia for its survival. It is difficult to be specific on figures because the Iraqis do not publish them. Informed sources suggest that a sum in the region of US$60 billion amounts to the external debt of Iraq. Much of this is owed to Arab allies and will never be repaid. It is estimated that US$30 billion of the US$60 billion debt will have to be repaid. The prospects for this depend on the price of oil and the ability of the Iraqis to get war damaged oil-fields back into production.

Oil Prices

Although Iraq is a member of OPEC it seems not to be restricted by OPEC guidelines on levels of production. Now that the War has stopped OPEC may tighten up on this. The main hinderance to Iraq in selling its oil is its ability to move it out of the country. This is a problem which it has been tackling for years with new pipelines through Turkey and Saudi Arabia. They also daily truck oil across the country and through Jordan to Aqaba. The absence of fighting in the Straits of Hormuz will also be of assistance in shipping oil through Persian Gulf ports.

The price of crude oil peaked at about US$20 per barrel during 1987. Since then the average price has been US$16 and the price trend is down. This month it fell below US$13 per barrel, the lowest since 1986. This is largely due to over-production by Saudi Arabia, primarily to fund the amount of money it has been pumping into Iraq. The OPEC pricing committee is meeting at the end of this month to review production and the price of oil.

If Iraq continues to ignore OPEC production levels and other Middle Eastern countries continued to produce oil at present levels to sustain their economies there is no doubt that the price of oil will remain at present or possibly more depressed levels.

Major Trading Partners

Appendix 1 shows the main suppliers to Iraq and gives some indication of the dramatic economic downturn caused by the Iran/Iraq War. In 1983, when the Iraqis got into difficulties with their foreign payments, the level of their imports dropped dramatically. However, from information obtained through various sources it appears that there was a further retrenchment in 1987. United Kingdom exports to Iraq amounted to approximately US$444 million or an estimated 10% or so of Iraq's non-military imports. Because of poor payment history the Japanese figure fell to about the same level. The United States figure was negligible prior to 1987 and that year it became Iraq's major supplier taking an estimated 17% of the market. In the event that The United States should stop supplying the Iraqis because of sanctions (dealt with later) or for any other reason it would have a major impact on the Iraqi economy.

Iran

Recent events inside Iran suggest that substantial progress has been made towards moderation. Other than Ayatollah Khomeni the key figures are Ali Akbar Al Rafsanjani speaker of Iran's parliament and Mr Hussein Moussavi the Prime Minister. The former is the moderate who persuaded Ayatollah Khomeni to end the war and is now much more powerful than Speaker Rafsanjani who is an entrenched radical. In a power play earlier this month Speaker Rafsanjani resigned as he did not get his own way in forming the cabinet. Ayatollah Khomeni stepped in and told Speaker Rafsanjani to stop complaining and "get on with the job". The new cabinet has a heavy bias towards the moderates. The radical members scraped in and a previous member, the Minister for Revolutionary Guards has been dropped. The Ministry for the Revolutionary Guards is itself under pressure particularly from a new Bill which proposes placing it under the control of the Ministry of Defence.

The overall position is very delicate and if the Iraqis try to humiliate the Iranians at the Geneva talks it is believed that the truce could end as Speaker Rafsanjani could not withstand the political fallout.

Kurds/Sanctions

20 million Kurds live in the mountainous region of Northern Iraq, Iran, Turkey and the USSR. For generations they have argued for autonomy and have been using force for the last 50 years or so. About one quarter of the Iraqi population of 17 million is Kurdish. At various stages of the war both sides formed alliances with the Kurds and played what became known as "the Kurdish card". By doing so a second front was opened to divert attention from the Shat Tal Arab waterway which was the main theatre of war. The Kurdish rebels are now a well trained and well armed force. They have been a problem for the Iraqis for many years and have had military assistance from both the Shah and Ayatollah Khomeni, the Russians, Americans, Syria and Israel. Anyone who wanted to destabilise Iraq helped them.

The Iraqis tackled the Kurds in earnest about 1975. Kurdish territory was depopulated by a forced migration and re-settlement policy. Kurdish strongholds — particularly around the Northern oilfields of Kirkuk — were re-populated with loyalist Arabs transplanted from the south. The Iraqis have used chemical gas against the Iranians and also against the Kurds in the past. It is now the contention of many groups that whilst world attention is being focused on the peace talks, Iraq is using this as an opportunity to exterminate the Kurds using, in particular, chemical warfare. Over

100,000 Kurds have fled to Turkey where they have been given some protection. As Turkey already has a Kurdish problem of its own this has not endeared Iraq to a major ally and trading partner. The European Parliament on the 15th September condemned Iraq for using chemical weapons to exterminate the Kurds. The United Kingdom government has said there is "compelling evidence" that the Iraqis are using poison gas in their fight.

These beliefs are strengthened by Iraq's refusal to allow United Nations or Red Cross personnel access to the war zone. this has given rise to calls for sanctions particularly within the USA where Bills have been both prepared and presented in the Senate and House of Representatives. If successful, (thought unlikely in present format) it would mean (*a*) that Iraq would not pay its outstandings to the USA if sanctions were imposed and (*b*) increase pressure on other trading partners to provide supplies in lieu of the Americans.

Hostilities

We believe that hostilities with Iran although currently the subject of a ceasefire have the potential to suddenly erupt, say in the event of the death of the Iranian leader Ayatollah Khomeni. The "Kurdish question" will continue to simmer in the guise of a protracted civil war. The use of mustard gas seems irrefutable and whilst the European Parliament and USA condemn the Iraqis for its use it appears that the United Kingdom and USA have had a hand in supplying the raw material. Its use does not cause concern to other Middle Eastern allies of Iraq and they have indicated their support, blaming much of the adverse publicity on Israel and it's US lobby. Israel is particularly worried about the shift of power in the Middle East where the Iran/Iraq conflict concentrated the Arab states on that war rather than on the PLO/Israeli/Arab conflict. The Middle Eastern States, particularly in the Gulf and Saudi Arabia are major buyers of arms from the United Kingdom and USA. This should influence a business decision on sanctions against Iraq but the power of the "moral majority" in the US cannot be ruled out particularly in an election year.

Ireland's Role in Iraq's Future

The current line of credit facilities available from Ireland while in relative terms is substantial in the overall context is quite small as far as Iraq is concerned. The USA has been increasing its support for Iraq substantially. Most Export Credit agencies are off cover having rescheduled or completed oil for debt deals and will only look at new cover on a case by case basis. If sanctions are imposed by the USA who have provided substantial food supplies there will be very strong pressure on us to increase our shipments of foodstuffs and pharmaceuticals.

Historically the level of business being done in the Iraqi market by Irish exporters was very small when compared with France, Germany and Japan. This was particularly so at the time Iraq ran into payment difficulties. When the other credit agencies suddenly found themselves with substantial commitments and debts that were not being paid on time we have a relatively low exposure. Although we have continued to cover the market for the last three to four years our total level of debt when compared to the others is relatively moderate. This leaves us in a very fortunate position because being one of the two/three countries granting facilities to Iraq we are virtually the only country the Iraqi's are paying. Our indebtedness is being serviced whereas most of the other countries are having to face rescheduling negotiations on an almost annual basis. The servicing of our debt by Iraq is only possible because the level of debt is manageable by them. They cannot, for example, service the West German debt in their present financial position. Because our total debt, although significant for us, is not large in Iraqi terms and is therefore being serviced.

Conclusion

As can be seen from the above resume of information it is extremely difficult if not impossible to forecast accurately the outcome to the situation in Iraq. However, it is our belief that the Minister for Trade and Marketing Mr Seamus Brennan TD on his visit will be under pressure to increase the availability of Export Credit cover. You are aware of our previously expressed views that we should have no Iraqi exposure, because it cannot be underwritten on a commercial basis. We would prefer to see the current exposure reduced and believe that any increase in cover or rollover of existing terms will not be in the best interest of the Scheme.

The present Iraqi outstandings to an economy of our size are more than sufficient bearing in mind the state of the Iraqi market. There is a strong likelihood that if there is an attempt to pull back from the existing level, the Iraqis would use this as an excuse to default. Our view is that it is best to test this position now at our existing high exposure rather than exacerbate the situation by extending existing or further limits in the hope of avoiding this outcome. Iraq continues to have difficulties in meeting its obligations notwithstanding the fact that credit terms of up to eighteen months have been negotiated with Irish suppliers of consumer goods. Our experience of late is that payment is made approximately six months after maturity of irrevocable letters of credit and that in the textile trade terms have not been met on major contracts in the past year which has resulted in one substantial claim on behalf of....... Although we have yet to experience eventual non-payment our concern on the business already transacted in Iraq has not been alleviated, particulary in view of the recent defaults. I attach a copy printout of the payment dates on the recent meat exports. At present we have received notification of payment on those drafts which are now overdue and total marginally less than £8m.

In summary, our recommendation remains unaltered from that previously advised and although difficult the Minister should strive to reduce our present exposure as soon as possible.

Should you require further detail on any of the above points please do not hesitate to contact me.

Yours sincerely
Robert D Frewen,
Manager
Credit and Guarantee Dept."

Mr Hanney and Mr Fanning of the Export Credit Section of the Department prepared a Memorandum on Export Credit Insurance For Iraq and which is dated the 21st October 1988 and submitted to the Minister for Industry & Commerce on that date.

This Memorandum reviewed the position with regard to Export Credit Insurance for Iraq and gave the following conclusion and made the following recommendations:—

"***Conclusions and Recommendations***

The economic outlook for Iraq is not good. Many bankers are convinced that the Iraqi authorities will have to reschedule again and some believe there may eventually be pressure for a multilateral rescheduling deal in either the Paris or London Clubs.

The Minister's existing liability in Iraq is very high relative to his overall liability under the Scheme and has resulted in what would be regarded by many as an unbalanced insurance portfolio. With debts amounting to some IR£65m falling due for payment under beef contracts in the period up to January 1989, (for which the Minister has an exposure of nearly IR£52m), it would be inappropriate to give any commitment to the Iraqis to increase the IR£150m ceiling. At the Joint Commission, the Minister might instead indicate the Irish Government's willingness to rollover the existing limit provided that payments are made promptly under existing contracts. It is accordingly recommended.

(i) That there be no increase in the IR£150m ceiling for the time being.

(ii) That the position be reviewed in the New Year in the light of payment performance under existing contracts and new demands for insurance cover from Irish exporters.

(iii) That the maximum credit terms of 18 months laid down by the Minister in 1987 stay in force and any Iraqi proposals for 2 years credit be refused.

(iv) That in the meantime spare capacity within the IR£150m ceiling be allocated to industrial/services exporters on a strict first come first served basis.

John Hanney John Fanning
Export Credit Section
21 October 1988"

On the morning of the 21st day of October 1988, Mr Connor of AIBP visited the Department of Industry & Commerce and was met by Mr Timbs, Mr Donnelly and Mr Walsh of that Department. During the course of a discussion, according to a minute of the said meeting made by Mr Donnelly on 28th October 1988, Mr Connor indicated that in addition to Export Credit Insurance cover already available to the company in Iraq, he would require additional cover for contracts valued at IR£325m for the remainder of 1988 and 1989. Mr Timbs replied that the limit on Export Credit Insurance for Iraq had almost been reached and increase in that limit was a matter for the Minister for Industry & Commerce and his Government colleagues.

On the afternoon of the 21st day of October the Minister for Industry & Commerce met the Secretary to the Department, Mr Joe Timbs and Mr Gerry Donnelly for the purpose of discussing the provision of Export Credit Insurance for Iraq.

The Memorandum dated the 21st day of October 1988 was produced to him and on the 24th day of October 1988 Mr Donnelly prepared a note of the discussions at that meeting.

The said note is as follows:—

"**NOTE OF DISCUSSIONS**

1. The Secretary, Mr Joe Timbs and Mr Gerry Donnelly met with the Minister on Friday 21 October, 1988 to discuss the provision of Export Credit Insurance for Iraq. The Minister was informed that existing exposure under the Export Credit Insurance Scheme was in the region of IR£300m. The limit on exports to Iraq was, as agreed by Government at their meeting on 7 September 1987 set at IR£150m. Existing exposure in Iraq amounts to IR£136m. In addition Anglo-Irish Beef Packers (AIBP) had sought additional cover on *contracts* valued at US$325m. for 1988/89. The other principal Irish beef exporters to the Iraqi market, Hibernia Meats had sought cover on two contracts valued at US$72m and IR£10m. respectively.

2. The Minister indicated that he had discussed the question of Export Credit Insurance for Iraq with the Government at their meeting on 8 June, 1988 at which the text of the Bill increasing the overall statutory limit had been approved. He said that at that meeting the government had agreed as follows,

   (*a*) further increases for Export Credit Insurance in Iraq should be at a discretion of the Minister for Industry and Commerce and,

   (*b*) that the provision of Export Credit Insurance for Iraq should be managed in the national interest so as to avoid damaging competition between exporters. (The effect of this was that Export Credit Insurance cover in Iraq would only be granted to existing exporters in the market i.e. AIBP and Hibernia.) This decision was to be communicated to Irish exporters by the Minister for Agriculture.

   The Minister was surprised to note that the decision at (a) above in particular had not been recorded in the formal Government decision resulting from the meeting on 2 June. He said that he would discuss the matter with Government at their next meeting on Tuesday 25 October and have the matter clarified.

3. The Minister decided that the following additional cover would be provided in the Iraqi market:

   (*a*) a roll-over of the existing cover held by AIBP (Liability under Scheme, IR£95.6m.) and Hibernia (Liability IR£23.1m.) as outstanding maturities were paid,

   (*b*) additional cover for AIBP and Hibernia up to a maximum liability under the Scheme of £80m and £20m respectively and,

   (*c*) additional cover for non-beef exporters up to a maximum liability under the Scheme of IR£20m subject to increase should demand necessitate such.

   The Minister also agreed that there should be no increase under any circumstances in the credit terms for exports to Iraq beyond the 18 months which applies at present.

G. Donnelly

24 October 1988
c.c. Secretary
Mr Ted O'Reilly, Assistant Secretary
Mr Joe Timbs
Mr John Fanning
Mr John Hanney"

Prior to this meeting, the Department officials concerned had not been informed by the Minister that it had been agreed at the Government meeting held on the 8th June 1988 that increases in the ceiling for Export Credit Insurance for Iraq should be at the discretion of the Minister for Industry and Commerce or that the provision of Export Credit Insurance for Iraq should be managed in the national interest so as to avoid damaging competition between exporters and that insurance cover for beef exports to Iraq should be confined to existing exporters in the market, that is Anglo Irish Beef Processors and Hibernia Meats or that this decision should be communicated to Irish beef exporters by the Minister for Agriculture.

Subsequent to this meeting of the 21st day of October 88, Mr Timbs contacted both AIBP and Hibernia Meats and "advised them that the Minister had given indications of additional cover which he was prepared to make available, £80m for AIBP and £20m for Hibernia.

There is no doubt whatsoever but that the Minister for Industry and Commerce, Mr Reynolds, believed on the 21st October 1988 that the Government had on the 8th day of June 1988 made the decisions set out at 2(*a*) and (*b*) of Mr Donnelly's meeting and expressed his surprise that such decisions had not been recorded, and stated that he would discuss the matter with Government at the next meeting which would be held on the 25th day of October and have the matter clarified.

Pending such clarification, the Minister indicated to the officials, the additional cover that would be allocated to the Iraqi market which is set out at Paragraph 3(*a*), (*b*) and (*c*).

The matter came before the Government on the 25th day of October 1988 and on that day the Secretary to the Government wrote to the Private Secretary to the Minister for Industry & Commerce:—

> "I am to inform you that, at a meeting held today, the Government decided that the Minister for Industry & Commerce might agree with the Minister for Finance a new limit for Export Credit Insurance for Iraq within the overall ceiling of £500,000,000. for Export Credit Insurance generally under the Insurance Acts, 1909 to 1988 in place of the existing limit of £150,000,000."

Mr Ray Burke TD, was the first member of the Government, at that time, to give evidence before the Tribunal and as stated by the Chief Justice in the course of his judgement in the case of *The Attorney General, Applicant* -v- *The Sole Member of the Tribunal of Inquiry into the Beef Processing Industry, The Honourable Liam Hamilton, Respondent.*

"The Respondent indicated an intention to ask questions of Mr Burke, who had been

> a member of the Government in June of 1988, concerning the details of discussions which took place at Government meetings at and around that time and, in particular, to inquire into any discussion which took place at a Government meeting held on the 8th June 1988 which related to decisions concerning the increase of Export Credit Insurance of beef exported to Iraq, and the confining of such insurance to two particular firms. The Respondent stated that it was in the public interest that he should so inquire as documentary evidence already produced to the Tribunal on behalf of the State had indicated an inconsistency between certain notes of the decisions made at that meeting, and a subsequent note purporting to constitute a recollection of the Minister for Industry and Commerce as to what was decided by the Government at that meeting. Counsel, on behalf of the Attorney General, objected to the asking of those questions on the grounds of specific instructions received by him, and submitted that having regard to the provisions of the Constitution discussions between Members of the Government meeting together for the purpose of making decisions were absolutely confidential and that the content of such discussions cannot be inquired into by the Tribunal.
>
> "The Respondent then, as appears from the transcript of the proceedings in the Tribunal, deferred the asking of any questions, though he made a ruling that he was entitled to ask them in order to give to the Attorney General an opportunity of applying to the High Court by way of judicial review for a resolution of the issues thus arising."

The Chief Justice concluded his judgement by saying:—

> "I would, therefore, conclude that the claim for confidentiality of the contents and details of discussions at meetings of the Government, made by the Attorney General in relation to the inquiry of this Tribunal is a valid claim. It extends to discussions and to their contents, but it does not, of course, extend to the decisions made and the documentary evidence of them, whether they are classified as formal or informal decisions. It is a constitutional right which, in my view, goes to the fundamental machinery of government, and is, therefore, not capable of being waived by any individual member of a government, nor in my view, are the details and contents of discussions at meetings of the Government capable of being made public, for the purpose of this Inquiry, by a decision of any succeeding Government.
>
> "I would, therefore, allow the appeal and grant to the Applicant a declaration in the terms of this judgement, it not being necessary, clearly, having regard to the attitude of the Learned Respondent, to make any form of order of prohibition."

The view of the Chief Justice was upheld by the majority of the Court.

As a result of this ruling, the Tribunal was limited in its inquiries to actual decisions made by the Government and documentation in regard to such decisions and was precluded from inquiring into "the contents and details of discussions at meetings of the Government."

This Ruling created certain difficulties for the Tribunal because documentation relating to discussions at Government had been received by the Tribunal some of which was referred to in evidence, prior to the judgement of the Supreme Court and the Tribunal is precluded from having regard thereto.

This Ruling has also inhibited the Ministers of the Government and in particular the then Minister for Industry & Commerce from dealing in evidence with the meetings of the Government held on the 8th day of June 1988 and the 25th day of October, 1988.

As appears from a minute dated the 2nd day of November 1988 from Mr Finbar Kelly, the Private Secretary to the Minister for Industry & Commerce to Mr Gerry Donnelly of the International Trade Division, the Minister for Industry & Commerce spoke to the Minister for Finance on that date concerning the recent Government decision in regard to Export Credit Insurance for exports to Iraq.

The note discloses that:—

> "The Minister for Finance indicated that this Department had not as yet made a submission to his Department on the matter. The Minister indicated that he will arrange to have this done immediately but that the Government decision was on the basis that an increase had been agreed, the only thing at issue at this point was the amount of the increase"

And the Minister asked if the submission could be submitted to Finance as a matter of urgency.

On this date the Minister for Trade and Marketing, Mr Brennan, who was leading the Irish delegation to the Fifth Joint Commission spoke to the Minister for Industry & Commerce and inquired what the position was as regards Export Credit Insurance cover for Iraq. He was informed that:

> "(1) The current limit on Export Credit Insurance for Iraq is IR£150m (the Department of Finance are opposing any increase in this limit).
>
> (2) There is *no policy* of confining Export Credit Insurance on beef exports to Iraq to particular companies.
>
> (3) Any exporter with a contract in Iraq will have an application for Export Credit Insurance considered in the normal way.
>
> (4) It was noted that the limit of exports to Iraq had almost been reached."

If there were to be any changes in regard to this policy, Minister Brennan was to be informed by Minister Reynolds."

No increase in the amount of cover for the exports to Iraq had been agreed between the Minister for Industry & Commerce and the Minister for Finance as required by the Government decision of the 25th October 1988, before the Fifth Session of the Irish/Iraq Joint Commission which was held in Baghdad from the 7th to the 9th of November 1988.

Difficult negotiations took place during the course of the said Commission with regard to Export Credit. The Iraqi side demanded that there be a significant improvement in the level of Export Credit facilities available from Ireland.

The Minister for Trade and Marketing, Mr Brennan contacted the Minister for Finance, Mr Mac Sharry by telephone and eventually the Irish side agreed to the inclusion of the following in the agreed minutes:

> "The Irish Government has decided to increase the overall export credit limit for Iraq from the present IR£150m ceiling by a significant and substantial amount in 1989."

On the 11th day of November Mr Timbs, at the direction of the Minister for Industry & Commerce wrote to the Secretary of the Department of Finance as follows:

> "I am directed by the Minister for Industry and Commerce to refer to the Government decision of 25th October 1988 (S.15005C) concerning Export Credit Insurance for Iraq. This Decision provides that the Minister for Industry and Commerce might agree with the Minister for Finance a new limit for Export Credit Insurance for Iraq in place of the existing limit of £150m.
>
> The Minister proposes that the ceiling on insured exports to Iraq should be increased from £150m to £270m. The aggregate limit under the Insurance Acts was raised from £300m to £500m in June 1988. Present exposure and commitments in Iraq are in the region of £130m.
>
> The requirement to raise the ceiling is partly based on the fact that we enjoy a special position in the Iraqi market in that our previous extensions of credit facilities to them, when others were less generous, are being suitably recognised. While there have been delays in payments from Iraq, these payments have eventually come through, despite the political and economic problems which have faced the country in the past. While other countries have been faced with requests from the Iraqis for rescheduling of debt this has never been suggested in the case of Ireland.
>
> The Minister believes that now is the time to capitalise on our previous commitments to this market and the goodwill generated therein. Since the ending of the war with Iran many Western countries are re-opening credit lines with Iraq. It is understood that the UK provided an additional credit line of £300m. for this country this week. The Minister believes that it is important that our position should not be undermined by countries who are only now prepared to underpin trade with Iraq. It is important to note that an export market of £270m p.a. would be our 8th largest export market.
>
> Insofar as the allocation of the proposed additional cover is concerned, the Minister would point out that not all applicants for Export Credit Insurance are successful in obtaining cover. The policy in this regard is to maximise the credit available for the best economic benefit of the State as a whole. It is clearly wasteful to expend this valuable facility in such a way that Irish companies compete against each other in foreign markets to the benefit of the buyer and the overall disadvantage of the State.
>
> The Minister is of the opinion that the foregoing considerations outweigh the fact that what is proposed would result in a significant proportion of the total risk permitted under the Insurance Act being concentrated in a single export destination and the agreement of the Minister for Finance is accordingly sought to this proposal."
>
> ---
>
> Joe Timbs
> Principal Officer,
> International Trade Division,
> November 1988"

By letter dated the 23rd day of November 1988, the agreement of the Minister for Finance for an increase of £100m in the ceiling for Iraq was conveyed to the Department of Industry & Commerce, as follows:—

> "Secretary
> Department of Industry and Commerce
>
> *Attention: Mr Joe Timbs*
>
> I am directed by the Minister for Finance to refer to your minute of 11 November proposing an increase of £120m in the ceiling for export credit for Iraq and seeking the agreement of the Minister for Finance to this increase.
>
> I am to convey the agreement of the Minister for Finance for an increase of £100m in the ceiling for Iraq. The Minister considers however that in view of the substantial increases in the ceiling which have taken place in recent years and the still uncertain situation in Iraq, there should be no further increases (other than that now agreed) in the limit for at least a year. The situation can then be reviewed in the light of ongoing trade developments and experience with the manner in which Iraq meets its commitments
>
> ______________________
>
> P.A. Howard

The agreement of the Minister for Finance for the increase in the ceiling for Iraq by £100m. rather than the £120m meant that the provisional allocation of cover made by the Minister for Industry and Commerce on the 21st day of October 1988 viz additional cover for AIBP and Hibernia up to a maximum liability under the scheme of £80m and £20m respectively and additional cover for non-beef exporters up to a maximum liability under the scheme of £20m, could not be provided in full.

A note in the handwriting of the Minister for Industry and Commerce appears to deal with this as it stated:—

> "Export Credit Limit of £100m needed. Can get by on that.
> And this will only represent cover for about 60% or so on contracts.
> Goodman can carry his balance as he did this year.
> Hibernia to get £20m.
> Leaving a further £10m. for small Cos."

This note, on the face of it, represented the view of the Minister for Industry & Commerce with regard to the proposed allocation of £100m cover, viz. £20m to Hibernia, £10m for small companies in the non-beef sector and the balance of £70m to "Goodman".

If this is correct, then there were no funds available within the increased ceiling of £250m for any allocation of Export Credit Insurance cover to any other beef exporting company.

Though the Minister for Finance had agreed to the increase in the ceiling for Iraq to £250m, Mr Howard of the Department of Finance had prepared a recommendation for the Minister for Finance on the 21st of November 1988, in which he stated inter alia that:—

"The overall ceiling on Export Credit was increased from £100m to £300m in 1981 and to £500m in June 1988. Within that overall ceiling, the limit for Iraq was first set at £12m in June 1983, then rose in stages to £70m in February 1986, then became £150m in 1987. We opposed successive increases in the Iraq ceiling which, because of the extremely volatile Iraq situation, we regarded as too much of a gamble with the Exchequer's resources.

Present commitments and exposure in Iraq are in the region of £130m. We understand that if the ceiling were to increase by £120m, most of that increase would be taken up very quickly.

While the ending of the Iran/Iraq war has undoubtedly made for a more secure destination for Irish exports, the present proposal would concentrate nearly 60% of Export Credit cover in one destination. This concentration of risk in one export destination is at variance with the normal commercial insurance practice of spreading risk and would leave the Exchequer extremely vulnerable if the situation in Iraq were to deteriorate. In the circumstances we recommend that you agree to an increase of only £50m in the ceiling for Iraq at this stage — the situation can be reviewed as necessary in the light of ongoing trade developments and experience with the manner in which Iraq meets its commitments."

On the 23rd day of November 1988, the then Minister for Industry & Commerce Albert Reynolds TD was appointed Minister for Finance and ceased to have responsibility for the allocation of Export Credit Insurance.

Mr Ray Burke TD was appointed Minister for Industry & Commerce and by virtue of such appointment became responsible for the allocation of Export Credit Insurance.

For the benefit of the Minister, a Memorandum on Export Credit Insurance cover for Iraq was prepared in the Department of Industry & Commerce and was submitted to him.

This Memorandum is dated the 12th day of December 1988 and because it represents in detail the views of the Department of Industry & Commerce on the position then existing with regard to the allocation of Export Credit Insurance, the Tribunal considers it desirable to set forth this Memorandum in detail.

"Memorandum

EXPORT CREDIT INSURANCE COVER FOR IRAQ

1. On 8th September, 1987 the Government decided (S. 15005C) to increase the ceiling on insured exports to Iraq to IR£150m. The decision to increase the ceiling was taken having regard to the considerable opportunities for Irish exporters, particularly beef exporters, in the Iraqi market. Total actual exposure at the time was in the region of IR£25m, so the Government decision in effect gave the go ahead for the Minister to take on additional liabilities of some IR£125m in respect of the Iraqi market.

2. ***Present Liabilities in Iraq***

   The table attached at Appendix 1 sets out the current position with regard to existing exposure and commitments in respect of export credit for Iraq. It will be noted that exposure (i.e. liabilities on insurance policies issued) amounts to IR£126.270m and commitments (potential liabilities on offers of cover) amounts to IR£15.565m giving a total contingent liability of IR£141.835m.

Included within this liability figure is a further IR£8m which has been set aside to provide cover on the new contract. The existing contract expires at the end of 1988. It is understood that PARC have now been successful in renewing their contract for the 1989-1991 period.

Export Credit Insurance cover for beef exports was written only for AIBP (exposure IR£76.75m) and Hibernia Meats (exposure IR£23m).

3. On Friday, 21 October, 1988 the Minister for Industry and Commerce, Mr Albert Reynolds discussed the question of new Export Credit Insurance for Iraq with the Secretary and officials from the Department's Export Credit Section. The Minister indicated that he had discussed the question of Export Credit Insurance for Iraq with the Government at their meeting on 8th June, 1988, in the context of a statutory increase in the Minister's overall liability in respect of export guarantees. He said that at that meeting the Government in addition to approving the text of the Bill to increase the statutory limit to IR£500m also agreed as follows:

   (*a*) Further increases for Export Credit Insurance in Iraq should be at the discretion of the Minister for Industry and Commerce.

   (*b*) That the provision of Export Credit Insurance for beef exports to Iraq should be managed in the national interest so as to avoid damaging competition between exporters. (The effect of this was that Export Credit Insurance cover in Iraq would only be granted to existing exporters in the market i.e. AIBP and Hibernia) and,

   (*c*) That the Minister for Agriculture was to advise beef exporters of the future position on export credit for beef exports to Iraq.

   The Minister was surprised to note that the decisions above, and that at (*a*) in particular, had not been recorded in the formal Government decision resulting from the meeting on 8th June, 1988. He said that he would discuss the matter with Government at their next meeting on Tuesday, 25th October and have the matter clarified.

4. However, in the light of the discretion given to him by Government, the Minister decided, on 21 October, that the following additional cover would be provided in the Iraqi market:

   (*a*) A rollover of the existing cover held by AIBP (liability under the Scheme, IR£76.75m) and Hibernia (liability IR£23m) as outstanding maturities were paid.

   (*b*) Additional cover for AIBP and Hibernia up to a maximum liability under the Scheme of IR£80m and IR£20m respectively and,

   (*c*) Additional cover for non-beef exporters up to a maximum liability under this Scheme of IR£20m subject to increase should demand necessitate such.

5. Following the Minister's decision, both AIBP and Hibernia were told informally of the new cover which would be available.

[On the instructions of the Minister IR£7m from Hibernia's allocation of IR£20m has already been made available and has been included in the commitments figure given in Appendix 1.]

6. At their meeting of 25 October, 1988 the Government, notwithstanding their apparent decision of 8 June 1988,

   "decided (S. 15005C) that the Minister for Industry and Commerce might agree with the Minister for Finance a new limit for Export Credit Insurance for Iraq within the overall ceiling of IR£500m for Export Credit Insurance generally under the Insurance Acts, 1909-1988, in place of the existing limit of IR£150m".

7. The 5th session of the Irish/Iraqi Joint Commission in which the Irish side was led by Minister Brennan, took place in Baghdad from 7-9 November, 1988. As anticipated, long and difficult negotiations took place on the export credit front. The Iraqi Side made it clear that their basic demand was for a significant improvement in the level of export credit facilities available from Ireland. As a result of telephone contracts between Minister Brennan and the then Minister for Finance, Mr Ray Mac Sharry, the Irish Side agreed to the inclusion of the following in the Agreed Minutes,

   "the Irish Government has decided to increase the overall export credit limit for Iraq from the present IR£150m ceiling by a significant and substantial amount in 1989".

8. At the direction of Minister Reynolds, this Department wrote to the Department of Finance on 11 November, 1988 seeking the approval of their Minister to increase the limit for Iraq from IR£150m to IR£270m i.e. an increase of IR£120m to facilitate the Minister's decisions of 21 October, 1988 regarding additional cover (see paragraph 4 above). The Department of Finance's reply of 23 November, 1988 conveys the agreement of the Minister for Finance to an increase of IR£100m with the proviso that there should be no further increase (other than that now agreed) in the Iraqi limit for at least a year.

9. Paragraph 4 ante outlines the previous Minister's decisions for additional cover for Iraq; i.e.

Company	*IR£m*
AIBP	80
Hibernia	20
Non-Beef	20
TOTAL	120

   As the approval from the Department of Finance is for an IR£100m increase only and on the basis of the previous Minister's decisions the most equitable solution would be for a one-sixth reduction all round which would result in the following level of cover being made available:

Company	*IR£m*
AIBP	66.66
Hibernia	16.67
Non-Beef	16.67
TOTAL	100.00

10. Apart from AIBP and Hibernia Meats, other Irish beef companies (Taher Meats, Agra-Trading and Halal Meats) sought cover for beef exports to Iraq over the past year. It was not possible to cater for their demands within the IR£150m ceiling. However, there has been a build-up of pressure for export credit for beef exports to Iraq in the very recent past with the following applications having been received:

Company	*Contract Value*	*Exposure*
AIBP	US$325m	IR£80m*
Hibernia	IR£50m	IR£20m*
Halal	IR£50m	IR£35m
Taher	IR£30m	IR£21m
Kerry Meats	US$8.25m	IR£3.7m
Agra Trading	US$40m	IR£18m
Kildare Chilling	US$12m	IR£5.5m
		IR£183.2m
TOTAL	IR£380m @ 70% = £266m (we have £100m)	

*The previous Minister decided that cover of IR£80m should be granted to AIBP and cover of IR£20m should be given to Hibernia on foot of their applications.

In addition the following non-beef applications have been received:

Company	*Contract Value*	*Exposure*
	Stg. £7m	IR£5.76m
	IR£2.5m	IR£1.75m
	IR£760,000	IR£532,000
TOTAL		IR£8.042m

Further evidence of the pressure from beef companies for cover in Iraq can be seen in the correspondence received from Taher Meats which was followed by a meeting with Minister Brennan on 2 November last. Furthermore, Minister Burke recently met both the Irish Farmers Association and representatives of Halal.

11. ***Role of the Department of Agriculture***

The Department of Agriculture were consulted on the allocation of insurance cover to the beef trade for exports to Iraq and replied on 6 October last that it would not be possible for their Department to make proposals on which Irish meat companies should be considered for cover nor on how any available cover should be allocated.

They have since repeated this general line but have confirmed that they recognise that the two exporters who currently enjoy cover in Iraq (AIBP and Hibernia) are reputable companies with a proven track record who are capable of maximising return to the industry in terms of exports to Iraq.

The Department of Agriculture, however, does not rule out the possibility that other beef exporters might in the future be able to establish a similar reputation and track record through the exploitation of markets other than Iraq.

The Department of Agriculture, therefore, would seem to agree with the policy of confining insurance cover to AIBP and Hibernia but are not prepared to offer us formal advice or to liaise with the trade on the issue. Nonetheless the Minister for Agriculture is responsible for the beef trade and it is suggested, that in the light of the sensitivity of the matter, Minister Burke might discuss with the Minister for Agriculture the subject of Export Credit Insurance for beef exports to Iraq.

12. It would appear from the applications which we have received that the main beef exporters are chasing the same Iraqi contracts — a deadline of *15 December* for bids has been mentioned. This is clearly not in the national interest. However, it is also clearly not the role of the Department of Industry and Commerce to "regulate" beef exports. The role of the Accounting Officer must be considered in that it may not be appropriate for him to oversee the allocation of such vast amounts of credit in an area where both he and his Department have no expertise and where the required expertise resides in another Department.

13. ***Recommendation***

This Minister should not act in this matter without the advice of the Minister for Agriculture. As in the case of IDA grant assistance for the food industry, it is the Department of Agriculture who have responsibility although overall responsibility for IDA resides here. The Department of Industry and Commerce/ICI can make its experience in insurance/credit underwriting available but this should be utilised only on the advice of the Minister for Agriculture.

12 December, 1988"

Iraq

Appendix 1

		IR£M
1. (*a*) Existing exposure — contingent liability	=	126.270
(*b*) Commitments — contingent liability	=	15.565
Total Contingent Liability	=	141.835

A. ***Breakdown of Existing Exposure***

Company	Contract Value Insured	Exch. Rate	Credit Period of (months)	Level of Indemnity (%)	Contingent Liability (IR£M)
	US$24m	Various	12	75	6.5
AIBP	DM 257.1m	2.6795	12	80	74.75*
AIBP 5% option	DM 12.86m	2.6795	12	80	3.84
Hibernia	US$ 46m	1.60	12	80	23.00
	Stg£3.164m	0.8471	18	70	2.61
	Stg£136,181	0.89	12	70	.107
	Stg£176,212	0.8471	18	70	.146
	Stg£94,000	0.8471	18	70	.077
	US$1,380,000	1.54	18	70	.627
AIBP	US$30m	1.4371	18	70	14.613
			TOTAL EXPOSURE	=	126.27

B. ***Breakdown of Commitments***

Company	Contract Value Insured	Exch. Rate	Credit Period of (months)	Level of Indemnity (%)	Contingent Liability (IR£M)
Hibernia	IR£10.00m	—	18	70	7.00
	Stg£200,000	0.85	12	70	.165
	—	—	—	—	8.00
	IR£570,000	—	18	70	.4
			TOTAL COMMITMENTS	=	15.565

*Original exposure IR£76.75m. Payment of IR£2m rec. in Oct. 1988."

As stated in this Memorandum, there were pending, in the Department of Industry & Commerce, applications for Export Credit Insurance in respect of beef exports to Iraq from AIBP, Hibernia, Halal, Taher Meats, Kerry Meats, Agra Trading and Kildare Chilling.

The total value of the contracts involved totalled approximately £380m and if cover was allocated in respect of all these alleged contracts at the rate of 70% of their value, this would have amounted to £266m, whereas there was only £100m available.

In addition there were three applications from non-beef exporting companies for Export Credit Insurance in respect of contracts with a total value of £12.76m which, if granted on the basis of 70% cover, would require an additional £8.042m.

This was the situation prevailing at the time Mr Burke, TD was appointed Minister for Industry & Commerce and the Minister responsible for the allocation of Export Credit Insurance in respect of what was regarded and described as Number 2 Account business.

On the 9th day of December 1988, Mr Burke, accompanied by Mr John Fanning of the Export Credit Section met with Mr Mohammed Rafique and Mr Seán Clarke, Chief Executive and Managing Director of Halal Meat Packers respectively in connection with an application for Export Credit Insurance for beef sales to Iraq.

Mr Rafique handed the Minister an application for Export Credit Insurance valued £30m to cover 10,000 tonnes of hindquarter boneless beef for export to Iraq. He did not at that stage have a confirmed order because the closing date for tenders in respect of contracts with the Iraqis was the 15th December 1988.

The Minister informed the representatives from Halal that he would examine the application and would respond by Wednesday 14th December 1988, the day before the expiry date for the submission of tenders.

By letter dated the 30th day of November 1988, Agra Trading Ltd had applied to the Insurance Corporation of Ireland for Export Credit Insurance cover in respect of a contract for 10,000 tonnes of beef for which they were tendering in Iraq and sought cover in the sum of $40m.

Representations on their behalf were also made to the Minister for Industry & Commerce.

As appears from a note dated the 14th day of December 1988 from Finbar Kelly, the Secretary to the Minister for Industry & Commerce to Mr O'Reilly, Assistant Secretary of the Department of Industry & Commerce, the Minister on that day had asked Mr Kelly to pass on the following message to Halal:—

> "The decisions have already been taken and that he was committed by the decisions of his predecessor".

This note also informed Mr O'Reilly that the message had been conveyed to Mr Clarke of Halal and that he also proposed to inform Mr Regan of Agra Trading of the position because Mr Regan "had direct contact with the Minister".

On the 12th day of December Mr Rafique had written to the Minister for Industry & Commerce in connection with a matter which he stated he had overlooked raising with the Minister during the meeting of the 9th of December 1988.

With his letter, Mr Rafique enclosed CBF export charts, showing the export of beef to Iraq during the period 1986 to August 1988 and commented thereon as follows:—

> "From the above table it can be seen that the total value of exports to Iraq over 1986, 1987 and the nine months of 1988 have a value to the Irish economy of £80,891,012.
>
> The credit terms made available during this period was 12 to 18 months. One would expect that all 1986 letters of credit would have been cashed at this stage. This would imply that the maximum exposure the Government would have in the form of ICI cover for beef exported to Iraq would be a maximum of £63.8m.
>
> The company was of the opinion that the Department of Industry, Commerce and Communications, through the ICI was underwriting cover to the tune of twice the amount outlined above for Iraq. If this is the case, it would appear that the ICI has been underwriting cover for export deals to Iraq done by Irish companies where the meat has been sourced outside Ireland, such as Germany and South America. This has been suggested to be the case to my company.
>
> I feel you can understand my resentment and disappointment if this is the case, while all the product I have in mind exporting to Iraq can be guaranteed to be West of Ireland origin."

In connection with this matter Mr Banks, Chief Executive of the Insurance Corporation of Ireland wrote to the Secretary of the Department of Industry & Commerce on the 22nd day of December 1988 in connection with certain problems that had arisen with regard to the Export Credit scheme.

Inter alia he stated that:—

> "I would hope that, before any decision is taken to extend further credit to Iraq, the following information available to ICI will be brought to the attention of the Minister:—
>
> 1. Iraq has in the past defaulted on its obligations to several countries and rescheduled its debts.
> 2. Of the present Iraqi exposure under the scheme of £122m, £51m is now seriously overdue and, if it remained unpaid, would require the Department to begin paying claims to exporters in March, 1989.
> 3. From an underwriting standpoint, it would generally be held to be inadvisable to increase the already high concentration of scheme risk in Iraq (44% of the total scheme portfolio).
> 4. The premium rate currently charged to participants for Iraq cover under this scheme is very low compared with

(*a*) the 15% to 22% we understand some Irish meat exporters have recently had to pay for insurance outside the scheme and

(*b*) the 6.25% charged for 12 month protection by ECGD in the UK.

Although the ceasefire in the war with Iran has allowed Iraq to renew efforts to revitalise its infrastructure, it still has major economic problems which will remain for the foreseeable future. If further facilities were granted to Iraq, we would be fearful that the size of the increased debt would increase the Iraqis ability to dictate their own repayment terms or to make further payments subject to yet more credit."

He went on however to deal with the point made by Mr Rafique in his letter dated the 12th day of December 1988.

"A further, possibly unfounded, anxiety relates to a disparity which has emerged between the official statistics for Irish exports to Iraq and the value of shipments recorded under the scheme. Exports per the Central Statistics Office for the period from 1.1.87 to 31.10.88 totalled £86m, but the scheme shipments for the same period were approximately £135m. If I have interpreted the trade figures correctly, they imply that considerable quantities of exports covered by Number 2 Account policies may not be of Irish origin. There may, however, be perfectly acceptable reasons for this which are known to you".

On the 14th day of December 1988, Mr Joe Shortall, Assistant Principal, Beef Division of the Department of Agriculture, telephoned Mr Timbs of the Department of Industry & Commerce to say that the Secretary of his Department had become aware that Agra Trading and Halal had sought Export Credit Insurance for Iraq and that the Secretary (Department of Agriculture) was anxious to point out that both of these companies were very reputable and should not be discriminated against in the matter of Export Credit Insurance.

Mr Timbs prepared a note thereof dated the 14th day of December and circulated same to the appropriate persons.

At that stage the Minister had 7 companies applying for cover.

He discussed the matter with the Taoiseach and the Minister for Agriculture & Food after a Cabinet meeting on the 16th day of December 1988.

On that evening, namely 16th December 1988, the Minister for Industry & Commerce instructed Mr Timbs to contact each of the said seven companies and inform them that:—

"In response to your application for Export Credit cover for the Iraqi market, the Minister wishes to inform you that on production of a signed/confirmed contract, he is prepared to consider your application as sympathetically as possible, within the overall limit of national cover established for that market which is limited. The Minister wishes to emphasise that the above should not be taken as a commitment to automatically grant the cover sought having regard to the constraint outlined above."

In response to queries, both AIBP and Hibernia were informed that this decision "did not necessarily override commitments previously given."

The terms of the telephonic communications were confirmed by Mr Timbs by letters dated the 19th day of December 1988 to the said seven companies, namely Anglo Irish Beef Processors International Ltd, Hibernia Meats International Ltd, Halal Meat Packers (Ireland) Ltd, Agra Trading Ltd, Taher Meats (Ireland) Ltd, Kerry Meat Products and Kildare Chilling Co. Ltd.

In reply to the said letter of the 19th day of December 1988, Hibernia Meats International Ltd wrote to Mr Timbs as follows:

> "Dear Joe
>
> Thank you for your letter dated 19th December, the contents of which I have noted.
>
> I do, however, wish to point out that on the 21/10/1988 I was verbally informed of a facility for Hibernia Meats International Ltd and associated companies, under Export Credit Insurance to Iraq of IR£20 million together with a rollover of the existing "46 million" Dollars for 1989.
>
> You will recall that this position was subsequently confirmed at a meeting with you in the Department on Wednesday the 23/11/1988 when arrangements were finalised for the first £10 million of the above facility to be put in place with the Insurance Corporation of Ireland.
>
> I trust that your Department will accept the position as outlined by me above.
>
> Yours sincerely
> Oliver Murphy, Managing Director"

AIBP verbally sought similar assurances but it was not until March 1989 that AIBP stated their position in writing. This was in response to a letter dated the 15th of March 1989 from Mr John Dully, Assistant Secretary, International Trade Division of the Department of Industry & Commerce.

This was a letter written on the 8th March 1989 to Mr Aidan Connor, Deputy Chief Executive AIBP, and was as follows:—

> "I am directed by the Minister for Industry & Commerce to refer to your recent application for Export Credit Insurance cover in respect of the supply of beef to Iraq.
>
> I am to inform you that your application remains the subject of consideration. Decisions thereon, which will be taken as early as possible, must have regard to the special difficulties of the Iraqi market at this time, in particular that country's payment position where considerable payments are now overdue.
>
> You will appreciate that it would be imprudent management of the Export Credit Insurance scheme with consequent injudicious use of taxpayers' money were the Minister to approve additional cover for the Iraqi market at this stage. The Minister believes that his primary responsibility in the matter is to ensure that existing debts are honoured by the Iraqis before he can assume new liability in that market.
>
> I am to inform you that cover has not been granted in respect of any application made in connection with the most recent Iraqi round of contracts."

Mr Connors reply was as follows:—

> "Thank you for your letter dated the 8th of March 1989.
>
> I am greatly concerned by the contents of your letter which is in effect the reneging by the Minister on an agreement made previously by your Department with our company. I demand an immediate explanation as to why this sudden change of mind has occurred.
>
> In order to set the record straight in this matter I list below in chronological sequence the events which details the background to the agreement ultimately given by the Minister's office in late October 1988.
>
> 1. On 27th November 1987 I wrote to your Department regarding insurance cover for Iraq on contracts which this company had signed at that time.
>
> 2. That application was followed by a series of meetings involving Mr L Goodman and myself from this company and a number of people in your Department, right up to Ministerial level. These meetings took place throughout the early part of 1988 and involved extensive negotiations on this matter.
>
> At all times in the negotiations it was stated that the limiting factor with regard to approving this Export Credit was the overall limit for the Export Credit scheme which was approved by the Oireachtas office. It was pointed out to us that the Department had put forward an increase in the overall level for approval by the Dail and on receipt of this, approval would be granted.
>
> 3. By telephone call in late October 1988, Mr Joe Timbs from your Department informed me that following the approval by the Dáil of the increased limit, the Department was now granting cover to AIBP in the amount of £100 million. This £100 million was to cover
>
>   (*a*) £20 million previously notified to us by Mr Joe Timbs on 10th February 1988 and which had been in abeyance pending the overall increase in the limit.
>
>   (*b*) a fresh £80 million against our application dated the 27 November 1987.
>
> In addition Mr Timbs also stated that the Department had a rollover of all existing cover to AIBP as and when payments were received from Iraq.
>
> The above represents a clear commitment and undertaking by the Department and your recent letter is a totally unacceptable repudiation of same. I demand an immediate restoration of the cover as promised by Mr Timbs.
>
> I look forward to your very early reply.
>
> Yours sincerely"

By letter dated 3rd day of January 1989 Mr Fitzpatrick, Managing Director of Taher Meats wrote to Mr Timbs informing him that the company had recently been awarded a contract to supply 5,000 tonnes of boneless beef to the State Establishment for Foodstuff Trading, Baghdad.

By letter dated the 12th January 1989, the full signed contract was forwarded to the Department.

This disclosed the agreement to supply and to purchase 5,000 tonnes at a rate of $3,280 per tonne.

By letter dated 4th January 1989, Oliver Murphy of Hibernia Meats International Ltd wrote to Mr Timbs as follows:—

> "Dear Joe
>
> I am pleased to inform you that we have signed a contract with the State Company for Foodstuffs Baghdad (Contract No. 9/89) for 20,000 metric tonnes, value $66 million dollars with 18 months credit to the purchaser. I will forward under separate cover a copy of the contract for your attention.
>
> I wish to have confirmation of Export Credit Insurance for the above in accordance with our telephone conversation of the 21/10/1988 and our subsequent meeting in the Department of the 23/11/1988 when the offer of £20 million together with a rollover of the existing 46 million dollars for 1989 was confirmed."

By letter dated 17th day of January 1989, a copy of the said contract was forwarded to Mr Timbs.

This contract was for the sale of 20,000 tonnes of boneless young bull meat at a rate of US$3,290 per tonne.

Halal had made an offer on the 15th day of December 1988 to sell to the State Company for Foodstuff Trading, 10,000 tonnes of hindquarter cuts.

On the 19th day of December 1988 they wrote to the Minister for Industry, Commerce & Communications informing him that the said company had made a counter-offer and requested negotiations.

They informed the Minister, in the said letter, that they were assuming that once they signed the contract, insurance cover would be made available. They stated that they were confident that once the contract was agreed and signed, cover would be made available to them and requested that if there were any doubts about that they should be informed in definite terms.

On this day Mr Fanning phoned Halal Meats and informed them that the Minister was not in a position to give a commitment to grant cover at that stage as the availability of insurance cover for the Iraq market was limited.

Mr Fanning informed Halal that the Minister would consider as sympathetically as possible the company's application once it had a signed contract.

Halal did not revert to the Department claiming to have signed contracts in Iraq.

Consequently, the only three companies who reverted to the Department, claiming to have signed contracts in Iraq were AIBP, Hibernia and Taher.

On the 5th day of January 1989, Laurence Goodman wrote to the Minister as follows:—

"Mr Ray Burke, TD
Minister for Industry & CommerceLeinster House
Kildare Street,
Dublin 2

Dear Minister

Further to our recent meeting I thought that it would be useful to give you a little background to the development of our business in Iraq and particularly in the light of developments that have occurred over Christmas.

Our company established the first sales for Irish Beef to Iraq after tremendous personal effort and financial commitment. This is more than nine years ago now and since then we have built up from a standing start a position of being the No. 1 supplier to that market with the largest market share and more particularly we have established a substantial premium for Irish beef above all other suppliers from any source.

It should be recognised that the above achievements have been against fierce competition from Argentina, Brazil, Uruguay, Australia and from the EEC particularly France and Germany. Irish suppliers have followed us to this market as they have to other markets over the past number of years but based on the difficulties there in terms of specification, scheduled daily deliveries, deliveries to war zones, etc., they have been unsuccessful to date.

Just prior to Christmas a new tender was called for additional supplies for 1989. On this occasion as well as having the normal competitors from other countries we had a selection of new potential Irish suppliers. Their comments were that life was being made much easier for them now based on the ending of the war and their view that if they were awarded a contract they would be in a strong position to get Export Credit Insurance.

The results of the efforts by the Irish competitors to date have been to accept a price of 3280 dollars per tonne or 570 dollars below our quoted price of 3850 dollars per tonne. They have also accepted 220 dollars below what we sold at last year, i.e. 3,500 dollars per tonne. To put this in context this is .17p per lb. lower than the price tendered by us and .7p lower than we supplied at last year, i.e. 1988. It should also be borne in mind that we are in a rising beef market.

The above horrifying facts were brought to my attention on Christmas Day and having invested nine years and considerable effort and expense in gaining the premier position in that country's market I felt obliged to cancel all my Christmas plans for myself and my family and to leave for Baghdad first thing on St. Stephen's morning together with a senior colleague. This may perhaps help to focus the significance I would place on Ireland losing its premium in the market over and above all other suppliers. My colleague and I spent five days there trying to pull things back on the rails but the Iraqi were very skilful in using the position to their advantage. The result is now that we have lost the premium and we have regained a share in the market but Irish meat as a result of our competitors' activities is now seen in a totally different light.

I am aware of the difficulties you and your senior civil servant colleagues have in administering an export credit scheme. However, I know you would all share with the view that the above happenings are unacceptable and even more so when caused by Irish companies which are non-Irish owned. I felt therefore it was important to bring this to your attention and that of your colleagues who put so much effort into trying to promote and increase our exports and our country's standing in the export markets abroad.

Perhaps I will have an opportunity of dropping in to see you and also the secretary of the department if you think it would be useful within the next week or so.

I would like to take this opportunity of wishing you and your family a very happy and peaceful 1989.

Yours sincerely

GOODMAN INTERNATIONAL LIMITED
L. J. Goodman
Chairman and Chief Executive"

On the 10th day of January 1989 Mr Donnelly of the Department of Industry & Commerce wrote to the Secretary, Department of Agriculture and Food, as follows:—

"Secretary
Department of Agriculture and Food
Agriculture House
Dublin 2

FOR THE ATTENTION OF MR SEAMUS HEALY,

EEC/FOREIGN TRADE DIVISION

I refer to recent discussions and correspondence, in particular your Department's minute of 4 January, 1989, concerning Export Credit Insurance for beef exports to Iraq.

This Department acknowledges and welcomes the proposal by the Department of Agriculture to assume a more active involvement in the appraisal and allocation of future cover for agricultural exports and in particular their willingness to decide on the apportionment of available cover among competing claimants in the beef sector.

While this Department is opposed to the establishment of any sort of Joint Committee for the purposes of determining allocation of cover we will revert to your Department in due course with our ideas as to the type of formal structure which should be established to facilitate the exercise.

In the meantime I would like to draw your attention to the fact that the Minister for Industry and Commerce has agreed to meet Mr Larry Goodman to discuss the subject of beef exports to Iraq and while no date for the meeting has yet been fixed I expect that it will take place within the next few days. The Minister has received the attached correspondence from Mr Goodman suggesting that his company have been undercut in the Iraqi market by other Irish exporters and arguing that this is an unacceptable position. The Minister would welcome the urgent views of the Department of Agriculture on this correspondence in order that he might be fully briefed on the matter for his meeting with Mr Goodman.

The urgent views of the Department of Agriculture are also requested on the differences between official CSO figures for beef exports to Iraq and the value of exports covered under this Department's Export Credit Insurance scheme. I should point out that the Scheme applies only to beef of Irish origin and it is a condition of individual insurance policies that exporters retain proof of the origin of beef covered under the policy. There is no question therefore of Export Credit Insurance being provided in respect of non-Irish beef. Nonetheless the differences between the figures are significant. For example official CSO figures show that in the period January to July, 1988 total Irish beef exports to Iraq amounted to IR£28m. In the same period, total shipments of beef to Iraq which were covered under our Export Credit Insurance scheme amounted to IR£54.1m. Similarly the official records show that in 1987 and up to the end of August 1988, Ireland exported beef to Iraq to a total value of IR£63.8m., whereas in the considerably shorter period from September 1987 to end of July 1988 we provided Export Credit Insurance for beef exports to Iraq valued at a total of IR£123.75m. The Minister would welcome the views of the Department of Agriculture on the foregoing and asks that they be submitted as urgently as possible again in anticipation of his proposed meeting with Mr Goodman.

Gerry Donnelly
10 January 1989."

By letter dated the 13th day of January 1989 Mr Nevin of the EC/Trade Division of the Department of Agriculture replied thereto as follows:—

"Secretary
Department of Industry and Commerce
Kildare Street
Dublin 2

For the attention of: Mr G Donnelly.

I refer to your letter of 10 January (plus enclosure), to Mr Seamus Healy in relation to beef exports to Iraq.

Regarding Mr L J Goodman's assertion that his firm's market prospects in Iraq are being seriously damaged because of substantial undercutting by Irish-based competitors this Department would obviously regret that the activities of a supplier to any particular market should prove injurious to his competitors. In the commercial situation, however, it is difficult to foresee that any specific action is open to us which would result in cessation of the activity complained of, especially if the possibility of voluntary agreement among the competing parties is discounted. As you know, we already seek through C.B.F. to ensure that Irish meat is marketed to optimum advantage on export markets. In the interest of the industry as a whole, therefore, and in the light of the current complaint we shall immediately renew our efforts in this regard.

Concerning the apparent discrepancies between the values attributed to our beef exports to Iraq in the official trade statistics published by the C.S.O. on the one hand and the value of exports covered under your Department's Export Credit Insurance scheme on the other, it is clear that resolution of this matter warrants the urgent attention of all the parties concerned. Accordingly, we would support convening an early meeting of the relevant officials from both our Departments together with C.S.O. and Revenue (Customs) personnel. Since the value of EC export refunds is not included in either the C.S.O. figures or those advanced by your Department, this may immediately be disregarded as an explanation for the differences and I note that you have already dismissed the possibility that non-Irish beef might have been insured. The only other possibility which immediately suggests itself is that because of the very considerable time lag which may exist between the period when an export contract is secured and insurance cover is applied for and the physical export of the beef quantities concerned, the C.S.O. export-values and the values insured under the export credit scheme would not necessarily balance over a given time span. This theory obviously needs to be researched more thoroughly than it is possible to do in the time available to us at present, but it represents one possibility at least and as such could merit further examination by the inter-Departmental meeting suggested above.

I note the point made in your final paragraph to the effect that in the 20 month period ended August 1988, Ireland's total beef exports to Iraq were valued at IR£63.8m according to the official trade statistics, whereas in the considerably shorter period from September 1987 through July 1988 Export Credit Insurance was provided for IR£123.75m worth of beef exports to that market. The tonnage involved in these periods (respectively 34,250 tonnes and 25,046 tonnes according to the C.S.O.) would indicate an average value — exclusive of refunds — of IR£1857 per tonne for the longer period and IR£1984 per tonne for the shorter on the basis of the official statistics but IR£4940 per tonne on the basis of the amounts insured. The results of my preliminary enquiries would suggest that the last mentioned figure is considerably inflated and that the values attributed by the C.S.O. are much closer to reality but again, this is something which the proposed meeting might address more fully.

Yours sincerely

B. Nevin
EC/Trade Division"

The Minister for Industry & Commerce met Messrs Goodman and Connor of Anglo Irish and Dilger of Food Industries on the 1st of February 1989, for the purpose of discussing the contents of Mr Goodman's letter dated the 5th day of January 1989.

Prior to the said meeting the Minister for Industry and Commerce Mr Burke, had received a Memorandum from Mr Donnelly of the Export Credit Section of the Department entitled "*Briefing Material for Minister — Export Credit Insurance for Iraq*" which is dated the 27th January, 1989 and is as follows:—

"BRIEFING MATERIAL FOR MINISTER

Export Credit Insurance FOR IRAQ

1. The Minister has agreed to meet Mr Larry Goodman to discuss the contents of his letter of 5 January, 1989 alleging that his company have been "under cut" by another Irish beef exporter in regard to a contract in Iraq and that this damaging

competition reflects poorly on the Irish beef trade as well as eliminating the premium which Irish Beef commands in the Iraqi market. Mr Goodman's letter is attached as Appendix 1.

2. The Department of Agriculture were asked for their views on Mr Goodman's letter. Their response is at Appendix 2.

3. It is clear that Mr Goodman would like the Minister to be selective in allocating available Export Credit Insurance cover to beef exporters for the Iraqi market. Ideally he would like to have a monopoly on such cover.

4. The current limit on export credit for Iraq is IR£150m. Current exposure and commitments amount to IR147.9m. Exposure on beef amounts to IR£115.19m and is divided between AIBP (Goodman) (IR£92.19m.) and Hibernia Meats (IR£23m).

5. Recent decisions have resulted in a further IR£100m being made available for the Iraqi market bringing the new limit on cover to IR£250m. Applications have been received from 7 beef exporters which would require cover under the scheme amounting to a total of IR£266m. The 7 applicants are AIBP, Hibernia Meats, Taher Meats, Halal Meats, Agra-Trading, Kerry Meats and Kildare Chilling. As a consequence of our inability to meet demand for cover for beef exports the Minister following discussion with his Government colleagues, informed the applicants on 19 December last that on production of a signed/confirmed contract he would be prepared to consider their applications as sympathetically as possible within the overall limit of national cover established for the Iraqi market. The Minister went on to emphasise that the above should not be taken as a commitment to automatically grant the cover sought having regard to the constraints outlined. Three of the companies involved (i.e. AIBP, Hibernia and Taher) have now reverted to the Department claiming to have signed contracts in Iraq. At the time of preparing this briefing the Goodman people have not actually submitted copies of the contracts to the Department. They say they will do so immediately the signed English versions are received.

6. AIBP say that they have a signed contract for the supply of beef worth US$50m. and expect to sign a further contract worth up to US$100m. in the coming weeks. The Minister in response might refer to his commitment to examine these applications sympathetically on the production of signed / confirmed contracts.

7. Mr Goodman will not be aware that the Department of Agriculture have agreed to decide on the allocation of available cover in Iraq to competing claimants in the beef sector. This is a logical development given that Department's primary responsibility for the beef trade and the great sensitivity involved in the allocation of export credit cover to competing exporters. However, nothing can happen in relation to the AIBP application until the signed contract(s) are received.

8. A separate submission has been made to the Minister recommending that, of the existing exposure of IR£150m. a total of IR£99.75m. would be "rolled over" specifically in respect of beef exports. This assumes, of course, that the Iraqis will pay existing amounts owing. In addition the submission proposes that a total of IR£83.33m. of the new cover being made available would be allocated specifically to beef. This would provide a potential allocation of cover to the beef sector alone

for 1989 of IR£183.08m. Remaining available cover would be allocated to the non-beef sector.

9. Total payments amounting to IR£59.7m. have fallen due for payment from Iraq and have not been received. State exposure on this amount is IR£47.7m. Of the total owing IR£42.7m. is due to Goodman and the State's exposure on this amount is IR£34.2m. The balance of the amount owing is due to Hibernia Meats.

10. The Minister might ask Mr Goodman what action is being taken in regard to the amounts overdue to his company. There is an obligation on the exporter to do all in his power to secure payment on insured contracts. The Minister can refer to the fact that Minister Brennan has written to his opposite number in Iraq and that it is proposed that senior officials of the Department should visit Iraq in the near future to seek to secure payment.

11. It is recommended that the Minister should avoid making any reference to the amount of new credit which it is proposed to extend to the Iraqis. The reason for this is that any disclosure of the amounts concerned to the beef exporters would likely get back to the Iraqi authorities. It is not in the best interest of securing payment on outstanding amounts that we display our hand to the Iraqis at this stage. It is also recommended that the Minister avoid any discussion other than on matters relating to the Goodman Group. For instance he should not disclose what other beef companies have applied for cover, what other companies are currently overdue money in Iraq or any other confidential information which represents a matter solely between the Department and the exporter concerned.

____

27 January 1989.

Mr Goodman reiterated and expanded on the complaints made by him in his letter dated the 9th day of January 1987 with regard to alleged undercutting by Taher Meats and Halal in respect of contracts for the export of beef to Iraq.

The Minister stated that he would arrange a meeting with the Minister for Agriculture with a view to taking decisions on market management.

At this meeting Mr Goodman also raised the question of outstanding difficulties in relation to a contract in respect of $155m which was the subject of the application made on the 27th November 1987 and in respect of which, £20m had already been put in place.

He stated that the balance of this contract had been substantially filled since then and it was critical that the necessary cover be put in place. The Department of Industry and Commerce's note of the discussion at such meeting is as follows:—

"1. The Minister met Messrs. Goodman and Connor of Anglo Irish and Dilger of Food Industries on 1 February. The undersigned was also present.

Regarding AIBP affairs, Mr Goodman made the following points:—

—his letter of 9 January detailed what he perceived to be serious developments in the Iraqi market which could imperil an attractive export outlet for Irish beef.

—his company spent many years developing this market, their efforts have been successful and a premium was being obtained for Irish beef.

—non Irish companies, i.e., Taher Meats and Halal were now killing the market by quoting at prices which were substantially lower than prices secured by AIBP in 1986/'87.

—Taher Meats are believed to have obtained a contract for 5000 tonnes at 7p. per pound below the AIBP price — this had the effect of not only creating pressure for AIBP but also for the Brazilians, who supplied to that market, the overall effect of which would be to significantly reduce the attractiveness of the market in Iraq.

—Additionally, these companies were also offering 18 months credit whereas AIBP had been successful in obtaining 12 months.

Mr Goodman agreed with the Minister that there was a need to manage the Iraqi market and suggested that there were many other markets which these companies could target.

As regards Hibernia, their French owner, CED, had obtained a contract in Iraq, part of which would be supplied by Hibernia and would benefit from Irish Export Credit Insurance. While acknowledging that portion of the contracts would be filled from Ireland, Mr Goodman objected that it was wrong that business written by a French company should benefit under the Irish scheme while, on the other hand, business which he would wish to supply from France could not benefit under the COFACE arrangements.

The Minister expressed his anxiety that every effort should be made to ensure that the Iraqi market remained a premium market for Irish exports and to this end, he would arrange a meeting with the Minister for Agriculture with a view to taking decisions on market management. Mr Goodman would be invited to such a meeting.

On the question of existing AIBP contracts for Iraq, Mr Goodman referred to the outstanding difficulties in relation to a contract in respect of $155m. The application was made in November, 1987 and cover in respect of £20m. had already been put in place. The balance of the contract had been substantially filled since then and it was critical that necessary cover be put in place. Mr Connor then left the meeting to discuss the details of this aspect with Mr Timbs.

3. The Minister referred to a Press query just received about a suggestion that this Department's Export Credit Scheme might have covered exports of mutton from Australia into the Iraqi market. Mr Goodman mentioned that members of the Press had been sniffing around with the story in the past while, but there was no substance to the point. Australian mutton was being supplied by Goodman to Iraq but there was no question of Irish Export Credit Insurance being used for this activity.

4. Mr Goodman referred again to a point he had raised at Malahide on 31 January regarding delays in processing export documentation which were causing delays both in actual exporting and in receipt of subsidies/release of bonds and guarantees, the cost of all of which was substantial. The basis of the complaint was complex but it would be explained in detail by Mr Connor of his company. I indicated that Mr Fisher of our Consultancy Unit would pursue this issue."

A Memorandum dealing with Export Credit Insurance cover for Iraq had been prepared for the benefit of the Minister on the 17th day of January 1989.

The terms of the Memorandum are important and are set forth in detail hereunder:—

"Memorandum

EXPORT CREDIT INSURANCE COVER FOR IRAQ

January, 1989.

1. Recent Memoranda on this subject have dealt primarily with the allocation of available cover to individual beef exporters for contracts in Iraq. The purpose of this Memorandum is to decide firstly what allocation should be made, if any, to non-beef exporters and secondly, at what rate the new cover should be made available.

2. The table attached at Appendix 1 sets out the current position with regard to existing exposure and commitments in respect of export credit for Iraq. Current exposure and commitments amount to a total contingent liability of IR£148.2m.

3. As a result of recent decisions a further IR£100m. of cover under the export credit scheme has now been made available for the Iraqi market. Applications have been received from 7 beef exporters which would require cover under the scheme amounting to a total of IR£266m. As a consequence of our inability to meet demand for cover for beef exports the Minister, following discussion with his Government colleagues, informed the 7 beef applicants that on production of a signed/confirmed contract he would be prepared to consider their applications as sympathetically as possible within the overall limit of national cover established for the Iraqi market. The Minister went on to emphasise that the above should not be taken as a commitment to automatically grant the cover sought having regard to the constraints outlined. Three of the seven Beef Companies have now reverted to the Department claiming to have signed contracts in Iraq. The Department of Agriculture have indicated their willingness to decide on the apportionment of available cover among competing claimants in the Beef Sector. Accordingly a submission to that Department has been prepared requesting such a decision. However, no decision has been made as to whether to allocate the full £100m. additional cover available to beef exporters or whether some small amount should be held back in respect of non-beef exports. This must be decided upon before deciding the submission to the Department of Agriculture.

4. The previous Minister for Industry and Commerce had sought an additional IR£120m. cover for Iraq to be allocated as follows:—

COMPANY	IR£
AIBP	80m.
Hibernia	20m.
Non-Beef	20m
Total	120m.

5. As the increase in cover approved amounted to only IR£100m. and on the basis of the previous Minister's decisions the most equitable solution would seem to be for a one-sixth reduction all round which would result in the following level of cover being made available:-

COMPANY	IR£m.
Beef Exporters	83.33m.
Non-Beef Exporters	16.67m.
Total	100.00m.

6. The previous Minister had also decided that the previous cover held by the two beef exporters in Iraq (AIBP and Hibernia) would be rolled over (i.e., reissued when repaid by the Iraqis) to those two companies in the same proportion as before. While the allocation is now a matter for the Department of Agriculture, it is proposed to adhere to the previous Minister's decision to roll-over the cover specifically for beef exports. This would make an additional IR£99.75m. available for beef exports for 1989 assuming of course that the Iraqis pay up on time.
7. When added to the suggested allocation of IR£83.33m. (Paragraph 5 above) the roll-over facility would result in a potential allocation to Beef exporters of IR£183.08m. in 1989. (The balance of the existing exposure would be "rolled-over" for non-beef exports).
8. Although most of the demand for export credit cover in Iraq comes from beef exporters there has been a smaller but nonetheless steady demand for cover from the non-beef sector covering such products as ....pharmaceuticals, refrigeration equipment and general industrial products. For example, we have an application on hands for a medium term contract from.............. for the construction of a water treatment plant in Iraq. Unless we make some allocation of available cover to such non-beef contracts it will not be possible to facilitate small companies to win valuable and in many cases strategic contracts in the Iraqi market. It is therefore recommended that cover be allocated to beef and non-beef sectors on the basis outlined in paragraph 5 and 6 above.
9. The second issue to be decided is the date at which the new cover would be released. It is proposed that the release of new cover should be tied to the repayment of our existing exposure by the Iraqis. For example, existing exposure is approximately IR£150m. The additional cover available is IR£100m. (Giving a new limit for Iraq of IR£250m). As a general rule it would seem prudent to decide that for every £15 repaid by the Iraqis, that £15 plus a further £10 would be reissued. This would encourage the Iraqis to pay existing debts promptly and mean a controlled and gradual build up of new liability in a way which can be closely monitored. It may not be possible to stick rigidly to this rate of release particularly in the Beef Sector. A lot will depend on how the available cover is allocated by the Department of Agriculture. For example we have existing exposure on Hibernia Meats of IR£23m. If it was decided to give that company a further £23m, release of this cover would probably have to be on a £1 for £1 basis.

10. It is proposed that the Minister agree to the general principle of the controlled release of the new cover having regard to repayment by the Iraqis of their outstanding liabilities as outlined in paragraph 8 above. In order to avoid having to revert to the Minister in an individual cases the Department should have discretion to vary the release rate having regard to prevailing circumstances.

11. **Summary of recommendations**

It is recommended as follows:—

(*a*) that the new credit facility for Iraq of £100m. should be allocated in the following proportions:—

Beef Exports	IR£83.33m.
Non-Beef	IR£16.67m.
Total	IR£100m.

(*b*) that cover currently held by (two) beef exporters in Iraq totalling IR£99.75m. be "rolled-over" *specifically* to cover new beef contracts and the balance of existing exposure "rolled-over" for non-beef exports.

(*c*) that the new cover of £100m. be released in a gradual manner having regard to repayment by the Iraqis of their existing debt.

---

12 January, 1989"

APPENDIX 1

"

IRAQ		IR£m.
1. (*a*) Existing exposure — contingent liability	=	138.007
(*b*) Commitment — contingent liability	=	10.205
Total Contingent liability	=	148,212.

A. ***Breakdown of Existing Exposure***

Company	Contract Value	Exchange Rate	Credit Period (months)	Level of Indemnity (%)	Contingent Liability (IR£M)
	US$66.6m	Various	12	75/80	19.0[2]
AIBP	DM 257.1m	2.6795	12	80	74.75[1]
AIBP (5% option)	DM12.86m	2.6795	12	80	3.84
Hibernia	US$46.m	1.60	12	80	23.00
	Stg£3.164m	0.8471	18	70	2.61
	Stg£136,181	0.89	12	70	0.107
	Stg£176,212	0.8471	18	70	0.146
	Stg£94,000	0.8471	18	70	0.077
	US$1.38m.	1.54	18	70	0.627
AIBP	DM 49.85m.	2.519	18	70	13.85
				Total Exposure	138.007

B. ***Breakdown of Commitments***

Company	Contract Value	Exchange Rate	Credit Period (months)	Level of Indemnity (%)	Contingent Liability (IR£M)
Hibernia	£IR10.00m	—	18	70	7.00
	Stg £200,000	0.85	12	70	0.165
	Stg £2.5m.	0.83	18	70	2.108
	IR£ 570,000	—	18	70	0.4
	IR£ 760,000	—	12	70	0.532
				Total Commitments	10.205

NOTES:

1. Original exposure IR£76.75m. Payment of IR£2m. received in Oct' 88. AIBP have been offered a roll-over of the IR76.75m cover as existing maturities under this cover are paid.

2. The figure of IR£19m. is the combined maximum exposure on the 1987/'88 and current (.......) contracts in the event of Iraqi default. This figure will decline as and when payments are received on the 1987/'88 contract. Next payment due 30 January 1989.

(9 January 1988)."

This Memorandum was discussed in detail on the 6th of February 1989 by the Minister for Industry and Commerce Mr Ray Burke, TD, with the Secretary, Mr Donlon, Mr Dully, Assistant Secretary and Mr Joseph Timbs.

Mr Timbs' note of the discussions was circulated to those, who had been present, as is set out hereunder:—

"**STRICTLY CONFIDENTIAL — CIRCULATION TO CC LIST ONLY**

Export Credit Insurance — IRAQ

1. On 6 February, 1989 the Minister discussed the question of Export Credit Insurance for Iraq on the basis of Department's Memorandum of 17th January, 1989 which had been submitted to him. The Secretary, Mr J. Dully, Assistant Secretary and the undersigned were present.

2. The position in relation to our exposure in Iraq was discussed and, in particular, the following were noted:

   (1) exposure stood at approximately IR£148m;

   (2) in November, 1988, the Department of Finance agreed that the overall limit for Iraq could be increased from £150m. to £250m;

   (3) delay, in payment from Iraq had now exceeded 5 months and while a small payment of just over £1m. had been received within the past week, the total amount overdue at this stage was in the region of £60m;

(4) while the previous Minister's decision of the 21st October, 1988 on the allocation of new cover could be justified *at that time,* circumstances had changed dramatically since then;

(5) Messrs Dully and Timbs had sought meetings in Iraq at senior level in the Ministries for Finance and Trade, in the Central and Rafadain banks and with the Iraqi Co-Chairman of the Joint Commission for the week commencing 13th February.

Having regard to the above, the Minister said that he was not prepared to make any decisions in relation to the issue of new cover or the rolling over of existing cover until the Iraqi payment position clarified it to his satisfaction. He could not issue any further cover in Iraq until such time as the Iraqis made payment. He agreed that circumstances had changed very substantially since the decision of the 21st October, 1988. He agreed that the question of allocation of cover for beef might be put formally to the Department of Agriculture along the terms of the draft letter shown to him subject to a considerable strengthening of the paragraph dealing with arrears of payment.

The Secretary stated that the Consultancy Unit were investigating the discrepancy between CSO figures and export credit figures on exports of beef to Iraq. Preliminary indications were that something may be amiss with the figures provided to ICI. The Minister directed that the investigation continue as a matter of urgency and with absolute confidentiality.

Joe Timbs
10 February 1989.

C.C. Secretary
Mr J. Dully, A/Sec.
Mr G. Donnelly."

As will be seen the said note payments from Iraq in the sum of £60m were overdue on the 10th day of February, 1989.

The Minister for Industry and Commerce had on the 21st day of October 1988 been forewarned of this possibility in the Memorandum from the Export Credit Section of that date which included, inter alia, the statement that

"With debts amounting to some IR£65m falling due for payment under beef contracts in the period up to January, 1989 (for which the Minister has an exposure of nearly IR£52m), it would be inappropriate to give any commitment to the Iraqis to increase the IR£150m ceiling"

and the recommendation that:—

(i) there be no increase in the £150m ceiling for the time being.

(ii) the position be reviewed in the New Year in the light of payment performance under existing contracts and new demands for insurance cover from Irish exporters.

As a result of such discussion, the Minister decided that he was not prepared to make any decision in relation to the issue of new cover or the rolling over of existing cover until the Iraqi payment position was to his satisfaction.

It had been the intention of the Minister for Industry & Commerce at this stage that there should be greater involvement by the Department of Agriculture in the allocation of cover of Export Credit Insurance for beef exports but his plans in this regard were not proceeded with in view of the decision to in effect suspend cover.

Though the Minister for Industry & Commerce had stated he was not prepared to make any decisions in relation to the issue of new cover or the rolling over of existing cover until the Iraqi payment position was clarified to his satisfaction, particulars of the contracts obtained by Anglo Irish Beef Packers, Hibernia Meats and Taher Meats were forwarded on the 10th February 1989 to the Department of Agriculture and their advice on suggested allocations in accordance with the Department's policy in relation to the development of export markets for beef.

By letter dated the 17th day of February 1989 the Department of Agriculture replied as follows:

"Mr J Timbs
Principal Officer
International Trade Division
Department of Industry & Commerce

Subject: Export Credit Insurance for Beef Exports to Iraq

I am directed by the Minister for Agriculture and Food to refer to your minute (ECI/INS-79) of 10 February 1989 plus enclosures and to earlier correspondence regarding the above.

The Minister for Agriculture and Food fully shares the concern of the Minister for Industry and Commerce regarding the extent of overdue payments owing from Iraq and accepts that neither roll-over of existing cover nor the advancement of new cover should be considered until the payment situation is satisfactorily resolved.

As regards the future allocation of insurance for beef exports to Iraq, we had understood that your Department would revert to the suggestion contained in this Department's minute of 4 January to the effect that an inter-Departmental structure should be established to decide on the issues involved (see minute of 10 January 1989 from Mr G Donnelly in which he undertook to revert in due course with your Department's "ideas as to the type of formal structure which should be established to facilitate the exercise"). We remain of the view that a group representing the relevant interests in both our Departments would afford the best possible course for dealing with this issue.

Pending finalisation of the definitive arrangements for dealing with future allocations of insurance and the outcome of the investigations being undertaken by your Department, following are this Department's preliminary observations on your minute and enclosures of 10 February:—

(*a*) It is noted that arising from the decision taken by the Minister for Industry and Commerce, up to IR£230m (subject to resolution of the payments situation already referred to), might in future be available to cover meat exports to Iraq. Assuming a 70% indemnity and an average price of IR£2135/t of beef (this represents the maximum average price of Irish beef exported to Iraq in the 1980s to date according to the official trade statistics), this amount would be sufficient to meet the State's liability in respect of about 154,000 tonnes. This quantity would equal about 150% of Ireland's total annual beef supplies to *all* North African and Middle Eastern destinations during each of the years 1987 and 1988 (the best years so far for these exports) and would represent more than eight times our highest annual export to date (18,547 tonnes in 1987 according to C.S.O. statistics) to Iraq alone. It is also worthy of note in the context of your current examination of beef exports to Iraq which are covered under the Export Credit Insurance scheme, that the average contract price for beef as revealed from those contracts which accompanied your 10 February minute is approximately IR£2205/t, which amount corresponds closely with the maximum average value of IR£2135/t indicated in the official trade statistics.

(*b*) some of the contract documents enclosed provide that supplies may be sourced in Europe, North America and/or South America.

(*c*) Contract No. 9/89 (Hibernia Meats International Limited) relates to the supply of 20,000 tons (sic) of boneless young bull meat, i.e. from animals of less than 2 years of age. Fulfilment of this contract would require the slaughter of about 170,000 animals; as Ireland's annual kill of young bulls rarely exceeds 3,000 head it would not be possible to source this contract from within this country alone.

(*d*) The firms named in your minute are well established trading companies; given the overall amount of cover available, apportionment amongst the competing applicants should not present insurmountable difficulties in the light of traditional volumes of Irish beef exports to Iraq.

In the light of the foregoing, it is suggested that the firms which have submitted contracts should at this stage be asked to specify what proportion of the contracts have been or are to be fulfilled with meat of Irish origin. The allocation of further cover can then be considered by whatever structure is agreed on between our Departments following completion of your Department's investigations.

As requested the papers which accompanied your minute of 10 February are returned herewith.

S. Healy
Principal Officer
EC/Trade Division
17 February 1989"

Though there were communications with and representations by Taher Meats, Hibernia Meats and AIBP, no new allocations of cover in respect of Export Credit Insurance in respect of beef exports to Iraq were made subsequent to the aforesaid decision of the Minister for Industry and Commerce.

Agra Trading Ltd on the 20th April 1989 inquired about the availability of cover and they were informed that no cover was available.

Having regard to the terms of the letter dated the 12th day of December 1988 from Mr Rafique to the Minister for Industry & Commerce and of the letter dated the 22nd day of December from Mr Banks, Chief Executive of the Insurance Corporation of Ireland to the Secretary of the Department of Industry & Commerce, the International Trade Division of the Department on the 11th day of January 1989 held a meeting with the Consultancy Unit to initiate a study into the difference which had come to light between official Central Statistics Office statistics of beef exports to Iraq and the level of insurance cover extended by Insurance Corporation of Ireland for beef exports to Iraq.

Recognising that not all the relevant data to resolve the issue was available within the Department of Industry & Commerce, a meeting was arranged between the Department of Industry & Commerce, the Department of Agriculture and the Central Statistics Office which meeting was held on the 20th day of January 1989.

The Consultancy Unit, under the direction of Mr Fisher of the Department of Industry & Commerce carried out an investigation of the matter. This Unit is a technical unit dealing largely with economic and financial issues, established in the early 1970s and composed of a number of professional accountants.

A number of meetings were held between members of the Consultancy Unit and representatives of AIBP and Hibernia Meats Ltd in connection with the matters being investigated by the said Consultancy Unit and there is no need for the Tribunal to refer to these discussions.

They prepared two interim reports dated respectively the 14th day of April 1989 and the 11th day of May 1989 and the final report was concluded on the 27th day of June 1989.

The principal and uncontested conclusions of the said report were that:

> "During the years 1987 and 1988 AIBP exported 60,730 tonnes of boneless frozen beef to Iraq of which 49,702 tonnes were insured or declared for insurance under the ICI Export Credit Insurance scheme.
>
> Of the insured tonnage 18,938 tonnes (38% of tonnage declared for insurance) were sourced outside the jurisdiction of the Irish Republic.
>
> Of the 18,938 tonnes sourced outside the Republic, virtually all such tonnage was processed outside the State.
>
> Of the total of 18,938 tonnes sourced outside the State, a significant volume of such tonnage was both slaughtered and processed in plants located in England, Scotland and Wales.

In the case of Hibernia Meats Ltd, a total of 14,866 tonnes was exported to Iraq and insured or declared for Export Credit Insurance during 1987 and 1988.

Of this 14,866 tonnes, 2,680 tonnes was sourced outside the State (18% of tonnage declared for insurance).

While most of this externally sourced tonnage of Hibernia Meats Ltd was processed in the North of Ireland, like AIBP a proportion was processed in the mainland UK."

The Unit also reached what they described as "subsidiary conclusions" as follows:—

"Examination of the documentary records and filing systems used by AIBP indicates a highly efficient system for keeping all the documentary records in relation to each shipment. The existence of this system and the manner whereby the UK official documentary records in relation to beef sourced in the UK were kept, indicates that external sourcing of beef for exports to Iraq was an obviously well established company policy of which the management of AIBP would have been fully aware.

Within the terms of the Export Credit Insurance policy, the company makes a written declaration to the effect that goods insured would be the produce of the Republic of Ireland, and that the company will retain evidence to prove that fact.At the moment AIBP is not retaining adequate evidence to the effect that beef exported under the jurisdiction of the Irish Customs & Excise authorities and insured under the Export Credit Insurance facility is not the produce of the Republic of Ireland.

There is a similar lack of adequate evidence in relation to the sourcing of beef produced and exported to Iraq by Hibernia Meat Ltd, which is also insured under the Export Credit Insurance facility.

If, for the future, beef exporters who are granted Export Credit Insurance facilities are required to prove that shipments of beef are the produce of the Republic of Ireland, then they will necessarily have to complete EEC type Health Certificates. At present, a less detailed Irish Department of Agriculture Health Certificate is used in connection with virtually all beef exports to Iraq.

Beef exports to Iraq by both companies which were sourced in the North of Ireland have documentary evidence which indicates precisely, both the slaughter house and the processing plant of such beef. The document is an EEC type Health Certificate.

It was noted that shipments of beef to Iraq by both companies were covered by Certificates of Origin indicating that the goods were of Irish origin. In respect of goods sourced in the mainland UK, such certificates were clearly incorrect."

From this report, it appears that:

"(*a*) of the tonnage exported to Iraq by AIBP during the years 1987 and 1988 and declared for and subject to Export Credit Insurance, 38% was sourced outside the jurisdiction of the Irish Republic.

(*b*) of the tonnage exported to Iraq by Hibernia Meats Ltd during the years 1987 and 1988 and declared for and subject to Export Credit Insurance, 18% was sourced outside the State."

Upon the completion of the report of the Consultancy Unit (the Fisher Report) the implications thereof were considered by the Export Credit Division and a Memorandum dealing with the implications thereof and setting forth the options open to the Minister was prepared and is dated the 11th day of August 1989. Annexed thereto is a Memorandum dated the 8th August 1989.

It is desirable that these memoranda should be set forth in full in this Report.

CONFIDENTIAL

"Export Credit Insurance FOR BEEF EXPORTS TO IRAQ
Options Arising from Statistical Investigation Carried out by Consultancy Unit

1. On 8th September, 1987, the Government decided to increase the ceiling on insured exports to Iraq from IR£70m. to IR£150m. Total liabilities in respect of insured exports to Iraq amounted to approximately IR£25m at the time so the Government decision in effect gave the Minister the go-ahead to assume additional liabilities of IR£125m on insured exports to Iraq. Following the Government decision, Export Credit Insurance was allocated to two beef processing companies in respect of contracted beef sales to Iraq in 1987 and 1988 amounting to IR£153.2m in total. The two companies involved were Anglo Irish Beef Packers Group Ltd (AIBP), part of the Goodman Group, and Dantean International Ltd, a company related to Hibernia Meats International Ltd and owned by Oliver Murphy.

2. The overall maximum liability of this insurance cover was IR£119.7m. However the companies did not utilise their full allocations and of the IR£153.2m contract values insured only IR£132m (liability IR£104m) was actually exported. Since cover was provided total payments of IR£46m have been received from Iraq with the result that the Minister's current liability stands at IR£67m approx.

3. All Export Credit Insurance policies state that they apply to shipments which relate to "the export from Ireland after the date of contract of goods produced or manufactured in Ireland" (Ireland being defined as exclusive of the Six Counties). Moreover the policies were issued on the basis of written proposals and declarations from the two companies which, inter alia, stated that the beef supplied under insurance cover was sourced and processed in the Republic of Ireland.

4. On 22nd December 1988, the Chief Executive of the Insurance Corporation of Ireland plc wrote to the Secretary of the Department drawing attention to a disparity which had emerged between official CSO statistics of beef exports to Iraq and the value of exports declared for insurance by AIBP and Dantean. Also in a letter to the Minister dated 12 December 1988, the Chief Executive of Halal Meat Packers, Sher Rafique, suggested that a substantial amount of beef exports to Iraq insured under the Export Credit Scheme was sourced outside of the Republic of Ireland.

5. Accordingly in January 1989 the Consultancy Unit were requested to carry out an investigation into the statistical discrepancy between official CSO statistics for beef exports to Iraq and the level of exports declared for insurance. As part of the investigation, it was necessary to carry out a detailed verification exercise covering some 7,000 separate documents and to consult with the Department of Agriculture, the CSO, ICI, and the two beef companies.

The Consultancy Unit has now completed its investigation and has delivered its findings in a report, copy attached, to the International Trade Division of the Department. The main conclusion of the report is that a significant proportion of beef supplied to Iraq in 1987/1988 under Export Credit Insurance cover was sourced from outside of the State. The amounts were 18,938 tonnes or 38% of tonnage declared for insurance in the case of AIBP and 2,680 tonnes or 18% of tonnage declared for insurance in the case of Dantean. In value terms the total amount sourced outside the State was IR£52.2m., approximately 35% of total beef sales declared for insurance.

6. Under normal insurance principles, a materially false or incorrect statement provided by the insured to the insurer renders the insurance contract null and void. This principle is enshrined in Article 4 of each of the insurance policies issued to AIBP and Dantean, which states:

   "The proposal made by the insured and the declaration contained in it shall be incorporated with this Policy as its basis.

   If any of the statements contained in the proposal and the declaration is untrue or incorrect in any respect, this Policy shall, unless the Company (i.e. ICI) otherwise elects in writing, be void".

7. In addition to Export Credit Insurance, the two beef companies obtained export credit finance amounting to IR£50.3m in total in respect of insured beef exports to Iraq. (IR£33.4m in the case of AIBP and IR£16.9m in the case of Dantean). This finance was made available on foot of State guarantees provided to the companies' financing banks under the Export Credit Finance Scheme. While there is no additional liability involved, the guarantees are unconditional and may be called on by the banks if payment is not received from Iraq within six months of the due date of payment, irrespective of the reason for non-payment.

8. The following options are available to the Minister arising from the conclusions of the Consultancy Unit's investigation.

   (1) Take no action unless and until, a claim is made by either company or by their financing banks.

   (2) Invalidate the entire insurance cover.

   (3) Disclaim liability in respect of all exports declared for insurance which were sourced outside the State.

   (4) Maintain insurance cover for all Republic of Ireland processed beef exported to Iraq plus a fixed percentage in respect of beef processed outside the Republic.

   (5) Substitute uninsured beef exports to Iraq in 1987/88 which were sourced in the State for insured beef exports to Iraq sourced outside the State.

   (6) Maintain insurance cover for all shipments notified by the two companies for insurance cover.

The implications of these options are described below:

Option 1

Take no action unless and until claim is made by either company or by their financing banks.

It is possible that no claim will be made in which case no action would be required under this option. If however a claim is made, a decision would then have to be taken as to whether to disclaim liability or not. This is not really an option but rather a postponement of a decision.

Option 2

Invalidate the entire insurance cover. This would mean disclaiming all liabilities under each facility (maximum liability of IR£119.7 in total) irrespective of whether shipments insured were of Republic of Ireland origin or not. The two companies would not be entitled to any refund of premium as a result of the invalidation of the insurance cover.

While this option would result in an immediate cancellation of *insurance* liabilities, the Minister would still be liable to the companies' financing banks for exports which are covered by unconditional guarantees issued under the Export Credit Finance Scheme and which are still outstanding from Iraq. The amounts in question are IR£32.1m in the case of AIBP and IR£6.2m in the case of Dantean. The guarantees however contain recourse agreements which enable the Minister/ICI to seek full reimbursement from the two companies in respect of claims paid to the banks under the guarantees which are not covered by the insurance. It is considered that the relevant recourse agreements should be invoked in such circumstances.

Option 3

Disclaim liability in respect of all exports declared for insurance which were sourced outside the State.

A total of IR£52.2m (liability IR£41.2m) of beef sales declared for insurance were sourced outside the State. Of this, IR£19m (liability IR£15m) has already been paid for, so that the net reduction in outstanding insurance liability under this option would be IR£26m.

However IR£16m of insured exports sourced outside the State are covered by bank guarantees and are still outstanding from Iraq. The Minister would remain liable to the banks for these amounts, and would have the same recourse options as mentioned above.

Option 4

Maintain insurance cover for all Republic of Ireland processed beef exported to Iraq plus a fixed percentage in respect of beef processed outside the State.

Under the Export Credit Scheme, insurance cover is sometimes granted to an exporter who has to source part of his contract outside of the State, the justification being the overall benefit to the Irish economy arising from the particular contract. This arrangement is designed to cater for exporters, particularly exporters of manufactured goods, who of necessity are obliged to source materials, components or finished goods outside the State due to unavailability within the State. This argument might not apply in the case of AIBP and Dantean. The percentage of non-Irish beef to be covered under this option would need to be determined.

Option 5

Substitute uninsured beef sales to Iraq in 1987/1988 which were sourced in the State for insured beef sales sourced outside the State.

This would apply to AIBP only as Dantean did not export beef to Iraq in the period without insurance cover. During 1987/1988 AIBP exported 11,028 tonnes (value IR£17.2m) to Iraq without insurance cover of which 6,094 tonnes (value IR£8.23m) were sourced in the Republic. The amount of uninsured sales sourced in the State would not be sufficient to offset the 18,938 tonnes (value IR£46.3m) of insured AIBP sales sourced outside the State.

APPENDIX 1

1. The limited slaughtering capacity was a most critical factor in aggravating the 1974/75 national beef crisis. Despite the fact that factories and individual farmers (those in winter feeding in the winter of 74/75) made a lot of money, the bulk of livestock producers lost heavily. This resulted in a loss of confidence in the industry from which it took several years to recover. The beef breeding herd dropped from over 700,000 to around 400,000 cows over the next few years. The beef breeding herd (suckler cow herd) never fully recovered from the crisis. This indicates the sensitivity of the cattle and beef industry to factors of this nature.
2. Ireland has a slaughtering capacity of around 70,000 cattle per week and its peak throughput is 50,000. In ordinary circumstances there is no shortage of slaughtering capacity. Associated facilities such as chilling capacity are more likely to be short under the normal range of circumstances.
3. Two firms, AIBP and UMP now control over 60% and one firm i.e. AIBP controls over 40% of the national slaughtering capacity. If the major firm involved in beef production were forced or decided to close down, the knock-on effects would be of major proportions. All of the plants of this firm would not be sold together and it is possible that all or most of them could be out of production for an extended period or indeed indefinitely. There is little doubt that in this eventuality there would be major repercussions for the beef industry, including farmers.
4. The pursuit of debts of the order mentioned to the full by the State could affect to varying degrees the following:

   —banks
   —the companies concerned
   —cattle producers
   —valuable markets
   —international trading reputation
   —Irish credibility in Brussels
   —employment
5. The current investigation by the Department of Agriculture and Food on APS beef certainly has already diminished confidence of the banking community in the industry. It will not be possible to restore this until the current investigations are completed. The banking credit facilities are of critical importance in the international trading arrangements of the beef industry.
6. AIBP are a major player in the international trading of beef. They have developed a number of markets in the UK, the rest of the EC and the Middle

East. As regards the APS scheme of 1989, under which 133,000 tonnes were taken in, AIBP provided 43,000 tonnes. While 69,000 of the total remains unsold AIBP has disposed of their quantities.

The second company involved in the Iraq export insurance guarantee appear to have some difficulties in the 1988 APS scheme. Further difficulties would probably create insurmountable obstacles for them.

7. While it is difficult to envisage closure on the scale referred to, the repercussions in the industry and related sectors would be seriously affected if some closures were to take place. Iraq is an important market in recent years with considerable potential in the years ahead. The knock-on effect in other markets would be significant if the reputation of important meat traders were seriously impaired.

8. Whatever actions are contemplated they must be taken in the context of the overall impact on the agriculture and food sector with due and full recognition of the possible impact of the current sampling on APS beef on the sector.

9. Fullest consideration should be given to the possible damage compared to possible losses that may be incurred due to guarantees given. Our view is that efforts to maximise payment of debts from sales of beef to Iraq should be sustained for whatever period is required. It is considered that this process should be fully explored and exhausted before considering other ways. The experience of An Bord Bainne, a major trader in the Middle East, is that Iraq, while slow to pay, always honours its debts.

10. The impact on farming prices of major factory closures would be the most serious outcome. Cattle slaughtering have been low so far this year and large numbers are available this autumn. The support mechanisms are now weaker, an APS in autumn will be smaller than last year, some APS from 1988 may overhang the market and the summer drought will affect fodder supplies. While the market position can be managed satisfactorily under stable conditions it will become rapidly unstable if the major beef production company with considerable international markets, closes or it is forced to reduce its operations. This would bring cattle prices into disarray if not total collapse.

11. Impact on international trade and effects on specific markets such as Iraq and other Middle East outlets to be considered. Total exports to the Middle East were 73,000 tonnes in 1986, 112,500 in 1987 and 98,000 in 1988. The figures for Iraq were 10,000 tonnes in 1986, 18,500 in 1987 and 28,700 in 1988.

12. Evidence that beef was sourced elsewhere would need to be incontrovertible. Rumours, without clear proof, would have serious consequences for the industry in trade and EC circles.

8 August 1989.
Department of Agriculture & Food"

Mr Desmond O'Malley, TD, who had been appointed Minister for Industry & Commerce on the 12th day of July 1989 met with Mr Timbs and Mr Donnelly on the 14th day of August 1989.

He had on that date received the Government Memorandum from the Department together with the documents annexed to it and informed Mr Timbs and Mr Donnelly that he was making the decision in principle to void the policies, that fraud should be assumed and the premiums not returned and instructed them to seek the advice of the Attorney General's office with regard to the issues highlighted in the said Memorandum.

On the 11th day of October 1989, Mr Frewen of the Insurance Corporation of Ireland wrote to Mr L J Goodman of Goodman Holdings in the following terms:—

> "Dear Mr Goodman
>
> On the 18th day of February 1988, we, as agents for the Minister for Industry & Commerce, entered into an Export Credit Insurance Comprehensive Shipments Policy Number 2436 with your company. Prior to the policy being issued, a Proposal with certain declarations and representations was made by your company, which were incorporated with the Policy as its basis in accordance with Article 4 thereof. It was proposed, declared and represented, inter alia, that the meat, the subject matter of the policy would be the produce of the "Republic of Ireland" and that evidence would be retained by your company to prove that fact. Further, it was an express and fundamental term of the policy that all meat the subject of any shipment covered by the policy was and would be meat produced within the State. Any liability on foot of the policy was conditional on this term being complied with.
>
> Extensive investigations having been carried out on behalf of our principal, the Minister for Industry & Commerce and his Department, it has been established, as you are aware, that substantial quantities of the meat, the subject matter of the Policy, were not produced in the State. At least 44% of the meat concerned was produced outside the State.
>
> Accordingly, we hereby give you notice on behalf of the Minister for Industry & Commerce that by reason of the matters aforesaid the Policy is, in accordance with Article 4 thereof, void.
>
> In the circumstances no liability will be accepted by the Minister.
>
> Yours faithfully"

A letter in somewhat similar terms was on the same day sent by the Insurance Corporation of Ireland to Dantean Holdings Ltd

Article 4 of the Export Credit Insurance Comprehensive Shipments Policy, issued by the Insurance Corporation of Ireland, provided that:—

> "The proposal made by the Insured and the Declaration contained in it shall be incorporated with this policy as its basis.
>
> "If any of the statements contained in the proposal and the Declaration is untrue or incorrect in any respect, this Policy shall, unless the Company otherwise elects in writing, be void."

The Insured were required and did declare that the meat, the subject matter of the policy, would be the produce of the Republic of Ireland and that evidence would be retained by the insured to prove that a fact.

Both AIBP and Hibernia Meats Ltd made declarations to that effect and the policies were issued on the basis inter alia of such declarations.

It is clear from the terms of the Report of the Consultancy Unit (the Fisher Report) that both AIBP and Hibernia Meats Ltd were in breach of the express terms of the said declarations because:—

(*a*) Of the tonnage of beef exported to Iraq by AIBP in 1987 and 1988, declared for and subject to Export Credit Insurance policies, 38% were sourced outside the jurisdiction of the State and

(*b*) Of the tonnage exported to Iraq by Hibernia Meats during the years 1987 and 1988, 18% were sourced outside the state.

It was because of the alleged breaches of these declarations, incorporated in the said policies of insurance, that the Minister for Industry & Commerce and the Insurance Corporation of Ireland plc, as his Agent, treated the said policies as void and repudiated liability on foot thereof.

While it had been ascertained from the findings of the Fisher Report that of the tonnage of beef exported to Iraq by AIBPI and Hibernia Meats Ltd during the years 1987 and 1988 and subject to Export Credit Insurance 38% of such tonnage exported by AIBPI and 18% of such tonnage exported by Hibernia Meats Ltd were sourced from outside the jurisdiction of the State, the Tribunal ascertained from the Department of Agriculture and Food that of the tonnage of beef exported to Iraq in 1987 and 1988, 84% of that exported by AIBPI and 75% of that exported by Hibernia Meats had been purchased from Intervention stocks held by the Irish Intervention Agency, the Minister for Agriculture and Food.

On the 12th day of October 1989, Anglo Irish Beef Processors International Ltd instituted proceedings in the High Court against the Minister for Industry & Commerce and the Insurance Corporation of Ireland plc.

In the course of the proceedings instituted by AIBP, they allege that the said purported repudiation of the policy was misconceived and for no legitimate reason or cause, was invalid and ineffective.

In their Reply to the Defence delivered on behalf of the Defendants, the Minister for Industry & Commerce and the Insurance Corporation of Ireland, they alleged that:

> "Insofar as the said representations were made, or the said warranties were given, they were made and given only in compliance with the standard form documentation required by the Defendants to be issued by any trader seeking export guarantee insurance cover from the State. Insofar as the said representations were made, or warranties given, the said were made and given by the Plaintiffs in ignorance of the said fact that the same were being exacted illegally on the part of the Defendants and in abuse of the powers of the State in that behalf."

All of the beef the subject matter of the said policies was exported by the Plaintiff from the State and was produced or originated either in a State or in a member State of the European Community.

"If the policies of export and guarantee insurance issued by and on behalf of the Defendants, contained provisions which purported to and were intended to have the effect of excluding from the cover thereby afforded, beef exported from the State, but which originated in, or was produced outside the State in another member state of the European Communities, such provisions were unlawful and unenforceable and are to be disregarded as being severed from such policy. In the alternative, such provisions fall to be construed as if the references therein to the State are intended to be references to member states of the European Communities and not exclusively to the State."

PARTICULARS

A provision or clause to the above effect in a system of export guarantee insurance operated by or on behalf of a Member State of the European Communities is contrary to the laws of the Communities in that it:—

"(1) Is inconsistent with and repugnant to the regulations establishing the common organisation of the market in beef and veal under the Common Agricultural Policy of the Community and in particular Regulation 805/68/EEC:

(2) Constitutes a quantitative restriction on imports to the State contrary to Article 30 of the Treaty establishing the European Economic Community and infringes the prohibition in Article 7 thereof:

(3) Constitutes an unlawful aid granted by the State which distorts competition by favouring those undertakings which produce beef within the State contrary to Article 92 of the said Treaty."

While there is no doubt but that the shipments of beef made to Iraq included beef sourced outside the State as set out in the Fisher Report and that such inclusion was contrary to the express declarations made by AIBP and Hibernia Meats Ltd, a fundamental issue raised in the proceedings pending in the High Court is whether:—

"The requirement of such a declaration was illegal and an abuse of the powers of the State having regard to the fact that all the beef, the subject matter of the relevant policies was exported from the State and was produced or originated either in the State or in a Member State of the European Communities on the grounds that such a requirement is contrary to the laws of the European Communities in that it:

(1) Is inconsistent with and repugnant to the regulations establishing the common organisation of the market in beef and veal under the Common Agricultural Policy of the Community and in particular Regulation 805/68/EEC:

(2) Constitutes a quantitive restriction on imports to the State contrary to Article 30 of the Treaty establishing the European Economic Community and infringes the prohibition in Article 7 thereof.

(3) Constitutes an unlawful aid granted by the State which distorts competition by favouring those undertakings which produce beef within the State contrary to Article 92 of the said Treaty."

This is an issue which will have to be determined by the High Court and/or the European Court of Justice and the Tribunal will express no view thereon and consequently will not refer to the evidence with regard thereto given during the hearing of the Tribunal.

Between the 14th day of August 1989 and the said 11th day of October 1989, the Minister and officials of his Department had a number of discussions with Mr Goodman and/or representatives of Anglo Irish Beef Processors International and with representatives of Hibernia Meats Ltd.

In view of the matters in issue in the proceedings instituted by Anglo Irish Beef Processors International, the Tribunal does not consider it desirable or necessary to deal with what transpired at these interviews.

In view of the controversy which arose, it is desirable to set forth in detail the nature of the claim made by Anglo Irish Beef Processors International in the said proceedings and this can best be done by setting forth in detail the Amended Statement of Claim delivered on their behalf on the 7th day of November 1990 which is as follows:—

"1. The Plaintiff was incorporated with limited liability under the provisions of the Companies Act 1963 and on the 18th day of September 1989 was re-registered as an unlimited company. The Plaintiff carries on in the State and abroad the business of slaughtering , processing and exporting beef and related products.

2. Under the provisions of the Insurance Acts 1953-1958 the First Named Defendant is authorised to make arrangements for giving to or for the benefit of persons carrying on the business in the State, guarantees in connection with, inter alia, the export of goods from the State for the purpose of encouraging such exports. Such arrangements are authorised so as to include agreements with insurance companies for the re-insurance of guarantees given by them.

3. The Second Named Defendant is established and authorised within the State to carry on business as an insurer including the business of export guarantee insurance and as such acts as the agent of the First Named Defendant in implementation of the arrangements aforesaid for the provision of export guarantees under the said Acts.

4. For a number of years prior to the year 1987 the Plaintiff had achieved considerable success in establishing a market for the export and sale of Irish beef in Iraq and had done so with the support and encouragement of the State and particularly that of the First Named Defendant.

5. ***The First Insurance Policy***

5.1 In the month of September 1987 the Plaintiff was successful in procuring the conclusion of a series of contracts with Iraqi import purchasers for the export and sale to them of a total of 40,000 tons of beef at a value of IR£97,028,215 for delivery between September 1987 and September 1988.

Date	Quantity Metric Tonnes	Particulars Purchaser	L/C No.	Value IR£
09.09.87	994.2437	The State Company for Foodstuff Trading	19546	2,422.444.52
25.09.87	2055.7416	,, ,,	,,	5,164,586.59
04.10.87	563.5074	,, ,,	,,	1,415,685.08
23.10.87	1014.2628	,, ,,	,,	2,548,106.22
29.10.87	1726.3092	,, ,,	,,	4,336,961.99
20.11.87	1186.7175	,, ,,	,,	2,981,359.71
30.11.87	1717.1665	,, ,,	,,	4,313,993.02
11.12.87	1436.3256	,, ,,	,,	3,608,443.68
31.12.87	1657.1441	,, ,,	,,	4,163,200.29
14.01.88	2026.0900	,, ,,	,,	5,090,093.54
27.01.88	622.4916	,, ,,	,,	3,203,726.29
27.01.88	665.4944	,, ,,	19677	
19.02.88	1706.9837	,, ,,	,,	4,288,411.03
27.02.88	1156.21065	,, ,,	,,	2,904,718.13
17.03.88	1868.4493	,, ,,	,,	4,694,056.89
31.03.88	831.2508	,, ,,	,,	2,088,329.90
14.04.88	1462.5676	,, ,,	,,	3,674,370.78
12.05.88	1476.9471	,, ,,	,,	3,710,496.03
18.05.89	1867.2501	,, ,,	,,	4,691,044.17
30.05.88	1297.4460	,, ,,	,,	3,259,540.05
01.06.88	1497.4531	,, ,,	,,	3,762,012.72
29.06.88	1169.1388	,, ,,	,,	2,937,197.19
	29999.19155		75,258,777.72	

Date	Quantity Metric Tonnes	Particulars Purchaser	L/C No.	Value IR£	IR£ C/Forward
09.09.87	515.5511	Iraqi Company for Agriculture Products Marketing	32088	1,295,206	75,258,777.72
25.09.87	143.9925	,, ,,	,,	361,749	
04.10.87	360.7275	,, ,,	,,	894,697	
29.10.87	121.2980	,, ,,	,,	304,734	
20.11.87	488.0317	,, ,,	,,	1,226,069	
11.12.87	516.8250	,, ,,	,,	1,298,406	
27.01.88	513.0268	,, ,,	,,	1,281,560	
27.02.88	468.82787	,, ,,	,,	1,177,824	
31.03.88	92.990	,, ,,	,,	233,616	
13.05.88	393.47360	,, ,,	,,	988,514	
26.07.88	1501.7029	,, ,,	,,	3,772,689	
					12,835,064
	5,116.44697			12,835,064	88,093,841.72

Date	Quantity Metric Tonnes	Particulars Purchaser	L/C No.	Value IR£	.£ C/Forward
09.09.87	480.9024	Iraqi Company for Agriculture Products Marketing	31905	828,452	88,093,841.72
25.09.87	216.4629	,, ,,	,,	372,901	
04.10.87		,, ,,	,,	2,053,819	
30.11.87	1195.18883	,, ,,	,,	484,958	
14.01.88	281.510	,, ,,	,,	358,651	
27.01.88	208.1909	,, ,,	,,	313,127	
19.02.88	182.8009	,, ,,	,,	248,646	
27.02.88	144.3347		,,	78,849	
	45.7708				4,739,403
				4,739,403	92,833,244.72

Date	Quantity Metric Tonnes	Particulars Purchaser	L/C No.	Value IR.£.	.£ C/Forward
31.03.88	912.2676	,, ,,	"	1,571,557	92,833,244.72
20.07.88	13.097	,, ,,	,,	22,562	
26.07.88	1509.7512	,, ,,	,,	2,600,852	
TOTAL =	5190.27223			4,194,971	4,194,971 IR£97,028,215

5.2 In order to ensure due receipt of payment of the sale proceeds of the said contracts, the Plaintiff applied to the second-named Defendant and on the 18th day of February 1988 received from the Second Named Defendant on behalf of the First Named Defendant a policy of Export Credit Insurance (No. EC 2436) in a maximum amount of IR£76,751,589.85 (being 80 per cent of total contract value) and paid to the Second Named Defendant a premium of IR£959,394.87 therefor.

5.3 Under the terms of the said policy the Second-Named Defendant in consideration of the said premium agreed to indemnify the Plaintiff on behalf of the First Named Defendant in respect of 80% of any loss which the Plaintiff might sustain in connection with any consignment to, of beef comprised in the contracts described in the said policy. The Plaintiff duly performed the said contracts and delivered the said consignments to the Iraqi purchasers thereof.

5.4 By a letter dated 11th October 1989, written on behalf of the second named Defendant on behalf of the first named Defendant as principal, the second named Defendant wrongfully and in breach of contract purported to avoid the said Export Credit Insurance policy No. 2436. The said purported repudiation of liability on foot of the said policy was misconceived and for no legitimate reason or cause and was invalid and ineffective.

6. ***The Second Policy***

6.1. In order to exploit further the markets which had been opened up for Irish goods and products in Iraq by the Plaintiff and with a view to encouraging trade between the State and Iraq under the terms of the Agreement on Economic, Scientific and Technological Co-operation of 1981 between Ireland and Iraq, the First Named Defendant sought and obtained from the Plaintiff assurances and commitments that the Plaintiff and other companies in the group of which the Plaintiff forms part would seek to procure further export contracts and engage in further trading and investment activities in Iraq.

6.2 Pursuant to the said Ireland/Iraq Agreement of 1981, meetings of the Irish-Iraqi Joint Commission on Economic, Scientific and Technological Co-operation took place during 1987 and 1988 and at a session of the said Joint Conference held in September 1987 the First Named Defendant undertook to ensure that Export Credit Insurance to an aggregate value of not less than IR£150 million would be made available by the State in respect of trade with Iraq including exports of beef.

6.3 Immediately following the said meeting of the Joint Commission, the Plaintiff was informed on behalf of the First Named Defendant that the said promise had been made and the Plaintiff was encouraged to seek further contracts for the export of beef to Iraq upon the basis that Export Credit Insurance therefore would be made available by the First Named Defendant through the agency of the Second-Named Defendant in due course to cover such contracts.

6.4 Acting upon the strength of the said information and the promise of the First Named Defendant that Export Credit Insurance would be provided as aforesaid, the Plaintiff company proceeded to seek and eventually obtain and concluded a series of further contracts for the export of beef to Iraq, the said contracts having an aggregate value of <u>IR£76,501,835.00</u>

Date	Quantity Metric Tonnes	Particulars Purchaser	L/C No.	Value IR£
30.05.88	602.6338	The State Company for Foodstuff Trading	19927	1,346,318
20.07.88	799.5664	,, ,,	,,	1,786,277
23.07.88	1054.7619	,, ,,	,,	
23.07.88	249.7691	,, ,,	19926)	2,914.396
23.08.88	595.2393	,, ,,	19927)	3,589,589
23.08.88	1011.5183	,, ,,	19926)	975,550
31.08.88	436.6718	,, ,,	,,	
29.09.88	40.0444	,, ,,	,,	
29.09.88	348.7119	,, ,,	20555	808,631
14.10.88	1010.8531	,, ,,	19926	
14.10.88	1446.7743	,, ,,	20555	5,490,480
27.10.88	1506.7092	,, ,,	,,	3,366,074
30.10.88	346.1555	,, ,,	,,	773,331
31.10.88	345.72644	,, ,,	19926	772,373
31.10.88	74.7540	,, ,,	,,	
31.10.88	641.037	,, ,,	20555	1,599,118
30.11.88	137.649	,, ,,	,,	307,516
31.12.88	389.688	,, ,,	,,	870,585
31.12.88	1119.0762	,, ,,	,,	2,500,080
31.12.88	589.4385	,, ,,	,,	1,316,839
	12.746.77814			28,417,157

6.5 Having procured contracts as aforesaid, the Plaintiff applied to the First Named Defendant to approve the issue by the Second Named Defendant on his behalf of corresponding policies of Export Credit Insurance to cover the said contracts. On the 10th day of February 1988 the Plaintiff was informed on behalf of the First Named Defendant that the First Named Defendant had agreed that Export Credit Insurance to a contract value of US£30 million had been allocated to the Plaintiff in respect of the said contracts subject only to the enactment of legislation amending the provisions of the said Insurance Acts so as to provide for an increase in the aggregate amount of the Minister's liability under the said arrangements.

6.6 On the 5th day of July 1988 the aggregate amount of the Minister's liability in respect of the said arrangements was increased to the sum of IR£500,000,000 on the passing of the Insurance (Export Guarantees) Act 1988.

6.7 In the month of December 1988 in response to a request on behalf of the First Named Defendant, the Plaintiff submitted details of copies of the said contracts to the Second Named Defendant and paid to the Second-Named Defendant a premium in a total amount of IR£194,303 on foot of the said Export Credit Insurance policy. The Plaintiff proceeded to perform the said contract and to deliver the said consignments to the Iraqi purchasers thereof.

6.8 Notwithstanding the First Named Defendant's promise and representation that Export Credit Insurance cover would be made available in respect of a contract concluded by the Plaintiff as aforesaid; notwithstanding the Plaintiff's reliance upon said promise and representation by acting to its detriment in concluding the contracts described in Paragraph 6.7 above; and notwithstanding the Plaintiff's payment of the said premium and its acceptance by the Second Named Defendant, the Second Named Defendant has wrongfully and in breach of contract neglected and refused to issue the said policy documents to the Plaintiff.

7. *Further Claims*

7.1 On or about the 7th/9th November 1988 a further session of the said Joint Commission on Economic, Scientific and Technological Co-operation took place between representatives of the State and representatives of the Government of Iraq. In advance of the said joint session the First Named Defendant procured from the Plaintiff and other companies in the Group of which the Plaintiff forms part, agreement that the Irish Delegation at the said Joint Commission might propose the establishment by the said Group of a joint venture for the construction of a modern beef-processing plant in Iraq and upon the basis that the Plaintiff's said Group would invest one half of the foreign currency portion of the costs of such plant and contribute to the management thereof and the technical training of personnel. In conjunction with the said proposal and as part of the arrangements proposed on behalf of the State at the said session of the Joint Commission, the Irish Delegation on behalf of the First Named Defendant undertook to increase further the aggregate value of the Export Credit Insurance available for Irish exports to Iraq "by a significant and substantial amount".

7.2 In consideration of the commitments undertaken by the Plaintiff and other companies in its Group towards the proposals put forward at the said session of the Joint Commission, the First Named Defendant undertook and promised to the

Plaintiff that further Export Credit Insurance would be made available through the Second-Named Defendant to cover additional beef export contracts to Iraq to a total value of IR£80 million.

7.3 Having procured promises and commitments from the Plaintiff as aforesaid and having induced the Plaintiff to organise and conduct its business arrangements on the basis that further substantial export credit guarantees cover would be made available to the Plaintiff the first named Defendant has caused the Plaintiff extensive loss and damage by:—

(*a*) breaking the said promise of further insurance cover;

(*b*) negligently and falsely publishing statements to the effect that the Plaintiff was in breach of the terms of its export guarantee contracts and that the Plaintiff had misused or abused the said guarantee arrangements

(*c*) by unnecessarily announcing in public an intention to attempt to repudiate contracts concluded on behalf of the State by the second named Defendant jeopardized the ability of the Plaintiff to obtain payment of sums due from the said Iraqi purchasers

(*d*) obstructing the Plaintiff in the conduct of its business by damaging its ability to attract and procure further valuable contracts in Iraq.

(*e*) procuring the reduction of or the withdrawal from the Plaintiff of the trading credit and bank support otherwise available to it

8. By reason of the matters aforesaid the Plaintiff has incurred loss and damage and is at risk of incurring further loss and damage in the event of payments due under the said contracts being defaulted upon by purchasers in Iraq.

THE PLAINTIFF CLAIMS:—

1. A declaration that the Second-Named Defendant (on behalf of the First Named Defendant) is bound to indemnify the Plaintiff in respect of losses incurred by the Plaintiff as a result of non-payment of sums due and owing to the Plaintiff by purchasers of consignments of beef under a series of contracts the due performance of which was insured under a policy of Export Credit Insurance issued by the Second Named Defendant on the 18th day of February 1988 under the serial No. EC 2436;
2. A declaration that the Second Named Defendant (on behalf of the First Named Defendant) has undertaken the Export Credit Insurance risk in respect of the contracts for the export and sale of beef described in paragraph 6.3 above;
3. A mandatory injunction directing the Second-Named Defendant to issue to the Plaintiff the policy document in respect of the insurance aforesaid;
4. Damages for breach of contract and negligence.
5. Further and other relief
6. Costs

IAN FINLAY B.L.
JOHN COOKE S.C.

To/

THE CHIEF STATE SOLICITOR
Dublin Castle
DUBLIN 2"

While the Tribunal is not concerned with the merits of the Plaintiff's claim in these proceedings, which will be determined in the first instance by the High Court, the amount of the claim and the potential liability of the State is relevant to the issues before the Tribunal.

As will be seen from the Statement of Claims, there are three separate items of claim viz

(i) A claim in respect of the loss sustained by the Plaintiff as a result of the voidance of the Policy of Insurance (No EC 2436) by the Minister for Industry and Commerce in or about the 11th day of October 1989, which loss is now stated to be £23,368.603.00,

(ii) A claim, under the heading "Further Claims" relating to the alleged promise made by the Minister for Industry and Commerce in October/November 1989 to cover in respect of additional beef exports to Iraq to the Plaintiff Company to a total value of £80m,

(iii) the other claim appears in the statement of claim under the heading "The Second Policy" and relates to the failure by the second-named Defendant to issue a policy in respect of promises alleged to have been made by and on behalf of the Minster for Industry and Commerce in the month of November 1987.

It is not clear from the Statement of Claims or the particulars given whether this particular aspect of the claim is limited to the $30m cover agreed to be given by the Minister for Industry and Commerce on the 13th day of November 1987 or whether it is a claim for 70% cover of the contracts valued at £76,501,835.00 referred to in paragraph 6.4 of the Statement of Claims.

Independent of the question of General Damages, which may be proved in the proceedings, the Plaintiffs' pecuniary loss claimed is either £115,262,332 if the claim under the alleged "Second Policy" relates to the $30m offer of cover or £159,113,616 if the claim under this heading relates to 70% cover of contracts valued at £76,501,835.

In either case, it is a very substantial claim and that is the only matter for determination by this Tribunal.

The Tribunal has during the course of this Report attempted to place in sequence the evidence both oral and documentary with regard to the operation and administration of the Export Credit Insurance Scheme during the relevant periods and fully appreciates that not all the evidence and documents are included within this Report.

It has however sought to detail within the confines of this Report all the relevant oral and documentary evidence to illustrate the manner in which the system was operated, the decisions made in relation thereto and the reasons for such decision.

The Tribunal appreciates that there is a considerable amount of repetition in the various memoranda dealt with in this Report but considered it appropriate that they should be printed in full rather than have edited extracts therefrom printed in this Report.

The abuses of the Export Credit Insurance Scheme alleged by Deputies O'Malley, Spring and Rabbitte were summarised by the Tribunal and set forth elsewhere in this Report and the issues raised thereby are:

1. (i) whether the decision made in 1987 to reinstate Export Credit Insurance was taken against the best professional advice and if so, was the decision made for improper reasons or motives.

   (ii) whether the decision made in September 1987 to increase the ceiling to £150m was made against the best professional advice and if so, was the decision made for improper reasons or motives.

   (iii) whether the decision made in November 1988 to increase the ceiling to £250m was made against the best professional advice and if so, was the decision made for improper reasons or motives.

2. (i) whether conscious decisions were taken to give one conglomerate (Goodman) more than 80% of the available cover for beef exports to Iraq and if so, did the grant of such cover disadvantage rivals and exporters in other products and if so, was the decision made for such purpose or for any other reason?

   (ii) whether the granting of E.C.I. was a political decision and depended on whether "you were a member of the club."

   (iii) whether Mr Goodman intervened with the then Taoiseach Charles J Haughey who then caused the Minister for Industry and Commerce to cancel the allocation of Export Credit Insurance to Halal

3. (i) whether, in 1987 and 1988 between one-fifth and one third of all Export Credit Insurance was given in respect of beef exports to Iraq and whether 80% thereof was given to the Goodman organisation and if so whether such provision amounted to an abuse of the scheme.

   (ii) whether the allocation of such insurance cover to two companies viz AIBP and Hibernia Meats Limited was an act of blatant favouritism.

   (iii) whether such allocation had the effect of strengthening further the already strong position of Goodman as the dominant group within the beef processing and allied trades contrary to the interests of farmers and employees and of exporters in other business sectors.

   (iv) whether such allocations were made because Goodman was extremely personally close to members of the Government.

4. (i) whether Export Credit Insurance cover was provided by the State in respect of the sale of beef to Iraq in 1987 and 1988 in excess of the amount actually exported.

(ii) if so, was such provision in breach of the terms of the Export Credit Insurance Scheme and did such provision constitute a substantial abuse amounting to a fraud on the taxpayer.

(iii) whether the scale of such alleged abuse and the potential liability of the State was unprecedented.

5. Whether very large quantities of non-Irish beef were included in shipments of beef to Iraq made by AIBP such shipments purporting to be covered by export insurance policies which were subject to the provision that the meat, the subject of the policy, would be the produce of the Republic of Ireland.

With regard to the issues at (4) and (5) above, the Tribunal is satisfied that:

(i) Very large quantities of beef, not sourced or produced within the State were included in shipments of beef to Iraq by AIBPI and Hibernia Meats Ltd during 1987 and 1988 : 18,938 tonnes representing 38% of total tonnage by AIBPI and 2680 tonnes representing 18% of total tonnage declared for insurance by Hibernia (Daintean).

(ii) Such beef was included in shipments purporting to be covered by Export Credit Insurance policies.

(iii) Such inclusion was contrary to the express terms of the Declarations made by AIBP and Hibernia Meats Ltd (Dantean)

(iv) Such inclusion constituted a substantial abuse of the express terms of the Scheme and the policies issued in pursuance thereof.

(v) Such abuse would not have led to a fraud on the taxpayer because of the insistence by the Minister for Industry and Commerce and his agent the Insurance Corporation of Ireland that proof of the origin of the beef to be exported be retained and such proof would have to be produced to the Insurance Corporation of Ireland before any claims on foot of the said policies would be paid. Such abuse however had the effect of tying up and rendering unavailable for allocation to other companies substantial amounts of Export Credit Insurance Cover, which would otherwise have been available for allocation.

(vi) Such abuse was substantial and if AIBP are successful in the proceedings hereinbefore referred to the potential liability of the State is very substantial. Such abuse and such potential liability may not be unprecedented but it certainly is not a regular occurrence.

With regard to the issues raised in the specific allegations made with regard to the management of the Export Credit Insurance Scheme, it is clearly established that the decisions made by the Minister for Industry & Commerce:

(1) In April 1987 to remove the suspension placed by the previous Minister for Industry & Commerce on the allocation of cover in respect of exports to Iraq,

(2) In September 1987 to secure the Government's approval to increase the ceiling for insured exports to Iraq to £150m, and

(3) In October 1988 to seek to secure the Government's approval to increase the ceiling for such exports to £270m, and

(4) To secure the approval of the Minister for Finance in November 1988 to an increase of £100m

were made against the advice made available to him by the Insurance Corporation of Ireland, the Department of Industry & Commerce and the Department of Finance and in each specific case were made as a result of an application for cover made on behalf of AIBP.

This advice was mainly based on commercial considerations and the nature and persistence of such advice has been illustrated in the evidence and documentation referred to herein.

The Taoiseach Albert Reynolds TD who was at the relevant times, Minister for Industry & Commerce, freely acknowledged in the course of his evidence that such advice was available to him and disregarded by him on the basis that the criteria involved in No. 2 account business, as all these transactions were, was whether cover should be granted in the national interest and not solely on a commercial basis.

As illustrated herein 50% of the entire amount available for Export Credit Insurance worldwide and in respect of all manufactured goods and services was between 1987 and 1988 allocated to exports to Iraq and of this allocation, 75% was allocated in respect of beef to Iraq, representing 37.5% of the entire amount available for all exports worldwide. Of this allocation 75.52% went to AIBP and 24.38% to Hibernia Meats.

These facts are clearly established and it is alleged that such allocations

(1) were made for political reasons and because Goodman was extremely personally close to members of the Government.

(2) constituted acts of blatant favouritism.

(3) disadvantaged rivals and exporters of other products.

(4) strengthened the position of the Goodman group within the beef processing and allied trades contrary to the interests of farmers and employees and of exporters in other business sectors.

The determination of where the "national interest" lay was a matter for decision by the Minister for Industry & Commerce.

On the 17th day of June 1983, the Government of the day had decided that the allocation of Export Credit Insurance in respect of exports to Iraq would no longer be made by the Insurance Corporation of Ireland on a purely commercial basis but in the case of all cover in excess of £250,000 would be made as a result of a decision by the Minister for Trade, Industry and Tourism (subsequently the Minister for Industry and Commerce) in the national interest.

Subsequent to that date, all allocations of cover in respect of exports to Iraq in excess of £250,000 and the terms and conditions of cover were made by the Minister for Industry and Commerce in "the national interest".

Such determination was the basis of the allocation of cover under the No. 2 account, from which all allocations of cover in respect of exports to Iraq were made.

Dealing with this question the Taoiseach and Minister for Industry and Commerce at the time said in evidence:

> "I want to make clear here, there are two separate accounts, the No. 1 Account is commercially managed by ICI and the Department and the second is the No. 2 Account to which no commercial criterion is applied, where it is the State that has to be covered, where it is the decision of the Minister for Industry & Commerce to decide that and I have already said that I took that decision. I don't need Government or anybody else's approval. I can go in and appraise it and keep them in contact but it is my responsibility and I take the decision and the bucks stops here."

Though the matter was purely a matter for him he explained in the course of his evidence to the Tribunal the context within which he made the decision to restore Export Credit for Iraq and the reasons for such decision. He stated that:—

> "The industrial policy, for which I was responsible, was to develop the maximum number of sustainable jobs in manufacturing industry and in international trade and services and, in short, this meant a policy of export led growth and import substitution and this required the maximisation of added value in manufacturing industry while retaining as much wealth as possible for further creating job development and the achievement of those objectives required concerted action surplus arrange at Government Departments and involving all the economic ministries."

> "£545 million was spent on encouraging and promoting industrial investment and development during that period. This money was actually spent in addition to a further £700 million on producing infrastructure."

> "The Agri-food sector was arguably the most important economic sector of all and offered the in-coming Government excellent opportunities for growth and development. This sector of the economy generated £266,800m of exports in 1986 and because of its highly indigenous nature and because it has a very low import content and negligible profit repatriation out of this country it contributed £2,401 million to our foreign currency earnings and this compares very favourably with the rest of the manufacturing industry which generated £7.56 billion in exports but only contributed £3.097 billion in foreign earnings."

> "In 1986 our total net foreign earnings from exports stood at 5.49 billion, 44% of which came from Agri-Food products. So, quite clearly, from what I am saying, the Agri-Food business is the best contributor to foreign earnings and it's the best contributor to wealth creation in this country."

> "Within the Agri-Food Sector the beef industry it offered great potential and I think that has been recognised for a long time back, it offered great potential. Just to give you an example, "There was a growth in exports from 712 million in 1986 to 790 million in 1987 and beef alone out of the Agri-Food Sector is the most indigenous of all our industries in terms of economic status and effect....."

> "In 1987, the first of a series of signals of uncertainty about the whole future and direction of the CAP support system. Intervention support had been the solution until

then. Intervention support, I have described on many times, as a lazy man's market but in December 1986 a new regime for beef had been agreed by the Council of Agriculture Ministers in response to a dramatically increased cost to the community of the CAP Guarantee supports, which then had reached 3.482 billion ECUs in 1986 and the situation in 1987 was made even gloomier by the high level of intervention stocks in Ireland which stood at 125,000 tonnes in 1986. So, starting into 1987 we had the uncertainty of the CAP, the new regime that was brought in in relation to intervention stocks, we had a depression in the cattle trade and, I think it has been said by the Agricultural Commissioner here last week, that if the cattle prices are down then every village and town in the country, and every part of this country, suffers as well."

"The establishment of plans and identification of an Irish food product on international markets is an expensive business, it takes a long time to do. If I may say so, I had personal experience, before I came into Government, I know exactly what's involved in international marketing. I know you have to have a consistent supply of raw material, I know you have to have a consistent supply and a reliable supply to your customer and I have said many times in my business life when asked what is the recipe for success in business and I have related it and with your permission Chairman, I could say it here again today. "Look after your customers, look after your workers and the profits will look after themselves".

"It must be realised that these problems were extremely complex and were further aggravated by the uncertainty about the direction of CAP. High intervention stocks, at that stage, were the order of the day and were over hanging the market and indeed they were limited third country market opportunities and by that I mean third countries, countries outside the EC, markets for commodities to off-load the glut of the Irish market at the end of the year, it was important to keep them in place and to keep them supplied so that we would have an outlet for our glut of cattle and beef at the end of the year."

"In 1987, the immediate problem remained one of stimulating commodities sales and thus relieving pressure on intervention with immediate positive consequence for trade confidence and prices in the market. First of all, put back the confidence into the supplier, into the people that were producing and keep them in the beef producing business. Export Credit Insurance was one of a range of instruments available to the Government to support the industry and export activity."

"Its use, in the case of the beef industry, in 1987 and 1988 must be viewed in the context of the overall £2,400 million that was actually spent. £2,400 million actually spent in Agriculture during that period. It gives you some idea, because otherwise you would have a catastrophic result for Irish farmers and it had to be tackled and tackled urgently, and how was it supposed to be done, and Export Credit Insurance was just one of the instruments available to the Government to do it and, furthermore, it must be understood that extending Export Credit Insurance cover in the case of the beef industry did not involve actual Government expenditure."

"All decisions made by me in relation to Export Credit Insurance were made in the context of commercial beef and the importance of stimulating exports, as I have already explained to this Tribunal. Because of the indigenous nature, the impact of additional exports would be extremely and very significant. Export sales of beef have a major effect on foreign exchange earnings, as I have already said, on farm incomes and employment within the processing industry. There are also significant spill-overs into other industries, such as sales to farm machinery, sales to suppliers, transport people, all of the people that are directly and indirectly involved in the food industry and such sales and the combination of such sales have a very substantial multiplier effect on the Irish industry as a whole."

As appears from this portion of the Taoiseach, Mr Reynolds' evidence, all decisions made by him in relation to Export Credit Insurance

> "were made by him in the context of commercial beef and the importance of stimulating exports."

He stated that what he meant by commercial beef was;

> "beef that is bought or cattle that is bought from farmers and killed within, I think, 90 to 100 days for Halal purposes if you were exporting that beef to Iraq."

The Taoiseach Mr Reynolds was clearly of the view that the national interest required that support, including Export Credit Insurance should be available and given to the export of "commercial beef", which would provide a much needed boost to the economy and that the provisions of such support was in accordance with Government policy as outlined by him.

In this context, it is relevant to refer to the letter dated the 27th February 1987 written by Mr Brian Britton of AIBPI to Mr Donlon, the Secretary to the Department of Industry and Commerce and dealing with "Trade with Iraq", wherein he had stated:

> "All the product being supplied is being processed and prepared in our own Group Factories, giving substantial employment in this country."

At no stage was it ever disclosed to the Minister for Industry and Commerce or the officials of his Department that the position was other than set forth in the said letter — it was never disclosed that 38% of the product being supplied would be sourced from outside the State and that of the tonnage of beef exported by AIBPI to Iraq during 1987 and 1988, 84% were purchased from the Intervention Stock held by the Irish Intervention Agency, a considerable portion of which was not prepared and processed in Goodman Group factories.

Though the Departmental advice, as outlined herein was that Export Credit Insurance should not be restored in respect of beef exports to Iraq in April 1987, he stated in evidence that

> "That was the advice in that particular document to me, was that none should be restored for the present. That was the advice given to me in that document by the Department of Industry & Commerce. Of course, in reaching my decision, I take that advice on board, but I also take into account my knowledge of the international market place, my knowledge about Iraq and my knowledge as to what is in the best national interests of this country. That's what my role and my responsibility is and in fact I made that assessment because I had just come back into Government having spent 4 years in opposition, having spent 4 years developing my own business, which

is very allied to this business that we are talking about, and I was very well up with what's happening in the world."

"And is the duty and responsibility of every Minister for Industry & Commerce to acquaint himself with what's happening and stay up with the reality so as to what's happening around the market place. That I did and I carried out my own valuation and that's what my decision was also based on, taking into account the advice of the civil servants."

At that time the Iraqis "had the benefit of the best technology from the west, they had war relief funds totalling about 30 million coming in from Saudi Arabia and from other Arab States in the Gulf to ensure that Iraq was able to stand up to the war. So, there was no question about the war depleting their resources. Iraq was then and is now a very very wealthy state. They are one of the oil rich countries of the Middle East. They have well over a hundred years of reserves of oil out there. So for anybody to suggest that this was a poor country, I am sorry, anyone who said that would be out of touch with the reality with what was the position of Iraq or indeed what still is the position of Iraq."

"We had a good trading relationship with them and there were considerable opportunities there and it was important that we would not lose them and that was one of the basic reasons why I restored Export Credit Insurance to Iraq."

This decision made by the Minister was to restore cover on a limited basis. As stated by him in evidence;

"First of all, I want everybody to recognise that the existing ceiling for cover in Iraq at that particular time, set by the previous government, was 70 million. The exposure as related to me by the Civil Servants at that time was 24 million, not including the 10 million that is normally allocated to the PARC Hospital Project. I increased the ceiling for Iraq, in my reinstatement decision up to and including PARC, to 45 million, leaving a balance of 30, 25 million un-allocated, and I allocated that on the basis of 6 million for beef and 5 million for smaller companies. Remember, apart from beef and the concentration at this Tribunal, and understandably so because that is the term of reference, the concentration is on beef but also remember there are quite a number of smaller companies around this country that have built up good business in Iraq, and indeed were starting to put on some bit of pressure to have their case examined too, and I allocated that case as 6 million for the beef industry and 5 million for smaller companies. And, in fact, in relation to the beef industry, I might say it was allocated on a basis of one third of the rest to be taken up by the company itself, a third by their bankers and a third by the State. I might say that was the most restrictive cover on Export Credit ever introduced in this country, and that was my original decision and my decision to reinstate export credit cover."

He further stated in evidence that:—

"It was quite clear at that stage that in relation to the 6 million cover I restored on the beef cover, I restored it to AIBP. Their payment and track record was unblemished at that stage except for an outstanding £300,000. They had already got in £4m for payment towards AIBP, so that left them £300,000 outstanding and I was told by the

> Department that Mr Britton had taken full responsibility for that £300,000 — which meant that they had a clean sheet".

No Export Credit Insurance for the export of beef to Iraq was taken up by AIBP or any beef exporter on foot of this decision but by letter dated the 31st day of August 1987 Goodman International sought 100% cover in respect of a contract valued at $134.500,000. This application was in respect of a contract signed by AIBPI on the 2nd day of July 1987.

The Taoiseach Mr Reynolds stated in evidence that;

> "The size of the contract was so economically significant for the Irish Beef industry and it would develop the industry along the lines that we wanted"

In a note prepared at the time by Mr O'Mahony of the Export Credit Section of the Department of Industry and Commerce for the Minister's information, Mr O'Mahony, having outlined the position with regard to "Export Credit Insurance for Iraq" stated "The Export Credit Section's view is that no further cover should be provided for Iraqi business"

Having decided in April 1987 that "the national interest" required that Export Credit Insurance be re-introduced in respect of exports to Iraq, but that the ceiling in respect thereof should be limited to £45m. and that the amount of cover to be given to AIBPI in respect of beef exports would be restricted to one-third of the value of the contract negotiated by them and within the limit stated by him, viz £6m the Minister for Industry decided on the 31st day of August 1987 to seek Government approval for the increase in the ceiling for exports to Iraq to £150m. and to radically alter the position adopted by him in April 1987 and this alteration was due to the application made by Goodman International (AIBPI) for Export Credit Insurance on the $134.5m. contract.

On the 23rd day of August 1987, while in hospital, the Minister had been shown a summary of the position with regard to the allocation of Export Credit Insurance to Iraq and the then existing potential State exposure of £34.74m. in respect thereof when the approved ceiling for such exposure was £35m. (exclusive of the £10m. allocated to PARC Hospital) and which summary contained the recommendation that no further cover should be provided for Iraq.

On the 31st day of August 1987, the Minister for Industry and Commerce, Mr Reynolds informed his acting Private Secretary, Mr McBride, that Anglo Irish Meats (AIBPI) would be making a submission for consideration by the Department and that he would make a decision following an examination of the Anglo Irish Meat proposals. He indicated that the AIBPI proposal would require Government approval and he wanted a memo for the Government meeting to be held on the 2nd day of September 1987, where the matter would be dealt with by the Taoiseach, Mr Charles J. Haughey, TD.

The AIBPI proposal was contained in the letter dated the 31st August 1987 and received in the Department on the 1st September. An Aide Memoire was prepared and submitted but was returned because it had not sought any particular decision.

The AIBPI proposal had sought 100% Export Credit Insurance cover for the supply of beef to Iraq with a total value of $134.5m.

Though, as stated by Mr Reynolds, "the size of the contract was so economically significant for the beef industry" and was the largest ever negotiated in respect of the export of beef, and Mr Reynolds was aware of the fact that the application was going to be made prior to the receipt of the application in the Department of Industry and Commerce, no satisfactory evidence was available to the Tribunal to establish the circumstances in which the Minister for Industry and Commerce was informed of the application prior to its receipt in the Department or of the necessity to have it dealt with at such speed, or why it was necessary to have the matter dealt with with such a degree of urgency that the Department of Finance and the Department of Agriculture and Food did not have an opportunity to express their observations on the matter in the Memorandum for Government.

Neither, the Minister for Industry and Commerce, Mr Reynolds, the Taoiseach, Charles J. Haughey, who was to deal with the matter in Cabinet, Mr Goodman, nor Mr Britton have any recollection of who informed Mr Reynolds that the application would be made.

While the size of the contract would have had significant benefit for the economy and, in particular, the agricultural economy, if, as the Minister for Industry and Commerce believed, "commercial beef" were to be exported in pursuance thereof, the grant of Export Credit Insurance cover in the amount sought could have had serious consequences for the Exchequer if the Iraqi authorities made default in payment, of which risk the Minister for Industry and Commerce had been advised and was aware.

The risk was referred to in the document which had been prepared by Mr Quigley of the Department of Finance for the advice of the Minister for Finance prior to the meeting of the Government on the 8th day of September, 1987 but not included in the Memorandum for Government which had not been received in the Department of Finance until the afternoon of the 7th of September, 1987.

The recommendations made by his officials to the Minister for Finance were as follows:—

2. "This Department recommends strongly that you oppose the Minister for Industry and Commerce's proposals, for the following reasons:—

3. the present "official" ceiling for cover for Iraq is £70m. This limit was decided upon by the Government in February 1986 and represented a major increase from the previous limit of £35m. However, the extent of cover has never remotely reached the £70 m "official" limit, as in the spring of 1986, the Iraqis started to default in their payments. Effective cover even now amounts to only £25m. In effect, what the Minister for Industry and Commerce is seeking is an increase in cover from *£25m to £150m.*

4. Even in normal circumstances, such an increase would be fraught with risk, as it would greatly increase the exposure of the Scheme and hence the Exchequer. In the present case, the risk is even greater. A very high proportion of the increased cover would relate to Iraq. As recent events in the Persian Gulf have illustrated, the Iraqi situation is extremely volatile. The Iraqis have to date, been erratic in

fulfilling their obligations. A deterioration in the country's military and economic position could lead to its defaulting on its foreign debts. If this were to happen (and if the Minister for Industry and Commerce's proposal to increase the ceiling of cover to £150m had already been accepted), *the Exchequer would be at a loss for a considerable sum possibly of the order of £120m.*

5. *In essence the Minister for Industry and Commerce's proposals are too much of a gamble with the Exchequer's resources.* You should seek to have the effective limit of cover for Iraq confined to £45m, (£20m above the present level of exposure) under the conditions as set out in paragraph 6 of the Memorandum. This roughly represents the limit beyond which the Exchequer should not go.

6. The second of the Minister for Industry and Commerce's proposals — to increase the ceiling for all markets from £300m to £500m — follows from the first. If the first proposal is not accepted by the Government, there would be no need for the second one."

These views were not forwarded to the Minister for Industry and Commerce for inclusion in the Memorandum for Government because time did not permit.

The Memorandum for Government submitted, prepared by officials of the Department of Industry and Commerce contained the arguments for and against the granting of the approval sought and are contained in Paragraph 9 of the said Memorandum as follows:—

"9. The question of increasing the ceiling for cover in Iraq must be viewed against present uncertainties in the Gulf region.

For

(*a*) While 3 claims have been paid in respect of default by Iraq the monies were subsequently received within a few months.
At present no claims have been paid where funds have not subsequently been recovered — but see Paragraph 5 above about the position between now and end year.

(*b*) A leading beef exporter has pointed out that the Iraqi market is a major and increasing market for Irish beef, despite major competition from Europe and that the beef contracts generate substantial foreign earnings by both Irish companies and the country.

(*c*) Premium income at 4% of contract values approximates £5.5m. representing over 70% of the cumulative deficit under the scheme as at 31/12/86.

Against

(*a*) The recent escalation of hostilities in the Gulf must further drain already strained Iraqi resources.

(*b*) Present applications plus existing commitments would mean that insurance for Iraq would constitute 40% of worldwide exposure.

(*c*) To provide cover under the conditions in Paragraph 6 would invariably involve additional requests from companies who have not approached us on the basis that they know we are effectively off cover.

(*d*) While it is difficult to ascertain precisely what other export credit agencies are doing credit lines are being reopened for the Iraqi market by many OECD countries on the basis of agreements with the Iraqis to reschedule existing debts e.g. the UK agreed credit lines of Stg £575m. up to y.e. 1987 are being renegotiated at present. In any event most OECD credit agencies operate an extremely restrictive cover policy or in some cases are totally off cover.

(*e*) If we were to substantially increase our credit line to Iraq and if their financial situation deteriorated further, we could be asked by the Iraqis to reschedule involving payment of claims to exporters, a moratorium on rescheduling repayments and payments spread over a number of years."

On the 8th day of September 1987 the Government decided:—

"1. that the ceiling for insured exports to Iraq should be raised from £70,000,000 to £150,000,000 and

2. that the question of increasing the ceiling for export credit insurance generally might be considered further at a later date, as and when the need arises;"

Consequently, while the ceiling for insured products to Iraq had been increased to £150m, the ceiling for Export Credit Insurance generally remained at £300m. This meant that, if the cover actually allocated to Iraq was granted, the amount available for cover in respect of goods exported to destinations other than Iraq was reduced from £230m. to £150m.

The Minister for Industry and Commerce had sought the approval of the Government to raise "the ceiling for insured exports to Iraq from the previous Government approved ceiling of £70m — to — £150m — the conditions of cover to be those as set out in Paragraph 6" which were;

"(*a*) 70% cover maximum on any contract;

(*b*) a maximum credit period of one year;

(*c*) a claims waiting period of 12 months as opposed to the normal 6;

(*d*) a minimum premium rate of 4% of full contract value as opposed to the usual .04% for good risk countries generally."

The cover as granted to AIBPI was in significantly better terms viz 80% cover in lieu of the "70% maximum cover : a claims waiting period of 6 months in lieu of 12 months and a premium of 1% in lieu of a minimum premium rate of 4% of full contract value and a further 2% in the event of a claim."

The same terms were subsequently offered to Hibernia Meats Ltd, Master Meat Packers Ltd and Halal.

As stated these terms were significantly better than those that had been offered to non-beef exporters pursuant to the decision made to re-introduce cover in respect of exports to Iraq, "the take up of which was very slow, partly due to the stringent conditions and partly due to not finalising contracts", and the conditions referred to in the Memorandum for Government significantly better than those envisaged and offered by the Minister for Industry and Commerce when he decided to reintroduce the scheme in April 1987 viz one-third cover on the amount of the Contract.

Though the advice remained the same and the situation in Iraq remained unaltered, the policy of restrictive cover as outlined by the Taoiseach Mr Reynolds, was altered. The ceiling was raised to £150m and less stringent conditions were imposed. These conditions were negotiated between Mr O'Reilly, Assistant Secretary, and Mr Brian Britton of AIBPI and subsequently approved by the Minister for Industry and Commerce.

The offers of Export Credit Insurance made subsequent to the increase in the ceiling to £150m, as outlined in this report, had by the 27th November 1987 reached the ceiling and no further offers of cover could be made because the effect of such offers had breached the statutory ceiling of £300m as outlined in this Report.

On the 8th day of June 1988 at the request of the Minister for Industry and Commerce the Government approved the text of The Insurance (Export Guarantees) Bill 1988 which was enacted into law on the 5th day of July 1988.

This Act increased the statutory limit to £500m.

On the 25th day of October 1988 the Secretary to the Government wrote to the Department of Industry and Commerce

> "I am to inform you that, at a meeting held today, the Government decided that the Minister for Industry & Commerce might agree with the Minister for Finance a new limit for export credit insurance for Iraq within the overall ceiling of £500,000,000 for export credit insurance generally under the Insurance Acts, 1909 to 1988, in place of the existing limit of £150,000,000."

By letter dated the 23rd day of November 1988 the Minister for Finance agreed to an increase of £100m in the ceiling for Iraq.

Between the 13th April 1987 and the 24th November, 1988 the authorised ceiling in respect of exports to Iraq had been increased from £70m to £250m, being an increase from 23% to 50% of available cover.

Such increases were authorised by the Government at the request of the Minister for Industry and Commerce, who at all times sought the approval of the Government for such increases and having obtained such approval, made his decisions in regard to the allocation of cover.

Though the advice available and given to the Minister for Industry and Commerce as outlined in this Report was against the re-introduction of the scheme for Export Credit Insurance in respect of exports to Iraq, the increase in the ceiling to £150m in the first instance, the Minister, with the approval of the Government, conceived it to be 'in the national interest' that the scheme should be re-introduced and the specified sums should be available for Export Credit Insurance in respect of exports to Iraq.

He considered that the risks of non-payment, of which he was advised were more than counterbalanced by the benefits which would accrue to the Irish economy by the development of exports, particularly of beef to Iraq.

His decision in this regard was based on the belief that the beef to be exported in pursuance of the contracts in respect of which Export Credit Insurance cover was granted was "commercial beef" as defined by him and would be sourced within the jurisdiction.

If such was the actuality and if payments were made in respect thereof by the Iraqi purchasing authorities, then the benefit to the Irish economy, and in particular the agricultural sector would have been substantial resulting in at least the stabilisation of cattle prices and more probably an increase in such prices thereby benefiting the farmers and thereby reducing the dependence of Ireland on the intervention system.

However the reality is that :—

(*a*) Of the tonnage of beef exported to Iraq in 1987 and 1988, by AIBPI and declared for and subject to Export Credit Insurance policies, 38% were sourced outside the jurisdiction of the State and

(*b*) Of the tonnage exported to Iraq by Hibernia Meats during the years 1987 and 1988, 18% were sourced outside the state.

and

"that of the tonnage of beef exported to Iraq in 1987 and 1988, 84% of that exported by AIBPI and 75% of that exported by Hibernia Meats had been purchased from Intervention stocks held by the Irish Intervention Agency, the Minister for Agriculture and Food".

The benefit accruing or likely to accrue to the Irish Economy from this situation would be minimal compared to the benefit which would accrue if the exports consisted of commercial beef and would not justify the risk involved in granting Export Credit Insurance in the amounts granted.

These facts were not known to the Minister for Industry and Commerce at the time that he made the decision to re-introduce the Scheme and sought increase in the ceiling for exports to Iraq or at the times that he authorised the granting of insurance cover in respect of the contracts to AIBPI and Hibernia Meats.

If he had been so aware it is unlikely that he would have granted cover in respect thereof as he, at all times believed that 'commercial beef' was being exported.

The then Taoiseach Charles J Haughey stated in evidence that

> "What I would say about that is that neither I nor indeed I would imagine any of my colleagues would ever have thought or visualised that anything other than Irish sourced beef would have been covered by our Export Credit Insurance Scheme."

It is clear from the evidence of the Taoiseach, Mr Reynolds, that he had no reason to believe either that beef was being purchased from outside the State or from Intervention stock, to substantially fulfil the contracts.

The evidence of the Department of Industry & Commerce officials is that at all times they believed that they were dealing with "commercial beef".

This was the position at the time of the decision made by Mr Reynolds to reintroduce Export Credit Insurance in respect of beef exports to Iraq in April 1987 and at the time that he approved the grant of cover to AIBPI in respect of the contract for the export of beef valued at $134.5m and at the time that he approved the issue of such cover to Hibernia Meats Ltd (Daintean) in respect of a contract valued at $46m, and to Master Meat Packers in respect of a contract valued at £10m.

Though Master Meat Packers Ltd did not take up the offer of cover made to them, the benefit thereof was subsequently transferred to Hibernia Meats Ltd.

These were the only offers of cover authorised by the Minister which were the subject of policies actually issued by the ICI in respect of beef exports to Iraq.

While neither the Minister nor the officials of his Department were aware of the fact that beef purchased from Intervention stocks was being used to substantially fulfil the requirements of these contracts, the position would have been completely different before the 21st day of October 1988 when the Minister informed the Secretary and other officials of his Department that he intended to increase the amount of cover available in respect of exports to Iraq by £120m and made provisional allocation of cover available to AIBPI and Hibernia Meats Ltd and other non-beef exporters if the contents of the briefing notes prepared by the CBF (The Irish Livestock and Meat Board) in anticipation of the Irish/Iraqi Joint Commission talks, due to be held in November 1988, had been brought to his and their attention.

This briefing note included the paragraph:—

> "In recent years the product supplied to Iraq has largely been from Intervention stocks with some APS. The market is mainly for frozen hindquarter boneless cuts. As the stocks of Intervention product decline, the market is likely to move towards APS and possibly forequarter cuts as prices rise. The type of beef should not be mentioned to the Iraqis. At present, Islamic slaughter is a requirement of the market."

This paragraph was taken out of the briefing document by Mr Shortall of the Department of Agriculture and the following paragraph substituted:—

> "The market is mainly for frozen hindquarters boneless cuts. In some cases the exporters have availed of the EEC aids to storage scheme prior to export. In view of rising price trends, there may be some move towards some forequarter cuts."

If this briefing document had been made available in its original form to the Department of Industry & Commerce officials as was the intention of the CBF, then they would have been aware, prior to the decisions made by the Minister on the 21st day of October 1988 that the major portion of the beef being exported to Iraq was from intervention stock and not commercial beef.

While it is undoubtedly clear that the decision made by the Minister to reintroduce Export Credit Insurance cover for exports to Iraq in April 1987, to secure the Government's approval of the increase in the ceiling for such exports to £150m in September 1987 and to secure the approval of the Minister for Finance in November 1988 to increase the ceiling of such cover to £250m, were made by the Minister for Industry & Commerce against the advice made available to him by the Insurance Corporation of Ireland, the Department of Industry & Commerce and the Department of Finance, such advice was based on commercial reasons, namely the real risk of default in payment by the Iraqi authorities, and the Minister considered that he was entitled to disregard such advice, if in his opinion, the "national interest" so required. For the reasons set forth by him in evidence, he conceived that the "national interest" so required and there is no evidence to suggest that he made his decisions other than in accord with his conception of the requirements of the "national interest", the determination of which on this issue was his responsibility. He had stated in evidence that:—

> "All decision made by me in relation to Export Credit Insurance were made in the context of commercial beef and the importance of stimulating exports, as I have already explained to this Tribunal. Because of the indigenous nature, the impact of additional exports would be extremely and very significant. Export sales of beef have a major effect on foreign exchange earnings, as I have already said, on farming incomes and employment within the processing industry. There are also significant spill-overs into other industries, such as sales to farm machinery, sales to suppliers, transport people, all of the people that are directly and indirectly involved in the food industry and such sales and the combination of such sales have a very substantial multiplier effect on the Irish industry as a whole."

However, Mr O'Reilly, Assistant Secretary in the Department of Industry and Commerce in the course of a minute written in March 1988 dealing with proposals to increase the statutory limit and the ceiling in respect of exports to Iraq stated:—

> "The real benefits of the business in Iraq are assumed to exist. I have never seen any analysis of them or whether such benefits might be obtained by exports to other country".

The decision made by the Minister for Industry and Commerce to seek the approval of the Government to increase the ceiling in respect of exports to Iraq to £150m, the Government's decision to give such approval on the 8th day of September 1987, and the Minister's decision, subsequent to the receipt of such approval, to approve of Export Credit Insurance to AIBPI in respect of 80% of $134.5, to Hibernia Meats Ltd, in respect of 70% of $46m. and to Master Meat Packers Ltd, in respect of 70% of £10m. involved a potential liability on the Exchequer of £98.65m. which amount does not include any liability in respect of the cover made to Halal and withdrawn, nor the offer of $30m. made to AIBPI in November 1987 which could increase such liability by £14.48m.

While the Minister for Industry and Commerce and the Government were entitled to make their respective decisions in "the national interest", the "national interest" would also appear to require that before exposing the State to a potential liability of well in excess of £100m a more detailed investigation or analysis of the benefits to the economy of such decisions which involved:— 

(i) the allocation of 50% of the amount of Export Credit Insurance cover available for all exports worldwide to one particular destination, and

(ii) such risk to the Exchequer if default in payment were made should have been carried out.

Such an investigation, if made, might and in all probability would have disclosed that a large portion of the beef to be exported was intended to be sourced outside the jurisdiction and an even larger proportion had been or was intended to be purchased from intervention stock and that the benefits to the Irish economy, arising from such exports, were illusory rather than real.

The Department of Agriculture and Food, as the Intervention Agency, were aware of the purchases of intervention beef from them by AIBPI and as the body responsible for the payment of the Export Refunds subsidy, were aware of the intended destination of such beef.

In addition AIBPI had in the course of a Memorandum submitted by them to the Department of Finance and dated the 2nd July 1987 in connection with a "Proposed Amendment to Section 84A of the Corporation Tax Act 1976" had stated that, inter alia

> "The sales by Anglo Irish include beef processed by Anglo Irish Beef Processors Ltd which is a fellow subsidiary of Anglo Irish and beef purchased from intervention stock which were processed by other beef processors within the State" and
>
> "The sale by Anglo-Irish of processed beef purchased from intervention will not qualify as manufactured goods under Section 39 and, in consequence, if the sales of this type of goods in any accounting period exceed 25% of all sales, then the entire borrowings would not qualify as Section 84 borrowing under the provisions of the Section 84A"

The entire of this Memorandum is printed in the Section of this Report dealing with Section 84 borrowings and is referred to here to illustrate that the Department of Finance were on notice that a considerable portion of sales by AIBPI consisted of beef purchased from Intervention and an inquiry from either Department would have ascertained the position.

The manner in which the allocations of Export Credit Insurance were made by the Minister for Industry & Commerce within the ceiling fixed by the Government was the subject of allegations made in Dáil Eireann and the issues raised by such allegations have been set forth in this report.

The factual position with regards to such allocations is as follows:—

(*a*) On the 12th day of February 1987, the then Minister for Industry and Commerce, Michael Noonan TD, decided, because of the general deteriorating financial and military situation in Iraq and the fact that there were a number of payments overdue from Iraq, not to offer any further Export Credit Insurance in respect of exports to Iraq until such times as the situation was seen to be sufficiently improved to enable such cover to be put in place;

(*b*) Prior to that date, application had been made on behalf of AIBP for Export Credit Insurance cover in respect of a contract for the supply of beef to Iraq for £34m and a number of meetings had been held between Mr Goodman and Mr Britton of AIBP and the Minister and officials of the Department of Industry & Commerce;

(*c*) AIBP were informed of the decision of the Minister for Industry & Commerce and this decision was confirmed by letter dated the 5th day of March 1987 to Mr Britton. This letter pointed out that "the question of further credit to Iraq was the subject of continuous review in the Department"

(*d*) On the 11th day of March 1987, Albert Reynolds TD was appointed Minister for Industry & Commerce;

(*e*) On the 9th day of April 1987, after a more formal meeting between members of the Government and Mr Goodman and Mr Britton of AIBP in connection with the IDA Development Plan, Mr Goodman spoke to the Minister for Industry & Commerce about the desirability of and necessity for the provision of Export Credit Insurance in respect of exports to Iraq.

(*f*) On the 13th day of April 1987, a copy of a note on Export Credit Insurance, prepared for the information of the Minister for Industry & Commerce, was given to him. This note referred to the fact that the outstanding liability in respect of Export Credit Insurance for exports to Iraq of £30m which existed at the time of the previous Minister's decision made in February 1985 and to the fact that payment of $1,994,685 had been paid to Dantean International Ltd and US$4m to Nenagh Chilled Meats Ltd subsequent to the making of that decision.

It pointed out that applications for Export Credit Insurance then current, amounted to approximately £30m and included an application from Anglo Irish Meats for £20m and from Dantean Meats for $2m.

(*g*) This note further contained the recommendation that the current policy in Iraq, i.e. no further cover, should be continued for the present.

(*h*) Between the 13th day of April 1987 and the 16th day of April 1987, the Minister for Industry & Commerce directed officials in his Department that Export Credit Insurance was to be made available for exports to Iraq on the following conditions:—

(*a*) that cover was to be made available up to a ceiling of £45m.

(*b*) that a pragmatic approach was to be adopted in the allocation of cover and that companies which already had payments overdue from Iraq should be given no further cover until amounts overdue had been cleared.

(*c*) that in the specific case of Anglo Irish Meats, cover should be offered to cover one third of their then proposed contract with the banks and Anglo Irish carrying one third each also.

(*d*) that a detailed approach be devised for allocating the cover being made available, taking into account the said directives.

(*i*) This offer of cover was not taken up by AIBP.

(*j*) On the 17th day of June 1987, Hibernia Meats Ltd sought from the Insurance Corporation of Ireland, cover for a two year period of contract valued at $46m which the company was negotiating with the Iraqi purchasing agencies.

(*k*) On the 2nd day of July 1987, Anglo Irish Beef Processors International Ltd negotiated a contract for the export of beef to Iraq with a total value of $134.5m.

(*l*) On the 31st day of August 1987 Mr Britton, the Deputy Chief Executive of Goodman International wrote to the Minister for Industry & Commerce applying for 100% Export Credit Insurance in respect of the said contract.

(*m*) The ceiling fixed by the previous Government on the 6th February 1986 in respect of Export Credit Insurance to Iraq was £70m and such ceiling would be breached if the applications made by Hibernia Meats Ltd and AIBP were acceded to.

(*n*) On the 8th day of September 1987, the Government decided that the ceiling for insured exports to Iraq should be raised from £70m to £150m and it is clear from all the evidence that the application for such decision made by the Minister for Industry and Commerce was to provide for the application of Goodman International (AIBPI) for cover in respect of the $134.5 contract.

(*o*) By letter dated the 8th day of October 1987, written by Mr O'Reilly, Assistant Secretary, at the direction of the Minister for Industry & Commerce, it was confirmed that the Minister had agreed that Export Credit Insurance would be provided, subject to the terms set forth in the said letter, in respect of 80% of the contract value of the said contract.

(*p*) On the 8th day of September 1987, subsequent to the meeting of the Government, at which the ceiling was increased from £70m to £150m, the Minister for Industry & Commerce met Mr Phelan of Master Meats and Mr Oliver Murphy of Hibernia Meats and after discussion with them, offered them £10m cover for each of their companies in respect of exports of beef to Iraq.

(*q*) On the 9th of September 1987 Hibernia Meats Ltd wrote to the Minister for Industry & Commerce, pointing out that in their correspondence with ICI as of June 1987, they had sought credit insurance cover for a total of $46m for a two year period.

(*r*) By letter dated the 23rd October 1987, the Department of Industry & Commerce informed Mr Murphy that the Minister for Industry & Commerce had decided that Export Credit Insurance cover would be available to Hibernia Meats for the full value of their contract in Iraq subject to the terms disclosed in the said letter.

(*s*) The allocation of the £10m cover to Master Meat Packers Ltd was confirmed by Mr Timbs on behalf of the Minister by letter dated the 22nd day of October 1987.

(*t*) On the 16th day of September 1987 Agra Trading Ltd had a meeting with officials of the Department of Industry & Commerce at which they requested Export Credit Insurance cover for a proposed contract of 5,000 tonnes of boneless beef valued at $17m with a credit period of 2 years and had been informed that their application would be facilitated in respect of a credit period of 18 months with 70% cover if they obtained a contract.

(*u*) On the 26th day of October 1987 Halal Meat Packers (Ballyhaunis) Ltd applied to the Insurance Corporation of Ireland for Export Credit Insurance cover in respect of a contract worth $25m in Iraq.

(*v*) On the 13th day of November 1987 the Minister for Industry & Commerce met Mr Larry Goodman. During the course of this meeting Mr Goodman informed the Minister that he had negotiated an extension of his contract with the Iraqi's and would require further cover. As at that time there was only $30m available within the ceiling, the Minister for Industry and Commerce agreed to the allocation of this amount to Goodman International (AIBPI).

At this meeting also Mr Goodman complained to the Minister about the activities of Halal and Agra Trading Ltd in Baghdad and alleged that they were engaged in price cutting.

(*w*) On the 13th day of November the Minister informed Mr Timbs that this additional Goodman contract would be covered on the usual terms on the following Monday 16th of November 1987.

(*x*) This brought the level of exposure up against the ceiling of £150m and meant that the Agra Trading Co. Ltd could not be accommodated. When this was pointed out to the Minister he stated that if Agra Trading Ltd obtained a contract, the question of the ceiling could then be considered.

(*y*) On the 17th day of November Halal informed the Insurance Corporation of Ireland that they were discussing an increased contract and as a consequence they sought an increase in the overall level of cover from $25m, which they had been offered, to $37.2m.

(*z*) As of the 17th day of November 1987 the position with regard to Export Credit Insurance for contracts in Iraq was:

(1) Existing exposure: £17.18m

Commitments: £133.48m

Total: £150.66m

This did not include the applications for cover from Agra Trading and the additional cover sought by Halal.

(*aa*) As of the 17th day of November 1987, concern was being expressed in the Department of Industry & Commerce and the Insurance Corporation of Ireland that, having regard to the then existing exposure and the commitments given with regard to the allocation of cover, that the statutory ceiling of £300m was in danger of being breached.

The then existing level of cover worldwide was approximately £200m and as pointed out by Mr Walsh in a minute dated the 25th day of November 1987, "if all the offers made were taken up, worldwide exposure will gradually increase to in excess of IR£350m, £150m of which will be in Iraq".

(*bb*) On the 27th day of November 1987, the Minister for Industry & Commerce decided to withdraw the offer of Export Credit Insurance made to Halal as it had not been accepted in the terms in which it was offered and because there was no evidence that they had a contract and instructed his Private Secretary to so inform Halal of his decision.

(*cc*) The position then was that of the £300m available worldwide for Export Credit Insurance, £150m (50%) had been allocated in respect of exports to Iraq.

Of this £150m, £84.9m was being allocated to AIBP and £28.52m (including £10m to Master Meat Packers) to Hibernia Meats Ltd

This meant that 75% of the amount available in respect of exports to Iraq, namely £150m, was allocated in respect of beef exported or to be exported by AIBP and Hibernia Meats Ltd This represented 37.5% of the entire amount available worldwide.

Of this allocation, 75.52% went to AIBP and 24.38% to Hibernia Meats Ltd

These percentages relate to the allocation of Export Credit Insurance cover made by the Minister for Industry & Commerce subsequent to his appointment to that office on the 11th day of March 1987 and are increased if regard is had to the exposure which existed at that time in regard to cover already granted namely £17.18m and the State's exposure in regard thereto. When this was done, the relevant percentages are: 63.92% to AIBP and 21.47% to Hibernia Meats Ltd of the total sum allocated and 74.85% and 25.15% respectively of the sum allocated for beef exports.

While the approved ceiling in respect of Export Credit Insurance for exports to Iraq was £150m, the overall statutory ceiling in respect of exports worldwide was £300m.

As of the 25th day of November 1987, the existing level of exposure worldwide was approximately £200m and having regard to the commitments to allocate cover given to AIBP of £69.42m, to Hibernia Meats Ltd of £28.42m and to non-beef exporting companies in the sum of £8.77m, which sums total £106.71m, there was no scope for the granting of any other Export Credit Insurance cover in respect of exports to Iraq unless the statutory ceiling was increased.

(*dd*) There was no scope within the existing ceiling to grant cover to Halal or Agra Trading Ltd if and when they or either of them produced contracts or indeed in respect of the cover in respect of the US$30m contract which the Minister for Industry & Commerce agreed to give to AIBP on the 16th day of November 1987 unless the statutory ceiling was increased.

This fact is confirmed by the minute faxed by Mr Timbs to Mr Aidan Connor on the 10th February 1988 in which he stated that:—

> "Further to our recent telephone conversation, I confirm that the Minister has agreed that a further US$30m contract value covered at either 80% or 70%, depending on the credit period has been allocated to AIBP International in respect of contracted beef sales to Iraq, subject to the enactment of amending legislation to increase the aggregate amount of the Minister's liability under the export credit scheme. As I mentioned to you, the US$30m contract value is available within our ceiling for Iraq but will not be available under our overall legislative ceiling until amending legislation has been enacted."

(*ee*) As of the 29th day of February 1988 there were pending in the Department of Industry & Commerce applications for Export Credit Insurance by:

1. AIBP in the sum of US$155m (including US$30m already referred to)

2. Agra Trading Ltd in the sum of US$17m.

3. Taher Meats in the sum of £11m.

4. Halal Meats Ltd in the sum of US$25m.

(*ff*) On the 31st day of March 1988 Mr Timbs of the Department of Industry & Commerce wrote to the Insurance Corporation of Ireland stating that "we are not open for business for meat contracts for Iraq and do not envisage being so in the future."

(*gg*) On the 8th day of June 1988 the Government approved of the text of the Insurance (Export Guarantees) Bill 1988 and authorised the Minister for Industry & Commerce to present the Bill to Dáil Eireann and have it circulated to Deputies.

(*hh*) The Insurance (Export Guarantees) Act 1988 was enacted by the legislature on the 5th day of July 1988.

(*ii*) On the 7th of July 1988 the Private Secretary to the Minister for Industry & Commerce wrote to Mr Fitzpatrick in the terms of the letter already referred to and stating that:

"The Minister regrets, therefore, that Export Credit Insurance cover cannot be made available for the Taher Meats contract or indeed any other major contract in Iraq for the foreseeable future."

Though the statutory ceiling was increased to £500m the Minister did not make or communicate to the officials of his Department any decision with regard to an increase in the ceiling for exports to Iraq until the 21st day of October 1988.

In view of the fact that the 5th Session of the Irish/Iraq Commission was due to take place in Baghdad on the 7th day of November 1988, the Export Credit section of the Department of Industry & Commerce had prepared a Memorandum dated the 21st day of October 1988 for submission to the Minister on that date.

On the morning of the 21st day of October 1988 Mr Connor of AIBP met Mr Timbs, Mr Donnelly and Mr Walsh of the Department of Industry & Commerce and informed them that in addition to the export credit cover already available to the company (AIBP) he would require additional cover for contracts valued at $325m for the remainder of 1988 and 1989.

During the course of a meeting with the Minister for Industry & Commerce on the afternoon of the 21st day of October 1988 at which the Secretary of the Department, Mr Donlon and Messrs Timbs and Donnelly were present, the Minister was informed that the then existing exposure in Iraq amounted to £136m and that AIBP had sought additional cover on contracts valued at $325m for 1988/89 and Hibernia Meats Ltd had sought cover on two contracts valued at $72m and £10m respectively.

The said officials were informed by the Minister for Industry & Commerce for the first time that it had been agreed at the Government meeting held on the 8th day of June 1988 that increases in the ceiling of Export Credit Insurance for Iraq should be at the discretion of the Minister for Industry & Commerce, that the provision of such insurance should be managed in the national interest to avoid damaging competition between exporters and that such decision should be communicated to the Irish beef exporters by the Minister for Agriculture.

When they informed the Minister that these decisions were not recorded in the communication of the Government decision made on the 8th day of June 1988, he expressed surprise and stated that he would have the matter clarified at the next meeting of the Government which was due to be held on the 25th day of October 1988.

Pending clarification of this matter, the Minister for Industry & Commerce decided that the following additional cover would be provided for the Iraqi market:—

(*a*) roll-over of the existing cover held by AIBP (liability under the Scheme is £95.6m) and Hibernia (liability IR£23.1m) as outstanding maturities were paid.

(*b*) additional cover for AIBP and Hibernia up to a maximum liability under the Scheme of £80m and £20m respectively and

(*c*) additional cover for non-beef exporters up to a maximum liability under the Scheme of £20m subject to increase should demand necessitate such.

If this decision had been implemented, the increased ceiling in respect of exports to Iraq would have been £270m representing 54% of the entire market worldwide and the total allocation of cover in respect of beef exports to Iraq to AIBP would have been £175.6m and to Hibernia Meats Ltd £43.1m representing 65% and 15.96% respectively.

Of the amount allocated or intended to be allocated in respect of beef exports to Iraq, i.e. £218.7m representing 81% of the total to be made available and of this amount 80.29% was to go to AIBP and 19.71% to Hibernia Meats Ltd.

On this basis no provision was made or could have been made for the allocation of cover in respect of beef exports to Iraq for any other company.

Subsequent to this meeting Mr Timbs contacted AIBP and Hibernia Meats on the 21st of October 1988 and the 22nd of October 1988 respectively and advised them that the Minister had given indications of additional cover which he was prepared to make available, £80m for AIBP and £20m for Hibernia.

On the 25th October 1988 the Government decided that the Minister for Industry & Commerce might agree with the Minister for Finance a new limit for Export Credit Insurance for Iraq within the overall ceiling of £500m for Export Credit Insurance generally.

On the 11th day of November 1988, Mr Timbs wrote on behalf of the Minister for Industry and Commerce to the Secretary of the Department of Finance seeking or proposing that the ceiling in respect of exports to Iraq be increased from £150m to £270m.

In the penultimate paragraph of the said letter he stated that:—

> "Insofar as the allocation of the proposed additional cover is concerned, the Minister would point out that not all applicants for Export Credit Insurance are successful in obtaining cover. The policy in this regard is to maximise the credit available for the best economic benefit of the State as a whole. It is clearly wasteful to expend this valuable facility in such a way that Irish companies compete against each other in foreign markets to the benefit of the buyer and the overall disadvantage of the State."

This clearly indicated the policy to be applied in the allocation of the proposed increase viz.

(i) that not all applicants for Export Credit Insurance would be successful;

(ii) to maximise the credit available for the best economic benefit of the State and

(iii) that it would be clearly wasteful to expend this valuable facility in such a way that Irish companies compete against each other in foreign markets to the benefit of the buyer and the overall disadvantage of the State.

The existence of this policy is confirmed in the memorandum on Export Credit Insurance for Iraq dated the 12th day of December 1988 which was prepared for the newly appointed Minister for Industry and Commerce, Mr Ray Burke TD and which has been printed in full in this Report and in relation to the policy of confining insurance cover to AIBP and Hibernia contained the statement that

> "The Department of Agriculture would seem to agree with the policy of confining insurance cover to AIBP and Hibernia but are not prepared to offer us formal advice or to liaise with the trade on the issue".

On the 23rd day of November 1988 the agreement of the Minister for Finance for an increase of £100m in the ceiling for Iraq was conveyed to the Department of Industry & Commerce and a note in the handwriting of the Minister for Industry and Commerce referred to in this Report showed his intended allocation of this £100m viz £70m to AIBPI £20m to Hibernia and leaving £10m for small companies.

Again on the basis of this allocation no provision was made for the allocation to any other beef exporting company of any cover under the Export Credit Insurance Scheme though on the 2nd day of November 1988 the Minister for Trade and Marketing, Mr Brennan TD, who was due to lead the Irish delegation to the Fifth Meeting of the Iraqi-Irish Joint Commission to be held in Baghdad was informed by the Minister for Industry and Commerce, inter alia, that

> "(1) The current limit on Export Credit Insurance for Iraq is IR£150m (the Department of Finance are opposing any increase in this limit).
>
> (2) There is *no policy* of confining Export Credit Insurance on beef exports to Iraq to particular companies.
>
> (3) Any exporter with a contract in Iraq will have an application for Export Credit Insurance considered in the normal way."

This statement is inconsistent with the declared intention of the Minister for Industry and Commerce on the 21st October, 1988 to the officials of his Department of his intention to increase the ceiling in respect of exports to Iraq and to allocate the amount of such increase viz £120m, as to £80m thereof to AIBPI, £20m to Hibernia and £20m to other non-beef exporters to Iraq and his intended allocation of the £100m by which he subsequently, with the consent of the Minister for Finance, increased the ceiling viz £70m to AIBPI, £20m to Hibernia Meats Ltd and £10m to other companies.

On the 23rd day of November 1988, the then Minister for Industry & Commerce was appointed Minister for Finance and Mr Ray Burke TD was appointed Minister for Industry & Commerce in his place.

No new allocations of cover were made subsequent to that date and the failure to grant the cover to AIBPI is, inter alia, the subject of proceedings in the High Court.

From this recital of the facts it is established that only two companies were issued with policies of insurance in respect of beef exports to Iraq viz AIBPI and Hibernia Meats Ltd, (Dantean) (to whom was transferred the benefit of the offer made to Master Meat Packers Ltd on the 8th of September, 1987) pursuant to the decision to increase the ceiling to £150m; that the offer made to Halal had been withdrawn, for the reasons given by the Minister for Industry and Commerce in his evidence; that subsequent applications made by Agra Trading Ltd, Taher Meats Ltd and Halal were refused on the basis that the ceiling had been reached and no cover was available within the ceiling of £150m. which ceiling was not increased until the 23rd day of November 1988 subsequent to a discussion between the Minister and officials of his Department on the question of Export Credit Insurance during the course of which he was informed of applications for Export Credit Insurance by AIBPI and Hibernia Meats which discussion took place on the 21st day of October 1988.

The Taoiseach, Mr Reynolds' account of this meeting is as follows:—

"240 Q. Where the question of the 325 million contract was discussed with AIBP in 1988, 1989

A. Yes, that would be between Department Officials and the representatives of the Company.

241 Q. It was one of the matters that was discussed by you with your officials at the meeting on the 21st?

A. Yes, the two matters being, one of them being AIBP and the other Hibernia.

242 Q. Did you decide, as set out at the bottom of page 304 Volume 15B,the note of the discussion

A. We are back to the note of the discussion. What part of this?

243 Q. The bottom of it. The Minister decided that the following additional cover would be provided in the Iraqi market.

A. Before we go any further, could I remind you as to the sequence of events? First of all, the meeting takes place on the basis of the two applications and the increase in export credit. Right? And I say to the officials concerned what my views are and what I believe is the appropriate increase for export credit for Iraq. And you have before you, I think I have already stated what my views were, and I think that you will understand as well as I understand that I cannot go any further than that, but the real decision of the 21st of October was that I would go back to government to clarify certain situations. That is the real decision that came out of the 21st of October. I went back to government on the first available opportunity, which was the 25th of October, and that decision was taken on the 25th of October, that the Minister for Industry and Commerce and the Minister for Finance would agree an appropriate ceiling for Iraq. Now, that is the real world. So, I think you and I know the problems we have because of the constraints that are put on us to get into any other aspects of it. My view was that the appropriate cover would be a 120 million increase and that the division, my view was that the division of that would be 80 for AIBP, 20 for Hibernia and 20 for others. And I wouldn't like anyone to think that, as far as I am concerned, that the small companies using the Iraqi market, they were equally as important as the larger ones because they had to make their own contributions, and they have an important role to play. That was my view, to clarify the situation. I told the officials I was going back to the government for decision clarification and decide on the 25th of October, which I did, and we all know where the events led from there on in, that we eventually made our submission to the Department of Finance which subsequently didn't accept the 120 million, which was my view starting off, but would accept 100 million to be divided, as was my view, 70/20/10. So, there was the position.

244 Q. Now, I think you also indicated that you were, paragraph 80, you were willing to roll over AIBP and Hibernia as a repayments

A. Yes, I said that on the basis that when money comes in money goes out. I stated my view on the new cover and roll over before I decided to go back to Government. Furthermore, I do not regard my statement to my officials, any of my officials, of what I suggested as of that time as in any way an irrevocable decision. Any cover would be provided, or cover to come would have been dependent on how the Iraqi payments situation developed, who got the contracts with the Iraqis. My intentions in this regard were quite clear and in clear terms claims, premiums, waiting periods those are the normal things that take place and my valuation of the Iraqi market as a risk when the proposed new cover or roll over cover would be put in place. In other words, in hindsight, what we have said is what happened. I would also have come off cover when the payment situation would have deteriorated. I have no doubt about that and no hesitation in saying that, and that would be my position. But, as we all know, that didn't happen until after it was gone. But I wanted to be clear that I would have taken the same decision about payments.

245 Q. As a result of the discussion with Mr Donnelly and Mr Timbs, the result of that was Mr Timbs made two telephone calls, one to AIBP and one to Hibernia?

A. That is correct. I have no recollection of telling Mr Timbs to communicate my view or my decision, and if you notice the way I have put italics on "Decision", because particular to that meeting that word is taken out of the way it was described in a subsequent Departmental Memorandum by Mr Donnelly who was at the meeting. I have no recollection whatever I said to Mr Timbs go ahead and tell them or indeed anybody else, because after all I had decided, I had decided to go back to government, so it wouldn't make a lot of sense to tell the companies in advance of a government decision what they were going to get, but I have no hesitation in saying my intentions were clear, and they would have been known to Mr Timbs and to anybody else at the meeting .... and in reading reports of the meeting afterwards it was described as "Informally telling them" by people who were at the meeting. So, my recollection seems to be borne out by that, but I can tell you straight up that I have no recollection of telling them and it wouldn't make sense that I would tell them because here I was going back to government to get a decision and get the Minister for Finance to make a decision at a certain level. Mr Timbs, I am not here to say what his evidence is, but certainly anybody at the meeting would be very clear about what my views were. But as to what, how you make or take decisions afterwards, I don't know.

A. On the 25th of October, the Government decided that the Minister for Industry and Commerce, as I have said, might agree with the Minister for Finance a new limit for export credit within the overall ceiling of £500 million for Export Credit Insurance generally under the Insurance Acts in place of the existing limit of £150 and that's what happened. The effect of the Government decision of October the 25th, was that pending agreement between the Minister for Finance and myself, it was not going to be possible to make any allocations of cover or any commitments to Iraq at the fifth Joint Commission. The Minister for Trade and Marketing, Mr Brennan was leading the Irish delegation to the fifth Joint Commission, spoke to me on November the 2nd 1987 and inquired what the position was as regards Export Credit Insurance for Iraq having regard to the aforesaid decision of October the 25th, the one that said a figure must be agreed between the Minister for Industry and Commerce and the Minister for Finance. I told him that the current limit on Export Credit Insurance for Iraq remained at 150 million and that it had almost been reached and I also told Mr Brennan that cover was not confined to particular companies, that any exporter with a contract would have an application considered. Of course, having regard to their track record, I did not envisage that any other beef exporter would actually secure a contract with Iraq. So, while the Minister for Finance indicated to me that he was disposed to an increase in the ceiling, he also indicated that his own Department viewed my proposed increase of £120 million as excessive. I was still very much of the view that a substantial increase in the ceiling was justified and that the least that would suffice was £100 million. The putting in place of this new cover would ultimately have depended on how the Iraqi repayments situation developed. Whether exporters could negotiate on an 18 month credit basis because Iraqis

> were looking for 2 years at this stage again or whether satisfactory terms of cover could be agreed. And on my on going view of Iraq as an assumable risk an increase in the ceiling of £100 million pounds would still have allowed £20 million for Hibernia, £10 million would have been available for the smaller beef companies and £70 million available for AIBP. However, had any exporter other than AIBP or Hibernia actually got a contract and after the experience that I have spoken about at length here since 1987, I did not believe that any of them would. I would not have felt obliged to confine cover to AIBP and Hibernia if such a situation had arisen. But in the any event, the Department of Finance did not communicate the Minister's agreement to an increase in the Iraqi ceiling of £100 million until November the 23rd, 1988. That being the same day that I ceased to be Minister for Industry and Commerce and in fact, moved over the following day to the Department of Finance to take up my duties as Minister for Finance."

From this evidence it is quite clear that on the 21st October 1988 that the Minister, Mr Reynolds, intended with Government approval, to increase the ceiling in respect of cover for exports to Iraq by £120m. and that, in his view, the division of such increased amount would be £80m. to AIBPI, £20m. to Hibernia and £20m. in respect of the smaller non-beef exporters to Iraq and in addition that there would be a "roll-over" of the allocations already made to these companies. As stated by him:—

> "I have no hesitation in saying that my intentions were clear and they would have been known to Mr Timbs and to anybody else at the meeting."

When the amount agreed with the Minister for Finance was limited to £100m. he expressed the view that the division would be £70m., £20m. and £10m.

The effect of the "roll-over" would be that as payments were made in respect of contracts already insured, further insurance would be granted in amounts equivalent to the payments.

As the amount of cover granted to AIBPI at this time was £69.42m. and to Hibernia was £28.52m., this allocation (if made) would increase AIBPI's cover to £139.42m. and Hibernia's cover to £48.52m. making a total, in respect of these two companies, of £187.94m. out of a total allocation of £250m. for Iraq and £500m. worldwide and no cover would be available for any other exporter of beef to Iraq.

The granting and intended granting of such cover and the amounts thereof, to the two companies named, was a cause of concern to the other meat exporters such as Halal, Agra Trading Ltd and Taher Meats Ltd, who felt that they were being discriminated against by the decisions of the Minister and letters were written by them protesting against such discrimination, copies of which letters are set forth in the course of this report.

In addition, Mr O'Reilly, Assistant Secretary to the Department had written a memo which was forwarded to the Minister on the 22nd day of March 1988 during the course of which he sought to highlight the problem and stated that:—

"On the question of giving further cover to meat exports.... to Iraq there are some basic considerations. It is necessary to try to ensure the best possible price for Irish meat but it is also necessary to be seen exercising equity in the allocations of cover. Within those factors there is the further consideration as to what extent the State should be prepared to go in supporting one individual entity. The outcome of continuing indefinitely is to increase the dominance of that entity with obvious consequences. Incidentally Goodman International have let over 100 people go at Bailieboro.

If it were decided to go that road there should be no difficulty about increasing the £150m. limit which would, as mentioned, only be operative after the legislation. This could be done, as it was before, with the approval of the Minister for Finance.

The critical issue is (i) whether to do so and (ii) if the decision on (i) is YES by what amount. It seems to me that it cannot be for AIBP alone. There are other applicants who say that they have contracts or that they have been invited to tender. On whatever additional amount of cover might be provided for AIBP in the event of extending the £150m limit it seems to me that, as their increased business is magnifying the State's risk, they would have to accept punishing terms. The entity operates on such a scale that the new business and the risks attaching to it should be borne in three equal segments, (i) by AIBP (they carried all their risk in 1985/1986), (ii) their banks and (iii) the State. As regards (iii) we could then negotiate terms that would have to be very stringent and would have to be more demanding than those in the existing bargain.

The same kind of terms would have to be required of other meat firms, who at present enjoy commitments on the same terms as AIBP, if they were to be given extended cover under new limits. On the other hand equity would seem to require that any cover commitment given under extended limits for Iraq should be on the same terms as that originally given in commitments to AIBP and others.

Because it is obviously wrong in terms of a balance in the total exposure I would be opposed to seeking extended cover for Iraq. The real benefits of the business in Iraq are assumed to exist; I have never seen any analysis of them in precise terms or whether such benefits might be obtained by exports to another country. One development is clear: the more contracts that Irish meat entities get in Iraq the more they will expect Export Credit Insurance cover and the more will the State's exposure in this obviously risky market be increased. Another obvious factor is the consideration whether Irish entities are getting the business because other countries do not provide insurance."

Dealing with his decision to grant export insurance cover to AIBPI and Hibernia Meats Ltd, the Taoiseach, Mr Reynolds in his evidence said that:

"At this stage those were the two companies that were in the market, those were the two companies who were in the market for years, those were the two companies who developed the market and there was no sign of anybody else around in the market at that time, and indeed there was a reference, at one stage, from the Department's officials and advisers, that it was quite clear that everybody looking for Export Credit could not possibly be satisfied."

and dealing with AIBPI went on to say:

"As far as I am concerned they were a company along with Hibernia Meats, who had pioneered the market and gone out to the market and identified the market and worked and developed the market and consequently their bona fides in having a contract in Iraq was not a concern. That is the criteria I would ask, what was the track record on which they were seeking cover.

.... Those who were in the market place and had established the market and had gone on to develop the market, they had established their credentials in relation to support for that market, and, in fact, they had done it under previous administrations and if one can look back over the years, those were the same two companies that the previous Government had supported in the Iraqi market. They were not selected as somebody might suggest. Some people suggested I select those two companies. Those two companies were being supported in the market. Those two companies by their own track record, had pre-selected themselves. It was not a concern because the people in the market had a track record".

While this was undoubtedly true, particularly where Goodman International (AIBPI) were concerned, Hibernia Meats had traded in Iraq through a French company CED Viandes who negotiated the contracts there.

AIBPI had exported beef to Iraq without the benefit of the Export Credit Insurance throughout the period of the Iran/Iraq war : had a satisfactory payment record in respect of such exports : had established a reputation for the supply of beef in accordance with contract : had established contacts with the relevant Iraqi purchasing authorities and had in July 1987 negotiated for and obtained the largest contract ever ($134.5m) for the export of Irish beef to Iraq and had an unanswerable case to be allocated Export Credit Insurance cover if such cover was available in respect of at least portion of the contract.

Having pioneered and established a market in Iraq, Mr Goodman was concerned to protect it and as stated by Mr Reynolds in his evidence:

"…. you can take it from me that every single opportunity both Mr Goodman or Mr Britton, or both, took every opportunity to look for the maximum amount of export credit wherever they could get it and they believed they were entitled to it all and that nobody else was entitled to any and they made no bones about it. The same with industrial grants, they looked for the maximum and canvassed for the maximum, and I don't think any of them would deny it and that is the role they have always carried."

Mr Haughey's evidence in this regard was of a similar vein.

At his meeting with the Minister for Industry and Commerce on the 13th day of November 1987 Mr Goodman according to the Minister's evidence had complained:

"Goodman had complained, during the course of this meeting that both Halal and Agra were causing him difficulties in Iraq by cutting prices. I take the view that such competition between Irish exporters can only be of benefit to foreign consumers. It is against, in my view, the national interest and the national economic interest to allow foreign consumers the benefits of lower prices."

In 1988, Halal, Agra Trading and Taher Meats were informed that no Export Credit Insurance was available because the limit had been reached and when in October 1988, the Minister decided to increase or seek to increase the ceiling that the portion of increase attributable to beef exports to Iraq would be divided between AIBPI and Hibernia in the proportion hereinbefore referred to for the reasons given by the Taoiseach Mr Reynolds in his evidence.

It is alleged that the allocations of Export Credit Insurance cover set forth in this Report:

(1) were made for political reasons and because Mr Goodman was extremely personally close to members of the Government;

(2) constituted acts of blatant favouritism;

(3) disadvantaged rival exporters and exporters of other products;

(4) strengthened the position of the Goodman Group within the beef processing and allied trades contrary to the interests of farmers and employees and of exporters in other sectors.

(5) that the Taoiseach Charles J Haughey caused the Minister for Industry & Commerce to cancel the allocation of Export Credit Insurance to Halal as a result of the intervention of Mr Laurence Goodman.

There is no doubt but that the allocation of Export Credit Insurance in the amounts which were allocated to AIBPI and Hibernia Meats Ltd with the consequent effect that no Export Credit Insurance cover was available to other exporters of beef to Iraq, placed other beef exporters at a considerable disadvantage when seeking to negotiate contracts for the export of beef to Iraq.

As appears from the reports of the 4th and 5th Irish-Iraqi Commission, the Iraqi authorities at all times sought a credit period in respect of such exporters starting at 12 months and finally reaching agreement on 18 months credit.

As a result of the size of the contracts involved, exporters who had not the security of a promise of Export Credit Insurance with the benefits of Short and Medium-Term Finance which was dependent thereon, were at a very considerable disadvantage in seeking to obtain such contracts.

The allocation of cover for beef exports to Iraq and the amount thereof left very little available for exporters of other products to Iraq, but no evidence was adduced to establish that any non-beef exporters were deprived of Export Credit Insurance in respect of non-beef exports to Iraq.

The cover and assurances in respect thereof, given by the Minister for Industry & Commerce to AIBPI undoubtedly strengthened the position already established by them in the market, and the failure to grant similar cover or assurances of cover to other potential beef exporters from Ireland undoubtedly placed them at a disadvantage and had the effect of further strengthening and protecting the interests of the Goodman Group in Iraq.

The Tribunal does not suggest or seek to imply that this was the intention or motive of the Minister for Industry and Commerce in making these decisions but rather was the effect of such decisions.

Before receiving an offer of or commitment to grant Export Credit Insurance, neither AIBPI nor Hibernia Meats Ltd (who at all times were partially fulfilling contracts on behalf of CED Viandes) were required by the Minister for Industry and Commerce to produce confirmation of an executed contract for the sale of beef to the Iraqi authorities whereas any other beef exporters were so required.

The Minister for Industry and Commerce's decision in this regard was made on the basis of what he described as "the track record" of these companies: they had previously shown their capacity to negotiate and fulfil contracts in that market whereas the other companies had not and this is particularly established in the case of AIBPI.

This undoubtedly gave an advantage to AIBPI and Hibernia Meats Ltd (who negotiated through their parent company, CED Viandes) who were able to negotiate and conclude their contracts with the Iraqi authorities in the reasonable expectation that they would be granted such Export Credit Insurance as would be available, whereas other companies, such as Halal, Agra Trading Limited, Taher Meats Ltd and other beef exporting companies would be expected to enter into contracts and assume the risks inherent in the fulfilment of such contracts without any guarantee other than that their applications for Export Credit Insurance would be considered.

It was alleged that these decisions were made for political reasons and because Mr Goodman was extremely personally close to members of the Government.

There is no evidence to suggest that either the Taoiseach at the time or the Minister for Industry & Commerce at the time was personally close to Mr Goodman or that Mr Goodman had any political associations with either of them or the Party that they represented.

Because of the position of Mr Goodman in the agricultural life of the country and because of the obvious concerns of the Taoiseach and the Minister for Industry & Commerce to develop the agri-food sector of the economy and exports of value added products, leading to job creation there is no doubt but that Mr Goodman had reasonably ready access to members of the Government, including the Taoiseach and the Minister for Industry & Commerce for the purpose of discussing his plans for the development of his companies and his exports. It is clear that he had similar access to the previous Taoiseach, Mr Fitzgerald and members of his Government.

Mr Goodman at all times availed of such access for the purpose of the development of his company and its exports to Iraq and pressed for the introduction of Export Credit Insurance and the grant of Insurance cover' in respect of his exports to Iraq and at all times, as stated by the Taoiseach and the Minister for Industry & Commerce, argued the case that his company, having developed the market should be entitled to the full support of the Government and that such Export Credit Insurance as was available should be granted to his companies and not to competitors, particularly those whom he stated to have been involved in price-cutting.

The views of the Goodman Group in this regard were clearly expressed in the paragraph excised from the controversial letter dated the 27th day of November 1987 at the request of Mr Timbs and which paragraph, representing, as it does, the clear view of the Group, warrants repetition:—

> "From a marketing perspective, it is imperative that the Iraqis see a united front from the sellers of Irish beef in order to preserve the price premium now clearly established. The Brazilian exporters openly compete with one another in Iraq and this fact has been exploited in full by the Iraqis as is evidenced by the successive reductions in selling price accepted by the Brazilians in recent tenders. For Ireland, a single voice is an essential marketing tool to prevent such an occurrence. Because of our history in the market, AIBP should be that voice and I would therefore request that your Department reject sundry applications for credit from various Irish suppliers in order to prevent a repetition of the Brazilian experience."

It would appear that the Minister for Industry & Commerce accepted the arguments put before him by Mr Goodman and without any independent appraisal but based on his experience, formed the view that it was against the national interest and the national economic interest to allow foreign consumers the benefit of lower prices which he feared would happen if Export Credit Insurance were granted to beef exporters other than AIBPI and Hibernia Meats Ltd and decided that Export Credit Insurance cover should only be granted to these two companies.

In forming this view, he considered that he was dealing with commercial beef, as already defined, and that price-cutting, if it existed, could have affected the price paid for cattle on the Irish market and lessened the benefit to the Irish economy. However 84% of the beef exported by AIBPI during 1987 and 1988 and 75% of the beef exported by Hibernia Meats Ltd consisted of beef purchased from Intervention Stock. This beef had been processed some considerable time before, the suppliers had been paid, the beef processed and sold into Intervention. The purchase of such beef and its export to Iraq conferred very little benefit to the Irish economy and the export of beef sourced outside the State (38% of the beef exported by AIBPI, and 18% of the beef exported by Hibernia Meats Ltd) conferred none.

While the decisions made by the Minister for Industry and Commerce with regard to the allocation of Export Credit Insurance in respect of exports to Iraq in 1987 and 1988 undoubtedly favoured AIBPI and Hibernia Meats Ltd, in the sense that they were the beneficiaries of such decisions, the decisions were made by him having regard to his conception of the requirements of the national interest and there is no evidence to suggest that his decisions were in any way based on improper motives, either political or personal.

The Tribunal has set forth all the relevant evidence with regard to the re-introduction of the Scheme of Export Credit Insurance in respect of exports to Iraq, the increase in the ceiling in respect thereof made by the Government Decision on the 8th day of September 1987, the allocations of cover made subsequent to that decision, the refusals to grant applications for cover under the Scheme, and reasons given for such refusals, the increase in the statutory ceiling from £300m to £500m by virtue of the provisions of the Insurance (Export Guarantees) Act 1988 enacted on the 5th day of July 1988, the subsequent increase in the ceiling to £250m in respect of exports to Iraq agreed between the Minister for Industry and Commerce and the Minister for Finance pursuant to the Government decision of the 25th October 1988 and the proposed allocation of the cover thereby granted and the reasons given by the Taoiseach and the then Minister for Industry and Commerce, Albert Reynolds TD for his decisions in regard thereto.

The Tribunal has set forth in detail the facts in relation to each of the allegations made in Dáil Eireann with regard to the administration of the Scheme and, the alleged abuses thereof, and the effect of the decisions made with regard to the administration of the Scheme.

There is no evidence to substantiate in any way the allegation made that the Taoiseach Charles J Haughey TD caused the then Minister for Industry and Commerce, Albert Reynolds TD to cancel the allocation of Export Credit Insurance to Halal. This allocation was withdrawn by Mr Reynolds TD for the reasons given by him in evidence and there is no evidence of any intervention by the then Taoiseach, Charles J Haughey TD in this matter.

The decisions made by the Minister for Industry and Commerce, Mr Reynolds TD, to:—

(i) Re-introduce Export Credit Insurance in April 1987 in respect of exports to Iraq on a restricted basis, subject to a limit of £45m. and to stringent conditions as outlined in this Report;

(ii) Seek the Governments approval to increase the ceiling on insured exports to Iraq from £70m. to £150m, which approval was granted by decision of the Government made on the 9th day of September 1987; and;

(iii) seek the Government's approval to increase the ceiling for insured exports to Iraq from £150m. to £270m. in October 1988, which ceiling was ultimately agreed between the Minister for Industry and Commerce and the Minister fro Finance as a result of the Government's decision made on the 25th day of October, 1988, in the sum of £250m.

were made by him against the professional advice available to him, which advice is set forth in detail in the course of this Report.

The Taoiseach, Albert Reynolds, TD, freely acknowledged, in the course of his evidence that such advice was available to him, was considered and disregarded by him on the basis that the criteria involved in the No. 2 account business, to which all these decisions related, was whether cover should be granted in the National Interest and not solely on a commercial basis and has given in evidence, as outlined in this Report, the factors which he took into account in determining the requirements of the national interest in relation to the decisions made by him.

The decision to increase, or authorise the increase in, the ceilings for insured exports to Iraq were made by the Government on the 8th day of September 1987 and on the 25th day of October 1988 and decisions with regard to the allocation or intended allocation of cover within such ceiling were made by the Minister for Industry and Commerce.

The necessity for the decision to increase the ceiling in respect of insured exports to Iraq was to enable consideration to be given to the application dated the 31st August 1987 from Goodman International for Export Credit Insurance for the supply of beef to Iraq with a total value of $134.5m.

The necessity for the request to increase the ceiling on the 25th October 1988 was to enable consideration to be given to the applications made by AIBPI for Export Credit Insurance for the supply of beef during 1988/'89 on contract value at $325m. and by Hibernia Meats Ltd on two contracts valued at $72m. and £10 respectively.

AIBPI had exported beef to Iraq without the benefit of the Export Credit Insurance throughout the period of the Iran/Iraq war : had a satisfactory payment record in respect of such exports : had established a reputation for the supply of beef in accordance with contract : had established contacts with the relevant Iraqi purchasing authorities and had in July 1987 negotiated for and obtained the largest contract ever ($134.5m) for the export of Irish beef to Iraq and had an unanswerable case to be considered for allocation of Export Credit Insurance cover if such cover was available in respect of at least portion of the contract.

The basis for these decisions was that they were in the "national interest" and the determination of the requirements of the national interest in these matters is a matter for the Government and the Minister for Industry and Commerce.

Section 2 of the Insurance Act 1953 as amended provides that:—

> "(1)(*a*) For the purposes of encouraging the exportation of goods and the provision of such services as are specified from time to time by order made by the Minister, the Minister, with the consent of the Minister for Finance, may make arrangements for giving to, or for the benefit of, persons carrying on a business or profession in the State guarantees in connection with the export, manufacture, treatment or distribution of goods, the provision of services or any other matter which appears to the Minister conducive to that purpose."

The purpose of providing Export Credit Insurance and other guarantees in connection therewith is clearly stated to be for the purposes of encouraging the exportation of goods and the provision of such services as specified from time to time by order of the Minister.

The encouragement of exports is clearly public policy within this State.

The amount available for export credit insurance was by virtue of the terms of the statutes limited and as consequence of such limitation, choices undoubtedly have to be made between different products, different destinations and between particular applicants within these categories. In making these choices strict criteria with regard to the economic benefit to the Exchequer should be applied.

By virtue of the terms of the judgement of the Supreme Court in the case of *the Attorney General* -v- *the Sole Member of the Tribunal* the Tribunal was precluded from inquiring into and reporting on the factors which influenced the Government in reaching its decision to increase such ceilings.

In the course of his judgment in that case the Chief Justice stated:

> "I would, therefore, conclude that the claim for confidentiality of the contents and details of discussions at meetings of the Government, made by the Attorney General in relation to the inquiry of this Tribunal is a valid claim. It extends to discussions and to their contents, but it does not, of course, extend to the decisions made and the documentary evidence of them, whether they are classified as formal or informal decisions. It is a constitutional right which, in my view, goes to the fundamental machinery of government, and is, therefore, not capable of being waived by any individual member of a government, nor in my view, are the details and contents of discussions at meetings of the Government capable of being made public, for the purpose of this Inquiry, by a decision of any succeeding Government."

Recommendation

The Tribunal recommends that:

(I) in regard to that portion of the Scheme which is operated by the Minister in the "national interest" that the Minister should make arrangements with the Minister for Finance for the

(*a*) establishment of procedures to govern the manner in which applications for such insurance should be made, specifying in particular:

(i) whether such applications should be made to the Insurance Corporation of Ireland, or other duly authorised Agent of the Minister, or to the Department of Industry and Commerce or other Department responsible under the Insurance Act 1955-1988 for the administration of the Scheme;

(ii) the information which should be contained in the application with regard to the nature and source of the product being exported in respect of which insurance cover is sought, the size of the contracts, the number of jobs involved and the importance of the contract to the applicant;

(iii) the conditions upon which such insurance would be granted dealing in particular with the extent of cover, the premium to be charged, the period of cover and the claims waiting period;

(iv) the criteria to be applied in the consideration of such applications, including the terms of economic benefit, the apportionment between export destinations, the contributions to the Exchequer, and the risk in the regard to repayment;

(*b*) all necessary information with regard to the foregoing should be available and made available to all potential exporters.

(*c*) that in the event of an application for such guarantees and insurance being refused on any ground, other than lack of availability of such insurance, the applicant should be notified of the reason for such refusal and be afforded the opportunity of making submissions to the Minister in regard to such refusal and the grounds therefor.

(*d*) that in the case of the export of food and dairy products the views of the Minister for Agriculture and Food be obtained.

(II) Having regard to the potential liability on the Exchequer if default in payment is made by the purchaser, an allocation of cover for an amount in excess of £3m should only be made by the Minister, with the specific consent of the Minister for Finance.

CHAPTER SEVEN

Industrial Development Authority

On the 12th day of June 1987 the Authority considered a proposal by Goodman International Limited to undertake a Five Year Development Plan 1987 — 1992 in respect of its Beef Operations in Ireland.

Having considered such proposal, the Authority recommended to the Government that the following facilities be approved for the project:—

(i) For Phase 1 (added value projects) a New Industry Grant of £16.77m. towards the cost of eligible fixed assets (at a number of locations) estimated at £80.5m. or 20.833% of approved eligible expenditure, whichever is the lesser.

(ii) For Phase 11 (expansion projects) a New Industry Grant of £8.23m. towards the cost of eligible fixed assets (at a number of locations) estimated at £39.5m. or 20.833% of approved eligible expenditure whichever is the lesser.

(iii) The purchase by the IDA of £5 million of redeemable preference shares (£2.5 m. to be purchased in year 1 of the project and £2.5m. to be purchased in year 2) to be redeemed by the company in equal amounts of £1million at the end of years 6 to 10 inclusive. The timing of the payment of the above amounts to be subject to the IDA Natural Resources Division being satisfied with the company's investment proposals for the immediately following period.

The Governments permission for such expenditure was necessary because of the provisions of Section 34 of the Industrial Development Act, 1986.

The approval sought was expressed to be subject to the following conditions:—

(*a*) Normal grant conditions including Grant Payments Department approval of fixed asset expenditure and the environmental aspects of the project.

(*b*) (i) Goodman International Ltd. proceeding with the three new Phase 1 plants only to the extent that the IDA in consultation with the Department of Agriculture is satisfied that the national herd will increase by up to an additional 150,000 cattle and up to an additional 250,000 sheep.

or

(ii) Goodman International Ltd. proceeding with Phase 11 on the basis of alternative proposals for securing the necessary raw material supply in a manner which would;

—not adversely affect raw material supplies for other existing processors

—and would result in additional added value to the satisfaction of the Authority.

(*c*) The cancellation of the unpaid grant balances previously approved for the Goodman International Limited companies listed below:

Year Approved	Grant Programme	Grant Approval	Grant Paid	Balance to be Cancelled
Anglo Irish (Bagenalstown)				
1972	Re-Equipment	31,300	30,300	1,000
1972	Training	8,500	2,988	5,512
1973	New Industry	228,000	63,140	164,860
Anglo Irish (Ravensdale)				
1981	Training	27,285	9,219	18,066
1976	New Industry	1,754,770	Nil	1,754,770
1974	Re-Equipment	128,750	98,400	30,350
Anglo Irish (Ardee)				
1978	New Industry	3,600.000	Ni1	3,600.000
Anglo Irish (Cahir)				
1975	Re-Equipment	126,000	125,000	1,000
1981	Training	154,350	22,628	131,722
1983	Research & Development	27,000	Nil	27,000

(*d*) A performance clause in standard form relating to jobs as outlined below included in the Grant Agreement together with a clawback clause which will provide that grants paid in year 1 — 5 will be repayable at the end of year 5 in proportion to the failure to achieve job targets and a similar clawback clause to operate at the end of year 8 in respect of years 6 — 8.

The Authority agreed that the annual job increase from a base of 783 would be an appropriate measure of performance on the basis that the Irish economy expenditures would develop in line with the growth in jobs: the relevant jobs to exclude those having existed in the previous 12 months in facilities taken over or replaced.

Performance Clause — Targetted Performance

	Year Number							
	Base	1	2	3	4	5	6	7
*Cumulation Permanent Jobs	783	833	913	1025	1148	1273	1411	1447
*Including Cumulative Additional for:								
—Phase I		40	90	162	215	270	341	344
—Phase II		10	40	80	150	220	287	320
Review dates 31 December	1988	1989	1990	1991	1992	1993	1994	
Grant Payments (Phase I & II)	£5m	£5m	£5m	£5m	£5m			

The Authority noted that the figures are based on the company investing £24m. in fixed assets each year and are subject to change depending on the progress of the project provided that any increase in grant payment will be matched by a pro rata increase in the targeted job figure.

The Authority noted that the Irish economy expenditure figures are expected to be as set out below:

Irish Economy Expenditure Figures: (Phases 1 & 11)

	Year Number							
	Base	1	2	3	4	5	6	7
IEE Build-up (£000)	31	3078	8331	14954	23295	31293	38165	38859
Incremental (IEE £000)	31	3047	5253	6623	8341	7998	6872	694

(*e*) The general development clause in the Grant Agreement providing for Irish economy expenditure to be substantially in line with the company's projections as set out above.

(*f*) A formal review of the company's overall development plan to be carried out each year by Natural Resources Division such review to take account of the company's performance against its projections and its overall progress towards full implementation of the development plan, as required for the purposes of the general development clause in the Grant Agreement.

(iv) Permission for appropriate re-allocation of grants between various locations as deemed necessary by Natural Resources Division of IDA to enable the company to carry out the project.

2. Recommends that the Government note that the financing of the project is based on the assumption of the availability of a loan facility of £120m. under the "Swap" currency Section 84 scheme (as available in line with current Central Bank and Revenue Commissioners regulation) and on the assumption that no liability to Capital Gains Tax in respect of such loan facility would arise in this case.

3. Recommends that the Government note that the cost to the Irish Exchequer of the S.84 Swap facility is likely to be less than £4m. per annum over the life of the loan. This would increase the cost per job of the project from £45,180 on the basis of the New Industry Grants and Preference Shares as at (i), (ii) and (iii) above to approximately £90,000 (assuming 66% Irish sourced S.84 facilities).

4. Recommends that the Government be asked to note that:

(*a*) the Authority will require that the Grant Agreement provide that Goodman International Ltd. will agree to allow IDA to undertake an independent evaluation of the overall marketing strategy which will confirm to the satisfaction of the Authority the key marketing elements of the proposal. In the event of such evaluation not confirming the marketing strategy, the Authority would reserve the right to review its assistance for the project accordingly.

(*b*) the Authority will require that Goodman International Ltd., Goodman Holdings Ltd. and Anglo Irish Beef Processors Ltd. will be parties to the Grant Agreement.

(*c*) The Authority will require that the Grant Agreement will provide that the promoters will use their best endeavours to ensure the early transfer of boning-out operations currently carried out under contract in the U.K. to Ireland.

(*d*) The Authority will require that the Grant Agreement will provide that the Audited Annual Accounts of

—Goodman International Ltd

—and Goodman Holdings Ltd

will be provided to IDA, for the duration of the Grant Agreement.

(*e*) No Training or R&D Grants are proposed in respect of the project.

5. Noted that the Board has agreed to the issue of a letter confirming that:

   (*a*) IDA will use its best endeavours to support Goodman International Limited in the arrangement of a loan facility of £120m. under the "Swap" currency Section 84 arrangement.

   (*b*) On foot of the negotiated package, the IDA will use its offices in strong support of the company, in consultation with the Department of Agriculture, in the company's application for FEOGA grants at the maximum level.

On the 16th day of June 1987 the Government agreed to support the development and approved the financing package on the basis set out in the proposal submitted by the Authority.

By letter dated the 17th June 1987 Mr Loughrey, Assistant Secretary in the Department of Agriculture and Food wrote to the Secretary of the Authority as follows:

ATTENTION: *Mr John Kerrigan*

Dear Sir,

"I am directed by the Minister for Agriculture and Food to refer to grant and other financing proposals for a major development by Goodman International Ltd. and to inform you that a meeting held yesterday, 16 June — the Government agreed to support the development and approved the financing package on the basis set out in the proposal submitted by the Authority.

"In particular the Government approved

(1) IDA capital grants totalling £25,000,000

(2) IDA redeemable preference shares of £5,000,000 to be taken up by the Authority and repayable at £1,000,000 per annum after five years, and

(3) current swap loans under Section 84 of the Corporation Tax Act, 1976, with an estimated capitalised value of £30,000,000.

Yours faithfully

____

John Loughrey
Assistant Secretary"

On the 18th day of June 1987 a Press Conference, presided over by the Taoiseach, was held to announce the plans for AIBP to expand their meat plants around the Country.

Present thereat were The Taoiseach and other members of the Government, representatives of the I.D.A, and Larry Goodman.

While the Government had approved the plan as put forward to it by the Authority, the terms of the Grant Agreement to be entered into between the IDA and the Goodman Group still had to be agreed and were the subject of continuing negotiation.

During the course of the said negotiations certain difficulties arose, particularly with regard to the performance clause referred to at (*b*) in the proposal submitted to the Government and approved by it on the 16th day of June 1987.

The Goodman Group objected to this condition being inserted in the Agreement.

On the 1st day of March 1988 Mr Lowery, the Executive Director of the IDA, gave to the Authority an up-to-date report on the negotiations with regard to the terms of the Grant Agreement and it was noted, by the Authority, that the performance and claw-back clauses were being objected to and the Authority agreed that the "Force Majeure" could apply to these clauses but that there was no scope for otherwise easing the requirement with regard to the insertion in the Agreement of both the performance clause and the clawback clause.

On the 2nd day of March 1988, Mr Aidan Connor of the Goodman Group, wrote to the IDA confirming the Goodman position in relation to the performance and repayment clauses in the draft Grant Agreement.

Mr Lowery replied to Mr Connor by letter dated the 4th March 1988, setting forth the position of the IDA, as follows:—

> "Dear Aidan
>
> Thank you for your letter of 2 March 1988 confirming the Goodman position in relation to the performance and repayment clauses in the draft Grant Agreement.
>
> The Authority's position is as follows:—
>
> First, the job creation targets in the Grant Agreement are the main basis on which the Authority and Government approved the financial support package for the project. The year by year performance review and the grant repayment provision as set out in the draft agreement are an integral part of the Government's decision to permit the Authority to grant aid the project.
>
> Second, it is unreasonable of the company to expect that £30m. would be paid out by the IDA without any reference as to whether or not our main objective for the project is being met as the project proceeds.
>
> Third, the job targets are the targets proposed by the company itself. They substantially lag the proposed payment of IDA money as follows:—

Period Ending 31 Dec	Cumulative Job Target	Cumulative IDA Payment IR£
1988	0	10*
1989	50	15
1990	130	20
1991	242	25
1992	365	30
1993	690	—
1994	628	—
1995	664	—

*Includes £5m Preference Shares.

Fourth, at our meeting here on Monday and Tuesday last we offered the following significant concessions in an attempt to meet the specific concerns outlined by the company.

(1) the performance and repayment clauses would be covered by "Force Majeure". This affords the company a mechanism of dealing with circumstances outside of the company's control which have a direct impact on the achievement of targets.

(2) the annual review mechanism puts in place a procedure for addressing circumstances causing deviations in any part of the programme and relaying them to the Authority itself for consideration.

In addition, I indicated that a decision to delay or reclaim grants would be a matter for the Authority itself and that the Authority would consider all relevant facts before deciding on a course of action.

Fifth, we find it impossible to understand why your objections to the performance and repayment clauses surfaced within the past two weeks. These provisions were in the draft agreement issued to the company in June 1987 and remained there while the agreement was negotiated paragraph by paragraph up to the final draft in November 1987. The negotiations with the IDA were concluded by a team from Goodman International up to Deputy Chief Executive level.

Finally, the IDA remains anxious that the project should go ahead as planned. It has expended considerable resources in facilitating the timely start up of the project. As part of the "package" and of our undertakings to you we worked directly with the banks to put in place a large tranche of Swap Section 84 funds specifically to fund the developments in the programme you negotiated with us. These are scarce funds which involve an Exchequer cost. (The use of these funds for projects outside the negotiated beef programme would be contrary to your commitments to us relating to the use of those funds and would obviously be a major issue.)

The Authority is committed to the total project as considered and approved by the Government and itself last year. As already indicated the performance and repayment clauses are part of a Cabinet decision.

I hope that upon further consideration you will agree that we have done everything within reason to meet the concerns raised by the company and that we can complete the Grant Agreement immediately."

Yours sincerely

____

Martin Lowery
Executive Director.

On the 7th day of March 1988, Mr Connor replied to Mr Lowery as follows:—

"Dear Martin,

I acknowledge receipt of your letter of the 4th inst., in relation to the Grant Agreement.

Your letter demands a response but firstly, it may be helpful to outline once again the basic premises which underlies all of our negotiations.

a) Goodman will build new facilities which will bring the Irish meat industry to the leading edge of meat technology.

b) The IDA will grant aid the building programme to the extent of 20.833% of eligible capital expenditure.

c) The payback for the IDA will be the creation of 664 new jobs in the company as a direct result of the capital expenditure.

To answer your points more fully:—

(1) The job targets were submitted by the Company. We stand over them absolutely. These targets were also agreed and accepted by the IDA, otherwise you would not have grant aided the project. We are confident that by the project's end in 1995 we will have lived up to our promises.

(2) The creation of new jobs in processing will lag behind the building programme and the expenditure by both Goodman and the IDA. This is commonsense. If we do not build the new facilities and operate them to maximum capacity, how can we create new employment?

(3) The programme calls for IR 120 million in capital expenditure over five years. By the time the IDA has committed its IR 25 million of grants, Goodman will have spent IR 95 million. Clearly, Goodman carries the greater risk.

(4) To give you greater comfort throughout the life of the project, we agreed to an annual review wherein the IDA would have access to all information it considers necessary to monitor not just this project but the entire operations of Goodman International.

(5) We also agreed to the ultimate comfort whereby we guarantee to repay grant money pro rata to any shortfall in the jobs target.

(6) Your "concessions" are neither significant nor helpful. In the case of "force majeure" we attempted to have a wide definition inserted in the agreement to cover unforseen circumstances. Your legal department emphatically rejected any attempt at such a definition insisting instead to rely on established precedent to define force majeure. This would be totally inadequate to cover our problems in this instance.

Furthermore, the Authority (under the agreement) reserves unilateral rights on all major points. There is no right granted to Goodman to dispute any decision of the Authority except under the lengthy arbitration procedures. All this would lead to would be more expensive delays which is precisely what we are seeking to avoid in the first instance.

(7) The very reason for not signing the draft agreement is that we are unhappy with its contents. At no stage did we acknowledge it as a "final" draft.

(8) You have not yet fulfilled your obligations in relation to Section 84 financing. Only about 50% of the total IR 120m. has been committed by the banks at this stage.

In summary, we remain ready and able to undertake the project and deliver on the targets. Our track record in the industry and our history of substantial investment in Ireland over 26 years proves our capability beyond doubt. What we are asking for is the IDA to agree to let us get on with the job with no undue delay or interference. If we fail to deliver on our promises, we will pay back any money to which we are not entitled. Nothing could be simpler or fairer to both parties.

We are prepared to sign the agreement as soon as this small change is incorporated in it. That decision rests with you.

Yours sincerely

AIDAN CONNOR
Deputy Chief Executive — International

The Tribunal has considered it necessary to print these letters in full, because they illustrate the point at which negotiations had been reached with regard to the terms of the Grant Agreement and illustrate, in particular, the clause which prevented agreement being reached by the parties as of the 7th of March 1988.

On the 8th day of March 1988, Mr O'hUiginn, Secretary to the Department of An Taoiseach, telephoned Mr Lowery in connection with the Goodman project.

He indicated that he was aware of the difficulties which had arisen in regard to the Grant Agreement with Goodman and he wondered why the IDA was taking such a hard line in requiring both :—

(*a*) Performance Clause;

(*b*) Clawback Clause.

He suggested that the IDA had sufficient protection if the payment of money were linked to fixed capital investment by Goodman on a year by year basis and that the "clawback" clause would come into play later on if the expected jobs had not been achieved.

He further suggested that payment on foot of fixed capital investment was the more normal basis for payment of IDA grants.

He was informed by Mr Lowery that the IDA had recently introduced the practice of linking payment, not only to fixed capital investment but also to actual job creation on a yearly basis.

Mr Lowery informed Mr O'hUiginn that the Authority could not change these conditions as they were an integral part of the Government's decision to permit the incentive package.

At Mr O'hUiginn's request, Mr Lowery, faxed him copies of the exchange of letters with the Goodman Group which set out the respective positions. This was done.

The Cabinet was then in session and Mr O'hUiginn, on his own initiative, caused the following memorandum to be sent to the Taoiseach in Cabinet.

"Subject: Goodman Project

"The IDA position is that they consider the annual job performance targets to be essential protection for their investment.

"On the other hand, they agree

(*a*) that the clawback provision which is covered by a Goodman International Company guarantee ultimately protects their money.

(*b*) that the draft agreement links that £5m. annual grants to annual investment in fixed assets of £24m. Goodman still accepts that the grant should be at the agreed rate of 20.833% of expenditure incurred.

Goodman, however, wants to change to a single 7 year clawback review from the 5 year review covering years 1-5 and the 8 year review covering years 6-8 which the original Government decision envisaged.

I would suggest that the Government could decide to relax the annual job targets while insisting that the overall target of 664 new permanent jobs be adhered to and that the original clawback reviews in the 5th and 8th years be retained.

Goodman might also be asked to give revised *non-binding* annual job targets which must add up to the original total of 664 which is basically what the Government decision envisaged.

I attach the most recent correspondence on the issue. The basic stance of Goodman is that withholding grants for the *building programme* will disrupt the project. They regard the project as an entirety and the clawback will protect the State investment if the job targets are not achieved."

Padraig O hUiginn
8th March, 1988

On receipt of this memorandum the Government decided that the Grant Agreement be amended to take account of the following:—

(1) the overall job targets were the essential job targets to be attained;

(2) in the event of these overall targets not being attained, the clawback provision would operate after the fifth and eighth years;

(3) the annual industrial grants should be related to the annual expenditure by the company in fixed assets, based on the principle that the grants should represent 20.833 per cent of the expenditure incurred; and

(4) Goodman International Limited would make its best endeavours to attain specified annual job targets consistent with the overall job targets to be attained.

This had the effect of removing the performance clause from the said agreement.

While the Government had so decided, Mr John Donlon, Secretary to the Department of Industry & Commerce, contacted Mr O'hUiginn, Secretary to the Department of An Taoiseach and informed him that the Government had no power to amend the Grant Agreement, that it was a matter for the IDA

After discussion with Mr O'hUiginn, Mr M. Nally, the Secretary to the Government amended the notification of the decision as follows:—

> "8 Marta, 1988
>
> An Rúnaí Príobháideach
> An tAire Talmhaíochta agus Bia
>
> I am to refer to the decision S. 25217 dated 16 June, 1987 concerning the provision of industrial grants to Goodman International Limited for the major development of its Irish meat operations and to inform you that, at a meeting held today, the Government decided that the decision should be interpreted as follows:—
>
> (1) the overall job targets were the essential job targets to be attained;
>
> (2) in the event of these overall targets not being attained the clawback provision would operate after the fifth and eight years;
>
> (3) the annual industrial grants should be related to the annual expenditure by the company in fixed assets based on the principle that the grants should represent 20.833 per cent of the expenditure incurred; and
>
> (4) Goodman International Limited would make its best endeavours to attain specified annual job targets consistent with the overall job targets to be attained.
>
> Dermot Nally
> Rúnaí Rialtais
>
> ---
>
> An Rúnaí Príobháideach
> An tAire Airgeadais
>
> Mar eolas don Aire."

It was on the basis of this communication that Mr John Loughrey, Assistant Secretary of the Department of Agriculture and Food wrote to the IDA on the 11th day of March 1988.

On the 15th day of March 1988, a special meeting of "the Authority" was held for the purpose of considering the decision of the Government as conveyed to them by the letter dated the 11th day of March from Mr Loughrey.

The Authority had obtained legal advice on the interpretation of the provisions of Section 35 of the Industrial Development Act, 1986 and were satisfied that the Government had the power to make the decision conveyed in the letter. They felt bound by it but being satisfied that the "clawback clause" provided adequate protection for the investment in the grant package they agreed to the deletion of the "performance" clause from the draft Agreement.

Their decision, in this regard, removed the last remaining obstacle to the signing of the Agreement which was formally signed on the 22nd day of March 1988.

In the course of the debate on the Taoiseach's motion to adjourn the Dáil on the 9th day of March 1989, Deputy Desmond said:—

> "Then in June 1987, the Government decided in an enormous P.R. exercise and, I believe, against the wishes of the IDA, to give £25million to Laurence Goodman including a plant in Tuam".

On the same occasion, Deputy MacGiolla said:—

> "This affair has also raised important questions about the extent to which one company, or indeed one person, can be allowed to control such a large part of one of our most important industries. The Goodman organisation is the very hub around which Fianna Fáil seem to have built their whole food development policy — beef, dairying and now sugar confectionery. Goodman has got the public backing of the Government as has been pointed out by Deputy O'Malley, of over £200 million two years ago, that is, in 1987. How much money he actually got we do not know, but certainly backing for £200 million or £250 million by the State gives him tremendous credit in raising finance wherever he wishes to go. It is understood that he has got IDA grants of up to probably £25 million. There is also some evidence which has been brought to my attention to suggest that the Taoiseach himself directly intervened with the IDA in some of these grants to get the IDA to drop their insistence on what is called the performance clause. The performance clause is required by the IDA in their contracts when issuing grants and this performance clause was dropped in the case of grants to the Goodman company. I do not know why that should be so. This is a hugh concern. It accounts for more that 42 per cent of the total beef exports from this county. Alone they now probably account for up to 6 per cent of our gross national product. They seem intent on gobbling up more of the food industry."

On the 24th day of May 1991 in the course of the motion establishing this Tribunal Deputy Bruton said:—

> "The first one was — and I stand over it — that when Mr Goodman was applying for assistance for his major Five-Year Plan for the beef industry, rather than that this examination be undertaken by the Industrial Development Authority in a normal unhurried way where such a large commitment of public funds would be examined carefully and dispassionately that particular grant package was rushed through by the IDA under political pressure and also rushed through the Department of Finance under similar political pressure, and the responsible authorities in the IDA were not able to assess that application properly because of political pressure.
>
> Furthermore, I assert, the fact that that particular application was approved in that way with the Taoiseach's own personal intervention made it more difficult for other competing firms in the beef industry to apply for funds because, given that there is only a limited number of cattle available for slaughtering, if one firm has been given the go-ahead for expansion along particular lines that, more or less, precludes the

creation of capacity in other firms. I contend that, whereas the IDA should have been allowed to assess that dispassionately, looking at not only all the applications before them but all the potential applications that they might receive in the future, as a result of the Taoiseach's almost childlike anxiety to be associated with good news, to appear at a press conference, this application was rushed through the IDA and rushed through the Department of Finance, and that the normal procedures, the normal controls which in my time as Minister for Industry and Commerce were always respected were not respected in this case."

In the course of the debate on the Companies (Amendment Bill 1990) Deputy Barrett said:—

"It is quite obvious that enormous political pressure from the highest possible level was brought to bear on the Goodman Group and the IDA to announce an expansion programme, the details of which has not been worked out, and which was launched in such a dramatic fashion solely as a PR exercise for the Taoiseach and his Government at the time."

In the events which have happened, the plan was never implemented but because of the allegations made in Dáil Eireann the Tribunal was obliged to inquire into the circumstances in which the plan was proposed, the manner in which it was approved by the Authority and the Government and ascertain whether there was any improper pressure imposed by the Government or any member thereof on either the Industrial Development Board or the Industrial Development Authority in the exercise of their statutory functions.

The gravamen of the allegations is that:—

"(i) the Authority did not and were not able to properly assess and evaluate the merits of the Five Year Development Plan submitted by Goodman International Limited;

(ii) the Board and the Authority were subjected to political pressure of such a degree that they were unable to examine carefully and dispassionately the grant package which involved a large commitment of public funds;

(iii) the normal controls were not respected and the Department of Finance was not afforded time to deal with the financial implications of the package;

(iv) the Taoiseach himself directly intervened to oblige "the Authority" to delete the "performance" clause from the Grant Agreement;

(v) the Goodman organisation was the hub around which Fianna Fáil seems to have developed their whole food development policy;

(vi) because of such political pressure, the Authority was precluded from assessing the implications of the plan on the industry generally having regard to the limited number of the cattle herd available for slaughtering;

(vii) the political pressure was brought to bear to rush the plan through the appropriate authorities so that it could be announced in a dramatic fashion as a Public Relations exercise for the Taoiseach and his Government at the time against the wishes of the Authority;

(viii) Against the wishes of the IDA the Government decided to give £25m. to Larry Goodman."

In addition, Deputy Rabbitte made the statement already referred to, that the Government, in the Finance Act, made a special arrangement to enable Mr Goodman to avail of high coupon Section 84 finance.

The Tribunal considered it necessary to examine the roles of the Industrial Development Authority and the Government in the grant making process under the provisions of the Industrial Development Act 1986.
The Industrial Development Authority was continued in being by the terms of Section 10 (1) of the Industrial Development Act 1986.

Section 10 (2) of the said Act provides that the Authority in the exercise of its powers and functions shall be responsible to the Minister, defined in the Act as "the Minister for Industry and Commerce".

Section 11 of the Act provides that subject to the provisions of the Act, the Authority shall be an autonomous body with the functions set forth in that Section.

These functions include the following:—

"(*a*) to act under the Minister as a body having national responsibility for the implementation of industrial development policies;

(*b*) to provide and administer such grants and other financial facilities for industry as may be authorised by the Oireachtas to provide and to administer;

(*c*) to initiate proposals and schemes for submission to the Minister for the creation and development of industry and the provision and maintenance of industrial employment;

(*d*) to provide, develop, construct, alter, adapt, maintain and administer industrial estates and factory buildings together with the associated facilities of such estates and buildings;

(*e*) to foster the national objective of regional industrial development;

(*f*) to survey possibilities of further industrial development and advise the Minister thereon;

(*g*) to advise the Minister on steps necessary and desirable for establishing new industry and for the expansion and modernisation of existing industry;

(*h*) to give on request advice and guidance to persons contemplating starting new industry or expanding existing industry;

Section 11(3) of the Act provides:—

"the Authority shall, in the exercise of its functions, act in accordance with policies set out for it from time to time by the Minister".

Section 13(1) of the Act provides that:—

> "The Minister may give the Authority such general policy directives as he considers appropriate having regard to the provisions of this Act".

This power is restricted to general policy directives because Section 13(2) of the Act provides that:—

> "a directive under subsection (1) shall not apply to any individual industrial undertaking or to giving preference to one area over others in regard to the location of an industrial undertaking otherwise than as part of a general review of industrial policy for the country as a whole indicated in the directive"

The Oireachtas desired to be informed of such directives and the manner in which they were implemented because Section 13(3) of the Act provided that:—

> "The Minister shall cause any directive given by him under subsection (1) to be laid before each House of the Oireachtas within twenty one days after it has been so given";

and

Section 13(4) of the Act provided that:

> "The Authority shall comply with any directive given to it under this section and shall set out the directive in its Annual Report and shall include in its Annual Report an account of the actions which "it has undertaken to give effect to the directive".

Section 14 of the Act deals with the financing of the workings of the Authority and provides that:

> "(i) In each financial year there may be paid by the Minister to the Authority out of money provided by the Oireachtas grants of such amounts as the Minister, with the consent of the Minister for Finance, may sanction to enable the Authority—
>
> (*a*) to meet its administration and general expenses, and
>
> (*b*) to discharge the obligations or liabilities incurred by the Authority under this Act or any repealed enactment or otherwise."

The aggregate amount of grants was limited to £700,000,000 and the aggregate amount of grants to enable it to meets its obligations under guarantees was limited to £125,000,000.

Section 21 to 32 of the Act deals with grant making powers of the Authority and Section 24 to 31 restrict the amount of such grants as may be made by the Authority without the prior permission of the Government.

With regard to grants made by the Authority under Sections 21, 22, 23 and 32, Section 34 of the Act provided that:—

> "Without the prior permission of the Government, the total amount of money granted in respect of a particular industrial undertaking shall not exceed £2,500,000."

Section 35 of the Act provides that:—

> "Where, under any of the preceding sections of this Part, the permission of the Government is required for the making of a grant or loan guarantee or the purchase of shares, the Government may, in lieu of granting such permission, grant permission to the Authority for the expenditure of a lower amount in respect of such grant, guarantee or purchase or may grant permission subject to such conditions as the Government may specify".

The Tribunal has set out the relevant provisions of the Act under the roles of the Authority, the Minister and the Government in the provision of grants and other financial facilities for industry as may be authorised by the Oireachtas.

By virtue of the foregoing provisions, the Oireachtas has provided that:—

(i) the Authority is an autonomous body though subject to the provisions of the Act;

(ii) one of its functions is to provide and administer such grants and other financial facilities as it may be authorised by the Oireachtas to provide and to administer;

(iii) the grants and other financial facilities which it may provide in accordance with the authority of the Oireachtas are those specified in detail in Sections 21 to 32 inclusive of the Act;

(iv) the only involvement of the Government in the grant making process is if the Authority requires permission to exceed the amounts specified in the different Sections of the Act in relation to grants, loan guarantees or the purchase of Shares;

(v) in such circumstances the Government may in lieu of granting the permission sought, grant permission to the Authority for the expenditure of a lower amount in respect of such grant, guarantee or purchase or may grant permission subject to such conditions as the Government may specify.

While the Authority is obliged, in the exercise of its functions to act in accordance with policies set out from time to time by the Minister, these policy directives which the Minister is empowered to give must relate to industrial policy for the country as a whole and cannot apply to any individual industrial undertaking.

In summary, the role of the Minister in the scheme envisaged by the Oireachtas is to lay down industrial policy for the country as a whole: to issue directives in regard to such policy to the Authority and to lay such directives before each House of the Oireachtas.

The role of the Authority is to administer the scheme subject to such directives if any in accordance with the scheme.

The role of the Government is as set out in Section 35 of the Act and only arises when the Authority seeks its permission to make a grant in excess of the amounts specified in the Act.

This summary of the manner in which the Oireachtas provided that industrial grants should be made may be regarded as extremely legalistic but the Tribunal considers that this is the clear legal position and that the whole purpose of the Act was to ensure that the Authority was the sole body with the power to make grants or provide other financial facilities in respect of any individual undertaking and that the Government had no role in regard thereto unless or until the Authority sought its permission to make a grant or provide financial facilities in excess of the specific amounts set forth in the different sections of the Act, in which circumstances the provisions of Section 35 of the Act applied.

The Oireachtas was concerned to ensure that the making of grants to individual undertakings was solely the prerogative of the Authority exercising its responsibilities under and in accordance with the provisions of the Act, and that the roles of the Minister and the Government in regard thereto were subject to the limitations set forth in the Act.

The Authority consists of nine Members, nominated by the Minister of Industry & Commerce with the consent of the Minister for Finance.

In the course of his evidence before this Tribunal Mr Haughey describes the IDA as "an instrument of Government" and that:

> "Basically it was set up by Government to achieve certain objectives and to suggest it should operate in some sort of remote distant region completely divorced from Government priorities or Government policy would be absurd."

The Tribunal was concerned to show by its recital of the relevant provisions of the Industrial Development Act 1986 that the Oireachtas did not intend that the Authority should be an instrument of the Government in the sense of being subject to the Government's direction with regard to any specific grants or any particular area but should operate as an independent and autonomous body in the exercise of its functions and should not be subject to or affected by the priorities or policies of any Government save such general policy directives as are dealt with in Section 13 of the Act: which policy directives are required to be laid before each House of the Oireachtas and cannot apply to any individual undertaking otherwise than as part of a general review of industrial policy for the country as a whole.

No such directives appear to have been laid before either House of the Oireachtas since the enactment of the Industrial Development Act, 1986.

The Authority then was completely free to consider the Five Year Development Plan hereinbefore referred to independently of either the priorities or policies of the Government.

In the course of his evidence before this Tribunal, Mr Padraic White, the Managing Director of the IDA at the relevant period stated:—

> "I think it's important to say the IDA Board and Authority act very much in an independent and extremely serious way. They take their decision-making responsibility seriously and the fact that the Board and Authority are the sole initiating group for grants is taken very seriously"

He also states that:—

> "I have very, very strong views on the integrity of the IDA on its right to make its own decision and I have protected that as Chief Executive".

In connection with negotiations with regard to the Goodman development plan he states:—

> "during these several negotiations there was nothing of a political pressure that I would regard as fundamentally objectionable or that caused me undue concern and I'd like to make that clear".

and

> "that it was perfectly open to the Minister (Mr Walsh), and perfectly natural for a Ministers portfolio, if he was specifically dedicated to food to want to develop the project as expeditiously as possible and his involvement in those two meetings on the 19th May, and on the 2nd June, 1987 when the negotiations had broken down, I regarded that as perfectly normal and I would have regarded the Minister wanting to know the status of the project as perfectly normal".

In reply to a question asked by the Tribunal as to whether he would regard as improper interference with the position of the IDA an attempt by a Minister, or anybody else, to influence the Authority with regard to the terms and conditions of any particular grant he stated:—

> "In the course of dialogue in politics and political representations, where politicians make representations you know, certainly if they say we should pay more attention to that or that the project deserves a bit more support and I have no problem with that. The division is if a politician or a minister says you must give that degree of support. Almost as an instruction. I mean in a heavy handed way and that line, you know, I am very happy to say I have never seen that line being crossed in my 10 years as chief executive never, and in fact, there is a very healthy respect by politicians of the integrity of the IDA and of its Board and they're very, they have never strayed over that line and in those set of negotiations in my judgement, they did not stray over that line and there was nothing that occurred that gave me personal concern that it was being too heavy handed or somebody was acting you know, outside of the bounds of normal reasonableness to have the project expedited."

In considering the allegations made with regard to the Five Year Development Plan approved by the Authority on the 12th day of June 1987 and approved by the Government on the 16th day of June 1987 it is important to note that this plan did not suddenly emerge on the change of Government in March 1987.

For a considerable time before that the plan had been the subject of negotiations between the IDA and Goodman International. It is not necessary to set forth in detail the progress of such negotiations.

It appears however from the evidence of Mr Sean Donnelly, Executive Director of the IDA who at all relevant times worked within the Natural Resources Division of the IDA with responsibility for the role of the IDA in relation to the Meat Industry.

He was responsible for the production of "A Strategy for the Development of the Agricultural Processing Industry in Ireland" published by the IDA in June 1982 and "A Future in Food" published by IDA in December 1987. Though not published until December 1987, this latter document had been approved by the IDA in March 1987 and subsequently endorsed by the Department of Agriculture.

Mr Donnelly had negotiated with the Goodman Group over a number of years with regard to numerous investments which qualified for grants.

From the mid-eighties it had been the policy of the IDA to encourage major Irish companies to draw up comprehensive development plans for a 3-5 year period which would enable the IDA to assess the Companies' various applications for IDA assistance in the context of these development plans.

In September 1986, the Goodman Group outlined to the IDA a number of investments which they intended to undertake over the following years.

During the course of discussions it was suggested by the IDA that the Group should draw up a comprehensive Five Year Group Development Plan rather than submitting a series of individual investments in an unstructured manner for IDA support.

As stated by Mr Goodman in a letter dated the 10th day of March 1987 to Mr Donnelly:—

> "At your express request in September last year, I prepared our strategy document on the expansion of the Irish Beef Sector for the five years 1987 to 1992. This was submitted, to you, in December 1986 and reconfirmed our plans to systematically upgrade the "5th Quarter" including blood, to edible status"

The plan as submitted in December 1986 was, according to Mr Donnelly not costed and in January 1987 there were four or five long meetings between members of the IDA executive staff and the Goodman Group with regard to the details of the Plan. The outline plan submitted by the Goodman Group was discussed at IDA Board meetings in December 1986 and January 1987.

The plan was costed in the sum of £260m being as to £120m thereof in respect of capital investment and £140m in additional working capital and non grant eligible expenditure.

The Goodman Group sought a grant of 75% of the capital expenditure portions of the investment, such 75% being a combination of IDA and FEOGA grants.

In previous projects the Goodman Group had been approved combined IDA/FEOGA grant assistance, at rates at between 45% and 50%.

It was made clear to the Goodman Group that the IDA could not meet the demand for 75% grants as the EC had placed a ceiling of 50% on the combined State and EC grants in the non disadvantaged parts of the country.

In February 1987, following a recommendation from the Department of Agriculture and Food and the IDA the EEC had approved a special package involving up to 75% grants and the whole country for the restructuring of the Irish pigmeat industry.

The Goodman Group sought a similar package in respect of the beef industry.

In March 1987, there was a change of Government and Mr Haughey TD was elected Taoiseach and inter alia, Mr Ray Mac Sharry TD was appointed Minister for Finance, Mr Albert Reynolds TD was appointed Minister for Industry & Commerce, Mr Michael O'Kennedy TD was appointed Minister for Agriculture and Food and Mr Joe Walsh TD was appointed Minister for Food.

Mr Walsh stated in evidence that:—

> "My responsibility, which was given to me by the incoming Government, was to accelerate economic activity in the general food area and, to that end, the new Government created a special office of food along the lines of the office of the Revenue Commissioners or the office of Public Works, which would be an autonomous unit within the Department of Agriculture & Food and my specific brief was to bring the Irish food industry up to international standards."

He set about, straight away, to implement that particular brief and in pursuance thereof met with the Industrial Development Authority.

The Government had held it more appropriate for the Natural Resources Division of the IDA to report directly to the Department of Agriculture & Food rather than hitherto sponsoring Department which was the Department of Industry & Commerce.

Shortly after his appointment the Minister for Food met with the IDA for the purpose of being given a presentation as to their then strategy for the development of the food industry and in the course of such presentation, the proposal which had been made to the IDA by the Goodman Group in December 1986 was brought to his attention.

He was informed that it had gone through a degree of evaluation and assessment and appeared to him to fit in well with the Programme for Government as being the best way forward for the Irish beef industry.

He stated that the IDA officials were enthusiastic about it and he considered that it was a project which was worth supporting and worth accelerating.

He informed the Taoiseach of the nature of the plan and as a result thereof, the Taoiseach arranged to have a round table meeting between the Ministers concerned and representatives of the Goodman Group.

On the 9th day of April 1987, the Taoiseach, accompanied by the Minister for Finance, the Minister for Industry & Commerce, the Minister for Agriculture and Food and the Minister for Food, met with Larry Goodman and Mr Brian Britton.

The purpose of this meeting was, in the words of Mr Haughey, "to explore, assess and find out all about this project and take a decision on it."

The nature of the scheme was outlined to the group of Ministers by Mr Goodman and Mr Britton and at the conclusion of the meeting, it was considered that the project was worthwhile and was exactly in line with the thinking of the Ministers and their approach.

They were also satisfied that it was exactly in line with the IDA's Five Year Plan and Mr Walsh, the Minister for Food, was given responsibility for promoting it and seeing it through and Mr Brian Britton was nominated as the representative of the Goodman Group with whom he should liaise and deal with in the event of difficulties arising.

On the 23rd day of April, 1987, a final copy of the plan was forwarded to the Department of Agriculture & Food and to the IDA and both Mr Goodman and Mr Britton met Mr Padraic White, the Managing Director of the IDA to brief him with regard to progress.

On receipt of this plan, Mr Loughrey, at the request of the Minister for Food, prepared an Aide Memoire for Government. The Aide Memoire was very supportive of the Plan and envisaged formal approval by the IDA Board though negotiations were still at a preliminary stage and the IDA had yet to make its proposal in relation to financial assistance.

This plan was much more detailed than the plan submitted to the IDA in December 1986 which had not included any costing.

This plan gave particulars of the cost of Capital Development as being £120m which was analysed as follows:—

Cost of Sites	£5,000,000
New Building Expenditure	£24,000,000
Modification to Existing Building	£6,000,000
Plant and Equipment	£85,000,000
Total Expenditure:	£120,000,000

In addition there were to be permanent funds increased hard core working capital in the sum of £140 million making a total development plan cost of £260 million, the Goodman Group providing in addition to their share of the cost of the Capital Development costs, the entire of the increased working capital costs, viz £140m.

The Goodman Group required that 75% of the Capital Development costs of the programme (£120m) be grant aided and the exclusion of food processing companies from the scope of Section 52 of the Finance Act 1986. Though this plan had not been submitted to the Board of the IDA or the Authority at this stage, the plan was placed before the Government which met on the 26th day of April 1987. The IDA had not been informed and were not aware of the fact that the proposal was being placed before Government.

This was a special meeting of the Government, taking place on a Sunday, at which all members of the Government and all the Junior Ministers were present for the purpose of evaluating a number of projects which were considered necessary to develop the economy, in particular with regard to job creation.

At this meeting held on the 26th day of April 1987, the Government decided that:—

"(1) the Ministers for Agriculture and Food, Finance and Industry and Commerce should make every effort to bring the project to a successful conclusion, in particular, by investigating the possibilities for meeting the financing requirements of the project in part by

(i) excluding the food processing industry from the scope of section 52 of the Finance Act, 1986 and/or adjusting the proposed FEOGA financing of the project, or

(ii) reversing the abolition of depreciation allowances on gross capital costs *in toto*, and

(iii) a revision of disadvantaged areas so as to include Louth; and

(2) the Minister for Agriculture and Food should submit definitive proposals for the project to Government in the normal way as soon as possible."

This decision required the three Ministers named therein to make every effort to bring the project by Goodman International for the development of the beef industry in Ireland referred to in the said aide-memoire to a successful conclusion and in particular to investigate the possibilities of meeting the financial requirements in the manner set forth.

With regard to (1) in the said decision, a submission in regard thereto had been made by Mr Britton of Goodman International to the Minister for Finance and the exclusion referred to was effected by Section 25 of the Finance Act 1987.

Irrespective of the merits or otherwise of the said development plan or of the desirability of implementing it as quickly as possible in the interests of job creation, this decision of the Government to require the said Ministers to make every effort to bring this project, which at that time had not been considered by the Board of the IDA, to a successful conclusion would appear to be contrary to the provisions of the Industrial Development Authority Act 1986 because at that time the plan submitted by Goodman International had not been formally considered by either the IDA Board or the Authority. While there had been discussions it appears from the evidence of Mr Sean Donnelly that the plan which had been placed before and considered by the Government on the 26th day of April 1987 was the first definitive document to be considered by the Authority and this was not done until the 28th day of April 1987, two days after the Government Meeting.

In view of the role envisaged for the Government in relation to the operations of the IDA as set forth in the provisions of the Industrial Development Authority Act 1986 referred to in this Report, the actions of the Government in deciding that three cabinet ministers viz the Ministers for Agriculture and Food, Finance and Industry and Commerce should make every effort to bring the project to a successful conclusion could be interpreted as a pre-emption by the Government of the role of "the Authority".

At the meeting, held on the 28th day of April, 1987 the Authority was fully informed by Mr Martin Lowery, Executive Director of the IDA of the nature of the discussions in progress with the Goodman Group and indicated to the Authority that a package of IDA Grant Assistance of about £30m would be required to bring the plan to fruition.

During the month of May 1987 there were intensive negotiations between the IDA and the Goodman Group with regard to the financing of the Capital programme provided for in the Plan viz. £120m.

At all times the Goodman Group sought 75% of the cost of this programme by way of combined IDA and FEOGA grants.

According to Mr Donnelly, who was at all times involved in the negotiations on behalf of the IDA, the IDA decided to pitch its opening offer at £13m in grants and £10m in preference shares and in addition suggested that the Goodman Group could adequately finance the Development Plan with this package if the Group used a form of low cost borrowing known as "High Coupon Section 84" loans. (Swap Currency).

The nature of these loans have been described in that portion of this Report dealing with Section 84 loans.

As the tax implications of this loan were complex the IDA undertook to use its best endeavours to obtain for the benefit of the Goodman Group such Section 84 Finance.

Goodman Group executives met the IDA on the 4th and 7th days of May 1987 to discuss various options for reducing the cost to the Goodman Group of the plan.

Negotiations between the parties broke down on the 15th day of May 1987 because of a fundamental disagreement on the financing of the programme, formal proposals in regard to which had been put forward by the IDA on the 12th day of May 1987.

The Goodman Group was prepared to pay £30m towards the cost of the capital investment and to provide the £140m being the increased working capital and required grant and assistance in the sum of £90m.

The IDA felt that this requirement was met by their offer in the following terms:—

IDA Grant	£13m
FEOGA Grant	£20m
Preference Shares	£10m
Section 84 loans	£47m
	£90m

but this offer was rejected by the Goodman Group.

On the following day Mr Britton met the Minister for Industry and Commerce, Mr Reynolds TD and the Minister for Food, Mr Walsh TD to inform them "where we were in relation to negotiations" and that the IDA package was unacceptable.

The Goodman Group were not convinced of the suitability or the benefits of Section 84 loans. A letter dated the 14th day of May 1987 written to Mr Donnelly set forth their position as follows:—

"Dear Sean 14 May 1987

PROPOSAL FOR GRANT AID

I refer to our recent meetings on the above matter and specifically to our meeting on Tuesday 12 May 1987 at which you outlined the suggested IDA package of grant aid for this Group's capital expenditure proposals.

Regrettably, I must inform you that your package, as it is currently structured is not acceptable to us for a number of reasons which are dealt with below:—

1. The package contains an offer of considerable substantial Section 84 borrowing (£50m) at a low interest cost. This is very undesirable for our point of view because:-

   (*a*) It negatively impacts our gearing ratio throughout the project with a consequent loss of flexibility to the Group.

   (*b*) Our Group structure cannot cater for large amounts of Section 84 borrowings.

   (*c*) Given our current tax profile, the true cost to the Group of taking on board Section 84 borrowings is very high, substantially in excess of the coupon rate mentioned in your proposal.

2. The question of equity participation is not acceptable. This point was emphasised at our last meeting.

I must stress that this decision was made only after intense deliberations amongst our own staff in consultation with a number of senior partners in Stokes Kennedy Crowley & Co. In this regard, I have asked Sean Mooney, tax partner, to write and speak to you directly on the technical aspects of our rejection.

For the record, I reiterate below our requirements to go forward with this plan which is a total aid package of £90m consisting of:—

1. A capital grant of 50% in designated areas and 75% in non-designated areas averaging out to a guaranteed grant rate of 62½% which is equivalent to £75m.

2. To make up the balance of £15m we are prepared to consider a number of options such as:—

   A property lease scheme under which we would build factories. The IDA would pay and we would lease for a nominal rent over a period of years with an option to purchase (again for a nominal sum) at some future date.

   A Scheme under which the IDA would subscribe for preference shares at a premium which would carry a nil coupon rate and would be redeemable at par after a period of years.

I would ask that you consider these options between now and our meeting tomorrow morning with a view to putting forward an acceptable proposal at that meeting.

If in the meantime, you have any queries, please do not hesitate to contact me.

Yours sincerely
BRIAN BRITTON
Deputy Chief Executive — Finance."

On the same day, Mr Mooney of Stokes Kennedy Crowley wrote to Mr Donnelly as follows:—

"Dear Mr Donnelly 14 May 1987

Re: Goodman International Ltd.

We refer to the proposal that part of the grant package to our clients should take the form of low interest Section 84 funds.

From a taxation view point, such a proposal would be unsuitable for a number of reasons. These are:—

1. Our clients exporting companies do not have a requirement for substantial Section 84 funding. They are funded principally by retained profits and by interest free borrowing from non exporting companies in the group.
2. The technical restrictions on the use of Section 84 finance imposed by Section 41 of the Finance Act, 1984 make it an unsuitable form of financing for use elsewhere in the group.
3. Our clients export companies do not own fixed assets. These are located elsewhere in the group where the capital allowances can be fully utilised. Incurring capital expenditure in the export companies would "waste" the allowances.

We trust these points are clear and the writer is available to elaborate on them more fully if required.

Yours sincerely
SEAN MOONEY"

Minister Walsh had been kept fully informed of the progress of the negotiations by Mr Donnelly of the IDA either through personal contact or through Mr Loughrey the Assistant Secretary to the Department.

Further meetings took place between the IDA and the Goodman Group on the 15th day of May 1987 but no agreement was reached.

Mr Sean Donnelly informed Mr Martin Lowery of the breakdown of the negotiations.

As he suspected that Mr Goodman would approach politicians with regard to the breakdown, Mr Lowery considered it necessary to inform the Minister for Industry and Commerce and the Taoiseach's Department of the position taken by the IDA which led to the breakdown.

At this point Mr Britton stated that the Goodman side, i.e. Mr Goodman and himself, reverted to lobbying.

On the 18th day of May 1987 Mr Goodman met the Taoiseach, Charles J Haughey TD and on the 19th, Mr Goodman and Mr Britton met Minister Walsh.

In his evidence before the Tribunal, Mr Britton said that the purpose of such meetings was to lobby for their assistance "to intercede with the IDA"

Minister Walsh and the Minister for Industry and Commerce, Mr Reynolds arranged a meeting with Mr White, Mr Donnelly and Mr Breen of the IDA, which meeting was held on the 19th May 1987.

It appears from the record of this meeting, the terms of which were confirmed by Minister Walsh during his evidence that a wide range of issues were covered during this meeting.

Both Ministers expressed the Government's desire to progress the project and to look for ways in which the "gap" between the IDA offer and the Group's demand might be filled.

A number of options were considered and the nature of the incentive package was discussed in detail on the basis of a split programme of expenditure.

During the course of such discussions, the Ministers encouraged an increase in the grant package from £13m to £20m.

The meeting concluded with agreement that the Ministers would contact Goodman International and request them to meet again with the IDA to progress negotiations along the lines agreed at this meeting.

On the following day, the 20th May, negotiations resumed as a result of a contact made by Minister Walsh with Mr Britton.

The IDA increased its grant offer to £20m. The Goodman Group however maintained that the project was not commercially viable if the after-grant cost to Goodman Group exceeded £30m and again referred to the pig meat industry where combined IDA and FEOGA grant of 75% were available.

Certain matters were eventually agreed and both Mr Donnelly and Mr Britton signed a hand written document incorporating particulars of such agreement but on the following day these terms were rejected by the Goodman Group.

On the following day Mr Goodman and Mr Britton met Minister Reynolds and Minister Walsh and told them that the IDA package was not viable and that there was no point in pursuing further negotiations.

The final offer made by the IDA was:

(i) A grant of £25m;

(ii) Redeemable preference shares 5m;

(iii) Support to FEOGA grant of 30m;

(iv) Support for £120m High Coupon Section 84 loans, the reduced interest payable on such loans would have a capital value of £30m

The package was not acceptable to the Goodman Group.

Minister Walsh intervened and arranged a meeting between the parties to be held on the 2nd June 1987.

Present at this meeting were the Minister for Food, Joe Walsh TD, Mr Lowery and a Project Executive from the IDA, John Loughrey, Assistant Secretary, and Vincent Keane from Department of Agriculture and Mr Larry Goodman, Mr Brian Britton and John O'Donnell of the Goodman Group.

At the meeting the Minister initiated discussions by indicating his desire that the project should go ahead but the real discussions took place between Mr Lowery of the IDA and Mr Larry Goodman.

Mr Lowery approached these discussions on the basis that the IDA offer of £25m in Grants and £5m purchase of redeemable preference shares would not be improved upon.

He emphasised the value of the other elements of the package which involved no cost to the IDA, the FEOGA grants and the High Coupon Section 84 loans.

The savings in interest on those loans were capitalised at £30m.

At this meeting Mr Goodman sought assurances with regard to:—

(i) the availability of FEOGA grants; and,

(ii) the availability of High Coupon Section 84 funding.

Mr Lowery stated that the IDA would support:—

(i) the Goodman Group in seeking to achieve the maximum levels of FEOGA support for the project; and,

(ii) the Goodman Group in seeking to secure the High Coupon Section 84 loans, but that the IDA would not underwrite the FEOGA grants or the securing of the Section 84 loans as it would be the responsibility of the Group to negotiate the said loans with the banks.

The Minister and the Department of Agriculture officials stated that they would support the application for the maximum level of FEOGA grants.

Mr Goodman then stated that he would accept the package.

The Minister and Mr Loughrey had played an active role in the evolution of the financial package necessary to implement the plan from the time of his instructions from the Taoiseach on the 9th day of April 1987 to promote it and see it through.

Neither his involvement or that of the Minister of Industry & Commerce, Albert Reynolds, was in anyway improper and did not constitute, in any way, an interference with the statutory role of "the Authority". The Development Plan, as submitted, could not be financed out of the resources available to "the Authority": it required the benefit of FEOGA grants in respect of which the Minister for Agriculture & Food was the designated authority under EEC Regulations and extra financing by way of Section 84 High Coupon loans.

The interests of the Government and the IDA coincided in respect of this plan and the involvement of the Minister in the negotiations having regard to all the necessary components of the financial package was necessary, justified and did not amount to political pressure of any kind.

Once the agreement had been reached, a detailed appraisal of the project was prepared by Mr Breen of the IDA

The Tribunal has read this appraisal and is satisfied that it contains all the necessary detail to enable the Board and "the Authority" to make a reasonable and full assessment of the plan.

Following the agreement reached on the 2nd June 1987, preparation of the Memorandum for Government started and on the 5th day of June 1987, Mr Loughrey wrote to the Secretary of the IDA stating that his Department considered that the proposals contained in the plan were generally in line with the Government's policy for the development of the meat sector.

Mr Loughrey stated in evidence that this letter was prompted by what he described as "the measured reservations" of the Beef Division of the Department of Agriculture regarding the effect of the Plan on other companies engaged in the industry if the proposed increase in cattle numbers didn't materialise or there was an increase in slaughtering capacity.

Following a meeting with the Chairman of "the Authority", Mr McCabe, Mr White decided to call a special meeting of "the Authority" for the 12th June 1987 to consider the project.

The Board of the IDA met on the 10th day of June 1987 and considered the plan.

After consideration the Board recommended to the Authority that the Government's approval be sought for the incentive package which had been agreed with the Goodman representatives.

Immediately following the decision of the IDA Board on the 10th day June 1987, Mr Loughrey at the request of Minister Walsh, began planning a press conference for the 18th June 1987.

Mr Walsh was not aware of the agreement made between successive Ministers for Industry and Commerce, Noonan and Reynolds, that there would be no announcement of IDA aided projects until the agreements had been signed.

On Thursday the 11th June 1987 sent a draft Memorandum for Government on the Plan to the Department of Finance for their observations.

Mr Molloy of the Department of Finance heard on the 12th June 1987 that a press conference was being arranged to announce the package and sent a note to the Minister for

Finance expressed concern about the project being "rushed to finality" and listed a number of "serious reservations", including the view that the Section 84 currency swap element of the package might not get Revenue clearance as being inconsistent with existing legislation and was not included in the cost per job.

On the 12th June 1987, Minister Walsh sent a confidential memorandum to Mr Goodman which was in the following terms:—

"1. Padraig White, Managing Director of the IDA, informed me today that, it was written into the proposal that the whole project was dependent on Section 84 financing coming into line as agreed.

2. Between Padraig White and Martin Lowery, they will negotiate with the top people in the banks to make sure that the financial package comes through as appropriate.

3. Mr White told me that the negotiations could best be done following Authority and Government approval"

A meeting of "the Authority" was convened for the evening of the 12th June 1987 for the specific purpose of considering the recommendation made by the Board.

At the outset of the meeting Mr McCabe satisfied himself that each member of "the Authority" had had sufficient time to consider the proposal.

Mr Lowery then presented the proposals to the Authority and he gave evidence before the Tribunal that "the Authority" had carried out a thorough examination of the proposals, sought clarification, where required of the details of the plan and proposed incentive package.

Having reviewed the details of the incentive package, the Authority decided to make the recommendations set forth at the beginning of this Chapter.

The Authority had decided to include in its recommendations, a recommendation that in addition to the year by year job creation or "performance" clause, which the IDA Board and executives had recommended, that a clawback clause should also be included and that the cost to the Exchequer of the High Coupon Section 84 loans should be quantified.

Before "the Authority" had met to consider this matter arrangements had been made to hold a press conference to announce the project and the media had become aware of some of the details of the plan.

This was a matter of concern to "the Authority" and grave dissatisfaction was expressed in regard thereto and was in breach of an agreement made between the Chairman of "the Authority" and the Minister for Industry & Commerce that there should be no announcement with regard to projects with which "the Authority" was concerned until after the Grant Agreement was signed by the parties.

However, this fact did not prevent "the Authority" from dealing with the proposal in regard to the project on its merits and "the Authority's" recommendations were made after consideration of such merits.

The minutes of the meeting of "the Authority" clearly stated that

> "no change in the conditions of approval should be agreed by the IDA without the express approval of "the Authority".

By letter dated the 15th day of June 1987 the Secretary to the Authority forwarded to Mr Loughrey the recommendations of the Authority to the Government, together with material to assist in the preparation of a Memorandum for Government.

The Memorandum for Government was prepared under the supervision of Mr Loughrey and completed on the 16th June 1987 when it was presented to Government pursuant to the Certificate of Urgency signed by Mr Loughrey. It had not been circulated to the members of the Government prior to its meeting.

On that date the Government agreed to support the development and approved the financing package.

The Press Conference to announce the package was held on the 18th June 1987.

There is no doubt but that the Government was extremely anxious to secure the implementation of this project and on the face of it, this project would appear to have been rushed through the IDA Board and "the Authority".

But this is not necessarily true and does not mean that the IDA Board and "the Authority" were deprived of the opportunity of objectively considering the plan.

In the first instance, the nature of the plan was in accordance with the policy which had evolved in the IDA from the early '80s, and was in accordance with the "Future in Food" published by the IDA in December 1987 but approved by "the Authority" in March 1987. The plan had been suggested by the IDA in September 1986 to the Goodman Group and originally submitted in outline to the IDA in December 1986: negotiations with regard thereto were held in January and February 1987 and the Authority of the IDA was kept informed of the developments of such negotiations and when the Board met to consider their recommendations on the 12th of June 1987 the members were satisfied that they had ample time in which to consider the merits of the plan and on the basis of such consideration made the recommendations aforesaid.

It is a matter for the Government to decide whether, before reaching any decision, they had sufficient information at their disposal and obviously in view of the decision reached by them on the 16th day of June 1987 they considered that they had.

After the Press Conference on the 18th of June 1987 negotiations took place between the IDA executives and members of the Goodman Group with regard to the details of the Grant Agreement.

Certain difficulties with regard to its terms arose from time to time and were dealt with.

On the 14th day of July 1987, Mr Britton met with Mr Donnelly to deal with arrangements not covered by the Grant and preference share legal agreements.

On the 15th day of July 1987 Mr Britton wrote to Mr Donnelly of the IDA enclosing a draft letter to be written by Mr Donnelly to Mr Goodman with regard to these terms.

This letter is as follows:—

"Mr L. Goodman
Chairman and Chief Executive
Anglo Irish Beef Processors Ltd and
Goodman International Ltd

Dear Larry

I refer to the package of assistance negotiated with the IDA in relation to your £120m. development program.

I set out below the commitments which the IDA have given as part of their support for the overall project.

The IDA will:—

1. ***S84 FUNDING***

(*a*) Obtain Revenue Commissioners' approval for the currency swap mechanism.

(*b*) Give full support to the legislative change required to allow Anglo Irish Beef Processors International Ltd to utilise the S84 finance (reference S84 Corporation Tax Act 1976).

(*c*) Obtain Central Bank approval for the currency swap mechanism.

(*d*) Obtain a written Government commitment that the availability of S84/currency swap finance to the Group will not be curtailed during the life of the project as a consequence of legislative changes.

(*e*) Position the project with the banks so that they recognise the national priority status given to the project by both Government and the IDA and the requirement of S84/currency swap funding amounting to a minimum of £120m p.a. (or equivalent) during the life of the project.

2. ***FEOGA***

(*a*) Provide the company with a letter of endorsement from the IDA Board confirming that they will use their good office to ensure that the FEOGA grant is achieved at maximum levels for the total project and for FEOGA purposes will give it priority status, in relation to other projects seeking FEOGA assistance.

(*b*) Seek in conjunction with Government (through the Department of Agriculture) a special FEOGA package for the beef industry.

(*c*) Obtain priority status from Government (through the Department of Agriculture) within the special FEOGA beef package for the Group's development program, in relation to other projects seeking FEOGA assistance under this package.

3. Effect a procedure for grant payments which allows payment immediately on receipt of monthly claims in advance of physical inspection.

4. Provide the Group with a letter confirming the flexibility of expenditure between locations.

5. Ensure that the draft amendments to the Grant Agreements incorporated in the second draft and in subsequent discussions with Natural Resources Division are incorporated into the final agreements.

Furthermore, it is understood that:—

(*a*) An integral part of the total package of assistance is the £1m "benefit in kind" from the Sugar Company should we use their site for the Tuam project.

(*b*) The maximum capital grant including FEOGA fallback offered to other meat companies during the life of the project will be 20.83%

(*c*) In relation to possible By Product projects (in the edible or inedible sectors) the IDA will ensure that there is no conflict of interest between these projects and our development plans leading to duplication of facilities.

Yours sincerely

SEAN DONNELLY

In this letter, Mr Britton is seeking a commitment from the IDA to:—

(1) use their good office to ensure that the FEOGA grant is achieved at maximum levels for the total project;

(2) for FEOGA purposes to give the project priority status in relation to other projects seeking FEOGA assistance;

(3) to seek a special FEOGA package for the beef industry;

(4) obtain priority status from Government (through the Department of Agriculture) within the special FEOGA beef package for the Group's development programme in relation to other projects seeking FEOGA assistance under this package; and

in relation to possible By Product projects the IDA, to ensure, that there is no conflict of interest between these projects and the development plans of the Goodman Group leading to duplication of facilities.

In this letter they were seeking a commitment, in effect, to give priority status to the Goodman 5-Year Development Plan over any other similar projects and in relation to possible By Product projects to ensure that grants would not be available in respect of such projects by other companies if they conflicted with the Goodman Group project.

On the 28th day of July 1987, Minister Walsh, wrote to Mr Goodman in relation to the package of assistance negotiated with the IDA in relation to the development programme.

This letter provided as follows:—

"Mr L. Goodman
Chairman and Chief Executive
Anglo irish Beef Processors
Goodman International Ltd
14 Castle Street
Ardee
Co. Louth

Dear Mr Goodman

I refer to the package of assistance negotiated with the IDA in relation to your £120m development programme.

I set out below the commitment which the Department of Agriculture and Food have given as part of their support for the overall project, specifically in relation to the FEOGA support for the package.

1. The Department of Agriculture and Food will seek a special FEOGA package for the Beef Industry.
2. In addition, the Department of Agriculture and Food will seek to obtain priority status for the development of the Irish Beef Industry within FEOGA
3. The Department of Agriculture and Food will strongly support the application by Goodman International for FEOGA Grants at the maximum level.

Yours sincerely
Joe Walsh TD, Minister for Food.

On the 6th day of August 1987, Mr Lowery wrote to Mr Goodman, as follows:—

"Mr Laurence Goodman
Chairman and Chief Executive
Goodman International ltd
Castle Street
Ardee
Co. Louth

Dear Mr Goodman

I refer to the discussions which have taken place between your company and the IDA in finalising the Grant Agreement in respect of your company's development programme.

The following outlines the additional action which IDA will carry out to complete all elements of the package.

1. *Section 84 Funding*

   a) The Authority has approved and recommended to Government the agreed funding / incentive package which includes the provision of the high coupon Section 84 Currency Swap facilities. A submission to obtain Revenue Commissioners approval to the proposed arrangements will be made by the IDA.

   b) The Authority will give full support to the legislative change required to allow Anglo Irish Beef Processors International Ltd to utilise the Section 84 finance.

   c) The Authority strongly supports and will assist in whatever way appropriate the company's request to the Central Bank for approval of the currency swap mechanism.

   d) The Authority supports strongly and will assist in whatever way appropriate the company's submission to Government, requesting assurances that the Section 84 Currency Swap facility will remain in place for the duration of the project (8 years).

   e) A presentation has been made by IDA to AIB, Bank of Ireland and Irish Intercontinental Bank at the most senior level setting out the National priority nature, importance and elements of the funding package (estimated at £120m. Currency Swap funding) to ensure a favourable attitude by the Banks to the project. The Authority will continue to assist the company in whatever way appropriate to achieve the company's objective.

2. *FEOGA*

   a) The Authority will use its offices in strong support of the company, in consultation with the Department of Agriculture and Food, in the company's application for FEOGA Grants at the maximum level.

   b) The Authority will make representations to the Department of Agriculture and Food with a view to securing a special FEOGA package for the beef industry. In addition, the Authority will make representations to the Department of Agriculture and Food with a view to obtaining priority status for the development of the Irish Beef Industry within FEOGA.

3. *a*) The provision of further assistance sought by G.I. by way of "benefit in kind" (estimated at £1m) for the Tuam project is a matter solely for negotiation between G.I. and the CSET.

   b) The Authority will endeavour to ensure that in relation to possible by-product projects (in the edible or inedible sectors) that there is no conflict of interest between these projects and the development plan for G.I.

Yours sincerely

____

Martin D. Lowery
Executive Director

In pursuance of the commitments contained in the said letter the IDA by letter dated the 17th day of August 1987 sought from the Revenue Commissioners an advance opinion on the efficacy of the High Coupon mechanism proposed and by letter dated the 8th September 1987 Mr Frank Cassells of the Revenue Commissioners gave the necessary confirmation.

This matter is dealt with in greater detail in the chapter of the Report dealing with "Section 84".

The legislative changes suggested by the Goodman Group involving an amendment to Section 84 of the Corporation Tax Act 1976 and the submission in regard thereto have also been dealt with in that chapter.

Despite these submissions no amendment was enacted as sought by the Goodman Group or to ensure that the High Coupon (Currency Swap) arrangement would be kept in place for the period of the Agreement (8 years).

A presentation was made by the IDA to the AIB, Bank of Ireland and the Irish Intercontinental Bank to ensure favourable attitude to the Banks to the proposal.

The Goodman Group proceeded to obtain and draw down such High Coupon Section 84 borrowings even before the Grant Agreement was signed.

Minister Walsh and Mr Loughrey met the EEC to press for support for the special beef package for Ireland and also progressed the FEOGA funding.

The exclusion of the food processing industry from the provisions of Section 52 of the Finance Act 1986 was effected by Section 25 of the Finance Act 1987.

As appears from the exchange of correspondence between Mr Aidan Connor of Goodman International and Mr Lowery of the IDA between 2nd March 1988 and 7th March 1988 hereinbefore set out, the only remaining obstacle to agreement on the terms of the Grant Agreement was the requirement by the IDA of the "performance clause" which "the Authority" insisted upon and which formed part of their proposals submitted to the Government on the 16th June 1987.

In the course of a memorandum of the 2nd March 1988 submitted to the Secretary of the Department of Agriculture and Food, Mr Loughrey stated:—

> "The Government decision of 16 June 1987 on the Goodman Programme approved in full the IDA's proposal that a performance clause in standard form relating to job targets from base year 1988 through to 1995 should be incorporated into the Grant Agreement. In essence this provides that grants paid in years 1 to 5 will be repayable at the end of year 5 in proportion to the failure to achieve job targets and a similar clawback clause to operate at the end of year 8 for years 6 to 8.
>
> The first draft Grant Agreement was issued on 22 June 1987. Negotiation went ahead with senior Goodman Executives, up to Deputy Chief Executive level, on a line by line, paragraph by paragraph basis throughout July and August. A final redraft was issued in August, 1987.

Arising from these negotiations a number of issues remained outstanding which did *not include* performance or clawback clauses. These outstanding items were resolved in November 1987 and the Company confirmed they were ready to sign the Grant Agreement.

The issue of the performance and clawback clause was first raised two weeks ago."

At a meeting of "the Authority" held on the 1st day of March 1988, Mr Lowery reported on the state of the negotiations between the parties and the relevant extract from the minutes of that meeting are:—

"(*a*) Noted that the Grant Agreement has not been signed and that the performance and clawback clauses were causing some difficulty and agreed that force majeure could apply to these clauses:

(*b*) Agreed that the agreements should be signed as a matter of urgency and that there was no scope for easing the clauses referred to at (*a*)."

It was, according to Mr White's evidence the "unanimous view of the members of "the Authority" that there was no scope for the further easing of the two clauses.

Mr Britton stated in evidence that this particular matter of the Performance Clause was regarded "as a matter for lobbying Mr Haughey's support" and agreed that Mr Goodman had gone to see Mr Haughey on the 4th day of March 1988 "to get him to get the IDA to see the wisdom of the Goodman stance".

Mr Haughey's recollection of that meeting was that:—

"It was almost certain that Mr Goodman came to me to tell me that the negotiations had broken down and it is almost certain that he would have mentioned the performance clause."

Mr Haughey's view of the performance clause was:—

"It seemed to us that Goodman was reasonable in not asking to be held to annual targets. He could be held to annual targets. He could be held up in any one of different sites and the idea that the scheme would abort if every annual target wasn't fully met seemed unreasonable."

On the morning of the 8th March 1988 Mr O'hUiginn the Secretary to the Department of the Taoiseach ascertained from Mr Lowery the position with regard to the breakdown of the negotiations and obtained a copy of the relevant correspondence.

Prior to ringing Mr Lowery Mr O'hUiginn was aware of the precise issue between the IDA and the Goodman Group.

This note was sent into Government by Mr O hUiginn without reference to the Ministers who would be involved such as the Minister for Agriculture and Food, the Minister for Industry & Commerce and the Minister for Finance.

On the basis of such memorandum the Government decided that the Grant Agreement be amended.

The decision made by the Government on the 8th day of March 1988 was that the Grant Agreement should be amended and the original letter dated the 8th March 1988 prepared and signed by the Secretary to the Government so records.

> S. 25217
>
> "8 Márta 1988
>
> An Rúnaí Príobháideach
> An tAire Talmhaíochta agus Bia
>
> I am to refer to the decision S. 25217 dated 16 June, 1987 concerning the provision of industrial grants to Goodman International Limited for the major development of its Irish meat operations and to inform you that, at a meeting held today, the Government decided that the Grant Agreement should be amended to take account of the following:
>
> (1) the overall job targets for the project were the essential job targets to be attained;
>
> (2) in the event of these overall targets not being attained, the clawback provision would operate after the fifth and eight years;
>
> (3) the annual industrial grants should be related to the annual expenditure by the company in fixed assets, based on the principle that the grants should represent 20.833 per cent of the expenditure incurred; and
>
> (4) Goodman International Limited would make its best endeavours to attain specified annual job targets consistent with the overall job targets to be attained.
>
> Rúnaí an Rialtais

It appears that on receipt of this letter Mr John Donlon, Secretary to the Department of Industry and Commerce, contacted Mr O'hUiginn and informed him that changes in the agreement are a matter for the IDA and in effect that the Government had no power to amend the agreement.

Mr O hUiginn then prepared a draft of the letter containing an interpretation of the decision which was ultimately circulated by Mr Nally and sent to the Department of Agriculture and Food.

It is quite clear that the Government had no power to amend the Grant Agreement on the 8th March 1988 and when Mr Donlon pointed this out, both Mr O'hUiginn and Mr Nally became aware of this and proceeded to amend the decision.

Mr Haughey, in evidence, said that he wasn't informed that the Government couldn't amend the agreement and that the matter was dealt with between the officials. He stated:

> "We took our decision at Government and that was the end of the matter as far as we were concerned. But, in all these circumstances, it is the duty of the officials,

secretaries of government departments, to clear up any of these sort of legal difficulties, which government decisions may give rise to."

Mr Loughrey was informed of the decision of the Government by letter from the Secretary to the Government and communicated this decision to "the Authority".

On the 10th March 1988 Mr White, who had been informed by Mr O hUiginn of the Government's decision, circulated members of "the Authority" with a memorandum setting forth the position as a result of the Government decision and recommended that the Authority agree to the modification of the terms of the Grant Agreement to incorporate the Government's decision.

The terms of the said memorandum are as follows:—

"1. The attached document includes the details of the Cabinet decision of the 8 March 1988 on the interpretation of their earlier decision of 16 June 1987 relating to IDA investment in the Goodman International beef programme.

2. The Cabinet decision of 8 March 1988 requires approval of the Authority if it is to be implemented.

   The Authority itself in its June 1987 decision on the proposals had indicated "that no change in the conditions of approval should be agreed without the express approval of the Authority".

3. In the circumstances, I recommend that the Authority agree to modification of the Grant Agreement to incorporate the recent Cabinet decision.

4. The Goodman Group are anxious to sign the Grant Agreement immediately in order to allow the proposed Goodman investment programme to proceed.

   In consultation with the Chairman, the following procedure for Authority consideration of the modification to the Grant Agreement is proposed.

   The documentation is being circulated on Thursday, 10 March. I will contact Authority members on Monday morning to ascertain their stance on the proposed modifications. If any Authority member cannot easily be reached, he might contact me.

   If an Authority member(s) feels that a meeting is necessary the Chairman provisionally proposes Tuesday, 15 March, 4.30 p.n. should such a meeting be deemed necessary.

   To facilitate members outside Dublin participating in such a meeting, the Chairman has indicated that arrangements could be made to have them linked to the meeting using a conference telephone speaker.

5. We will let the Authority members know by Monday afternoon of the need or otherwise for the Tuesday meeting.

Padraic A. White
Managing Director
10th March 1988

Mr White stated that his recommendations were based on:

(1) the fact of the Government decision, and

(2) the totality of the remaining agreement gave to the IDA substantial power to implement/protect the integrity of the Agreement.

As appears from the said memorandum it was not his intention to have a meeting of the Authority if all the members of the Authority so agreed but to get the agreement of the members on the telephone. Such a course was not acceptable to Mr Brendan Dowling. He informed Mr White that he would resign from the Authority if there was no meeting. Mr Dowling told Mr White that since the Authority had originally agreed that any changes in the terms of the agreement had to be approved by the Authority, then an Authority meeting was required.

He asked Mr White to get legal advice on the status of the decision, i.e. whether it amounted to a direction or a suggestion. If a suggestion, then the Authority should reiterate the views of earlier Authority meetings. He was of the view that the wording of the original agreement was unambiguous and not capable of being reinterpreted; it was only capable of being changed.

The meeting called at Mr Dowling's insistence went ahead on the 15th. The Secretary of the IDA had informally asked Mr John Darley, the IDA's in-house lawyer, to attend the meeting and to do a note on the effect of the Government decision.

According to Mr Dowling, Mr Darley's written opinion dated 14th March said that the Government seemed to have had the right to remove the Performance Clause under the 1986 Act. Mr McCabe said he was highly disappointed at the Darley view. The minutes of the Authority meeting of the 15th indicates the Authority's agreement that the Government's decision of the 8th March was deemed to be a decision under Section 35 of the Industrial Development Act, 1986. Mr Dowling said the minute was "formally noting or putting into effect a decision made elsewhere."

Mr Lowery, in the course of his evidence stated:—

> "The position is very clear. The Authority had taken its own position on the performance and clawback clause. The Government took a decision in relation to the performance clause which obviously was against the wishes of the Authority. The Authority none the less accepted the Government's decision and went ahead and implemented it and it accepted the Government's decision readily in the context of having examined the extent to which it still had adequate cover for the grant to be paid out".

The relevant portion of the minutes of this meeting of the Authority reads as follows:—

> "*Goodman International Group*
>
> "The Authority considered a document (Ref. 236/1) which had been circulated beforehand, setting out changes resulting from the Government's decision of 8th March, 1988. Also considered were a letter from the Department of Agriculture and Food of 11th March, 1988 advising IDA of the Government's decision and

a draft of the revised Grant Agreement between IDA and Goodman International Ltd. which was amended to incorporate changes resulting from the Government decision.

The Authority:

(i) Agreed that the conditions set out in its earlier decisions on the project be amended to take into account changes resulting from the Cabinet's decision as set out in the letter of the 11th March 1988 received from the Department of Agriculture and Food.

(ii) Agreed that the Grant Agreement with the Company be modified to reflect the revised conditions.

(iii) Agreed that the amendment of the conditions as at (i) and (ii) above had been agreed by the Authority on the basis that the Government decision incorporated in the letter of the 11th March 1988 from the Department of Agriculture and Food was deemed to be a decision under Section 35 of the Industrial Development Act, 1986".

Mr Joe McCabe was the Chairman of the Industrial Development Authority and in his evidence with regard to the performance clause confirmed that at a meeting of "the Authority" held on the 1st March 1988, the Authority agreed that there was no scope for easing the "performance clause" and the "clawback clause" and that the meeting of the Authority held on the 15th March 1988 was held at the insistence of Mr Brendan Dowling a member of "the Authority" and to consider the effect of the Government decision made on the 10th day of March 1988 and communicated to the Secretary to the Authority by letter dated the 11th day of March 1988.

He stated that "the Authority" had sought and obtained legal advice from their solicitor to the effect that the Government was entitled to make the decision that they had made on the 8th day of March 1988.

With regard to such advice, Mr McCabe stated that "he was highly disappointed to hear it but the Government had overruled the IDA"

The "Authority" had on the 12th June 1987 included in their Recommendations to Government the inclusion of the "clawback" clause in addition to the "performance clause" which had been recommended by the IDA executives and Board and though negotiations on the terms of the Grant Agreement had broken down on the necessity for the "performance clause", "the authority" at their meeting on the 1st day of March 1988, though aware of the breakdown of the negotiations, had refused to delete it.

It is obvious that "the Authority" would not have deleted the clause were it not for the intervention of the Government.

"The Authority" did not seek such intervention and the intervention by the Government, whether it was originally initiated by the Secretary of the Department of the Taoiseach on his own initiative or not, must have been based on a proposal by the then Taoiseach because Mr O'hUiginn's memorandum was addressed to him and no copy documents were provided for the other members of the Government.

The Revised Grant Agreement (Schedule 1) was signed on the 22nd March, 1988.

Prior to the signing of the agreement the following steps had been taken in regard to the implementation of the plan:—

(i) The Food Processing Industry had been excluded from the provisions of Section 52 of the Finance Act 1986.

(ii) The IDA had obtained the opinion of the Revenue Commissioners with regard to the application of the High Coupon Section 84 (Currency Swap)

(iii) The IDA had made supportive submissions with regard to such borrowings to a number of Banks and the following facilities had been put in place by 13 October 1987:—

£15m	drawn down as of that date
£64.5m	to be drawn down by 31.10.1987
£14m	to be drawn down by 31.12.1987
£22m	to be drawn down by 31.3.1988

All high coupon at interest rates ranging from .68% to 3.0% (positive)

(iv) The Goodman Group had made a submission with regard to the amendment of Section 84.

(v) Representations had been made to FEOGA.

A review of the agreement took place on the 5th day of September 1988 by which time FEOGA grant approval for that portion of the plan which related to the plants at Tuam and Dublin.

The first Annual Review under the Agreement took place on the 21st April 1989. The Goodman Group executives reported to the IDA with regard to progress made on procurement initiatives, rationalisation of slaughtering capacity and the Group's market development and informed the IDA that there would be no progress on the capital investment programme until the following matters in respect of which they alleged that they had been given assurances from the IDA, the Department of Agriculture and the Government:—

"2 (*a*) Government assurances sought in relation to the availability of Section 84 Currency Swap Finance for the life of the development plan (8 years).

(*b*) A change in the Finance Act to allow Anglo Irish Beef Processors International Ltd. to utilise Section 84 Finance.

(*c*) A request for the provision of a special FEOGA package for the beef industry.

(*d*) The re-designation of Co. Louth in relation to FEOGA grant monies."

At its meeting on the 28th April 1989, "the Authority" considered a report on the Review of the National Resources Division.

This Report indicated, inter alia, that at that stage the Goodman Group had drawn down an estimated £75m. under the Section 84 currency swap mechanism, FEOGA approval had been obtained in June 1988 for two projects under plan in respect of Tuam and Cloghran (Dublin) and a decision had been sought by the Department of Agriculture from the EEC in respect of the re-designation of Co Louth as a disadvantaged area.

It appears from the minutes of this meeting of the Authority that

> "3. Considered that the items set out at 2(*a*) — (*d*) were never conditions related to the Beef Development Plan and decided that they could not be considered as reasons for lack of progress."

Having regard to:—

1) The fact that "the Authority" in its recommendations to the Government, recommended that:—

> "the Government note that the financing of the project is based on the assumption of the availability of a loan facility of £120m. under the Swap Currency Section 84 scheme and on the assumption that no liability to Capital Gains Tax in respect of such loan facility would arise in this case".

and noted that:—

> "on foot of the negotiated package, the IDA will use its offices in strong support of the Company, in consultation with the Department for FEOGA grants at the highest level."

2) The commitment given by the Department of Agriculture & Food and confirmed by letter dated the 28th July 1987 to:—

> "(*a*) seek a special FEOGA package for the Beef Industry:
>
> (*b*) obtain priority status for the development of the Irish Beef Industry within FEOGA;
>
> (*c*) to strongly support the application by Goodman International for FEOGA grants at the maximum level: and

3) the additional action to be carried out by the IDA to complete all elements of the package as outlined in Mr Lowery's letter to Mr Goodman dated the 6th day of August 1987 including inter alia.

> "1. *Section 84 Funding*
>
> *a*) The Authority has approved and recommended to Government the agreed funding / incentive package which includes the provision of the high coupon Section 84 Currency Swap facilities. A submission to obtain Revenue Commissioners approval to the proposed arrangements will be made by the IDA.
>
> *b*) The Authority will give full support to the legislative change required to allow Anglo Irish Beef Processors International Ltd to utilise the Section 84 Finance.
>
> *c*) The Authority strongly supports and will assist in whatever way appropriate the company's request to the Central Bank for approval of the currency swap mechanism.

d) The Authority supports strongly and will assist in whatever way appropriate the company's submission to Government, requesting assurances that the Section 84 Currency Swap facility will remain in place for the duration of the project (8 years).

e) A presentation has been made by IDA to AIB, Bank of Ireland and Irish Intercontinental Bank at the most senior level setting out the National priority nature, importance and elements of the funding package (estimated at £120m Currency Swap funding) to ensure a favourable attitude by the Banks to the project. The Authority will continue to assist the company in whatever way appropriate to achieve the company's objective.

2. *FEOGA*

a) The Authority will use its offices in strong support of the company, in consultation with the Department of Agriculture and Food, in the company's application for FEOGA Grants at the maximum level.

b) The Authority will make representations to the Department of Agriculture and Food with a view to securing a special FEOGA package for the beef industry. In addition, the Authority will make representations to the Department of Agriculture and Food with a view to obtaining priority status for the development of the Irish Beef Industry within FEOGA.

3. *a*) The provision of further assistance sought by G.I. by way of "benefit in kind" (estimated at £1m) for the Tuam project is a matter solely for negotiation between G.I. and the CSET.

b) The Authority will endeavour to ensure that in relation to possible by-product projects (in the edible or inedible sectors) that there is no conflict of interest between these projects and the development plans for G.I."

It is difficult to accept the view of the Authority that these items never related to the Beef Development Plan.

The Government had approved the financial package as follows:—

(1) IDA Grants totalling £25m.

(2) Redeemable preference shares of £5m. to be taken up by the Authority and repayable at £1m. per annum after 5 years, and

(3) currency swap loans under Section 84 of the Corporation Tax Act 1976 with an estimated capitalised value of £30,000,000.

As illustrated in the chapter of this Report dealing with Section 84 the provision of (3) above, was necessitated by the fact that in negotiations between the IDA and the Goodman Group with regard to the financing of the capital investment portion of the Development Plan, the Goodman Group at all times, requested that 75% of the cost thereof would be grant assisted and insisted that the project would not be commercially viable unless this level of grant aid was provided.

Combined IDA and FEOGA grants were limited to 50% of the capital cost of the project viz £60m. and it was eventually agreed that the balance of £30m. could be provided by interest savings on borrowings under the Currency Swap Section 84 scheme.

Inherent in this arrangement was

(i) the requirement that this facility should continue during the period of the project and the period of the borrowings (8 years); and

(ii) the necessity of establishing that AIBP carried on a "specified trade" within the meaning of Section 84A of the Corporation Tax Act 1976 as introduced by Section 41 of the Finance Act 1984.

As appears from the submission made on the 20th January 1988 by AIBPI to the Minister for Finance proposing an amendment of Section 84A of the Corporation Tax Act 1976.

"(i) Anglo Irish Beef Processors International Limited (Anglo-Irish) sells processed beef on the export market by means of sale by wholesale. Anglo Irish carried on this activity prior to 1 January 1981 and, in consequence, is entitled to claim, export sales relief on the profits derived from that activity

(ii) The sales by Anglo Irish include beef processed by Anglo Irish Beef Processors Limited which is a fellow subsidiary of Anglo Irish and beef purchased from intervention stock which were processed by other beef processors within the State".

As it appears from the submission that the sale by Anglo Irish of processed beef purchased from intervention, which had been processed by other processors other than AIBP or an associated company, would not qualify as manufactured goods within the Act and if the sale of this type of goods in any accounting period exceed 25% of all sales, then the entire borrowings would not qualify as Section 84 borrowings. Both these requirements were of considerable importance to the Goodman Group and they justifiably in the opinion of the Tribunal regarded them as part of the agreed package and pressed for their implementation.

During the course of evidence before this Tribunal, in relation to the Export Credit Insurance issue, it was ascertained that 84% of the beef exported to Iraq, between September, 1987 and December 1988 was beef purchased by the Goodman Group from intervention.

Having regard to such a high level of purchases from intervention for export the Tribunal sought to ascertain from the group's auditors, Stokes Kennedy Crowley, whether they had satisfied themselves and the manner in which they so satisfied themselves that at least 75% of the goods sold by it consisted of beef processed by its associated companies.

Mr Mooney of SKC informed the Tribunal that:—

"(i) under his instructions, the audit staff carried out a series of tests to establish the source of beef purchased and sold by Anglo Irish Beef Processors International Ltd.;

(ii) AIBPI was the only company within the group that availed of Section 84 borrowings;

(iii) the proportion of total sales by AIBPI of beef purchased from AIBP (an associated Company) during the years 1987 to 1990 were as follows:—

1987....................71.3%

1988....................87%

1989....................83%

1990....................97%

(iv) In 1987, direct purchases from AIBP by AIBPI were 71.3%, leaving a shortage of 3.7% which was sufficient to deprive them of the benefits of the Section 84 scheme unless it could be established that at least 3.7% of the beef purchased by AIBPI from the Intervention Authority was beef which had been processed by AIBP and placed in intervention by them.

(v) SKC satisfied themselves that the 75% requirement was reached by the addition of product purchased from intervention which had been processed by AIBP."

It was for the purpose of giving to AIBPI greater flexibility in the purchase for export of beef processed by processors other than its associated companies, that the amendment was sought.

Between May 1989 and May 1990 there was an exchange of correspondence between the IDA and the Goodman Group with regard to various matters.

The Annual Review meeting was due to be held on the 23rd May, 1990 and by letter dated the 21st day of May 1990 Mr Britton stated:—

> "As you are aware, following a series of meetings with the IDA last year, our Group decided to place the Development Plan on hold as circumstances did not exist which would have allowed the Plan to proceed in the manner and with the support originally envisaged and deemed essential by the promoters."

The Annual Review meeting was held on the 23rd May, 1990 with no change in the position.

The Executives of the IDA reported to "the Authority" and having considered the report and the contents of the letter dated the 21st May 1990 directed Mr Donnelly to write to the Goodman Group informing them of the attitude of "the Authority".

By letter dated the 1st June 1990, Mr Donnelly wrote as follows:—

"IDA IRELAND

PRIVATE & CONFIDENTIAL

Mr Larry Goodman
Goodman International
14 Castle Street
Ardee
COUNTY LOUTH

01 June 1990

Dear Larry

Re: *Five Year Beef Development Plan:*

The Authority has considered the contents of your letter of 21 May 1990 in conjunction with the views expressed at the Review Meeting on 23 May 1990. From these considerations, it would appear that you decided last year to postpone indefinitely the implementation of your Group Development Plan without regard to the provisions of the Grant Agreement.

As you know, the Authority was not informed of that decision when it was made and your letter of the 21 May was the first indication received by the Authority of your decision.

In our view, that decision was, in legal effect, a repudiation of the Grant Agreement dated 22 March 1988 and the Authority regards it as such. As a result the Grant Agreement has no longer any effect and the Authority has no outstanding obligations to you by virtue of that Agreement.

As you are aware, the Department of Agriculture & Food is reviewing the Beef Industry and I share the view that a National Beef Plan is necessary. I welcome your interest in the promotion of a National Beef Plan and your expressed intention of initiating discussions with the Department of Agriculture & Food in that connection.

You advised me that you wished to have the discussions with the Department of Agriculture & Food and to reflect further on your future plans.

I am happy to consider any new Corporate Plan that results from your discussions and reflection. However, any assistance which the Authority might provide towards a new Corporate Plan would be separate from, and not a continuation of the Grant Agreement of 22 March 1988.

Yours sincerely
Sean Donnelly
Executive Director"

This letter finally signalled the end of the Five Year Beef Development Plan that had been submitted by Goodman Group on the 23rd day of April 1987; considered by the Government on the 28th day of April 1987 and "the Authority" on the 26th day of April 1987, finally approved by "the Authority" on the 12th day of June 1987 and by the Government on the 16th day of June 1987, announced at a Press Conference on the 18th day of June 1987 and amended by the deletion of the "performance" clause by "the Authority" on the 15th day of March 1988 pursuant to a decision of the Government taken at its meeting on the 8th day of March 1988 and in respect of which the Grant Agreement was finally signed on the 22nd day of March 1988.

On the basis of the facts outlined in this Report, it is clear that

(i) the Five Year Development Plan 1987-1992 in respect of its Beef operations in Ireland was produced by Goodman International Limited at the request of the Industrial Development Authority (IDA) and with its encouragement and there has not been established any basis for the allegation that "the Authority" did not and were not able to properly assess and evaluate the merits of the plan;

(ii) the concept inherent in the plan had the full support of the IDA as it was in accord with their development policy;

(iii) the plan was also in accordance with the policy of the Government in regard to the development of the food industry and job creation;

(iv) when the Government became aware of the plan and the negotiations in regard thereto being carried out between the IDA and the Goodman Group it decided to support the concept of such plan and to encourage and assist the parties in the negotiations;

(v) the support given and assistance provided did not mean that a similar plan put forward by another beef processor would not receive similar support from either the Government or the IDA;

(vi) the support and assistance given by the Government and the Ministers thereof prior to the announcement of the plan did not in the words of Mr White the then Managing Director of the IDA, in any way amount to "political pressure";

(vi) at no stage did the Government decide that it would rely solely on the Goodman Group to develop the beef industry;

(vii) at no stage did the Government decide against the wishes of the IDA to give a grant of £25m to the Group;

(viii) at no stage did the entire or any member of the Board of the IDA threaten to resign over a grant to expand an industry that had a surplus processing capacity;

(ix) on the contrary the proposed plan had the full support of the Board of the IDA and "the authority", and neither the Board nor "the Authority" was precluded from assessing the implications of the plan on the cattle industry because of political pressure;

(x) The Press Conference held to announce the agreement between the IDA and Goodman International was undoubtedly held prematurely and not in accordance with the wishes of the IDA and Goodman International but at the instigation of the Minister for Food, who was unaware of the agreement made between the

Chairman of "the Authority" and the Minister for Industry and Commerce that there should be no announcement of any plan until the Grant Agreement had been signed but there is nothing unusual in a Government or a Minister being anxious to announce good news and seeking to derive political benefit from such announcement;

(xi) in view of the failure on the part of the Goodman group to proceed with the plan, no grants were paid in respect of any development under the plan.

There is no doubt whatsoever but that the Government on the 8th day of March 1988 wrongfully and in excess of their powers under the provisions of Section 35 of the Industrial Development Act 1986, directed 'the Authority' to remove 'the performance' clause from the Grant Agreement being negotiated between the IDA and the Goodman Group and that this direction was made either at the instigation of the then Taoiseach or the Secretary to his Department.

CHAPTER EIGHT

Section 84

In the course of the debate on the Companies (Amendment) Bill 1990, Deputy Rabbitte stated on the 28th August 1990:—

> "I am now stating in this House that I have information which suggests that Mr Goodman proceeded to draw down much of the £170 million package of Section 84 loans. I am stating that it is my information that those exceptional credit lines were manifestly not used for the purpose for which they were approved: rather that Mr Goodman used these facilities to fund imprudent and speculative investments outside the State that had nothing to do with the beef industry and that in that process the Exchequer was effectively defrauded of substantial revenue. These are serious charges which I am asking the Minister for Industry and Commerce to address"

On the 15th May 1991 in Dáil Eireann he stated:—

> "We also know that in the Finance Act the Government made a special arrangement to enable Mr Goodman to avail of High Coupon Finance for the Schemes that I referred to earlier. Since this finance could only be drawn down as working capital and since at least some of it was used outside the State to fund speculative ventures that had nothing to do with the development of Agriculture or the reasons for which it was authorised. I would ask the Minister why was there never any prosecution. Surely this is tantamount to tax evasion."

In October 1990 the then Worker's Party Deputies, of which Deputy Rabbitte was one, had tabled a motion in Dáil Eireann seeking the establishment of a public inquiry for the purpose of inquiring, inter alia,

(*a*) Whether or not money borrowed by the Goodman Group under favourable tax terms for the expansion of the beef processing industry was used for the purposes for which it was acquired.

In the course of his evidence before this Tribunal, Deputy Rabbitte stated that his information which provided the basis for the foregoing quoted statements and the allegations contained therein came from the unnamed banking source referred to already and claimed privilege in respect of the name of such source.

The Tribunal was obliged to inquire into these allegations without the assistance of such "banking source".

The reference to "The money borrowed by the Goodman Group under favourable tax terms for the expansion of the beef processing industry" is to borrowings to which the provisions of Section 84 of the Corporation Tax Act, 1976 as amended relate.

These allegations cannot be considered "in vacuo" and their consideration requires an understanding of the operation of what are described as "Section 84 Loans" and "High Coupon Section 84 Finance".

It appears that the purpose of Section 84A of the Corporation Tax Act 1976 as inserted by Section 41 of the Finance Act 1984 was to prevent perceived abuses of the Corporation Profits Tax/Income Tax regime for companies whereby money could be taken out of companies in a tax effective way.

Section 84 sought to treat many of these tax effective disbursements from companies as distributions of profits and consequently not deductible or allowable as expenses for tax purposes.

One of the payments contemplated by Section 84 as a distribution was interest on a loan where the level of interest is dependent on the Company's financial performance.

Such interest is paid to the lending bank but being treated as a distribution was not subject to the payment of Corporation Tax by the Bank. Normally the Bank would pay Corporation Tax at the rate of 40%.

Because of this saving of Corporation Tax, the legislation was used to enable the lending institution, the Bank, to share the benefit of its tax saving with the Corporate Borrower by lending at a reduced interest rate where the particular circumstances so allowed.

These circumstances so allowed when a Company was only liable for 10% manufacturing rate or a nil rate as a result of export sales relief.

Section 84 of the Corporation Tax Act of 1976 was amended by Section 41 of the Finance Act 1984.

The effect of this amendment was to control the operation of the Section 84 finance so that it became a form of industrial incentive: it limited the meaning of the term "distribution" thereby narrowing the range of Corporate borrowers who were entitled to avail of Section 84 Finance.

The criteria laid down were that:

(i) The borrower carried on a "specified trade" as defined by sub-Section (3) and (5) of 84A in the State

and

(ii) The interest, if it were not a distribution would be treated as a trading expense.

A specified trade is a trade which consists wholly or mainly of the manufacture of goods which are sold by a Company which are manufactured by a fellow subsidiary (provided the share capital in both companies have at least 90% common ownership shall be deemed to have been manufactured by the company selling them).

A trade will only consist wholly or mainly of the manufacture of goods if the total amount received from the sale of qualifying goods is not less than 75% of the total amount receivable from all sales.

Although the operation of Section 84 finance as an industrial incentive required the lender to share the tax saving with the corporate borrower, there was no statutory obligation to do so: it was a matter for negotiation between borrower and lender.

Mr Cassels of the Revenue Commissioners said that as a result of the amendment they were administering a tax incentive provision that they had hitherto considered offensive and they applied the provisions very strictly.

The operation of the "High Coupon Section 84" scheme also known as "the Swap Currency Section 84" was explained by Mr Sean Donnelly of the Industrial Development Authority who stated that:—

(1) the object of this type of finance is to maximise the tax saving on the loan by borrowing in a weak currency carrying a high rate of interest.

(2) the money so borrowed is then converted into the currency required.

(3) the high rate of interest provided is converted into a distribution by the Section 84 mechanism results in an increased tax free payment to the lender.

(4) this usually results in a lower rate of interest on the borrowings for the company borrowing.

(5) as the loan is to be repaid in the currency in which it was borrowed, financial instruments are employed by the borrower to secure a foreign exchange gain so as to offset the higher interest to be paid in the weak currency.

The importance of the question of Section 84 and "High Coupon Section 84" borrowings in the negotiations between the IDA and Goodman International with regard to the "5-year Development Plan" will be dealt with in the chapter of this Report dealing with the IDA.

At this stage, the Tribunal is dealing with the allegations

(1) that borrowings made by the Goodman Group under these headings were used to fund speculative ventures, and

(2) that in the Finance Act, the Government made a special arrangement to enable Mr Goodman to avail of High Coupon finance (in respect of Section 84 loans) to fund speculative ventures abroad.

It appears from the evidence adduced before the Tribunal that:—

(1) Section 84 funds were drawn down on an annual basis by Anglo Irish Beef Processors International Ltd (AIBP) for the years 1986 to 1990 inclusive.

(2) In the period between June 1987 when the terms of the agreement for the Beef Development Plan were agreed and the 22nd March 1988 when the agreement was signed, approximately £106m was drawn down.

(3) In the accounting period ending the 31st December 1987, AIBP paid £.7m by way of Section 84 interest and in the period 31 December 1988, £10.9m to various financial institutions.

(4) On the appointment of the Examiner to the Goodman Group of Companies by the High Court pursuant to the provisions of the Companies (Amendment) Act 1990 on the 29th August 1990 the Section 84 loans included in the Statement of Affairs were as follows:—

Anglo Irish Beef Processors International Limited
Section 84 loans included in the Statement of Affairs at 29 August 1990

Bank	*Balance Currency*	*Ir£ Equivalent Ir£*
Allied Irish Banks	Ir£3,400,00	3,400,000*
Banque National de Paris	Ir£1,500,000	1,500,000
Irish Intercontinental Bank Mocnico	Nz$15,463,000 /Us$10,000,000	5,833,285
Contiguous	Nz$7,500,000 /Dm8,485,588	3,162,370
KBL Investments	Nz$14,970,000 /Dm16,886,301	6,293,110
Bank of Ireland	Dm38,333,969	14,286,129
	Us$9,218,040	5,377,145
Ulster Investment Bank	Stg 6,500,000	7,331,378
	Total Ir£ Equivalent	47,183,417

*Included is a Section 84 loan of Ir£400,000 which was borrowed by AIBP Carlow Exports. This amount was also included in the Statement of Affairs of AIBP Financial Services at 29 August 1990.

(5) In the course of the settlement negotiations between the Examiner, Stokes Kennedy Crowley and the Revenue officials dealt with in the Chapter on Tax Avoidance the question of Section 84 borrowings was raised in the context of the suggestion that there had been substantial amounts of money paid out by the Group for investments which would not have qualified for Section 84 relief, and in particular shares in Berisford International plc and Unigate plc.

(6) The Revenue officials accepted assurances from SKC that the usages to which the money borrowed was put by AIBP qualified for Section 84 purposes.

(7) In the course of his evidence to the Tribunal, Mr Peter Fitzpatrick, the Examiner appointed by the High Court stated that:

(*a*) the Revenue authorities confirmed to him that there was no claim in respect of Section 84 financing.

(*b*) the source of the funds for the purchase of the shares in Berisford International plc and Unigate plc was a matter of concern to a number of the Creditor Banks with whom he was negotiating a scheme of arrangements.

(*c*) he prepared a schedule of the sources of the funds for each of these two investments.

(*d*) the total cost of the investment in Berisford International plc was £174.3m sterling including the initial acquisition costs, and margin deposits and carrying costs.

(*e*) he was satisfied from his inquiries that £90m sterling being portion of the initial acquisition cost which was £98.7m sterling was funded by a syndicate headed by Bank Nationale de Paris and carrying costs of £54m sterling was sourced from a spread of Banks, none of which had an Irish operation.

(*f*) to him the source of the balance of £30.3m sterling comprising the balance of £8.7m sterling towards the original acquisition costs and £21.6m sterling in respect of carrying costs was unclear.

(*g*) the total cost of the investment in Unigate plc was £110.9m sterling including initial acquisition costs and margin calls.

(*h*) the purchase consideration of £62.6m sterling had been sourced through Bank Paribas and margin calls of £33m sterling were sourced from a list of Banks none of which had an Irish place of business through which they could avail of Section 84.

(*i*) he did not undertake any examinations to trace how the Section 84 loans were used but was satisfied that the amounts to which he referred as being sourced by Banks with no operations in the State were not loans to which Section 84 applied.

(8) In the course of his evidence to the Tribunal Mr Mooney of SKC stated that:

(i) AIBP paid the £30.3m sterling about which Mr Fitzpatrick was unclear in respect of the shares in Berisford International plc, between September 1988 and August 1990.

(ii) AIBP paid the £15.3m sterling about which Mr Fitzpatrick was unclear in respect of the shares Unigate plc during the period November 1988 to August 1990.

(iii) AIBP never borrowed Section 84 funds.

(iv) AIBPI was the only company within the group which availed of Section 84 finance.

Mr Mooney produced the accounts of AIBPI for each of the years ending 31.12.1986, ‘87, ‘88, ‘89 and ‘90.

These were analysed by both Mr Mooney and Mr O'Donghaile of the Revenue Commissioners and such analysis, together with the evidence of Mr Fitzpatrick satisfied the Tribunal that the Section 84 funds borrowed by AIBPI were used for working capital purposes and no portion thereof was used for any other purpose including the acquisition of shares in Berisford International plc and Unigate plc.

In the course of his speech on the 15th day of May 1991 Deputy Rabbitte had stated:—

> "We also know that in the Finance Act the Government made a special arrangement to enable Mr Goodman to avail of High Coupon Finance for the schemes that I referred to earlier".

In the course of his speech on the Companies (Amendment) Bill 1990 Deputy Spring had stated, in the course of detailing the support given to the Goodman Group by the Fianna Fáil Government, that:—

> "Support included changes in the tax laws, to enable a substantial amount of Mr Goodman's income from beef processing to be taxed at the 10 per cent manufacturing rate. Further changes included provisions which made Section 84 financing for Mr Goodman more advantageous."

In these extracts, there are contained three instances of alleged changes in the tax laws enacted for the benefit of the Goodman Group of Companies viz.

(i) in the Finance Act, the Government made a special arrangement to enable Mr Goodman to avail of High Coupon Finance.

(ii) provisions were made to make Section 84 financing more advantageous to Mr Goodman.

(iii) changes were made in the tax laws to enable a substantial amount of Mr Goodman's income to be taxed at the 10% manufacturing rate.

It appears from the Memorandum for Government dated the 16th June 1987 and prepared by the office of the Minister for Agriculture and Food seeking the approval for grant aid by the Industrial Development Authority towards the cost of a major capital development by Goodman International Ltd. of its Irish meat operations that the total plan would cost an estimated £261m over an eight year period of which the capital expenditure would amount to £120m.

It was proposed that the capital expenditure would be funded as follows:—

(*a*) IDA Capital Grants		£25m
(*b*) IDA Redeemable Preference Shares		£5m
(*c*) Capital Value of Section 84 loans		£30m
(*d*) FEOGA aid sought		£30m
(*e*) Shareholder's Funds		£30m
		£120m
Working Capital provided by Shareholder		£141m
	TOTAL:	£261m

The memorandum further stated that:

3. "The Section 84 currency swap arrangements will require both Central Bank and Revenue clearance. They comprise an essential element of the financial package for the promoters and any diminution of the expected interest savings would undermine the proposed project financing. There is, however, no commitment by the IDA to make good any reduction in such interest savings. An estimated two thirds of the £30m capital value of the Section 84 arrangements will in practice be met by the Exchequer by way of tax relief for domestic financial institutions; the remaining one third approximately will use overseas tax capacity."

12(ii) of the memorandum provided that:

The Government is asked to note:

(ii) "that the financing of the project is based on the assumption of the availability of a loan facility of £120m under the "Swap" currency Section 84 scheme (as available in line with current Central Bank and Revenue Commissioners procedures) and on the assumption that no liability to Capital Gains Tax in respect of such loan facility would arise in this case."

The Minister for Finance's reply to this statement is contained in the memorandum and is as follows:—

"The Minister for Finance points out that the claim in Para. 12.2 of the Memo that the Section 84 currency swap arrangement is "available in line with current Central Bank and Revenue procedures". There are no such arrangements in existence. The Revenue Commissioners have ruled that the one other case presented to them to date was in breach of the law. With regard to capital gains tax the Minister cannot say whether a C.G.T. liability will arise in the absence of details."

On the 16th June 1987 the Government agreed to support the development and approved the financing package as outlined in the memorandum involving:—

(1) IDA capital grants totalling £25,000,000,

(2) IDA redeemable preference shares of £5,000,000 to be taken up by the Authority and repayable at £1,000,000 per annum after 5 years, and

(3) Currency swap loans under Section 84 of the Corporation Act Tax 1976, with an estimated capitalised value of £30,000,000.

It is clear from the account of the negotiations between the IDA and the Goodman Group that the Goodman Group at the early stage of such negotiations wanted a grant package of 75% of the capital costs of the scheme and only accepted the agreement with regard to the swap currency Section 84 finance with reluctance.

The initial response of the Goodman Group to the suggestion that portion of the development could be financed by such borrowing was that it would not work because of

(i) the Group's existing structure.

(ii) the fact that the definition of "specified trade" would not allow the Group to avail fully of the provisions of Section 84.

(iii) the requirement that 75% of the goods sold by the exporting Company AIBP had to be manufactured by a company with which there was a 90% joint ownership relation.

(iv) the fact that AIBPI sold beef purchased from intervention in addition to that purchased from AIBP and provision for this situation would require an amendment of Section 84.

The IDA indicated that it was for the Group to seek such amendment from the Department of Finance but that it would support any change that was necessary in order to realise the Beef Development Plan.

On the 2nd July 1987 Goodman International submitted to the Deptartment of Finance that Section 84 of the Corporation Tax Act, 1976 be amended. The terms of the submission are set out hereunder:

PROPOSED AMENDMENT TO SECTION 84 A OF THE C. TAX ACT 1976 TO FACILITATE THE BORROWING OF SECTION 84 FINANCE BY ANGLO IRISH BEEF PROCESSORS INTERNATIONAL LIMITED

The suggested addition to Sub-section 3 of Section 84 A of the CT Act 1976 would be as follows:

(*d*) A trade which consists wholly or mainly of either or both of—

(i) the manufacture of goods within the meaning of paragraph (*a*), and

(ii) the selling by wholesale of goods manufactured within the State which by virtue of Sub-section (3) of Section 54 are brought within the definition of "goods" in Sub-section (1) of that Section.

Commentary

(1) Anglo Irish Beef Processors International Limited (Anglo Irish) sells processed beef on the export market by means of sale by wholesale. Anglo Irish carried on this activity prior to 1 January 1981 and, in consequence, is entitled to claim export sales relief on the profits derived from that activity.

(2) The sales by Anglo Irish include beef processed by Anglo Irish Beef Processors Limited which is a fellow subsidiary of Anglo Irish and beef purchased from intervention stock which were processed by other beef processors within the State.

(3) Anglo Irish now wishes to borrow substantial Section 84 funds to finance its export sales and debtors. Section 84 A of the CT Act 1976 as introduced by Section 41 of the Finance Act, 1984 requires that the borrower carry on a "specified trade" within the meaning of that Section. A specified trade means a trade which consists wholly or mainly of:

(*a*) The manufacture of goods (including activities) which would qualify as the manufacture of goods under section 39 of the Finance Act, 1980 were the borrower to make a claim for relief under that Section. Section 39, 1 (A) and (B) provides that goods sold by a company which are being manufactured by its fellow subsidiary (providing the share capital in both companies have at least 90% common ownership) shall be deemed to have been manufactured by the company selling them. In consequence, the sale by Anglo Irish of the beef processed within the Group is deemed to be a sale of manufactured goods for the above-mentioned purposes. However, paragraph 5 of Section 84 A states that a trade will only consist wholly or mainly of the manufacturer of goods if the total amount received from the sale of qualifying goods is not less than 75% of the total amount receivable from all sales. The sale by Anglo Irish of processed beef purchased from intervention will not qualify as manufactured goods under Section 39 and, in consequence, if the sales of this type of goods in any accounting period exceed 25% of all sales, then the entire borrowings would not qualify as Section 84 borrowing under the provisions of Section 84A.

(4) We have reviewed the issue as to whether classifying Anglo Irish as a trading house under Section 29 of the Finance Act, 1987 would bring the sale of goods purchased from intervention within Section 39 of the Finance Act, 1980 while still not preventing Anglo Irish from claiming export sales relief on the profits deriving from those sales. We have come to the conclusion that this is not possible because in order to be defined as "export goods" within the meaning of Section 29, it would be necessary for Anglo to be claiming the 10% reduced rate of corporation tax, rather than export sales relief, which is the relief which we understand will continue to be claimed by Anglo.

(5) In order to deal with the above-mentioned problem, we suggest a commitment be obtained for the Finance Act, 1988 to include the amendment as outlined above.

2 July 1987

The reasons for seeking the said amendment as set out in the said submission were that:

(i) under the existing legislation a company to claim relief under the Section must either manufacture the goods sold or they must be manufactured by a fellow subsidiary (provided that the share capital in both companies have at least 90%),

(ii) if the company however purchased goods manufactured by other companies for the purpose of export, then the entire borrowings would not qualify as Section 84 borrowing under the provisions of Section 84A unless if such purchases exceeded 25% of the entire goods sold.

As it was the practice of Anglo Irish to purchase beef from intervention for the purpose of export, the amendment was sought.

The submission suggested that the proposed amendment be included in the Finance Act 1988.

On the 20th day of January 1988 Laurence Goodman wrote to the Minister for Finance as follows:—

"20 January 1988 Our ref: LJG/GB

Mr Ray McSharry TD,
Minister for Finance
Government Buildings
Merrion Street
Dublin 2

Re *Section 84 Financing*

Dear Minister

You will be aware of the Goodman Group Development Programme announced last September. In our negotiations on the programme with the IDA and the Department of Agriculture and Food we received an assurance that they would pursue with your Department the question of a minor amendment to Section 84 of the Corporation Tax Act 1976.

The amendment is of a technical nature and its effect is to enable us to use Section 84 borrowings more fully in funding our export business. I enclose a memorandum prepared by Stokes Kennedy Crowley & Co., detailing the proposed amendments and setting out a more complete commentary on its effects.

The Minister for Agriculture and Food may have been in contact with officials of your Department on this matter already. I should be grateful if you would confirm that the Finance Bill will contain an amendment along the lines suggested in the enclosed draft.

I look forward to hearing from you.

Yours sincerely
GOODMAN INTERNATIONAL LIMITED
LAURENCE J. GOODMAN
Chairman and Chief Executive.
14 Castle Street, Ardee, Co. Louth, Ireland"

A submission from Mr D Quigley, of Department of Finance to the Minister dealing with Section 84 loans and dated the 11th March 1988 included the following:

"The *second* Section 84-related item concerns a letter from Goodman International requesting an amendment to the Section 84 provisions to allow a particular company in their group to take a substantial Section 84 loan to finance its export sales and its debts. This company sells processed beef on the export market on a wholesale basis and gets export sales relief on the profits from these sales. The beef is obtained partly from another company in the group which processes it and partly from intervention. If the intervention portion exceeds 25 per cent of the total sales of the exporting company in any accounting period, then the company will not qualify for a Section 84 loan because of the requirement that at least 75 per cent of the sales in this instance has to come from the associated company which processed the beef. Goodman International state that they want to use Section 84 loans more fully in funding their export business and they have suggested an amendment to cover this intervention situation.

The amendment suggested by them would allow a non-manufacturing company to qualify for a Section 84 loan if it sells goods by wholesale on the export market and such goods are manufactured in the State.

This amendment would extend the scope for Section 84 loans and we would be strongly opposed to it: the intention, as already decided by the Government, is to consider moving towards greater restriction."

The proposed amendment was not included in the Finance Bill 1988 and on the 28th September 1988 Brian Britton, Deputy Chief Executive of Goodman International wrote to Mr Sean Donnelly, who had been seconded from the IDA to the Taoiseach's office as follows:—

"Goodman International

28 September 1988

Mr Sean Donnelly
The Taoiseach's Office
Dail Eireann
Kildare Street
Dublin 2

Dear Sean

Further to our telephone conversation of yesterday, I set out briefly the amendment which was supposed to be in the Finance Bill.

Section 84 is available only to a company which carries on a *specified trade*. A specified trade is one where not less than 75% of sales consist of the sale of goods manufactured by the company or by Group companies.

If in any accounting period non-Group processed sales by AIBP International exceed 25% of total sales then AIBP International Limited is not carrying on a specified trade and cannot avail of Section 84. Therefore, the bank would be taxable on any interest received from AIBP International. The bank could then presumably call on AIBP International to pay further interest to cover its tax exposure.

Our proposed amendment was that AIBP International Limited could avail of Section 84 if it exported by wholesale Irish processed beef, notwithstanding that the beef might have been processed by an unconnected company.

I enclose a copy of a detailed submission which we made to the Department of Finance on 2 July, 1987 regarding this matter and which will have to be reactivated. I note that you will be speaking initially to the IDA and thereafter to the Department of Finance. If I do not hear from you within the next 10 days I will contact you to see what progress has been made on the matter and what further action is required.

Kind regards.

Yours sincerely
GOODMAN INTERNATIONAL LIMITED
BRIAN BRITTON
Deputy Chief Executive — Finance"

The question of this proposed amendment was again the subject of a submission dated the 21st March 1989 in connection with the Finance Bill 1989 which having reviewed the position concluded as follows:—

> ".......the same reasons that applied last year are still valid and we see no reason to change our recommendation."

The Minister for Finance agreed with this recommendation and no legislative changes were introduced.

Consequently the statements made by both deputies with regard to amendments of the provisions of the Finance Acts with regard to Section 84 borrowings to make them more advantageous to the Goodman Group are incorrect.

Mr Liam Murphy, Principal Officer, Budget Section of the Department of Finance stated in evidence before the Tribunal that there was no change in the Section 84 legislation contained in either the 1987 or 1988 Finance Acts and while there were changes in the 1989, 1990, 1991 and 1992 Finance Acts, such changes were designed to restrict the application of Section 84 and not to extend it.

While this proposed amendment was never enacted into legislation it would appear however that Goodman International during the course of negotiations with regard to the Development Plan sought that such an amendment would be made.

In the course of the First Annual Review of the Plan by the IDA and the Goodman Group, on the 21/4/1989, Mr Britton informed the IDA representatives that there would be no progress on the capital investment programme until certain matters, in respect of which Mr Britton said they had received assurances from the IDA, the Department of Agriculture and the Government, were forthcoming.

These included:—

(i) an assurance that Section 84 currency swap finance would not be curtailed during the eight year life of the Plan:—

(ii) a Government undertaking to obtain the legislative change to allow IABPI utilise Section 84 Finance more fully.

It would appear that the Goodman Group alleged that agreement had been reached on these proposals and that such agreement would be confirmed in writing and a draft letter containing such assurances was submitted through the Office of the Minister for Food to the Department of the Taoiseach on the 28th July 1987.

The terms of the draft letter were:

> "Dear Mr Goodman
>
> I refer to the package of assistance negotiated with the IDA in relation to your £120m development programme.
>
> I set out below the commitments which the Government have given as part of their support for the overall project in relation to Section 84 currency swap finance.

(1) I confirm on behalf of the Government, that the availability of Section 84 currency swap finance to the Group, Goodman International Ltd., will not be curtailed during the 8 year life of the project as a consequence of legislative changes.

(2) I undertake on behalf of the Government to obtain the following legislative change required to allow Anglo-Irish Beef Processors International Ltd. utilise the Section 84 finance. An addition to Sub-section 3 of the Section 84 (A) of the CT Act, 1976 would be as follows:—

"A trade which consists wholly or mainly of either or both of—

(i) the manufacture of goods within the meaning of paragraph (*a*), and

(ii) the selling by wholesale of goods manufactured within the State which by virtue of Sub-section (3) of Section 84 are brought within the definition of "goods" in Sub-section (1) of that Section."

Yours sincerely,"

A letter in the terms sought was not sent but on the 28th day of July 1987 the Minister for Food wrote:

"Dear Mr Goodman

I refer to the package of assistance negotiated with the IDA in relation to your £120m development programme.

I set out below the commitment which the Department of Agriculture and Food have given as part of their support for the overall project, specifically in relation to the FEOGA support for the package.

1. The Department of Agriculture and Food will seek a special FEOGA package for the Beef Industry.
2. In addition, the Department of Agriculture and Food will seek to obtain priority status for the development of the Irish Beef Industry within FEOGA
3. The Department of Agriculture and Food will strongly support the application by Goodman International for FEOGA grants at the maximum level."

However Mr Lowery of the IDA wrote to Mr Goodman on the 6/8/1987 and in connection with Section 84 funding stated:—

"1 *Section 84 funding.*

a) The Authority has approved and recommended to Government the agreed funding/incentive package which includes the provision of High Coupon Section 84 currency swap facilities. A submission to obtain Revenue Commissioners approval to the proposed arrangements will be made by the IDA.

b) The Authority will give full support to the legislative change required to allow Anglo Irish Beef Processors International Ltd to utilise the Section 84 Finance.

c) The Authority strongly supports and will assist in whatever way appropriate the Company's request to the Central Bank for approval of the currency swap mechanism.

d) The Authority supports strongly and will assist in whatever way appropriate the Company's submission to Government requesting assurances that the Section 84 currency swap facility will remain in place for the duration of the project.........."

The action of the Taoiseach in refusing to approve of the terms of the letter drafted by Minister Walsh containing commitments on behalf of the Government indicates that the Government was not in any way committing itself to the introduction of the legislative changes sought by the Goodman Group and did not in fact introduce such legislation.

On the 16th day of November 1987 Mr Britton wrote to the Minister for Food as follows:—

"Mr Joe Walsh TD
Minister for Food
Department of Agriculture and Food
Kildare Street
Dublin 2

Re: *Section 84 Currency Swap Finance*

Dear Minister

Further to commitments given by the Government during our negotiations on the Group's Development Plan earlier this year, wherein these commitments would be confirmed to us, in writing, I would now like to request formally such confirmation in writing.

Suggested wording was submitted through you to the Taoiseach's Department on 28 July, 1987. It was then suggested that Mr Padraic O'hUiginn, Secretary to the Government in the Taoiseach's Office would be contacting us to discuss the matter. We have had no contact from him other than a request also through you for a memorandum summarising the proposed change to the legislation together with an explanation on the change. This was submitted, once again via yourself, on 12 August 1987.

Following a further request from you, John Loughrey wrote to us on 11 September, 1987 indicating that the proposed change in the legislation raised by us could only be considered in the context of next year's Finance Bill, and that your Department was pursuing this matter with the Department of Finance.

We have raised the matter on a number of occasions with you subsequently but to date we have not received the required commitment from the Government in writing. The matter was also raised by Larry Goodman in a meeting with the Taoiseach this weekend.

I enclose, for your convenience, a copy of our required commitment letter and memorandum summarising the proposed change to the legislation, together with an explanation on the change.

I would be available to meet with you at any time to progress this matter.

Yours sincerely
GOODMAN INTERNATIONAL LIMITED
BRIAN BRITTON
Deputy Chief Executive — Finance"

Because of the amount of High Coupon Section 84 Finance required it was extremely important that it be known in advance the tax implication thereof.

By letter dated the 17th day of August 1987, the IDA sought from the Revenue Commissioners, an advance opinion on the efficiency of the proposed High Coupon mechanism applicable to the facts set out in the letter and the letter stated that the funds would be borrowed and used by AIBPI to fund working capital commitments.

The IDA sought confirmation

(*a*) that the interest payable on such borrowings would be treated as a qualifying distribution of a specified trade and

(*b*) that any exchange gain on the disposal of currency would be treated as income attributable to the sale of goods for the purposes of Section 58(4) of the Corporation Tax Act 1976.

By letter dated the 8th September 1987, Mr Frank Cassells on behalf of the Revenue Commissioners, gave the necessary confirmation based on the hypothetical fact set out in the IDA letter of the 17/8/1987.

In the course of his evidence Mr Cassells stated that

(i) the opinion as expressed by him was based on the existing law and practice.

(ii) the opinion was not influenced in any way by the context of the Beef Development Plan.

(iii) the opinion was based solely on the facts outlined in the letter.

(iv) in the case of Section 84 relief the Revenue construed the Act very strictly.

(v) it was the first occasion upon which the IDA had sought such an opinion.

(vi) the key issue was the second question viz the status for tax purposes of the exchange gain.

(vii) the Revenue Commissioners were forced to the conclusion that having regard to the wording of the Export Sales Relief provisions the gain could not be severed from the Company's trading income and as such was a receipt of export sales relieved trading and accordingly not liable to tax.

He particularly reiterated the Revenue Commissioners position viz that the opinion given was one given on a particular statement of facts : that if the actual facts differed from those stated, the opinion would have no validity and the fact that the particular funds were to be used in furtherance of the Beef Development Plan would not be a factor in considering the application of the relief.

The IDA acted quite properly in seeking the opinion of the Revenue Commissioners and the Revenue Commissioners acted independently in giving the opinion sought on the basis of the existing law and without reference to the Beef Development Plan.

With regard to the allegation made by Deputy Spring that "support included changes in the tax laws to enable a substantial amount of Mr Goodman's income from beef processing to be taxed at the 10 per cent manufacturing rate" Mr Liam Murphy stated that:—

(i) the 10% manufacturing rate had always applied to meat processing

(ii) An amendment was considered necessary to deal with the situation created by a number of decisions by the Courts which gave the benefit of this rate to activities such as banana ripening, coal grading, milk pasteurisation and grain drying which were not regarded as genuine manufacturing activities

(iii) the Finance Act 1990 introduced provisions to limit the definition of manufacturing and thereby deprive such activities of the benefit of the 10% rate of tax

(iv) In order to ensure that genuine manufacturing activities were not inadvertently excluded from the benefit of the 10% rate of tax, Section 41 of the Finance Act 1990 specifically provided that meat processing fish processing and certain other activities would continue to be regarded as manufacturing.

To suggest that such amendment was introduced for the benefit of Mr Goodman or his companies was obviously incorrect.

Section 52 of the Finance Act 1986 provided that capital allowance for plant and machinery would be reduced by the amount of any IDA or other grants paid from State sources.

On the 15th day of April 1987 Mr Brian Britton wrote to the Minister for Finance Mr Ray McSharry TD submitting a proposal on behalf of Goodman International for the amendment of Section 52 of the Finance Act 1986 as regards the Food Sector.

This submission entitled "Capital Allowances on Grant Aided Assets for the Food Industry" proposed that

"1. Companies in the Food Sector purchasing plant and machinery (for their own use) should be exempt from Section 52 of the Finance Act 1986

2. The section would still apply to leased assets and to capital expenditure in the non-food sectors"

and expressed the belief that "the major effects on the economy as a result of this proposal would be

(1) Increased investment in food processing technology by the Food Processing Sector.

(2) A significant increase in export earnings as our food industry increases its raw material base to accommodate its market lead development in export sales.

(3) The development would shift the emphasis away from commodity trading thereby offering more stable returns to the farmer and creating an environment whereby

he can expand his activities without fear of widely fluctuating prices or over-reliance on price support mechanisms.

The submission contained detailed arguments in support of the case for such amendment. The necessity for such amendment had been indicated by the Goodman Group to Mr John Loughrey of the IDA who prepared a briefing on the Development for the Government Meeting to be held on the 26th April 1987.

The Minister for Industry and Commerce Mr Reynolds had written to the Minister for Finance in connection with this matter on the 16th of April 1987 as follows:—

"OIFIG AN AIRE TIONSCAIL AGUS TRACHTALA
(Office of the Minister for Industry and Commerce)
BAILE ATHA CLIATH 2
(Dublin 2)

"16 April 1987

Mr Ray MacSharry TD
Minister for Finance
Government Buildings
Dublin 2

Dear Ray

I refer to our discussions concerning tax-based leasing.

I consider that the changes made in the Finance Act 1986 in relation to such leasing were retrogressive.
I propose to discuss this matter with you in the near future with a view to devising appropriate remedial provisions for inclusion in the year's Finance Act.

Yours sincerely

Albert Reynolds TD
Minister for Industry and Commerce"

On receipt of the submission the Minister for Finance sought from Mr Maurice O'Connell a note on the subject, which he obtained on the 23rd April 1987.

In this note he expressed strong reservations about the amendment and indeed drafted a letter to Mr Britton for signature by the Minister for Finance in the following terms:—

OFIG AN AIRE TIONSCAIL AGUS TRACHTALA
(Office of the Minister for Industry and Commerce)
BAILE ATHA CLIATH 2
(Dublin 2)

April 1987

Mr Brian Britton
Deputy Chief Executive — Finance
Goodman International Limited
14 Castle Street
Ardee
Co Louth

Dear Mr Britton

Thank you for you letter of 15 April 1987 about section 52 of the Finance Act 1986.

This section, as you know, provides that capital allowances will be given on the basis of the expenditure actually incurred by the claimant. This is entirely appropriate and, in fact, capital allowances here in Ireland remain quite generous by international standards. 100% allowances in the first year are, for example, no longer available in the UK or the US.

The previous arrangements which obtained in relation to capital allowances, i.e. calculation of the value of allowance without regard to grants paid from State sources, represented an entirely unacceptable *double* charge on the Exchequer. These arrangements involved, in effect, extending a full measure of relief from taxation on expenditure which the Exchequer itself had incurred..

In these circumstances, and having regard to the extremely tight budgetary circumstances this year, I regret I am not in a position to contemplate amending section 52 as proposed by you.

Yours sincerely

____

Ray MacSharry TD
Minister for Finance."

The letter was neither signed nor issued.

At its meeting on the 26th April 1987, the Government decided in relation to the Beef Development Plan that:—

> "The Minister's for Agriculture and Food, Finance and Industry and Commerce should make every effort to bring the project to a successful conclusion, in particular, by investigating the possibilities for meeting the financing requirements of the project in part by

(i) excluding the food processing industry from the scope of Section 52 of the Finance, Act 1986 and/or adjusting the proposed FEOGA financing of the project, or

(ii) reversing the abolition of depreciation allowances on gross capital costs in toto, and

(iii) a revision of disadvantaged areas so as to include County Louth".

The exclusion referred to at (1) was effected by Section 25 of the 1987 Finance Act in respect of food processing companies which purchased their own plant and machinery.

Mr Murphy stated in evidence that the amendment was limited in its effect on the Exchequer and no tax benefit would accrue to a Company for a particular project if for whatever reason the project did not go ahead and no State grants issued for plant and machinery.

As Goodman International did not, in the events which happened, received any State Grants for plant and machinery they did not benefit from the amendment.

If the Development Plan had gone ahead and Goodman International had received the projected grants for plant and machinery, Goodman International would have benefitted by a further £6m approximately.

So far as the Tribunal has been able to ascertain, this is the only amendment to tax laws introduced by the Government at the request of the Goodman Group.

The request for the amendment was made not only in its own interests but in the interests of the food processing industry as a whole but there is no doubt but that the amendment was made at the request of Goodman International and in the context of the Beef Development Plan which the Government was actively encouraging.

CHAPTER NINE

Tax Evasion and Avoidance

TAX EVASION

In the course of the ITV programme, Mr McGuinness had stated that:—

> "The company had a wide scheme of under the counter payments. Cheques were made out against bogus invoices, endorsed by Goodman Employees and cashed at local branches of the Allied Irish Bank. These cheques were payable quarterly in March, June, September and December of each year. They were paid to everyone in the company from the floor up and amounted approximately to 3 million pounds per year".

In addition to dealing with this allegation, the Tribunal finds it convenient in the interest of brevity and to avoid repetition, to deal with the allegations made in Dail Eireann with regard to "Under the Counter" payments to employees. These allegations were made by Deputy Pat Rabbitte on the 28th day of August 1991 and the 15th day of May 1991 and can be summarised as follows:—

> "(1) That because of Goodman's political connections the Revenue Commissioners turned a blind eye to the type of "remuneration packages" enjoyed by Senior Executives and to a non return of PAYE and PRSI to the Exchequer because of the operation of the contract system for large proportion of the Goodman work force.
>
> (2) A great many of Goodman workers were on the dole and were being paid "under the counter".
>
> While both allegations are undoubtedly serious, the first one is particularly serious because it challenges the independence of the Revenue Commissioners and its freedom from political interference."

In the course of his speech in Dáil Eireann on the 15th day of May 1991, Deputy Dick Spring stated:

> "Mr Goodman, at his press conference last night referred to various tax schemes in operation for the employees of his company. He referred to them as "bona fide" schemes and advised the media that they had been brought to the attention of the Revenue Commissioners and that the situation was now fully regularised.
>
> I understand that what happened in this case is that following intensive negotiations with the Revenue Commissioners, the Commissioners agreed not to take proceedings against Mr Goodman or his company in respect of large scale tax evasion practices going back over many years. In return, Goodman International paid the Revenue Commissioners £4 million in respect of all outstanding liabilities and penalties. That, I am told, is by far the largest settlement of its kind in the history of this State. Quite frankly, I find it very difficult to understand how anyone can effectively admit to tax evasion on that scale and still escape scot-free from any kind of prosecution."

In the course of his speech in Dáil Eireann on the 15th day of May, 1991 Deputy Bruton stated:—

> "A very serious allegation was made on the programme with regard to taxation. It was alleged that under-the-table payments were being made to employees of the Goodman Group at a rate of £3 million per year. Presumably, these payments were made in a way that the payment of income tax was avoided. At his press conference yesterday Mr Goodman said this matter was first brought to his attention last August when the exáminer was appointed. He also said that meetings followed with the Revenue Commissioners after which, to quote Mr Goodman, "everything was regularised". What does this mean? Ordinary individuals who evade tax cannot simply go to the Revenue Commissioners, have a few meetings with them and then have "everything regularised". This is not available to me, or to any other taxpayer. Is it possible to have things quietly regularised if one is a big company but not possible if one is a small company or a private individual? Do we have equality before the law with regard to tax matters? This matter must be sorted out in this public inquiry, otherwise those on PAYE will feel that there is one tax law for the rich and another for everybody else."

Deputy Bruton on the 24th day of May, 1991 stated that:

> "the writing off of £4m in taxes in respect of under-the-counter payments to Goodman employees was a wrong judgment on the part of the Revenue Commissioners."

In support of Mr McGuinness statement on the ITV programme, ITV in the course of their submission in writing to the Tribunal had enclosed a number of documents which had been referred to in the programme namely a set of bogus livestock purchase remittances together with photo-copies of cheques drawn against them; bogus purchase remittance documents without corresponding cheques; a document relating to bogus haulage services and two documents showing hand-written calculations of unreported payments to be made to employees. Eighteen in number receiving a total of £12,135 to be paid by bank drafts and such payments to be recorded as a payment to a named haulier, (Keenan Transport).

With his written submission Deputy Rabbitte forwarded to the Tribunal documents which he had received from an anonymous source on AIBP note paper relating to two payments totalling £8,280 and £3,278.06 respectively made to workers and two invoices for these amounts alleged to have been paid to a named haulier and a copy cheque for £3,750 Stg (cash) drawn on AIB Newry. On receipt of these documents and having considered the implications of same, the Tribunal caused a number of enquiries to be made in relation thereto.

In the first instance, the attention of the Revenue Commissioners was drawn to the allegation and to the system whereby such payments were concealed in the returns / records of the company. In the course of a meeting held on the 7th day of August 1991 between Solicitor and Counsel to the Tribunal and Messrs S Moriarty Assistant Secretary and P O'Duinn, Inspector of Taxes attached to the Investigation Branch of the Revenue Commissioners the Tribunal sought their assistance in determining the truth or otherwise of the allegations into which the Tribunal was obliged to inquire. The Revenue Commissioners through the officers of the Special Investigation Branch agreed to provide this assistance and to fulfil their own responsibilities in this regard once the fact had been brought to their attention.

The Tribunal wishes at this stage to acknowledge the assistance and support which it received from the Officers of the Revenue Commissioner in particular Mr O'Donghaile and Mr O'Duinn and to state at the earliest opportunity in this Report that there was no basis for Deputy Rabbitte's allegation that the Revenue Commissioners had turned "a blind eye", whether because of political connections or otherwise to the activities of the Goodman Group in relation to Tax Evasion.

On the 2nd and 3rd days of September 1991 the Tribunal wrote to the Secretary of Anglo-Irish Beef Processors, Ferry Bank, Waterford and the Secretary AIBP Ravensdale, and approximately 50 (fifty) employees of the said companies named in the documents given to the Tribunal by Mr McGuinness and Deputy Rabbitte enclosing copies thereof and seeking information with regard thereto including the full names and addresses of the persons named in the bogus invoices. Copies of all these letters were sent to the solicitors to the Goodman Group of Companies who represented not only the said companies but the employees thereof. By way of illustration of the matters in respect of which information was sought the Tribunal refers to one such letter to an employee/accountant of the Goodman Group which was as follows:—

" 2 September '91

Liam Coleman, Esquire
7, Rockcourt,
Blackrock
Co. Louth

Re: Tribunal of Inquiry — Beef Processing Industry

Dear Mr Coleman

The Government of Ireland by Resolution passed by Dail Eireann on the 24th day of May, 1991 and by Seanad Eireann on the 29th day of May, 1991, established a Tribunal of Inquiry, which Tribunal of Inquiry was appointed by Warrant of the Minister for Agriculture and Food dated the 31st day of May, 1991.

The Terms of Reference of the Tribunal are as follows:—

1. To inquire into the following definite matters of urgent public importance:
   1. Allegations regarding illegal activities, fraud and malpractice in and in connection with the beef processing industry made or referred to (*a*) in Dail Eireann and (*b*) in a television programme transmitted by ITV on May 13th, 1991.
   2. Any matters connected with or relevant to the matters aforesaid which the Tribunal considers it necessary to investigate in connection with its inquiries into the matters mentioned at 1. above.
2. To make such recommendations (if any) as the Tribunal having regard to its findings thinks proper.

The Tribunal as part of its inquiries into the matters referred to above is investigating matters in relation to the payment of employees both prior to and at the time of the appointment of the Examiner.

It has been suggested that you as the Internal Auditor for the Company is the person best able to assist the Tribunal in relation to matters following part of its inquiry. We would appreciate at this time whether you would confirm and let us know the following matters:—

1. Are you the Internal Auditor for the Group?
2. What is your function as Internal Auditor?
3. Confirm that you are in a position to assist the Tribunal in relation to all matters affecting employees paid within the Group.
4. Confirm how all employees were paid.
5. Confirm how senior management were paid.
6. What cash payments were made by the Group to its employees in any position.
7. What cash payments were made to Senior Executives or senior employees in any position.
8. How were cash payments by the Group to employees disguised.
9. Confirm that some cash payments were disguised by creating false invoices made payable to either:—
   (*a*) existing or non-existing haulage companies;
   (*b*) additional payments to existing or non-existing farmers or dealers in the form of invoices;
   (*c*) invoices made out to non-existing persons.
10. Confirm that when these invoices were made out the cheques were then made payable for the invoices cashed at the bank and the cash was either received or alternatively converted into drafts.

The Tribunal would appreciate if you would detail all cash payments made by the Group to its employees together with their names and addresses from 1st January, 1985.

This is an information request and therefore should be replied to by you to the Tribunal. You are free to discuss the matter with your solicitor and may make a statement through your solicitor should you so require.

The Tribunal would appreciate an immediate response to this letter and we await hearing from you.

We are sending a copy of this letter to the Company's Solicitors, Messrs A & L Goodbody for their information.

Yours faithfully
Mr Justice Liam Hamilton
President of the High Court
Sole Member of the Tribunal of Inquiry."

On the 9th day of September 1991, Messrs. A & L Goodbody wrote to the Tribunal in the following terms:—

"Dear Miss Loughlin,

We have received copies of approximately 50 letters addressed to employees.

We have consulted with Counsel.

We wish to state that these letters taken as a whole constitute an attack upon and a violation of our rights under the decision in In Re Haughey [1971] IR.

We have clearly indicated in previous correspondence our clients' wish to reserve their position, in accordance with their rights under In Re Haughey, and we would refer also to the express statement of their entitlement to that effect in the Tribunal's letter of the 26th July.

Taking as an example the letter of 2nd September, addressed to Mr Liam Coleman, this letter is an attempt to elicit evidence on company affairs, from a company employee. It bears directly on our rights under In Re Haughey. Our clients rights under that decision are effectively eliminated if the interrogation of company employees in the manner of Mr Coleman's letter continues.

It is , to borrow a phrase from the judgement of C. J. O'Dalaigh, another case of 'clocha ceangailte is madrai scaoilte'. The protections of In Re Haughey are meaningless and worthless, and they simply disappear if the Tribunal interrogates company employees on company affairs in the manner of your letter of 2nd September. You will be aware that we still have not been told the allegations which are supported by evidence and are therefore the subject of ongoing enquiry by the Tribunal.

We would submit and are advised that the Tribunal has no power to conduct this sort of interrogation of the witnesses of an accused party, and should not persist in doing so after we have clearly stated our position.

In addition to the foregoing, the format of some of the questions give rise to other serious concerns.

Take question 8 in Mr Coleman's letter:

(8) How were cash payments by the Group to employees disguised?

This seems to involve a prejudgment or determination of a serious issue without our clients being heard.

We refer also to a copy letter which you have sent to Mr J Peters dated 4th September and which asserts that a copy is being "sent to your employer's solicitor at the insistence of their solicitors A & L Goodbody".

The entitlement of our clients as de facto co-accused, to see letters making allegations against their employees, arises from the Constitution of Ireland and not from the insistence of A & L Goodbody or our Counsel or anybody else

It is for the Tribunal to identify and vindicate the constitutional rights of parties with whom they deal. If the matter is one of constitutional right (and we are advised decisively that it is) then to characterise it as arising merely from some insistence on our part, lends an unfair gloss to the letter."

Because of the failure to secure answers to the questions and information sought in the aforesaid letters, which was readily available to the Goodman Group of Companies for the reasons set forth in this said letter dated 9th September 1991 the Tribunal was obliged to continue with its inquiries and devote considerable time thereto.

On the 26th August 1991, the Tribunal wrote to George McMillen, Esquire, General Haulier, Crossmaglen, Co. Armagh being one of the hauliers named in the documentation furnished to the Tribunal in the following terms:-

"

26th August 1991.

George McMillen, Esq.,
General Haulier
Crossmaglen
Co. Armagh

RE: Tribunal of Inquiry — Beef Processing Industry

Dear Sir

The Government of Ireland by Resolution passed by Dail Eireann on the 24th day of May, 1991 and by Seanad Eireann on the 29th day of May, 1991, established a Tribunal of Inquiry, which Tribunal of Inquiry was appointed by Warrant of the Minister for Agriculture and Food dated the 31st day of May, 1991.

The Terms of Reference of the Tribunal are as follows:—

1. To inquire into the following definite matters of urgent public importance:

   1. Allegations regarding illegal activities, fraud and malpractice in and in connection with the beef processing industry made or referred to (*a*) in Dail Eireann and (*b*) in a television programme transmitted by ITV on May 13th, 1991.

   2. Any matters connected with or relevant to the matters aforesaid which the Tribunal considers it necessary to investigate in connection with its inquiries into the matters mentioned at 1. above.

2 To make such recommendations (if any) as the Tribunal having regard to its findings thinks proper.

The Tribunal considers that you may be in a position to be of assistance to the Tribunal to help it in considering the matters raised under its Terms of Reference.

In particular, the Tribunal has been furnished with an original invoice purporting to be from your company dated 21st day of March, 1986, made out to AIBP Newry in respect of refrigerated transport for January / February 1986, for the sum of £9,524 Sterling. We enclose herewith copy of document.

We would appreciate the following information:—

1. Confirmation that you received the sum of £9,524 Sterling in respect of this invoice.
2. Copy of the receipt issued by you for the sum of £9,524 Sterling in respect of this invoice.
3. Full details of how the sum of £9,524 Sterling was made up for refrigerated transport between January and February of 1986 indicating:—

   (*a*) the nature of the vehicle, and/or the container;

   (*b*) the serial number or registration number of same;

   (*c*) the destination of each load.

As this is a request for information the Tribunal would appreciate an urgent and prompt response. However, if you have any difficulty in relation to understanding the context of same do not hesitate to contact your own solicitor and should he wish, he can contact us with a view to clarifying. However, we would appreciate an urgent reply and await hearing from you by return.

Yours faithfully

____

Mr Justice Liam Hamilton
President of the High Court
Sole Member of the Tribunal of Inquiry."

GEORGE McMillen
General Haulier
Crossmaglen Co. Armagh

M. AIBP Newry
Newry
Co. Down

Date: 21 March 1986.

	Description	Amount
21 March	Refrigerated Transport for January/February 1986	£9,524 Stg.

On the 5th September 1991, this letter was returned to the Tribunal office in Dublin Castle by the Royal Mail, Northern Ireland with the following comment; "Incomplete Address"

The Tribunal, on receipt of the return of this letter, wrote to Anglo Irish Beef Packers (Newry) Ltd., on the 12th September 1991, at the same time sending a copy to Messrs A & L Goodbody, Solicitors, in the following terms:—

"Secretary
Anglo Irish Beef Packers (Newry) Ltd.,
Warrenpoint Road,
Newry
Northern Ireland VT3 42PD. 12 September '91

Re: Tribunal of Inquiry — Beef Processing Industry

Dear Sir,

The Tribunal has been furnished with an invoice (copy enclosed herewith). The invoice is in respect of work done by George McMillen. General Haulier, Crossmaglen, Co. Armagh for the sum of £9,524 sterling in respect of work done prior to 21st March, 1986.

The Tribunal wrote to George McMillen, in respect of the invoice and the envelope has been returned by the Royal Mail as "address incomplete".

The Tribunal would appreciate if you would arrange to make available the full address of George McMillen to enable him to deal with queries being raised by the Tribunal.

In respect of the invoice the Tribunal would further appreciate if you would give the following information:—

1. Full details of the work allegedly done by George McMillen to earn the sum claimed.
2. Copy of the cheque (front and back) given to George McMillen to pay for the sum.

It is alleged that this document is a document prepared for the purposes of obtaining money from the bank to enable the Company to pay its workers in cash. It is this allegation the Tribunal is inquiring into but at this time requires the information referred to above to enable it to inquire further into the allegation.

The Tribunal would appreciate an urgent and early reply for this information and are sending a copy of this letter to your solicitors Messrs A & L Goodbody.

Yours faithfully
Mr Justice Liam Hamilton
President of the High Court
Sole Member of the Tribunal of Inquiry."

No response was received to this letter from either the company or their solicitors.

At the same time on the 12th September 1991 the Tribunal wrote to the Royal Ulster Constabulary in Crossmaglen and in its material part wrote as follows:—

"The Tribunal as part of its inquiry into matters covered by its Terms of Reference has been furnished with an invoice purporting to be an invoice of one George McMillen, General Haulier, Crossmaglen, Co. Armagh, it is dated 21st March, 1986 and we enclose a copy herewith.

The Tribunal wrote a letter to the person at the address on the invoice and it was returned by the Royal Mail as "address incomplete". The Tribunal would appreciate your assistance in confirming the address of the person named above or confirmation that he did not and does not exist.

The Tribunal would appreciate an early response and thanks you for your co-operation in anticipation."

The Royal Ulster Constabulary responded by a statement of the 18th October 1991, from Sergeant D. Meeke in the following material terms:—

"I am aware of an enquiry from the Tribunal of Inquiry, Tribunal Office, Upper Yard, Dublin Castle, Dublin 2., into the whereabouts of a George McMillen, General Haulier, Crossmaglen. I have been attached to Crossmaglen RUC Station since May 1988 and at no time during my service in Crossmaglen RUC Station have I become aware of a George McMillen, General Haulier, Crossmaglen. I have carried out an exhaustive search of records held at Crossmaglen RUC Station and can find no evidence to suggest that this man or indeed the Company exists."

On the 17th September 1991 Mr S Mooney of Stokes, Kennedy Crowley Accountants, contacted Mr O'Donghaile of the Revenue Commissioners and in an interview noted by Mr O'Donghaile and subsequently given in evidence before the Tribunal Mr Mooney said that:—

"in the course of discussions during the examination process in Autumn 1990 he had advised us that the company had paid amounts to executives and employees (loans/dividends) which had not been subjected to tax. He said that the Board of Goodman Group had now become aware that there were further payments, in excess of £2m. made to employees which had not been subjected to tax. The Board had contacted him and he had advised that the Revenue should be informed. He said that his information, was that the Board had not been aware of the payments. He was not giving a preliminary outline of the position. He emphasised that he had not been aware of these payments during our discussion last year.

He said that the payments, which were in excess of £2m. were particularly prevalent in the Dundalk operation but would have existed in other plants to a lesser extent. He said that the payments were made to various categories of shop floor employees and casuals. He stated that there was a system of bogus/fictitious invoices relating to hauliers or farmers in operation and that as far as he knew the payments were cloaked by these. The payments had been made over 5 or 6 years. He said staff in SKC's were presently working on the matter and attempting to quantify the precise amounts of this remuneration which had not been subjected to PAYE deductions.

He hoped to have a fuller picture in about two weeks when he would contact me.

I stated that such a system of under the counter payments could not have been put in place and maintained over a long period without there having been knowledge of it at a high level in the Goodman organisation. I stated that what had occurred appeared to be evasion and I said that the question of culpability would have to be investigated. I advised that this in practice meant that apart from quantifying the tax loss the question of penalty and/or prosecution would require to be considered. He enquired whether we would require to talk to senior Goodman employees. He mentioned John McDonnell the Financial Controller. I said that we would obviously require to interview people who could give full information on the matter, and this could include Mr Goodman. He suggested that particular Plant Managers or Plant Accountants would probably have had knowledge of the system.

I advised that I would record his comments on the matter.

P.S. O'Donghaile
Principal Inspector"

The Tribunal was made aware of the above interview by the State Solicitor representing the State authorities before the Tribunal.

The Tribunal immediately wrote to A & L Goodbody on the 19th September in the following terms:—

" 19 September '91

A & L Goodbody
Solicitors,
1 Earlsfort Centre
Hatch Street
Dublin 2.

Re: Tribunal of Inquiry — Beef Processing Industry

Dear Sirs,

The Tribunal has been furnished with a note of a 'phonecall dated 17th September 1991 and a note of interview dated 17th September 1991 (copy enclosed) signed by Mr P S O'Donghaile, Principal Inspector of the Revenue Commissioners. The notes speak for themselves.

The Tribunal is concerned with all of the matters contained within the notes but particularly is concerned with the Board of Goodman Group has now become aware that there were further payments in excess £2 million made to employees which had not been subjected to tax. The memo continues "he said that the payments which were in excess of £2 million were particularly in the Dundalk operation but would have existed in other plants to a lesser extent. He said that the payments were made to various categories of shop floor employees and casuals. He stated that there was a system of bogus and fictitious invoices relating to hauliers or farmers in operation and that as far as he knew the payments were cloaked by these. The payments had been made over five or six years".

The Tribunal, at this time, requires a full explanation in relation to the matters referred to above and in that connection also requires confirmation:—

1. That John McDonnell is the Financial Controller of the Group who is in the position to give the full information in relation to these matters.
2. The names and addresses of the plant managers or plant accountants of all of the Group's subsidiaries and companies to enable the Tribunal to write to them.

The Tribunal would appreciate an immediate response.

Yours faithfully,

____

Mr Justice Liam Hamilton
President of the High Court,
Sole Member of the Tribunal of Inquiry.

The following response was received:—

" Our Ref: CMP 23rd September, 1991

Goodman International
Re: Tribunal Inquiry into the Beef Processing Industry

____

Dear Mr Justice Hamilton

We refer to your letter of the 19th of September, 1991, and in particular to your request for a full explanation in relation to the matters referred to in your letter.

It will be apparent to the Tribunal that the matters referred to in your letter relate to the alleged possible commission of a criminal offence or offences. Indeed, it is apparent from the note of interview signed by a Principal Inspector of the Revenue Commissioners, which is enclosed with your letter, that the Revenue Commissioners have already stated that prosecution requires to be considered in relation to the matters in question.

We would refer to the summary of Submissions in relation to the scope of the Tribunal of Inquiry, which has been furnished to you by us, and in particular to the Submission that the Tribunal cannot investigate or make findings of fact in respect of allegations which constitute a criminal offence or the major factual components of a criminal offence. While we wish to continue to co-operate with the Tribunal in every respect which is consistent with those Submissions, we regret that the particular Submission to which we have referred precludes us at this point in time from furnishing to the Tribunal the explanation sought in the letter of the 19th of September, 1991.

In this regard, we wish the Tribunal to be aware that the matters in question have come to light as a result of the company's own investigation, that these matters have been brought to the attention of the Revenue Commissioners by the company's auditors, and that these matters are the subject of on-going discussions with the Revenue at this time.

Yours faithfully,
A & L Goodbody"

While this letter states "that the matters in question have come to light as a result of the Company's own investigation" it is fair to comment that the contact with the Revenue Commissioners was not made until after the letters from the Tribunal.

Mr McGuinness gave evidence before the Tribunal between the 22nd and 29th November 1991, in support of his allegations and stated:—

"(1) When interviewed by Mr Brian Britton, the Financial Controller of the Group, he was offered a starting salary of £12,000 which was to include a tax free component of £3,000, payable quarterly in cash.

(2) In September 1984 he became aware that it was standard practice to make quarterly tax free payments to employees and that he was instructed by Mr Nobbie Quinn, Manager of the Plant, and Mr John O'Donnell, the Accountant with responsibility therefor to prepare the necessary documentation to conceal such payments and to record them as livestock purchases.

(3) He queried such practice with Mr Brian Britton and received his approval to carry out such practice.

(4) That from the first quarter of 1985 he was responsible for such untaxed payments and the concealment thereof by recording them as payments to hauliers and livestock purchases both in Newry and in Waterford on his appointment there.

(5) He was made aware from his discussion with other factory accountants and with Mr David Murphy of the Head Office of the Group of Companies that it was a company wide practice in Northern Ireland and in the Republic of Ireland to make untaxed payments to employees by way of payment of production bonus in cash, by payment of some overtime payments in cash and by payment of portion of their wages by cheque and portion by cash.

(6) That such payments appear in the weekly Profit and Loss Accounts prepared by Plant Accountants and submitted to Head Office at Ravensdale but not shown to the Group's Auditors."

The Tribunal does not consider it necessary to deal in detail with the evidence of Mr McGuinness in regard to this allegation because the position with regard to tax evasion is much more efficiently and comprehensively dealt with in the course of the evidence of Messrs O'Duinn and O Donghaile of the Revenue Commissioners.
The cross-examination by counsel on behalf of Stokes, Kennedy and Crowley of Mr McGuinness is however relevant because Counsel appearing for Stokes, Kennedy and Crowley, the Auditors to the Goodman Group availed of the opportunity in the course of his cross-examination to suggest and state that during the course of the audit of the accounts of the Cahir Plant for the year ended 31st of December 1986, in March 1987, Mr John King of Stokes, Kennedy and Crowley had discovered

> "in the course of testing firstly one invoice which appeared odd and on the investigation found a number of invoices which were in fact transpired to be quite odd because they were supposed to relate to Northern Ireland hauliers. They were typed on sheets of paper, they transpired not to relate to any existing hauliers in Northern Ireland although they purported to and he found eventually that approximately £840,000 was represented by these phoney invoices."

This amount of information was assembled by the 13th March 1987 by the audit team from Stokes, Kennedy and Crowley.

It was further suggested by Counsel that on the 20th day of March 1987 that the said Audit Team:—

> "they were quite fortuitously late at the plant in Cahir, working late at the plant in Cahir, and they noticed that at that point in the day, there were a number of workers at the wages office who were being paid in cash and this in conjunction with what they had found about the hauliers, the phony invoices for the hauliers, this information was conveyed back to Dublin to Mr Niall O'Carroll one of the senior people in Stokes, Kennedy Crowley at that time......"

The above matters were confirmed by Mr John King in his evidence, to the Tribunal. He told the Tribunal that in the week of the 13th March, 1987 while carrying out the audit of the accounts for the year ended the 31st December 1986, he came across an invoice from a haulier that appeared unusual. It was typed on a plain piece of paper. It had a vague address. It did not disclose a business telephone number. All the figures disclosed on the invoice including the total were in a round sum. It did not have a V.A.T. number. Mr King and his team found invoices for seven other hauliers of shipments and they had balances on them. All the invoices were from Northern Ireland hauliers. The invoices totalled approximately £840,000 in total.

Mr King, in trying to establish the validity of these invoices examined:—

(i) the Northern Ireland telephone book;

(ii) the client's haulage book to trace a reference to the hauliers or their destination;

(iii) the year end creditors and listing to see if the people stated on the invoice were listed therein.

Mr King found no evidence of the hauliers listed in the above.

Mr King, then obtained and looked at copies of the cheques that had been made payable to these hauliers and noted that they had all been presented for payment at a local bank in Cahir.

Mr King then approached the plant accountant, Mr James Geoghegan, who initially was reluctant to discuss the matter and referred him to the plant manager who equally was reluctant to discuss the matter. Mr King, again approached Mr Geoghegan, who accepted that there was no documentation to support the invoices and offered various untruthful explanations concerning the existence of the hauliers.

Mr King asked him;

> "if Goodman International Head Office was aware of these hauliers and he said that they were".

At the time, the £840,000 represented approximately 40% of the total haulage charges for the year ended the 31st December 1986 and it was clear that on a year to year basis the figure for the haulage charge in 1985 compared with 1986 and some of the names on the bogus haulage invoices were similar, raising a clear suspicion that similar practices occurred in the previous year.

When Mr King returned to Dublin at the weekend, he contacted and spoke to the Audit Manager, Mr M Buttanshaw and the Audit Partner, Mr N O'Carroll. They were shocked and told Mr King that they would contact the client and try and establish what was going on. Mr King sent a memorandum in the following terms to both Mr Buttanshaw Audit Manager and Mr N O'Carroll, Audit Partner of Stokes, Kennedy and Crowley in the following terms:—

"To: Michael Buttanshaw/Niall O'Carroll.
From: Mr King. Date: 13 March 1987.

RE: *AIBP Cahir-Haulage Costs — Irregularities:*

As part of our regular testing of the client's system, an invoice came to light which appeared unusual. It related to a Northern haulier but was typed on a plain piece of paper, with a vague address, no telephone number and all the figures included, per the various destinations, were in round sum amounts. We perused this particular haulier and found the other invoices to be of a similar nature. On further investigation we found seven other hauliers whose invoices were the same — see list of names attached, all from the NI. Indeed, it appeared that all invoices were typed on the same typewriter. In all, a total of approximately £840,000 seems to have been paid to these hauliers.

We examined the NI phone book at the address given on the invoices and they were not listed in the phone book.

We then tried to trace a sample of names to the haulage book — none of the names existed in the haulage book and it was not possible to tie in the location.

We requested a sample of returned paid cheques — see attached. While the hauliers named agreed we noticed that they were all cashed/presented in Cahir.

At no time during the audit did we see any of the named haulier vehicles. There was not evidence of the names in the year end creditors listing and none of them were included in the accruals listing which included a large amount of hauliers.

Of all the hauliers used, these eight hauliers between them account for approximately:—

840,000 = 40% of the total charge

2,116,832

This seems excessive.

See attached details of discussion with James Geoghegan.

As a result of this discussion, I feel that those payments of £840,000 do not relate to haulage, but are some other payment.

It must be noted that the carriage cost on this year appears reasonable on last year. Therefore, it would appear that similar payments of this nature were being made in 1985. Indeed, the names on the attached list appear in the 1985 audit files.

No further work will be done, until the matter has been discussed.
John King"

Mr Niall O'Carroll, in his evidence, to the Tribunal said:—

> "When Mr King's memorandum was brought to my attention, I immediately thought that it might be a case of embezzlement and it appeared to be a local problem of significant proportions."

Mr O'Carroll immediately contacted Mr Britton and explained his concerns. Mr Britton indicated "he would investigate and come back to me."

Before Mr Britton had come back to Mr O'Carroll and before Mr O'Carroll had a meeting with his audit team Mr King made his second discovery on the 20th March 1987 which he memoed to Mr Niall O'Carroll on the 21st as follows:—

> "MEMO
>
> To: NIALL O'CARROLL/MICHAEL BUTTANSHAW
> From: John King.
>
> Re: *CAHIR*
>
> Date: 21 March 1987.
>
> As we were leaving the client's premises on the 20.3.'87 we noticed an unusual amount of activity near the wages office. It appeared that a number of employees were being paid.
>
> However, as we were leaving we noticed that the employees would appear to have been paid in cash. We felt that this was strange as we were made to understand that *all* employees were paid by cheque"

On the 24th/25th March 1987 a review meeting of the audit team was held where the discoveries were discussed. It was the view of the audit team that the payments to the hauliers were being made to employees. It was understood at that meeting that Mr O'Carroll was meeting Mr Brian Britton to discuss the matter.

Mr King had no further subsequent involvement in the matter.

> "Stokes Kennedy Crowley had carried out audits similar to the Cahir Audit in the other Goodman plants and particularly in Dundalk, Dublin and Bagenalstown, but these audits did not reveal any problems and particularly the problem similar to the one discovered by Mr King in Cahir, although an admission was shortly to be made by the Goodman Group".

Mr O'Carroll contacted Mr Britton and explained that in the light of Mr King's new revelations that it appeared that now they were dealing with tax evasion and fraud rather than embezzlement of company funds. When he informed Mr Britton of these matters:

> "Mr Britton heard what I said and he said that he would extend his investigations accordingly and come back to me."

Mr Britton made no other comment at that time.

Mr O'Carroll felt it prudent to draw Mr Britton's attention to Section 94 of the Finance Act 1983 which deals with Revenue Offenses and sets out the onus and obligations and responsibilities and penalties for those involved in Revenue offences.

Mr O'Carroll stated in evidence that:—

> "As auditors, our first responsibility was to bring the matter to the attention of the shareholders and bring it out in full to the shareholders and to senior management.
>
> The further responsibilities are set out at Part 5 of our Institute's Guidelines and I discussed their implications with my partners and particularly the partner in charge of professional standards."

The Tribunal considers it appropriate to set out here in detail both Section 94 of the Finance Act 1983 and Section P, Part 5 of the Miscellaneous Legal, Ethical and Practical Guidance considered by Mr O'Carroll and Stokes Kennedy Crowley and relevant to the issues here.

"Revenue Offences. **94.**—(I) In this Part— 40

"the Acts" means—

(*a*) the Customs Acts

(*b*) the statutes relating to the duties of excise and to the management of those duties,

(*c*) the Tax Acts,

(*d*) the Capital Gains Tax Acts 45

(*e*) the Value-Added Tax Act, 1972, and the enactments amending or extending that Act,

(*f*) the Capital Acquisitions Tax Act, 1976, and the enactments amending or extending that Act,

(*g*) the statutes relating to stamp duty and to the management of that duty, and

(*h*) Part VI,

and any instruments made thereunder and any instruments made under any other enactment and relating to tax;

"tax" means any tax, duty, levy or charge under the care and management of the Revenue Commissioners.

(2) A person shall, without prejudice to any other penalty to which he may be liable, be guilty of an offence under this section if, after the date of the passing of this Act, he—

(*a*) knowingly or wilfully delivers any incorrect return, statement or accounts or knowingly or wilfully furnishes any incorrect information in connection with any tax, (*b*) knowingly aids, abets, assists, incites or induces another person to make or deliver knowingly or wilfully any incorrect return, statement or accounts in connection with any tax,

(*c*) claims or obtains relief or exemption from, or repayment of, any tax, being a relief, exemption or repayment to which, to his knowledge, he is not entitled,

(*d*) knowingly or wilfully issues or produces any incorrect invoice, receipt, instrument or other document in connection with any tax,

(*e*) knowingly or wilfully fails to comply with any provision of the Acts requiring—

(i) the furnishing of a return of income, profits or gains, or of sources of income, profits or gains, for the purposes of any tax,

(ii) the furnishing of any other return, certificate, notification, particulars, or any statement or evidence, for the purposes of any tax,

(iii) the keeping or retention of books, records, accounts or other documents, for the purposes of any tax, or

(iv) the production of books, records, accounts or other documents, when so requested, for the purposes of any tax.

(*f*) fails to remit any income tax payable pursuant to Chapter IV of Part V of the Income Tax Act, 1967, and the regulations thereunder, or section 7 of the Finance Act, 1968, and the said regulations, or value-added tax within the time specified in that behalf in relation to income tax or value-added tax, as the case may be, by the Acts, or

(*g*) obstructs or interferes with any officer of the Revenue Commissioners, or any other person, in the exercise or performance of powers or duties under the Acts for the purposes of any tax."

Section 94, subsection 3 and subsection 4 of the Finance Act 1983 deal with penalties in respect of the foregoing offences. The discoveries made by Mr King and the audit team with regard to the bogus invoices disclosed prima facie evidence of offences contrary to the provisions of Section 94 of the Finance Act 1983.

Part 5 of the Miscellaneous Legal, Ethical and Practical Guidance issued by the Institute of Chartered Accountants in Ireland referred to by Mr O'Carroll provides as follows:—

9. WITHDRAWAL OF SERVICES

A Member must cease to act for a client if he knows that

(i) his client intends to do an unlawful act, or

(ii) his client has committed an unlawful act and that such unlawful act would compromise the Member.

If, in the above circumstances, a Member has been communicating with a third party (for example, the revenue authorities) on the client's behalf and such third party is affected by the illegality, the third party should be notified of the withdrawal of the Member's services but should not be informed of the reason why.

Explanation

A Member may consider it preferable that he continue acting for a defaulting client rather than that the client retain a less scrupulous accountant. In such a case, he should use his best endeavours to persuade the client to desist from the unlawful act or, if it has been committed, to disclose his illegality and make the necessary restitution, if appropriate. If such advice is not heeded, he must withdraw.

If the Member is a Company Auditor and the Directors commit an illegality, he may retain his position so long as he complies with his duty of disclosure to the shareholders (his clients). See paragraph 13.

A Member is not obliged to withdraw his services if the unlawful act does not concern the Member's work (for example, if the Member knew that the client was making fraudulent tax returns on his personal accounts and the Member was dealing only with the client's business accounts or vice-versa, or if the client was making fraudulent tax returns when the Member was doing work not involving any taxation matter).

If a client dispenses with the Member's services before he has completed his work and reported on the accounts, no further legal duty rests on the Member and he is not obliged to and must not give any information to the revenue authorities.

If a Member discovers that he has submitted fraudulent or negligent tax returns, in the past, because he was misled by his client, he must advise the client to disclose the inaccuracies to the revenue authorities and if this advice is not heeded he must inform them and his submitted returns can no longer be relied upon and that he is withdrawing from this client's services.

In other words, the Member himself must not be a party to his client's fraud on the revenue authorities. If the fraud concerns returns which the Member has not prepared or which the Member has not submitted, then the duties set out in this paragraph do not apply.

Circumstances vary and it is not always that a client fully appreciates the seriousness of his offence or the consequences which may ensue, and in particular he may not realise that if there is no disclosure and the revenue authorities later discover a fraud there will be a greater likelihood of a criminal prosecution (with the possibility of imprisonment on conviction) than where a suitable monetary settlement is offered on the client's own disclosure.

The client may also not realise that if a Member is obliged to withdraw and so notifies the revenue authorities, this may well result in the revenue authorities starting enquiries which led to the discovery of fraud.

12. AUDITOR'S DUTY TO EXAMINE

An auditor of a company must use all reasonable care and skill in his examination mindful of his duty to report as indicated in paragraph 13 on unlawful acts insofar as they come to light in the course of his duty and affect the accounts.

Explanation

When a Member is appointed an auditor of a company pursuant to the Companies Act 1963 or the Companies Act (Northern Ireland) 1960 his duties are governed by the relevant Act. Each Act endeavours to guarantee the independence of an auditor (e.g. special notice is required to remove him from office, and he may not be an employee of the company). Apart from the minimum requirements of the relevant Act, an auditor should not do anything which would compromise or be seen to compromise his independence or integrity.

The Law on Companies is concerned that reliance can be placed on the Auditor's Report (as provided respectively in section 163 and section 156 and the Seventh Schedules of the said Acts). Section 163 and section 156 should also be consulted for the powers of inspection etc. conferred on the Auditor. The veracity of published accounts and the Auditor's Report are the linch-pin of the system of protection of investors and creditors through disclosure.

An auditor will be liable in negligence if he fails to use that skill and care which a reasonable, competent auditor would use. His duty is not to confine himself merely to the task of verifying the arithmetical accuracy of the accounts but to inquire into their substantial accuracy and to ascertain that they have been properly compiled so as to contain a true and fair view of the state of the company's affairs. Legal opinion now holds that he should not rely wholly on the honesty and accuracy of officials or employees.

However, an auditor is not a detective or bound to approach his work with suspicion or with a foregone conclusion that there is something wrong. Provided the auditor takes reasonable care, makes all necessary investigations and there is nothing calculated to excite his suspicion, he is entitled to rely on the representation of the company's officers. But if his suspicion is aroused, he must ask for explanations and investigate the matter fully until he is satisfied."

13. DUTY TO SHAREHOLDERS OF A COMPANY

"If an auditor discovers an act he believes to be illegal or questionable, he must report to, and obtain consideration of that act from, the appropriate level of authority within the entity. In certain cases, this may necessitate his reporting in such a manner as to bring the matter to the notice of the shareholders.

Explanation

An auditor's client is the company and not the directors of the company. His duty is to the company. Therefore, there is no breach of confidence if he discloses some wrongdoing to the shareholders. (Gower in his textbook on Modern Company Law suggests that an auditor's duty is to each individual member of the company. If this be so, any such individual could sue the Auditor for damages in negligence).

If an auditor knows that some material unlawful act or default in relation to the company has been committed by the management, he must report this to the directors and, where appropriate, ensure that it is reported to the members of the company. If his efforts to have such matters reported at a general meeting are frustrated, he must still ensure that the company members are informed (for example, if necessary, by circulating them personally by post). The fact that there has been no deliberate misconduct by the directors does not relieve him of this duty nor can he evade this duty to report by asking not to be re-appointed.

However, where the Directors have acted in some illegal manner (for example, by defrauding the revenue authorities) the Member should give the advice to the Directors suggested in paragraph 9 above, even though, properly speaking, the client is the shareholder.

14 AUDITOR'S REPORT

When an auditor signs and issues his report (qualified or not) he should ensure that it is not misleading or ambiguous.

Explanation

This is to be read with paragraph 13 above. The report must be unambiguous even if the result will be to disclose to the shareholders, and others who may read the accounts that an offence has been committed."

24 SUSPICION OF A CLIENT'S CONDUCT

A Member should not continue acting for his client while suspicious of his conduct.

Explanation

Mere suspicion of an unlawful act is not sufficient to make a Member liable to criminal prosecution or to be obliged to withdraw his services (paragraphs 2 and 9 above). On the other hand, he must not bury his head but must interrogate his client and make such appropriate inquiries and investigations as will either exonerate his client or confirm his suspicions. From the moment the Member becomes suspicious to the time the matter is concluded, the Member should keep a careful record of all his conversations etc. so that he can exonerate himself at a later date, should this be necessary"

An investigation was carried out by the management of the Company at this time March/April 1987 to determine the amount of such payments for the year ended 31/12/1986 and they represented to the said auditors that the amount paid in that year without deductions of tax was £1.927m and their representation in that regard was accepted by the said Auditors.

On the 7th April 1987 there had been a management meeting in Ravensdale, which was recorded in a document dated the 29th April 1987. It was on Stokes Kennedy Crowley headed working paper.

Management meeting held in Ravensdale, April 7, 1987.

Present: Larry Goodman
Peter Goodman
Gerry Thornton
Brian Britton

ATTENDANCE: Aidan Connor

I reported to the meeting, that arising from a review of the consolidated group accounts for the year ended 30 December 1986, it was evident that a number of plant managers had "jumped the gun" by paying out money under the new scheme which is currently under discussion with SKC but is still not finalised at this time. In total the payments amounted to £1.92m. and the plants involved were Cahir (£872k) principally and to a lesser extent Dundalk, (£259k) Dublin (£316k) and Bagenalstown (£450k).

L. G. instructed PG/GT to meet with the relevant Meat Plant Managers immediately and to forcefully reprimand them for exceeding their authority in this matter. The scheme will be finalised in next few weeks but it was bad management practice that we had introduced it at some plants pre-emptively. All payments under the scheme are to cease at once until the final details are agreed with SKC.

L.G. who instructed BB/AC to meet with SKC immediately:

(i) finalising the scheme details as soon as possible;

(ii) agreeing with SKC the appropriate accounting treatment for the year end accounts.

"Follow up reports are to be directly with LG (not me).

Mr O'Carroll met with Brian Britton, Aidan Connor on the 9th of April 1987. Mr O'Carroll's evidence in regard to this meeting was as follows:

"They told me that they had had a full investigation carried out, that the sum of all payments at all locations — there were three other locations as well as Cahir — amounted to 1.927 million in total payments to the end of 1986. These payments had been made, without authorisation, by certain local plant managers jumping the gun, as they put it, by making payments to employees in anticipation of a new employee bonus scheme which had been under review and discussion throughout 1985 and 1986. They had instructed all such unauthorised payments should be ceased immediately pending implementation of a proper scheme. It was further represented to me, at the time, that all such advances and payments, whether paid on foot of the bogus haulage invoice or otherwise, would be repaid and I sought and obtained guarantees to that effect from third parties or from Mr Goodman's holding company. I subsequently confirmed the representations made which were handwritten with Brian Britton in correspondence."

Mr O'Carroll continued:—

"I had in my possession — and I was concerned that there should be a formal written position in relation to this — I had in my possession a hand written memo, an extract of a management meeting written by Aidan Connor and I was concerned that this should be clearly seen to represent everybody's understanding of the position. I took it on myself then to write ultimately to Mr Britton confirming all the details as I understand them and asking him if he would come back to me confirming that these were correctly understood".

This letter was written to Brian Britton on the 5th May 1987.

"Dear Brian

I refer to our discussions in connection with the audited accounts for Goodman International and its subsidiaries for the year ended 31 December 1986.

In particular I refer to the advances made to employees by various Plant Managers. I understand that the sums involved for each location for all years in aggregate amount to:—

	£'000's
Cahir	872
Dundalk	259
Dublin	316
Bagenalstown	450
	1,927

In the light of our discussions it is clear that these sums represent advances to various employees made by Managers in each location and that these do not have the approval of the Board. I confirm that it is our understanding that these advances will be repaid immediately and that the repayment has been guaranteed by Goodman Holdings. On this basis it would appear that the payments are not part of the employment remuneration for each individual and, therefore, not subject to the PAYE regulations.

I would be grateful if you would confirm:—

a. That the above is the correct outline of the position.

b. to the best of your knowledge and belief there are no other payments which have been made to employees on which PAYE regulations have not been applied.

Kind regards
Yours sincerely

____

Niall D. O'Carroll"

It was not responded to.

According to Mr O'Carroll he wrote in September 1987 to Mr Goodman in similar terms asking him to get Mr Britton to respond to the letter. Again, there was no response.

Further confirmation of Mr McGuinness' evidence in this regard was provided by the evidence of Mr Mooney, Tax Consultant with Stokes, Kennedy and Crowley, that when this discovery was drawn to the attention of Senior Executives of the Goodman Group, it was admitted that certain employees in addition to their normal wages, which were subject to the deduction of PAYE and PRSI, were being paid sums of money without deduction of tax. Stokes, Kennedy and Crowley understood from such enquiries that the method used was

"—to draw cheques in respect of bogus invoices payable to the names on the bogus invoices,

—to have such cheques cashed by the plant accountant or other employees at the local bank despite the names on the cheques and the fact that some cheques were payable to well known and reputable hauliers and were restrictively crossed, and

—to make cash payments to a considerable number of employees.

At the end of 1986 the Employer Incentive Scheme was still the subject matter of discussions between the tax personnel in SKC and the top management within the Goodman Group. Mr Goodman had not given the formal imprimatur for them to go ahead."

Mr O'Carroll in evidence:—

"From about the 25th March to the 25th April, I was in touch with the Goodman Group about once a day. This involved contact with Brian Britton, the Group Financial Director, Aidan Connor, the Group Financial Controller, at the time and occasionally Larry Goodman.

In the first week of April 1987, it was clear that we couldn't sign off the audit while this outstanding matter was not resolved.

Following the meetings and representations with Mr Britton I discussed the matter with Mr Reid, one of SKC's tax partners, who had had an involvement with the Goodman Group from time to time. I learnt then that Mr Goodman, who had become aware of and to whom I had spoken about the problem, had met Mr Reid and he had discussed the implementation of the scheme and asked Mr Reid to implement the final details and instruct his staff on how to do that as quickly as possible. None of the schemes proposed through 1985 / 1986 to the Goodman Group involved:—

(i) the recording of bogus invoices for hauliers;

(ii) the use of invoices to cattle dealers which were not valid;

(iii) the use of expense slip which were not valid expenses."

The Tribunal has already referred to the handwritten document signed by A. Connor and dated the 29/4/'87 containing a record of the management meeting held on the 7/4/'87 at which were present Mr Larry Goodman, Mr Peter Goodman, Mr Brian Britton and Mr G Thornton and to the 2nd instruction of Larry Goodman's to Brian Britton and Aidan Connor namely:—

"agreeing with SKC the appropriate accounting treatment for the year end accounts".

Mr O'Carroll clarified the meaning of the second recommendation as being:—

"At this stage we had one outstanding point in relation to the group audit and that was 1.927. of bogus invoices. The one reality, in terms of accounting treatment, was that they were not haulage invoices which is how they were booked. So we had to decide what, if they weren't haulage invoices, were they. I consulted at length with my tax partners and professional standards partners and we formed the view that because the payments had not been authorised they could not be deemed anything other than unauthorised advances taken by employees. We, therefore, put them in a suspense account in debtors and sought and obtained guarantees that they would be repaid. They were subsequently repaid. This happened around the middle of April.

229 Q. On the 9th of April, you had a meeting with Mr Britton and Mr Connor where they tell you that these were effectively unauthorised payments by the managers jumping the gun or were advances or loans.

A. Correct.

230 Q. Subsequent on that then, are you saying that you then, with your tax people, decided to treat them as advances?

A. Yes, I'm the audit partner, I don't pretend to know all the nuances of the PAYE/PRSI regulations, so I had to seek expert advice as to whether this explanation was sustainable.

231 Q. Which explanation?

A. That these were not the haulage payments, that they were nothing more than unspecified advances to employees and, therefore, were not caught by the PAYE/PRSI regulations.

Though Mr John King of SKC and the Audit team lead by him had discovered during the course of the audit of AIBP Cahir Exports Ltd for the year ended 31st December 1985 Bogus Records of Payments alleged to have been paid to haulier's in the sum of £840,000 and had in his memorandum dated the 13th day of March 1987 stated that:—

> "It must be noted that the carriage cost on this year appears reasonable on last year. Therefore it would appear that similar payments of this nature were being made in 1988. Indeed the names on the attached list (bogus hauliers) appear in the 1985 audit files"

no further examination of the account for the year ended the 31st of December 1985 was carried out by Stokes, Kennedy and Crowley.

Stokes, Kennedy and Crowley accepted the representations made to them by Mr Brian Britton which were;

(1) that the amount involved for all years ended the 31st December 1986 was £1.927m in respect of the following locations and companies:—

Cahir (AIBP Cahir Exports Ltd)	£872,000
Dundalk AIBP	£259,000
Dublin	£316,000
Bagenalstown AIBP Carlow Exports Ltd	£450,000

(ii) that these payments represented advances made to various employees by Managers in each location

(iii) that such payments were made without the approval of the Board, and

(iv) that such advances would be repaid immediately and that such repayments would be guaranteed by Goodman Holdings Ltd.

At that stage, all the work necessary for the completion of the accounts for the year ended 31-12-1986 had been completed by the various audit teams employed by SKC but one outstanding issue remained to be dealt with.

That issue was the manner in which these cash payments which had been made to employees without deduction of PAYE and PRSI payments should be dealt with in the accounts of the respective Companies.

As stated by Mr O'Carroll:

> "At that stage we had one outstanding point in relation to the Group Audit and that was £1,927m. of bogus invoices. The one reality in terms of accountancy treatment was that they were not haulage invoices which is how they were booked. So we had to decide what, if they weren't haulage invoices, were they."

Mr O'Carroll stated that he consulted at length with his tax partners and his professional standards partners and they formed the view that

> "because the payments had not been authorised they could not be deemed anything other than unauthorised advances taken by employees"

and that

> "We therefore put them in a suspense account in debtors and sought and obtained guarantees that they would be repaid"

He stated that the payments were

> "nothing more than unauthorised advances to employees and therefore were not caught by the PAYE/PRSI Regulations."

When Mr Mooney, another tax expert in SKC gave evidence he described these payments as a "loan to the employees", which as stated, by Mr Carroll was a different concept to an "unauthorised, specified advance payment" which he regarded them as.

The accounts for the year ended the 31-12-1986 were signed on the 29th March 1987 but not released until the end of April 1987 after the guarantees with regard to repayment were signed.

The guarantees were all dated the 8th April 1987 and were in the following terms:—

GOODMAN HOLDINGS LIMITED
14 Castle Street, Ardee, Co. Louth

"The Secretary
AIBP Cahir Exports Ltd.,
14 Castle Street
Ardee
Co. Louth

Dear Sir,

We, Goodman Holdings Ltd., in accordance with powers contained in Clause No. 2 (*g*) of the Memorandum of Association of the Company, hereby guarantee repayment of loans made by you to employees totalling £872,000..

We undertake that we will pay such amounts on demand, if an whenever the employees, or any of them, make default in repayment of such loans.

Yours faithfully

____

Laurence J. Goodman
Chairman and Chief Executive."

There is no reality in the representations made by Mr Britton to Mr O'Carroll of SKC that these payments made to employees without deduction of PAYE and PRSI contributions were either unauthorised or represented loans to those employees without the approval of the Board of the Goodman group of companies or that they would ever have to be repaid to the companies by their employees and there is no evidence that they were ever repaid.

These payments were and were intended to be payments to certain employees free of deduction of income tax and PRSI contributions and should have been declared as such and the appropriate tax paid and PRSI contributions in respect thereof made.

The evidence of Mr McGuinness in this regard has already been quoted in this Report but bears repetition at this stage.

(1) When interviewed by Mr Brian Britton, the Financial Controller of the Group, he was offered a starting salary of £12,000 which was to include a tax free component of £3,000, payable quarterly in cash.

(2) In September 1984 he became aware that it was standard practice to make quarterly tax free payments to employees and that he was instructed by Mr Nobbie Quinn, Manager of the Plant, and Mr John O'Donnell, the Accountant with responsibility therefor to prepare the necessary documentation to conceal such payments and to record them as livestock purchases.

(3) He queried such practice with Mr Brian Britton and received his approval to carry out such practice.

(4) That from the first quarter of 1985 he was responsible for such untaxed payments and the concealment thereof by recording them as payments to hauliers and livestock purchases both in Newry and in Waterford on his appointment there.

(5) He was made aware from his discussion with other factory accountants and with Mr David Murphy of the Head Office of the Group of Companies that it was a company wide practice in Northern Ireland and in the Republic of Ireland to make untaxed payments to employees by way of payment of production bonus in cash, by payment of some overtime payments in cash and by payment of portion of their wages by cheque and portion by cash.

(6) That such payments appear in the weekly Profit and Loss Accounts prepared by Plant Accountants and submitted to Head Office at Ravensdale but not shown to the Group's Auditors.

It is quite clear from this evidence, which the Tribunal accepts, that the system of making payments to certain employees, without the deduction of tax and PRSI contributions, was widespread throughout the Goodman Group of Companies, that such payments appeared in the Weekly Profit and Loss Accounts prepared by the Plant Accountants and submitted to the Head Office of the Group in Ravendale but such profit and loss accounts were not disclosed to the Company's Auditors and such payments were concealed by the provision of fictitious invoices allegedly in respect of payments to haulage contractors, to farmers in respect of livestock purchased and in respect of fictitious expenses.

When this practice was discovered in the Cahir Plant by the Audit Team lead by Mr King of SKC, Mr O'Carroll of SKC quite properly drew the attention of the top management of the Group to the practice, though the Tribunal is satisfied that they were at all times fully aware of the practice, and on the 9th April 1987 was informed by Mr Britton and Mr Connor that

> "they had instructed that all such unauthorised payments should be ceased immediately pending implementation of a proper scheme".

However as hereinafter appears, the practice did not cease but continued during the years 1987, 1988, 1989 and 1990 until the appointment of the Examiner in respect of the Group in August of that year.

The Tribunal has stated that the position with regard to tax evasion by this group of companies was comprehensively dealt with in the evidence of Mr O'Donghaile and Mr O'Duinn of the Investigation Branch of the Revenue Commissioners. Their evidence was detailed, complementary and of considerable assistance to the Tribunal.

The Tribunal does not consider it necessary to review in detail such evidence within the confines of this Report and will merely deal with the important aspects thereof.

This evidence established:—

(1) Between 1985 and 1990 Tax Inspectors had visited Goodman plants on about 90 occasions, generally in connection with VAT payments but occasionally in connection with PAYE but no evidence was found of "under the counter" payments to employees.

(2) The reason why such payments were not discovered according to Mr O'Donghaile was because of the system which "was very well and professionally put together" and "had been organised by a large organisation and it had been organised by

professionals....chartered accountants were involved in this and they had put a lot of thought into it."

(3) That the records as disclosed did not show or identify "under the counter" payments.

(4) As a result of a meeting between Mr Moriarty and Mr O'Duinn with Solicitor and Counsel to the Tribunal on the 7th August 1991 and the information and copy invoices given to him, he decided to make inquiries to establish the truth or otherwise of the allegations made about the invoices.

(5) On the 14th August 1991 Mr O'Duinn, accompanied by J Flynn, Higher Grade Inspector, Investigation Branch and Mr S Bell, Inspector of Dundalk VAT, Mr O'Duinn visited AIBP International at Ravensdale, Dundalk. The records of the company for 1989 (expenditure) had been requested by him in advance of such visit and these records were made available to him.

(6) The volume of books and records available for just one year was very large, purchases listing of over 1,000 pages, approximately 10,000 Purchase Invoices and cheque payment books totalling over 100 pages.

(7) To overcome the problems associated with such volume, the Inspectors focused on areas of likely irregularities such as payments to hauliers, having regard to the information received by him from the Tribunal.

(8) Mr O'Duinn noticed details of transactions with 6 individual hauliers, with Northern Ireland addresses to whom £150,000 had been paid in 1989.

Most invoices could not then be traced for these transactions but the company representative promised to locate and forward them together with paid cheques relating thereto.

(9) Mr O'Duinn formed the opinion that the few invoices produced to him did not appear to be genuine.

(10) Mr O'Duinn also noticed payments to a company called Spitfire Ltd which supplied labour services to AIBP International at a value of £25,000 per month from May 1989 onwards.

(11) In the weeks that followed the visit the requested invoices and paid cheques were forwarded to Mr O'Duinn.

(12) The invoices were all of a simple duplicate type with no addresses, telephone numbers or VAT numbers shown. The pay cheques indicated that they were by and large cashed on the date of issue at the bank on which they were drawn.

(13) Mr O'Duinn concluded at the time that in 1989 AIBP International had created funds and paid out at least £180,000 in untaxed remuneration.

(14) On the 15th day of August 1991, Mr O'Duinn, accompanied by Mr Flynn, called to meet two livestock dealers who supplied cattle to AIBP Waterford. These were livestock dealers named in the documents supplied by the Tribunal and, having examined these dealers records and from interviews with them, Mr O'Duinn was satisfied that they did not receive the sums supposedly paid to them.

(15) Mr O'Duinn contemplated visits to AIBP Waterford and a return visit to AIBP International and other AIBP plants, but before he had made arrangements for such visits, he was advised by Mr O'Donghaile, Principal Inspector in mid-September 1991, that the Goodman Group were now admitting to untaxed remuneration payments.

(16) On the 17/9/91 Mr Mooney of Stokes Kennedy Crowley, the Auditor of the Goodman Group of companies visited Mr Ó Donghaile and informed him that:

(*a*) the Board of Goodman International had recently become aware that payments, which had not been subject to tax, had been paid to employees of the Group;

(*b*) the Board was anxious to advise the Revenue Commissioners;

(*c*) he had advised this course of action;

(*d*) he thought the amount involved was in excess of £2m but work was going on to establish the correct amount;

(*e*) the payments were made to employees in all plants, going back over several years;

(*f*) such payments had been cloaked by fictitious invoices;

(*g*) he had not been aware of such payments when he had discussed the Examinership settlement in October 1990."

(17) In the course of a meeting held on the 10th October 1991, Mr Mooney informed Mr Ó Donghaile:

(*a*) that Goodman International had advised his firm, SKC, that in the period 1/1/1987 to 31/12/1990 between £4m and £4.5m had been paid to employees, without being subject to tax;

(*b*) that investigations were still continuing to ascertain the current amount;

(*c*) that these payments were in addition to the dividends/loans that had been previously advised during the course of the negotiations for the settlement of outstanding or claimed liabilities during the Examinership's proceedings;

(*d*) in response to a statement by Mr Ó Donghaile that if under the counter payments were made after the 1/1/1987, it was reasonable to assume that they were also paid before that date, he said that, he would examine that position;

(*e*) that in the course of preparation of the 1990 accounts, it had been discovered that dividends in excess of the amount disclosed to the Revenue Commissioners in the course of the negotiations prior to the said settlements and which were subject thereto, were paid to employees but the amount thereof was not yet ascertained."

(18) On the 7th day of November 1991, Mr Mooney and Mr Fagan of SKC and Mr J O'Donnell and Mr J O'Loughlin of Goodman International met with Mr O'Donghaile and Mr Ó'Duinn.

Mr O'Donnell informed them that:

(*a*) they now had a more complete picture of the under the counter situation;

(*b*) a total of £4.7 million had been paid to employees in this manner between 1/1/1987 and 31/12/1990;

(*c*) since 1990, there had been no under-the-counter payments;

(*d*) since the inception of the dividend scheme (mid 1987) dividend payments of £2.9m had been paid whereas £1.7m only had been disclosed to the Revenue Commissioners in October 1990.

(19) The following issues requiring attention were raised by Mr Ó'Donghaile and Mr Ó'Duinn:—

(*a*) they requested a detailed analysis of the payments on an annualised and plant by plant basis to enable the Revenue Commissioners to examine all books and records;

(*b*) they required particulars of pre 1/1/1987 payments;

(*c*) they required particulars relating to a company called Spitfire Limited;

(*d*) the question of VAT claimed on fictitious invoices;

(20) By letter dated the 14th day of November, Goodman International forwarded to the Revenue Commissioners:

(i) annualised details in respect of each plant showing under the counter payments in the sum of £4.7m in the period 1.1.1987 to 31.12.1990;

(ii) details of wages and dividends paid out of 8 employee companies viz:

Fleggburgh Ltd
RedRobin Ltd
Castlerigg Limited
Panache Limited
Cottesmore Ltd
Wistaston
Armcliffe Ltd
Nailsworth Ltd

At no stage had there been any disclosure by either Mr O'Donnell or Mr Mooney to Mr O Donghaile of the Revenue Commissioners with regard to the payments made to employees without being subjected to tax or PRSI contributions during the year ended 31-12-1986.

(21) At the Tribunal hearing on the 26th day of November 1991, Mr O'Donghaile learnt for the first time that under the counter payments had been made in Cahir in the period prior to 1/1/1987, that such payments had been discovered during the course of preparatory work for an audit and as a result thereof obtained a copy of the submission of SKC to the Tribunal, a copy of which had been served on the Chief State Solicitor, which dealt with the discovery of under the counter payments of £1.927m to employees of AIBP prior to 1/1/1987.

(22) On the basis of the information then available to the Revenue Commissioner PAYE estimates were raised on the 20/12/1991 and were appealed by the Group.

(23) Investigations of records, visits to Plants and negotiations continued and between 17.1.1992 and 15.7.1992, 17 submissions were received from Goodman International in relation to untaxed payments.

These submissions were comprehensive identifying amounts and dates of fictitious invoices and other expense cheques and employees who had received payments.

(24) On the 30th day of April 1992 Mr Mooney of SKC and Mr O'Donnell met with Mr Ó'Donghaile and Mr Ó'Duinn to discuss progress and it was ascertained that the under-the-counter payments would be about £5.5m.

Mr Ó'Donghaile informed them that the undisclosed dividends of £1.2m would be treated as untaxed remunerations.

(25) On the 27/7/1992 a meeting was held between Mr Ó'Donghaile, Mr O'Duinn and Mr Flynn of the Revenue Commissioners and Mr Mooney of S.K.C and Mr John O'Donnell of the Goodman Group.

Mr Mooney and Mr O'Donnell were informed that the Revenue Commissioners were of the view that the following matters required to be included as untaxed remuneration:

(*a*) £5.5m under the counter payments as per Goodman International submission.

(*b*) £1.2m additional dividends.

(*c*) Possible additions re outstanding queries.

(*d*) pre-1987 untaxed payments (£1.927m as per SKC and that a tax rate of about 55% would be applied and that interest at the rate of 1.25% per month would also be a factor.

They requested that outstanding queries be dealt with without delay.

That was the position when Mr Ó'Donghaile and Mr Ó'Duinn gave evidence before the Tribunal on the 8th and 9th days of September 1992.

While considerable work had been done in an effort to finalise the matter, negotiations were not complete and were continuing at that time.

Prior to the completion of this Report the Tribunal sought information from the Revenue Authorities as to the then current position with regard to the negotiations with the Goodman Group of Companies referred to in the evidence before the Tribunal and referred to in this Report and ascertained that

(i) efforts to reach a final settlement of outstanding payroll taxes with Goodman International and related meat processing companies continued after the Tribunal ceased its public sittings;

(ii) from June 1993 onwards, these negotiations took place in the context of the provisions of the Waiver of Certain Tax Interest and Penalties Act 1993 (Tax Amnesty Scheme) which enabled companies and individuals to settle their payroll liabilities up to the 5/4/1991 without the imposition of interest and penalties.

(iii) in January 1994 an offer of about £4.1m was submitted by the Goodman Group to settle outstanding pay roll taxes for the period up to 5/4/1991;

(iv) this offer of £4.1m comprised a direct payment of £3.654m and a set-off of about £.4m available to the Group, being DIRT credits;

(v) the precise amount of a small part of the DIRT credits has yet to be determined;

(vi) while the Revenue authorities have not yet issued a formal written acceptance of this proposal, the Tribunal has been informed that they are likely to do so;

(vii) the £4.1m relates to payroll tax evasion/avoidance investigated by the Revenue authorities as a result of the allegations inquired into by the Tribunal as is distinct from the sum of £4.53m previously paid by the Group as a result of the agreement negotiated in October/November 1990 during the period of the Examiner's appointment and confirmed by the letter dated the 14th day of February 1991 hereinafter referred to;

(viii) by virtue of the provisions of the Waiver of Certain Tax Interest and Penalties Act 1993, Companies and individuals were enabled to settle their pay roll liabilities up to 5/4/1991 without imposition of interest and penalties.

Were it not for the provision of this Act, the statutory rate of interest (1.25% per month on underpayments) would have applied to the underpayments in the period under investigation 1983-1990 inclusive.

In addition Penalties would have been incurred in each of the PAYE locations of the Company where there had been failures to remit the correct monthly taxes with the P30 returns and to submit correct end of the year returns for each of the registered locations.

On the basis of the evidence before it, the Tribunal was satisfied, and could not have been other than satisfied by the entire of the evidence it heard and the admissions made on behalf of Goodman International to the Revenue Commissioners as outlined above that:

(i) the allegations made with regard to under-the counter payments have been fully substantiated;

(ii) there was a deliberate policy in the Goodman Group of companies to evade payments of Income Tax by way of under-the-counter payments to employees;

(iii) the making of such payments was concealed in the records of the company by recording of fictitious payments to hauliers and farmers;

(iv) the records of the company were misleading and calculated to deceive the Revenue Authorities in the event of an investigation and did so deceive them;

(v) the records submitted by the Goodman Group of companies were misleading and intended to be so;

(vi) the reason why the Revenue Authorities were deceived and the ordinary investigations carried out by the Revenue Authorities did not disclose the payments made without deduction of income tax or PRSI contributions was because such payments were cloaked by fictitious invoices allegedly showing payments to hauliers, farmers and fictitious expenses;

(vii) the system of concealment was common in all relevant plants, was known to the top management of the group, undoubtedly authorised by them and in the words of Mr O Donghaile of the Revenue Investigation Branch "was very well and professionally put together and had been organised by a large organisation and it had been organised by professionals".

While the Tribunal is satisfied that the Companies' Auditors acted in accordance with their responsibilities as set out in Paragraph 14 of the Miscellaneous Legal, Ethical and Practical Guidance issued by the Institute of Chartered Accountants in Ireland which provided that:—

> "If an auditor discovers an act he believes to be illegal or questionable, he must report to, and obtain consideration of that act from the appropriate level of authority within the entity. In certain cases, this may necessitate his reporting in such a manner as to bring the matter to the notice of the shareholders".

the Tribunal considers that in the case of tax evasion the obligations placed on an auditor should not be limited to reporting such tax evasion to the "appropriate level of authority within the entity or bringing the matter to the notice of the shareholders but should be extended to oblige them to report such evasion to the Revenue Commissioners and recommends that a provision which would have that effect, be included in the next Finance Bill to be placed before the Houses of the Oireachtas.

TAX AVOIDANCE

In the course of the evidence with regard to under-the counter payments, reference was made to tax-free dividends paid by companies of which employees were shareholders, to such employees.

The efficacy of such companies as a tax-avoidance scheme was disputed by the Revenue Commissioners in October 1990 during the course of negotiations to determine and settle the tax liability of the Goodman Group of Companies which eventually led to the settlement made between the Goodman Group of Companies and the Revenue Commissioners and the Examiner appointed by the High Court, Peter Fitzpatrick.

These statements and allegations when taken with Deputy Rabbitte's allegation that the Revenue Commissioners, because of Goodman's political connections, turned a blind eye to the type of remuneration packages enjoyed by senior executives and a non-return of PAYE and PRSI for many workers because of the operation of a contract system for a large proportion of the Goodman force constituted a serious attack on the independence and integrity of the Revenue Commissioners.

Because of such challenge to the independence, integrity and capacity of the Revenue Commissioners the Tribunal was obliged to and did carry out an independent and careful inquiry into the circumstances of the settlement reached between Goodman International and the Revenue Commissioners and confirmed by the Examiner, appointed by the High Court on the 29th day of August 1990, of Goodman International and its related companies on the 14th February 1991 to which these statements and allegations relate.

The Tribunal does not consider it necessary to set out in this Report the nature and extent of such inquiries other than to state that they were detailed and exhaustive involving many consultations by Solicitor and Counsel to the Tribunal, searches of Companies' records, correspondence with auditors and examination of accounts and other relevant matters.

The Tribunal is satisfied that it had as a result of such inquiries all information necessary to deal with this question, which is and was of such public importance involving as it did the independence and integrity of the Revenue Commissioners.
Before dealing with the circumstances in which the settlement was reached, it is desirable to set forth in detail the particulars of the settlement reached and which are contained in the letter dated the 14th February 1991 which is reproduced hereunder:—

"**GOODMAN INTERNATIONAL** 14th February 1991

Mr P Donnelly
Office of the Revenue Commissioners
Setanta Centre
Nassau Street
Dublin 2

Dear Sir,

The following sets out the terms of the agreement reached between the company and the Revenue Commissioner:-

1. The following payments will be made to the Revenue Commissioners. £2 million will be paid within 7 days of the signing of the Support Agreement between the 33 Banks and the company. £1.9 million will be paid on the first anniversary of the first payment or £1 million will be paid on the first anniversary of the first payment and a further £1 million can be paid on the second anniversary of the first payment.
2. In the context of the Companies (Amendment) Act, 1990, the payments are in full and final satisfaction of:-

   (*a*) all liability to corporation tax in respect of the fifty six companies set out in Appendix A attached to this letter for all accounting periods ending on or before 31st December, 1989; and

   (*b*) all liability to corporation tax in respect of the two companies set out in Appendix B for all accounting periods ending on or before 31st December, 1988; and

   (*c*) all liability to PAYE and PRSI of the fifty eight companies up to 30th September, 1990 save for PAYE and PRSI arising on current returns which may not yet be filed with the Revenue.

3. It is projected that tax adjusted losses will arise in the accounting period 31st December, 1990 in respect of the companies in Appendix B.

   The two companies in Appendix B (AIBP and AIBP International) shall claim relief under Section 16(2) CTA 1976 against profits arising in the accounting period ended 31st December, 1989.

On the understanding that the profits arising in these two companies will be fully relieved by the aforesaid losses, the Revenue undertakes to await the precise outcome of the periods ended 31st December, 1990 (but not later than the 30th September, 1991) before pursuing any possible corporation tax liability which may arise in respect of the accounting period to 31st December, 1989.

The companies shall submit final audited financial statements, tax-computations and outstanding returns for the accounting period ended 31st December, 1989 as soon as is practicable.

4. Notwithstanding paragraph 3, AIBP International will be entitled to claim exports sales relief for the accounting period to 31st December, 1989 as the facts permit. However, this is agreed only on the basis that the reserves for that period will be applied to satisfy the Section 84 CTA 1976 requirements of the interest paid in that year on the section 84 loans and are not otherwise available, and a corresponding amount of the loss in 1990 will not be available to be used in any other manner. On this basis and notwithstanding the Section 16(2) claim referred to at paragraph 3 above, it is agreed that a Nil tax credit will apply to the interest paid in the year to 31st December 1989 on the Section 84 loans.

5. This agreement is conditional on none of the fifty eight companies claiming relief under Section 18, Corporation Tax Act, 1976 ("CTA 1976") in respect of the years 31st December, 1986, 31st December 1987 and 31st December 1988.

6. The Revenue will withdraw their appeal against the decision of the Circuit Court Judge given in the case of Anglo Irish Meats Limited in respect of the accounting periods ending 31st December 1977, 31st December 1978, 31st December, 1979, 31st December 1980 and 31st December 1981.

7. The company waives the right to costs awarded to it by the Circuit Court Judge in the case of Anglo Irish Meats Limited.

8. The Revenue will not dispute the commencement date for export sales relief for AIBP Cahir Exports (formerly Cahir Meats Limited) and will amend the assessments under appeal in accordance with the computations submitted.

9. As at 31st December 1989 the export sales relieved reserves available for distribution with 100% export sales relief applying in the following companies is as set out in this paragraph:—

	IR£
Silvercrest Foods Exports Ltd.	2,789,623
Munster Proteins Exports Ltd.	3,524,086
AIBP Carlow Exports	2,563,608
AIBP Nenagh Exports Ltd.	5,807,352
Anglo Irish Beef Processors International Ltd.	50,798,862
AIBP Cahir Exports Ltd.	1,497,677

10. All corporation tax outstanding for all accounting periods ending on or before 31st December, 1988 for the fifty eight companies detailed in the schedules shall be discharged.

11. The existing position in relation to Section 84 interest is not to be regarded as affected by this agreement.

12. With effect from 1st November, 1990 all emoluments to employees will be paid under deduction of PAYE and PRSI and all deductions will be properly accounted for to the Revenue.
13. With effect from 1st November 1990 no further dividends from export sales relieved reserves shall be paid to the following companies:—

    Cottesmore Limited
    Armcliffe Limited
    Panache Limited
    Wistaston Limited
    Fleggburg Limited
    RedRobin Limited
    Castlerigg Limited
    Nailsworth Limited
14. The Revenue will not seek to assess employees of the fifty eight companies in respect of emoluments paid to them without deduction of PAYE and PRSI prior to 30th September 1990.
15. All Group relief that is available for claim or surrender, for the accounting periods ended 31st December, 1986, to 31st December, 1989, in the fifty eight companies, is deemed to have been allowed, in arriving at the corporation tax liabilities on which this agreement is based.

    I would be obliged if you would confirm that this letter reflects the agreement reached with the Revenue Commissioners.

    Yours faithfully
    LAURENCE GOODMAN
    CHIEF EXECUTIVE

I confirm my agreement to the contents of this letter.

PETER FITZPATRICK
EXAMINER

The Revenue Commissioners confirm their agreement to the contents of this letter.

Signed: P.S. O'Donghaile
Inspector of Taxes"

The employee companies referred to in the aforesaid agreement were incorporated subsequent to the discovery by SKC of the under-the-counter payments in Cahir in the year prior to the 1.1.1987 and the admission to SKC of the payments of £1.927m to employees during that year which were paid free of tax.

This situation was discussed between SKC and the Group's Senior Management Committee on the 9th April 1987, when it was agreed that:

(i) All further such payments should cease pending the implementation of the employee scheme which was then under discussion with SKC and

(ii) that the employees would repay any advances they had received on foot of bogus haulage payments or otherwise.

An indemnity in regard to (ii) was obtained from Goodman Holdings in regard to such repayment before the Accounts were signed and released.

The employee companies were sought to be put in place in 1987.

As appears from SKC's submission to the Tribunal and confirmed by the evidence of Mr Mooney of S.K.C the following arrangements were to be put in place:—

"1. Independent companies were to be set up at each location and the operatives at that location were to be shareholders, holding preference shares. The ordinary shares in the company at each location were to be held by the local plant manager and plant accountant.

2. Employees at each location were to cease employment by Anglo Irish Beef Processors and were to be employed by the new (employee) company.

3. An agreement was to be reached between the employee company and AIBP whereby the employee company would exclusively provide labour services to the plant. This labour would be supplied at an agreed rate and payments to the employee company would be made on the basis of invoices. The amount on the invoice was to be calculated by reference to the number of production workers at the agreed rate. The employee company in each case was to be registered for PAYE/PRSI and VAT; and wage payments were to be made by these companies to the employees under PAYE/PRSI in the normal way.

4. In addition, the employee companies were to take shares in a subsidiary of Goodman International, AIBP Northern Limited. This company in turn had shares in AIBP Carlow Exports Limited, a company obtaining benefit from export sales relief. Dividends were to be paid out of AIBP Carlow Exports to AIBP Northern. AIBP Northern was to declare and pay dividends to the various employee companies.

   The employee companies were to use the export sales relief dividends to make dividend payments to preference shareholders. It was not intended that these dividend payments would replace the normal wages.

   As the payments of dividends were to be based on productivity as a participation in the success of the Group, advances by way of loan were made to employees weekly at the same time as their wages were being paid. The extent of this advance was to be calculated by reference to productivity targets reached.

   The employee companies were to declare dividends to preference shareholders on a periodic basis. The dividends were to be used to repay the advances.

There was to be no other relationship between Anglo Irish Beef Processors (i.e. the trading company) and the employee company except as described above. It was intended, however, that the local accountant/plant manager would administer the employee company."

In pursuance of this proposal, SKC supplied the following stock companies:

Company	Location	Date of incorporation
Fleggburg Limited	AIBP Dundalk	14th August 1987
RedRobin Ltd.	AIBP Dublin	03rd June 1987
Castlerigg Ltd	AIBP Longford	28th August 1987
Panache Ltd	AIBP Carlow	03rd June 1987
Cottesmore Ltd	AIBP Cahir	30th April 1987
Wistaston Ltd	AIBP Waterford	04th June 1987
Armcliffe Ltd	Munster Proteins	13th August 1987
Nailsworth Ltd	AIBP Nenagh	29th June 1987

The Revenue Commissioners had during their ordinary investigations become aware of the existence of such companies and were in the course of consideration of the tax implications thereof.

On the 29th day of August 1990 Mr Peter Fitzpatrick was by Order of the High Court appointed Examiner pursuant to the provisions of the Companies Amendment Act 1990 of Goodman International and 25 of its related companies.

Between the 31st day of August 1990 and the 4th day of October 1990 he was by further orders of the High Court appointed Examiner of 35 other related companies.

He was Examiner to the said Group of Companies at the time of the agreement which was the subject of the statements made by Deputies Spring and Bruton and by Senator Raftery and which was formally signed on the 14th day of February 1991.

The terms thereof are set out in a letter dated the 14th day of February 1991 from Goodman International to Mr P Donnelly of the Office of the Revenue Commissioners signed by Laurence Goodman as Chief Executive and confirmed by Mr Fitzpatrick as Examiner and by Mr O'Dongaile (Donnelly) as Senior Inspector of Taxes.

The circumstances with regard to this settlement as appears from the evidence were that:

(i) the provisions of the Companies Amendment Act 1990 dealing with the appointment and powers of the Examiner could detrimentally affect the pre-existing rights of the Revenue Commissioners.

(ii) Mr O'Donghaile was nominated by his superior and Mr Moriarty to look after the Revenue's interest in relation to the examinership of the Goodman Group of Companies.

(iii) The Revenue Authorities were concerned lest there might be a tax liability on some of the Companies comprised in the Goodman Group arising out of certain issues which were the subject of correspondence between the Revenue Commissioners and the tax agents of the companies Stokes, Kennedy & Crowley.

(iv) These issues related to the questions of Corporation Tax, export sales relief and potential liability in relation to dividend schemes.

(v) On the 25th day of September 1990, Mr O'Donghaile wrote to Mr Fitzpatrick informing him of the position and informing him that liabilities were likely to arise in respect of the following:—

(i) Corporation Tax on various companies engaged in the processing and sale (including export) of meat and meat products.

(ii) Payroll taxes on employer companies within the group: where payments made to the employees had not been fully subjected to the PAYE and PRSI deductions."

(vi) As a result of this letter a meeting was arranged between Mr Ó'Donghaile and Miss Walsh of Coopers and Lybrand, who was acting on behalf of the Examiner and held on the 4th day of October 1990.

Mr Ó'Donghaile advised her:

(i) that the Revenue Commissioners were at that stage concerned about two aspects of the potential liability of the Goodman Group of companies in respect of Corporation Tax viz. that effective transfer pricing was being practised by the Companies and that the exports in the export sales relief companies were not totally sourced within the State, and

(ii) that they (the Revenue Authorities) had become aware of the existence of a series of dividend companies, about which they were not satisfied and proposed to investigate in detail.

(iii) there was a possibility that there may be other types of payments made by the Company upon which tax had not been remitted to the Revenue Authorities.

By virtue of the terms of Section 66 of the Corporation Act 1976, dividends paid out of profits which are free of liability for Tax because of Export Sales Relief are not subject to tax when paid to Irish residents.

Section 54 of the Finance Act 1974 was enacted specifically to tax dividends paid out of export sales relieved profits to persons who were employees as well as shareholders where the Revenue authorities were of the view that the employees independently of such dividends were not receiving adequate remuneration for services rendered.

In effect the Revenue authorities had to be satisfied that independent of the dividends paid by the Company, the wages of the employees were adequate for the services rendered and that the payment of dividends free of tax was not by way of compensation for such services.

As a result of such meeting, the examiner became aware that the main concerns of the Revenue Commissioners were as follows:-

"(*a*) **Export Sales Relief** — transfer pricing policies. The Inspector (Mr O'Donghaile) was of the view that those companies within the group which availed of Export Sales Relief had being making excessive profits, probably as a result of transfer pricing between group companies which have the effect of maximising profits in those companies which were not liable to tax.

(*b*) **Export Sales Relief** — Trading in non-Irish Meat.
The Inspector of Taxes was aware of the claims being made by the Department of Industry and Commerce in relation to the cancellation of the groups Export Credit Insurance Cover that the group had allegedly being exporting non-Irish Meat. The Inspector was of the view that, if these allegations were proven to be correct, then there could possibly be an impact on the level of Exports Sales Relief claimed by the group companies.

(*c*) **Payroll Taxes.** The Inspector of Taxes referred to seven companies in which ordinary employees of the group were preference shareholders and who received approximately one third of their normal remuneration by way of dividends rather than by direct payments which would have been subject to PAYE/PRSI deductions.

These companies were as follows:—

Cottesmore Limited.
Fleggburg Limited.
Panache Limited.
RedRobin Limited.
Castlerigg Limited.
Armcliffe Limited.
Wistaston Limited.

The concern of the Inspector of Taxes was that the dividends payments were in effect disguised remuneration which ought to have been subject to taxation at source. The Inspector was also concerned that he thought payments from contractors had been made by the group without deduction of PAYE.

(*d*) **Payments to higher paid employees.** The Inspector indicated that he had concerns in relation to Directors and higher paid employees returns for tax purposes in that there appeared to be a understatement of remuneration."

The Examiner's concern at that point in time was directed to the preparation of the first Examiner's report which was required by Sections 15 and 16 of the Companies Amendment Act 1990.

He stated in evidence that in order to report effectively and accurately he was required to ascertain the full indebtedness of the Goodman Group of Companies, including tax liability and considered it desirable that the tax agents of the Group, SKC should deal with the negotiations with the Revenue Commissioners and this was agreed by them.

Thereafter the negotiations with the Revenue Commissioners were conducted by SKC though a representative attended such negotiations on behalf of the Examiner.

Evidence was adduced before the Tribunal dealing with meetings on the 22nd day of October 1990, the 25th day of October 1990, the 26th day of October 1990 and on the 5th day of November 1990 and the issues dealt with during the course of such meetings and the conflicting views thereon expressed during the course of such meetings and negotiations.

The issues which were dealt with during the course of such negotiations were those previously referred to viz. Corporation Tax, tax on the remuneration of the Executives of the Group of Companies and the liability for tax on dividends by the employee Companies.

The issue with regard to Corporation Tax and the liability therefor was based on the view that the Companies within the Group which availed of Export Sales Relief had been making excessive profits as a result of transfer pricing between group companies which had the effect of maximising profits in those companies which were not liable to tax.

This view was strenuously opposed by the Companies and to establish same would have involved the Revenue Authorities in lengthy investigation and inevitably litigation.

The Revenue Authorities also considered that if non-Irish beef had been exported by the Companies and Export Sales Relief had been claimed thereon, there could be a possible impact on the level of the relief claimed by the company.

On the question of Executive Remuneration it was disclosed that £3.8m had been paid to Executives without deduction of tax in the period 1986 to 1989, allegedly by way of loans pending the intended introduction of a scheme which would, it was anticipated, free such payments from liability for tax.

The efficacy of the operations of the employee companies as a tax avoidance scheme was challenged by the Revenue Commissioners and strenuously supported by the tax agents of Goodman International.

In respect of liability for corporation tax, Mr O'Donghaile suggested a figure of £5.6m though he admitted that the liability was arguable, £1.75m in respect of executives remuneration and £.95m in respect of the employee companies, making a total of £8.30m.

All these figures and liability in respect thereof were challenged by the Company's representative and it was apparent that if compromise were not reached that a long and tedious investigation and appeals procedure would have to be undertaken.

It was in the interests of both parties that a settlement be reached: the Company because the early determination of the tax liabilities of the Companies under examination by the High Court was essential to the Examiner's report: the Revenue Authorities because of the doubt with regard to the legal basis for their claims and the uncertainty with regard to the solvency of the Company, the effect of the Examiner's Report and their desire to be dealt with other than under any scheme or arrangement put forward by the Examiner and approved by the High Court.

After further negotiation the settlement outlined in the letter of the 14th day of February 1991 was reached.

The significance of the terms of Paragraph 9 of the said letter dated the 14th day of February 1991 is that the reserves of export sales relieved profits available for distribution free of tax by companies named therein was reduced by £20m as a result of the insistence by Mr O'Donghaile that the basis on which sales relieved profits were calculated resulted in excessive reserves being available for distribution.

The settlement did not purport to deal with under the counter payments or the activities of any companies which had not been disclosed to the Revenue Commissioners at that point of time.

The under the counter payments made to employees prior to the establishment of the said companies were not disclosed to the Revenue Commissioners, the amounts paid to the employee companies were understated by the sum of £1.2m as disclosed in the Chapter on Tax Evasion and the existence of Spitfire Ltd as an employee company was not disclosed.

The question of Section 84 finance was discussed but not dealt with in the settlement because it did not affect the tax liability of the Companies.

Mr Peter Fitzpatrick who as the Examiner appointed by the High Court and who had performed a clearly independent function in these negotiations stated:

"(i) That the settlement confirmed by him was an attractive settlement from the point of view of the Revenue Commissioners.

(ii) That there was considerable argument on both sides with regard to the Corporation tax issue and the PAYE issue and the position was by no means clear.

(iii) That the determination of the issues in dispute could have led to lengthy litigation in the Courts.

(iv) That having regard to the time limits set forth in the Companies Amendment Act 1990 it was in the interests of both parties to effect a settlement.

(v) That the settlement was not in respect of all outstanding liabilities and penalties did not include any matters which were not within the knowledge of the Revenue Commissioners."

The Tribunal agrees unreservedly with this view which is consistent with all the evidence heard by the Tribunal and is satisfied that in agreeing to the settlement the Revenue Commissioners did not make a wrong judgment, that there was a real issue as to the extent of the Goodman Group's liability in respect of income tax and Corporation Profits tax and that the settlement reached was a fair and reasonable one having regard to the circumstances disclosed by the Company and its Auditors to the Revenue Commissioners.

The settlement did not cover or relate to any liabilities in respect of payments which had not been disclosed to the Revenue Commissioners and as appears from the Report of the Tribunal dealing with tax evasion, Goodman International and associated companies remained liable for Income Tax and Pay Related Social Insurance in respect of such payments.

In the course of the negotiations which led to the Agreement dated the 14th February, 1991, there was no admission made by or on behalf of Goodman International of tax evasion and the Revenue Commissioners representatives, who were not aware of the tax evasion practised in the Goodman organisation, believed that they were dealing with the efficiency of the Tax Avoidance Scheme which had been put in place in 1987, when agreeing to the Settlement.

Consequently, while there was large scale tax evasion being practised in the Goodman organisation over many years, as alleged by Deputy Spring, there was not any agreement by the Revenue Commissioners "not to take proceedings against Mr Goodman or his company in respect of large scale tax evasion practices going back over many years".

The Revenue Commissioners were not aware of such tax evasion practices at the time and did not become so aware until informed by the Tribunal.

CHAPTER TEN

CBF

On the 25th day of October 1990 in the course of a debate in Dáil Eireann on a statement made by the Minister for Agriculture and Food, Deputy Rabbitte made a speech, in which he referred to many matters which were the subject of inquiries by the Tribunal. For this reason, it is desirable that a substantial portion thereof be reproduced in this Report because it places in context many of the allegations made and required to be dealt with by the Tribunal.

In the course of his speech, he stated that:—

> "The cattle and beef sector accounts for 36 per cent of gross agricultural output and contributes almost 10 per cent to GDP. When export refunds are included, export earnings amount to 7 per cent of total exports. Some progress has been made towards increasing value-added exports in recent years in so much as the export of cattle on the hoof has been reduced from 46 per cent of the total destined for export in 1969 to just over 10 per cent in 1989. The scale of the beef industry is therefore very significant sustaining almost 100,000 farmers in the production of cattle and employing an estimated 5,000 workers in meat processing.
>
> Unfortunately the grand scale £260 million development plan for the industry promoted by the Government in 1987-88 has served only to undermine the confidence of companies outside of the Goodman Group and to facilitate the threatened demise of the Goodman Group itself. I explained to the Dáil on 28 August 1990, *inter alia*, how Mr Goodman grossly abused the unprecedentedly generous package of section 84 loans to sow the seeds of his own disaster. I have often been misquoted since then on this point, but the record shows I made the following statement:—
>
> > *I am now stating in this House that I have information which suggests that Mr Goodman proceeded to draw down much of the £170 million package of section 84 loans... that those exceptional credit lines were manifestly not used for the purpose for which they were approved; rather that Mr Goodman used these facilities to fund imprudent and speculative ventures outside the State that had nothing to do with the beef industry and that in that process the Exchequer was effectively defrauded of substantial revenue.*

I have other questions for the Minister for Agriculture and Food and I am sorry he has left the House. I have received authoritative information that a major competitor of the Goodman Group, having secured substantial markets in Iraq in 1987, was extended the protection and benefit of the newly-restored export credit insurance scheme. That competitor I name as Hal-Al which has about 17 per cent of the beef kill as compared to Goodman's 35 per cent.

When Mr Larry Goodman quickly learned of Hal-Al becoming a beneficiary under the ECIS he immediately intervened with the Taoiseach, who caused the then Minister for Industry and Commerce, Mr Albert Reynolds, to cancel the protection for Hal-Al. Next day an official in Minister Reynolds' office — Mr Michael Kelly — telephoned Hal-Al to convey the bad news. Unfortunately, he said, the earlier decision had been an error and the facility had been entirely used up.

Mr Goodman has always ruthlessly dealt with competitors who stood in the way of his ambitions to get a stranglehold on the market. In this case he used his close political contacts with the Taoiseach, Charles Haughey, TD, and the then Minister for Industry and Commerce, Albert Reynolds, TD, to strike a telling blow against Hal-Al. The effect of the decision was that the future of the Hal-Al plants hung in the balance for some time as the main shareholder calculated whether he could, or should, continue to do business in a country where politics and business worked hand in glove to distort the market.

When I posed such a set of circumstances to the Taoiseach at Question Time yesterday, he seemed to deny any such intervention. In the public interest I should like him to clarify his role without delay. Is he saying that the Goodman intervention took place at the level of the Minister for Industry and Commerce or the then Minister for Finance, or did the Minister for Industry and Commerce take the decision to withdraw cover from Hal-Al on his own initiative? My authoritative information suggests that the intervention took place at the level of the Taoiseach.

I have received separate reliable information which directly implicates the Taoiseach in a separate matter involving Mr Goodman exerting political muscle. In the preparation of the original 1988 Estimates it was decided, either in Cabinet or by a group of Ministers, to accede to a request from CBF — the Meat Marketing Board — to increase their financial assistance so that they could work at securing new markets for Irish meat products. That substantial increase sought by CBF was agreed. When the Goodman organisation heard of the decision, Mr Goodman again intervened with Mr Haughey, who instructed the Minister for Agriculture, Mr O'Kennedy, who I am sorry has left the House, not only to reverse the decision but to reduce dramatically the CBF allocation. The record shows that CBF were allocated £965,000 in 1987 and only £515,000 in 1988, a dramatic reduction of 47 per cent. Once again Mr Goodman's political connections had been used to shut out the prospects of expanding competitors and had ensured that he maintained his pre-eminent position in the industry.

We are not yet in a position to evaluate the damage done to agriculture, to our reputation as a trading nation, to our standing in the international financial community and to our economy. But it is clear that one man's greed and self-aggrandisement allied to the blinkered and, apparently, inexplicable blanket approval of the Government of the day has brought one of our major industries to the brink of destruction.

It is clear that powerful men who aspire to embodying the spirit of the nation have combined to use power as if the livelihood of others were mere playthings and with little regard to the damage likely to be inflicted on our international reputation. While this oligarchical structure was being put in place, those outside the Goodman Group had few friends.

When Hal-Al eventually persuaded the then officers of the IFA to bring the facts and figures to Minister Reynolds, they had scarcely time to get back to the Irish Farm Centre before Mr Goodman paid them an unscheduled visit. I cannot say whether he parked his helicopter nearby in Baldonnel on that particular day, but I do know that Mr Tom Clinton, then IFA President, and his colleagues, quickly withdrew their espousal of the Hal-Al cause when Mr Goodman threatened to refuse to collect the IFA levy for the farm bosses.

It is all so reminiscent of a similar visitation by Mr Goodman to Liberty Hall when Mr Goodman was on the acquisition trail and cutting a swathe through traditional union conditions. As he sacked workers right, left and centre and rehired some of them as sub-contractors, some trade union bosses felt similarly constrained in their response by financial considerations, since the cost of the superstructure at Liberty Hall makes costs at Bluebell look like a local regional office.

The other source of protest, the largest opposition party, was also uncommonly acquiescent. Not all of this was due to the Tallaght Strategy. Although Goodman was the dominant beneficiary of ECIS, Hibernian Meats also benefited — quite properly, as far as I know — and blood being thicker than water, Fine Gael did not complain too loudly. Another consideration in the minds of party managers in Fine Gael was that they had also become entangled in the Goodman financial web. Mr Goodman had an ecumenical approach to the purchase of influence and, after Fine Gael used the structure of the IFA President, Tom Clinton, as a go-between to mend bridges between Goodman and Fine Gael, party coffers received an immediate injection of £60,000 in 1988.

All of this confirms me in my view that it is imperative that there be a public inquiry held into the entire ramifications of the Goodman organisation so that our international reputation and Irish agriculture in general can recover from the damage suffered."

This speech contained and referred to many serious allegations against Mr Larry Goodman, the then Taoiseach Charles J. Haughey, the then Minister for Industry and Commerce, and now Taoiseach Albert Reynolds, the Minister for Agriculture and Food Michael O'Kennedy, the then President of the Irish Farmers' Association Mr Tom Clinton and the Fine Gael Party.

These allegations can be summarised as follows:—

(i) Deputy Rabbitte had information that suggested that Mr Goodman proceeded to draw down much of the £170m package of Section 84 loans: that those lines of credit were not used for the purpose for which they were approved and that the Exchequer was effectively defrauded of substantial revenue.

(ii) When Hal-Al, a major competitor of the Goodman Group, was extended the benefit of the protection of Export Credit Insurance in respect of substantial markets in Iraq in 1987, Deputy Rabbitte had "authoritative information" to suggest that Mr Larry Goodman used his close political contacts with the then Taoiseach and the then Minister for Industry and Commerce to strike a telling blow against Hal-Al by having the grant of Export Credit Insurance withdrawn and to do this he (Larry Goodman) immediately intervened with the then Taoiseach who caused the then Minister for Industry and Commerce to cancel the Export Credit Insurance.

(iii) When the Goodman organisation learned of a decision made either in Cabinet or by a group of Ministers in the course of preparing the Estimates for 1988 to accede to a request from CBF — the meat marketing board — to increase their financial assistance to secure new markets for Irish meat products, Mr Goodman intervened with Mr Haughey who instructed the Minister for Agriculture not only to reverse the decision to increase the CBF allocation but to reduce same.

(iv) When the I.F.A. espoused the cause of Hal-Al by making representations on its behalf, Mr Goodman threatened to refuse to collect the I.F.A. levy for the farm bosses and they quickly withdrew their support.

(v) Mr Goodman had an ecumenical approach to the purchase of influence and the Fine Gael party coffers received an immediate injection of £60,000 in 1988.

These allegations are of the most serious nature, involving as they do allegations of fraud, corruption, the exercise of improper influence and bribery. Each of them necessitated careful and thorough examination and investigation by the Tribunal because they were properly regarded by both Houses of the Oireachtas as of urgent public importance.

These allegations are dealt with in different chapters of the Report, but the allegations of interfering with the work of CBF by reduction of its grant at the instigation of Mr Goodman is of particular importance with regard to the role of the CBF in the beef industry.

In order to understand the nature and effect of the allegation involving the CBF, it is necessary to outline briefly its role in the beef industry and in view of the many issues involving the sale and export of beef to Iraq and other Third Countries with which the Tribunal had to deal, the Tribunal sought and obtained a submission from CBF, which is set forth hereunder:—

"1. **INTRODUCTION**

1.1 CBF was established by Government for the development of markets for livestock, meat and meat products in partnership with Irish industry.

1.2 Its primary role in International Markets is to promote Ireland and the Irish meat industry as a major source of quality product, raised in a grass-based environment, supplied by a modern processing industry.

1.3 During the period 1984/87 CBF activities included:

—Developing and maintaining contacts with buying agencies

—Fostering Irish trade co-ordination

—Providing market information to exporters and industry generally

—Participating at important trade fairs

—Participating in Government visits/officials talks.

1.4 The Middle East was the major growth region for beef in the 80s — the direct result of improving prosperity due to higher oil revenues. Between 1980 and 1985/86 total imports into the three most important markets in the region — Iran, Iraq and Egypt — grew by more than 200,000 tonnes."

"1.5 The main features of these three markets were:

Central Buying Agencies
Large Volume Contracts
Price sensitive
General high risk due to
- instability in the region
- oil dependent economies.

1.6 Compared with Europe and North America where Irish exporters had traded historically, trading conditions in the Middle East contrasted sharply. For example, in Europe and the United States, there are many buyers and more broadly based stable economies.

1.7 The main CBF market development work for the region was carried out in the first half of the 1980s. By the second half of the 1980's the key contacts had been fairly well established and CBF activity tended to be confined to market visits, trade fairs and participation in the Government-to-Government talks.

2. DEVELOPMENT OF POLICY

2.1 The overall CBF objective was to maximise Irish earnings from these markets. How this was to be achieved in practice changed over time and from market to market. In the early stages, only a few companies decided to become involved and clearly, they were the effective vehicles for opening up these markets; others appeared to take the view that it was essentially a short-term business. From relatively small beginnings, business expanded considerably.

2.2 CBF policy evolved in response to the changing market situation and trading conditions. At no time was CBF dealing with a static situation, either from the point of view of the Third Country Purchasers, or the Irish Exporters. Consistent with its central objective, CBF sought to consolidate and then expand these markets, where possible. We felt that Irish exporters competing individually, could have been at a disadvantage dealing with Central Buying Agencies and likely to end up in unnecessary price cutting to the detriment of the Irish beef producer. We believed that a "unified selling approach" would have been mutually beneficial and, furthermore, could have facilitated the assimilation of smaller exporters, who otherwise would not realistically have been able to participate in this business.

2.3 From the early to mid 1980s CBF made a number of attempts to achieve co-ordination and, while there was some success in this regard, the necessary overall consensus did not materialise. The efforts to introduce an appropriate structure at industry/national level took different forms over the period, namely:—

Loose Consortium/Co-Ordinating Role
CBF Trading Company
Test Marketing
Formal Industry Consortium service by CBF

2.4 While CBF continued to promote co-ordination and maximum involvement by the Irish meat industry, it also had to deal with commercial opportunities as they arose in the absence of an agreed industry-wide approach by the Irish exporters. For example, in 1983 and in 1984 CBF co-ordinated negotiations between two Irish exporters and the Iranian Government Trading Corporation resulting in contracts for both years.

2.5 Because of the size of the contracts and the small number of Irish suppliers involved (the 1984 agreement involved only AIBP and Purcells) some members of the Board of CBF were concerned that the development of these contracts should be more widely available to the industry.

2.6 In the following year, February 1985, the Board of CBF came to consider its position in relation to the renewal of the Iranian contract for that year. Despite approaches by CBF the existing suppliers did not agree to participate in a more broadly-based approach allowing other Irish suppliers to be involved in the negotiations and, in the circumstances, CBF felt it could no longer participate in the negotiations for this 1985 contract. An extract from the Board Minutes of the February 1985 meeting records as follows:

> "The Chairman reported on discussions with Anglo Irish and Purcells regarding the new Iranian contract. While CBF's obligation to service the whole industry had been stressed, the existing suppliers had not agreed to participate in a more broadly based approach to the negotiation of the new contract. Accordingly, CBF took the view that since it had played a major role in the development of this business by the current suppliers, it was not necessary, in the circumstances, to provide them with further assistance and also that, in the absence of a more broadly based approach, it could not adopt what could be construed as a selective, or partisan position".

2.7 CBF went through considerable change during 1985. The then Managing Director, Dr O'Sullivan, resigned and the Market Development Director, Mr Paddy Moore, became Acting Chief Executive in September, 1985. Later, in January, 1986, he was appointed as Chief Executive of CBF. He adopted as one of his priorities, a review of policy and practice in relation to CBF's services to exporters. This review was undertaken as part of the implementation of the CBF 5 Year Plan for the years 1986 to 1990.

2.8 A number of Statutory Committees were set up including a sub-Committee called the Meat Exports Advisory Committee, which was set up in January 1986 to assist the Board in the formulation of its export marketing policy. The Committee consisted of a Chairman, who was a member of the CBF Board, six industry members and the CEO of CBF. Three of the industry representatives were nominated by the Irish Meat Processors Association (IMPA), and the other three were nominated by the Irish Meat Industries Association (IMIA).

The IMPA nominations were from the following companies:

Meadow Meats
Master Meat Packers
UMP

The IMIA nominations were from the following companies:

AIBP
Tara Meats
IMIA Executive

2.9 The Meat Exports Advisory Committee agreed a CBF Policy Document entitled "Business Development Programme" which was approved by the Board of CBF on 28 May, 1986. (A copy of the Business Development Programme is appended hereto). Thereafter, CBF Policy in Third Countries was carried out in accordance with this Programme.

3. TRADE SERVICES TO EXPORTERS

3.1 CBF policy in servicing exporters to these markets, and underpinning the policy document referred to above, was that all potential exporters should be treated equally, but that CBF should not cut across existing business.

3.2 Given the competitive nature of the industry and the degree of rivalry between Irish firms however it can be appreciated that the application of this policy on a day-to-day basis was never an easy matter.

3.3 The essential elements of CBF trade services to exporters to these markets were:—

—Circulating tenders to companies already in the marketplace and to companies showing interest in entering the market

—Providing market information to all exporters on request, including names of appropriate contacts

—Mounting meat exhibits at trade fairs and offering all exporters attending the fairs full access to them.

4. HISTORY OF TRADE CO-OPERATION & CBF SERVICES

4.1 CBF was first involved in attempting to co-ordinate contracts for both Libya and Egypt in 1981/82.

4.2 In February/March, 1983, CBF assisted in the negotiation of the first contract with Iran. This was for 15,000 tonnes of beef and was secured by AIBP when Dublin Meat Packers withdrew.

4.3 In 1984 CBF were again involved in co-ordinating contract negotiations with Iran where AIBP and Purcell Meats signed substantial contracts.

4.4 In December 1984, CBF having advised all interested exporters about the Iraqi tender for 1985 invited them to a meeting to discuss a co-ordinated industry approach. The following companies attended:

AIBP
IMP
Purcell Meats
Horgan Meats
Kildare Chilling
Slaney.

4.5 After a protracted meeting when it became clear that it would not be possible, due to the lack of appropriate response by the members in the industry to organise a joint bid approach, CBF reluctantly had to advise the companies to make their own responses.

4.6 In January, 1985, and before the start of negotiations, CBF withdrew from the joint selling approach to Iran because the existing suppliers (AIBP and Purcell Meats) were unwilling to allow any other companies to become involved in negotiations for the contract.

4.7 In May 1985, CBF was part of the Government/Trade Delegation to Egypt which resulted in the signing of a protocol for the supply of 30,000 tonnes of beef.

The following companies were involved in the contract:

AIBP
Purcell Meats
Horgan Meats
Agra Trading
Kildare Chilling.

CBF requested an allocation on behalf of Irish companies not present at the negotiations and was ultimately allocated 2,000 tonnes.

4.8 In 1990, following price under-cutting by Irish exporters in the negotiation of the 1989 contract with Iran, CBF concluded a "voluntary supply agreement" with a group of six exporters for the supply of beef to Iran.

The companies were:

AIBP
UMP
Hibernia Meats
KMP
Kepak
Dawn Meats.

A share of the total contract was to be reserved for allocation to other Irish suppliers.

4.9 In addition to the above CBF:

—organised a presence at the major Trade Fairs including hosting stands at the annual Tehran and Baghdad Trade Fairs each year

—undertook regular market visits

—passed tenders from the proposed customers to exporters having an interest in supplying the relevant market

—carried out market surveys and provided reports and market information to exporters as requested

—sent representatives to Joint Commission meetings as they arose.

5. **Export Credit Insurance**

5.1 CBF from time to time briefed the Department of Agriculture on the requirements of the market. In general, CBF supported the view that export credit insurance was necessary to enable firms to trade with Iraq and supported the view that such insurance should be made available.

5.2 CBF was not consulted by ICI or the Department of Industry & Commerce about the allocation of cover given to individual exporters.

5.3 Exporters contacting CBF regarding export credit insurance were referred to the Department of Industry & Commerce or the ICI."

The reasons for the reduction in the grants to CBF in 1987 and 1988 are clearly set forth in the documentation presented to the Tribunal by the Deptartment of Agriculture and confirmed by the evidence, in particular of Mr Gerry Dromey, who was a Principal Officer in Beef 11 Division of the Department of Agriculture between 1984 and 1989.

It appears from an examination of such documentation and consideration of such evidence, that the position in regard thereto was as follows:—

(1) The Grant in Aid to CBF was £787,000 in 1985 and £1,097,000 in 1986.

(2) In February 1987 the then Coalition Government approved funding for CBF for 1987 in the sum of £1.125m.

(3) On the 27th March 1987 the new Government, as part of general cutbacks in Government expenditure decided to reduce the Grant in Aid from £1.125m to £0.965m, a reduction of £160,000.

(4) On the 13th day of May 1987, the Taoiseach wrote to the Minister for Agriculture & Food, a letter in the following terms:—

"Dear Minister

It is imperative that we carry further the progress we have made so far this year in getting public expenditure under control. Unless we achieve further significant cuts in expenditure, the growth in public sector debt will continue to be a burden on the economy inhibiting economic growth and employment and making it impossible for us to get development under way.

We must begin to identify the specific programmes and expenditures for further cuts *now* if we want to get results for the remainder of 1987 and for 1988.

I am anxious to get this process underway as soon as possible. I therefore ask you to submit to me, and to the Minister for Finance, a paper, by Friday 22 May at the latest, identifying the proposed reductions to expenditure. The proposals must have the effect of achieving a significant reduction on your Department's present level of spending. They may cover capital as well as current expenditure. Your paper should state whether legislation, or other important preparatory steps, would be required in order to bring them into effect, and the timetable you would envisage for taking these steps. It should also cost the proposals made, showing the possible longer term, as well as 1987 and 1988 savings.

In arriving at your proposals, all options should be considered, including the elimination or reduction of particular schemes and programmes, rooting out overlaps and duplications between organisations, the merger of organisations, the closure of institutions which may have outlived their usefulness, the scaling down of the operations of organisations and institutions and the disposal of physical assets which are no longer productively used. A radical approach should be adopted and no expenditure should be regarded as sacrosanct and immune to elimination or reduction. We do not want a series of justifications of the status quo or special pleadings.

I am depending on you to make it clear to officials that their full co-operation in and commitment to this exercise is required and that the Government expect worthwhile results to emerge.

Following the 22 May, the Secretary, Department of Finance, will head a team of Finance officials to meet each Accounting Officer to review each Group of Votes to identify the savings that can be made. The proposals identified will come before Government for decision. The timetable I want to hold to is that these decisions be made on a weekly basis from end-May and that we be in a position to have the full programme of reductions agreed by end August — early September.

Yours sincerely
Charlie Haughey
Taoiseach."

(5) On the 5th day of June 1987, the Minister for Agriculture & Food wrote to the Taoiseach enclosing a list of possible savings measures on the Department's expenditure and enclosed an explanatory note in respect of each measure.

(6) The note in respect of the reduction in the grant to CBF was in the following terms:—

"*Measure* *Reduce Grant-in Aid to CBF*

1. CBF's total budget this year is £3.178m of which £0.965 will be provided by the State Grant-in-Aid and £2.213m by a statutory levy on producers. The latter is at the rate of £1 per head of cattle (10p per sheep) and is collected on all slaughtered and exported animals.

2. It should be possible for the State to reduce its contribution to the body so that within three years it will be largely funded by the industry it serves.

3. The aim would be to increase the statutory levy by Order (with Dail and Seanad approval) so as to reduce this year's grant by £150,000. Further increases in the levy could be made in 1988 and 1989 so as to reduce dependence on the State grant.

4. This measure could only be considered if somewhat similar action were being taken in regard to CTT.

(7) On the 30th July 1987 the Government decided to reduce the Grant in Aid to CBF to effect savings of £450,000 in 1988 and £965,000 in 1989.

The effect of this decision was to reduce the CBF Grant for 1988 to £515,000 and effectively to abolish it in 1989.

(8) In June 1987 the Minister for Agriculture & Food proposed that CBF's role should be extended to pig meat.

(9) In January 1988 the Government agreed to the proposal from the Department of Agriculture & Food that the role of CBF should be extended to cover pig meat initially and later to other meats if necessary.

It also agreed that the Grant in Aid for CBF should be increased by the transfer from CBT to CBF of an amount to be agreed between the Ministers for Agriculture & Food, Finance and Industry & Commerce.

After discussion the amount agreed was £400,000, bringing the CBF Grant in Aid to £0.915m for 1988.

In July 1988, the Government decided to phase out the pig meat grant for CBF over two years, and fix the grant at £0.250m for 1989 and £.150m for 1990.

In September 1988 the Minister for Agriculture & Food submitted an Aide Memoire to the Government seeking a review of the proposed CBF Grant in Aid for 1989 and 1990; he sought to have the Grant in Aid retained at the 1988 level, namely, £0.915m for 1989 and 1990 in order to enable CBF to complete its 5 year plan which was designed to increase added value, employment and a significant market-led expansion in sales to other EC member states.

The Grant in Aid for 1989 was subsequently revised upwards on 5th October 1988 to £0.5m as part of the Government decision on the estimates for 1989.

Mr Haughey described as a fabrication the allegation that he had discussed the reduction in the grant to CBF with Mr Goodman, who confirmed in evidence that he had no such discussion.

The Tribunal also heard the evidence of Mr Paddy Moore, the Chief Executive of CBF and the then Minister for Agriculture & Food.

It is clear from this evidence that the reason for the reduction in the grant to CBF was that it was part of the overall plan of the Government which came into power in March 1987 to curtail public expenditure, that expenditure cuts were sought by the then Taoiseach from each Department of State, that the suggested cut in the grant to CBF emanated from the Department of Agriculture and was dealt with in the usual manner when Estimates are being prepared by Government.

There is no suggestion that the then Taoiseach was particularly involved other than in requiring expenditure cuts from all Departments.

Deputy Rabbitte stated that he bases his allegation on information supplied to him from " a banking source" but claimed privilege when refusing to give to the Tribunal the name of such source.

The Tribunal is satisfied that the allegations that Mr Goodman intervened with the Taoiseach to have the grant to CBF reduced, and this serious allegation made by Deputy Rabbitte, is baseless.

CHAPTER ELEVEN

ICC

In the course of a speech made by him in Dáil Eireann on 28th August, 1990, Deputy Rabbitte stated that:-

> "Knowing the inside political track has enabled Goodman to get access to exceptional lines of credit and to benefit from risky but profitable Middle East contracts, confident in the knowledge that he is guaranteed by the Government so long as Fianna Fáil remain in power. This did not start in 1987. For example, some years ago there are good grounds for believing that Mr Goodman was allowed to cherry pick the best of the ICC property portfolio because he had the inside political track before any other body became aware."

This is a further allegation of the use of alleged political influence by Mr Goodman with Fianna Fáil which the Tribunal was obliged to expend considerable effort and time in its investigation.

It was obliged to contact the Industrial Credit Corporation and to obtain from them a detailed submission with regard to their property portfolio, Corporate Plan indicating property disposal policy, copies of correspondence/reports leading up to decision to accept Goodman's offer for the Airways Properties and a table summarising details of the Airways Properties including a note on acquisition cost and estimated market value at 31st October 1987 and the Board report dated 16th August 1988 approving of the acceptance of the Goodman offer.

In addition Mr Leo Roche, Manager, Property Investments of the Industrial Credit Corporation, prepared a detailed statement which he repeated under oath in evidence.

The conclusion of the Report and confirmed by him in evidence was:—

> "In relation to this transaction, ICC at all times had regard to commercial criteria only. The property which was acquired by Goodman was available to the market and/or other interested parties during 1988. It was sold to Goodman at a profit of £1.42m over cost and at a profit of £.625m in excess of open market valuation. There was absolutely no question of Goodman being allowed to "cherry pick" the best of the ICC portfolio. The information supplied herewith supports these facts."

As the Tribunal has stated the said information consisted of the matters to which it has already referred. Again, the Tribunal is satisfied that there was absolutely no basis for this allegation.

CHAPTER TWELVE

Ray Mac Sharry

The male presenter on the ITV programme in the course thereof, stated:

> "The current European Commissioner for Agriculture is Ireland's Ray Mac Sharry. He frequently takes the opportunity to speak out against fraud within the subsidy system. He declined to take part in this programme. His staff told World in Action that he was "too busy". Commissioner Mac Sharry knows Larry Goodman well. He too started life selling cattle. He is a former Irish Minister for Agriculture. One of Mac Sharry's sons works as a Sales Manager for Goodman International. The Commissioner's son-in-law is part of the team which audits Goodman's accounts and also a cousin of Goodman's former Deputy Chief Executive, Brian Britton. Mac Sharry himself has appeared in Goodman's own promotional literature. He helped him celebrate a contract with a German supermarket back in 1981.
>
> "But Mac Sharry's most controversial connection with Goodman came last year. Goodman had secretly borrowed £500 million from 33 international banks which he couldn't repay. The Dutch bank, Ambro, was threatening with liquidation. The Bank held off after a request by the Dutch Agriculture Minister, Herik Braks. Braks had previously received a telephone call from Ray Mac Sharry asking him for help. In the breathing space, Prime Minister Charles Haughey recalled the Irish Government from its holidays and pushed through emergency legislation to prevent Goodman's companies from going under. Mac Sharry insists he did not intervene on Goodman's behalf."

The programme then televised an extract from a speech made by Commissioner Mac Sharry in the European Parliament on October 11th 1990 and showed him as stating:—

> "I made no request to Mr Braks to intervene vis a vis any particular bank in favour of Goodman International. It would be presumptuous of me, anyway, to think that my initiatives would be instrumental in keeping this firm in business. I doubt it. However, if it did help, I am rather pleased."

This was represented on the programme as a continuous statement by Commissioner Mac Sharry.

As it was represented, it appeared to, at the same time, contain a denial that he had requested Mr Braks to intervene and at the same time expressing pleasure if his intervention had helped.

One John Tomlinson, an English Member of the European Parliament was shown on the programme as stating:—

> "I am still persuaded that Mr Braks was invited to intervene specifically in favour of Goodman International at the behest of the Commissioner and that does not seem to me to be a proper course of action. The Commission's responsibility is not to be there seeking to protect the commercial or the financial interest of a particular trader. Now, if Mr Goodman had got a particular banking problem with Dutch banks, it is his commercial responsibility to deal with that and not the Commissioner's responsibility to deal with it."

These allegations are serious, involving as they do an allegation that Commissioner Mac Sharry had abused his position as the Commissioner for Agriculture of the European Economic Community to intervene on behalf of a particular trader, Goodman International and used his status as such Commissioner to influence the outcome of such intervention.

Commissioner Mac Sharry, in the course of his evidence before this Tribunal, denied that he had made any such request to Minister Braks to intervene on behalf of Goodman International with any Dutch bank and has stated that he was not aware of the Dutch bank alleged to have been a creditor of the Goodman Group.

The Tribunal accepted his evidence in this regard and that would have been the end of the matter as far as the Tribunal was concerned were it not for the evidence with regard to the manner in which the producers of the television programme distorted and edited the video which showed Commissioner Mac Sharry's statement to the European Parliament on the 11th day of October 1990.

The actual video of what Commissioner Mac Sharry stated to the European Parliament was shown to the Tribunal and, when compared with the purported extract therefrom shown on the ITV programme, it was manifest that the excerpt which was shown on the ITV programme was carefully and, in the opinion of the Tribunal, unethically, edited to present a distorted picture of what he had actually said.

During the course of his speech to the European Parliament, Commissioner Mac Sharry made the following points:—

(1) Mr Braks was one of a number of people, including representatives of Governments and other organisations, to whom he spoke about the Community's beef industry as he saw it developing in August 1990.

(2) He referred to steps taken by him to ensure that for the purposes of maintaining stability and confidence in the industry, all concerned in the industry should be aware that the Community would continue to ensure a guaranteed outlet for virtually all steer production.

(3) He made it clear that he had not asked Mr Braks to intervene with any particular bank in favour of Goodman International and pointed out that at the time he was speaking to Mr Braks, that he did not even know the identities of the Dutch bankers involved with Goodman International.

(4) He said that it would be presumptuous of him to think that the initiatives which he did take with, among others, Mr Braks, were instrumental in keeping Goodman International in business.

(5) He said that if his initiatives had helped, he was pleased that the ongoing difficulties in the market confirmed him in his view that the loss or temporary closure of the outlets provided by Goodman International would have been very detrimental to the interests of producers of the trade and of the employment of thousands of people.

The producers of the ITV programme deleted from the video of the European Parliament his explanation of the initiatives he had taken. While it did show his denial of having asked Mr Braks to intervene with Goodman's Dutch banks, they deleted his statement that he did not know who Goodman's Dutch banks were at the time he was speaking to Mr Braks.

By broadcasting that portion of his speech in which he stated that he was pleased if his initiatives worked, the programme thereby gave the impression that while at one stage denying that he had asked Mr Braks to intervene with Goodman's Dutch banks, he was at the same time expressing a pleasure that such initiative on his behalf may have been successful.

The ITV programme did not state that the extract from Commissioner Mac Sharry's speech had been edited, but represented it as being a continuous extract from his speech, thereby deliberately distorting its effect.

CHAPTER THIRTEEN

Greenore

During the course of his speech in Dáil Eireann on the 15th May 1991, Deputy Rabbitte stated, inter alia:

> "It has been suggested that Goodman was subjected to a lesser degree of Customs' inspections than other commercial operations (especially in regard to container loads going North) and that he was able to virtually close off the port of Greenore to other people when he was exporting meat".

That statement made by Deputy Rabbitte appeared to imply that the Customs officials responsible for inspection of containers going north or the supervision of the port of Greenore discriminated in favour of the Goodman Group with regard to inspection of container loads.

When the Tribunal was about to deal with this allegation, Mr White, who appeared for Deputy Rabbitte, stated that:

> "My client wants to make clear that this was never his intention, nor did he feel that he was, at any stage, making an allegation that any action of the regulatory authorities was deliberate."

No evidence was adduced before this Tribunal, which, in any way, sought to support this allegation. In fact, all the evidence was to the contrary.

Mr Seamus Ó Conchuir, a Higher Officer of Customs & Excise, stationed at Greenore Port/Road Station since September 1980, gave evidence with regard to exports both through the Road Station and the Port in respect of the period 1985 to 1990.

Dealing with the Road Station, he stated that during that period approximately 175 containers in total passed through the station, 148 of those were the property of AIBP and 27 the property of other traders.

He stated that none of these containers were destined for non-EEC countries and consequently did not attract Export Refunds.

Consequently, no examinations were carried out at exportation of either the containers the property of AIBP or of the other traders.

He stated that his function of examination at the Road Station arose in circumstances only where there were Export Refunds being claimed in respect of the exportation and that would be outside the EEC.

With regard to exportation through the port, Mr Ó Conchuir gave to the Tribunal details of shipments for a period from 1986 to the date upon which he gave evidence before the Tribunal on the 4th November 1992.

He stated that during that period, approximately 150 shipments of beef totalling approximately 240,000 tonnes was made through the port in addition to 20 shipments of live cattle.

All these shipments attracted Export Refunds and the traders who used the port were AIBP, Agra Trading, Halal, UMP Midleton, Liffey Meats, Slaney Meats, Horgan Meats Charleville, Kildare Chilling, Rangeland, WD Meats (Northern Ireland), KMP Co-op, Hibernia, United Meat Packers, Master Trade Longford, Taher Meats, Dawn Meats, Meadow Meats, Master Meats, Cahir Meats and Blanchvac.

AIBP were by far the major exporter of the beef but were not involved in live cattle exports.

He stated that all exporters were treated equally and that there was no discrimination in favour of AIBP Ltd or any other exporter.

All examinations were carried out on a random and selective basis and in all cases were in the presence and with the assistance of Department of Agriculture officials who attended at the port during loading of beef on ships.

He stated that during this period no irregularities or evidence of fraud were detected by him or any other officer from examinations of beef and scrutiny of export documentation presented by AIBP Ltd.

He stated that there were of course minor errors and discrepancies not of a fraudulent nature, discovered by him in export documents which were fully investigated and corrected to his satisfaction.

It appears further that from time to time, when AIBP were loading cartons of beef for export on ships, that security personnel were engaged at night time to protect the trucks, containers and cartons, which were at the port awaiting loading. There is no evidence to suggest that anybody who had lawful access to the port was in any way inhibited in the exercise of that right of access.

Mr Aodoghan O'Rahilly, who was the Chief Executive of the company Greenore Ferries Services Ltd., the proprietor of the port, gave evidence that there was no substance in the allegation that the Goodman Group was able to or did in effect, secure the closing of the port to any other person or persons.

Other evidence was given by Customs & Excise personnel, involved in the operation of the port and border stations, that there was no truth whatsoever in the allegation that the Goodman Group was subjected to a lesser degree of control than any other exporter.

The Tribunal accepts their evidence in this regard and there was no basis whatsoever for the suggestion that the Goodman companies were subjected to a lesser degree of Customs inspections than other commercial operations.

CHAPTER FOURTEEN

Carousel Operation

In the course of a speech made in Dáil Eireann on the 15th day of May 1991, Deputy Rabbitte stated:

DAIL EXTRACT

"It has been suggested to me that there was an extraordinary incident in 1988 which would have been normally expected to lead to immediate prosecution, but in respect of which no action was taken. My information is that in 1988 a container load of boneless beef left a Goodman plant near Wexford. The lorry, carrying 500 sides of beef, crossed by ferry to Britain, travelled up the mainland and crossed to Northern Ireland at Larne. The lorry was intercepted by our customs officers on an unapproved road near Castleblayney on its way back to the Republic. The driver, whose name I have, explained to the customs officers that he was on his way to a Goodman plant near Enniskillen, but had got lost.

It was, of course, ludicrous to suggest that anyone would transport a container of meat from Wexford to Northern Ireland via Britain. The customs officers concluded that they had intercepted a "carousel" operation. This is a scam whereby meat is exported, on which export refunds are paid, secretly imported and then re-exported to claim yet more export refunds. To the astonishment of the customs officers, no action was taken against any of those involved. I have the name of the lorry driver, which I am prepared to give the Minister privately."

—15 May 1991, Col 1261.

The effect of this statement is:—

(i) that it has been suggested to Deputy Rabbitte that officers of Customs & Excise had intercepted a "Carousel Operation" being carried out by the Goodman Organisation in 1988;

(ii) A "carousel" operation is a "scam whereby meat is exported, on which export refunds are paid, secretly imported and then re-exported to claim yet more export refunds"

(iii) normally the discovery of such incident would have led to immediate prosecution but no action was taken and that the officers of Customs & Excise were astonished that no action was taken.

Deputy Rabbitte asked the Minister to tell the House why no action was taken.

His statement implies that a decision was taken other than by the Customs & Excise officers who had investigated the matter not to take action on foot of the results of the investigation.

In view of the seriousness of the allegation involving, as it did, an allegation that the Goodman Organisation was attempting to defraud the Export Refund subsidy system and an allegation that no action was taken by the appropriate State Authority by way of prosecution when the Customs officers had concluded that they had interrupted such an attempt, the Tribunal examined, in detail, the facts relating to the incident referred to.

The evidence in relation to this matter was given by:—

(*a*) Brian Maguire, an Assistant Officer of Customs & Excise;

(*b*) Eoin Prunty, an officer of Customs & Excise;

(*c*) Sean Brosnan, Higher Officer of Customs & Excise;

(*d*) Nollaig O'Broin, Higher Officer of Customs & Excise:

(*e*) John McBennett, driver;

(*f*) Peter Maguire, Manager AIBP plant at Cahir;

(*g*) Larry Goodman;

(*h*) Pat Rabbitte, TD,:

The relevant facts as disclosed in such evidence are that:—

(1) on the 23rd day of March, 1988, Brian Maguire, who was stationed at Ballybay, Co. Monaghan, was on mobile patrol duty overlooking the concession road between Castleblayney and Dundalk;

(2) he had all traffic coming from the direction of Crossmaglen, Co. Armagh under observation;

(3) he observed a blue and white tractor unit and container bearing the name "Creggan Transport" entering Culloville, County Armagh from the direction of Crossmaglen, County Armagh and heading towards Castleblayney, Co. Monaghan.

(4) he informed his colleague Mr Eoin Prunty of this fact by radio contact.

(5) Mr Prunty proceeded to intercept this vehicle and observed it entering the State at Ballincarrybridge which is an unapproved Land Frontier crossing point between Crossmaglen and Castleblayney.

(6) Mr Prunty kept the vehicle under continuous observation until it was stopped at Anahale, Castleblayney.

(7) the registration number of the vehicle was KIB 5243;

(8) the seals in the doors of the Container were intact and bore the Northern Ireland Department of Agriculture Markings "D.A.N.I./222187" and "D.A.N.I. 222188";

(9) the container contained 567 boneless beef flanks bearing markings "Ireland EEC 300";

(10) the driver was John McBennett of Creggan Transport, an employee of Creggan Transport of Newry Road, Crossmaglen, Co. Armagh, who informed Mr Prunty that the beef was owned by AIBP Cahir Exports Limited.

(11) Mr McBennett claimed that he had driven the trailer unit and container from Dundalk and had travelled the main Dundalk/Castleblayney Road;

(12) Mr McBennett accompanied Mr Prunty to the Castleblayney Road Station where, in reply to further questioning he stated that:—

(*a*) when stopped he was on his way to Lough Egish to collect more meat there;

(*b*) Another driver had collected the container of meat at AIBP Cahir Exports Ltd., Cahir and had packed the container at Castleblayney Road near Dundalk for collection by him;

(*c*) he was unable to produce any documentation with regard to the contents of the container.

(13) documentation found in the cab of the tractor unit related to a consignment of fore and hindquarters of beef consigned from AIBP Cahir Exports Ltd to Fermanagh Meats Ltd., to Mid-Cornwall Meats;

(14) when questioned about this transaction, Mr McBennett produced an invoice from Mid-Cornwall Meats showing that the meat referred to in such documentation had, in fact, been delivered by him;

(15) Mr McBennett stated that after delivery, he had returned to Northern Ireland via Stranraer;

(16) Mr Prunty informed Mr McBennett that he was seizing the conveyance and goods on the grounds that the goods were illegally imported into the State from Northern Ireland without payment of import duties payable thereon pending the decision of the Revenue Commissioners in the matter;

(17) the conveyance and goods were removed to the official compound at Clones, Co. Monaghan;

(18) on the 24/3/1988 at Clones the District Veterinary officer certified the contents of the container as chilled boneless beef flanks and in payment of a deposit of £10,000 the tractor unit and container was released to Mr McBennett;

(19) the goods were liable to Monetary Compensation Amount charges on importation from Northern Ireland into the State;

(20) the file, in the matter, was then referred to the Investigation Branch of Customs & Excise for investigation.

In reply to Mr White, appearing for Deputy Rabbitte, Mr Prunty stated that when he intercepted the vehicle he was dealing with a straight smuggling transaction from Northern Ireland into the State but it was considered, at the time, that there may have been a more serious C.A.P. fraud so the file was sent to the investigation branch for investigation.

Mr Sean Brosnan, is a Higher Officer of Customs & Excise stationed at Brook Buildings, Ardee.

When informed, by Mr Prunty, of the incident he went to Castleblayney and was fully informed by Mr Prunty of the circumstances leading to the seizure of the tractor and container including the contents.

He then contacted Mr Garvey, who was the Customs Control Officer; and requested him to go to Cahir as soon as possible and to conduct an investigation there.

At. 6.30 p.m. on the 23rd March 1988 Mr Brosnan was telephoned, at Clones, by Mr Kirwan, the processing manager at AIBP, Cahir, who informed him that:—

(i) the container of beef could not have come from Northern Ireland;

(ii) it had left Cahir for Dundalk that morning and couldn't possibly have been or have been detected coming from Northern Ireland.

When asked by Mr Brosnan for the name of the driver, the lorry number, the time of collection and any other details to verify Mr Kirwan's claim he was not in a position to supply any of the information sought.

On the following day Mr Kirwan again contacted Mr Brosnan and informed him that the container may have been collected on the evening of the 22nd March 1988 but despite request, Mr Kirwan was not in a position to give any of the details sought from him the previous evening.

He informed Mr Brosnan that the driver was a resident of Northern Ireland and may have entered Northern Ireland to change his clothes. Mr Brosnan knew that this was untrue as he knew Mr McBennett was a resident of Castleblayney.

When asked by Mr Brosnan for an explanation of the Department of Agriculture Northern Ireland seals on the container he was unable to give any explanation.

During the course of the conversation Mr Brosnan informed Mr Kirwan that he was satisfied that the container had been smuggled in from Northern Ireland on the afternoon of the 23rd March 1988 and would release the container on payment of a deposit of £10,000.

Mr McBennett stated in evidence that:—

(i) in 1988 he was employed as a driver by Creggan Transport and had been employed by them for about 3 years before the incident and during that period had transported many loads of beef on behalf of AIBP;

(ii) on the 19th day of March 1988 he collected a container of beef from the AIBP plant in Cahir for transport to Mid-Cornwall Meats in Truro;

(iii) he received all the necessary documentation for production to the Customs & Excise authorities in Rosslare;

(iv) he delivered the said load of meat to its destination: went to Nottingham and re-loaded the container with plaster for British Gypsum in Kingscourt: returned to Northern Ireland via Larne: cleared the load through the Customs at Newry/Dundalk and left the container/trailer in Dundalk, returning to the Creggan Transport yard in Newry with the tractor unit;

(v) when he arrived there he attached the tractor unit to a container AN334 in the yard;

(vi) his boss had gone to a funeral and was not available to tell him where to deliver the load, which he knew to be meat because it was refrigerated but had no idea where it came from or where it was to go;

(vii) he decided to go to his home in Castleblayney for lunch and drive off for this purpose in the Tractor with the container load of meat;

(viii) he was in the process of so doing when he was intercepted;

(ix) when intercepted he panicked as he knew that he crossed the border; he told the Customs officers the first thing that came into his head; he didn't know the meat had come from Cahir and it was just co-incidence that he said it had.

The Customs & Excise file in connection with this incident was forwarded to the Revenue Commissioners and on the 3rd day of August 1988 was assigned to the CAP Unit of the Investigation Branch.

On the 4th day of October 1988, Mr Nollaig Ó Broin, a Higher Officer in Customs & Excise and attached to the Investigation Branch contacted the Investigation Unit of H.M. Customs & Excise, Belfast to ascertain how a container with Department of Agriculture Northern Ireland seals on it came to be intercepted on an unapproved road between Dundalk and Castleblayney.

By letter dated the 19th day of October 1988, the Northern Ireland authorities replied as follows:—

HM CUSTOMS AND EXCISE

Mr B Bolger — Your Ref. 242/88 NOB
Surveyor
SIB Dublin — Our Reference 147/88
18 Lansdowne Road
Dublin 4 — Date 19 October 1988

Dear Brian

RE: SEIZURE OF 567 CHILLED BEEF SKIRTS IN VEHICLE KIB 5243 23.3.88 AT CASTLEBLAYNEY

From inquiries made through DANI it has been established that the 2 seals DANI 222187 and DANI 222188 were put on a Creggan Transport Vehicle, container number AN 334 on 23.3.88 at ABP, Fermanagh Meats, Enniskillen.

According to the Meat Plant Veterinary Officer he rejected the consignment which came from Cahir Meats as being unsound. The Veterinary Officer has no record other than the use of the seals. He cannot recall whether or not he endorsed the Republic of Ireland Health Certificate, however he gave it to the Creggan Transport driver and told him to take the meat back to where it came from. We have been unable to trace the import.

As Creggan Transport's base is in Crossmaglen it is quite possible that the vehicle seized, KIB 5243, was not the import vehicle.

Perhaps your Department of Agriculture could trace the export more easily through their Claims Section and withhold/reclaim any relevant payments.

Regards,

H L. SNODDEN
SURVEYOR
BELFAST INVESTIGATION

On the 30th of March 1988, Mr O'Broin visited the AIBP premises at Cahir and interviewed Mr Tim Kirwan, the processing manager at the plant and who had telephoned Mr Prunty at the time of the incident.

In addition, Mr O'Broin examined the company's books and accounts for a month prior to and a month subsequent to the seizure on the 23rd day of March 1988.

As a result of such examination and an examination of the Department of Agriculture records he ascertained that:—

(i) that two particular tractor units with containers of meat had left Cahir on the 19th day of March 1988,

(ii) one of these units Registration No. LIJ 3550 and Container No. GT2, containing 9990 kilogrammes of separated beef hinds and 5742 beef hindquarter flanks was destined for the AIBP plant in Fermanagh;

(iii) The second unit, Registration No. KIB 5243 with container CT4, containing 105 hinds and 225 forequarters of chilled beef, though the consignee was described as Fermanagh Meats was destined for export via Rosslare to Mid-Cornwall Meats in the U.K.

The unit destined for AIBP plant in Fermanagh was cleared through Customs at Castleblayney road station on the 19th day of March 1988.

The unit destined for Mid-Cornwall Meats travelled via Rosslare and the meat contained in the container was delivered there.

Upon release from Customs upon payment of the £10,000 deposit hereinbefore referred to, the meat contained in the container seized, which was part of the shipment made to Fermanagh Meats and rejected, was send to Rangeland Meats Ltd., of Tullynahinera, Castleblayney in the County of Monaghan.

On the 11th day of April, 1989 Mr O'Broin visited Rangeland Meats and from inquiries and an examination of this Company's records verified that:—

(i) vehicle KIB 5243 had delivered the beef to the premises;

and,

(ii) Rangeland Meats had, in fact, paid AIBP Cahir for the meat contained in such load.

On that date Mr Ó Broin also interviewed Mr Hand, the District Veterinary Officer in Ballybay who had issued the certificate with regard to the meat before its release from Customs Control and satisfied himself with regard to minor discrepancies on the Certificate viz the description of flanks as shirts and a wrongful estimate of the weight of the beef and returned to Dublin.

Mr O'Broin examined the export documentation, the SAD documentation, in respect of each vehicle from which it appeared that:

(i) Vehicle No. LIJ 3550, being the vehicle destined for Fermanagh, was cleared through Customs at Castleblayney at 6.30 p.m. on the 19th March 1988; and

(ii) Vehicle No. KIB 5243, being the vehicle destined for Mid-Cornwall Meats left Rosslare on the M/V David for Wales on the 19th March 1988.

On the 13th day of April, 1989 Mr O'Broin returned to AIBP Cahir and interviewed both Mr Peter Maguire, the Manager of the Plant and Mr Kirwan the processing manager.

On this occasion, they acknowledged that they had been aware of what had happened i.e. that the meat, which had been seized, had been rejected by the Veterinary authorities in Fermanagh, that they had requested the haulier to hold the goods in their yard in Creggan while they made arrangements to have the meat transported to AIBP, Newry, that for some reason or other the driver had brought the goods south of the border, that when the goods were seized, they were apprehensive about the unfavourable publicity that might ensue and in an effort to distance themselves and the Company from such unfavourable publicity Mr Kirwan gave wrong information to the Customs officers.

When questioned about the fact that the consignee of the meat destined for Mid-Cornwall Meats was shown to be Fermanagh Meats they explained that all exports to Northern Ireland or the U.K. are shown as consignee to AIBP Fermanagh in order to centralise and facilitate the receipt of Monetary Compensation Amounts, levies and export refunds.

On the question of MCA payable Mr O'Broin stated that:—

(i) the normal documentation for the claim for such payment was handed in at the Castleblayney station on the 19th March 1988;

(ii) the said documentation was transferred to the FEOGA Section of the Department of Agriculture on the 8th April 1988;

(iii) payment was made to Cahir Exports on the 15th of June 1988;

(iv) A refund of the amount paid in respect of the goods seized viz £731 was made in February 1990.

The Tribunal is satisfied that the claim for this amount, which was made before the goods were rejected was neither the reason for or motive behind this incident.

Having thoroughly investigated this incident, Mr O'Broin, whose primary purpose in carrying out this investigation was to ascertain whether or not there was a "carousel" in the sense already described, concluded that:

(i) the meat seized by Mr Prunty and the customs patrol at Castleblayney on the 23/3/1988 was part of the meat exported in Vehicle No. LIJ 3550 through Castleblayney Road Station and not part of the meat exported on the same day to Mid-Cornwall Meats through Rosslare;

(ii) the documents found in Cab Registration No. KIB 5243 related to the meat exported through Rosslare and not to the meat seized;

(iii) the incident was not part of a "carousel" operation.

Once Mr O'Broin was satisfied that the incident was not part of "a carousel operation" he considered that the matter had been satisfactorily dealt with in the first instance. There had been an illegal importation of meat into the State, it and the vehicle had been seized and subsequently released upon a deposit of £10,000 being paid, which sum was paid by the haulier and he recommended that this sum be retained by the Revenue Commissioners, which they did.

Mr O'Broin was satisfied as a result of these inquiries that AIBP were not aware of the importation of the meat until after it had been seized and were not involved in it.

The Tribunal is satisfied that the incident referred to in Deputy Rabbitte's speech was not part of a carousel operation.

The presence in the trailer of documentation relating not to the meat in the container attached thereto but to the meat exported to Mid-Cornwall Meats, led to a reasonable suspicion on the part of the Customs authorities that it might have been part of such an operation.

This suspicion led to the very thorough investigation outlined herein by Mr O'Broin, whose investigation established the facts set forth above.

The Tribunal is satisfied that this investigation was not in any way hindered by any person in authority, political or otherwise, and was a completely independent investigation and the recommendations made by Mr O'Broin were based on such investigation and on no other basis.

The Tribunal is however, concerned by the lack of co-operation with the Customs & Excise authorities shown by Mr Kirwan, the processing manager at the AIBP Plant in Cahir and his attempts to mislead the officers of Customs & Excise who were dealing with this incident and who only admitted the circumstances of the incident to Mr O'Broin when the facts had been established by him by independent investigation.

CHAPTER FIFTEEN

Department of Agriculture, Food & Forestry

The Department of Agriculture acts as the "Intervention Agency" on behalf of the European Commission under the European Communities Act 1972.

In its capacity as such intervention agency, the Department of Agriculture operates the following EC schemes in the beef sector:

Intervention Purchases, Storage and Sales
Export Refunds
Aids to Private Storage (APS)
Beef carcase classification scheme
Monetary Compensation amounts (MCAs)

In view of the nature of the allegations being inquired into by the Tribunal, it is desirable to summarise very briefly the workings of the said schemes.

Intervention is one of the main Community market support mechanisms under the Common Agricultural Policy. The main regulations governing same are Commission Regulations 805/68, 859/89, 2173/79, 2539/84, 985/81, 2824/85, 2182/77 and 2220/85.

Only steer beef from Carcase Classification Categories U3, U4, R3, R4 and 03 are eligible for Intervention in Ireland (Annex to Commission Regulation 859/89).

Intervention beef can be deboned prior to placing it in storage (Article 18 of 859/89).

Deboning is carried out only by those meat export plants and approved cutting plants which have facilities to carry out deboning in accordance with EC regulations and the Department of Agriculture's specifications (Article 19 of 859/89).

Deboning is carried out on a contract basis between the Minister for Agriculture & Food and the contractor (Article 19 of 859/89).

Boning must be carried out in accordance with a strict specification (Article 19.2 of 859/89)

In all, eleven different cuts of meat are produced and the deboning contract lays down the cutting and trimming requirements, as well as the wrapping, boxing and labelling requirements.

Under Paragraph 3(*e*) of the boning contract all meat derived from the deboning, trimming, packing and freezing operations is the property of the Minister.

In addition to the eleven cuts listed in the Schedule to the Boning Contract, the Contract provides that identifiable pieces of meat which are removed when preparing the cuts should be packed in boxes of plate and flank meat.

The contractor is entitled to keep all bones, fat and trimmings resulting from the boning operation.

The Schedule to the boning contract provides that boned cuts from the hindquarter, excluding the plate and flank, may not have a fat cover of more than 1 centimetre and the visible fat must not exceed 10% of the product. The weights of the visible fat on plates and flanks and briskets must not exceed 30% of the product.

Article 2 of Commission Regulation No. 230/79 provides that:—

> "The maximum tolerance referred to in Article 3.(2)(*c*) of Regulation (EEC) No 2305/70 shall be 1%. It shall apply to the difference between the unwrapped weight of the product recorded when it is taken over and the wrapped weight recorded when it is removed from storage.
>
> "This limit shall apply to boned meat produced by boning fore and hindquarters taken over, allowance being made for a weight loss of 32% as a result of the boning.
>
> "In the deboning contract, the Department of Agriculture have set a minimum yield of red meat to be achieved by the contractor of 68%."

The deboning contractors are paid on the basis of a rate per kilo which is reduced where the deboning contractor fails to meet the 68% yield. No additional fee is paid for yields in excess of 68%.

The beef is presented in half carcase form in accordance with the standard carcase dressing specification.

The following conditions must be met:—

The beef must come from one of the 5 eligible categories.

It must be stamped with the classification grade with indelible ink.

The animals must have been slaughtered not more than 6 days previously (the day of slaughter not being counted).

The beef may be quartered prior to weighing but the two quarters must be presented.

The beef is weighed-in in chilled bone-in form. At this point the product becomes the property of the Intervention Agency viz the Department of Agriculture.

The weight, together with the carcase number and classification grade are recorded on an IB4 Form (Intervention Beef Purchase Record — Boneless Beef).

The supervising AO is required to check weigh at least 10% of the quarters. The relevant carcases are encircled and initialled on the IB4 Form.

When a form is completed, the top copy is removed by the official and the factory retains the bottom copy.

Weights and classification details recorded on the IB4 form are required to be cross-checked against particulars shown on the daily classification sheets (Kill Sheet). This sheet is completed at the earlier stage of classification and records the carcase number, hot weight and classification.

The IB4 Certification form is then signed.

The quarters are required to be deboned under the supervision of officials. Meat cuts produced are wrapped, boxed, weighed and labelled.

During deboning officials are required to examine at least 5% of the boxes produced, subject to a minimum of 20 and the following details are required to be checked to ensure that:—

(1) The number of catch weight cuts marked on the boxes are actually in the boxes.

(2) The weights are correct.

(3) The markings on the boxes, including the lot number, are present in the appropriate places and are correct and legible.

(4) The beef has been properly cut, wrapped and packed in accordance with the specifications.

Each day a random sample of boxes of forequarters, plates and flanks are required to be selected for defatting analysis.

Details of each box produced are entered on an IB7 Form — Boneless Beef Analysis Schedule. This form is signed by the factory representative and an Official.

In addition, each day the boxes of at least one particular cut are selected by the Veterinary staff at random and the following check is required to be carried out:

1. That the correct weight of meat in each box of that cut has been recorded on the IB7.
2. That the cuts are properly wrapped and prepared (a 20% sample of boxes is required to be reopened).
3. That the number of cuts marked at both ends is actually present in the box.
4. Reconciliation of the total number of cuts in all the boxes of the catch weight cuts the number of quarter/sides actually deboned.

The accuracy of the scales is required to be checked each day using a known weight.

An IB6 form is required to be completed in the Veterinary Office by the official there. This form summarises the details of the IB7 form, i.e. production from the deboning and after comparison with the weights recorded on the IB4 form, shows the yield of meat achieved and placed in storage.

The top copies of the IB4 Purchase Record and the IB4 Certification sheet together with the IB6 and IB7 forms are returned by the official to the Department of Agriculture.

Checks are required to be carried out in the Headquarters of the Department of Agriculture on the documents associated with each intervention purchase.

SALE OF STOCK FROM INTERVENTION

The EC Commission in consultation with the Beef Management Committee authorises regular sales of beef from intervention stock.

The decision to sell beef from storage is based on the availability of beef in both the Community and world markets.

Beef is sold for a variety of uses including:—

(1) For general export outside the EEC.

(2) For a special export to specific destinations.

When the Commission, in consultation with the Beef Management Committee, agrees to sell intervention beef, a regulation is published giving effect to the sale, including the quantities to be sold, prices, terms and conditions etc.

Where beef is sold under a sale with specific end-use requirement, the Department of Agriculture is responsible for ensuring compliance with these requirements. Where beef is sold for export, the normal Customs' documentation is required as proof of export from the Community. When Intervention beef is sold, the contracts for sale are completed between the purchasers and the Minister for Agriculture & Food and the purchasers pay for the beef before taking it over.

A Removal Order is completed showing the purchaser's name, description of product, weight etc.

Beef sold for export must be removed from store and exported in the same state. There are exceptions to this rule. The Department can give permission to replace torn or soiled wrapping and also in most sales of boneless product for export purchasers, may cut and/or repack product (Commission Regulation No. 2824/85).

The conditions governing this procedure required that the beef continues to be fully identifiable.

The prior approval of the Department of Agriculture is required for this operation and the work undertaken must be done under the supervision of Department Officials. When the repackaging is completed the Department Official prepares a "repackaging certificate" certifying that beef has been repackaged. The Official then prepares a further Removal Order certifying the removal of the product, the container number and the relevant seal number.

AIDS TO PRIVATE STORAGE (APS)

This scheme is a market support measure which can be introduced when cattle prices are low and the market has a surplus of beef.

The scheme operates on the basis of removing the product off the market for a limited period and is normally introduced in the Autumn period when very large numbers of summer fattened cattle come on the market.

Under APS the beef remains the property of the trader and the trader retains complete discretion on the disposal method when a minimum storage period has elapsed.

The main regulations governing APS are Commission Regulations Nos. 1091/80, 2965/89 and 2220/85.

Under this scheme beef producers enter into contract with the Minister for Agriculture and Food to place beef in store for a minimum specified storage period.

At the end of the said period the amount of aid, which is fixed in advance, is paid to the storers. The aid can also in certain circumstances be paid in advance subject to the provision of securities fixed by regulation.

One of the features of the APS scheme is that it allows traders to qualify for storage payments while the product can be simultaneously placed under a Customs export warehousing regime (Article 5 of 3445/90). This permits a product to be stored for up to six months prior to the product actually leaving the store. It also provides for the advance payment of the appropriate level of export refund subject again to the provision of the relevant level of guarantee security.

The Department of Agriculture supervises all APS beef production, deboning, transporting and storage operations. Beef in private storage must be produced and stored in EC approved slaughter premises and stores which have a permanent presence of Department Officials.

The operator must lodge a security with his contract application to ensure compliance with the conditions of the contract. This security is returned to the operator upon the proper completion of the storage contract.

A formal contract is entered into between the intervention agency and the operator. An individual identification number is assigned to each contract and all product produced under the contract and all the associated documentation must bear this number. The minimum number contract quantity is 15 tonnes. There is no maximum quantity. If an operator withdraws his application for a contract the security is forfeited.

The contract imposed the following obligations on the contractor;

(i) the production and storage of the agreed quantity of beef within the time limits set and in accordance with the conditions set out in the regulations and the Departments instructions.

(ii) Leaving the product in store for a minimum storage period and not altering or exchanging the beef whilst in store.

(iii) Properly completing all supporting documentation.

(iv) Permitting the Department at all times to check that all the obligations laid down in the contract are being observed.

The APS Regulations specifically require that all meats placed in private storage must be classified in accordance with the Community Carcase Classification System.

However all grades are eligible; the Classification Officer affixes labels to the sides giving the classification and these labels must remain on the carcases until the product is either deboned or exported.

The beef is taken under APS control in chilled bone-in form. Before the product is physically placed in store it can be cut/deboned. There are conditions which must be met when APS beef is deboned (Article 4 of 3445/90 and Article 4 of 2965/89).

The decision to debone the beef is at the discretion of the contractor.
When beef is deboned the rules are:—

(1) The beef must be weighed into the boning hall in a chilled state.

(2) Only meat produced from the carcases weighed in under a contract may be deboned and stored under that contract.

(3) All the meat produced from the deboning operation must be stored.

(4) A minimum deboning yield of 67% must be achieved, otherwise no aid is payable. The yield is the ratio of the weight of meat obtained to the weight of the original bone-in product; a maximum yield of 75% is also set in the Regulation. At the 75% yield level, full aid is payable on the beef. If the yield falls below 75% a proportional deduction in aid is made. No additional meat can be employed to bring up the yield.

(5) The large tendons, cartilages, pieces of fat, lean trimmings and other scraps left over from deboning cannot be placed in storage.

Beef must be placed in storage in accordance with criteria set by the Department of Agriculture.

These requirements are to ensure that

(1) the beef is readily identifiable by contract and

(2) it can be physically inspected to confirm its presence and to ensure that it has not been altered in any way.

(3) an adhesive label in the form as described by the Department which indicates that the product is under APS control and the contract number is given.

APS meat in a cold store must bear a label on each side which shows

(1) The APS contract number.

(2) The date into storage.

(3) The number of boxes or quarters on the pallet.

(4) The type of cut.

(5) The relevant store intake docket number.

Pallets contain meat from one contract only. These labels and marks must remain on the product throughout the period it is under APS control.

The deboning operation is supervised by officials. The beef is boned to the contractors own specification because at all times it remains his property.

Export Refunds

Market prices in the EEC are significantly higher than those on the world market. To facilitate the trade of community products on world markets subsidies known as export refunds are paid to the exporters. These export refunds compensate for the difference between prices on the world market and prices in the community.

In the beef sector alone Ireland makes payments of around £250 million annually in export refunds, although within that average the amount expended can vary widely from year to year.

The rates of refund and the countries to which they apply are fixed regularly by the EC Commission in conjunction with the Beef Management Committee. The rates are fixed by the adoption of a Commission Regulation which is published in the Official Journal of the Communities.

The Granting of Export Refunds is subject to a number of regulations made
by the Council of the EEC and the Commission and the most important of these are Commission Regulations 3665/87, Regulation 565/80, Regulation 2220/85, 32/82 and Regulation 1964/82.

The calculation of refund depends on three factors, destination, type of beef being exported and date of export.

In general different rates of refund are payable according to whether the beef is:

fresh/chilled or frozen,
male or female,
presented as sides, hindquarters or forequarters,
bone-in or boneless.

The rate payable reflects the perceived value of the product. Consequently fresh beef attracts a higher refund than frozen. Beef from male animals attracts a higher rate than beef from female cattle and the rates payable on hindquarter beef exceed those for fore quarter beef and the rates for boneless beef are higher than those for bone-in beef.

The classification of goods for export refund purposes is carried out in accordance with general interpretative rules and in Ireland the Custom Authorities are the competent authorities to interpret the export refund nomenclature classification.

There is no differentiation of refund according to the carcase classification of the animal.

The beef must fall within a category defined as eligible for refunds and it must be exported to a third country to which refunds are fixed.

In addition to these there are other criteria which must be met:

The beef must be of sound and fair marketable quality and fit for human consumption.
It must be placed under Customs control with a view to export.
It must leave the Community within prescribed time limits.
It must be placed on the market of an eligible non EC country within a prescribed time limit.
Proof of export and import must be supplied within prescribed time limits.

Placing of Beef under Customs Control

Beef may be placed under Customs Control prior to export either in a Custom warehouse or in the traders own premises. It can also be placed under control at the point of export.

The majority of beef exporters avail of the Customs warehousing procedure which is detailed in Council Regulation No. 565/80.

It provides in Article 5 for advanced payment of the refund subject to provision of a security as soon as the trader has placed the beef in an approved warehouse under Customs control. The beef can remain in the warehouse for up to six months and upon removal the trader has a further 60 days to effect export from the EC.

Prior to placing exports under Custom control, the exporter must first obtain the "approved consignor" status from Customs and Excise.

To obtain this approval exporters must lodge a number of undertakings with Customs and the bona fides of the exporter, the premises and undertakings are vetted by Customs.

Approved consignors are obliged to maintain records of a high standard open to inspection by Customs.

Customs are presented with:

(*a*) An AP form (advanced payment declaration.)
This form is in two parts and gives full details of the beef, that is quantity, product tax and intended destination.

(*b*) C and E 977 form.
This is a register of CAP goods placed under control prior to export. In the case of boneless beef this form must be supported by boning hall production sheets.

The entries on these forms are verified by Customs and Excise. If they are satisfied a copy of the AP form is signed and stamped and despatched to the Department of Agriculture for calculation and issue of payment.

When the warehousing period is completed the beef is placed under Customs Export Control in which the exporter has 60 days to effect physical export of the product.

The exporter must send in in advance for Customs a C & E 978 form which is a declaration of CAP products for export.

A declaration and control (D and C) form must also be completed and presented.

This gives full particulars of the beef (quantities, production description, codes etc) and is cross referenced through serial numbers to the relevant AP forms.

A Control Form T5 is required when the beef transits through another Member State en-route to the third country. This form is cross referenced with the relevant D and C Forms. It travels with the consignment and is surrendered to the Customs Authorities of the Member States from which the beef exits from the Community.

These Customs authorities complete the declaration of the T5 that the beef has left the Community and the endorsed T5 is returned to Irish Customs who in turn forward same to the Department of Agriculture.

The single administrative document (SAD) which is the normal Customs export declaration must be lodged with Customs.

It is clear that the Customs and Excise authorities, as the competent authorities to interpret the export refund nomenclature classification have an important role in the control of exports and in the working and control of the export refund system.

This control is exercised through a combination of physical and documentary checks which are carried out at meat plants and cold stores throughout the country and at export locations.

Customs and Excise authorities do not maintain a permanent presence at meat plants or cold stores. The control is exercised by means of visits and spot checks at these premises. They do however have a full-time presence at ports and other export points.

The role of the Department of Agriculture, the controls exercised by them and the action taken by the Department when the irregularities are uncovered are set forth in detail in a previous chapter.

CHAPTER SIXTEEN

1986 APS Scheme and Waterford/Ballymun

The 1986 APS Scheme was governed by Regulation EEC No. 2651/86 which set out the specific details and conditions of the scheme.

The 1986 Scheme began on 1st September 1986 and traders were entitled to submit applications to store beef under the terms of that Scheme until the 10th October 1986.

Waterford

Anglo Irish Beef Processors Ltd (hereinafter referred to as AIBP was an "approved consignor" and so approved by the Revenue Commissioners, whose beef was processed at the premises of AIBP (Carlow) Exports Ltd at Christendom, Ferrybank, Waterford and stored in premises, the property of Autozero Ltd, trading as Waterford Cold Store of the same address. These latter premises were approved by the Revenue Commissioners as a CAP Warehouse.

Beef was placed under Customs control at this CAP warehouse in accordance with the APS Scheme in 1986. 70,000 cartons of beef, produced at AIBP Waterford were stored in the Waterford Cold Store and simultaneously placed under Customs Control and export refunds in the sum of £2,289,000 were advance paid against securities lodged.

On the 26th day of November 1986 Richard B. Hanrahan, an Officer of Customs & Excise visited the Waterford Cold Store to check a bond belonging to AIBP. The bond selected for examination was Bond No. 716 which related to the production for the 25th of November 1986 and related to 464 cartons of boneless beef.

He had obtained from Anglo Irish Beef Processors the production records required to be kept and made available for this bond.

He attempted to compare the individual weights on the cartons with the individual weights on the production records and had great difficulty in so doing and eventually abandoned the attempt.

On the following day he informed Sean P O'hOdhráin, Higher Officer of Customs & Excise of his difficulty.

On the 12th day of September, 1986, Sean R O'Briain, an Officer of Customs & Excise, during the course of a visit to Autozero Ltd., examined Bond No. 695 which related to production on the 18th day of November 1986 and noted that the total of the weights marked on the cartons of shins did not agree with the stated weight.

He informed S P O'hOdhráin, Higher Officer, of the situation.

Mr O'hOdhráin, then immediately went to Autozero Ltd and obtained the production records for Bond No. 695 from AIBP.

They checked the weights on the cartons and the bonded weights. This disclosed an over-declaration of 58.3 kilograms.

On the 15th day of December 1986 Mr O'hOdhráin and Mr O'Briain informed Mr O'Braonáin, who was the Surveyor of the Customs & Excise District of Waterford of their suspicions that there were serious discrepancies between the actual weight of cartons being placed under Customs control under Article 5 of Commission Regulation (EEC) No. 565/80 and that declared to be in the cartons.

Having discussed the matter in some detail Mr O'Braonáin instructed Mr O'hOdhráin and Mr O'Briain to extend their checking and report to him.

Under Mr O'hOdhráin's supervision, the Waterford Customs & Excise staff continued checks of AIBP beef products in Autozero Ltd until the 23rd day of December 1986.

During that period they check weighed 4,000 cartons to ensure the accuracy of the weights on the cartons.

On the 24th day of December 1986 Mr O'Braonáin and Mr O'hOdhráin met with Mr Gerry Thornton, representing AIBP in the AIBP offices in Waterford. Mr Thornton had become aware of the investigation being carried out by Customs & Excise officials. He admitted to Mr O'Braonáin and Mr O'hOdhráin that there was a problem with the weights bonded. He informed them that the Boning Hall at AIBP (Carlow) Export Ltd was leased to Mr Eamon Mackle and that everyone in the Boning Hall was employed by Mr Mackle, who was boning the beef for Anglo Irish Beef Processors Ltd on a contract basis.

He stated that AIBP placed the onus on Mr Mackle to achieve specific yields from the cattle slaughtered, but that they did not conduct any check on Mr Mackle's operation to verify that the yields claimed by Mr Mackle were accurate.

He informed them that he had been informed on the previous day by Mr Mackle that he was having difficulty achieving yields and that he had increased the weight on production records to give the impression that the required yields had been achieved.

Mr Thornton was informed that the investigation would continue until the full extent of the over-declarations were quantified in line with normal Customs' procedures.

The Waterford Customs & Excise Staff under the supervision of Mr O'Braonáin and Mr O'hOdhráin continued their investigation on the 5th, 6th and 7th days of January 1987 and during that period examined 11,719 cartons during loading for export.

Details of the discrepancies found on the examinations on that date were submitted to the Surveyor of the Waterford Customs & Excise.

Prior to the 19th day of January 1987, it was decided that due to the scale of the investigation required and the demands on the Customs & Excise staff levels, AIBP should be requested to provide personnel for the purpose of listing the carton weights.

AIBP agreed to this proposal and provided 10 of their employees for this purpose.

These employees were required to list the weight of each carton on the pallets and the listing was spot checked by Customs & Excise officials before the beef was loaded for export.

In the early afternoon of the 19th January 1987 it was observed by Mr O'Briain and another officer that the weights written on one carton on a pallet appeared to have been altered by changing 21 kilograms to 27 kilograms. The carton was weighed and revealed the correct weight to be 21 kilograms. When Mr O'hOdhráin and Mr O'Braonáin learnt of this, they immediately terminated all involvement by the employees of AIBP in the listing of carton weights.

All cartons due for export on the 19th day of January 1987 were subsequently examined and check weighed and it was discovered that the weights of 17 cartons had been similarly altered. In the case of 15 of these cartons the weight had been changed from 21 kilograms to 27 kilograms, in one from 20 kilograms to 26 kilograms and in the other the weight had been increased by 6 kilograms.

The discrepancies in the weights discovered during the examination of the beef being exported on the 19th day of January 1987 were communicated to the relevant Customs & Excise authorities.

On the 3rd day of February 1987, 14 cartons of boned beef (plate cuts) from different APS contracts were thawed for the purpose of examining the beef therein.

On the 9th day of February 1987, in the presence of Mr O'Braonáin, Mr Dermot Ryan of the Department of Agriculture, Mr James Fairbairn and Mr Don O'Brien of AIBP, the thawed beef was examined.

In the case of 6 cartons examined, a sizeable amount of trimmings had been packed within each cut contained in the cartons. Trimmings are not eligible for APS purposes nor are they entitled to export refunds.

On the 9th day of February 1987 a further 25 cartons were selected for thawing from three further APS contracts.

On the 13th day of February 1987, again in the presence of Mr O'Braonáin and Mr O'hOdhráin and Mr Fairbairn, Mr O'Brien and Mr Ryan, the 25 cartons in question were examined.

It was found that of the 25 cartons, 20 contained a sizeable amount of trimmings within the internal cuts.

The trimmings found in 10 individual cartons were then weighed, and of the total weight of the 10 cartons — 246.7 kilograms — the weight of trimmings was found to be 38.08 kilograms, representing 15.43% thereof.

Mr Fairbairn conceded that there were indeed trimmings included in the cartons in question and alleged that the trimmings had been included without the knowledge of AIBP.

As it was eventually agreed by AIBP that the findings of trimmings in 27 out of 39 cuts of plate/flank was representative of the entire 1986 production of plate/flank, it was not considered necessary to thaw out any further cartons of Plate and Flank.

This was eventually confirmed by letter in writing from Mr Fairbairn, dated 31st July 1987 wherein he stated that:—

> "We herein confirm that on foot of your findings we would accept the result as regards defects and the degree thereof to be representative of the plate cuts presently held in warehouse at Waterford cold store".

From the 22nd January 1987 until 16th February 1987 the investigation was carried on by members of the Customs & Excise staff assisted by staff from the Department of Agriculture & Food.

A quantity in excess of 60,000 cartons were examined during this investigation.

This investigation disclosed that the net weight marked on cartons were correct but discrepancies by way of inflated weights to relevant C&E 977 forms were found in respect of 57,767 cartons of beef under Customs' control and to 1,308 cartons of shin beef, which were not placed under Customs' control but which were part of APS contracts.

Of the 57,767 cartons of beef under Customs' control, 45,789 cartons were APS beef and 11,978 cartons were non-APS beef.

The computation of the discrepancies led to the conclusion that 30,607.9 kilograms of non-existent beef was placed under Customs' control and that there was a shortage of 1,031.4

kilograms in the shin beef which was part of the APS contracts but not under Customs' control.

In addition it was concluded that in the 6,800 cartons of plate and flank, there was included an average of 3.8 kilograms per carton of trimmings which was not eligible for either APS or Export Refunds.

Ballymun

95,977 cartons of beef of a total weight of 2,639,003.7 kilograms was placed under Customs' control at a CAP warehouse at Autozero Ltd., Cabra in the City of Dublin and a number of other cold stores under the said 1986 APS scheme and simultaneously placed under Customs' control.

Export Refunds in the sum of £3,972,000 were advance paid against securities lodged.

On the 4th day of March 1987 Thomas MacCraith, a Higher Officer of Customs & Excise carried out a full examination of 182 cartons of beef placed under Customs' control by AIBP which had been produced at their plant at Cloghran in the County of Dublin and stored at Autozero Ltd., Cabra, Dublin. From such examination he established an over-declaration of weights.

On the 9th and 10th day of March 1987 he carried out full examination of 973 cartons of beef produced by Anglo Irish Beef Processors Ltd and stored at Hibernia cold store, Sallins in the County of Kildare.

Again he established from such an examination an over-declaration of weights.

Between March 1987 and June 1987, Mr MacCraith supervised the examination and checking of a further 64,326 cartons of beef placed under Customs' control by AIBP and stored in various cold stores.

All the beef referred to was boned out at AIBP at Cloghran in the County of Dublin.

This investigation which was a very detailed and intensive one, disclosed that of the 95,977 cartons placed under Customs' control, 33,100 had been exported prior to the investigation.

The cartons still in stock were stored in the following cold stores:—

(1) Anglo Irish, Cloghran, Co Dublin

(2) Autozero, Cabra, Dublin

(3) Transfreeze, Santry, Dublin

(4) Eirfreeze, East Wall, Dublin

(5) Irish Cold store, Tallaght, Dublin

(6) QK, Naas, Co Kildare

(7) Hibernia Meats, Sallins, Co Kildare

(8) Norish Food City, Castleblayney, Co Monaghan

(9) Norish, Kilkenny

(10) Cahir Cold Store, Cahir, Co Tipperary

(11) Hanley's, Rooskey, Co Roscommon.

The physical check of the cartons in the Dublin Cold Stores was carried out by CAP Control Station staff, assisted at various stages by other officers provided for this purpose.

The physical check of the cartons in cold stores located outside the Dublin area was carried out by the local staff and a report of their findings forwarded to Head Office. The Department of Agriculture also supplied a number of Agricultural Officers to assist in this task during part of this investigation.

The findings were correlated in Head Office and the full quantification thereof completed of all beef produced by AIBP at their plant in Cloghran Co Dublin during the period October '86 to February '87.

Most of this beef (95,977 cartons) was produced in the period October/November 1986 under APS contract and the remaining 1,827 cartons were produced after the APS system ended.

Only 62,877 cartons of APS beef was examined as the balance had been exported by the time the irregularity was detected.

This investigation disclosed that in the 62,877 cartons of beef examined, there was an over-declaration of 70,285.4 kilograms, representing a percentage of 4.2.

Applying that to the cartons of beef which had been exported prior to the beginning of the investigation, and assuming as the Customs & Excise did that similar proportionate discrepancies would exist in regard thereto, it was assumed that there was an over-declaration of 36,525.44 kilograms, representing a total over-declaration of weights in the amount of 106.8 tonnes.

In the course of the examination of meat produced by AIBP at their plant at Cloghran, there was no evidence of the inclusion of any trimmings in the cartons of beef placed under Customs' control and the irregularities discovered related solely to over-declarations of weights in the production records.

Despite intensive investigations, no irregularities were discovered as a result of the examination of the records of any plant operated by AIBP other than the irregularities discovered in respect of the production at Waterford and at Cloghran. Despite an extensive scrutiny, the records held by AIBP at Ravensdale, the headquarters of AIBP, did not disclose any of the irregularities discovered in Dublin and Waterford.

The actual physical examination of cartons took place in two phases:—

Phase 1

Cartons were examined as they were being taken out of cold store for delivery in containers to the place of exportation. This phase took place in two stages, one involving the shipment of 6 container loads in early March 1987 and the second stage, a complete shipload in April 1987.

Phase 2

AIBP agreed to have the remaining cartons still in store after Phase 1 taken out onto the loading bays for physical examination and then re-deposited in the cold store. This exercise took place during May and June 1987.

When the amount over-declared in Waterford, namely 30,607.9 kilograms, is taken into account, the total over-declaration of weights in respect of beef processing at the Waterford and Cloghran plants amounts to 137,418.74 kilograms, i.e., 137.418 tonnes.

In respect of beef processed at Cloghran, a total of 82 cartons of "plate and flank" were selected at random and thawed out for examination to ascertain if trimmings had been included. No irregularities were found.

All records relating to the production of beef for AIBP International Ltd. for the period under investigation were called for and produced from AIBP Ltd., Nenagh and AIBP Ltd., Cahir.

A detailed examination of these records was carried out and did not reveal any irregularities.

A detailed examination of the records of AIBP held at their office at Ravensdale did not disclose any information in relation to the weight irregularities at Dublin (Cloghran) and Waterford factories.

The only irregularities occurred at Waterford and Cloghran where the boning was carried out on behalf of AIBP Ltd. by Daltina Traders Ltd.

Daltina Traders Ltd was a company incorporated in October 1985 by Eamon Mackle for the purpose of carrying out deboning operations for AIBP at Cloghran.

Prior thereto, these operations had been carried out by a company called Dubned Exports Ltd., of which Liam Marks was a director and which employed Mr Mackle as manager.

By agreement between Mr Mackle and Mr Marks, Daltina Traders Ltd took over the work previously being carried out by Dubned Exports Ltd at Cloghran and thereafter AIBP were invoiced by Daltina Traders Ltd with the costs thereof. In or about the month of August 1986, Mr Mackle, on behalf of Daltina Traders Ltd agreed with AIBP to set up a deboning operation at AIBP premises at Waterford, to supply the necessary boners and be responsible therefor.

Their arrangement with regard to Cloghran was to continue and Mr Mackle appointed Mr Marks to manage and supervise same as he (Mr Mackle) would be required to spend most of his time in Waterford.

All these agreements and arrangements were verbal.

The deboning of beef in respect of the 1986 scheme at Waterford and Cloghran (Ballymun) was carried out by employees of Daltina Traders Ltd and the boning hall production records were required to be kept by their employees, Joe Devlin in Waterford and Ray Watson in Cloghran.

These records provided the basis for the boning hall production sheet which were required to support the C&E 977 form which were prepared and submitted by AIBP staff.

The declarations in respect thereof were made by AIBP.

At all times the management of AIBP disavowed all knowledge of the irregularities discovered during the investigation and attributed sole responsibility therefor to Daltina Traders Ltd who carried out the deboning operations and provided the details of the production.

On the 13th April 1987, Mr Healy of the CAP Division of the Revenue Commissioners, wrote to Mr D Russell of the FEOGA division of the Department of Agriculture as follows:—

> "I am directed by the Revenue Commissioners to refer to the beef irregularity found in Anglo Irish bondings in Waterford and to state that as far as the Commissioners are concerned, there is no reason for prolonging their investigation, and if you have no objection they would regard the inquiry as concluded at this stage. If Anglo Irish now proceeded to export the beef, except the plates and flanks, the exportation would be entitled to be documented in the ordinary way.
>
> "The remaining problems concerning the plates and flanks cannot finally be decided until such time as both your Department and Anglo Irish clarify your respective requirements.
>
> "For example, if you were to decide to refuse any payment for these cuts and Anglo Irish were to acquiesce, then they could be exported. If Anglo Irish were not to agree, then we would have to carry out further sampling.
>
> "It is envisaged that your Department would deal with the financial aspects of the remaining cuts."

While the investigation in respect of Waterford was, at that stage, completed, the investigation in respect of the APS scheme at Cloghran was still in progress.

Though the investigation was completed in respect of Waterford, one serious problem remained outstanding with regard to the results thereof.

This related to the boxes of plate and flank which had been placed under Customs' control.

A sampling of these boxes had been thawed out and the presence of trimmings, which were neither eligible for export refund or APS had been discovered. Unless AIBP accepted the findings in respect of those cartons as being representative of the entire of the boxes of plate and flank, the thawing of a considerable number of other boxes or cartons would be required.

A number of meetings were held between representatives of the CAP Division of the Revenue Commissioners, the Department of Agriculture and AIBP for the purpose of discussing this problem and securing a solution thereto.

AIBP were anxious to have the matter dealt with as quickly as possible as the delay was in effect costing them money because shipments of the beef could not be undertaken.

Eventually by letter dated the 31st day of July 1987 to the Revenue Commissioners, AIBP confirmed that

> ".....on foot of your findings we would accept the result as regards defects and the degree thereof to be representative of the "plate cuts" presently held in warehouse at Waterford Cold store."

On receipt of this letter the Revenue Commissioners wrote to the Department of Agriculture, stating that in view of the said admission they were regarding the investigation of the plate/flank cuts as concluded.

The report of the Customs & Excise Investigation Branch into the irregularities found at Waterford and Cloghran was completed and signed on the 4th day of September 1987.

This report was submitted to the Surveyor of the Investigation Branch and on the 5th day of October was forwarded to Mr Michael Murphy of the Finance Division, Department of Agriculture.

The terms of the report were considered by the Finance Division of the Department of Agriculture who prepared a memorandum dealing with sanctions to be imposed in respect of APS and export refund payments, which report was completed and dated on the 18th day of January 1988.

The principal recommendations contained therein were:—

(1) To sanction Goodman International for £1.111m including £430,000 in forfeited securities

(2) Report the case to Brussels as an irregularity and seek bi-lateral discussions with the Commission to get their opinion on the sanction proposed.

(3) Refer the case to the Garda Fraud Squad for investigation.

The matter was reported to the European Commission in Brussels by the Department of Agriculture on the 29th day of January 1988.

The Commission requested a copy of the Customs report and an edited version thereof, from which the names and locations had been deleted, was provided by the Department of Agriculture.

The deletion of the names and locations was in accordance with standard practice and EEC Regulation.

A meeting to discuss the implications of the said report was held with the EEC Commissions legal service in Brussels on the 18th day of February 1988.

Present at the meeting were four representatives from the Commission and Mr Donal Russell, Principal Officer, Finance Division, Department of Agriculture, Mr John Hickey, Assistant Principal, Refunds Section, Department of Agriculture, Mr Maurice Mullen, Assistant Principal Officer, APS Section, Department of Agriculture and Mr Murphy, Higher Officer, Customs & Excise.

The options open to the Department of Agriculture under the regulations were discussed and there was informal agreement that:—

(*a*) No refunds should be paid on the excess weight.

(*b*) No refunds should be paid on the plate cuts, and

(*c*) A reduced rate of APS payment should apply.

Subsequent to this meeting various calculations on the amounts to be disallowed were made and eventually the letter dated the 16th day of January 1989 was written by Ms Harvey, Principal Officer of the FEOGA Division, to the Chief Executive of Anglo Irish Beef Processors International Ltd. which set forth the sanctions to be imposed and the principles upon which such sanctions were applied.

This letter was as follows:—

"16 January 1989

Chief Executive
Anglo Irish Beef Processors (International) Ltd
Ravensdale
Dundalk
Co Louth

Sir

I refer to previous contacts between officers of this Department and representatives of your group of companies in connection with the findings of the investigation carried out in 1986/87 by the Customs Authorities into the beef production operations of the group over the period September 1986 to February 1987 inclusive. The investigation arose as a result of suspected misdeclarations by the group on export refund documentation.

The investigation found that, in respect of beef produced at the group's plants at Cloghran, Co Dublin and Christendom, Co Waterford over the period concerned, there was an overstatement of the weights actually achieved and that at the Waterford plant trimmings were included in cartons of beef declared as containing plate cuts only. These finding mean that incorrect declarations have been made by the group both in respect of export refund claims and in respect of claims for aid under the Private Storage Aid Scheme (APS) for beef produced under the terms of Regulation 989/68 (Council), 1091/80 (Commission) and 2651/84 (Commission).

Arising from the findings of the investigation and following consultation with the Commission services the Department proposes to apply sanctions based on the following principles viz.

Export Refunds

1. In respect of weight overdeclarations recovery of the advances already made or, as appropriate, forfeiture of entitlement to the refunds not yet paid will be confined to the extent of the weights overdeclared as indicated in the Customs reports. The recovery/forfeiture will be applied to all bondings from Cloghran/Waterford production over the period concerned. The regulatory financial penalty of 20% will also be applied.

2. In respect of trimmings, as the regulatory provisions that each piece of meat in the cartons concerned be individually wrapped were not met the entire contents of each of the cartons concerned are ineligible for export refunds. The results of the sample of plate production cartons examined by the Customs Authorities will be applied to the total Waterford plate production bonded and recovery of refunds already advance paid or, as appropriate, forfeiture of entitlement to refunds already advance paid or, as appropriate, forfeiture of entitlement to refunds not yet paid will be made in respect of that quantity. The regulatory 20% financial penalty will also be applied.

APS

3. Each contract concerned will be adjusted for the overdeclared weight and, where appropriate, for the exclusion of trimmings which are ineligible for APS aid. Where the resultant adjusted yield meets the regulatory 67% minimum the recovery of aid already advance paid or, as appropriate, the adjustment of entitlement to aid not yet paid will be confined to the extent of the weights overdeclared and of any ineligible product. Where the resultant adjusted yield is less than 67% aid on such contracts is not payable and any advance payments made will be recovered to the extent of the amount advanced + 20%.

4. Where the obligations imposed by Article 3 of Regulation 1091/80 have not been met the Department is obliged to invoke the provisions of Article 5 of that regulation. Having regard to the extent and gravity of the breach the Department is declaring forfeit in their entireties the contract securities for the contracts concerned.

The financial consequences of the misdeclarations are calculated as follows on the basis of the above and of the information at present available in the Department:

Export Refunds	*IR£*
1. Amounts already advanced which must be recovered, including 20% penalty.	
(*a*) Weight overdeclaration	283,535
(*b*) Trimmings	312,467

2. Amounts not yet paid and for which entitlement is forfeit, including 20% penalty.	
(*a*) Weight overdeclaration	9,275
(*b*) Trimmings	10,508
3. *APS*	
1. Amounts already advanced which must be recovered (including 20% penalty)	33,553
2. Amounts not yet paid and for which entitlement is forfeit.	43,086
3. Securities forfeit	392,442
Total	1,084,866

As already stated, these amounts are indicative and some may require adjustment when the outstanding information comes to hand. Contact will be made with the group in the very near future in respect of each APS contract and export refund bonding.

As you are aware the Commission Services have been formally advised of this matter in accordance with the provisions of Regulation 283/72. The papers have also been referred to the Garda Authorities.

Finally, I am to inform you that the Department views with concern the misdeclarations made by the group and requests an explanation of how the misdeclarations occurred together with an undertaking that such occurrences will not recur.

Yours faithfully, (M Harvey), Principal, FEOGA Division."

There was subsequent correspondence and meetings between officials of the Department of Agriculture and Mr Fairbairn with regard to the basis of the said sanctions in which he challenged the securities forfeiture figures, the effect of trims in the cartons and extrapolation across product which had left store prior to investigation and the imposition of penalties.

By letter dated the 23rd May 1989 and 31st day of May 1989 the following amounts were sought from AIBP: £392,441.67 being securities forfeited and £567,624.00 being refund of export refunds paid.

In addition AIBP lost entitlement to about £63,000 in respect of APS and refunds claimed but not paid.

By letter dated the 4th day of February 1988, Mr D Russell, Principal of the Finance Division of the Department of Agriculture wrote to the Garda Commissioner enclosing a memorandum on excess claims made by Goodman International (Beef Processors and Exporters) under EC schemes of Aids for Private Storage and Export Refunds in the beef sector and requested that the matter be investigated by the Garda Siochána. Two Fraud Squad Officers were assigned to this case, Sgt. F. Murphy and D. Garda Meagher.

D. Garda Meagher received a copy of the said letter and memorandum in March 1988 and contacted Mr Donal Russell and arranged a meeting with him which took place on the 29th of March 1988. During this meeting, Mr Russell outlined what had been said in the memorandum and advised D. Garda Meagher to contact Mr Dave Murphy of the Customs & Excise in connection with the investigation.

D. Garda Meagher contacted Mr Murphy on the 2nd day of May 1988 and asked him for a copy of the Customs' report and was informed that before same could be released, a letter would be required from the Garda authorities requesting same.

On the 16th day of May 1988, a letter was sent by Superintendent Casey to the Surveyor of the Department of Customs requesting a copy of the Customs & Excise report on the investigation.

On the 18th October 1988 D. Garda Meagher contacted Mr Murphy who informed him that he had not received the said letter.

Between the 16th of May 1988 and the 18th of October 1988, D. Garda Meagher made a number of efforts to contact Mr Murphy by phone but was not successful in so doing. Later that day D Garda Meagher brought a copy of the letter to Mr Murphy's office and on the 2nd day of December 1988, Mr Murphy telephoned and said the Customs' report was available.

It was collected by D. Garda Meagher on that date.

Upon receipt of the said report a detailed investigation into the irregularities was begun.

On the 11th day of January 1991, a report on the investigation was submitted to the Chief State Solicitor by Det. Supt. O'Donoghue.

The report had been prepared by D. Garda John Hayes as Det. Sgt. Meagher, who had been in charge of the investigation, had retired before the matter was finalised.

The file was submitted to the Director of Public Prosecutions and by letter dated the 16th day of May 1991, the Senior Legal Assistant directed that no prosecution was warranted by the available evidence.

In the course of his letter dated the 16th day of May 1991 the Senior Legal Assistant in the Office of the Director of Public Prosecutions stated that:

> "Whatever hope there might have been of bringing home criminal responsibility for such activities was effectively eliminated by the inordinate delay in completing the investigations and in particular in referring this matter to the Garda Siochana."

The speech made (on the 15/5/1991 in Dail Eireann) by the then Deputy and now Tanaiste, Dick Spring, was extremely critical of the role of the Minister of Agriculture and Food and the Department of Agriculture in regard to these matters and alleged that;

(i) The Department of Agriculture and Food did not diligently assist the Garda Fraud Squad in relation to the Waterford and Ballymun investigations

(ii) Though a memorandum from the Department of Agriculture and Food requested the Garda Fraud Squad to investigate serious irregularities in Goodman's

Waterford and Ballymun plants the Department, despite numerous requests, failed to release their file to the Fraud Squad until December 1988

(iii) the essential matters to be inquired into were in the case of the Waterford Plant, the false altering of weights and cartons, both before the Customs investigation and during it, and the inclusion of beef trimmings in the cartons to maximise the weight, and in the case of the of the Ballymun plant, the false altering of case weights and documentations

(iv) No investigation appears to have been carried out by the Fraud Squad in relation to the Waterford Plant although a file was submitted to the office of the D.P.P. in respect of the Ballymun allegations

(v) Notwithstanding their knowledge of the irregularities at Waterford and Ballymun and the prosecution of Mr Nobby Quinn in relation to the bogus African stamps the Department (and the Minister) was prepared to release bank guarantees up to £20 million (frozen because of the irregularities at Waterford) as part of the overall deal (in the Examinership) last Autumn."

These were serious allegations and extremely critical of the Department of Agriculture and the Tribunal considers it necessary to set forth the position in regard thereto.

The position in relation to the allegation to the release of the bank guarantees referred to the last allegation is that:—

(1) When the irregularities in respect of AIBP production at Waterford and Ballymun came to the notice of the Department of Agriculture, all export refund securities which related to the Company's production at these plants was frozen, at the Department's initiative and in accordance with standard practice until such time as the extent of the irregularity and the level of financial correction could be established.

(2) Securities in respect of other production in related export documentation were also frozen as a consequence. The amount frozen was approximately £20m.

(3) The level of financial correction for export refunds was established following consultation with the Commission services as to correction methodology at £567,624 and this amount was paid by AIBP.

(4) Upon receipt of payment work resumed on the processing for release of all export refund securities still held arising out of the Waterford and Ballymun investigation.

(5) The processing operation was extremely complicated and cumbersome as it covered not only the issues arising from the investigation but also all the normal checks as to whether all other requirements of the export refund regulations were met.

(6) The Examiner to the Goodman Group of companies was not appointed until the 29th day of August, 1990 and the relevant bank guarantees had been released prior to such appointment.

(7) The release of these securities was unconnected in any way with the Examinership of the Goodman Group of Companies.

The Tribunal has heard detailed evidence from the Officers of the Customs & Excise who carried out the said investigations and from the Department of Agriculture Officials who assisted therein.

The investigation carried out by the Customs & Excise authorities, with the assistance of the Department of Agriculture Officials concerned was a detailed and exhaustive one, involving the weighing of all the cartons of beef, thawing of a selected proportion thereof, a careful, detailed and exhaustive examination of the boning hall production records and the comparison of such records and the weights shown thereon with the weights shown on the cartons of beef referrable thereto and the interviewing of many witnesses.

The investigation began in December 1986 and did not conclude until September 1987 when the report on the investigation was completed and signed on the 4th day of September 1987.

This investigation and the evidence before this Tribunal clearly established that:—

(1) AIBP availed of the benefits of the 1986 APS scheme.

(2) In pursuance of the said scheme 70,000 cartons of beef, processed by AIBP at their plant in Waterford and 95,977 cartons of beef, processed by AIBP at their plant in Cloghran (Ballymun) were placed under Customs' control by AIBP and bonded for export to Third Countries.

(3) AIBP thereby became entitled to APS storage payments and advance payment of export refunds.

(4) That in both plants the beef, the subject of the APS scheme and such bonding was deboned by employees of Daltina Traders Ltd in pursuance of a verbal agreement made by Mr Eamon Mackle of Daltina Traders and AIBP

(5) That it was the responsibility of Daltina Traders Ltd. to keep records of the Boning Hall production.

(6) That these records provide the basis for the boning hall production sheets which were required to support the C&E 977 form.

(7) That a comparison between the weights shown on the Boning Hall production sheets with the weights shown on the cartons to which they related showed a sustained and regular over-declaration of weights in the boning hall production sheets.

(8) That the weights shown on the cartons were correct and the over-declaration of weights related to the boning hall production sheets.

(9) That the over-declarations of weights amounted in Waterford to 30,607.9 kilograms and in Cloghran (Ballymun) an estimated 106,800 kilograms making a total of 137.418 metric tonnes.

(10) That in Waterford, though not in Cloghran, trimmings which were not eligible for APS storage payments or export refund payments were found to be included in the cartons of plate and flank.

(11) That AIBP at all times denied any knowledge of authorisation of such irregularities and attributed responsibility therefor to Daltina Traders Ltd.

(12) That all declarations and claims in respect of this beef were made by AIBP.

(13) The effect of these declarations and claims would have resulted in substantial payments to AIBP in excess of their proper entitlement under the APS scheme and Export Refund scheme as set forth in the letters dated 16th January 1989, the 23rd May 1989 and the 31st day of May 1989.

The detailed investigation carried out by the Investigation Branch of the Customs & Excise and the subsequent investigation by the Garda authorities were unable to establish whether or not AIBP were knowingly involved in the over-declaration of weights and the insertion of trimmings in the cartons of plate and flank in Waterford.

In the course of his evidence, Mr Murphy, who had signed the report of the investigation on the 4th day of September 1987, stated that:—

1. The full extent of the fraud only became evident at a late stage of the investigation prior to the completion of the report.
2. "If the Gardaí were to be contacted, that would be the time to contact them."
3. The Customs Authorities tended to take the view that they were responsible for the investigation of the irregularities but the party which was most injured was the party which was paying out the funds, which was the Department of Agriculture, and any action which would be taken by them would be in respect of that offence which would be more serious in these circumstances than the Customs offence.

The investigation report was submitted by Mr Murphy to his authorities and on the 5th day of October 1987 was forwarded by Mr Healy of the CAP Division of the Office of the Revenue Commissioners to Mr Michael Murphy of the Finance Division of the Department of Agriculture.

The information in all of the reports and the support documents were then examined in detail by the Finance Division of the Department of Agriculture and this involved lengthy discussions with the Beef Division and the Export Refund Payments Division.

It was the responsibility of the Department of Agriculture to determine the possible corrections to payments and sanctions for Waterford and Ballymun beef which should be imposed in respect of the irregular declarations and claims. The various options open to the Department of Agriculture were considered in detail and on the 18th day of January 1988 a detailed memorandum on sanctions to be imposed in respect of APS and Export Refund payments was completed.

This memorandum recommended inter alia:

"(1) To sanction Goodman International for £1.111m including £430,000 in forfeited securities.

(2) To report the case to Brussels as an irregularity and seek bilateral discussions with the Commission to get their opinion on the sanction proposed.

(3) Refer to the Garda Fraud Squad for investigation."

This memorandum was submitted inter alia to the Secretary of the Department and it was decided to report the matter to the European Commission in Brussels and refer the matter to the Garda Fraud Squad for investigation.

As already stated, the matter was reported to the European Commission in Brussels by the Department of Agriculture on the 29th day of January 1988 and to the Garda Commissioner on the 4th day of February 1988.

Enclosed with the letter dated the 4th day of February 1988 was a memorandum prepared by the Department of Agriculture on the excess claims made under the EEC schemes of Aids for Private Storage and Export Refunds in the beef sector.

Mr Donal Russell of the Finance Division of Agriculture was contacted by and arranged a meeting in regard to the matter with D. Garda Meagher, which meeting was held on the 29th March 1988.

In the course of this meeting, Mr Russell explained the contents of the memorandum to D. Garda Meagher and advised him to contact Mr Dave Murphy of the Customs & Excise in connection with the investigation.

D. Garda Meagher contacted Mr Murphy on the 2nd of May 1988 and asked him for a copy of the Customs report. He was informed by him that before the same could be released, a letter would be required from Garda authorities requesting same.

On the 16th May 1988 a letter was sent by Supt. Casey to the Superintendent of Customs requesting a copy of the Customs & Excise report on the investigation.

The letter containing the written request does not appear to have reached the Revenue Commissioners and it was not until the 18th October 1988 that D. Sgt. Meagher gave a copy of the letter to Mr Murphy.

This file was not received until the 2nd day of December 1988.

While Deputy Spring was justified in his criticism of the delay in making the Customs & Excise file available to the Garda Siochana, he was incorrect in attributing the blame in respect thereof to the Department of Agriculture and in his allegation that they did not diligently assist the Garda Fraud Squad in relation to the Waterford and Ballymun investigations.

Such a delay as occurred was due to the attitude of the Customs & Excise authorities in requiring a written application for such records and the mislaying of the letter requesting same. The delay between the receipt by the Department of Agriculture of the Customs & Excise Investigation Branch report on the 5th day of October 1987 and the reference of the matter to the Garda Commissioner on the 4th day of February 1988 was not unreasonable, having regard to the complexity of the matter and the decisions to be made by the Department of Agriculture in regard thereto.

The Tribunal is satisfied that there is no basis for the allegation that the Department of Agriculture did not diligently assist the Garda Fraud Squad in relation to the Waterford and Ballymun investigations.

The Tribunal is further satisfied that there was nothing sinister in the delay of the Customs & Excise authorities making the file available to the Garda Fraud Squad.

On receipt of the file, the matter was considered by the Garda authorities and, having regard to the nature and complexity thereof and the voluminous documentation which required to be investigated, and the resources available to them they decided that the proper approach in dealing with the matter was to carefully and thoroughly investigate one APS contract from beginning to end.

In pursuance of this decision, which was a reasonable one, they concentrated on an APS contract in respect of beef processed at Cloghran.

This investigation provided the basis of their report to the Director of Public Prosecutions.

While it is true to state, as Deputy Spring stated, that no investigation was carried out by the Fraud Squad, in relation to the Waterford plant, this was the reason therefor.

On the 15th day of May 1991 Deputy Pat Rabbitte stated in the Dáil that Mr Goodman's denial of the responsibility for Waterford and his blaming a sub-contractor, Mr Marks, stands contradicted by the evidence of Mr McGuinness who said that there was "a high level plan to obstruct the investigation at Waterford."

This was an obvious reference to the statement made by Patrick McGuinness on the ITV programme that a plan was agreed at senior management level, within the Goodman Group, with Customs people at their Head Office to contain the damage resulting from the investigation, which plan involves selecting samples of good meat without trimmings for investigation by the Customs & Excise authorities.

In this connection Mr McGuinness had stated on the ITV programme that:—

> "There was a massive panic within the company and a plan was put forward as to how the damage might be limited. The plan was basically agreed between our people and the Customs people at their Head Office, that a certain sample of good product would be selected for thawing out and for investigation. This was a deliberate scheme to contain the damage because of the explosive nature of the investigation" ".....was agreed at senior management level."

In addition the presenter of the said programme stated that:—

> "World in Action had a document stated to be the Master Plan which showed the locations where the boxes of good meat were supposed to be opened by Customs officials. The plan failed because local (Waterford) Customs agents became suspicious and kicked up a fuss."

The question of the circumstances with regard to the creation of this alleged Master Plan will be considered by the Tribunal at a later stage.

Irrespective of the existence of such plan or the circumstances under which it was drawn up, the Tribunal is satisfied on the basis of the evidence heard by it, that at no stage was it produced to the Customs & Excise authorities or any official thereof or that there was any agreement in respect thereof and the allegation is completely and utterly without foundation.

During the course of the investigations at Waterford, AIBP had been requested by the Customs & Excise authorities to provide personnel to assist in the listing of the weight of each carton on the pallets.

On the 19th day of January 1987, ten employees of AIBP were assisting Customs & Excise in the course of the examination.

In the early afternoon of that date, it was observed by Mr M.C. O'Briain that the weights written on one carton on a pallet had been altered by changing 21 kilograms to 27 kilograms.

It was suspected that this alteration had been made by a member of the staff employed by AIBP and their assistance was immediately terminated.

A check carried out that evening showed that the weights of 17 cartons had been similarly altered and the weights thereon increased by 6 kilograms per carton.

No such alteration of weights had appeared on any of the cartons previously examined by the Customs & Excise staff, nor did such alteration appear on any of the cartons subsequently examined by them.

Though it is strenuously denied by AIBP that such alterations were made by any of their personnel, it is a reasonable and indeed inevitable inference to be drawn from the fact that the only day upon which an alteration of the weights on the cartons was discovered was on the day that personnel employed by AIBP were assisting in the checking of the weights on the cartons.

This inference is further confirmed by the evidence of Patrick McGuinness in relation to a conversation which he had in early January with Mr Eoin Lambe, who was employed by the International Division of the Goodman Group and who had been sent to assist in Waterford during the investigation. He stated that:—

> "Eoin had been based in Waterford on behalf of the International Division and he had stated that he had been part of a team who had been re-numbering boxes of APS meat in the cold store, the basic purpose being to confuse anybody who attempted then to compare the nett weight on the box to the document weight. He said that it had been discovered by the Customs officials and had been told to leave the cold store".

The Tribunal is satisfied that there was an attempt by the employees of the International Division of AIBP, who were purporting to assist the officers of Customs and Excise in the course of their investigations to conceal the extent of the over declarations of weights by altering the weights shown on the cartons.

From the outset of the investigations AIBP management personnel denied any knowledge of or involvement in the irregularities being discovered and attributed responsibility therefor to Daltina Traders Ltd.

Gerry Thornton did so at Waterford on the 24th December 1986, Jim Fairbairn did so at Waterford on the 13th February 1987, Larry Goodman did so in the course of a discussion with the Secretary to the Department of Agriculture & Food, James O'Mahony on the 5th March 1987, Larry Goodman, Gerry Thornton and Jim Fairbairn did so during the course of a meeting with the Chairman of the Revenue Commissioners and others on the 20th day of March 1987 and Colm O'Loughlin the manager at Cloghran did so on 23/3/1987.

In the course of his evidence before this Tribunal, Mr Eamon Mackle, Daltina Traders Ltd confirmed that he had never received any requests or orders from either Larry Goodman or Peter Goodman or any plant manager or representative of Goodman International to carry out any fraudulent practice either directly or indirectly or by hint or suggestion.

This evidence is confirmed by the evidence with regard to the explosive reaction of Mr Gerry Thornton when on 23rd December 1986, he learnt of the irregularities being investigated in Waterford. He phoned Mr Mackle and asked him to attend a meeting at Ravensdale the following morning.

This meeting was attended by Mr Thornton, Mr McGuinness, Mr Fairbairn and Mr Nobby Quinn.

At that meeting Mr Mackle was accused of falsifying the weights and was required to accompany Mr Thornton and Mr McGuinness to Waterford and his contract was terminated.

Mr Mackle attributed blame for the over-declaration of the weights to the checker employed by him, Joe Devlin.

Despite its best efforts, the Tribunal was unable to secure the attendance of Mr Devlin as he was resident outside the jurisdiction and refused to attend.

Mr Marks was the manager employed by Daltina Traders Ltd., at Cloghran/Ballymun at the time of the APS contracts which were investigated by the Customs & Excise authorities.

He stated in evidence that he did not suggest or authorise any alteration of production sheets and was unaware of such alteration.

An employee of Daltina Traders Ltd., Mr Ray Watson, was responsible for the checking of the weights and the preparation of the production sheets in the boning hall. These production sheets were handed in by him to the office of AIBP which was adjacent to the boning hall.

This office was occupied by a number of girls, in particular Ms Angela Magee and Ms. Imelda Murray, who were responsible for the documentation with regard to the APS scheme.

These girls believed that a yield of 78% should be achieved and when the yield was below that, Mr Watson suggested to them that they add weights to the production sheets, which they did. Mr Watson, in evidence, said that the only persons who knew about this was himself and the two girls in the office.

The only difference between the evidence of Mr Watson and the two girls, Imelda Murray and Angela Magee, is that they say that they were told by Mr Watson to alter the production sheets for the purpose of producing the yields expected, whereas he states that he merely suggested the alteration of such weights and the girls then made the necessary alterations and produced the APS yield sheets showing what they considered to be the expected percentage.

These yield sheets were forwarded each day to Ravensdale.

Both Imelda Murray and Angela Magee were, in the opinion of the Tribunal, truthful witnesses and their evidence is accepted by the Tribunal.

Both swore that they never received any instructions from Colm O'Loughlin, the manager of the plant, or any representative of AIBP to make the alterations which were clearly established in evidence or that they were aware of the fact that such alterations had been made.

While AIBP can and should be criticised for failing to exercise any reasonable degree of supervision over the activities of their staff, such as Angela Magee and Imelda Murray in the preparation of important documents with relation to the APS scheme and the Export Refund Scheme, the Tribunal is satisfied that the AIBP management personnel were not aware of either the over-declaration of weights in the boning hall production sheets and the APS yield sheets or of the presence of trimmings in the cartons of plate and flank in Waterford until the matter was drawn to their attention by the officers of Customs & Excise carrying out the investigation.

The Tribunal is, however, satisfied that there was an attempt by personnel employed by the International Division of AIBP to conceal the extent of the over-declaration of weights by increasing the weights on some of the cartons placed in private storage in the cold store.

CHAPTER SEVENTEEN

The Eirfreeze Investigation

At all times material to this inquiry Eirfreeze Limited were the owners and occupier of the cold store situated at North Wall in the City of Dublin. The said company was one of the Goodman Group of companies.

Deputy MacGiolla alleged in Dail Eireann on the 9th day of March 1989 that:—

> "The Eirfreeze plant located in the North Wall was shutdown at 6.00pm or 7.00pm on Saturday, the 4th day of March 1989, by inspectors from the Department of Agriculture and Food because of very serious illegal activities by a team acting on behalf of one of the Goodman companies — changing labels and dates of slaughter on meat".

On the 15th day of May 1991 Deputy MacGiolla alleged that:—

> "On the 10th day of March 1989 (the day after Deputy MacGoilla statement to the Dail) the Goodman PR company accused Deputy MacGoilla of seriously damaging the reputation of Goodman International and the whole meat industry, denied that the Eirfreeze plant had been shut down and stated that the charges made by Deputy MacGoilla were utterly false".

He further alleged, in Dail Eireann, on the same day that;

> "At the hearing of the prosecution against Eirfreeze at the District Court on the 30th of July 1990, Defence Counsel on behalf of the company pleaded guilty on two charges relating to the illegal labelling of meat carcases"
>
> and further alleged that
>
> "It was stated in court that the Eirfreeze plant was shutdown on Saturday night 4th of March 1989 and Department inspectors took away 63 carcases on which they found false CU2 labels which indicated the meat was from steers of good conformation and of low fat, in other words, a high quality product, which was at variance with the original grading by the Department Official".

He further alleged in Dail Eireann on the 15th day of May 1991 that;

> "The Department of Agriculture and Food and the Prosecuting Counsel seemed very reluctant to pursue the charges against Eirfreeze (and AIBP) with any vigour, particularly with the issue of fraud and forgery and withdrew one charge against Eirfeeze and two charges against AIBP".

The Tribunal carried out a detailed inquiry into the incident which occurred at Eirfreeze on the 4th of March 1989, the subsequent investigations carried out by the Customs and Excise officials, by the Department of Agriculture Officials and to the prosecution instituted by the Minister for Agriculture and Food arising out of such investigations.

This inquiry disclosed that between the 28th day of February 1989 and the 3rd day of March 1989 13 container loads of bone-in forequarter beef arrived in the Eirfreeze cold store at the North Wall in the City of Dublin.

An examination of the "Meat Inspection Certification for the Movement of Meat/Meat products between Approved Establishments in Ireland" disclosed that the containers came from 4 different cold stores as follows:—

Store	Number of Containers	Number of Quarters
Cahir Cold store	3	668
Lyonara	2	292
QK Naas	5	1037
Jenkinsons	3	630
Total	13	2627

In addition there were six containers of beef in South Bank Quay containing 1,149 frozen bone-in hindquarter beef which had been loaded at the premises of AIBP Nenagh, and due for export to Morocco, the consignee being the Office de Commercialisation et D'Exportation at 45 Avenue des F.A.R., Casablanca, Morocco.

Mr Patrick Gregan was a Veterinary Inspector with responsibility for operations at Eirfreeze Limited. He stated that from his records 12 consignments had arrived at Eirfreeze during the days prior to the 4th of March 1989 and the ultimate destination of the meat contained in the containers was Morocco and that they were brought to Eirfreeze for a repacking operation which was a customer requirement. The rebagging was perfectly in order as was all the relevant documentation provided that the repacking operation was carried out under the supervision of the Department of Agriculture official at the plant.

The nature of the rebagging consisted of the removal of a plastic liner from the carcase and when this was removed a stockinet cover was placed over the carcase and the carcase with the stockinet cover would be placed in a polypropylene bag and sewn across with a sewing machine which was electrically operated.

Mr Pat Birdy of the International Division of AIBP said that the customer did not want the inner plastic bag so the inner plastic bag had to be removed and replaced by a stockinet and outer bag.

This operation was carried out at Eirfreeze under the supervision of the Department of Agriculture official for approximately 3 days and on the Saturday the 4th day of March 1989 six container loads of beef still required to be rebagged. Mr Gregan was aware that the rebagging had to be completed before Monday the 6th day of March 1989.

Departmental supervision had stopped at 4.00pm approximately on the 4th day of March 1989.

Mr Gabriel Daly was a Higher Agricultural Officer of the Department of Agriculture and Food and was on duty in Eirfreeze on the 4th day of March 1989 and supervised the work there from 8.30am until 4.10pm.

He left then because in his opinion, the work had ceased and all the racks of beef had been moved back into the store and the majority of people had left.

He stated that he received no indication that it was the intention to resume work at a later stage.

At 4.30pm. on that afternoon, Harbour Sergeant Kinlan, as a result of information received from Harbour Constable Mates, drove to Molloy & Sherry's yard, which adjoined the Eirfreeze Cold Store, and saw four (4) men sitting in a Northern Ireland registered car.

He spoke to the Molloy & Sherry watchman, John O'Connor, who in turn spoke to the men, who stated that they were waiting for their boss to begin work on a container in the yard for AIBP. About 15 minutes later Peter O'Reilly arrived with two other men and he informed Sergeant Kinlan that he worked for Anglo Irish Beef Processors and that he was in charge of the men, numbering 12, who were required, he said, to check some containers and Sergeant Kinlan, though suspicious, left the yard. He returned at 5.20pm. accompanied by Sergeant Curtin of the Harbour Police and having entered the yard met Peter O'Reilly who appeared to be returning from Eirfreeze. Mr O'Reilly informed Sergeant Kinlan that the men were working in Molloy & Sherry yard and when asked to show them he said that they also had some work to do in Eirfreeze where they were at that point in time.

He informed Sergeant Kinlan that a consignment of beef came over the border for export and it had the wrong covering on it and they were replacing the coverings with woven plastic sacks to meet their customer requirements and the meat was due for export on Monday the 6th to Morocco and that it was a rushed job to complete the order.

Sergeant Curtin and Sergeant Kinlan drove to the Department of Agriculture opposite North Wall but there was no-one in the office and at approximately 6.05 pm rang the office of the Department of Agriculture in Kildare Street and reported the incident.

At approximately 6.30pm. Mr Doug Smyth, a Veterinary Inspector employed by the Department of Agriculture and who at the particular time was acting Deputy Portal Inspector entered the office of the Harbour Police Control office, informed them that he was from the Department of Agriculture and that he would require assistance to enter the Eirfreeze Cold Store.

Sergeant Kinlan contacted the Gardaí by 'phone and at 6.45pm. members of the Gardaí, Harbour Police and Mr Smyth entered the premises of Eirfreeze Ltd. Mr Smyth, had been in the process of visiting the "NV Bison" berthed in Dublin Port when he was stopped by Harbour Sergeant Curtin, who asked him to report to Sergeant Kinlan at the Harbour Police Headquarters. After inspecting sides of beef which hung from rails at ground level on the premises Mr Smyth was satisfied that all was in order and the party left the premises. In the meantime, the official in Agriculture House who had been contacted by Harbour Sergeant Kinlan contacted Patrick Gregan and requested him to investigate the complaint accompanied by a Department of Agriculture officer.

Mr Gregan contacted Mr Mellett, SAO, and they met at the Harbour Police check point on the docks at 7.30pm.

Accompanied by members of the Garda Siochana and the Harbour Police, Mr Gregan and Mr Mellett entered the Eirfreeze premises at 7.45pm. They observed a team of men re-wrapping forequarter carcases. The loading assembly area was covered with pallets of hanging forequarter carcases and Mr Gregan estimated that about 20 men were involved in the re-wrapping operation.

Sergeant Gabriel McIntyre stated that there were approximately 20 men working at various tasks, most being engaged in removing plastic covering from the frozen sides of animals and replacing it with a type of muslin cloth and sealing the ends of same.

Mr Gregan rubbed the marking on one side of the beef and his hands were stained with fresh ink marking.

When challenged by Mr Gregan, Mr O'Reilly denied that any stamps were being applied. Both Mr Gregan and Mr Mellett examined various pallets and Mr Gregan ordered Mr O'Reilly to cease all work immediately. Mr Gregan spoke to Mr Pat Birdy of AIBP, who was on the 'phone from Dundalk and informed him that he had stopped the work on the floor and his reasons for such action.

Mr Birdy stated that there must have been a break down in communications in not informing the Department that they were continuing to work in the evening and that they were under pressure to have the consignment ready for shipping out on Monday afternoon. Mr Gregan informed Mr Birdy of his suspicions that stamps were being used and Mr Birdy denied that any stamps were being applied.

The pallets were then fork lifted off the floor into the freezer units where they were being stored and all work ceased. Leaving Mr Mellett in charge Mr Gregan left the premises at 8.45pm. At 10.30pm. Mr Gregan met Mr O'Hagan, Senior Veterinary Inspector and reported the situation as he had found it at the Cold Store and the action he had taken.

At. 11.50pm. Mr O'Hagan, who had contacted Mr John Ferris S.S.V.I., of the Department of Agriculture telephoned Mr Gregan with instructions to arrange for the withdrawal of Department of Agriculture attendance at the Eirfreeze Cold Store the following morning.

At 9.30am. on the 5th day of March 1989, Mr Gregan attended the Eirfreeze Cold Store and informed Mr P. O'Neill that he was withdrawing Department of Agriculture attendance at the Cold Store under instruction and that the export status of the meat involved in the previous evenings investigations would have to be reassessed and Veterinary Certification would be held up until that was clarified.

On Monday the 6th day of March 1989, Mr Sean O'Connor, D.D.V.S., received a request from AIBP to enable normal commercial activities to be carried out at Eirfreeze Cold Stores. This request was granted and an Agricultural Officer, Mr John Kelly, was instructed to return and resume his duties in relation to ordinary daily commercial functions. This did not in any way interfere with the investigation being carried out by the Customs & Excise officials and Department of Agriculture Officials.

A meeting took place, that morning, in the Veterinary office at Eirfreeze between Customs' officers, Mr J. Naughton, Mr Ben Clarke, Mr Tom McGrath, Mr B. Murphy and Mr D. Kelly, representing the Customs' Authorities and Mr Benny Bennett, S.V.I., and Mr Gregan from the Department of Agriculture.

At this meeting it was decided to examine the pallets from Saturday night and they were taken out from the freezers. The polypropylene sacks and stockinet were removed and the carcases were examined in detail.

Mr Murphy and Mr Kelly examined the frozen forequarters which had been taken out of cold storage at Eirfreeze on the 4th day of March 1989, during the period when there was no supervision by the Department of Agriculture Official. These were divided into two lots, namely those in respect of which the packings had been replaced and those which had not been dealt with by the time the operation was stopped by the officials of the Department of Agriculture.

Mr Murphy and his colleague Mr O'Kelly examined firstly the lot of frozen beef forequarters which had been re-bagged on the 4th of March 1989. Each forequarter was examined individually and the stamps and other markings on it noted. Each forequarter had the following stamps: "CU2", "EEC Ireland 290", (E) and (M).

The "CU2" stamp referred to appeared to be of more recent origin than the other stamps and when Mr Murphy touched the "CU2" stamps he found them still to be wet. He and Mr Kelly then proceeded to examine the other lot namely the forequarters which had been taken out of cold storage at the same time but which was still in its original packing. All packing was removed from these forequarters before examination, each forequarter was examined individually and the stamps and other markings noted.

Each forequarter had the following stamps: "EEC Ireland 290", (E), and (M). There was no "CU2" stamp on any of these quarters.

In the presence of Mr Murphy, Mr Patrick Birdy, a member of the International Division of AIBP with responsibility for transport, who was present at the time, and in reply to questions by Mr Bennett S.V.I., alleged that all stamps were applied at the time of production and denied that the "CU2" stamp or any stamps had been applied to the carcases while they were being re-bagged in Eirfreeze or that any marks had been removed from the carcases.

On the 6th and 7th days of March 1989, Mr Murphy and Mr Kelly, continued the examination of forequarters forming the balance of the consignment. In all, they examined 745 forequarters. 126 had been re-bagged without official supervision and these all bore the "CU2" stamps, the remaining 619 had not and bore no grading mark.

On the 8th day of March, 1989, he returned to Eirfreeze to continue the examination but was informed that no examinations were to be carried out as the Department of Agriculture had decided that no further handling of this consignment would be permitted until further notice as some "intervention beef frozen" forequarters had been found amongst the beef forequarters under control.

It was subsequently ascertained that this intervention beef had been removed from the cold store in error and was not part of the consignment under investigation.

On Tuesday the 7th day of March, 1989, Mr Gregan received a phone call from Mr O'Hagan and as a result of that he sent a Senior Agricultural Officer, Mr Danny Gavigan to the South Bank as the Customs wished to open the six containers at the South Bank.

The seals were broken, the containers were inspected and "CU2" stamps were observed on the outer carcases in the containers. The movement certificates to the South Bank and the meat inspection certificates were issued in Nenagh.

On the 24th day of March 1993, the 184th day of oral hearings before this Tribunal, Mr Patrick Birdy of the International Division of AIBP with responsibility for transport, stated that a CU2 stamp was being placed on a limited number of pieces for each container and that this was being done at Eirfreeze on the occasion of the investigation.

If this statement or admission had been made by Mr Birdy, during the course of his telephonic conversation with Mr Gregan on the night of the 4th day of March 1989 or on the 6th day of March 1989 to Mr Bennet or the other officials present, a considerable amount of time spent by officials of the Department of Agriculture and officers of Customs & Excise, in investigating this matter and of the time spent by the Tribunal in hearing evidence in regard thereto would have been unnecessary.

In the course of his evidence before the Tribunal, he gave as his reason for not admitting to the application of the CU2 stamp to the quarters at Eirfreeze that he "knew that there would be a fuss over it and was waiting to see which way the fox was going to jump" and that he "was keeping his powder dry by saying nothing".

On the 10th and 15th day of March 1989, Mr Nicholas Finnerty of the Beef Carcase Classification Unit of the Department of Agriculture visited Eirfreeze to conduct an examination of the forequarters.

He examined approximately 100 forequarters and all of them had "slices cut off the shoulder area" and in his opinion, the removal of these slices happened at the point of slaughter.

He found a fresh CU2 stamp on all of the carcases examined except for one which had a CU3 stamp.

On comparison with an official CU3 stamp, Mr Finnerty found that the outline was bolder and thicker than the official stamp impression.

In his opinion the stamps on the carcases other than the CU2 stamp were authentic official stamps, the normal stamps applied on a slaughter line.

It appears from the evidence of Mr Maurice Mullen that the Department of Agriculture do not have a CU2 stamp because only steer beef from carcase classification categories U3, U4, R3, R4 and 03 are eligible for intervention in Ireland and these are the only official stamps of the Department of Agriculture.

On the 10th of March a meeting was held at Agriculture House at which Mr Donal Creedon, the Secretary to the Department was present. Other persons present were: Mr Gregan, Mr Sean O'Connor, Mr Gerry Twomey and Mr Power, Assistant Secretary to the Department.

The problem which was discussed at this meeting related in particular to the 6 containers at the South Docks herebefore referred to and the decision to be made was whether they would be retained or released for export.

It appears from Mr Creedon's evidence that, after long and vigorous discussion, to which everybody contributed, they weighed the pros and cons with regard to the decision.

They had two options, one to hold them or the other to release them.

On the question of releasing them they took into consideration the fact that there was no public money involved, that the beef was perfectly good beef and that the Veterinary Officers who were concerned in providing veterinary certificates had no difficulty in so doing and that the containers were not required for any investigative purposes as it was in the opinion of the Secretary that there was adequate evidence available with regard to the alleged irregularities in the product detained in Eirfreeze Ltd stores under the control of the Department of Agriculture. Mr Creedon, having considered all these matters decided to release the containers for export.

The Tribunal is satisfied that Mr Creedon properly exercised his judgement in this regard having considered all the relevant matters.

Mr Creedon stated that, as Secretary to the Department, he advised the Minister for Agriculture & Food, Mr Michael O'Kennedy TD of the events as they occurred and kept him fully informed, but the Minister did not interfere in any way with the investigation or steps to be taken as a result thereof.

In his evidence Mr O'Kennedy stated that he wasn't involved at any time in any decision arising out of the Eirfreeze investigation nor was he at any time involved in the conduct of the prosecution in that case.

The Tribunal accepts the evidence of the Secretary to the Department of Agriculture & Food and the Minister for Agriculture & Food and is satisfied therefrom that the Minister was not involved in any way in either the investigations or the prosecution in respect of matters arising therefrom.

In pursuance of the Secretary's decision, the 6 containers referred to were released with the exception of 64 forequarters held at Eirfreeze.

The allegation made by Deputy MacGiolla that "the Eirfreeze plant located in the North Wall was shut down at 6 pm or 7 pm on Saturday the 4th day of March 1989, by inspectors from the Department of Agriculture & Food because of very serious illegal activities by a team acting on behalf of one of the Goodman companies changing labels and dates of slaughter on meat", is substantially established.

The Tribunal has referred to the evidence of Mr Gregan and Mr Mellett that at approximately 8.45 pm on the 4th of March they directed that all work cease and the decision of Mr O'Hagan, communicated to Mr Gregan at 11.50 pm on that night, to arrange for the withdrawal of Department of Agriculture attendance at the Eirfreeze cold store the following morning, which was done.

On the 6th of March 1989 a request from AIBP to resume normal commercial activities in the store was acceded to.

While it could be argued that the premises were not shut down as such, the effect of the directions of Mr Gregan was that all work necessitating the presence of Department of Agriculture & Food officials was stopped.

With regard to the allegation made by Deputy MacGiolla that the Dept of Agriculture & Food and the prosecuting counsel seemed very reluctant to pursue the charges against Eirfreeze and AIBP with any vigour particularly with the issue of fraud and forgery and their withdrawal of one charge against Eirfreeze and two charges against AIBP, the facts established in the evidence before this Tribunal were that:—

Mr O'Keeffe who was the Higher Executive Officer attached to the Department of Agriculture with responsibility for the administrative side of the classification systems and other matters relating to classification and for prosecutions under the Carcase Classification Regulations received a copy of the report in relation to the events of the 4th March 1989 and wrote to the Attorney General's office on the 13th March 1989 for advice in respect of the bringing of prosecutions under the Carcase Classification Regulations.

A reply was received from the Attorney General's Office on the 21st March and summons were issued against four different companies, namely, Lyonara Ltd., Lyonara Cold Store, AIBP and Eirfreeze Ltd.

There was not available any evidence which would justify a prosecution for fraud or forgery and the prosecutions related to breaches of the European Communities Carcase Classification Regulations.

Deputy MacGiolla was present at the hearing of the prosecutions against Eirfreeze Ltd and AIBP in the District Court on the 30th day of July 1990.

Mr Joseph Matthews, a practising barrister of twenty years experience had been instructed by the Chief State Solicitor to appear on behalf of the Minister for Agriculture & Food in the said prosecutions.

Mr Matthews gave detailed evidence with regard to what transpired at the said hearing, of the fact that he had prescribed the necessary proofs and that the witnesses directed by him were present in Court and that adequate consultation had taken place at the office of the Chief State Solicitor prior to the hearing.

He gave evidence with regard to the difficulties in the case, procedural and otherwise.

Counsel for the Defence had indicated during the course of submissions to the Court in the morning that there were fundamental flaws in the summonses relating to both his clients, namely, AIBP and Eirfreeze Ltd.

During the luncheon recess Mr Matthews pointed out the difficulties to the legal officer of the Attorney General, Mr Alkin.

As a result of these difficulties, Mr Matthews with the approval of Mr Alkin agreed to accept pleas of guilty to two of the charges set out in the summonses by Eirfreeze Ltd and to the striking out of the summonses against AIBP.

The President of the District Court imposed fines of £200 on each of the summonses to which Eirfreeze Ltd had pleaded guilty.

The Tribunal is satisfied that there is no substance in the allegation that the prosecutions were not conducted with vigour. It was a matter for Counsel engaged on behalf of the Minister for Agriculture to reach a realistic assessment of the difficulties in the case and the Tribunal is satisfied that he exercised that responsibility in a responsible manner.

With regard to the CU2 stamp applied to the carcases in Eirfreeze on the 4th March 1989, Mr Birdy stated that the application of such stamps was not a requirement of the consignee, the Moroccan company, but the requirement of a third party described by Mr Birdy as an "ambulance chaser, a fall back client, who would purchase the beef if for any reason the containers of beef were not accepted into Morocco."

He stated that in November the contract had been running for quite a while and that in that month they sent out some containers which for one reason or another were not accepted in Morocco. The specific reason was a temperature failure.

He stated that he was approached by a gentleman who is called an ambulance chaser: that there are people who look out for distressed cargoes and made enquiries in connection with the product.

Mr Birdy stated that he brought him to Nenagh where they were loading similar product and he said that he could access that beef to another market in the same refund zone, which was very important moneywise, but that it would facilitate its entrance if his back door CU was used.

This "ambulance chaser" provided Mr Birdy with the stamp to be applied to a number of carcases situate at the back doors of the containers. He stated that they would facilitate its entry to the alternative market.

The Tribunal does not accept Mr Birdy's evidence in this regard but is of the view that the application of the CU2 stamp to the carcases was an attempt to show compliance with the terms of the contract which required that approximately 25% of the merchandise would be of Grade U and that 45% thereof should have a fat cover of 2.

The Tribunal is further satisfied that, while the Department of Agriculture had been notified that work would be carried out in Eirfreeze Ltd on Saturday the 4th of March and Sunday 5th March, that at 4 p.m. in the afternoon of the 4th March, a situation was deliberately created to give the impression that work had ceased for the day when it was the intention to resume work at a later stage in the absence of the Agriculture official, when the stamps could be applied without the risk of detection in the absence of supervision by the Agricultural officer.

If it were otherwise, there was no necessity for Mr O'Reilly to inform Harbour Sergeant Kinlan that they were only checking some containers and in attempting to conceal the fact that they were due to work in Eirfreeze and not in the Molloy and Sherry's yard.

CHAPTER EIGHTEEN

AIBP (Nenagh) Jerry O'Callaghan

It appears from the official reports of proceedings in Dáil Eireann on the 12th day of April 1989, Deputy Desmond had asked the Minister for Agriculture & Food if he was aware of an allegation, notified to his Department, that a journalist saw the removal of and changing of stamps, dressings and labels, on beef carcases in a plant (details supplied) on 12/13th January 1989 and if he would state the outcome of his investigations into this matter.

The Minister for Agriculture & Food replied as follows:

> "I am informed by my Department that a journalist submitted a number of queries to an official of the Department in relation to the alleged incident. Inquiries, which have not yet been concluded, are underway in the Department, arising from the queries submitted"

The purpose of Deputy Desmond in tabling the said Parliamentary Question was to seek confirmation that the journalist in question had notified the Department of Agriculture & Food of his allegations and to inquire as to the response of the Minister.

The journalist in question was Jerry O'Callaghan of RTE and he had informed Deputy Desmond that he notified the Department of Agriculture & Food of what he had seen on the night in question.

For reasons which will appear, the Tribunal considers it desirable to deal with this matter at this stage of its Report.

At this time, Mr O'Callaghan was employed by RTE as a journalist and was involved in the preparation and presentation of programmes for the "Today Tonight" series of current affairs programmes.

Mr O'Callaghan had graduated from UCD with a degree in Agricultural Science and after spending a number of years teaching, he did a Master's degree in Agricultural Science. Between 1970 and 1977 he worked in the Department of Agriculture. In 1977 he joined the staff of RTE and originally worked on and presented the "Landmark" programme which was a weekly programme dealing with farming.

In 1986 he joined the team responsible for the "Today Tonight" programme and was responsible for a number of programmes about farming and about the meat business.

In January 1989 he was attempting to investigate allegations that he had heard about stamp changing and malpractices, in particular in the Goodman Organisation.

In pursuance of such investigations he visited the AIBP plant in Nenagh on four different occasions, namely the nights of the 12th January 1989, 13th January 1989, 1st March 1989 and the 13th March 1989.

The only visit which is of importance to the Tribunal is that of the visit on the night of the 12th January 1989 because that is the only night upon which he was able to gain admission into the plant.

The relevant portion of Mr O'Callaghan's evidence to the Tribunal was that on this night of the 12th January 1989, he secured admission to the plant at about midnight and that:

1. Before he gained admission to the plant, he saw a number of cars, some with Northern Ireland registration, parked in the carpark; two large containers parked in the vicinity of the entrance to the plant. The names on the containers were: (*a*) Molloy & Sherry and (*b*) Sealand.

2. After he secured admission to the plant he saw quarters/carcases being taken from the Molloy & Sherry container and being placed on a moving rail along which carcases are moved or pushed which ran around the hall; that there were about 6 or 7 men, dressed in overalls or wearing white coats, working in the hall; that the carcases or quarters taken out of the Molloy & Sherry container were covered; that the first thing done with these carcases was the removal of the covering therefrom; that once the covering was removed, one man operating an angle grinder, a circular disc, shaved stamps off the carcases and may in addition have removed dirty or damaged parts of the carcases; a man, other than the man removing the coverings from the carcases, then washed the carcases and another man placed a stamp with a different number to that which had been removed; that the carcase/quarter was then covered up again and placed in a container.

The Tribunal carried out detailed inquiries into this incident and heard evidence with regard thereto from James Monaghan, Factory Manager, AIBP Nenagh, Edward Kelly, the Agricultural Officer on duty at the plant that night, Frank Brislane, Supervisory Agricultural Officer and Joseph Mangan, Veterinary Inspector responsible for the AIBP plant at Nenagh.

This evidence established to the satisfaction of the Tribunal that:

(1) In October/November 1988, the International Division of the Goodman Group of companies had requested Mr Monaghan to make the plant available for an extensive re-packaging operation to be carried out by them.

(2) He agreed, subject to the condition that the carrying out of such operation would not in any way interfere with the day to day operations of the plant.

(3) This meant that it could only be carried out after the completion of the ordinary day's work in the plant because, as stated by Mr Monaghan, it was not desirable to have frozen meat coming into the factory while there was fresh beef in the loading bay or in transit from the chills to the boning hall.

(4) None of the usual factory operatives using the plant were engaged in this re-packing or re-packaging operation.

(5) The re-packaging operation was carried out by operatives employed by the International Division of the Goodman Group of companies.

(6) Mr Monaghan had no involvement in the operation other than making the plant available outside the normal working hours of the plant.

(7) The Department of Agriculture was at all times notified by the International Division of the intention to re-package the carcases/quarters.

(8) On the night in question Mr Kelly was the Agricultural Officer on duty.

(9) Mr Kelly had taken up duty at the plant at 9 a.m. on the morning of the 12th January 1989 and remained on duty until 5 a.m. on the 13th day of January 1989 and resumed work again at 9 a.m. on that day.

(10) The frozen meat being re-packaged was meat intended for export to Morocco, in accordance with the terms of the contract referred to in the chapter dealing with the Eirfreeze incident.

(11) Before the meat was unloaded from the containers, Mr Kelly checked the seal number thereon with the movement certificate and only then broke the seal on the container.

(12) The re-packaging then began and on completion thereof Mr Kelly prepared the documentation to cover the transfer of the meat from one container into another, including a Movement Certificate, a Release Form and the Veterinary Inspector would provide a Meat Inspection Certificate.

(13) All the relevant documentation was prepared by Mr Kelly.

(14) Due to the necessity to prepare such documentation Mr Kelly was not at all times in the hall and only visited it from time to time.

The Tribunal accepts the evidence of Mr O'Callaghan with regard to what he saw happening in the Boning Hall and having regard to the fact that the meat being re-packaged was intended for Morocco. The Tribunal is satisfied that what Mr O'Callaghan described was an operation similar to that being carried out at Eirfreeze when it was interrupted by the arrival of the Officers referred to elsewhere in this Report. The carcases/quarters were being packaged in accordance with the customer requirements as described by Mr Birdy the Transport Manager of the International Division of the Goodman Group.

Having regard to the evidence that the six containers on the South Quay referred to in the previous chapter were loaded in Nenagh and had "CU2" stamps thereon the Tribunal is satisified that in addition to the legitimate and authorised re-packaging, stamps as described by Mr O'Callaghan were applied to carcases in Nenagh.

CHAPTER NINETEEN

1988 APS Scheme

The Tribunal was obliged to inquire into the operation of the 1988 APS scheme which had been introduced under Commission Regulation 2675/88 and the investigation carried out by the Customs & Excise authorities and the Department of Agriculture & Food into such operation.

This Regulation set forth in detail the terms and conditions applicable to the scheme and provided, inter alia, that:

"(1) Provisions should be laid down enabling the applicants to benefit from an advance payment of the aid subject to a security;

(2) In view of the exceptional circumstances in the beef market and in order to encourage operators to make use of private storage it should be provided that, for a limited period, products under private storage contract should be able at the same time to be placed under the system laid down in Article 5(1) of Council Regulation (EEC) No. 565/80 of 4th March 1980 on the advance payment of Export Refunds in respect of Agricultural products;

(3) In the case of boning, if the quantity actually stored does not exceed 67 kilograms of bone meat per 100 kilograms of unboned meat employed, private storage shall not be paid.

(4) If the quantity actually stored exceeds 67 kilograms but is lower than 75 kilograms of boned meat per 100 kilograms of unboned meat employed, the aid referred to shall be reduced proportionally.

(5) No aid shall be granted in the case of boning for quantities in excess of 75 kilograms of boned meat per 100 kilograms of unboned meat employed.

(6) The large tendons, cartilages, pieces of fat and other scraps left over from cutting for boning may not be stored."

The necessity for such inquiry by the Tribunal arose because:

1. of the allegation made in the ITV programme that:

"A proposed major European investigation into Goodman's organisation did not happen after assurances were received from the Irish authorities that they themselves had a wide-ranging investigation into Goodman in hand; but that there is no evidence of any such investigation."

and

2. the allegations made in Dáil Eireann:

   (*a*) By Deputy Dick Spring on the 26th April 1989 that the Department gave advance notice of inspections at meat plants and in particular at Foynes on the 15th and 16th April 1989 and that almost all the samples taken there had trimmings in them or were otherwise suspicious.

   (*b*) By Deputy Barry Desmond on the 9th March 1989 that there had been a major investigation into the Charleville plant of the Halal associated United Meat Packers export company in relation to export refunds.

   (*c*) By Deputy Brendan McGahon on the 24th May 1991 that:

   Four foreign owned Irish based companies like Halal, Hibernia Meats, Agra and Taher had not been mentioned in Dáil Eireann when Larry Goodman and his companies were vilified in a manner that no other public person had been vilified in Dáil Eireann in the 70 years of this State.

   That fines initially totalling over £20m for irregularities uncovered in the operation on the APS scheme were reduced to £3.6m by the Minister for Agriculture & Food. Four of the five companies were fined for serious and deliberate breaches of regulations and the fifth and smallest fine was £90,000 for technical breach and this was Anglo Irish.

The allegation made in the ITV programme implied that there was collusion between "the Irish authorities" and the Goodman organisation to thwart a major European investigation by the appropriate authority of the European Commission into the Goodman organisation and was part of the general approach in the said programme that the Goodman organisation was protected and favoured by the Irish Government, because of political connections therewith.

Because of the seriousness of this allegation the Tribunal sought at the earliest possible opportunity the assistance of the European Commission in determining its truth or otherwise.

On the 16th day of July 1991, the Tribunal wrote to Monsieur Michel Jacquot who was the Director of FEOGA which is the French acronym for the European Agricultural Guarantee and Guidance Fund (EAGGF).

In this letter Monsieur Jacquot was informed of the terms of reference of the Tribunal, the nature of the specific allegations made and his assistance invoked.

In reply, Monsieur Jacquot informed the Tribunal that:

"The allegations as conveyed in your letter 16th July and equally in the Statement of Allegations is without foundation (in respect of the services of EAGGF)."

Though this reply satisfactorily established the falsity of the allegation the Tribunal was obliged to hear evidence in regard thereto and as a result of such evidence it was clearly established to be baseless and false. Such was the cogency of the evidence in this regard that the allegation was withdrawn by Ms Susan O'Keeffe, the journalist who had researched and played a large part in the preparation of the programme.

The allegations made in Dáil Eireann arise out of the investigations carried out by the Customs and Excise authorities and the Department of Agriculture into the operations of the said scheme and it is necessary to deal in a general way with these investigations because of their individual importance and of their relevance to the question of the adequacy of the controls exercised in the operation of the scheme, the adequacy of such controls in the operation of the 1986 scheme having been queried by the Comptroller and Auditor General and the European Commission.

The Department of Agriculture is responsible for the supervision of all APS production, deboning, transporting and storage operations and ensuring compliance with all regulations relevant thereto. Under the 1988 scheme all contracted beef had to be placed in store prior to the 23rd December 1988.

The following table illustrates the traders who availed of the scheme, the number of contracts, the tonnage contracted and the percentage share of contracted production.

Contractor	No. of Contracts	Tonnage Contracted	Share of Total APS contracted Production %
1. AIBP	1,173	42,383	31.9
2. Agra	433	25,978.5	19.5
3. UMP	237	19,927	15.0
4. Hibernia	191	18,335	13.8
5. Taher	86	9,598	7.2
6. Kildare	246	7,770	5.8
7. Liffey	34	3,395	2.6
8. Slaney	27	2,234	1.7
9. KMP	6	1,000	
10. Kepak	10	720	
11. DJS	9	700	2.5
12. J. Doherty	9	700	
13. Meadow	5	210	
14. Rangeland	2	60	
Totals	2,468	133,010.5	

Almost all the beef placed in store in pursuance of the said contracts and scheme was "bonded" which is a Customs term to describe the procedure already referred to in the course of this Report, under form Customs & Excise 977 (C&E 977) headed "Register of CAP goods placed under control prior to date of export."

The purpose of this procedure is to fix a date prior to the date of exportation as the operative date for the rate of CAP refund/charge applicable.

As the provisions of Article 5 of Council Regulation (EEC) No. 565/80 were applicable to beef stored in accordance with the provisions of Commission Regulation 2675/88, the approved consignor was entitled to claim advance payment of the export refund involved at the time of placement of the beef under Customs control.

As previously pointed out, the Customs & Excise authorities are the competent authorities to interpret the Export Refund nomenclature classification and have an important role in the control of exports and in the working and control of the Export Refund system.

This control is exercised through a combination of physical and documentary checks which it carried out at meat plants and cold stores throughout the country and at export locations.

As will be seen from the table produced above, 133,010.5 tonnes of beef were placed in storage under the 1988 APS scheme. The evidence established that, of this 133,010.5 tonnes, some 90,000 tonnes of beef were deboned under Article 4 of Regulation 2675/88, producing approximately 67,000 tonnes of boneless beef, giving an average yield of 74.44%.

The amount of beef stored under this APS scheme was the highest ever experienced in Ireland.

Given the large quantities involved and in consequence the large sums payable in Export Refunds and APS aid, it was felt in the Department of Agriculture that some additional checking measures were desirable to enable the Department to be fully satisfied that the requirements of both the APS scheme and the Export Refund scheme were met.

In addition, this decision was influenced by the outcome of the investigations conducted on product produced under the 1986 scheme already dealt with in this Report and concern was also expressed at the initial findings of a Customs investigation undertaken in the early months of 1989.

This investigation was undertaken by Seán P. O'h-Odhráin, a Higher Officer of Customs & Excise stationed at Waterford with responsibility for Autozero Ltd trading as Waterford Coldstores, Christendom, Ferrybank in Waterford.

United Meat Packers (Exports) Ltd. of Clare Road, Ballyhaunis in the County of Mayo, an approved CAP consignor, stored 68,467 cartons (weighing approximately 1,650 tonnes) of beef, which had been boned at United Meat Packers, Charleville Ltd., Charleville in the County of Cork, at Autozero Ltd.

During late December 1988 and early January 1989, Mr O'h-Odhráin had discussed with Mr Stapleton, a Higher Officer of Customs & Excise in Ballina in the County of Mayo, the rate of yield achieved by United Meat Packers (Exports) Ltd.

Examination of the yields for all APS contracts stored at Autozero Ltd showed that, of the 27 contracts stored in Waterford, United Meat Packers had achieved a yield in excess of 76% in 6 contracts, in excess of 75% in 5 contracts and in excess of 74% in all other contracts.

These yields were considered to be exceptionally high by Mr O'h-Odhráin who had the benefit of the experience of the investigations carried out by him at Waterford in respect of the 1986 APS scheme.

He decided to thaw out cartons of flank beef from two APS contracts initially, for the purpose of conducting an effective internal examination.

On the 23rd January 1989 he selected for thawing 39 cartons of flanks ex APS Contract No. 2675/B/10202 and 12 cartons of flank ex Contract No. 2675/B/10207.

Of the 39 cartons taken from the first contract, 20 cartons were found to contain pieces of beef which were not individually wrapped and six cartons were found to contain trimmings in varying amounts.

Of the 12 cartons examined from the second contract, 1 carton was found to contain trimmings, 2 further cartons contained pieces of beef which were not individually wrapped.

As he was not satisfied with the purported explanation of these matters given by UMP, on the 8th February 1989 he selected for thawing a further 396 cartons of flank beef from the remaining 25 APS hindquarter bonds and 80 cartons of forequarter beef from 5 forequarter bonds from APS contracts.

This involved the thawing out of 476 cartons. As a result of what was discovered on the examination of the contents of these cartons a further 352 cartons of forequarter beef were selected for thawing on the 6/3/1989 and examined on the 13/3/1989.

Overall, 879 cartons were thawed and the contents examined.

This examination disclosed that of the 447 cartons of hindquarter beef examined:

173 cartons contained trimmings of an average weight of 4.57 kilograms per carton,

110 cartons contained beef not individually wrapped,

and

Of the 432 cartons of forequarter beef examined:

51 cartons contained trimmings of an average weight of .95 kilograms per carton,

22 cartons contained pieces of beef not individually wrapped,

3 cartons contained shin beef.

The presence of trimmings and shin beef and the failure to individually wrap pieces of meat was in breach of the regulations.

In addition to the thawing and examinations carried out on the aforesaid 879 cartons, two whole bondings comprising 2,478 cartons were check weighed and compared with the weights as declared when the goods were placed under Customs' control and found to be in agreement.

Preliminary reports on the said findings were sent to the Paying Agency (the Department of Agriculture & Food) on February the 8th 1989, on March 1st and March the 16th 1989 and the final report was sent on the 19th of April 1989.

In the course of a reply to certain queries raised by the Comptroller & Auditor General on the 13th of July 1988 with regard to the operation of the 1986 Scheme, the Secretary to the Department of Agriculture had, on the 27th day of October 1988, stated, inter alia, that the Department's checking and supervisory procedures are kept under constant review and that checks would be carried out on foot of selected contracts after the product had been placed in private storage.

From the experience gained in the AIBP Waterford / Ballymun Investigation it was decided that two forms of post vacuum checks would be undertaken on the 1988 APS Scheme product. Namely:—

(1) Weighing of randomly selected contracts or cuts within contracts, and

(2) Examination of boxes of meat, including thawing of plate and flank and forequarter boxes.

The weighing of product under selective contracts was undertaken in December 1988 and this check did not reveal any discrepancies between weights declared and actually entered into store.

In the early months of 1989 an examination into the documentary details of all contracts under this scheme was undertaken. The largest participants were identified, in particular those that had placed beef in store in boneless form and contracts with high yields were noted.

The Department of Agriculture, following consultations with the Revenue Commissioners, decided to conduct an in depth investigation into meat stocks held in warehouse storage by the major exporters.

This investigation took the form of a joint Customs & Excise / Department of Agriculture sampling programme.

The Tribunal has heard detailed evidence from the Department of Agriculture officials and the officers of Customs & Excise engaged in this joint sampling exercise but considers it unnecessary to review same during the course of this Report and will limit this Report to:—

(1) the nature and extent of the investigations;

(2) the findings as a result of such investigations;

(3) the sanctions imposed on companies as a result of such findings; and

(4) the manner in which such sanctions were determined and the reasons therefor.

This Joint Sampling Programme conducted by the Customs and Excise and the Department of Agriculture extended over the six month period, April to September 1989, and proceeded in three phases.

As appears from the report of the Customs & Excise authorities made on the 20th day of December 1989 and verified in evidence during the course of the Hearings before the Tribunal the facts in relation to the nature and extent of the investigations are as follows:—

Phase 1, commenced in April 1989 and concluded in May 1989;
Samples were drawn from the following seven traders:

Liffey Meats;
Halal (United Meat Packers);
Anglo Irish Beef Processors;
Agra Trading;
Taher Meats;
Kildare Meats;
Hibernia Meats;

These traders together process almost 98% of total national beef production.

The sampling selections made, while equally valid for both APS and Export Refund purposes, focused on those APS beef contracts where suspiciously high yields had shown up in the Department of Agriculture Analysis of Yields.

For security and operational reasons all the samples were brought to a central coldstore — National Cold Storage, Tallaght and Irish Cold Stores, Tallaght — for thawing and subsequent examination.

Initially the samples selected consisted of some two hundred (200) cartons per producer plant. However, in one case it was necessary to draw a large sample of boxed beef at the ship side at the Port of Foynes, Co. Limerick, which had been declared for export by Halal (United Meat Packers) and was in the course of shipment.

The necessity arose because this particular consignment, under export movement, contained many of the Customs' bonds and beef cuts which the Customs & Excise wished to sample and a total of 1,692 cartons were drawn.

Due to the level of infringements found in cartons containing plate & flank it was decided to extend the sampling operation and to include smaller traders, not already sampled.

This was done in Phase 2 of the Joint Sampling operation, which commenced in June 1989 and concluded in July 1989.

It increased the sampling of plate and flank products stored by the seven traders listed above and extended the exercise to smaller traders.

It also involved a further sampling of forequarters from one large producer, namely AIBP. The selection basis in Phase 2 was orientated more towards the Customs bonding lots held in warehouse in order to extend the sampling coverage for Export Refund purposes.

Thawing and examination of the sampled beef was conducted in the premises of National Cold Storage, Tallaght and Irish Coldstores, Tallaght.

Following the findings in Phases 1 and 2 it was clear that as regards the scale of infringements found, the most serious problems lay with the plate and flank cuts declared by the following four traders:—

Agra Trading;
Hibernia Meats;
Halal (United Meat Packers);
Taher Meats.

The Department of Agriculture discussed the question of the representativeness of the sample findings with these four traders but no agreement could be reached. It was then decided to conduct a final sampling phase to complete the sampling coverage, and to answer the traders' complaint that the sampling was not sufficiently representative of overall plate and flank production.

The selection was made centrally in conjunction with the Department of Agriculture, account being taken of production plants, coldstores and bonding lots which had not been sampled previously.

This third phase commenced in August 1989 and concluded in early September 1989. The specified samples were drawn by local Customs' Control Officers and were thawed and examined in suitable local premises countrywide.

Sampling Procedure

The sampling phases outlined above were part of one overall operation. The main difference between the phases lay in the manner of the selection of samples.

The selection procedure employed in each phase was as follows:—

In Phase 1 (April-May 1989) the selection of samples was made by the Department of Agriculture.

In Phase 2 (June-July 1989) the selection of samples was made by local Customs & Excise Control Officers. Samples were drawn from the broadest possible range of bondings.

In Phase 3 (August-September 1989) the selections of the bondings from which samples were to be drawn were made centrally by Customs & Excise in consultation with the Department of Agriculture.

The samples were then drawn by local Customs' Control Officers. In Phase 3 sampling was concentrated on the products of four companies. Namely:—

Agra Trading;
Hibernia Meats;
Halal (United Meat Packers); and;
Taher Meats.

This concentration on the said four companies was as a result of the high level of infringements detected during the course of the investigations in Phases 1 and 2.

In all phases the thawing and examination operations were conducted as follows:

Most selected samples were brought in sealed refrigerated lorries to the premises of National Cold Storage (NCS), Tallaght and Irish Coldstores (ICS), Tallaght.

Some samples, however, were examined in suitable local premises.

All samples were placed in freezers and sealed with both Department of Agriculture and Customs & Excise seals. When meat was due for examination, it was removed from the freezers and placed in a separate room where a controlled thawing process began.

The thawing room was secured under Customs & Excise / Department of Agriculture seal throughout the process.

The examinations were carried out in a room approved by Veterinary staff of the Department of Agriculture. The boxes of meat were brought into the examination room on pallets. Each box was examined externally and the following information noted on examination record sheets:—

(1) Veterinary Seal Number;
(2) APS Contract Number;
(3) Deboning Certificate Number;
(4) Date Code;
(5) Type of meat cut;
(6) Net Weight.

Each box was weighed and the weight recorded on the examination record sheets. The contents of the box were then examined jointly by the Customs' officer in attendance and by the Veterinary Inspector assisted by Agricultural Officers of the Department of Agriculture.

Those boxes whose contents were found regular as declared were re-palleted and returned to the freezer.

In all cases where boxes were found to contain infringements, the nature of the infringements were recorded on the examination record sheets. These boxes were then clearly marked, stored on separate pallets and returned to the freezer.

At the end of each day's examination, Customs staff compared their examination record sheets for accuracy with the sheets used by the Department of Agriculture staff.

The Customs' officer countersigned "Agriculture" work sheets and copies of these were given by Department of Agriculture to a representative of the trader concerned whose goods had been examined. Normally, trader representatives were in attendance for the duration of the examination but in some cases, however, traders declined to send representatives.

The Tribunal is satisfied, from the evidence adduced before it, that the Joint Sampling Investigation carried out by the Department of Agriculture and the Customs & Excise Authorities was detailed, thorough and exhaustive.

During the week beginning the 10th day of April 1989 and ending on Friday the 14th day of April 1989 an EC Commission FEOGA Audit visited Ireland for the purpose of conducting a FEOGA Article 9 Inquiry.

As appears from the evidence of Monsieur Jacquot, this visit was part of an overall investigation by the FEOGA Audit Unit into the operation of the APS Scheme in the several Member States including Ireland.

During the course of the week the FEOGA Audit team consisting of four members, visited Autozero Cold Store, Cabra, National Coldstore, Tallaght, Norish Coldstore, Rangeland Meats, Kepak, Clonee, Hibernia Coldstores, Sallins, Roscrea Coldstore, Autozero, Waterford, National Coldstore, Tallaght and examined in each of these coldstores the departmental and coldstore records and carried out a physical examination of product selected by the FEOGA Audit Team. The said team expressed themselves to be very happy with the system in place for the control and operation of the APS Scheme.

In addition to visiting these coldstores the Audit team carried out an examination of records corresponding to product and records which they had examined locally in the coldstores and found that all records in the Headquarters of the Department of Agriculture, reconciled with the Audit's teams local findings.

It was agreed that the result of the sampling operation then in vogue would be provided to the Commission as soon as possible.

In the course of his evidence Monsieur Jacquot stated that:

> "when we actually began our inquiry, we discovered that there was already an inquiry in progress in Ireland and decided that we should really join forces." He stated that "the team felt that there was no reason to disperse our efforts because neither the Commission nor the Irish Authorities had enormous numbers of staff and consequently "it was decided that we should join hands with the Irish Authorities and proceed with the inquiry jointly".

This clearly establishes the fact that there was no effort by the Irish Authorities to interfere, in any way, with the inquiry being carried out by the FEOGA Unit.

The summary of the findings in relation to the aforesaid investigation was submitted to the Department of Agriculture which collated and analyzed the findings.

A report was prepared by Mr Mullen and Ms. Cannon and established the following facts:—

(1) All of the products selected for sampling were physically present and available in its designated coldstore;

(2) All samples were produced by the coldstore operators on request and there were no indications, from the exercise, that the product contracted and declared for APS and refund purposes respectively had not been placed in store.

(3) The initial selection of samples had particular regard to the high yields achieved from the boning. It was expected in planning the sampling programme that this would be a key indicator of possible infringements, particularly the inclusion of ineligible pieces. It was concluded from the results of the examination that high yields were not, in themselves, indicative of such infringements. In the case of a number of contracts with exceptionally high yields no infringements were uncovered.

All of the meat examined was found to comply with the relevant community and national hygiene/quality requirements.

No systematic pattern of weight inflation was observed. Such discrepancies as were uncovered were of a very minor nature, such differences of plus or minus 0.1 kilograms or 0.2 kilograms between declared weight and weight recorded on examination.

No infringements were observed in respect of cuts other than plate and flank. The description and details of the product on the labelling of boxes were found to be in order.

The only irregularities ascertained during the course of the said investigation related to the inclusion of trimmings in some of the cartons examined and the failure, in some cases, to individually wrap meat as required by the regulations.

The findings of the investigation into UMP production from their Charleville plant were incorporated into the Joint Sampling Programme results and financial corrections.

The UMP investigation, which was undertaken in January 1989, and dealt with earlier, showed the inclusion of varying levels of non-individually wrapped pieces and trimmings in plate and flank boxes. In addition, similar findings observed, but to a much lesser degree, in boxes of chuck and blade and brisket examined.

The Sampling Programme had covered products from nine contractors, namely:—

Agra Meats Ltd.;
Hibernia Meats Ltd.;
United Meat Packers (Halal);
Taher Meats;
Anglo Irish Beef Processors Ltd.;
Kildare Meats;
KMP;
Liffey Meats; and
Slaney Meats.

In the case of products sampled from Liffey Meats and Slaney Meats all samples were found to be in full compliance with the regulatory requirements.

In the case of KMP and Kildare Meats, a small number of boxes sampled were found to contain one or two pieces of meat not individually wrapped.

In the case of AIBP from which by far the greatest number of samples were examined, all samples were found to be in full compliance with the regulatory requirements, except that the method of wrapping of individual pieces of meat was not in compliance with the regulations.

The system of wrapping used, was by way of a single long sheet of wrapping folded in such manner as to separate each piece completely covering all pieces with wrapping and preventing them from touching which system of wrapping was not regarded as in compliance with the regulation.

At worst it was clearly a technical breach of the regulation.

—All cuts of meat were physically present:

—All weights were correct:

—All hygiene and quality requirements were complied with:

there was no inclusion of trimmings and all other regulations were complied with.

The findings in respect of the other four companies can be summarised briefly as follows:

Agra Trading Ltd.

This company contracted to place 26,978 tonnes in private storage under the 1988 scheme. All of the beef was deboned.

The total quantity of plate and flank meat placed in store was 3,719 tonnes and this was produced in 16 different deboning plants under 433 contracts.

The examination of this company's product involved some 3,836 boxes from 127 contracts. Product from almost all of the plants was examined.

Out of the 3,836 boxes examined 3,197 (84%) were found to be clear and in compliance with the regulations.

Hibernia Meats Ltd.

This company contracted to store 18,335 tonnes of APS beef under 190 contracts. Overall 12,774 tonnes of beef covered by 138 contracts were deboned in 9 separate deboning plants. The company produced a total of 1,668 tonnes of plate and flank meat.

There were 2,623 boxes of plate and flank meat examined, 1,829 of these (70%) were found to be in full compliance with regulations. Trimmings were found in some 647 boxes totalling 2,193 kilograms.

There were also 226 boxes found with pieces of meat not individually wrapped, weighing 728 kilograms.

United Meat Packers (Halal)

This company placed 14,949 tonnes of boneless beef in store involving 237 contracts. This included 4,189 tonnes of plate and flank. There were 5 different deboning plants used by the company.

With the exception of product from Charleville, very little trimmings were found, in all 40 kilograms between the other 4 production plants.

All weights were found to be in order.

Pieces of meat not individually wrapped were found in each of the production plants used.

In respect of the meat produced at Charleville, some 1,258 tonnes of plate and flank meat under 76 contracts was produced in this plant.

A total of 1,961 boxes were examined.

58 of the 76 contracts were sampled and 14 of the contracts examined were found to be clear. The examination disclosed significant levels of trimmings. The total quantity of trimmings found was 839 kilograms and were found in 29 of the 58 contracts examined.

Non-individually wrapped pieces of meat were found in 236 of the 1,961 boxes examined, weighing a total of 1,009 kilograms.

Taher

Under the 1988 APS scheme, Taher Meats Ltd contracted to store 9,598 tonnes of beef under 85 contracts. All the meat was deboned in a total of 7 production units. The company produced a total of 1,606 tonnes of plate and flank meat.

Meat from 40 of the 85 contracts was examined involving a total of 1,097 boxes. Some 898 (82%) of these were found to be clear.

The remaining 199 boxes included 44 with trimmings weighing a total of 54.6 kilograms and the rest with pieces not individually wrapped.

Sanctions

On the 25th day of January 1990, the Department of Agriculture & Food sent to the Director of FEOGA the following:—

"(1) A report of the Beef Sampling operation on product produced under the 1988 APS Scheme.

(2) An assessment of report on Beef Sampling operation and consideration of penalties.

(3) Results of the Beef Sampling operation in respect of each contractor. "

The memorandum in relation to the assessment of the report of the operation and consideration of penalties and stated inter alia, that:

"Penalties are considered in respect of the following:—

Failure to individually wrap pieces of meat;

Inclusion of trimmings;

Failure to individually wrap pieces of meat

It is considered that Export Refunds (including regulatory penalty) should be disallowed on the quantity of pieces found not to be individually wrapped and that this disallowance/forfeiture should be extrapolated across the plate and flank production of the companies in question.

There is an argument that the application of penalties on the above basis is unduly harsh. Indeed the trade will doubtless argue that the breach of the regulations is not fundamental but more technical in nature and will point to the fact that they did not gain financially from it. Technical or not, the regulatory requirement is to wrap the meat pieces individually. It is clear that this could have been done in the cases under consideration. Since to be eligible for special refunds the meat must be individually wrapped, and since the regulation does not confer entitlement to refunds in any other circumstances, the conclusion is that the penalty proposed is the only one available.

Inclusion of trimmings

It is considered that Export Refunds and APS should be disallowed on the quantity of trimmings uncovered and that this disallowance should be extrapolated according to the criteria established by the Department of Agriculture. The regulatory 20% penalties on Export Refunds and APS (where appropriate) would also apply. This negates any financial benefit which would have accrued from including such product.

In order to differentiate between product which would have been refund eligible (had it been wrapped) and the product which should have been excluded, it is considered that a penalty reflecting more than the potential financial gain may be appropriate in the case of the latter.

This could be dealt with by applying forfeitures on the APS contract securities. In this context it is accepted that there is a need to distinguish between contracts with a low level of inclusions and contracts with more serious levels. Account can be taken of such differences by applying varying levels of forfeiture on the APS contract.

Financial consequences of suggested penalties.

The net effect of disallowing Export Refunds in APS as outlined in the report was that the following penalties would have been imposed.

Agra Trading	£214,000
Hibernia	£249,000
UMP	£830,000
Taher Meats	£60,000"

No penalty was suggested in regard to AIBP.

A meeting was held to discuss the report, its findings and suggested penalties, between representatives of the Department of Agriculture and officials of FEOGA at Brussels on the 1st day of February 1990.

During such discussions the FEOGA representatives stated that the method of wrapping employed by AIBP was in breach of Regulation 1964/82.

On the 6th March 1990 Michel Jacquot, the Director of FEOGA wrote to the Department of Agriculture and the Revenue Commissioners confirming the position taken by his officials on matters raised during the discussions on the 1st day of February 1990.

In the course of his letter he stated that:—

> "The sampling exercise employed a valid methodology, was properly supervised throughout and the prime documentation is to hand. Consequently, Commission services consider it to be perfectly reasonable to extrapolate on the basis of those results across both sampled and unsampled lots/contracts under private storage measures. Remarks below are relevant to the method of extrapolation employed in the report presented to my services on 1st February.
>
> Recovery of amounts unduly paid should be effected (with regulatory penalties) as a result of isolated and minor infringements. However such infringements, if they are both minor and isolated need not be considered in deciding on the method of extrapolation.
>
> There is no contradiction between the conclusion under the second indent and the interpretation presented at the Beef Management Committee. The meaning of individual wrapping in Regulation 1964/82 is clear, i.e. the wrapping of each boneless cut should be unique, peculiar and uncombinate, conditions which are not satisfied in the circumstances reported. Consequently the Intervention Agency should proceed, in the case of the one company quoted (namely AIBP) to the recovery of amounts unduly paid in this instance together with appropriate regulatory penalties.
>
> Extrapolation, where it is employed by the Irish authorities, should be the application in a reasonable fashion of the symptoms of the part (i.e. results of the sample) to the whole. The extent and depth to which the said application is effected is in direct relation to the findings, their relative gravity and, in these circumstances, the level of confidence possible in the control arrangements.

The following observations thus arise:—

- —The main results: 3,107 boxes ineligible from 10,534 in the principal tranches are indeed grave. It is quite proper therefrom to examine the situation for individual deboning plants to establish the scale of the irregular practice in more detail. No benefit however of any doubt can be extended to the control arrangements then in operation;
- —The basis for extrapolation in the report presented i.e. measurement of percentage of trimmings, bears little relation to the principal irregular practice uncovered. A carton containing ineligible scraps is to be automatically excluded from EAGGF aid; the weight or any other element in the said scraps is not directly relevant; extrapolation therefore should be built around the exclusion of cartons.
- —It has been noted that in the documents conveyed to the Commission, extrapolation has been confined to plate and flank production of contractants. EAGGF reserves its position in respect of such a restriction. A reserve equally applies in relation to

any treatment of excessively high yields which prompted the Customs' initiative in the first place."

This letter and the proposals therein contained with regard, in particular to the basis for extrapolation which the Commission sought, namely that it be built around the exclusion of all cartons in which trimmings or meat not individually wrapped were found, were considerably harsher than those proposed by the Department of Agriculture, namely that the extrapolation should be based, not on the exclusion of the entire carton, but on the exclusion of a percentage thereof, based on the percentage of trimmings or unwrapped individual cuts of meat found in such cartons.

In effect, FEOGA sought to exclude the entire carton from the benefit of payments either under the APS scheme or the Export Refunds system, whereas the Department of Agriculture & Food sought to limit the exclusion of such cartons based on the percentage of trimmings and meat not individually wrapped found in the cartons.

It is clear from the evidence that, if the system proposed by FEOGA was adopted, the sanctions to be imposed on the four companies concerned, namely Agra, Hibernia, UMP and Taher, would have been far in excess of those ultimately imposed.

AIBP were not involved in this dispute as their concern was limited to the interpretation placed by FEOGA on the provisions of Regulation 1964/82 which they regarded as incorrect.

A number of meetings were held between representatives of the Department of Agriculture and representatives of FEOGA and a number of letters were exchanged between the parties in an effort to reach agreement on the appropriate correction methodology.

At all times the Department of Agriculture considered the original approach by FEOGA to be too severe, having regard to the nature and extent of the infringements and the financial effect of financial corrections based on such approach on the four companies concerned and argued for a less severe approach. Eventually, agreement was reached by the Department of Agriculture and the Commission of the European Communities on the general framework for imposition of financial corrections.

In the course of a letter dated the 22nd day of March 1991, written by Mr Michael Dowling, Secretary to the Department of Agriculture, to Mr Guy Legras, Director General, the guidelines proposed by the Commission were accepted.
These guidelines were:—

"That all cartons found to contain trimmings be excluded from APS and Export Refunds.

"That all cartons found to contain non-individually wrapped pieces be excluded from Export Refunds.

"That the sampling results be extrapolated across the total plate and flank production of the production units concerned, with separate calculations for each production unit.

"That the extrapolation method for APS be based on the percentage by weight of trimmings found relative to the weight of the cartons sampled.

"That the extrapolation method for Export Refunds be based on the percentage by weight of trimmings and non-individually wrapped pieces found relative to the weight of the cartons sampled.

"That where the weight of trimmings in any carton is greater than or equal to 3 kgs, the weight of the entire carton be included in the extrapolation calculations.

"That where cod fat has been uncovered the weight of the entire carton be included in the extrapolation exercise.

"That where the weights of the infractions have not been recorded the entire weights of cartons be included in the extrapolation calculation, except where the evidence shows that the failure to record weights was not indicative of major infractions (e.g. AIBP case).

"That where trimmings and non-individually wrapped meat have been uncovered in the same carton and a global weight only recorded, each non-individually wrapped piece be accorded a weight of .5 kgs, where possible, and that the balance be treated as trimmings."

On the basis of this approach, the penalties imposed on each of the four companies concerned were:—

Agra	£529,817.45
Hibernia Meats	£1,525,748.93
United Meat Packers (Halal)	£1,418,148.55
Taher Meats	£92,613.00

A penalty of £90,228.78 was imposed on AIBP in respect of the contested breach of the regulations in relation to the individual wrapping of meat.

The determination of the appropriate penalties in cases of the irregularities or breaches of the Community regulations in regard to APS and Export Refund schemes is not a matter for the Tribunal.

Anybody aggrieved by such determination has a remedy in the appropriate court.

The Tribunal is satisfied, however, that the representatives of the Department of Agriculture acted with propriety in seeking a fair and equitable level of penalties having regard to the nature of the infringements discovered and breaches of regulations disclosed during the investigation.

The statement made by Deputy McGahon in Dáil Eireann on the 24th day of May 1991 that "fines initially totalling over £20m for irregularities uncovered in the operation of the APS scheme were reduced to £3.6m last week by the Minister" is incorrect.

No fines were reduced by the Minister for Agriculture & Food.

It may well be that if the original approach of FEOGA had been adopted, the penalties imposed would have been higher and might have reached £20m but a more reasonable approach was subsequently adopted as set out above.

The Minister for Agriculture & Food was not involved in any way in the determination of the penalties and did not reduce any penalties, which had been determined by the Department of Agriculture in consultation with the appropriate authorities of the Commission.

Four of the five companies were indeed fined for serious and deliberate breach of the regulations as alleged by him and a fine imposed on AIBP was for an alleged technical breach of regulations.

As stated by Deputy Desmond on the 9th day of March 1989, there was a major investigation into the Charleville plant of the Halal associated United Meat Packers Exports Co. in relation to Export Refunds.

Deputy Spring had alleged that the Department gave advance notice of inspection of meat plants and in particular at Foynes on the 15th and 16th day of April 1989.

The circumstances in regard to this incident were outlined by Mr Mullen of the Department of Agriculture during the course of his evidence and Deputy Spring accepted his evidence in that regard and withdrew the allegation.

Samples were taken in Foynes and, while Deputy Spring alleged that they had trimmings in them or were otherwise suspicious, the examination of the cartons taken in Foynes did not disclose the presence of trimmings but disclosed the presence of meat which was not individually wrapped as required by the regulations.

The cartons taken at Foynes did not appear to have originated in the Charleville plant, where the cartons contained a considerable amount of trimmings.

CHAPTER TWENTY

Defatting Analysis

The Tribunal has dealt in the chapter dealing with Rathkeale with the examination of a random sample of cartons of Intervention Beef produced from Rathkeale and selected for defatting analysis which disclosed that of the 25 cartons of fore-quarter examined the fat content was 12.4%, being 24% above the permitted level of 10% and of the 28 boxes of Plate and Flank examined, the average fat content was 32.68% being approximately 9% more than that permitted. The significance of these percentages is that the percentage of visual fat on the forequarter may not exceed 10% and on the plate and flank may not exceed 30% and (Article 21.4 Commission Regulation (EEC) No. 859/89).

The detailed results of these tests are as follows:—

Premises:- TALLAGHT COLD STORE Ex AIBP RATHKEALE

Weight	Cut	Fat Weight	Lean Weight	Percentage	Production
26.30	Forequarter	2.65	23.15	9.36	23.01.91
27.10	" "	4.90	21.30	17.81	" "
27.00	" "	3.55	22.40	12.90	" "
27.40	" "	1.75	25.10	6.30	" "
26.65	" "	4.35	21.40	15.80	" "
27.20	" "	2.90	22.75	10.50	" "
27.80	" "	3.10	23.25	11.27	" "
24.90	" "	1.40	23.10	5.05	22.11.90
24.35	" "	3.65	20.10	13.27	07.11.90
24.15	" "	3.45	20.20	12.54	08.11.90
25.40	" "	3.75	21.15	13.63	08.11.90
24.90	" "	3.05	21.25	11.00	28.11.90
24.20	" "	3.00	20.60	10.90	21.11.90
24.25	" "	4.40	19.90	16.00	28.11.90
25.20	" "	3.40	21.10	12.36	21.11.90
25.80	" "	3.70	21.70	13.45	09.11.90
26.00	" "	3.60	22.30	13.00	09.11.90
25.80	" "	3.55	21.75	12.90	08.11.90
25.55	" "	3.45	21.85	12.54	21.11.90
25.30	" "	1.95	23.65	7.09	26.11.90
25.50	" "	6.60	18.50	24.00	22.11.90
24.85	" "	4.40	20.05	16.00	26.11.90
24.00	" "	3.05	20.90	11.09	22.11.90
23.45	" "	0.95	22.50	3.45	22.11.90
26.60	" "	5.05	21.15	18.36	23.01.91
27.70	Plate/Flank	9.34	17.45	33.96	23.01.91
27.75	" "	9.35	17.35	34.00	" "
27.40	" "	8.25	18.55	30.00	" "
27.60	" "	8.80	18.60	32.00	" "
27.40	" "	6.75	20.05	24.54	" "
27.90	" "	6.65	20.50	24.18	" "
26.45	" "	10.10	16.10	36.72	" "
26.30	" "	9.55	16.05	34.72	" "
26.70	" "	6.90	19.50	25.00	" "
27.20	" "	11.60	15.05	42.18	" "
28.40	" "	8.75	19.20	31.81	" "
26.55	" "	10.60	15.45	38.54	" "
26.55	" "	9.10	17.05	33.09	" "
26.90	" "	9.15	17.20	33.27	" "
27.30	" "	7.30	19.05	26.54	" "

Premises:- TALLAGHT COLD STORE Ex AIBP RATHKEALE

Weight	Cut	Fat Weight	Lean Weight	Percentage	Production
26.60	Plate/Flank	8.00	18.20	29.09	7.11.90
26.00	" "	10.00	15.95	36.36	7.11.90
25.80	" "	6.65	18.15	24.18	9.11.90
26.50	" "	11.00	15.05	40.00	9.11.90
26.90	" "	5.60	21.00	20.36	7.11.90
26.90	" "	8.10	15.60	29.45	9.11.90
26.85	" "	8.75	17.60	31.81	9.11.90
26.25	" "	7.70	17.20	28.00	9.11.90
27.55	" "	8.25	18.05	30.00	9.11.90
26.50	" "	9.55	16.65	34.72	9.11.90
26.35	" "	7.65	17.60	27.81	9.11.90
24.50	" "	9.70	14.30	35.27	9.11.90
27.80	" "	10.45	16.55	38.00	9.11.90

This defatting analysis took place between the 11th and 27th November, 1991 at Tallaght Cold Stores.

Between the 6/2/92 and 11/3/92 the Controls Enquiry Team under the supervision of Mr Matthews V.I and Mr Gregan V.I carried out a further defatting analysis of cartons of forequarter and plate and flank boxes produced at various AIBP plants vis Nenagh, Longford, Cahir, Waterford, Cloghran, Bagnelstown and Carrigans.

A summary of the results of such de-fatting analysis in respect of each plant is set out hereunder:-

Defatting analysis at Tallaght Cold Store - Summary of results

AIBP NENAGH EEC 290

FOREQUARTER

Date of Production	22/3/1991	20/6/1991	22/6/1991	4/7/1991	19/7/1991
% fat	17.27	10.00	6.91	15.45	17.82
	10.36	10.73	10.36	11.64	6.91
	18.18	11.82	6.36	11.64	12.00
	9.27	11.27	9.09	8.73	16.36
	16.00	14.36	12.00	12.73	12.54
Daily average	14.22	11.64	8.94	12.04	13.13
Plant average		11.99			

PLATE AND FLANK

Date of Production	22/3/1991	20/6/1991	22/6/1991	4/7/1991	19/7/1991
% fat	34.00	18.18	40.36	29.82	24.45
	29.82	26.18	20.36	25.09	23.27
Daily average	31.92	22.18	30.36	27.45	24.36
Plant average		27.25			

Defatting analysis at Tallaght Cold Store - Summary of results

AIBP LONGFORD EEC 352

FOREQUARTER

Date of Production	16/1/1991	25/1/1991	8/2/1991	26/2/1991
% fat	9.10	12.54	18.73	23.27
	8.54	12.18	12.36	15.09
	12.18	14.36	9.45	8.18
	8.73	7.27	12.91	9.63
	11.64	10.91	13.82	12.18
	12.18	10.36	7.45	6.36
	4.73	7.09	17.64	10.91
	13.09	7.27	7.82	9.09
	5.64	12.54	16.00	18.73
	7.45	9.09	17.27	7.09
	13.82	6.18	17.82	13.82
	6.91	10.18	8.91	9.09
	5.45	12.54	15.09	10.18
	8.18	7.64	10.54	5.09
	12.73	16.00	8.18	9.64
Daily average	9.36	10.41	12.93	11.22
Plant average		10.98		

PLATE AND FLANK

Date of Production	25/1/1991	28/1/1991	8/2/1991	26/2/1991
% fat	24.54	27.64	26.54	28.36
	27.81	27.45	31.09	25.27
Daily average	26.17	27.54	28.81	26.81
Plant average		27.33		

Defatting analysis at Tallaght Cold Store - Summary of results

AIBP CAHIR EEC 300

FOREQUARTER

Date of Production	8/1/1991	30/1/1991	25/2/1991
	1.82	8.18	10.00
	8.36	8.18	6.18
	12.36	9.27	6.36
	7.27	13.82	10.73
% fat	4.36	8.54	7.27
	11.09	9.64	14.54
	4.36	5.45	7.45
	11.45	15.09	4.18
	5.64	6.90	5.45
	8.18	5.09	12.54
	4.91	5.27	6.73
	6.54	17.45	6.00
	4.18	4.91	5.45
	3.45	5.64	8.18
	7.89	9.64	19.82
	2.18	10.73	19.64
	6.36	4.00	11.45
	8.00	9.09	17.45
	11.45	6.91	5.09
	10.36	7.09	15.45
		6.91	5.27
		6.36	8.54
		18.18	13.82
		12.73	7.82
		9.45	9.64
		8.36	19.09
		21.45	13.45
		5.64	18.91
		14.00	22.54
		18.00	3.80
Daily average	6.97	9.73	10.77
Plant average		9.43	

PLATE AND FLANK

Date of Production	30/1/1991	25/2/1991	27/3/1991	26/4/1991
	22.36	32.54	27.27	26.90
	26.73	28.00	27.09	18.00
Daily average	24.55	30.27	27.18	22.45
Plant average		26.11		

Defatting analysis at Tallaght Cold Store - Summary of results

AIBP WATERFORD EEC 344

FOREQUARTER

Date of Production	**8/3/1991**	**19/2/1991**	**21/2/1991**	**5/3/1991**	**19/3/1991**
	8.73	19.82	6.73	6.36	6.00
	4.73	14.73	2.91	18.73	12.91
	4.54	20.54	3.64	17.45	8.36
	3.09	6.91	7.82	8.18	15.45
	13.27	14.73	8.18	6.18	25.63
	6.00	6.18	7.82	4.54	9.09
		11.64	9.64	7.27	16.36
		10.18	10.18	11.82	22.91
		6.73	7.45	17.27	25.09
		17.64	11.82	13.45	16.73
		22.36			19.27
		11.82			21.27
% Fat		4.54			30.36
		10.54			17.27
		17.09			14.54
		11.27			20.90
		9.45			17.45
		9.64			24.54
		9.27			15.45
		3.45			16.91
		19.64			15.27
		11.45			12.00
		16.00			20.18
		12.00			14.91
		9.82			6.91
		15.09			8.54
		11.09			16.18
		18.73			2.36
		10.36			2.36
		4.36			12.54
		10.18			6.54
		12.73			10.36
		12.18			11.45
		6.36			17.27
					8.18
					22.36
					8.54
Daily average	6.73	12.01	7.62	11.13	14.93
Plant Average		12.25			

PLATE AND FLANK

Date of Production	**8/3/1991**	**19/3/1991**
% fat	30.73	28.54
	16.00	24.54
	14.18	33.82
	23.82	30.73
Daily average	21.18	29.41
Plant average		

Defatting analysis at Tallaght Cold Store - Summary of results
AIBP CLOGHRAN EEC 333

FOREQUARTER

Date of Production	11/12/1990	6/3/1991	14/3/1991
% fat	4.54	9.64	6.54
	9.82	9.09	6.36
	7.64	5.64	10.91
	9.27	8.36	11.64
	3.82	8.18	17.09
Daily average	7.02	8.18	10.51
Plant average		8.57	

PLATE AND FLANK

Date of Production	6/12/1990	11/12/1990	1/3/1990	6/3/1990	14/3/1991
% fat	26.36	20.00	28.18	19.27	19.27
	31.64	26.36	28.18	26.73	33.82
Daily average	29.00	23.18	28.18	23.00	26.54
Plant average			25.98		

Defatting analysis at Tallaght Cold Store - Summary of results

AIBP BAGENALSTOWN EEC 303

FOREQUARTER

Date of Production	5/3/1991	11/3/1991	14/3/1991	28/3/1991
% fat	8.18	9.45	7.45	7.27
	12.73	12.90	6.36	8.18
	5.82	8.00	10.18	6.18
	11.45	10.54	10.18	6.54
	11.09	8.73	7.09	9.27
Daily average	9.85	9.92	8.25	7.49
Plant average			8.88	

PLATE AND FLANK

Date of Production	5/3/1991	11/3/1991	14/3/1991	28/3/1991
% fat	27.27	28.90	37.64	27.81
	32.18	20.36	38.90	23.64
Daily average	29.72	24.63	38.27	25.72
Plant average		29.59		

Defatting analysis at Tallaght Cold Store - Summary of results

AIBP CARRIGANS EEC 292

FOREQUARTER

Date of Production	12/11/1990	15/11/1990	22/11/1990	29/11/1990
% fat	8.36	11.63	14.54	6.54
	8.54	7.82	8.00	11.45
	10.18	10.00	8.91	7.27
	4.91	12.18	15.82	7.45
	8.91	10.18	6.36	11.09
Daily average	8.18	10.36	10.73	8.76
Plant average		9.51		

PLATE AND FLANK

Date of Production	**12/11/1990**	**13/11/1990**	**20/11/1990**	**30/11/1990**
% fat	24.90	28.18	28.18	26.54
	20.36	17.82	30.54	30.54
Daily average	22.63	23.00	29.36	28.54
Plant average		25.88		

These results showed an excessive amount of fat in the cartons of forequarter examined

(i) in Nenagh, where the amounts varied between 18.18% and 6.91% and represented a plant average of 11.99%

(ii) in Longford, where the amounts varied between 23.27% and 5.09%, and represented a plant average of 10.98%

(iii) in Waterford, where the amounts varied between 30.36% and 2.36%, and represented a plant average of 12.25%.

The results in respect of the AIBP Plants at Cahir, Cloghran, Bagenalstown and Carrigans were satisfactory, the plant averages being within the prescribed limits.

The foregoing tests were carried out at Tallaght Cold Stores between the 6th day of February 1992 and the 11th day of March 1992.

Because of these results, it was decided by the Beef Control Division of the Department of Agriculture to carry out, under the supervision of Mr David Lynch VI, a detailed defatting analysis of samples of intervention beef produced at 39 plants in the case of forequarter beef and 38 plants in the case of 'plate and flank'.

This examination and analysis was carried out in the QK premises at Naas between the 8th July 1992 and the 11th of August 1992.

While a random selection of cartons was taken from 39 different plants a considerably greater number of cartons were taken from AIBP plants than from the other meat plants which resulted in a more detailed examination and analysis of beef produced by AIBP than by any other Company.

Detailed records of the results of such examination were produced to the Tribunal and the said results are fairly summarised in the following table which shows the plant, the number of cartons examined in both the forequarter and the plate and flank, the fat content in the cartons of forequarter and the fat content in the cartons of plate and flank.

CUMULATIVE RESULTS BASED ON TEST 1, 2 & 3

		% FAT FQ	BOXES			% FAT P & F	BOXES
1	AIBP DUBLIN	7.01	110	1	SLANEY	18.84	18
2	DAWN CX	7.21	25	2	DAWN CX	19.29	10
3	AIBP BAGENALSTOWN	7.73	28	3	OXFLEISCH	20.10	08
4	OXFLEISCH	7.99	25	4	AIBP DUBLIN	21.91	50
5	AIBP CAHIR	8.50	140	5	LIFFEY	23.13	08
6	AIBP NENAGH	8.78	107	6	UMP B'DEREEN	23.46	10
7	AIBP CARRIGANS	9.29	20	7	AIBP CAHIR	23.49	42
8	LIFFEY	9.46	25	8	AIBP WATERFORD	23.66	30
9	UMP B'DEREEN	9.93	25	9	KEPAK CLONEE	23.84	10
10	KEPAK CLONEE	9.97	25	10	MASTER M CLONMEL	24.20	22
11	MASTER M CLONMEL	10.01	25	11	AIBP NENAGH	24.50	52
12	AIBP LONGFORD	10.05	143	12	FRESHLANDS	24.51	10
13	FRESHLANDS	10.12	25	13	SALLINS	24.53	16
14	SALLINS	10.13	55	14	MEADOW R'DOWNEY	24.64	16
15	MEADOW R'DOWNEY	10.14	25	15	KMP MIDDLETON	25.08	14
16	SLANEY	10.16	45	16	AIBP LONGFORD	25.31	62
17	KILDARE CHILLING	10.24	45	17	AGRA CORK	25.34	10
18	AGRA CORK	10.39	25	18	EUROWEST	25.51	24
19	UMP SLIGO	10.40	25	19	UMP SLIGO	25.56	10
20	KMP MIDDLETON	10.44	30	20	AIBP CARRIGANS	25.66	08
21	TARA TALLAGHT	10.66	25	21	KILDARE CHILLING	25.77	18
22	MEADOW W'FORD	10.76	25	22	TARA TALLAGHT	25.95	10
23	AIBP RATHKEALE	11.06	118	23	MEADOW W'FORD	26.64	10
24	AIBP WATERFORD	11.11	172	24	ASHBOURNE	26.76	10
25	EUROWEST	11.11	50	25	RANGELAND	26.89	10
26	ASHBOURNE	11.57	25	26	WESTERN DROMAD	27.12	10
27	RANGELAND	11.68	25	27	MEADOW CLONES	27.59	10
28	WESTERN DROMAD	11.68	25	28	CONTINENTIAL MEATS	28.49	10
29	MEADOW CLONES	11.88	25	29	BLANCHVAC TALLAGHT	28.96	10
30	CONTINENTAL MEATS	12.18	10	30	AIBP RATHKEALE	29.29	62
31	BLANCHVAC TALLAGHT	12.30	25	31	AIBP BAGIN	29.37	08
32	UMP B'HAUNIS	13.28	25	32	UMP B'HAUNIS	29.37	08
33	HIBERNIA ATHY	13.90	90	33	MASTER BANDON	29.73	10
34	MASTER BANDON	13.96	25	34	UMP C'VILLE	30.14	10
35	UMP C'VILLE	14.53	25	35	KEPAK B'MAHON	30.91	12
36	KEPAK B'MAHON	15.33	25	36	HIBERNIA ATHY	30.99	12
37	TARA KILBEGGAN	16.89	25	37	TARA KILBEGGAN	33.66	10

These tables show that in the case of cartons of plate and flank, where a fat content of 30% is permitted no AIBP plant exceeded the permitted level and indeed were well within it.

Only 4 of the 37 plants exceeded the permitted level, three marginally and one by a excessive amount.

With regard to the cartons of forequarter beef, where a fat content of 10% is permitted, five of the eight AIBP plants were well below the 10%, one was .05% above it and two, Rathkeale and Waterford exceeded it by 1.06% and 1.11% respectively.

The record of the AIBP plants compare very favourably with the records of the other companies.

CHAPTER TWENTY-ONE

PART I

Intervention Scheme AIBP at Rathkeale Boning Hall

For the proper understanding of the nature of the inquiry carried out by the Tribunal into the workings of the Intervention Scheme at the Rathkeale Processing Plant, owned and operated by AIBP, it is necessary to review briefly the manner in which the Intervention System is required to operate.

Intervention beef is placed in storage and may be deboned prior to being so placed. The de-boning may be carried out only at the meat export plants and specialist cutting plants which have facilities to carry out the de-boning in accordance with EC Regulations and the Department of Agriculture specifications.

The AIBP plants at Rathkeale had such facilities and the boning hall was duly licensed by the Department of Agriculture and Food.

Only steer beef from carcase classification categories U3, U4, R3, R4 and O3 are eligible for intervention in Ireland.

All carcases produced in Meat Export Plants in Ireland are classified at the weighing scales on the factory slaughter-line by Classification Officers of the Department of Agriculture and Food. Each carcase is individually weighed within one hour of slaughter to give hot carcase weight. Meat factories deduct 2% from such weight to allow for loss of weight in chilling. The weight is recorded on labels fixed to the carcase/half carcases by factory staff and this weight is recorded in a daily classification sheet (kill sheet). A label is placed on the carcase by the classification officer giving details of sex, conformation and fat. This label cannot be removed at anytime prior to exportation or boning.

If a factory indicates that it wishes to sell some carcases into intervention these carcases must be stamped with the appropriate classification by way of a purple dye.

This intervention stamp is in addition to the classification label attached to the carcase. The beef is presented in half carcase form and must have been slaughtered not more than six days previously. The intervention beef, which has been presented in half carcase form is de-boned pursuant to the terms of a Contract made between the Minister for Agriculture and Food and the Contractor, which Contract is subject to the Schedule annexed to it. This Contract provides, inter-alia:

> "all meat derived from the deboning, trimming, packing and freezing operations is the property of the Minister (other than the materials referred to in 3(*d*)) and, unless rejected by an Authorised Officer, shall be produced to the Minister on completion of each day's operations by the Contractor and any material failure to do so shall be regarded as a fundamental breach of the terms and Conditions of the Contract and the Schedule attached thereto".

Paragraph 3(*d*) of the said Contract provides that:—

> "All bones, fat and certain small trimmings as defined in the Schedule attached to this Contract resulting from operations carried out under the terms and conditions of this Contract and the Schedule attached thereto shall become the property of the Contractor".

Paragraph 4 of the said Contract deals with the payment by the Minister of a uniform rate of fee per kilogram net of intervention bone-in beef in respect of all operations carried out in accordance with the terms and conditions of the Contract.

Paragraph 5 of the said Contract however provides that:—

> "The fee referred to under paragraph 4(*a*) shall be reduced for each production in respect of each 0.1% by which the yield of boxed beef (net weight) falls below 68% of the weight of the bone-in beef from which it is derived. The amount of such deduction shall be fixed by the Minister from time to time by reference to the actual value of the beef involved".

From this it is quite clear that in order to qualify for the maximum payment in respect of the deboning and allied operations the deboners must achieve a yield of 68% and that the fee is reduced proportionally if that yield is not obtained.

The Intervention beef is weighed into Intervention in chilled bone-in form. Once it is weighed the beef becomes the property of the Minister for Agriculture and Food as the Intervention Agency.

The factory operative at the time of the weigh-in is obliged to enter on the IB4 form, the weight, carcase number and classification grade of each quarter beefweighed in. The supervising Agricultural Officer is required to check weigh at least 10% of the quarters and circle and initial the carcases checked by him on the IB4 form. When the IB4 form is completed the original should be taken by the Agricultural Officer and a copy retained by the factory operative. The particulars shown on the IB4 form should be cross-checked against the particulars on the daily classification sheet (kill sheet).

After weigh-in the beef is deboned in accordance with the specification contained in the Contract and the Schedule annexed thereto. Once the beef is deboned it is packed in cartons by factory operatives.

Agricultural Officers (AOs) are required to examine at least 5% of the cartons produced subject to a minimum per day of 20 such cartons. Details of each box produced are entered in the IB7 form, which is signed by a factory representative and the appropriate official of the Department of Agriculture. By way of further check the boxes of at least one particular cut are required to be selected and checked to ensure that the weights are correct, that the cuts are actually present and properly wrapped and prepared.

An IB6 form is then completed in the Veterinary Office of the Department of Agriculture and signed by the official and a factory representative. This form shows the yield of beef placed into intervention. These documents are, together with the purchase agreement and other documents which may relate to the transfer of meat to the cold store, sent to Head Office of the Department of Agriculture where they are checked to ensure that the carcase was slaughtered within six days of presentation for storage, that the number of pieces produced from deboning corresponded with the number of quarters deboned; that the documents have been properly completed and stamped and the details on the documentation correspond. When the Department of Agriculture Officials are satisfied in regard to these matters payment is authorised.

The procedures outlined above were designed to provide a system which adequately ensure that the Intervention Agency received the beef that it had contracted to purchase and that the beef had been deboned in accordance with the specifications annexed to the Contract.

However in the course of his evidence before this Tribunal, Mr John Lynch, who had been employed by AIBP as Boning Hall Manager at Rathkeale during the period April 1988 to January 1991 when he was dismissed, presented a completely different picture of the way that this system was operated in the Rathkeale plant, particularly during the period August 1990 to January 1991 and produced documentary evidence in regard thereto. He stated that;

"(1) Though the specification required that cod fat should be removed from carcases being weighed in for Intervention purposes, such carcases with cod fat attached were weighed inwithout objection from Department of Agriculture Officials.

(2) Such weighing was done openly even in the presence of Officials.

(3) The yield of meat produced from deboning the carcases was considerably in excess of the 68% referred to in the deboning contract and varied between 74% and 78%.

(4) That this yield was substantially achieved by failure to trim the carcases in accordance with specification and by leaving excess fat on some of the cuts of meat.

(5) The yield in excess of 68% was taken from intervention and transferred to commercial stock to be later sold by the company.

(6) To provide for the possibility of too much meat being transferred from intervention stock to commercial stock on any particular day and thereby failing to return a yield of 68%, the company had in reserve what he described as a) "buffer stock of frozen meat" which could be transferred into intervention if required to make up the yield.

(7) When this occurred, meat was taken from this stock, reboxed into intervention boxes, the current date placed on the intervention box which was then placed with the current intervention boxes.

(8) This practice began when Mr Tony Butler Plant Manager recruited boners for a second line in the boning hall who achieved the higher yield by not properly trimming the fat and Mr Lynch, after discussion with Mr Butler adopted a similar practice and gave instructions to the boners not to trim the fat.

(9) That the completed IB4 forms and IB7 forms were delivered to the office of the boning hall Accountant but were not given to the Department of Agriculture Officials at the end of the day.

(10) The IB4 forms and the IB7 forms were, when necessary, altered to ensure that the inclusion of cod fat did not appear from a comparison between the weights on the IB4 form and the kill sheets and that the yield achieved as shown on the IB7 forms was in the region of 68% and not the actual yield achieved.

(11) The yield actually achieved was shown in daily job costing sheets which were produced for internal company information and differed considerably from that shown on the IB6 returned to the Department.

(12) This difference is illustrated by a number of boning hall job costing sheets produced by Mr Lynch. The following is a sample of the boning hall job costing sheets produced by Mr Lynch to the Tribunal.

BONING HALL JOB COSTING: AIBP RATHKEALE

Job. No.	:	50103S001	Date:10/12/90	Start Time	:	
Customer	:	Intervention Job		Finish Time	:	
Cut Type	:	S001		Intake Wat.	:	23381.80
Intake Value	:	0.00		Intake Price	:	0.00 p/LB

Product No.	Description	Trays /Boxes	Net Wt.	Act % Yield	Target % Yield	Stock Per Lb	Stock Value
AP20124	LMC VP	1	20.50	0.09	0.00	117.00	52.88
AP20152	CHUCK TENDER V	5	127.50	0.55	0.00	113.00	317.63
AP20809	LT-90 VAC-PAC	5	103.60	0.44	0.00	59.00	134.76
FZ21007	CTT(FZN) -Canner	11	302.50	1.29	0.00	40.00	266.76
FZ21214	PIW FZN	1	23.50	0.10	0.00	70.00	36.27
FZ21811	FQ85 - FZN	1	27.50	0.12	0.00	83.00	50.32
FZ21812	PF85 - FZN	39	1072.50	4.59	0.00	50.00	1182.23
IN21401	Interv. PF	112	3077.50	13.16	13.80	0.00	0.00
IN21402	Interv. SS	51	1256.40	5.37	4.40	0.00	0.00
IN21403	Interv. BR	38	964.70	4.13	3.40	0.00	0.00
IN21404	Interv. CR	17	409.60	1.75	1.90	0.00	0.00
IN21405	Interv. IN	68	1449.10	6.20	6.00	0.00	0.00
IN21406	Interv. OU	67	1344.20	5.75	5.70	0.00	0.00
IN21407	Interv. KN	33	918.90	3.93	3.70	0.00	0.00
IN21408	Interv. R	37	844.90	3.61	3.60	0.00	0.00
IN21409	Interv. FL	11	284.10	1.22	1.20	0.00	0.00
IN21410	Interv. SL	33	844.00	3.61	3.50	0.00	0.00
IN21411	Interv. FQ	170	4685.90	20.04	20.80	0.00	0.00
TE20124	TE299-LMC	22	445.10	1.90	0.00	172.09	1688.68
SUBTOTAL A: Red Meat..		722	18202.00	77.85	68.00	9.29	3729.53
FR29924 FAT		0	570.50	2.44	6.00	3.50	44.02
FR29925 BONES		0	4609.30	19.71	21.00	-1.81	-183.93
bone			0.00	0.00	0.00	0.00	0.00
SUBTOTAL B: Fat/Bone		0	5179.80	22.15	27.00	-1.23	-139.91
TOTAL		722	23381.80	100.00	95.00	8.07	3589.62

Side Price A	: -0.17p	Boning Charge	:3021.09
Side Price B	: 7.24p	Meat Intake Value	: 0.00
Side Price C	: 5.09p	Labour	:1407.00
Yield Gain	:£3729.52	Packaging	: 656.27
Yield Gain	: 7.24 p per LB	Freezing/Storage	:1050.32
Net Margin	: 5.09 p per LB	Overheads	: 837.50
		Allowance	: 33.50
		Handling	: 0.00
		Net Margin	:2626.12

This sheet illustrates the job costing of an intervention job performed in Rathkeale on the 10th day of December 1990 and illustrates in respect of this particular job carried on in the boning hall on that day.

(1) That the yield of meat achieved on that day was 77.85%.

(2) That the net weight of such meat was 18,202kg.

(3) That of that meat 2,122.7kg (11.66%) was transferred from intervention stock to commercial stock.

(4) That of 722 boxes/trays produced, 85 (11.77%) were transferred to commercial stock.

(5) The value of such stock transferred was stated to be 3,729.53.

(6) That the cuts of meat transferred to commercial stock were clearly stated.

In addition a yield/loss analysis was carried out and such analysis of the above showed a gain to the company of £3,729.52 being the value of the beef illegally transferred to stock. A copy of such yield/gain loss analysis is set forth hereunder:—

YIELD GAIN/LOSS ANALYSIS

Product No.	Description	£	£/LB
AP20124	LMC VP	52.88	0.10
AP20152	CHUCK TENDER VP	317.63	0.62
AP20809	LT-90 VAC-PAC	134.76	0.26
FZ21007	CTT(FZN)-Cannery	266.76	0.52
FZ21214	PIW FZN	36.27	0.07
FZ21811	FQ85 - FZN	50.32	0.10
FZ21812	PF85 - FZN	1182.23	2.29
IN21401	Interv. PF	0.00	0.00
IN21402	Interv. SS	0.00	0.00
IN21403	Interv. BR	0.00	0.00
IN21404	Interv. CR	0.00	0.00
IN21405	Interv. IN	0.00	0.00
IN21406	Interv. OU	0.00	
IN21407	Interv. KN	0.00	0.00
IN21408	Interv. R	0.00	0.00
IN21409	Interv. FI	0.00	0.00
IN21410	Interv. SL	0.00	0.00
IN21411	Interv. FQ	0.00	0.00
TE20124	TE299-LMC	1688.68	3.28
	TOTAL	3729.52	7.24

In addition to this job costing sheet, Mr Lynch produced ten other job costing sheets in respect of ten other intervention jobs performed in Rathkeale between the 10th day of December 1990 and the 13th day of December, 1990.
The copies of the eleven job costing sheets showed that of the 8772 trays or boxes of meat produced in the intervention deboning in this period of four days, 987 trays or boxes of meat were transferred to the commercial stock of the Company, representing 11.25% of the number of boxes and trays produced and an average of 90 trays or boxes per job.

It further appeared, from the said documents, that the Company valued such meat wrongfully transferred to Company stock in the sum of £40,551.25.

Each job costing sheet provided the basis of a weekly job cost summary sent to AIBP Headquarters at Ravensdale on the Tuesday of each week.

A number of copies of these cost summaries were produced by Mr Lynch including a summary in respect of the week ending the 16th day of December 1990 incorporating the particulars in respect of the job carried out on the 10th day of December as shown above. Information contained in such cost summary is very detailed and shows in respect of each job, the job number, the customer, the number of cattle, the cut type, the purchase, the transfer to stock, the fat and trim, bone, the expenses, overheads, the net margin, the net margin per lb, the intake weight, the target yield, the actual yield and a gain/loss occurring from such actual yield.

This weekly job costs summary is produced hereunder:-

WEEKLY JOB COST SUMMARY
WEEK NUMBER : 50
LOCATION : AIBP RATHKEALE
WEEK ENDING : 16/12/90

JOB NO.	DATE	CUSTOMER	NO. OF CATTLE	CUT TYPE	PURCHASE	TRANSFER TO STOCK	FAT & TRIM	BONE	BONING CHARGE	GROSS MARGIN	LABOUR	PACKING	FREEZING & STORAGE	OVERHEAD	ALLOWANCE	HANDLING	NET MARGIN	NET MARGIN P/LB	INTAKE WEIGHT KGs	TARGET YIELD %	ACTUAL YIELD %	YIELD GAIN/LOSS	PRICE P/LB
50101S002	10/12/90	TESCO HFRS SIDES	20.00	S002	11055	11947	5	0	0	897	430	445	56	250	160	0	-445	-3.42	5899	73.39	67.13	-1635	85.00
50102S001	10/12/90	Intervention Job	49.50	S001	0	1244	-174	0	2391	3461	1040	443	810	619	25	0	524	1.29	18506	68.00	73.70	1244	0.00
50103S001	10/12/90	Intervention Job	67.00	S001	0	3730	-140	0	3021	6611	1407	656	1050	838	34	0	2626	5.09	23382	68.00	77.85	3730	0.00
50107S001	11/12/90	Intervention Job	135.00	S001	0	6873	-392	0	6237	12718	2835	1316	2150	1688	68	0	4661	4.38	48271	68.00	76.77	6873	0.00
50110H083	11/12/90	TESCO - 3 RIB ST CUT HIND	15.00	H032	9548	10644	-15	0	0	1081	323	301	47	188	120	0	103	1.08	4331	75.86	70.87	-2014	100.00
50111S002	12/12/90	TESCO HFRS SIDES	30.00	S002	18399	19744	0	0	0	1345	645	693	144	375	240	0	-752	-3.48	9818	73.39	67.13	-2860	85.00
50112S001	12/12/90	Intervention Job	100.00	S001	0	4403	-210	0	4467	8669	2100	998	1498	1250	50	0	2773	3.64	34571	68.00	76.01	4403	0.00
50113S001	12/12/90	Intervention Job	25.00	S001	0	959	-51	0	1073	1980	525	224	370	313	13	0	536	2.93	8307	68.00	76.49	958	0.00
50116H072	13/12/90	TESCO 7 RIB HINDS	37.50	H072	32334	35858	4	0	0	3528	919	1039	190	469	300	0	612	1.93	14379	69.00	73.73	4347	102.00
50117S001	13/12/90	Intervention Job	50.00	S001	0	1286	-134	0	2306	3458	1050	479	749	625	25	0	530	1.35	17848	68.00	73.83	1286	0.00
50121S001	13/12/90	Intervention Job	49.00	S001	0	1781	-158	0	2117	3740	1054	455	690	613	25	0	904	2.50	16383	68.00	73.49	1781	0.00
50122H072	14/12/90	TESCO 7 RIB HINDS	36.00	H072	29611	32283	-98	0	0	2574	846	1001	168	450	288	0	-179	-0.63	12897	69.00	69.88	4019	104.14
50204S001	10/12/90	Intervention Job	75.00	S001	0	4726	-203	0	3507	8031	2400	809	1194	938	38	0	2653	4.43	27142	68.00	78.24	4726	0.00
50205S013	10/12/90	CANNERS - BLUE COW SIDES	25.50	S013	5446	7003	-64	0	0	1493	485	126	213	319	38	0	313	2.70	5256	69.09	67.49	-424	47.00
50206H083	10/12/90	8-rib PISTOLA COW HINDS	18.50	HO83	12148	13237	-33	0	0	1056	463	304	55	231	28	0	-25	-0.20	5623	78.25	71.32	1161	98.00
50208H083	11/12/90	8-rib PISTOLA COW HINDS	17.25	H083	11320	12868	-21	0	0	1527	431	297	52	216	26	0	506	4.38	5240	78.25	76.04	1615	98.00
50209S001	11/12/90	Intervention Job	90.00	S001	0	5701	-256	0	4053	9498	2880	930	1382	1125	45	0	3136	4.53	31370	68.00	77.25	5701	0.00
50214S001	12/12/90	Intervention Job	104.50	S001	0	5836	-292	0	4656	10199	3344	1127	1565	1306	52	0	2805	3.53	36033	68.00	76.81	5836	0.00
50215H083	12/12/90	8-rib PISTOLA COW HINDS	18.50	H083	11464	12558	-22	0	0	1072	463	311	42	231	28	0	-3	-0.03	5591	78.25	73.31	550	93.00
50218S001	13/12/90	Intervention Job	69.00	S001	0	4013	-188	0	3081	6906	2208	740	1046	863	35	0	2015	3.83	23843	68.00	77.40	4013	0.00
50219S013	13/12/90	CANNERS-BLUE COW SIDES	28.50	S013	5465	7091	-68	0	0	1558	513	79	94	356	43	0	474	4.07	5274	69.09	66.45	-341	47.00
50220HO83	13/12/90	8-rib PISTOLA COW HINDS	21.75	HO83	10706	11970	-22	0	0	1243	544	315	56	272	33	0	24	0.19	5851	78.25	72.22	-595	83.00
		Total for Week	1082.50		157496	215755	-2523	0	36909	92645	26905	13088	13621	13535	1714	0	23791		365815			44374	

This summary contains particulars from 11 intervention jobs carried out at AIBP Rathkeale during the week ending the 16th day of December 1990. In respect of each intervention job the target yield was shown as 68% and the actual yields achieved are shown considerably in excess of that percentage, averaging 76% approximately.

In addition it is clearly shown that out of the meat produced for intervention, meat valued £40,512 was transferred to commercial stock. The amount of fat and trim involved in such transfer is being shown as being valued £2,189 leaving what could be clearly seen from an examination of such job costs summary sheet a transfer of meat valued at £38,323, which was the property of the Minister for Agriculture and Food.

These and other documents and an explanation thereof had been made available to the Solicitor and Counsel to the Tribunal during the course of an interview which took place at Dublin Castle between them and Mr John Lynch on the 27th day of September 1991. The Tribunal had been informed by a representative of Deputy Spring that Mr John Lynch, who had formerly been manager of the boning hall of AIBP factory at Rathkeale, had information which, it was considered, would be of assistance to the Tribunal. As a result of such communication the Tribunal wrote to Mr Lynch on the 23rd day of September 1991 and an arrangement was made to interview Mr Lynch on the 27th day of September 1991. The information and documents given to the Solicitor and Counsel to the Tribunal was of such a nature that the Tribunal requested the Department of Agriculture and Food to compare the documentation provided by Mr Lynch with the corresponding records kept by the Department with regard to the intervention jobs that were referred to, including the IB4s, the IB7 and the IB6.

As a result of such request the said documents were compared by Mr Maurice Mullen and Ms Brid Cannon with the corresponding intervention documents submitted to the Department of Agriculture and Food. As a result of their findings on such comparison it was decided by officials in the Department of Agriculture and Food that Control Enquiry Team (CET) be established and that this team would visit the AIBP plant at Rathkeale on the 2nd day of October 1991 and examine intervention production there.

On the 2nd of October 1991 they visited Rathkeale where they found that:—

(1) Pieces of meat, the property of the Minister, were found;

(2) Fat levels greater than the maximum permitted was found in boxes of forequarter beef;

(3) Boxes from the previous day's production (1st October 1991) were found in a Marshalling area close to the chills; and

(4) other product from production on that day (1 October 1991) had already been moved to a separate cold store.

On the following day the team visited Limerick Coldstore for the purpose of examining Rathkeale production there. In addition visits were made to the AIBP premises at Ravensdale, Ardee, Ballymun and Rathkeale. During the course of such visits certain production records were taken into possession and placed in the custody of the Gardai. The relevant records were compared with the corresponding intervention records kept by the Department of Agriculture and Food by Mr Seamus Fogarty, an Assistant Principal

Officer in that section of the Department that dealt with Intervention Purchase and Storage.

Mr Fogarty compared all of the documents which related to boning hall yields for intervention beef for AIBP with the corresponding intervention beef purchase schedules held by the Department of Agriculture and Food. The basis of his examination was to compare what was produced on the day in the boning hall, according to the company's records, with what was declared by the company to the Department. He examined the daily job costing sheets which referred to intervention jobs carried out in the boning hall in AIBP Rathkeale. The content of these job costing sheets have been illustrated during the course of this Report.

Mr Fogarty's evidence can be briefly summarised as follows:—

(1) that he examined 34 daily job costing sheets referring to intervention production dates in the AIBP plant at Rathkeale between the 17th of July 1990 and 7th of February 1991.

(2) that from records retained in the Department, intervention production occurred on dates other than those shown on the production sheets examined by him but he was unable to carry out an examination in respect of these dates because the daily job costing sheets in respect of such dates kept by AIBP were not available.

(3) the examination of the said 34 daily job costing sheets disclosed that 129975.53kgs of intervention beef was misappropriated by AIBP Rathkeale over the said 34 occasions.

(4) that in the said daily job costing sheets the company valued this beef at £257,511.58.

(5) that the intervention loss of the value of the beef which must be repaid to the EC is £3.04584 per kilo up to 30th of September 1990 and £3.14735 per kilo until the 30th of September 1991.

(6) the estimated value of the misappropriated beef was then £405,742.84.

In addition to the said 34 daily costing sheets, Mr Fogarty examined weekly boning hall job costings discovered in Ravensdale but relating to the production in Rathkeale between January 1991 and August 1991 and compared them with the corresponding Intervention Purchase Schedules retained in the Department.

Having compared the documents and information contained therein, he estimated that the total weight of the meat appropriated to commercial use by the company, as appeared from an examination of the said weekly job cost summaries, to be 158,050.71 kgs.

In reaching this figure he stated that he had granted, to the company, an allowance in respect of trims of 1.24%.

In addition to the Daily Job Costing Sheets already referred to, Mr Fogarty also examined a number of Daily Boning Hall Costing Sheets obtained by the Garda Siochana from Mr Tony Butler, the Plant Manager at AIBP, Rathkeale premises, but was in a position to only compare 5 (five) of them with the Department records.

Such examination disclosed that on the days covered by such sheets, the misappropriation of 1285.1 kgs of meat which was valued at IR£3,986.79.

In total Mr Fogarty estimated that the total loss to the Department of Agriculture & Food as a result of the misappropriation of beef disclosed in the aforesaid documents as IR£907,170.65.

These figures were challenged by Counsel for AIBP, who at all times maintained that the company was entitled to retain possession of all meat produced in the deboning in excess of the yield of 68%, suggested that there would be no loss due to the retention of meat if credit was given for the meat added back to make up the yield of 68% as disclosed in the documents and if an allowance of 6.2% was made in respect of trims.

While this suggestion may be mathematically accurate, it bears no relation to the evidence given with regard to the nature of the misappropriation of meat at Rathkeale during this period.

The evidence from Mr Lynch, supported by the documents which he produced, related not to the transfer of trimmings to company stock, which the company was entitled to do but to the transfer of identifiable cuts of meat, which the company was not entitled to do.

Mr Lynch's evidence in this regard was supported, not only by the documentation, which he produced but by the evidence of a number of workers in the boning hall in the AIBP plant at Rathkeale.

The Tribunal does not consider it necessary to refer in detail to the entire of such evidence but by way of illustration will refer to the evidence of:—

(1) James Higgins, who was a boner employed in Rathkeale during the period June 1988 to June 1991 and stated that at one period or another, during the course of his employment in Rathkeale, he was asked by his supervisors to take the briskets or sirloin from the intervention beef and trim it to commercial specifications, when it would be packed in boxes for commercial use. The supervisors named by him were Larry Kelly and John Lynch in the earlier period of his employment and Martin Finucane and John Raftery in the latter period;

(2) James Leahy, who was employed at AIBP Rathkeale during the period February 1991 to May 1991 and stated that he was instructed by John Raftery to take beef, including fillets and other cuts from the Intervention line and place it on separate trays and the meat would subsequently be vac-packed for commercial use;

(3) Peter O'Connor, who was a Boner employed at AIBP, Rathkeale, from June 1988 to January 1991 and who stated that all the meat produced as a result of Intervention deboning did not go into intervention:

    That on occasions he would be told by his supervisor, Larry Kelly, that certain cuts would be needed for commercial cuts and that they should be trimmed for such purpose; such cuts included silver-sides, knuckles, top-sides: That this occurred on a daily basis when intervention work was being carried out and that the said cuts were taken from intervention and transferred for commercial use.

(4) Liam Quirk, who started work as a trainee Boner and Trimmer in the boning hall in AIBP in Rathkeale in October 1988 and continued working there until May 1992 and stated that during the course of his employment he received instructions from his supervisor to take cuts from intervention boning, such as briskets, sirloin and fillets and trim them for commercial purposes: That these cuts would subsequently be vac-packed in the evening for commercial purposes and that this practice of taking beef from intervention continued on a regular basis during the years 1991 and 1992.

From a consideration of this and other evidence, the Tribunal is satisfied that there was a clear, definite and deliberate policy by the management staff employed by AIBP at the Rathkeale Plant, including the Plant Manager, the Boning Hall Accountant, the Boning Hall Manager and supervisors to misappropriate intervention beef, the property of the Minister for Agriculture & Food and apply it for the commercial purposes of the Company.

The consistency and deliberate nature of such policy is clearly shown in the evidence of Mr Brian Kennedy.

Mr Kennedy was first employed as a Production Clerk in the boning hall at Rathkeale.

His duties included recording the weights and specifications for each job contract boned on the day.

In respect of intervention contracts these weights were recorded in the "IB7" forms and in respect of commercial contracts in what he described as "business carbon copy books".

When he had completed the "IB7" forms he handed them to Sean O'Shea.

In addition to this work, he was also responsible for stock control and was familiar with the buffer stock maintained by the Company in the cold store, which consisted of pieces of meat taken from intervention boning or other cuts from commercial jobs which were not needed and would take cuts from such stock and place them in intervention boxes to make up any deficit he might discover.

After a year working in the boning hall, Mr Kennedy was promoted to the accounting office.

His particular job there was to complete the monthly accounts which involved the costing of each particular job and the determination of the profit or loss from each job.

In the preparation of such accounts, regard had to be taken to the amount of beef taken from intervention and used for commercial purposes.

Mr Kennedy kept a record of meat taken from intervention and applied for commercial use.

This information was recorded in pages of the carbon copy book which were given to the Accountant, Mr Goodwin.

This procedure was followed all the time he was there, from August 1990 to January 1992. The cuts transferred were LMCs, foreribs and most of the other cuts.

The cuts were transferred from intervention to commercial use and were stacked on pallets of 30 trays in the cold stores.

It was Mr Kennedy's job to record the number of trays being stacked on such pallets in his books and each day to give the page from such books to Mr O'Shea who passed them on to Mr Goodwin. There was one page for the contents of each pallet.

His function in relation to the monitoring of the beef taken out of intervention and transferred to commercial stock was to record the weight of each individual box and have regard to such in regard to the meat of inferior quality being put back in.

An indication of the extent of the transfers from intervention to commercial use is given by his evidence that each pallet contained 30 trays, each pallet was allocated a separate page in his book, the book contained 50 pages and during the busy periods he would use 5 books per week.

As the books were completed, they were stacked in the store room.

The top sheet of each page had been given in to the Accounts office and were filed there. These provided a complete record of the daily transfers of meat from intervention to commercial use.

In October 1991, as the search for documents was being carried out by the Department of Agriculture & Food officials, he received instructions from Mr Goodwin, the Accountant;

(i) to locate the said carbon copy books from the store; and

(ii) to examine them for incriminating evidence with regard to transfer from intervention to commercial.

He complied with such instruction and gave the books to Mr Goodwin.

In relation to the pages containing this information, which were filed in the Accountant's office, he received instructions to destroy them. He, with two other employees, destroyed these sheets by shredding and flushing them down the toilet.

This was a clear and successful attempt by Mr Goodwin to destroy the records kept by him in his office which would establish the full extent of the quantities of beef transferred from intervention to Company stock to be utilised for commercial purposes.

As already stated the weekly job costs summary is forwarded each Tuesday morning to Mr David Murphy the Group Accountant for the Meat Division at Ravensdale and as such reports to Mr Gerry Thornton who is the Deputy Chief Executive of this Division.

Mr Murphy stated in evidence that the only documentation received by him from the different plants, including Rathkeale, was the weekly boning hall profit and loss summary report and the weekly job costs summary, and that his real interest was in the boning hall profit and loss summary report which provided the basis for the spreadsheet which he prepared for Mr Thornton.

He stated that he was not aware of what was happening at the Rathkeale plant and that a review of any of the documentation coming to him from that plant did not indicate that there was anything untoward going on there.

Mr Gerry Thornton, the Deputy Chief Executive of the Meat Division, with overall responsibility for the operations of 35 operating units on behalf of the Goodman Group stated that neither he nor any member of his staff at Ravensdale were aware of any irregularities in the operation of the plant at Rathkeale and that the documentation submitted to his Division and furnished for review each Tuesday would not disclose that the alleged activity was taking place.

In the course of his evidence before this Tribunal, Mr Larry Goodman stated that he had no knowledge of what was going on at Rathkeale and would neither have approved of or condoned such practices.

There is no evidence in the proceedings before the Tribunal to contradict the evidence in this regard of Mr David Murphy, Mr Gerry Thornton and Mr Larry Goodman.

Mr Tony Butler, the Plant Manager at Rathkeale, Mr Larry Kelly the Supervisor in the Boning Hall and Mr Sean Goodwin, the Boning Hall Accountant all appeared before the Tribunal and refused to answer any questions in relation to these matters claiming privilege against self incrimination. Having regard to the Garda investigation, the Tribunal accepted the claim made by them.

In addition to his evidence with regard to the transfer to intervention meat into commercial stock, Mr Lynch stated that non-intervention animals were from time to time substituted for intervention animals. He stated that this was achieved in the following ways:—

Cattle of intervention quality were declared for intervention and so stamped and recorded in the kill sheet. Such cattle were not then weighed in for intervention but for commercial deboning. The numbers of such carcases would not be relevant in respect of carcases being boned for commercial purposes and would be allocated to non-intervention quality carcases. The non intervention carcases would be brought into the boning hall with legitimate intervention carcases but would not be weighed in and recorded on the IB4s. The said carcases would be deboned mainly in the morning before the Agricultural Officers arrived. The meat and cuts produced as a result of such deboning would be incorporated into the intervention contract by including particulars thereof in the IB4 forms produced in the office, the carcase numbers of the carcases which had been boned for commercial purposes would be allocated to such non-intervention carcases. This meant that carcases which were not eligible for intervention were being placed in intervention in lieu of carcases which were.

He stated that this practice occurred on most days during the period between September 1990 and January 1991 and the documentation in regard thereto was prepared by a clerk in the boning hall office.

Mr Lynch's evidence in this regard is accepted by the Tribunal as correct, having regard to the fact that his evidence with regard to the transfer of intervention meat to the company for commercial use and the substitution therefor, when necessary to achieve the yield of 68%, of meat taken from the "buffer stock" has been established beyond all reasonable doubt.

It is a cause for legitimate public concern to ascertain how the activities as described herein and constituting a flagrant abuse of the Intervention System, one of the market support schemes under the Common Agricultural Policy, was allowed to continue for such an extended period particularly having regard to the heavy responsibility placed on the staff employed by the Department of Agriculture and Food at the Plant to enforce the procedures outlined herein which were designed to ensure that the Department of Agriculture and Food, as the Intervention Agency, received the beef that it had contracted to purchase and was only discovered as a result of the information given by Mr Lynch to the Tribunal and made available to the Department of Agriculture and Food and the investigations carried out by them and the Garda Síochana on receipt of this information. During the course of his cross-examination by Counsel on behalf of the State Authorities Mr Lynch stated that;

1. After he had received instructions from Tony Butler, the Plant Manager, not to adhere to the intervention specification by not trimming the fat off the meat and he had so instructed deboners, this boning and failure to remove the fat was done openly in the presence of the Agricultural Officers.
2. That the Agricultural Officers had ample opportunity for observing the manner in which the meat was being trimmed, that they observed the operation on a daily basis, that it was obvious that the meat was not being trimmed according to specification and that they never once interfered.
3. That the boners carrying out the work in such a manner were not concerned with the presence of the Agricultural Officers and did not fear that they might be caught trimming in breach of specification.
4. That he inferred from their failure to interfere in any way with the operations of the boning hall that an arrangement had been made with regard thereto between the said Tony Butler and Mr Denis Carroll, the Senior Agricultural Officer employed by the Department of Agriculture and Food in the plant.
5. That each of the Agricultural Officers concerned namely Messrs. Lyons, Buckley, Vaughan, O'Rourke and Carroll were fully aware of the fact that the meat was being boned in breach of specification.
6. That the Agricultural Officers permitted carcases with cod fat attached thereto to be weighed in for intervention purposes.
7. That the Agricultural Officers did not check weigh 10% of the carcases as required by the regulations.
8. That the Agricultural Officers did not check 5% of the boxes of meat after they had been packed.
9. The fat analysis was not done daily and that only on one occasion between September 1990 and January 1991 did he see a fat analysis being carried out.

10. That if documentation in relation to daily fat analysis was sent to the Department of Agriculture and Food, the contents thereof were not based on any physical examination.

Mr Denis Carroll against whom some of these allegations were made, was a Senior Agricultural Officer, stationed at the AIBP plant at Rathkeale from January 1988. His function was to deploy the Agricultural Officers available to him and to cover the lairage, the deboning hall, the slaughtering line, the cold store, the despatch area in the boning hall, the loading bay, the Cannery and the offal area, and to attend to the office duties necessitated by all the documentation that had to be checked and prepared in respect of intervention purchases. He was also responsible for the port at Foynes and occasionally would have to make an Agricultural Officer available for work there. At the relevant time namely from July 1990 on he had only 4 Agricultural Officers available to him. It is perfectly obvious having heard evidence with regard to the nature of their duties that the staff available to Mr Carroll was completely inadequate to enable proper supervision of the operations of the plant which they were obliged to supervise.

As stated in evidence by Mr Carroll "people would have double duties around the place, especially in the busy time of the year". It was obvious to Mr Carroll who requested additional staff from time to time and was informed that they were unavailable, that such was the position. In addition to his other duties, Mr Carroll was responsible for checking the documents required to be completed by the factory operatives and certified by them and the preparations of the documents required by the Department of Agriculture and Food in regard to all intervention jobs.

The documents which Mr Carroll was required to submit to the Department of Agriculture and Food within 3 days of the deboning operation were:-

(1) The Purchase Agreement on meat for deboning, which is signed by the duly authorised officer of the company and the duly authorised official of the Department of Agriculture and Food.

(2) IB4 — boneless beef purchase record and certification sheet, which is required to be signed by a representative of the factory.

(3) IB7 deboning yield record, which is completed by a factory representative and signed by him.

(4) IB6 — deboned yield summary, which must be signed by both a representative of the company and the Agriculture Officer.

(5) IB8 — boneless beef transfer form.

(6) IB9 — boneless beef transport record.

During the course of his evidence Mr Lynch had stated in respect of one document purporting to have been signed by Mr Kelly, the supervisor in the boning hall, that this was not his signature.

These documents provide the basis upon which the Department of Agriculture and Food authorise payment for the beef and the deboning charges.

Mr Carroll stated in evidence that in respect of all the documentation submitted to the Department of Agriculture and Food in respect of all intervention purchases made in Rathkeale during the period January 1990 and October 1991, Mr Larry Kelly the company representative whose name appears on all such documentation, including the Tender for Intervention Purchases of Beef, the certification of the IB4, the IB6 and IB7, did not sign any of them and that he, Mr Carroll, had, with Mr Kelly's knowledge, signed his name to such documentation.

His explanation of such action is that due to pressure of other duties, the documentation would be prepared in the evening and when completed, Mr Kelly would have left the plant.

Mr Carroll was obliged to check the particulars shown on the IB4 forms and the IB7 forms and to prepare the IB6 forms on the basis of the information contained in these forms.

The IB4 was required to be completed in duplicate by the factory operatives weighing in the carcases of beef and on completion the top copy should be immediately handed to the Department official. In Rathkeale this was not done and the copies were handed into the office of the boning hall Accountant. The IB7 form is required to be completed by the factory operative and signed by him. This was not the practice in Rathkeale and the IB7 was presented with no signature of a factory official thereon.

Mr Carroll stated in evidence that;

(1) He checked the particulars shown on the IB forms with the particulars shown on the kill sheet to ensure that the weights were correct.

(2) Having checked them he would be obliged to have "a fair share of them" rewritten for various reasons such as

a) Their legibility and condition.

b) The inclusion in one form of different grades of carcases, whereas the regulations required that different grades be shown on different forms.

c) The reduction of weights because of the wrongful inclusion of cod fat in the intervention beef.

d) The presence of obvious clerical errors.

(3) The rewriting of the IB4 forms would be done by Mr Owen Maher in the boning hall Accountant's office and rechecked in the Agriculture office to ensure that there had been no fundamental alterations in the IB4 forms.

(4) Having rechecked the rewritten IB4 forms, any markings on the original IB4s were transferred by him to the said rewritten forms, such marking would include the circles placed on carcase numbers which had been checked by an Agricultural Officer and initialled by him.

(5) The IB7 forms were also checked by Mr Carroll and would be sent back to Mr Maher in the boning hall Accountants office to be rewritten, if mistakes were discovered in the IB7 forms such mistakes included a wrong carton or a wrong cut of meat being included in the IB forms.

The main reason for the necessity to have the IB4 forms rewritten was due to the practice of AIBP Management in weighing in carcases with cod fat attached thereto. This practice seems to have been known to all the relevant officials of the Department of Agriculture and Food on duty there including Mr Joe Mangan, the Veterinary Inspector, Mr Denis Carroll, Senior Agricultural Officer, Leonard Buckley, Higher Agricultural Officer, John Vaughan, Agricultural Officer and the other Agricultural Officers.

Mr Mangan discussed the problem with the Plant Manager, Mr Tony Butler on a number of occasions but in spite of the admonition of Mr Mangan and the penalty being imposed of one kilo per side the plant continued to weigh in the carcases with the cod fat attached thereto. A decision had been made by Mr Carroll with the approval of Mr Mangan to deduct a kilo from the weight of each carcase side with cod fat on it. The Agricultural Officer responsible in the boning hall marked the IB4 forms which required to be rewritten with an X.

In spite of this the practice continued but at no stage did Mr Mangan report this practice to his superiors or to Head Office. Mr Butler continued with the practice and accepted the deduction of 1 kilo when it was imposed. This meant that there was an excess of fat going into the boning hall, which should have been accounted for but was wrongfully included in intervention boxed meat, enabling some intervention meat to be transferred to company stock. If it had not been placed in intervention stock, it would have been transferred to the company.

However the Tribunal has examined 51 daily boning hall costing sheets and these disclose what fat was actually taken by the company. These sheets disclosed that while the target yield in respect of fat for each intervention job was 6%, the yields achieved varied from 0.25% to 3.76% (excluding two jobs where the 6% yield was achieved). The average yield of fat transferred was 2.11%. On the basis that the proper yield was 6%, this meant that the balance of the fat representing an average of 3.89% of the weight of the carcase was wrongfully included in the intervention boxes.

The Department of Agriculture and Food regulations require that the de-fatting tests be carried out each day in the plant by a factory operative under the supervision of an Agricultural Officer and that the results of such tests be forwarded weekly to the Department. The records produced by the Department of Agriculture and Food and produced in evidence in respect of the period 4th of August 1990 to 2nd of February 1991 showed that the returns made to the Department did not disclose the presence of excess fat in any of the boxes of meat examined on any day during this period. The permitted fat content was 10% in respect of the forequarter and 30% on the plate and flank and the results of the tests returned to the Department showed fat content of between 9.3% and 9.4% on the forequarter and 29.1% and 29.6% on the plate and flank, all within the permitted levels.

Mr Lynch had stated that all these returns were falsified because no tests were carried out on a daily basis. The Tribunal is satisfied however from the evidence from Mr Mangan, Mr Carroll, Mr Buckley and Mr Vaughan that tests were carried out but that the results of the tests as transmitted to the Department did not present a true and accurate result of the tests carried out. The accepted practice in regard to such tests was that if the test on the box of meat examined disclosed the presence of fat in excess of the permitted level, the company would be given an opportunity of having a second test sometime later and that if this second test was satisfactory the average would be taken and returned to the Department.

On occasions, a third test might have to be taken to ensure that the average weight returned was within the acceptable levels. This practice obviously gave to the company an opportunity to rectify any irregularity disclosed on the initial test. The results of the test were given to Mr Carroll the Senior Agricultural Officer, who prepared the relevant documentation for signature, by Mr Mangan, the Veterinary Officer.

During the course of an examination of intervention beef produced at AIBP Rathkeale carried out under the supervision of Mr B J Bennett, Area Superintending Veterinary Inspector, at Tallaght Coldstores, the days production of January 1991 together with other products from various days in 1990 and 1991 were examined.

A random sample of boxes of forequarter and plate and flank were selected for defatting analysis. The defatting was performed under the supervision of either the Veterinary Inspector or Agriculture Officer by Mr Larry Dunne and staff employees of AIBP who did not disagree with the findings. In 25 boxes of forequarter examined the average fat content was 12.4%, being 24% above the permitted level, and in 28 boxes of plate and flank examined the average fat content was 32.68%, being approximately 9% more than that permitted. These results are completely inconsistent with the returns made to the Department but do not establish that the returns made from Rathkeale were "falsified" in the sense used by Mr Lynch but rather the undesirability of failing to return the findings as a result of the first tests carried out and of giving to the company the opportunity of rectifying the situation before a second or third test is carried out.

It is unreal to expect that the company would not have availed of an opportunity to put the matter in order.

Having heard the evidence of Mr Mangan, the Veterinary Inspector, Mr Denis Carroll the Senior Agricultural Officer and the other Agricultural Officers employed by the Department of Agriculture and Food and the AIBP plant at Rathkeale, the Tribunal is satisfied that there was no agreement between them and the representatives of the company with regard to the irregular activities which occurred there. The situation which arose in the boning hall in Rathkeale, which in the words of Mr Mangan was catastrophic when he first went there arose because of;

(1) The huge volume of beef being processed in the boning hall, both for intervention and commercial purposes.

(2) That Mr Carroll and the limited number of AOs were unable to exercise adequately the supervision and fulfil their responsibilities with regard thereto imposed by the Department of Agriculture and Food.

(3) That irrespective of the volume of beef being processed in the boning hall, there was never more than one Agricultural Officer present and indeed he may not have been present during the entire of the time.

(4) This Agricultural Officer was expected to supervise the weigh-in of the sides of the beef, check weigh at least 10% thereof, to initial such checking of the IB4s,

> to supervise the boning and trimming of the meat, the packing of the separate cuts (11) into different boxes, the weighing thereof and the recording of such weights on the cartons and to physically check 5% thereof with a minimum of 20 cartons per day and to ensure that the intervention meat was handed up to the Department at the end of each day and placed in storage.

The Tribunal is satisfied that it was physically impossible having regard to the nature of such work and the hours which he was expected to work for a single Agricultural Officer to adequately perform these duties. It is quite clear from the evidence of Mr Mangan, Mr Power and Mr O'Neill that the supervision of the boning hall particularly during the busy season (September to December) was minimal and totally inadequate. Extra assistance had been sought from the Department of Agriculture and Food from time to time but was not made available. The failure to make adequate staff available is not the fault of the employees on the ground in the factory but rests on the Department of Agriculture and Food.

PART II

Intervention Rathkeale (Cannery)

On the 5th day of March 1991 the Council of the European Communities adopted Regulation No. 598/91, one of the purposes of which was to enable the community to transfer agriculture products available as a result of Intervention free of charge to the "Soviet Union" in response to specific requests.

The then Soviet Union requested this aid by supply of canned processed beef and the Commission decided that this request should be granted by releasing a sufficient quantity of intervention beef for that purpose.

The Commission Regulation EEC No. 1582/91 laid down certain detailed rules for the application of Regulation No. 598/91 for the supply of canned beef intended for the people of the "Soviet Union".

This regulation, provided, inter alia, that the following intervention beef should be made available for the manufacture of the product to be delivered.

(1) Forequarters bone-in

(2) Hindquarters bone-in

(3) Boneless cuts, except fillets and striploins, having being trimmed to an intervention specification of a minimum of 90% visible lean.

The regulation provided for:—

(*a*) A tender procedure, and that the intervention beef would be delivered free of charge and that the Community would pay the costs of processing and canning such product and transport costs incurred in connection therewith.

The product specifications for the canned beef was:—

(i) Composition of the content of each can before cooking:

(*a*) 80% in form of trimmed beef:
A minimum of 90% of this meat must have been diced into pieces of at least 1.5 centimetres and at most 3.0 centimetres on each face.

(*b*) 18.5% water;

(*c*) 1.5% salt.

These specifications were designed to ensure that intervention beef and intervention beef only would be included in the cans of beef to be delivered to the people of the Soviet Union.

On the 18th day of June, 1991, Shannon Meats submitted tenders (16) for a total of 1,600 tonnes of Intervention beef and were successful with all of them.

This was confirmed on the 27th day of June, 1991 by fax from the Department of Agriculture & Food.

The tender submitted by Shannon Meats included undertakings:—

(1) to produce through deboning, trimming and cutting operations the highest possible quantity of eligible meat to be canned;

(2) to trim, debone and can the beef in accordance with the conditions and specifications of the regulation and to keep the canned beef in store in easily identifiable lots, each lot comprising uniform can size;

(3) to start delivery of the canned beef to the designated organisation free at the loading bay of their warehouse not later than the 21st September 1991.

It appears from the evidence of Ms. Ní Dhuinn, an Assistant Principal Officer employed in the intervention operation Beef Sales Section of the Department of Agriculture & Food that:—

(i) She confirmed to Shannon Meats that they had been successful in their tender and requested that they confirm that they were still interested in fulfilling the contract, which confirmation was forthcoming on the 28th day of June 1991;

(ii) Shannon Meats required the beef in forequarter and hindquarter cuts;

(iii) Between the 8th of July 1991 and the 18th of September 1991 the beef was released from cold stores and delivered to the premises of Shannon Meats at the AIBP plant in Rathkeale, the necessary securities having been provided.

As the beef was processed and delivered to the transport company designated by the authority entitled to receive possession of the canned beef, written documentation was sent to Ms. Ní Dhuinn and particulars thereof sent by her to the EEC Commission.

(iv) This documentation included:

(1) A certificate headed: "Certificate of Processing in Ireland of Frozen Beef from Intervention Stock".

"Canned Intervention beef for the People of the USSR;"

(2) This certificate included particulars of the frozen intervention beef and the processed product and contained a certificate by the Veterinary Inspector that all the details were correct and that he was satisfied, in line with his report, that all the beef submitted for canning has been canned in accordance with the specifications in Annex 1 and that the beef as delivered to the designated organisation fully corresponded to the beef taken over.

Accompanying this document was a short certificate or statement from the Veterinary Officer that the contents of the named contract was taken in at the premises and was processed and exported under Department of Agriculture & Food supervision.

(v) The removal order and the takeover certificate by the transport company (Irish Transport International) would be provided;

(vi) All the documentation, in respect of these tenders, was forwarded to the Department of Agriculture & Food and checked by Ms. Ní Dhuinn. They were all found to be in order and showed that, between the 10th day of July 1991 and the 18th September 1991, all the beef which had been taken over by Shannon Meats (1598.8658 tonnes of beef) had been correctly processed to give 1762.0416 tonnes of canned product.

All the details on these certificates were cross-checked by her against the Department's removal orders, the management Goods Outward dockets from the cold stores releasing the beef and against the takeover certificates issued by the transport company, and all were found to agree.

As everything appeared to be in order instructions were given to pay Shannon Meats £1,254,014.12 for their processing costs and £805.59 for storage against the invoices which had been submitted by them.

Before the cheques issued, however, the allegations made by Mr John Lynch had been brought to the attention of the Department of Agriculture & Food and as a result of their investigations the payment thereof was frozen.

Mr Lynch had had a meeting with Mr Mullen and Ms. Bríd Cannon of the Department of Agriculture & Food on the 12th of October 1991. During the course of that meeting Mr Lynch had alleged misconduct by AIBP relative to the processing at Rathkeale of the intervention beef for the Russian Food Aid Programme.

He had stated that most of the hindquarter beef supplied to the company for processing into cans was not, in fact, so processed and that the product was re-boxed and sold off commercially, inferior forequarter meat and heart meat being substituted in the processing operation.

Because of these allegations, which are in addition to the allegations made by him in respect of the Boning Hall operations, a visit of the Control Enquiry Team was organised for the 15th of October 1991.

The Control Inquiry Team consisting of Mr Mullen, Mr Fogarty, Mr Mullins, Mr Darcy and Ms Hogan, accompanied by members of the Gardai visited the plant.

In her evidence Ms. Cannon stated that she concentrated on the records held in the Cannery Office.

She requested from Mr Doherty and Mr Denis Murphy the Daily Production Records for the Cannery in respect of the period May to August 1991.

These were produced and dealt exclusively with the USSR contract.

She examined the company's records in the office and identified five (5) files which she believed were relevant.

Mr Mullen identified a diary maintained by the company in the Cannery Supervisor's Office and took possession thereof.

According to her evidence Ms. Cannon examined these files in the Askeaton Garda Office.

She stated, in evidence, that the Daily Production Diary produced by the company showed that between the 1st of July 1991 and the 30th of September 1991 a total of 6,382 cartons of, insides, outsides, knuckles and rumps were used in the canning operations, whereas a total of 17,107, cartons of insides, outsides, knuckles, and rumps were released to Shannon Meats, Rathkeale from intervention and free of charge, for processing for the USSR into cans.

An examination of the said Production Diary also showed that such insides, outsides, knuckles and rumps were used in canning on particular dates only and the records were not in accordance with the information entered on the certificates of processing submitted by Shannon Meats to the Department.

The examination of the Production Diary also showed the almost daily usage of hearts in the canning operation.

Mr Mullen went to inspect the Cannery and in the Cannery Supervisor's Office he found the 1991 Production Diary kept by one, Mr Michael Dunne.

In addition he examined the cold store and he noticed a number of brown boxes with only the cut and weight indicated on them.

He instructed that they be removed from the cold store to the marshalling area outside the store and examined a number of the boxes.

He instructed that the store be sealed at night and that an Agricultural Officer be placed on duty.

An explanation of these matters was sought from Mr Butler, the Plant Manager at a meeting attended by Mr Mullen and Ms Cannon and other officers of the Department but apart from offering full co-operation with regard to the investigations being carried out, he offered no explanation of the matters raised by Mr Mullen and stated that Mr M Doherty was the manager responsible for the cannery.

On examination of the cold store, Mr Mullen found 182 boxes of beef, with only the weight and cut recorded on them.

These boxes were subsequently taken possession of and examined under supervision at Tallaght cold store and a detailed check of the Department's records made by Ms Ni Dhuinn, disclosed that 52 of these 182 cartons had been released to Shannon Meats free of charge in accordance with regulation 1582/91 and which should have been included in the canned meat sent to the Soviet Union.

The production records also disclosed the use of hearts in the preparation of the meat for canning.

A number of employees of AIBP at the cannery during June/July gave evidence before the Tribunal, which was uncontradicted, in relation to the trimming of hearts, the inclusion of hearts with beef in the mix for the cans, the inclusion of fresh meat from the boning hall and instructions from management to conceal the hearts if the Agricultural Officers were present or likely to be present but it is only necessary to deal with the evidence of Mr Michael Dunne who was employed by Shannon Meats in Rathkeale as a Clerical Officer in the cannery at the relevant times.

It was his responsibility to check in the meat being used for the fulfilment of the "Russian contract".

He kept a detailed personal diary which disclosed the meat and hearts used in the fulfilment of the contract.

Each day he transferred the entries from this personal diary to the production sheets of the plant.

This diary disclosed the amount of meat, the amount of hearts, and the mix sought by the company, namely, 180 kilograms of meat to 20 kilograms of hearts.

This diary was produced in evidence and had been analysed by Ms Ní Dhuinn of the Department of Agriculture and her evidence was that the contents of this diary disclosed that:

(1) Of 4,216 boxes of rumps made available, free of charge, from the intervention Agency, only 613 were used, leaving a shortfall of 3,603.

(2) Of 4,672 boxes of outsides provided, only 472 boxes were used, leaving a shortfall of 4,200.

(3) Of 4,251 boxes of insides provided, only 1,898 were used, leaving a shortfall of 2,353.

(4) Of 3,968 boxes of knuckles provided, 3,399 were used, leaving a shortfall of 569.

It further appeared that 45,306 boxes of forequarters were used, whereas only 43,664 were released from intervention, a surplus of 1,642.

The diary also showed the use of 2,367 boxes of forequarters which she understood to be fresh meat from the boning hall, 895 boxes of fresh heart and 2,366 boxes of frozen hearts.

Only frozen boneless beef as released from intervention by the Department of Agriculture & Food was to be used in processing into the cans.

These records show that 7,625 boxes of meat of various cuts supplied by the intervention authority were not used.
The Daily Production Diary prepared by the company during the relevant period in connection with the canning of the beef for the Soviet Union showed that 10,725 cartons of beef supplied from intervention, free of charge, to be processed and canned was not utilised for this purpose.

The records produced by the company to the Department of Agriculture and Food showed that all of the beef taken over by Shannon Meats from intervention 1598.8658 tonnes of beef had been processed in accordance with the regulations to give 1762.0416 tonnes of canned product.

The records were produced by Mr Sean Hartnett, Manager of the Cannery, to Mr Joe Mangan the Veterinary Inspector who signed the Certificate of Processing, the particulars on which had been inserted by the company's employees. A copy of this certificate is produced overleaf for the purpose of illustrating the detail contained therein.

AN ROINN TALMHAIOCHTA, ARAS TALMHAIOCHTA, BAILE ATHA CLIATH
DEPARTMENT OF AGRICULTURE, AGRICULTURE HOUSE, DUBLIN 2.

TEL 789011
TELEX 24280
REF-

Certificate of Processing
in Ireland of Frozen Beef
from Intervention Stock

Canned Intervention Beef for the People of the USSR - Regulation 1582/91

Sale Contract Ref:	REX M 1
Removal Order No:	3079; 3080
Name & Address of Processor	Shannon Meats Rathkeale Co Limerick
System of Processing:	per Specification in Regulation 1582/91

Details of Frozen Intervention Beef					**Processed Product**		
Date of Intake	Cut	No. of cartons	Weight Kgs	Date of Canning	Trade Description	Weight Kgs.	Beef Content Salt Content
10-7-91	Fqs	1454	39,985-0	15-7-91 16-7-91	Stewed Steak	19996.2 28717.2	Beef 80.0%
					Sub total	48713.4	Salt 1.5%
					Less other R.O. Nos	(4729.9)	Water 18.5%
TOTAL		1454	39,985-0			43983.5	

I certify that the details are correct and that the beef as above has been transferred to Shannon Transport and Warehouse, Ballysimon Road, Limerick.

Signed: J. Mangan VI IRELAND JM 354 EEC (stamped)

Date: 2/8/91

Mr Mangan accepted, without investigation, the accuracy of these details.

At a later date, the Certificate required to be signed by Mr Mangan was enlarged to provide as follows:—

> "I certify that the details are correct and that I am satisfied in line with my report that all the beef submitted for canning has been canned in accordance with the specifications in Annex 1 and that the beef as delivered to the designated organisation fully corresponds to the beef taken over."

This form of certificate was first used on the 14th day of August 1991.

The report referred to therein consisted of the following statement signed by Mr Mangan:

> "I certify that the contents of Contract No. Rexm $\frac{1}{4}$, quantity 100 tonnes was taken in at the above premises and was processed and exported under Department of Agriculture supervision."

All Certificates of Processing signed by Mr Mangan from the 14th day of August 1991 were accompanied by a similar report.

While Mr Mangan gave evidence that Mr Hartnett had made a number of complaints to him about the quality of the intervention beef supplied for the purpose of the contract, particularly the amount of fat contained in the cartons, there had been no rejection of the meat supplied.

Mr Hartnett and Mr M Doherty, when called before the Tribunal and afforded an opportunity of explaining these matters, refused to answer any question claiming the privilege of refusing to answer on the grounds of self incrimination.

Mr Laurence Goodman, Mr Gerry Thornton and Mr David Murphy again stated that they were unaware of any such practices being carried out at the cannery and that such activities did not appear from the records in respect thereof sent by the factory to headquarters at Ravensdale.

It is clear from the uncontradicted evidence given before the Tribunal that there was a deliberate policy on the part of the management of Shannon Meats:

(i) Not to comply with the known specifications of the Regulations 598/91 and 1582/91.

(ii) Not to comply with their undertaking to produce the highest quantity of eligible meat to be canned and to can the beef in accordance with the conditions and specifications of the regulations.

(iii) To include frozen hearts, fresh hearts and fresh meat obtained by deboning cattle of inferior quality in the mix of meat going into cans in breach of the said regulations which provided that the meat supplied from intervention, free of charge, only would be used.

(iv) To misappropriate and use for commercial purposes the intervention beef supplied to the factory for the purpose of providing canned beef to the Soviet Union not so used.

A total of 10,725 cartons of beef so supplied is unaccounted for.

Article 9 of EEC Regulation 1582/91 provided that:

(1) Intervention Agencies shall be responsible for the supervision of all movements and operations related to the beef concerned until the time where the canned beef is taken over by the designated organisation indicated in the takeover certificate laid down in Annex 11.

(2) Supervision must include:

(*a*) permanent physical control to verify *that all the meat taken over from intervention stores* and trimmed to the necessary extent is used for the manufacture of canned beef in accordance with the specifications laid down in Annex 1.

(*b*) when actual delivery takes place, physical control to verify that the canned beef produced and stored fully corresponds with the canned beef to be delivered.

(3) In respect of each delivery contract a report shall be made stating the findings of the supervision referred to in paragraph 1. Where those findings are considered as satisfactory by the official responsible he shall issue the appropriate certification to the successful tenderer.

By notice of these regulations the Department of Agriculture & Food was:

(1) responsible for the supervision of all movements of the beef,

(2) responsible for the supervision of all operations relating to the beef, and,

(3) obliged to maintain permanent physical control to verify that all meat taken over from intervention stores,

(i) was trimmed to the necessary extent, and

(ii) used for the manufacture of canned meat in accordance with the specifications.

In respect of each delivery contract a report was required to be produced stating the findings of the supervisor and only when such findings are considered as satisfactory by the official responsible was he entitled to issue the appropriate certification.

The Tribunal is satisfied from the evidence that the employees of the Department of Agriculture adequately supervised the delivery of the beef to the cold store but failed to exercise adequate control of the movements of the beef within the plant.

The evidence clearly established that any supervision carried out was merely by way of spot checks. This was because of the inadequacy of the number of staff and the other duties which they were required to perform.

The Department of Agriculture and Food did not provide an adequate number of staff to enable them to fulfil the responsibilities imposed on them by the EEC regulations.

There was no permanent physical presence to supervise the transfer of meat to the cannery, to supervise and be in a position to verify that all the meat taken over was used for the manufacture of canned meat.

It is obvious that if such presence was there as required and effectively exercised that the irregularities disclosed in this Report could not have occurred.

The failure to provide such supervision meant that there was no official available to make a report stating the findings of such supervision to the official responsible for issuing the appropriate certificate as required by the Regulations (Article 9 (3)) and this fact was well known in the Department of Agriculture and Food.

The processing certificates used by Mr Mangan were designed by Ms Ní Dhuinn of the Department of Agriculture & Food and the wording of the Certificate at the bottom was agreed between Mr Mangan and Ms Ní Dhuinn. The alteration of the wording of the Certificate between the 2nd and 14th August 1991 was obviously designed to show compliance with the requirement of Article 9 (3).

No report was made to Mr Mangan by any official supervising the canning process.

The report or certificate accompanying the appropriate certification was prepared for Mr Mangan's signature by Mr Hartnett or another employee of the company but not by any official of the Department of Agriculture.

In all these matters the Department of Agriculture & Food as the Intervention Agency, was in clear breach of the obligations imposed on it by the aforesaid regulations and their failure to fulfil these obligations resulted in the misappropriation by the company of intervention beef supplied free of charge and intended to be used in the canning operation and the use of hearts and substituted meat for such purpose is in clear breach of the regulations and specifications.

There is no evidence to suggest that the AIBP Management at Ravensdale were aware of the fraudulent activities being carried out by the management of the Plant at Rathkeale with regard to the misappropriation from the boning hall of beef intended to be placed in public storage and the misappropriation of intervention beef supplied free of charge to the Cannery for inclusion in the Russian contract and the records furnished weekly to AIBP management at Ravensdale did not disclose such misappropriation.

CHAPTER TWENTY-TWO

Intervention Doherty Meats (AIBP) Carrigans

This plant was taken over by AIBP in June 1989. Subsequent to the takeover, APS and Intervention boning took place in the plant in October, November and December 1989.

After this period the plant was closed and remained closed until September 1990. In September 1990 it was opened for boning of beef slaughtered by other plants but slaughtering in the Carrigan's plant began in October 1990.

Mr Dermot Butler was the Veterinary Inspector in charge in Doherty Meats, Carrigans and had been employed in that capacity by the Department of Agriculture & Food from May 1987.

Mr Hugh MacCloskey was the Supervisory Agricultural Officer with responsibility for staff supervision and the implementation of the standard controls at the EEC licensed slaughter house/deboning/cold store premises.

He stated in evidence that when deboning commenced in September 1990, there was only a small amount of boning being carried out and the paper work in connection therewith was completed reasonably accurately and punctually.

He went on to say, however, that during October, slaughtering commenced and the level of boning increased and they soon found that they had great difficulty in having the paperwork completed accurately and this led to a huge backlog of intervention documents that could not be processed by the Agriculture Office because the paperwork coming from the factory was being completed inaccurately. This situation arose because of the huge increase in the level of deboning because of the fact that slaughtering had begun in October 1990.

During the period from the beginning of October to December 1990, a considerable number of irregularities were discovered.

It is quite clear that both Mr Butler and Mr McCloskey were most assiduous in the fulfilment of their responsibilities and took particular care in the execution of their duties.

The irregularities discovered by Mr McCloskey and Mr Butler occurred between the 6th October 1990 and the 6th December 1990, were recorded and reports in respect thereof were forwarded to the Department of Agriculture.

Mr John Ferris is the Senior Superintendent Veterinary Inspector and the reports were considered by him and he considered that the events referred to therein were extremely suspicious in that it appeared that:—

(1) Over-weighing of bone-in quarters brought in from outside plants took place when the Department staff did not recheck or were liable not to recheck all of weights themselves.

(2) Upgraded carcases were finding their way into the boning hall during intervention deboning.

(3) Overweighing of Carrigans own bone-in intervention quarters onto the IB4 as shown by the reweighing of 10 hindquarters on the 29th November 1990.

(4) The possible creaming off of surplus intervention forequarter beef as shown by the 11 cartons found in the corner of the cold store from the production of the 8.11.1990.

(5) The possible replacing "new for old" intervention forequarter beef with 1989 APS forequarter beef.

(6) The general careless manner in which the paperwork was completed with numerous mistakes all adds to the confusion and makes cross checking more difficult. There are numerous examples of incorrect grades and wrong weights being recorded.

In all the incidents, explanations were presented by management staff, namely that the errors occurred because of the huge volume of work being carried out, the inexperience of the staff and operatives, who were not fully conversant with the procedures, the inclusion of non-intervention grade carcases was due to inexperience of the operatives, the use of boxes with wrong dates in respect of packaging and so on.

Many of the irregularities were dealt with at "floor level" and suitable penalties imposed.

Mr Ferris stated in evidence that he was satisfied that all these incidents were properly investigated by Mr Butler and Mr McCloskey went on to say that:—

> "The big difficulty here was the fact that the paperwork was — particularly in relation to the product in cold store — was way behind. Nothing had been made out correctly

> since the start of intervention at the end of September and up to my visit. The cold store was full. The company was looking for payment and our people couldn't agree to signing the documentation because they weren't in a position to check the contents of the cold store. In fact, when I got there myself, I couldn't even gain entry past the door. So eventually management agreed that there would be something done. I met a Mr Finbar McDonald and it all started to fall into place within the succeeding weeks and I told them that day we would be taking the intervention store and delisting it for storage of intervention, which we subsequently did, and that unless the full intervention operation was tidied up very quickly in regard to all of these type of incidents taking place, that we would be considering and I would be recommending they would be delisted. So within the next week, or after a week, the whole thing started to come together very quickly."

Having considered all the reports and the explanations given by management, Mr Ferris stated that:—

> "The conclusions that I drew were that there was a lot of various things happening that shouldn't be happening. Insofar as there seemed to be upgraded carcases and on occasions they would be finding their way into intervention, and also there seemed to be differences in weighing and there was bad paper work and the explanations couldn't be given for intervention boxes and the eleven boxes that were missing and all this took a long time to come forward. So we had to move fairly quickly on this whole issue, once it was reported to us in Agriculture House and at the same day I arrived in Carrigans, I had also arranged that a team would be sent to QK (Cold store) in Naas to take out 7 containers and that was done. They took out 7 containers to check the carton numbers because the documentation that we got in Carrigans for the days productions that were involved with those 7 containters showed that we were 50 or 52 cartons short. In fact when we did a complete check down, we got 71 too many or something like that."

In reply to Counsel on behalf of the Goodman Group, he stated that "there was nothing but bad housekeeping problems, serious housekeeping problems but nothing worse than that".

When a check was done about mid-January, the cold store records were found to agree with the Department of Agriculture records.

Mr Kenneth Robinson the Manager of the plant stated that most of the incidents occurred in the early months after the reopening of the plant during which they were training people in key practices and accepted that errors were made for which the plant was penalised.

Mr Butler gave evidence with regard to an investigation carried out by him into the deboning carried out on the 11th day of April 1991 which led him to the belief that 18 sides of beef were wrongfully recorded in the IB4s of that date. He stated that he had checked the carcases in the chill room on the previous evening, had recorded the carcase numbers and on subsequently checking the carcase numbers on the IB4s for the deboning on the 15th April 1991 discovered that they included numbers of carcases not seen by him in the chill on the previous evening.

He suspected that these carcase numbers were the carcase numbers of sides which had been exported the previous day.

This allegation is disputed by management and the Tribunal considers the evidence on this incident to be inconclusive.

CHAPTER TWENTY-THREE

Intervention Waterford

The manner in which the Intervention system operates has hopefully been adequately explained in the previous chapters of this Report.

The manner in which it can be and was manipulated to the advantage of the processor can best be illustrated by the evidence of Miss Fionnuala Fenton who was employed by AIBP as a clerk in the office of the accountant at the premises of AIBP in Waterford, and from September 1986 was the clerk who dealt with the assembly of the documentation required to be prepared by the company's operators for submission in the first instance to the Agricultural Office in the plant, and then for transfer by them to Head Office of the Department of Agriculture.

This documentation included the IB4 forms, which record the weight of the sides of beef weighed in for intervention purposes and the IB7 forms which contain particulars of the beef to be placed in storage.

A comparison of the particulars shown on each of these two forms provides the basis for the determination of the yield actually achieved in the deboning process.

It is essential to the operation of the Intervention system that the particulars shown on these forms be accurate as it is on the basis of the particulars contained in these forms that payment in respect of the beef taken into intervention is made and the fees paid by the Department in respect of the deboning contract are determined.

The original of the IB4 should be taken by the Agricultural Officer and a copy retained by the factory operative. Miss Fenton's evidence was that:

(i) From day one she was instructed that the target yield was 68.5%.

(ii) In March 1987 she was instructed by Mr McGuinness that she was to contact him if the yield was not being achieved.

(iii) Between March 1987 until Mr McGuinness left in October 1987 she participated in the alteration of the IB4 forms in order to ensure that the target yield of 68.5% or an approximation thereof was achieved.

(iv) This yield was achieved by altering the weights shown in the IB4 forms.

(v) These IB4 forms had not been taken into the possession of the Agricultural Office on completion but delivered to Miss Fenton's office and retained by her until the corresponding IB7 forms were delivered to her.

(vi) She then calculated the yield which had been obtained.

(vii) If the yield was in excess of 68.5% she calculated the additional weight that would have to be added to the weights shown on the IB4 forms to achieve the desired yield.

(viii) If the yield achieved was below 68%, she calculated the weight which would have to be deducted from the weights shown on the IB4 form so that a yield of 68% would be shown.

(ix) When she had calculated the adjustments that needed to be made in the IB4 forms, she would contact the factory operative who had completed the IB4 forms, usually Mr Ken Brennan, who would come to her office, where she had blank IB4 forms.

(x) In her office, Mr Brennan would re-write the IB4 forms in accordance with her directions to show the appropriate weights. She never discussed the reason for the alterations with Mr Brennan but gave him instructions as to what weights should be altered.

(xi) When Mr Brennan had completed the re-writing of the IB4 forms, certain further adjustments had to be made.

(xii) The IB4 forms contained circles and initials placed by Agricultural Officers on the original IB4 forms indicating the carcases check weighed by them and these had to be reproduced on the newly written IB4s.

(xiii) Miss Fenton reproduced these markings by photocopying the originals and tracing these markings into the re-written IB4s.

(xiv) The re-written IB4 forms would then be given to the Agricultural Officers, together with the IB7 forms.

The documents so presented were on their face in order and no examination thereof would disclose the irregularities outlined above, their nature or extent.

The system as outlined above showed a deliberate policy on the part of employees of AIBP to conceal from the Department of Agriculture, as the intervention authority the yield being achieved as a result of deboning of intervention beef, to alter upwards the weights of the beef actually weighed in and secure payment in respect of the increased quantity to which they were not entitled, and where necessary to reduce the weights to show the yield which would entitle them to maximum payment in respect of charges for deboning and amounted to a deliberate fraud on the Intervention authority.

In her evidence, Miss Fenton stated that:

"(i) this practice was introduced by Mr Patrick McGuinness;

(ii) that in implementing it she was following his instructions;

(iii) at the beginning he did the calculations and instructed Mr Ken Brennan as to the alterations to be made;

(iv) as she became more efficient she did it on her own initiative, and

(v) when Mr McGuinness was transferred from the factory in October 1987, the practice was terminated on the instructions of his replacement Nick O'Connor."

Mr Patrick McGuinness denied that he introduced this practice and his evidence in this regard will be considered later in this Report.

Mr Ken Brennan was the factory operative responsible for weighing in the quarters of beef into the boning hall during the relevant period.

He stated in evidence that:—

(i) he was employed in the AIBP factory in Waterford in 1987 as a general worker in charge of weighing the beef into the boning hall;

(ii) as he weighed in the beef into the boning hall his duties included completing IB4 forms;

(iii) he was instructed as to his duties in regard thereto by Mr Sean Robinson, who was the manager of the boning hall;

(iv) he was instructed by him to add two kilos to the actual weight of each quarter weighed if he could do so in the absence of the Agricultural office in charge of the boning hall;

(v) he did so whenever he got the opportunity;

(vi) he was frequently asked by the supervising manager if he was adding the weights;

(vii) if he was unable to add on weights due to the presence of the Agricultural Officer, he so informed the Supervisor who would come back later and take the IB4 forms to the office to be re-written;

(viii) when this occurred he would go to the office and re-write the forms;

(ix) when he was doing this Mr Robinson was present on a few occasions;

(x) when he was doing this, Miss Fenton was always present and the door of the office was locked;

(xi) when the Agricultural Officer was present at the weigh-in he would initial the weights and circle the number of the relevant carcase;

(xii) when Mr Robinson was replaced by Mr Salmon as the manager in charge of the boning hall he was frequently asked by Mr Salmon if he was upping the weights and if he replied no, Mr Salmon would arrange for him to go to the office and re-write them with Miss Fenton;

(xiii) Mr Gerard Kelly, who had come from Bagenalstown and had responsibilities in the boning hall also was aware of the upping of the weights;

(xiv) on only one occasion did the factory manager, Mr John Connolly ask him about upping the weights;

(xv) the practice of increasing or upping the weights continued during the entire period he was employed there;

(xvi) he was supplied with books of IB4 forms by the Agricultural Officer;

(xvii) completed IB4s were not immediately given to the Agricultural Officer but handed into the office of AIBP or to Miss Fenton;

(xviii) having completed the re-writing of the forms, he would crumple or dirty the re-written forms so as to give the impression that they were the original forms;

(xix) he did not know what became of the original forms;

While Mr Brennan mentioned Messrs Robinson, Salmon, Kelly and Connolly as being aware of and/or concerned with the increasing of the weights he never mentioned in the course of his evidence the name of Patrick McGuinness or suggested that he was in any way involved in giving him instructions with regardthereto nor was it ever suggested to him in the course of his cross-examination by Counsel for AIBP that Mr McGuinness was so involved.

A number of the original forms had been produced and made available to the Tribunal through Mr McGuinness and ITV and these had been compared with the re-written forms as submitted to the Department of Agriculture.

The IB4 forms produced by Mr McGuinness were the original ones written by Mr Brennan and the corresponding ones produced by the Department of Agriculture were those re-written by him.

In addition, the Department of Agriculture & Food produced IB4 forms submitted by AIBP from Waterford between the period March 1987 and June 1988.

Because of the allegations made with regard to the alteration of weights on and the re-writing of some of these forms and the forgery of initials of Agricultural Officers thereon, the Tribunal considered it necessary to have such documentation examined in detail by a member of the Garda Technical Bureau.

With the consent of the Commissioner of the Garda Síochána, the Tribunal was extremely fortunate to secure the services of Detective Sergeant John P Lynch of the Document Section of the Garda Technical Bureau who attended at the Offices of the Tribunal and carried out a detailed, careful and meticulous examination of the many documents involved, prepared reports thereon which were circulated to all interested parties and gave evidence in respect thereof.

In view of the admissions made in regard thereto by Miss Fionnuala Fenton and Mr Ken Brennan, the Tribunal does not consider it necessary to detail his evidence in this Report but wishes to acknowledge the contribution made by him to the resolution of this issue and to express its gratitude to Det. Sergeant Lynch for the many hours devoted by him to the examination of such documents.

Mr Brennan was led in detail through a number of these IB4 forms produced by Granada and compared them with the relevant corresponding IB4 forms submitted to the Department of Agriculture & Food and pointed out the difference between the weights recorded on the said documents.

These weights showed increases of between 3 kilos per quarter and 6 kilos per quarter.

He further confirmed that he had examined all the forms produced by Mr McGuinness and the Department of Agriculture and confirmed that they were written by him apart from the odd one written by Joe Devlin who stood in for him from time to time at the weighing in.

During the course of his cross-examination, Mr Gleeson, on behalf of AIBP pointed out and established that in 17% of the forms, the alterations showed not an increase, but a deduction.
This is in accordance with the evidence given by Miss Fenton that the alteration of the IB4 forms frequently involved a reduction in the weights shown thereon in order to show the appropriate yield.

Mr Gleeson queried the illogicality of increasing the weights in the first instance and then subsequently having to reduce them and sought an explanation therefor from Mr Brennan, which he was unable to give because, as stated by him and subsequently by Miss Fenton, the alterations in weights were done at the direction of Miss Fenton, who decided what weights had to be increased and what weights had to be decreased depending on the yield actually achieved.

Irrespective of the apparent illogicality of the procedure of adding on weights on the IB4 form and subsequently re-writing them, the Tribunal accepts the evidence with regard thereto of Mr Brennan, whom the Tribunal accepts as an outstandingly truthful and honest witness, that he added where possible 2 kilos to each side and his evidence and that of Miss Fenton established the rewriting of the IB4 forms and the reasons therefor viz to show the appropriate yield of 68% to 68.5%.

The Tribunal sees no illogicality in this since it is accepted that the largest yield obtained was to be shown as between 68% and 68.5%..

If the carcases are properly trimmed in accordance with the Intervention specifications it may at times, depending on the quality of the animal, be difficult or indeed impossible to achieve this yield.

In such circumstances that difficulty would be compounded if the weigh-in weight had been increased beyond the achieved weight.

Until the IB7 was produced and the yield calculated no decision could be made as to whether the added-on weight could be returned on the IB4 form. If the yield was below 68% adjustment downwards would then have to be made.

The weights added on as a result of the re-writing varying between 3 kilos and 6 kilos per side were in addition to the 2 kilos already added on by Mr Brennan at the weigh in and clearly established that yields considerably in excess of 68% to 68.5% were achieved in many cases 83% and apparently not achieved in 17% of the cases referred to.

Miss Fenton stated that the practice of re-writing IB4 forms ceased with the departure of Mr McGuinness in October 1987.

While Mr Brennan did not deal specifically with this question he did state that the practice of increasing the weights on the IB4s continued until he left in May 1988.

The continuance of these practices cannot be established from an examination of the documentation submitted to the Department of Agriculture because the documents appear to be in order and any irregularity in respect of weights shown on such documentation can only be established by physical examination of the meat contained in the boxes and the weighing thereof or by evidence from the persons who prepared such documentation unless an irregularity is discovered as a result of the vigilance of the Agricultural Officer in charge of the boning hall in the course of his duties.

On the 19th day of February 1988, Mr John Comerford was on duty in the Boning Hall and Supervising the weighing-in of beef into the boning hall. As Mr Ken Brennan was absent ill on that day, the weighing-in was being performed on that day by an inexperienced operative who included in a single IB4 form carcases of different grades whereas the practice was to record different grades on different forms.

Mr Comerford allowed him to continue with this practice until the end of the production but checked the weighing in of 50% of the sides and circled and initialled the carcases check-weighed.

He then informed the operative that he should re-write the IB4 forms and segregate the different grades.

The operative later in the day returned the new IB4 forms to Mr Comerford.

Mr Comerford asked for the original forms so that he could transfer therefrom the recording of his check of the weights.

When he received the original forms back, he checked the carcase number and weights and discovered that out of 34 sides of beef, the weight of 22 had been systematically increased by 3 kilos per side.

Mr Comerford reported the matter to Mr Michael Staff, the Senior Agricultural Officer, who reported to Mr Kiersey, Veterinary Inspector.

By way of explanation Mr Gilbert, the Boning Hall Manager, wrote as follows:

"26th February 1988

AIBP WATERFORD

Christendom
Ferrybank
Waterford

Mr Michael Staff, S.A.O.
Veterinary Office
AIBP Waterford
Christendom
Ferrybank
WATERFORD

Dear Michael

With reference to weight discrepancies from our Intervention Production of Friday 19th February 1988. I found from my initial investigation that our Check Weighman, Mr Ken Brennan was away sick on the day in question. The position was therefore covered by an inexperienced employee. This inexperience clearly demonstrated by the inaccurate recording of Grades on Form IB4 i.e. The 'Original'. On realising his mistake he then re-recorded on new sheets and attempted to segregate the grades, in doing so he also transposed a number of weights incorrectly.

This incident is totally unacceptable and I assure you those involved have been severely reprimanded. Unfortunately, our line management failed to ensure that this employee was fully briefed on the correct procedure and assure you that in future only qualified and thoroughly briefed personnel will undertake such duties.

Finally, please accept our apologies and assurance that this type of incident will not re-occur.

Yours faithfully,
W V Gilbert
BONING HALL MANAGER

c.c. J. Connolly — Factory Manager, N. O'Connor — Accountant.

This irregularity was reported to Head Office and the explanation accepted.

While it is understandable to accept that an inexperienced operative would mix up the grades on the IB4 form, which was what attracted the attention of Mr Comerford, different considerations must apply to the question of consistently increasing the weights on the IB4 forms of 22 sides by 3 kilos per side.

The factory operative, inexperienced though he may have been must either have known that it was the practice to increase the weights or have been instructed to do so when afforded the opportunity of re-writing the IB4 forms.

The Agricultural Officer could not have been expected to know that this was not an isolated incident. Neither the Department of Agriculture or its representatives in the boning hall or plant were aware of the practice outlined by Mr Ken Brennan and by Miss Fenton of consistently increasing weights on the IB4s when the opportunity presented itself so to do, in the absence of the Agricultural Officer, and of the re-writing, when necessary to show the appropriate yields, of the IB4s.

The discovery of this incident by Mr Comerford on the 19th day of February 1988 provides corroboration of the evidence of Mr Ken Brennan that the practice of increasing the weights on the IB4 forms continued during the entire of the period he was employed by AIBP at their plant in Waterford and that these practices did not cease when Mr Patrick McGuinness was transferred from the plant in October 1987.

PATRICK McGUINNESS

Mr Patrick McGuinness was the accountant employed by AIBP in Waterford during the period September 1986 to October 1987. He was employed by the Meat Division of this company.

He was a Chartered Accountant who had completed his articles in 1984. Subsequent to this, he was employed by the Goodman Group as an Accountant at the Newry plant where he remained until August 1986, when he was transferred to the Waterford plant, taking up duties there in September 1986.

He remained there until October 1987 when he was transferred to the Meat Division in Dundalk where he worked on a number of projects until he emigrated to Canada in June 1988.

As Financial Accountant his responsibilities involved the preparation of accounts, weekly and financial accounts, monthly accounts, annual statements, approval of payroll and purchase of livestock and generally the administration of the office in the plant in Waterford.

Mr McGuinness was the main source of much of the information upon which the ITV programme was based and he appeared on the programme.

His allegations made in that programme were as set out in Chapter 3 which dealt with the content of the ITV programme and included allegations:

"(i) that the Intervention scheme under which subsidies are paid by the EEC was abused by falsification of documents, by use of bogus stamps to alter the classification of animals, switching of meat going into or in intervention storage and the substitution therefor of inferior product, and falsification of weights shown on cartons of beef;

(ii) that the Export Refund subsidy system was abused by failing to comply with the contractual requirements of Middle East customers, the unauthorised use of Islamic stamps, in the possession of the company to show compliance with this requirement, and reboxing of meat purchased from the Intervention Agency for the purpose of misleading customers;

(iii) that the 1986 APS scheme was abused in the AIBP plant in Waterford by falsification of weights and additions of poor quality meat and by attempting to conceal the extent thereof by altering case weight at the Cold Stores and the preparation of a master plan agreed between senior management of the Goodman Group and the Customs authorities at their Head Office in Dublin to contain the damage to the reputation of the Group because of the explosive nature of the investigation;

(iv) the Irish Tax system was abused by:

(*a*) a company wide system of under the counter payments to employees which were not returned to the Revenue authorities for payment of Income Tax or PRSI, and such payments were concealed in the company's records by bogus records of payments to hauliers and farmers and amounted to approximately £3m annually;

(v) that these abuses were institutionalised within all the factories;

(vi) that Larry Goodman set the tone;

(vii) that the Intervention system was vital to the Goodman companies because at the end of the day that is where the profits came from;

(viii) that, in connection with Iraqi contracts, there were many different sources for meat, there was fresh meat, there would have been intervention meat, there would have been frozen meat, it may have been Halal slaughtered, it could have been bull, it could have been anything;

(ix) that there was, in the company a feeling that "we were invincible, we had the right connections in the right places that could control any investigation that would be put in place";

Mr McGuinness gave evidence before this Tribunal over a period of six days between the 22nd day of November 1991 and the 29th day of November 1991.

During the course of these six days, he gave evidence in support of his statements on the programme and was intensively cross-examined by Counsel for the Goodman Group of companies, Counsel for Mr Larry Goodman, Counsel for the State authorities and all the other interested parties.

On the 14th day of January 1992 Counsel on behalf of the Goodman Group of companies took the unusual course of making submissions with regard to the credibility of Mr McGuinness though unfortunately the evidence before the Tribunal had not concluded at that stage.

During the course of his submission on that day he submitted that:

"(i) it was important to consider whether Mr McGuinness transpired to be a sturdy and dependable corner-stone or foundation stone for theReport or whether any report based or dependent on his testimony would contain within itself the seeds of its own potential destruction.

(ii) a report based on Mr McGuinness' testimony was open to the risk of foundering upon the mire of his misunderstanding and misinterpretation.

(iii) Mr McGuinness was a witness whose testimony the Tribunal was compelled to reject.

(iv) he was a sub-class of witness, an involved witness whose evidence has to be treated with circumspection by the Tribunal.

(v) that his evidence had to be tested and evaluated carefully."

In his submissions he outlined a number of inconsistencies both in his evidence and with statements made by him, allegations made and withdrawn by him: statements which were obviously incorrect; statements which showed a misunderstanding of various EEC regulations and of the workings of the Goodman Group of companies.

He then submitted that;

> "the extent to which the Tribunal would base any adverse findings on the evidence of Mr McGuinness, that part of the Report would be based on the evidence of an unreliable witness."

The question of the reliability or otherwise of the evidence of any witness given before the Tribunal is a matter for the Tribunal which gave careful consideration, as it was bound to do, to its assessment of all the witnesses.

The Tribunal's duty is to find the facts and report thereon to Dáil Eireann by presentation of the Report to the Minister for Agriculture and Food.

It is not practicable for the Tribunal to enclose within the confines of this Report even a summary of all the evidence given before it, including the cross-examination by Counsel representing the interests of all the parties before it.

However the evidence of Mr McGuinness is of particular importance because of the nature of the allegations made by him and the probability that the ITV programme broadcast on the 11th of May 1991 would not have been made without his assistance and the probability that the resolution setting up this Tribunal would not have been passed by Dáil Eireann on the 24th day of May 1991 were it not for the publicity generated by the said programme.

Rather than deal in detail with the entire of Mr McGuinness' evidence, the Tribunal will in the first instance deal with those portions thereof about which there is no controversy or which has been confirmed and corroborated by independent evidence accepted by the Tribunal. It does so in the interest of brevity, to avoid repetition and to provide a platform for the assessment of the contradicted evidence of Mr McGuinness.

The Tribunal will deal with these under the following headings:

(*a*) *1986 A.P.S. scheme (Waterford)*

The facts in relation to the over declarations of weight of the beef stored in pursuance of this scheme on the supporting documentation and the inclusion of ineligible meat in the boxes thereof are well established and dealt with in this Report in the Chapter entitled 1986 APS Scheme Waterford. In respect thereof the only matters to be dealt with by the Tribunal relates to the question of the alleged "Master Plan".

In this connection Mr McGuinness had stated:

"... (iii) that the 1986 APS scheme was abused in the AIBP plant in Waterford by falsification of weights and additions of poor quality meat and by attempting to conceal the extent thereof by altering case weight at Cold Stores and the preparation of a master plan agreed between senior management of the Goodman Group and the Customs authorities at their Head Office in Dublin to contain the damage to the reputation of the Group because of the explosive nature of the investigation."

The facts in relation to the over declarations of weights of the beef stored in pursuance of this scheme on the supporting documentation and the inclusion of ineligible meat in the boxes thereof are well established and dealt with in this Report in the chapter entitled 1986 APS Scheme Waterford. The Tribunal has, in the course of its Report on the 1986 APS Scheme, dealt with and found that there had been an attempt by personnel employed by the International Division of AIBP to conceal the extent of the over declaration of weights by altering, by way of increase the weights shown on some of the cartons placed in private storage in the cold store.

The only matters to be dealt with by the Tribunal relates to the question of the alleged "Master Plan" which is alleged to have been agreed between the top management of the Goodman Group and the Customs and Excise authorities at their head office in Dublin the existence of which plan having been strenuously denied by all witnesses other than Mr McGuinness.

Mr McGuinness stated in evidence that:

(i) a few weeks after the first incident viz the incident on the 19th day of January, 1987 when it was suspected that the weights shown on the cartons were altered by employees of AIBP he was informed by Gerry Thornton and Jim Fairbairn that meat was going to be taken from the Cahir Plant to Waterford.

(ii) this meat was to replace the contents of some of the APS boxes that were in the cold store.

(iii) two small loads of meat came from the Cahir plant.

(iv) the drivers and personnel accompanying these loads stopped at a public house on the Cahir Road slightly outside Waterford.

(v) he and Jim Fairbairn met them there and told them to remain there until the cold store in Waterford was empty.

(vi) he and Jim Fairbairn returned to Waterford where he met Gerry Thornton.

(vii) Gerry Thornton instructed him to go to the cold store with the meat and to write up the necessary documentation.

(viii) Gerry Thornton then informed him that he was returning to Dundalk as it wouldn't be appropriate for him to be there.

(ix) he and Jim Fairbairn returned to the public house.

(x) approximately ten people had come from Cahir, including a member of what he described as the A Team, Peter O'Reilly, David Dunne, Larry Dunne and Owen Lambe.

(xi) they stayed in the public house and checked to make sure that there were no Customs officials present.

(xii) when they were informed that the coast was clear, they got into their cars, travelled to the cold store and were admitted by the cold store supervisor.

(xiii) the cold store employees had set up a series of tables to assist in the re-boxing.

(xiv) there were two fork lift operators, employees of the cold store, for the purpose of removing the pallets from the cold store, place them at the tables and then return the pallets to the cold store in the same position from which they were removed.

(xv) when the pallets of cartons of meat were removed from the cold store a number of cartons or boxes were removed from the pallet, the cartons opened, the meat taken out and replaced by meat taken from the cartons brought from Cahir, the cartons resealed and the two strappings put over the box.

(xvi) this carton was then replaced in the exact same position it had held in the pallet.

(xvii) this procedure was followed in respect of all the cartons which were changed.

(xviii) He (McGuinness) was in the room recording all the details in respect of the cartons changed and their location.

(xix) Mr Fairbairn had indicated the type of information he would need to be recorded on the document and Mr McGuinness prepared such documentation and recorded the information.

This documentation is recorded in the document, a copy of which was produced by Mr McGuinness, and which is known as "The Master Plan".

A copy is printed hereunder.

Cont No	**Pallet No**	**Vet Seal No 1**	**Vet Seal No 2**	**Vet Seal No 2**	**Map**
5462	P40	15985 20.5	15990 26.3	No Vet No. 24.0	P40 **1** **2** **3** 2 on top of 1 + 3
4897	P50	85161 21.2	85172 21.4	85164 22.5	C B D **3** **2** **1** P50
4960	P25	89620 25.3	89596 20.7	89604 24.3	P25 **1** **2** **3** C
5385	P17	93043 21.5	93045 23.6	93044 20.0	**3** **2** **1** P17
5498	P9	62063 30.8	62089 28.1	62064 24.1	**1** **3** **2**
4715	P28	98167 24.9	98097 24.8	98137 29.4	P28 **1** B **2** **3**
5152	P24	40855 21.8	39149 23.8	40868 26.2	**3** **1** **2** **2** 2nd down under **1**
5381	No Custom No/W.C.S No. is EM 1983	48281 23.4	48262 26.5	48767 21.8	**1** **2** **3** No. **3** 2nd down in pallet

Cont No	Pallet No	Vet Seal No 1	Vet Seal No 2	Vet Seal No 2	Map
5470	P. 16	25429 25.7	- 28.5	- 27.5	**3** **2** **1** P. 16
4748	P. 31	102429 21.6	102426 26.7	102427 24.1	P. 31 **2** A **1** **3** B C
5451	P. 33	84950 24.8	84942 23.2	84933 26.0	P. 33 **3** **2** **1**
5484	P3 R 21/ /87	21247 31.9	21251 23.0	21257 21.5	P3 **2** **1** 3 under 1
5445	P9	12634 19.8	12638 23.4 (B)	12637 24.7	P9 **1** **2** **3** **1** on top of **2** + **3**
5521	P2	61249 21.8 Cancelled as Pallet is "P"	-- —	-- —	
5428	P2	22738 25.1	22744 23.9	22749 28.6	P2 **1** **2** **3**
5460	P10	61702 26.3	61703 27.7	61711 26.3	**3** **1** **2** P10
5461	P27	49245 23.4	49258 (c) 27.1	49250 (A)(D)23.7	P27 D **1** **3** **2C** **A**

Cont No	Pallet No	Vet Seal No 1	Vet Seal No 2	Vet Seal No 2	Map
5499	Wooden Pallet 6 Bx on Pallet.	20136 26.0	20138 19.9	20141 26.2	
5436	Wooden Pallet as above 6 Bx on Pallet. Same pallet as 5499	25809 25.1	25816 25.3	25790 21.6	
5380	Wooden Pallet	45237 21.1	43329 21.3	45248 21.6	
4713	Wooden Pallet	70795 22.6	70811 24.7	70729 23.1	
4714	Wooden Pallet	69968 23.3	69855 23.1	69854 25.8	
5434	P1	21660 28.3	21652 28.2	21661 22.5	P1 **2** **3** **1** **3** 2nd down
5492	P48	17590 25.0	14.837 25.0	117582 26.3	P.48 **3** **2** **1**
4865	P25	101134 26.4	101140 21.6	101106 24.60	P25 **1** **3** **2** **2** + **3** 2nd Row under **1**

Cont No	Pallet No	Vet Seal No 1	Vet Seal No 2	Vet Seal No 2	Map
5502	No Custom number on Pallet 10 Bx on Pallet Netting on Pallet	20818 19.7	20447 22.6	- 25.1	3 1 2 1 on Bottom 2 2nd from Bottom 3 3rd from Bottom over No 2
4908	3 Bx No Pallet Number	68814 22.0	68896 25.5	68806 25.80	

Mr McGuinness stated that:—

(*a*) He gave the original of this document to Mr Fairbairn and photocopies thereof to the cold store supervisor, to Gerry Thornton and to John Connolly, the Plant Manager at Waterford.

(*b*) The purpose of preparing this plan was to provide a good sample of meat which could be checked by the Customs' officials and which would be indicated to them as suitable for thawing out and if this was successful, would indicate that the APS regulations had been complied with.

The effect of this evidence of Mr McGuinness is that Mr Fairbairn, either on his own or in conjunction with Mr Gerry Thornton, decided to attempt to deceive the Customs' officials who were carrying out a detailed investigation into the operation of the 1986 APS contract in the AIBP Waterford plant, where it had been established that there was an over-declaration of weights on the cartons of meat stored under the scheme and it was suspected that the cartons of meat so stored or a substantial portion thereof including trimmings which were not eligible for APS, by substituting a number of cartons of meat obtained from the AIBP Cahir plant for cartons which had been placed in storage by AIBP Waterford and by hoping or intending to persuade or direct the Customs & Excise officials to an examination of these boxes.

The original allegation made and presented on the ITV programme was that this plan had been agreed between top management of the Goodman Group and top officials in the office of Customs & Excise in Dublin but the evidence clearly established that there was no such agreement and Mr McGuinness withdrew that portion of the allegation but insisted that the plan was conceived, was implemented to the extent described by him, namely, the substitution of meat obtained from Cahir for meat contained in the cartons which had been stored and that he was not aware of what transpired subsequent to this action.

All of the persons alleged by Mr McGuinness to have been involved in this operation and plan have strenuously denied such involvement and denied that they were in any way party thereto or involved in any discussions with Mr McGuinness about such a plan.

Mr Fairbairn stated in evidence that he first became aware of the problem at the AIBP plant in Waterford after the Christmas break at the end of December 86 or the early portion of January 1987.

He had a discussion with Gerry Thornton who informed him of the problem with regard to the discrepancy in the weights declared on the cartons but stated that there was no discussion about trimmings.

As the meat stored under the APS scheme was under the control of the International Division he travelled to Waterford to deal with the problem.

He stated that he did not discuss with Mr McGuinness the question of the substitution of cartons or the bringing of cartons of meat from Cahir.

He stated that he had no discussion at all with Mr McGuinness during this period as Mr McGuinness was employed by the Meat Division and had "no brief at all with our activity in the International Division".

He stated that he had no discussion, good, bad or indifferent with Mr McGuinness about the plan or any aspect of it.

Mr Thornton was equally vehement in his denial of any involvement in the alleged plan.

He stated that he did not instruct Mr McGuinness to prepare a Master Plan and went on to say that he would not have the knowledge of the essential elements that would have been necessary to complete such a plan, that he had only been in Waterford by helicopter on one day during this period, namely the 3rd day of February 1987 in the company of Israeli clients with whom he spent most of the day.

He stated that from inquiries made by him that no meat had been transferred from Cahir, and that he was not aware of any agreement or attempt to conclude an agreement with anybody at Head Office with regard to this plan.

Mr Michael Mullowney was the Operations Manager at the Waterford Cold store/Autozero, situated at Ferrybank in Waterford.

He outlined to the Tribunal the nature of the cold store which consisted of a 6 acre site, 17,500 feet of space, divided into 11 cold rooms, and that in the centre there was an area for distributing between the cold rooms.

He described the manner in which beef was stacked in the cold rooms, that there was a central aisle in each cold room and that the beef was stored on either side of the aisle in rows, being stacked on steel pallets that are 48" × 40" pallets.

He described the special trucks available for moving pallets which are what he described as "reach trucks", usually with tilting forks.

For the purpose of identifying product, tags were placed on pallets and a room plan kept of where the product is stored and that this room plan is necessary before a customer product can be located.

Dealing with the security of the premises, he stated that during a working day, the premises are occupied by the management and staff of Waterford Cold store and that at night time or outside working hours and weekends, a security company is engaged.

In this way, the cold store is protected 24 hours a day.

Entry to the cold store outside ordinary working hours would not be permitted by the security men on duty without prior authorization and that he, Mr Mullowney, was the person who was authorised to give such authorization.

With regard to location of product, he stated that it was the company's practice to "record the product in" and various information relating to it like contract number, lot number, serial number, quantity, any other information that the customer required or that the regulations required would be recorded in their stock.

He stated that if the Customs decided from the documentation to examine a portion of a consignment that the cold store would be able to move the appropriate product out into the examination hall.

With regard to the evidence of Mr McGuinness and the allegations with regard to the changing of meat in the cold store, he stated that he did not permit Mr McGuinness or any other person into the cold store at night, and at no stage did he approve, authorise, or become aware of any person coming into the premises at night and carrying out the operation as described by Mr McGuinness in his evidence.

He stated that for AIBP's representatives or employees to locate a specific product they would have to have the assistance of the staff in the cold store, but he did say that there would be limited information available to them to enable them to locate contracts but they wouldn't be able to locate specific products within the product.

This would appear to indicate that the representatives of AIBP would be aware generally of the location in the cold store of the beef, the subject of particular contracts.

He stated that he saw a copy of the plan before giving evidence at the Tribunal and stated that as a "stock location instrument is virtually useless".

Mr John Connolly, the Manager of the plant at Waterford denied that he was given a copy of the plan or that he was in any way aware of an attempt to deceive the Customs & Excise officers.

The weight and indeed vehemence of such denials presented the Tribunal with very considerable difficulty, in view of the very detailed evidence of Mr McGuinness.

It was because of this difficulty that the Tribunal decided to print in detail in this Report the plan which showed the APS Contract Number, the Pallet Number on which the beef was stored, the number of the Department of Agriculture seal on the relevant cartons and under the heading "Map", a rough outline of the position of each of the boxes on the pallets.

The APS numbers and the Department of Agriculture seal numbers are correct and it is impossible for the Tribunal to envisage any circumstances, other than those described by Mr McGuinness in which this record could have been compiled.

This record was prepared in the cold store by Mr McGuinness and consisted of a record of the substitution of meat which had been brought from Cahir for meat taken from cartons in the cold store, the numbers of the APS contracts to which they related, the Veterinary Seal number of each box and a rough map of the position of the boxes on the pallets.

The purpose of this exercise was to attempt to deceive the Customs and Excise authorities by in some way directing or persuading them to examine these boxes.

The plan did not succeed nor is there any evidence to suggest that it was sought to be implemented. But that does not affect the fact that the plan was conceived, that it was sought to be implemented by the substitution of meat in the cartons in the cold store, that the substituted cartons and their location in the cold store, were recorded by Mr McGuinness and copies of the record were given to Mr Fairbairn, Mr Thornton, Mr Connolly and the supervisor of the cold store as alleged by Mr McGuinness.

This document was a record of the movement of cartons of meat from the cold store and the replacement of the meat therein taken contemporaneously by Mr McGuinness. Its detail has contributed to the certainty that Mr McGuinness' account of the substitution of meat in the cold store at Waterford is true and accurate and his evidence in this regard is accepted by the Tribunal.

There is another fact that contributed to the certainty of the accuracy of Mr McGuinness' evidence.

He had stated that Mr Thornton was in Waterford at the time but had to return to Dundalk by helicopter.

Mr Thornton had stated in evidence that the only day during this period that he had travelled to Waterford by helicopter was the 3/2/1987.

It was on this day that Mr O hOdhrain of the Customs and Excise had withdrawn from the cold store a number of cartons of meat which he proposed to thaw out and examine the contents in detail.

This was the first indication given by Customs and Excise that they proposed to adopt this course as hitherto they had only been concerned with the over declaration of weights. In the course of this statement and evidence Mr Thornton stated that on the 3rd of February 1987 they "were not aware that the Customs and Excise officers intended to thaw out the product" and that:

> "While we knew that the Customs and Excise investigation was going to be a full investigation, we were not aware and could not have anticipated that they intended to actually thaw out the product. They had never done that before. Our understanding at the time was the investigation was limited to a reconciliation of the actual weights as against the weights submitted by the employees of Daltina".

While that may have been the understanding of AIBP prior to the 3rd day of February, 1987, they must have become aware, as a result of the action taken by Mr O hOdhrain on the 3rd February in withdrawing cartons from the cold store for the purpose of thawing out and examination, that the investigation was no longer and would not be confined to a "reconciliation of the actual weights as against the weights submitted by the employees of Daltina", but was being extended to the contents of the boxes.

It is inconceivable that the actions of the Customs and Excise officers and the implications thereof would not have been discussed between Mr Fairbairn and Mr Thornton and some action considered in regard thereto.

Such discussion lead to the formulation of the plan described by Mr McGuinness. It was a plan of action conceived in panic and subsequently found to be impractical.

The Tribunal accepts the statement, allowing for some exaggeration, made by Mr McGuinness on the ITV programme that

> "There was a massive panic within the company and a plan was put forward as to how the damage should be limited."

but it completely rejects his following statement that

> "The plan was basically agreed between our people and the Customs people at their Head Office."

This latter statement was without foundation.

It is true that Mr Larry Goodman wrote to Mr Pairceir, the Chairman of the Revenue Commissioners but that was not until the 18th March 1987 and after he had been to see Mr O'Mahony, the Secretary of the Department of Agriculture , on the 5th March 1987.

During the course of this meeting Mr Goodman informed Mr O'Mahony of the sub-contracting arrangements at Waterford and Ballymun and the problems with regard to inflated weights and inclusion of trimmings and made it quite clear that any monies found due to the Department of Agriculture would be paid.

He was obviously more concerned about possible damage to the reputation of his companies and asked that he should be given an opportunity to meet officials to see whether damage to his company abroad could be prevented.

He was informed by Mr O'Mahony that the Department of Agriculture was awaiting a report on the matter from the Customs and Excise authorities.

On the 18th March 1987 Mr Goodman wrote to Mr Pairceir as follows:—

"Dear Mr Pairceír

On the advice of CAP Division, and in the absence of Mr Sanfey and Mr Curran, and based on the urgency and delicate nature of the matter, I have taken the liberty of writing to you to request a meeting to discuss a major problem that has arisen in respect of Export Refunds under Reg. EEC 2730/79.

At the meeting we would like to discuss possible solutions to the problem, and to explain the circumstances under which it arose. The report of the customs investigation is with CAP Division.

An early resolution is imperative as the beef involved is urgently required to fill contracts including the Egyptian Government Protocol Contract. Also, apart from the potentially serious financial implication for our company, untold damage will be done long term to the good name and business relationships enjoyed by us with retailers and the major supermarkets of Europe.

Due to the delicate and most urgent nature of the matter we would respectfully request a meeting at 11.00 a.m. on Thursday, 19th March. We will contact your office to confirm this time or any time more convenient to you on that day.

Yours sincerely

LAURENCE J GOODMAN
Chairman & Chief Executive."

As a result of this letter a meeting was held on the 20/3/1987 in the office of Mr Pairceir, the Chairman of the Revenue Commissioners, at which were present Messrs Goodman, Fairbairn and Thornton (representing AIBP) and Mr Pairceir, Mr J Hallissey and Mr C Healy of the Revenue Commissioners.

This was not in any way a meeting designed to curtail the investigations of the Revenue Commissioners but to discuss ways in which they could be expedited because at that time the beef was required for fulfilment of contracts and the Tribunal is satisfied that there is no basis for the statement that there was ever any discussion with the Customs and Excise authorities with regard to the limitation of their investigation or agreement in regard thereto.

(*b*) *Falsification of IB4 forms by increasing the weights on and re-writing thereof.*

In this connection Mr McGuinness had stated that "the intervention scheme under which subsidies are paid by the EEC was abused by falsification of documents"

The practice in the Waterford Plant in regard thereto is set forth in detail in the evidence of Mr Ken Brennan and Miss Fionnuala Fenton and the only matters in dispute are:

(i) whether this practice was introduced by Mr McGuinness.

(ii) whether it was continued after his departure in October 1987 and

(iii) whether it was "institutionalised" throughout all the plants owned and operated by the Goodman Group of companies.

The purpose of this practice was to satisfy from documentation the Department of Agriculture that a yield of 68% was being achieved in the deboning operation and that the meat to which they were entitled was being stored in Intervention.

During the course of the hearings of the Tribunal, controversy developed as to the effect of the deboning contract and the entitlement of the Department of Agriculture thereunder and because of Mr McGuinness' evidence in regard to the question of the 68% it is appropriate that it be dealt with at this stage of the Report.

(*c*) 68%

In the Chapter dealing with AIBP Rathkeale, the Tribunal has referred to the deboning contract entered into between the Minister for Agriculture & Food and the contractor Anglo Irish Beef Processors Ltd and in particular to the condition which provides that:—

> "All meat derived from the deboning, trimming, packing and freezing operations is the property of the Minister other than all bones, fat and certain small trimmings ......which become the property of the Contractor."

The only reference to 68% is in the paragraph dealing with payment for deboning and is the percentage yield to be achieved to qualify for the agreed rate of remuneration. A lesser yield involved a penalty by way of reduced payment: a higher yield did not confer any benefit on the deboning contractor.

Mr McGuinness, in the course of his evidence to this Tribunal, stated that:—

"(1) based on his knowledge of boning operations during the course of his employment in Newry and Waterford that yields of 71% to 75% were achievable:

(2) the variation in the rates depended on:—

(*a*) the quality of the animal;

and;

(*b*) the experience and ability of the persons carrying out the deboning operation.

(3) when these yields in excess of 68% were achieved, steps were taken to ensure that only a yield of 68% or slightly in excess of this figure was returned to the Department of Agriculture ;

(4) it was explained to him, by the Manager of AIBP plant in Newry, Mr Nobby Quinn, that this could be achieved in either of two ways:—

(i) by increasing the weigh-in weights of the quarters of beef in the IB4 forms as the quarters were being weighed in the boning hall,

(ii) if the yield being achieved in the deb,oning process was in excess of 68% by the removal of meat and its transfer to the Company's own stock.

(5) at times it would be necessary to re-write the IB4 forms to show the target yield of 68%;

(6) though the specifications require that the quarters are deboned in such a way as to provide the eleven different cuts of meat set forth in the said specifications

particular cuts can be deboned to a tighter standard and pieces of such cuts can be removed from them and treated as "own stock";

(7) transfers to "own stock" were recorded on a weekly basis in the stock books of the Company at a standard price and is effectively recorded in the boning hall weekly accounts as revenue."

The Tribunal has dealt in detail with the practice of increasing weights on the IB4s and re-writing of IB4s in Waterford and the removal of intervention beef and its transfer to the Company's own store in the AIBP plant in Rathkeale.

This practice of transferring to the Company's own stock meat surplus to the yield of 68% was also prevalent in the AIBP plant in Cloghran/Ballymun.

Miss Imelda Murray, who was employed as a clerk in the AIBP plant at Ballymun/Cloghran, Co Dublin, gave evidence that she dealt with intervention records and documentation on behalf of her employers, AIBP, during the years 1987 and 1988.

Her immediate superior was the Boning Hall Manager, Mr Edward Burns. She stated that;

"she was given the IB4s for the purpose of totalling the amount of beef weighed in the boning hall, that once this weight was ascertained, she was in a position todetermine the number of cartons of beef that would be required to achieve the yield of 68% , that she informed the Boning Hall Manager of this number and that any additional cartons, which were not required to achieve the yield were transferred into the Company's own stock and particulars thereof were recorded on an internal yield sheet which was kept at Cloghran."

Though, cross-examined in detail, she was quite adamant that the boxes of meat, to which she referred, were not trimmings.

This practice undoubtedly provides the explanation for the incident which occurred in Cloghran on the 10th day of August 1987 and which was described in evidence by Mr Patrick Connolly, Veterinary Surgeon attached to the AIBP plant at Cloghran.

He stated that:—

"What happened was in checking the IB7 forms against the boxes we found that we had more boxes than we had reported on the IB7s and we reported the matter to the Intervention agency and we rejected the production"

and

"The IB4s and the IB7s matched up. The yield was not a problem. It was simply the extra boxes were left over."

It is a fair and reasonable inference to draw from these facts that the extra cartons were intended to be transferred to the Company's own stock but were counted by the Agricultural Officer before they were so transferred.

These incidents illustrate that the practices described by Mr McGuinness were followed in at least some of the plants owned by AIBP.

At this stage it is necessary for a proper understanding of this issue to consider the regulations in regard thereto.

The EEC Commission recognised that allowances would have to be made for the loss of meat in the deboning process due to the necessary adherence to the specifications annexed to the deboning contract.

Council Regulation 3492/90 sets forth the factors to be taken into account for the financing of intervention measures and provides for the establishment of tolerances for quantity losses arising from the preservation or processing of agricultural products in intervention.

Article 2. of Commission Regulation (EEC) No. 230/79 provides that:

> "The maximum tolerance referred to in Article 3 (2) (*c*) of Regulation (EEC) No. 2305/70 shall be 1%. It shall apply to the difference between the unwrapped weight of the product recorded when it is taken over and the wrapped weight recorded when it is removed from storage.
>
> This limit shall apply to boned meat produced by boning fore and hindquarters taken over, allowance being made for a weight loss of 32% as a result of the boning."

Article 2.2. of Commission Regulation (EEC) No. 147/91 of 22nd January, 1991 provides that:

> "the percentages for allowable losses during processing are hereby fixed as follows:—
> de-boning of beef — 32%"

It is clear from these Regulations that the weight loss allowed by the EC Commission for the purpose of deboning intervention beef is 32% and 0.6% in respect of storage.

The import of these Regulations is that the Intervention Agencies have to refund to the EEC Commission the value of any weight losses in excess of these tolerances viz 32% in respect of boning and 0.6% in respect of storage.

Consequently the Department of Agriculture , acting as the Intervention Agency, made provision in the deboning contract for the cost of weight losses in excess of the 32% to be passed on to the deboning contractor by providing for the deboning allowance to be reduced in cases where the yield fails to reach 68%.

The terms of this deboning contract came into effect from March 1985.

A short history of the minimum yield requirement is necessary for a proper understanding of the issue involved in this Report.

When Ireland joined the EEC Community, the relevant Regulation was EEC Regulation 221/72 and this provided for a weight-loss of 36% as a result of deboning, thus requiring a minimum yield of 64%. By virtue of the provision of Commission Regulation (EEC) 3180/74, the minimum yield required was increased to 68% with retroactive effect to the 14th of May 1973.

At the end of that year it was established that average yield achieved nationally in Ireland was 63.6%. The retroactive application of the maximum weight-loss tolerance resulted in the Irish Intervention Agency becoming liable to the EEC Commission in the sum of approximately £2 million pounds in respect of the short fall in the boneless yield requirement.

Representations were made to the Commission that the 68% minimum yield required was not achievable in Irish circumstances and the Department of Agriculture carried out controlled deboning trials. These revealed that at that time 64% was the highest yield which could be achieved from Irish cattle deboned in accordance with the Intervention specifications.

Following consultation with the EEC Commission it was agreed that the regulation could be interpreted as allowing the weight of the boxes to be included in the yield of boneless beef and that the 1% weight loss tolerance during storage provided for in the Regulation could be allowed at the deboning stage. This interpretation enabled the minimum yield of 68% to be achieved on paper and no money had to be refunded to the EEC Commission. This interpretation however was not justified on any reasonable consideration of the regulations involving, as it did, the transfer of the 1% storage allowance to the deboning allowance and the inclusion of the weight of the cartons as meat.

It is however a tribute to the negotiating skills of the Department of Agriculture officials who thereby avoided liability to refund to the Commission approximately £2m in respect of the shortfall in the boneless yield requirement.

The average boneless yields increased from an average of 64.2% in October 1975 to 65.47% in July 1976. In May 1981 the Department decided to increase the minimum yield requirement per hindquarters from 65.4% to 66%. The Meat Plants had considerable difficulty in meeting the new yield requirements. In the year May 1981 to April 1982 the average yields for all plants were 65.83%. In 1983 average yields increased to 66.3%.

In 1984 the EEC Commission not surprisingly reversed its previous decision to allow the weight of the carton and the 1% tolerance to be used to offset weight-loss resulting from deboning. The effect of this decision was that the effective minimum yield required was increased to 68%. The meat factories were unhappy with the new yield requirement of 68% and the accompanying clawback on the deboning allowance consequent on its implementation.

In view of the concern expressed by the meat industry the Department carried out further controlled deboning trials in April 1984 to determine the levels of yield achievable in the context of the existing intervention deboning specifications. The results were as follows:—

CLASS	YIELDS %
U3	69.02
U4	66.0
R3	67.2
R4	65.7
O3	65.1

The Department however insisted that the minimum yield requirement from April 1984 be 68%. They had no alternative but to require this yield in view of the relevant EEC Regulations. In the years following this increase, yields returned to the Department of Agriculture increased gradually from an average of 66.3% in 1983 to 68.14% in 1985.

Two factors appear to have contributed to this development namely:—

(1) On 1st of July 1984 a Commission Regulation required cod fat and top side fat be removed from the carcase prior to weighing over into intervention and this would have the effect of increasing the boneless yield by 0.8%.

(2) Class Steer 04, the fattest class at that time eligible for intervention, was excluded from intervention from 9th of April 1984 and the effect of this was to increase the average boneless yield. In the years following 1985 the average boneless yield increased marginally from 68.14% in 1985 to 68.5% in 1991.

These yields did not include the "bones, fat and certain small trimmings" which under the terms of the deboning contract became the property of the contractor.

Figures obtained from the EC Commission on yields returned in the Member States where deboning takes place, on 1991 were:

Denmark	70.00%
Italy	68.10%
U.K.	70.00%
France	70.60%
Ireland	68.71%

The eleven cuts which the deboning contractor was obliged to return to the Minister were as follows:— fillets, striploin, insides, outsides, knuckles, cube rolls, rumps, shin and shank, brisket, plate and flank and forequarter.

In addition to these particular and named cuts, it was further provided in the Schedule to the deboning contract that "identifiable pieces of meat — e.g., chain of fillet, chain of striploin, cap of knuckle and cap of cube roll, which are removed when preparing cuts should be packed with plate and flank as appropriate.

These identifiable pieces of meat which are removed when preparing the cuts required to be placed in intervention are distinguishable from the "fragments of muscle and fatty tissue and other tissue resulting from the cutting and deboning of meat" which may be retained by the Contractor.

By letter, dated the 11th May 1989, the Department of Agriculture wrote as follows to all plants involved in deboning intervention beef.

> "*Attempted Misappropriation of various pieces of Intervention Boneless Beef by certain meat plants.*
>
> Following a series of random surprise inspections of deboning operations during the latter part of March when intake of intervention beef was at its peak it became apparent that certain meat plants were attempting to misappropriate significant amounts of beef for their own use. These attempts generally involved the removal of chains of fillets and striploins as well as the rump tails. In the case of chains such pieces of meat should of course be packed with plate and flank and in the case of rump tails the meat should be left attached to the flank during the deboning process so that this cut does not arise as an individual piece. The attempted misappropriation usually involved a low key insidious removal of anything between 8 to 10 lbs of quality plate and flank meat from each carcase. It is obvious that if these practices went unchallenged that large amounts of beef could be taken at the Intervention Agency's expense.
>
> In this connection your attention is drawn to the terms of the Contract for Deboning of Intervention Beef. The Schedule, paragraph 4 xii concerning "Wrapping and Packing", specifies that "Identifiable pieces of meat e.g. chain of fillet, chain of striploin, cap of knuckle and cap of cube roll which are removed when preparing cuts shall be packed with plate and flank as appropriate. In paragraph 3 (*e*) of the contract it is specified that "*All meat* derived from the deboning, trimming, packing and freezing operations is the property of the Minister (other than the materials referred to in 3 (*d*) and, unless rejected by an authorised officer, shall be produced to the Minister on completion of each day's operations by the Contractor and any material failure to do so shall be regarded as a *fundamental breach* of the terms and conditions of the contract and the Schedule attached thereto". Paragraph 6 of the Contract spells out the options open to the Minister or an authorised officer if the terms of the Contract are not complied with in a satisfactory manner in accordance with the provisions laid down. Your particular attention is drawn to the provisions at (*c*) and (*d*) under this paragraph which allow among other things for rejection of all or part of the boxed beef produced under Contract and for the suspension of deboning of intervention beef without notice.
>
> Breaches of the carcase dressing specification have also been noted and in this regard particular attention is drawn to the need to have carcases presented without cod fat and without fat on the inside of topside. Failure to observe this requirement may be penalised as above.
>
> The purpose of this communication is to remind you that the contractual requirements concerning not alone the matters referred to but all others must be rigorously imposed and that failure to comply may meet with automatic imposition of the penalties provided for and in particular those mentioned above.
>
> A. McNamara
> Beef Intervention Agency."

A copy of this letter/circular was sent to each deboning plant and was forwarded to the Veterinary Officers in each meat factory together with a covering letter in the following terms:—

> "Department of Agriculture and Food, Agriculture House, Dublin 2.
>
> Vet-Officer in Charge
> Each Meat Factory
>
> Your attention is directed to a circular issued on 11 May 1989 to all plants involved in Deboning of Intervention Beef (copy to Vet Inspector in Charge) regarding the attempted misappropriation of various pieces of Intervention Boneless Beef by certain meat plants.
>
> As a follow up to the circular you are requested to ensure that all the terms of the Contract for Deboning of Intervention Beef outlined therein are complied with in accordance with the provisions laid down. You should also ensure that carcases are presented without cod fat and without fat on the inside of the topside.
>
> Supervision of the Boning Hall should include surprise random checks by veterinary staff at least once per day and random examination of boxes before and after freezing. The quantities before boning should be compared with the amount of boneless meat produced and the bones fat and trimmings. This could be done by selecting quarters and retaining bone, fat and trimmings.
>
> Surprise checks should also be made by Veterinary Inspectors as part of their ongoing duties. Senior management of the Veterinary Inspectorate will also carry out random examination from time to time.
>
> A. McNamara
> Beef Division
>
> 22 June 1989"

This circular dated the 11th May 1989 was sent to all plants involved in deboning intervention beef specifically drew the attention of the plants to the requirements of the Schedule to the deboning contract that "identifiable pieces of meat, e.g. chain of fillet, chain of striploin, cap of knuckle which are removed when preparing cuts shall be packed with plate and flank as appropriate" and to the provisions of Par. 3 (*e*) of the contract which provided that "All meat derived from the deboning, trimming, packing and freezing operations shall be the property of the Minister (other than the materials referred to in 3(*d*)).

As referred to earlier in this Report, Par. 3(*d*) of the Deboning Contract provides that:—

> "All bones, fat and certain *small trimmings* as defined in the Schedule attached to this Contract resulting from operations carried out under the terms and conditions of this Contract and the Schedule attached hereto shall become the property of the Contractor."

This Contract and the Schedule annexed thereto appears to distinguish between "identifiable pieces of meat" which are removed by the deboner when preparing the eleven different cuts hereinbefore referred to, which are the property of the Minister and required to be packed with the "plate and flank" and the certain small trimmings, which together with the bones and fat, which are the property of the Contractor.

It appears from the letter dated the 11th day of May 1989 and the evidence of Mr Ferris of the Department of Agriculture in regard thereto that:

> "it had become apparent that certain meat plants were attempting to misappropriate significant amounts of beef for their own use"
>
> and
>
> "the attempted misappropriation usually involved a low key insidious removal of anything between 8 to 10 kilos of quality plate and flank from each carcase".

In the letter from Mr McNamara of the Beef Division of the Department of Agriculture to the Veterinary Officers in charge of each meat factory, it was emphasised that:—

> "Supervision of the Boning Hall should include surprise random examination of boxes before and after freezing. The quantities before boning should be compared with the amount of boneless meat produced and the boxes of fat and trimmings."

The Goodman Group of companies engaged in the deboning of beef for intervention storage maintain that, by virtue of the provisions of the said Contract and the custom of the trade, their obligations under the contract are merely to trim the beef and produce the eleven cuts in accordance with specification and produce a yield of 68% of red meat from the quarters weighed into Intervention and that the balance of the meat is trimmings which they are entitled to retain, to transfer not into intervention storage, but to their own stock and to dispose of the same commercially.

Mr Gerry Thornton is the Deputy Chief Executive of the Meat Division of the Goodman Group of companies with considerable experience having joined AIBP in 1970.

Though a Chartered Accountant by profession, he had experience as manager of the abattoir in Bagenalstown and was manager of the Plant in Cahir from June 1983 to the end of 1985, when in the course of a restructuring of the organisation of the Group he was appointed Deputy Chief Executive of the Meat Division.

When dealing with the issue of the 68% yield he stated that:—

> "(i) When he was Manager at Bagenalstown and Cahir he had some involvement with the Department of Agriculture in connection with contracts for the deboning of intervention products.
>
> (ii) Prior to 1984, the yield which the Department required was 66% and the Department were happy to receive this yield provided that the cuts were to specification.
>
> (iii) In 1984 the yield required was increased to 68% and the Department was happy to accept this yield of meat, provided that the cuts were to specification.
>
> (iv) Since the advent of the Tribunal, the Department appeared to be taking a different approach to the application and construction of the Contract and that the Department is now emphasising that all the production of red meat, including trimmings, apart from certain small trimmings, are for the Account of the Department.
>
> (v) This approach has not been his experience.

(vi) In his experience over the years the practice of his company has been to return to the Department a yield as close as possible to 68%, while holding to and not breaking the specification.

(vii) To achieve that yield, the Group's production people at the various plants would monitor deboning operations during the course of each day and would, where necessary adjust the deboning to ensure that the Department get their minimum requirement, while at the same time ensuring that the cut specification is maintained.

(viii) Since the introduction of the deboning contract, the Group has consistently returned as close as possible to the stipulated yield.

(ix) The Department of Agriculture over the past decades have been aware of this and have accepted it without inquiry in the knowledge that "trims" represent a compensation to the processor for the unfavourable deboning allowance applicable in Ireland.

(x) The Department of Agriculture at all times received exactly what they are entitled to and what they expected."

The basic thrust of Mr McGuinness' evidence referred to herein is that yields in excess of 68% were achievable and when a yield in excess of 68% was achieved or likely to be achieved, corrective measures were taken to ensure that only a yield of 68% or one slightly in excess of that percentage was returned to the Department of Agriculture.

The evidence of Mr McGuinness in that regard was substantially confirmed by the evidence of Mr Thornton that in his experience the practice of the Goodman Group plants over the years had been to return to the Department a yield as close as possible to the 68% while holding to and not breaking the specification set out in the deboning contract and in order to achieve that yield the deboning operations would be monitored during the course of each day and where necessary the system of deboning would be adjusted to ensure that the Department got their minimum yield.

The Tribunal has heard evidence from a number of Plant Managers with regard to the deboning of Intervention Beef and has examined in detail a number of job costing sheets in relation to the daily production of intervention beef of a type similar to the one set forth in the chapter of this Report dealing with Rathkeale and a considerable number of the Weekly Job Costing Summaries sent in by each plant to the headquarters of the Meat Division in Ravensdale and their explanations in respect thereof.

Despite the detailed examination thereof and the detailed evidence in regard thereto, it is not necessary to refer to them in detail in the course of this Report because it is clear from the examination thereof that they show that they merely returned to the Minister for Agriculture and Food the minimum yield of red meat viz 68% or close thereto: that they retained as property of the Company the balance of meat obtained as a result of the trimming of the various eleven cuts in accordance with specification, and that they transferred same to their own Company Stock to be disposed of commercially.

Generally these costing sheets and weekly summaries provided in respect of intervention jobs showed the yield achieved as substantially more than 68% and that the increased yield achieved was transferred to the Company's stock and not into intervention, providing substantial corroboration of Mr McGuinness' evidence that yields of 71% to 75% were easily achievable.

The Tribunal in the Chapter of its Report dealing with Rathkeale reproduced a copy of the Weekly Job Costs Summary.

Similar type weekly cost summary sheets were used at this time in the AIBP plant in Cloghran/Ballymun, AIBP plant in Nenagh and the AIBP plant in Bagenalstown.

While these job costing sheets contain a column for showing the target yield in respect of Intervention jobs, Rathkeale appears to be the only plant which inserted a figure in such column.

The Weekly Job Cost Summary for the week ending the 21st April 1991 submitted by AIBP Dublin is reproduced hereunder:-

Weekly Job Cost Summary
Week Number: 16
Location: AIBP Dublin (PAD Room)
Week ending: 21/04/1991

Job No.	Date	Customer	No. of Cattle	Cut Type	Purchase	Transfer to stock	Fat & Trim	Bone	Boning Charge	Gross Margin	Labour	Packing	Freezing & Storage	Overhead	Allowance	Handling	Net Margin	Net Margin P/LB	Intake Weight Kgs	Target Yield %	Actual Yield %	Yield Gain/Loss	Price P/Lb
16101S001	22/04/91	Intervention Job	50.00	S001	0	882	84	-120	2287	3133	1043	362	405	700	125	0	498	0.01	17697	0.00	76.85	882	0.00
16102S001	24/4/91	Intervention Job	39.50	S001	0	706	47	-109	1679	2323	824	268	288	650	99	0	194	0.01	12993	0.00	74.53	706	0.0
16103S001	24/4/91	Intervention Job	36.50	S001	0	1116	42	-99	1649	2708	761	265	281	350	91	0	960	0.04	12763	0.00	78.24	1116	0.00
16104S001	25/4/91	Intervention Job	73.50	S001	0	1611	83	-188	3009	4515	1533	498	513	650	184	0	1137	0.02	23290	0.00	75.15	1611	0.00
16105S001	26/04/91	Intervention Job	72.50	S001	0	1705	123	-209	3627	5246	1512	556	621	650	181	0	1726	0.03	28070	0.00	75.66	1705	0.00
		Total for Week	272.00		0	6020	379	-725	12251	17925	5673	1949	2108	3000	680	0	4515		94813	0.00		6020	

While this summary does not specify the target yield, which was obviously 68% as all the jobs shown were Intervention jobs, it does show the actual yields as 76.83%, 74.33%, 76.24%, 75.15% and 75.66% respectively, the transfers to Company's stock as 882, 706, 1116, 1611 and 1705 respectively and the fat and trim as 84, 47, 42, 83 and 123.

A random inspection of the other job costing summaries establish that the yields achieved and shown were not unusual at this time at the Dublin/Ballymun plant of AIBP.

Five intervention jobs carried out in the week ending 21/4/1991 show yields of 76.85%, 74.33%, 76.24% 75.15% and 75.66% respectively: three intervention jobs carried out in the week ending 26/4/1991 show yields of 72.49%, 76.36% and 70.40% respectively: four intervention jobs carried out in the week ending the 16/6/91 showed yields of 74.73%, 72.67%, 72.92% and 72.94%.

The weekly job cost summary for the week ending 17/5/1991 forwarded from the AIBP plant at Nenagh show that three intervention jobs were carried out in the plant on the 17/5/1991 and the relevant extracts from such summary are set out overleaf.

Weekly Job Cost Summary
Week Number: 19
Location: Nenagh Co. Tipperary
Week-ending 17/5/91

Job No.	Date	Customer	No. of Cattle	Cut Type	Purchase	Transfer to stock	Fat & Trim	Bone	Boning Charge	Gross Margin	Labour	Packing	Freezing & Storage	Overhead	Allowance	Handling	Net Margin	Net Margin P/LD	Intake Weight Kgs	Target Yield %	Actual Yield %	Yield Gain/Loss	Price P/Lb
19101S001	17/05/91	Intervention Job	150.00	5001	0	4092	115	-303	6775	10679	2310	1223	1119	1500	0	0	4527	3.92	52432	0.00	76.09	4092	0
19102S001	17/5/91	Intervention Job	35.00	5001	0	2855	71	-158	3974	6742	1493	737	669	855	0	2	2988	4.41	30758	0.00	74.03	2355	0
19110S001	17/5/91	Intervention Job	121.50	5001	0	6184	99	-241	5563	11664	2378	1317	875	1215	0	14	5806	4.12	43052	0.00	76.59	6184	0

Again the target yield is not inserted but the yields achieved are shown as 76.09%, 75.03% and 76.39% respectively, the transfer to stock is shown as 4092, 2855 and 6184 respectively and the fat and trim as 115, 71 and 99 respectively.

A random inspection of the other weekly job cost summaries from Nenagh show that in respect of two intervention jobs carried out on 19/7/1991 the yields were 74.37% and 74.44%, in respect of one job on the 26/7/1991 the yield was 75.28% and in respect of five jobs on the 16/8/1991 the yields were 75.09%, 76.28%, 76.84%, 73.62% and 74.13% respectively.

Four intervention jobs carried out in the week ending the 24/2/1991 however show yields of 70.14%, 70.44%, 70.76% and 70.98%.

The Company's records are detailed and well kept.

The Company recognised that the eleven cuts specified in the deboning contract must be trimmed in accordance with the specifications and were the property of the Minister for Agriculture but maintained that if this were done and a yield of 68% achieved as a result thereof, the remainder were "trimmings" which they were entitled to retain.

It is clear from the evidence of the Plant Managers that in the deboning of beef for intervention that their target yield was, as stated by Mr Thornton, 68% and that the deboning was carefully monitored to ensure that that yield was achieved.

This was done in the first instance by the blending of carcases which are weighed into the Boning Hall.

This procedure was described by Mr Colin Duffy, Manager of the AIBP plant at Cahir who stated in evidence that:

> "(i) they endeavoured with the mix of cattle they had to acquire a yield in or around 68 to 69%;
>
> (ii) they were fairly specialised in this and they would simply calculate the weights of their intake for the following days or the day after from their carcase sheets;
>
> (iii) the production man would go and see the carcases and their grades and that it was his job to blend the carcases that went to the Boning Hall so as to obtain a good average grade of weight at the intake point;"

This practice of blending the carcases was also followed at the AIBP at Waterford and at the AIBP Plant at Nenagh because as stated by Mr Monaghan, Manager of the AIBP Plant, Nenagh "there are a number of grades of cattle that won't make the yield of 68% and comply with the specification. So, therefore you have to blend the cattle or carcases."

This practice depended on the variety of the cattle available for intervention and was not practicable in some plants because of the lack of variety in the cattle.

Having regard to the different grades of cattle eligible for intervention, it was a perfectly legitimate practice to blend the different grades to acquire an average yield.

All plants monitored the yield being obtained in the boning hall during the course of the boning operation to see whether the yield was being achieved and depending on the result of such monitoring the extent of the trimming of the cuts of meat was varied.

The purpose of this variation was either:

(i) to leave more meat on the cuts if the yield being achieved was less than 68%

or

(ii) to take more meat from the cuts if the yield being achieved was more than 68%.

Obviously the trimming of more meat from the cuts was to ensure that no more than the minimum yield was made available to the Minister and this involved taking more meat and fat from the cuts than is required by the specification.

It appears from the evidence that the trims are placed in trays and are then transferred to a table or trimming station within the boning hall where the trims are trimmed again to the specifications of customers of the plant. As stated in evidence by Mr George Mullan, Manager, Longford Plant, they would be further cut for the particular customers specification at the stage where the trims would come down the production line and would be transferred to a trimming station within the Boning hall where they would have personnel involved in Quality Control and preparation of trims.

There is no doubt but that these trimmings included identifiable pieces of meat which are removed when the cuts are being prepared and trimmed. They are variously described in the job costing sheets as LMCVP, chuck tender V.P., L.T.90 Vac Pac.

Having been trimmed to various customers' specifications or requirements many of the cuts and trimmings are vacuum packed before being transferred to the Company's stock.

Mr Patrick Connolly, the Veterinary Surgeon at Cloghran also described an incident which occurred on the 24th day of February 1988. He stated:—

> "On that occasion Mr John Mitchell A.O. was on duty in the boning hall and he found the tare on the weighing scales had been changed since he last checked it and as a result the weights entered in the IB4 were incorrect. In addition to that he found that they were packing away commercial cuts of beef out of intervention and packing them separately and as a result we decided that we would reject the day's production."

He stated that:—

> "We are not talking about trims." "This particular cut of meat was taken out of the forequarter" "They were known as LMCs leg of mutton cut."
>
> "These cuts of meat which were taken out were being packed in commercial boxes and this was being done in the boning hall."

It is denied by each of the Plant Managers that in the plants managed by them that there was any alteration of the weights on the IB4s produced from their plants or that there was any transfer of boxes of Intervention Beef to the Company's own stock. They allege that the boxes of meat which they transferred to Company stock consisted of boxes of trimmings to which they were entitled.

The Tribunal accepts their evidence that it was not the practice to increase the weights shown on the IB4's. Such a practice would, if carried out, have made it more difficult to achieve the yields, which they were obliged to return to the Department of Agriculture, and lessen the percentage of meat to be transferred to the Company's own stock.

In addition, the Defatting Analysis carried out by the Department of Agriculture which has been dealt with in this Report established that there was no excess fat included with the cartons of the specified cuts that they were obliged to produce in accordance with the terms of the deboning contract.

At all times, the Company and the Managers of the plants maintained that under the terms of this contract they were only obliged to debone the carcasses/quarters, prepare the cuts in accordance with the specifications and to achieve a yield of 68% or close thereto and that they were entitled to retain as trimmings any meat yield in excess of such 68%.

This claim made by the Goodman Group of companies is not accepted by the Department of Agriculture.

Mr John Ferris is the Senior Superintending Veterinary Inspector in the Department of Agriculture and has been involved in the operation of the beef intervention system at plant, regional and national level since its introduction in 1974.

He was a member of the Group involved in the preparation and drafting of the deboning contract and had discussions with the beef processing trade with regard to the terms thereof and particularly with regard to the technical aspects set forth in the Schedule.

In dealing with the question of the 68% yield and in reply to questions put to him by Counsel for the Goodman Group of companies, he stated that:—

"(i) Since he became involved back in 1974 it was always the understanding at every level, that everything above the minimum yield came to the Department.

(ii) Until it was raised during the hearings of the Tribunal, he was unaware of any claim to the contrary within the industry.

(iii) The Department policy was that if the deboning and trimming is carried out exactly in accordance with the specification, any of the excess meat that would arise or should arise, should go into intervention.

(iv) Neither he nor anybody else could for one moment suggest that yields should be minimised in order to improve the quality of the cut with the subsequent loss of that yield to the Intervention Agency."

He further stated that:—

"(v) In March, 1989, there had been concern within the Department that they might not be getting all the bits and pieces to which they were entitled and he decided that he would visit a number of plants on a completely unannounced basis.

(vi) He visited nine plants, some of which were working and some of which had completed their intervention boning.

(vii) Every plant that was engaged in intervention work was packing off pieces or trims which should go into intervention boxes, such as chains of fillet, chains of striploin, caps of cuberolls and caps of knuckles.

(viii) He reported such activity to head office and this report led to the letter dated the 11 May 1989 hereinbefore set forth."

The letter dated the 11th May 1989 to the boning plants and the letter dated the 22nd June 1989 to the Veterinary Officers in charge of each meat factory represented the view of the Department on these practices.

In his evidence Mr Ferris also

"(*a*) described "Chuck Tender" as a single piece of muscle taken out of the forequarter and lying at the front end of the scapula which is contained in the forequarter, looking very like a fillet and incapable of being described as trimmings.

(*b*) described the "Leg of Mutton" cut as being part of the forequarter, not usually isolated individually and is a very large piece of meat.

(*c*) stated that both the "Chuck Tender" and "Leg of Mutton" cut should be included in the intervention cartons."

In the course of his evidence Mr Ferris dealt with supervised tests carried out by him to establish the percentage of the weight of a carcass/quarter which would be regarded as trimmings available after deboning in accordance with specifications and these varied between 4.1% and 5.3%, an average in the region of 4.5%.

Mr Richard Healy, an Assistant Principal Officer in the Department of Agriculture stated in evidence that the focus of the Regulations is that the Minister gets all of the meat to which he is entitled and if the yield falls below 68% there is a claw back applied to the deboning allowance in order to recoup the liabilities the Minister would have to the EEC Commission.

The Tribunal has received from the Department of Agriculture and considered the details thereon, a table showing the average boneless yields returned by all plants who carried out intervention deboning during the years 1983 to 1991.

This table, without going into the details thereof clearly show that in the years 1987 to 1991, 92.66% returned a yield of 68% or slightly in excess thereof, 4.66% a yield of 69% or slightly in excess thereof, 2% a yield of 67% or slightly in excess thereof and .67%, a yield of 70%.

This clearly establishes that during the entire of this period the returns from all plants from all companies engaged in the trade showed a yield of approximately 68% or slightly in excess of this percentage.

In the course of his cross-examination by Counsel for the Goodman Group of companies with regard to these tables Mr S Fogarty, a Principal Officer in the Department of Agriculture, agreed that during this period that there had never been any controversy between the Department and any processor for failing to return enough meat once a yield of 68% had been achieved.

It appears from the evidence of Mr McInerney of the Finance Division of the Department of Agriculture and Food that the Commission of the European Communities in calculating the losses resulting from deboning restricted weight losses to 32% and in their Aide Memoir dealing with the 1990 Financial Year stated:

> "In accordance with the second paragraph of Article 2 of Regulation (EEC) No. 230/79 weight losses due to boning operations are restricted to 32% of the carcase weight. It follows that the production yield must reach *at least* 68% of the quantities treated."

It is clear from this that the view of the EC Commission that 68% was a minimum requirement and not just the yield which should be returned or which they expected to receive.

In spite of this, Mr Buckley an Agricultural Officer employed at the AIBP Plant in Rathkeale stated "once we got our 68% everybody was happy."

Mr Vaughan, an Agricultural Officer employed at the AIBP plant in Rathkeale and who had previously worked in Nenagh, stated in the course of his evidence that:—

> "I wouldn't be aware of what percentage of trimmings would be left over after a day's production. Once I got my 69% or from 68% to 69% I was happy: we were satisfied. So if they are 3 or 4 per cent over, it was none of my business."

It is clear from the foregoing that:

(i) No queries were raised by the Department of Agriculture once the documentation, returned to it in pursuance of the regulations showed a yield of 68%.

(ii) The officials in the plants were satisfied once a yield of 68% was achieved and did not ascertain or in any way query the extent of the trimmings obtained as a result of the deboning.

The Goodman Group of companies maintain that this attitude by the Department was consistent with their interpretation of the deboning contract and amounts to acquiescence in the practice followed by them in all plants of deboning and trimming for the purpose of obtaining a minimum yield of 68% and their entitlement to anything in excess thereof. They maintain this attitude in spite of the terms of the letter dated the 11th day of May 1989 sent to all meat plants.

The terms of the deboning contract were criticised by Counsel on behalf of the Goodman Group of companies as being ambiguous and bureaucratic and non-commercially viable.

Dr. Cento Veljanovski was a witness made available to the Tribunal by the Goodman Group of companies.

He was a very well qualified witness who had a consulting practice in Economic and Regulatory Analysis, was a Senior Research Fellow of Law and Economics at the Institute of Economic Affairs in London and also advised on Economic reform to the Republic of Macedonia, Yugoslavia.

He stated in evidence:—

"(i) that he had carried out an evaluation of the Irish Intervention Deboning Contract and that from his review of the deboning contract itself and the submissions to this Tribunal from the Department of Agriculture , it would appear that there is a divergence in the views of how deboning contracts should be interpreted.

(ii) that the Intervention Agency, the Department of Agriculture , claimed, under the contract, that it retained ownership of all meat derived from the carcase and that the objective of the contract was to maximise the value of deboned beef.

(iii) that an alternative interpretation of the contract is that the Agency only seeks to achieve a yield of 68% and permits the deboning contractor to legitimately substitute a substantial proportion of fat for small meat, trimmings.

Inter alia, he stated that:—

"1. The deboners' remuneration is based on two components, a rate per kilo of deboned beef with severe penalties for not achieving a 68% yield, and the value of the trimmings whose ownership is transferred to the deboning contractor.

2. This structure of payments is clearly designed to provide powerful incentives to achieve a 68% delivered yield. There are several penalties for shortfalls below 68%, such that at a 64% yield the deboning contractor receives no payment.

3. There are no incentives to deliver in excess of 68%, even though this involves progressively greater effort and expense on the deboner's part.

4. The transfer of ownership of "certain small trimmings" provides the bonus necessary to cut trimmings in excess of 68% by conveying the value in excess of this residual to the deboning contractor.

He further stated that:

"This system of fee per kilo, penalty for shortfalls and a bonus of certain small trimmings, after satisfying the quality specifications in the deboning contract, displays the characteristics of an economically efficient contract designed to maximise, not the amount of delivered meat, but the attainment of the 68% yield of a quality consistent with the boning contract."

That there was no clear definition in the contract of "certain small trimmings."

The contract permits substitution of fat for meat in the definition of acceptable delivered boxed beef.

The past actions of the Agency show that they were not concerned with maximising delivered meat.

The commercial value of the contract depends on the fee and the value of trimmings.

A deboner tendering for a contract would take the value of the trimmings and the boning fee into account. If the Intervention Agency wanted to maximise yield, it would have instituted a system of bonus payments for a yield in excess of 68%."

Having completed his examination of the contract, he concluded that:

"(*a*) The contract is structured to provide a strong incentive for yields of 68%.

(*b*) It does not provide incentives for yields over 68%.

(*c*) It does not suggest that the Agency is concerned to achieve a yield of greater than 68%.

(*d*) Nowhere in the contract or the past actions of the Intervention Agency is there any indication that it sought to maximise the yield from deboned beef.

(*e*) The contractual terms permit a significant amount of fat to be provided under the contract and for the ownership of trimmings to be transferred to the deboner.

(*f*) 68% is the norm across Europe and the European experience is that the Intervention Agencies only want 68%.

(*g*) There is a reasonable expectation, a reasonable reliance on the practice of the Agencies only demanding 68%."

Basically the issue, as stated by Dr. Veljanovski and outlined in this Report, is whether the Minister for Agriculture and Food is entitled under the deboning contract to retain ownership of all meat derived from the carcases/quarters weighed into the boning hall with the exception of bone, fat and certain small trimmings or whether the Minister is only entitled to 68% of such meat.

In spite of the criticisms made of the deboning contract and the drafting thereof it is perfectly clear on the salient terms thereof viz "that all meat derived from the deboning, trimming, packing and freezing operations is the property of the Minister".

The only limitation on that statement is that contained in Para. 3(d) of the Contract, which as already stated provided that:—

"All bones, fat and certain small trimmings as defined in the Schedule attached to this Contract resulting from operations carried out under the terms and conditions of this Contract and the Schedule attached thereto shall become the property of the Contractor."

While "certain small trimmings" are not defined in the Schedule, the Schedule sets out in detail the manner in which the quarters are to be cut and trimmed and the trimmings which are to be removed.

There is no doubt but that the Contractor is entitled to small trimmings remaining after the deboning provided such deboning and cutting is carried out in accordance with specifications but such trimmings cannot include extra trimmings or cuts of meat obtained as a result of trimming done with the specific purpose of limiting to 68% the yield to the Minister.

Under the contract, the Minister is entitled to the entire of the yield and trimming designed to limit that yield is not in accordance with the clear and unambiguous terms of the deboning contract and any meat derived as a result of such trimming was and remained the property of the Minister and should not have been transferred to the Company's own stock. The action of the plants in so doing was in compliance with the policy of AIBP as enunciated by Mr Gerry Thornton in his evidence but in breach of the terms of the Contract and EEC Regulations.

There is no evidence that the Department of Agriculture were aware of or acquiesced in this practice; in fact Mr Ferris stated that the first he ever heard of it or the claim of the companies to be entitled to anything in excess of 68% was during the course of the hearings before this Tribunal. The documentation returned to the Department viz the IB4s, the IB7s and the IB6s would not disclose this practice.

The Department of Agriculture is responsible for the supervision of the deboning operation.

Article 20 of Commission Regulation EEC 859/89 provides that:—

> "1. Intervention Agencies shall be responsible for the supervision of operations referred to in Article 19.
>
> Supervision must include either permanent physical control or an unannounced inspection of the boning operation not less than once per day and random examinations of the cartons of cuts before and after freezing in such a way that a comparison of the quantities before boning with the quantities boned on one hand and the bones, fat and trimmings on the other hand."

There is no evidence from the Department of Agriculture of a random examination of the cartons of cuts *before* and *after* freezing in such a way that would enable a comparison to be made between the quantities before boning with the quantities boned on the one hand and the bones, fat and trimmings on the other hand. This, in spite of the contents of the letter sent by Mr McNamara of the Beef Division to the Veterinary Office in charge of each Meat Factory in which he stated that:

The comparison of the IB4 and the IB7 to produce the IB6 does not appear to comply with this requirement as the IB7 is produced before freezing and the bones, fat and trimmings do not appear to be weighed in the presence of a representative of the Department of Agriculture.

It appears from the evidence of the Plant Managers that the trims are transferred to a trimming table in the boning hall and openly dealt with there.

There is no evidence that the trimmings were ever weighed by any Agricultural Officer or that there was any examination or inspection of any kind of the trimmings and pieces of meat that were being transferred to the Company's own stock.

There is no doubt whatsoever but that:—

(i) it was the deliberate practice and policy of the management of AIBP in the State engaged in Intervention operations including deboning operations in accordance with the Deboning Contracts entered into between the Minister for Agriculture and Food and AIBP to debone quarters of beef taken and weighed into Intervention in such a manner as to ensure that the yield of the meat placed in Intervention storage did not substantially exceed 68% of the weight of the quarter.

(ii) the balance of the meat remaining after such yield had been achieved was regarded as the property of the company and dealt with accordingly by them by way of further trimming vac packing or other compliance with commercial customer requirements.

(iii) the practice and policy was in clear breach of the said deboning contract and whatever misapprehension they might or might not have had about their entitlements thereunder that misapprehension was clearly dissipated by the terms of the letter dated the 11th day of May 1989 from Mr McNamara of the Beef Division to all plants involved in deboning intervention beef.

(iv) In spite of the terms of this letter the practice and policy continued and was sought to be justified before this Tribunal on the basis of the terms of the contract and on the alleged acquiescence of the Department of Agriculture officials in the practice.

Consequently it appears that the obligation imposed on the Department of Agriculture and Food to be responsible for the supervision of the deboning operation was purported to have been fulfilled by "permanent physical control".

According to Mr Ferris, this was exercised by one man and it is clearly not possible with one man in the boning hall to cover all the requirements.

There can be no doubt about this because the regulations place an onus on the Department to check virtually everything, the weighing in of the quarters, the weighing of the cartons, supervision between these two areas to ensure that the specifications were being kept and that there was no slippage and in Mr Ferris' view would require a minimum of three Agricultural Officers rather than the one supplied.

He also considered it necessary that there should be cover for lunch breaks, tea breaks and other occasions.

(*d*) Switching of meat going into or in intervention storage and the substitution therefor of inferior product.

Mr McGuinness had alleged that "the intervention scheme... was abused .... by the switching of meat going into or in intervention storage and the substitution therefor of inferior product"

The Tribunal has dealt in detail with this practice at the AIBP plant in Rathkeale in the course of the chapter entitled "Intervention Rathkeale".

In the course of his evidence Mr McGuinness stated that the Goodman Group engaged in the switching of beef presented for storage under the APS scheme.

He stated that he was aware that non-steer carcases were regularly substituted for steer carcases into the APS scheme.

In support of this contention he produced a document which contained a hand written calculation of weight differentials for a number of supply contracts being handled by the Waterford Plant in September 1986 and October 1986.

This document had been prepared at Mr McGuinness' request by a clerk in the AIBP offices at Waterford, a Paul Shevlin, and showed differentials in the cold weight and the intake weight in a number of carcases and is hereinafter referred to as the "Shevlin document".

In some cases the intake weight exceeds the dead cold weight of the carcases and is indicative of an overstatement of the intake weight and in the cases where the intake weight is considerably less, this indicates a switching of carcases.

During the proceedings before the Tribunal, there was considerable controversy over the effect of the records contained in this document and the statement of Mr McGuinness that "there is no innocent explanation for a net weight gain".

It appeared from the evidence that the weight of an animal is recorded immediately after slaughter and this weight is known as the "Hot weight".

The farmer is not paid for his animal on the basis of this weight but on the basis of what is known as "the cold weight" which is by agreement artificially calculated as 98% of the "Hot weight" i.e. 2% less than the Hot Weight.

Depending on the time spent in the chill and other factors including moisture content there may not always be a difference of 2% between the actual cold weight and the hot weight of an actual carcase and the weight loss may be as low as 1% or 1.2%.

Trimming and removal of fat may further increase the difference between the intake weight and the "Hot Weight" and the artificially calculated "Cold Weight" and according to the evidence of Mr Maurice Mullen of the Department of Agriculture a weight loss of 3. 5% would be well within the parameters.

The in-take weight of a carcase must always be less than the "hot weight" of the same carcase because chilling and trimming reduces the weight of the carcase.

This takes place before the carcases are weighed into intervention or for storage under the APS schemes. Any document which shows an increase over the hot weight must be incorrect.

Any excessive loss of weight, shown on the relevant documentation of a carcase as between the Hot Weight and the intake weight, can only be explained either by:

(i) the incorrect recording of the number of the carcase on the relevant documentation or

(ii) the fact that there has been a switching of the carcase after weighing to ascertain the hot weight and its being weighed in for Intervention or APS purposes.

In respect of the 1986 APS scheme AIBP had ten contracts of storage involving 792 tons of beef and the documentation in respect of these contracts was examined and analysed by officials of the Department of Agriculture under the supervision of Mr Mullen at the request of the Tribunal for the purpose of comparing the Hot Weight and the weights recorded on AIBP documentation.

A summary of such findings is set forth hereunder.

Contract No. *2651/BM*	Hot Weight (Kgs.)	Cold Weight (Kgs.)	% Difference
4717	58229	56868.8	–2.35%
4861	49240	46566.5	–5.43%
4885	102862.5	99649.5	–3.12%
4990	100076.5	95290	–4.78%
5145	97985	94361.5	–3.7%
5146	93454.5	89631	–4.09%
5380	88574	85697	–3.25%
5382	52929	51930	–1.89%
5386	88340.5	85712	–2.98%

While the overall picture as disclosed above was satisfactory and within the accepted parameters, the comparison and analysis of the documentation disclosed a wide range of discrepancies with wide variations up and downwards for individual sides and quarters.

In addition there were a number of instances where the same carcase number listed from the Daily Classification Sheet accounted for the equivalent of more than two sides in the APS 1 and APS 4 forms and the Intervention IB1 and IB4 forms and specific instances were identified where carcases from female animals were included on APS forms in respect of male contracts.

In addition the recorded intake weight of some carcases was so much below the recorded hot weights of the same carcases that the difference could not be explained by the normal wastage due to the chilling and trimming process after the carcases had been weighed and indicated the switching of carcases if the numbers of the carcases shown on the relevant forms were correct.

It is conceded that at the beginning of the operation of the APS scheme in 1986, the system of numbering the carcases was rather chaotic and led to confusion and that it was some weeks before a more satisfactory system was introduced.

The system as first introduced was that the carcases would be numbered in sequence from a Monday to Friday and on the following Monday the sequence would start again.

As the scheme was a ten day one viz there could be ten days between the date of slaughter and the placing in storage, there was a risk or danger of two carcases having the same serial number and this undoubtedly led to some confusion.

The system of numbering the carcases was related to a fortnightly sequence and this removed the possibility of two carcases having the same serial number.

It appears from the analysis referred to and the evidence of Mr Mullen that the number of discrepancies in the documentation lessened thereafter.

The fact that the "Shevlin Document" and the analysis of the documentation dealt with by Mr Mullen disclosed evidence of a practice of over-declaration of weights on the APS and IB4 forms was strongly contested by Counsel for the Goodman Group and the evidence adduced in regard thereto.

While the Tribunal has considered this evidence in detail, it does not consider it necessary to set it out in the course of this Report because the "Shevlin document" was produced in the first instance at Mr McGuinness' request in order to enable him to obtain some indication of the extent of the over-declaration of weights which could be ascertained by a comparison of the weights on the relevant documentation if such a comparison was carried out by Department of Agriculture officials, and was then produced before this Tribunal to provide corroboration of his evidence before this Tribunal with regard to the over-declaration of weights and the switching of carcases.

His evidence with regard to the over-declaration of the weights of carcasses on admission to the Boning Hall at Waterford in respect of APS and Intervention contracts has been substantiated. In regard to the APS contracts in Waterford by the evidence of the Customs and Excise officials and Department of Agriculture officials more particularly dealt with in the chapter of this Report dealing with the Waterford/Ballymun 1986 APS investigation and in regard to the over-declaration of weights on the IB4 forms in the intervention process by the evidence of Mr Ken Brennan and Miss Fionnuala Fenton already referred to and by the comparison of the weights shown on the IB4 forms retained by Mr McGuinness and those submitted by the AIBP company to the Department of Agriculture for payments.

Consequently there is no need for the Tribunal to rely on the Shevlin Document to provide corroboration for Mr McGuinness' evidence in this regard.

The "Shevlin Document" and the analysis of the documentation carried out by Mr Mullen is inconclusive on the question whether the weights recorded therein provide corroboration of his evidence with regard to the switching of carcases by the substitution of poorer quality and lighter carcases for other carcases.

The discrepancies in weight shown in respect of some of the carcases are indicative either of such switching or incorrect recording of the numbers of the carcases being compared.

It is conceded by the Department of Agriculture that the system of recording could have led to confusion in this respect and that the incorrect numbering of carcases could provide an explanation of the discrepancies both in regard to the weight difference and the number of quarters shown per carcase.

Consequently the Tribunal is satisfied that the "Shevlin Document" and the evidence of Mr Mullen on this issue do not of themselves provide corroboration of Mr McGuinness' evidence with regard to the switching of carcases for Intervention purposes but neither are they inconsistent therewith.

However the Tribunal accepts Mr McGuinness' evidence with regard to this.

In this regard he stated that:—

> "what I saw was the process of cutting off the original stamp, the replacement putting on a new stamp, and also you have to go a step further and you have to take off the killing docket details because the kill docket details, they were applied to the fore-quarter and hind quarter and they show the kill number and the grade and they have to be taken off as well and effectively a new one has to be put on."

It appears from his evidence that Mr McGuinness witnessed the process of switching carcases subsequent to slaughtering and grading and prior to being weighed in at the boning hall, that the process involved the removal of the original grading mark on the carcase and the substitution of a higher grading mark and the removal of the label attached to the forequarters and hindquarters containing the kill number and the grade and the substitution therefore of a new label.

This process is similar in practically every detail to the practice described by Mr Lynch as being followed in the AIBP Plant in Rathkeale.

In addition it appears from the evidence of Mr Hughes, who was the Veterinary Inspector in charge of AIBP plant at Bagenalstown, Co. Carlow that between May and June 1984 that 149 ineligible carcases were deboned into intervention.

Mr Colm O'Loughlin was at that time Manager of this plant and Mr Hughes stated in evidence that he kept Mr O'Loughlin informed of the discoveries but "ineligible carcases found their way into intervention until the first of June that year"

The fact of the inclusion of such ineligible carcases was reported to Head Office of the Department of Agriculture, and at a meeting held in Agriculture House in October 1984, Mr O'Loughlin informed the Department officials that there had been a strike at the plant in February / March of that year, that one or two of the employees were disappointed with the result thereof and alleged that they, in an attempt to destroy the smooth running of the plant or to discredit Mr O'Loughlin, had decided to put the ineligible carcases into intervention.

Evidently the local Garda Sergeant had warned Mr O'Loughlin that these two problem employees were likely to do something within the plant that might upset the smooth running of the plant.

The Department officials noted the explanation and AIBP were penalised to the extent of £26,700 being the value of such carcases.

In the AIBP plant at Cloghran, John Mitchell, an Agricultural Officer, discovered and rejected four ineligible beef hindquarters being weighted into the Boning Hall on the 2nd December 1987 and Pat Rooney an Agricultural Officer, discovered and rejected eight ineligible hindquarters in the Boning Hall there on the 16th March 1990.

Viewed on their own each of these incidents may not seem important but taken collectively and involving as they do there different plants viz Rathkeale, Bagenalstown and Cloghran/Ballymun, they provided some corroboration of the evidence of Mr McGuinness with regard to incidents in Waterford.

(*e*) Abuse of Export Refund subsidy system by failing to comply with the contractual requirements of Middle East customers, the unauthorised use of Islamic stamps and the reboxing of meat for the purpose of misleading customers and that in connection with Iraqi contracts the inclusion of meat from many sources, fresh meat, frozen meat, intervention beef etc.

The use of intervention beef purchased from the Intervention Authorities for export to Third Countries, the repackaging and/or reboxing of such meat and the removal of markings therefore do not in any way constitute an abuse of the Export Refund subsidy system.

The Department of Agriculture , as the Intervention Authority, was at all times aware of the fact that the Intervention beef purchased from it was destined for export to Iraq, authorised the repackaging or reboxing of the meat and authorised the payment of the appropriate rate of export refunds.

In the ITV programme the presenter thereof had alleged that boxes of old frozen meat from European stores were brought by the truck load to the Goodman owned Ulster Cold stores, Craigavon Northern Ireland. Then a transformation took place. For a solid eighteen months old frozen meat was turned into new.

A former employee at this plant Mr Thomas Ruddy stated that "all of it was reboxed as killed within the last week or two". Despite it's best efforts the Tribunal was unable to secure the attendance before it of Mr Ruddy, who was resident outside the jurisdiction and not answerable to a sub poena.

Though at this stage of its Report, the Tribunal is dealing with the allegations made on the ITV programme, particularly by Mr McGuinness, the Tribunal considers it desirable, in the interests of avoiding repetition to refer to the allegations made in Dail Eireann by Deputy MacGiolla on the 9th day of March, 1989 and by Deputy Pat Rabbitte on the 15th and 24th day of May 1991.

These allegations can be summarised as allegations that the Goodman Group of companies were abusing the system under which subsidies were paid by having the labels on meat changed in different parts of the country by a team moving about to do this job on behalf of Goodman Companies: by the maintenance of an entire production line in Nenagh designed for taking stamps from frozen carcases and re-stamping and re-labelling them and by carrying out repackaging and restamping operations in Goodman plants in operations heavily subsidised by the Irish Taxpayer thereby putting Ireland's reputation for quality at risk.

In the course of that portion of the ITV programme dealing with exports of beef to Iraq, Mr McGuinness had stated :—

> "There was many different sources for the meat, there was fresh Irish meat, there would have been intervention meat, there would have been frozen meat, it may have been Halal slaughtered, it may not have been Halal slaughtered, it could have been cow, it could have been bull, it could have been anything,"
>
> and
>
> "the whole system was that you switch product to show what the customer wanted. If that meant re-boxing, you re-boxed the meat, to show what the company thought he was getting. "

In the course of his evidence, before this Tribunal, Mr McGuinness stated:—

"(1) That he was aware of the fact that re-boxing had been taking place at the Cold Store, Craigavon, from discussions with the plant accountant, the members of what he describes as the "A Team" and from his own observations, on about two occasions in 1986, when he saw the re-boxing line in operation;

(2) Intervention meat was re-boxed for Iraqi contracts on a 24 hour basis and several shifts were employed during the day;

(3) The intervention meat would be taken out of the box, the original box discarded, a new box would be prepared and the meat would be placed in that box;

(4) The box would be re-strapped, re-palleted and returned to the cold store;

(5) The new boxes were Goodman International type boxes."

In a follow-up programme broadcast by ITV on the 22nd day of July 1991 Mr Brendan Solan, a former worker at the AIBP plant in Cahir, was interviewed for the purpose of confirming that the practices outlined in the original programme were carried out in the plant, in which he was employed, namely Cahir.

On the said programme he stated that re-stamping of meat did occur at Cahir and that:—

> "for about at least two weeks there were guys who came down from headquarters in Dundalk, and they would come into the factory with their own stamps, knives - everything else, and there were sides of meat brought up to the Loading Bays. They would take them off, start cutting off whatever markings were on them and putting their own marking back on the sides of meat and this went on for about two weeks".

He stated that unfit meat was repackaged as new and that about 5% of it was in an absolutely appalling condition.

In the course of his evidence before the Tribunal, Mr Solan, who was employed in the AIBP Plant at Cahir from May to October 1988 described in detail the reboxing of beef at this plant during this period. The Tribunal is however satisfied from the evidence adduced before it that the reboxing of the beef referred to by Mr Solan was the reboxing of intervention beef for the purpose of export, was duly authorised in accordance with the relevant regulations and notice of such re-boxing had been given to the Department of Agriculture.

A considerable quantity of evidence was adduced before the Tribunal, and a considerable amount of the time of the Tribunal was taken up by the Tribunal in hearing such evidence with regard to the reboxing of meat taken from Intervention, the relabelling of cartons into which it was placed and the removal of stamps from such meat in different plants not only during the normal working hours of such plants but late at night and over week-ends, not only by the usual operatives in such plants but by a team of operatives from outside such plants, with the inherent implication that such reboxing, such relabelling and removal of stamps was illegal and contrary to regulation.

The reboxing of meat purchased from intervention is authorised under certain circumstances by the provisions of EEC Regulation 2824/85.

This regulation lays down detailed rules for the sale of certain frozen boned beef which is held by the intervention agencies of the member states and which is to be exported either in the same state or after cutting and or repacking.
The regulation provides that the trader must state, in the purchase application or tender, whether the meat will be exported in the same state or after cutting and or re-packing.

The regulation further provides that:

(i) cutting and/or re-packing should take place only with the authorisation of the competent authorities who may only authorise the cutting and/or repacking of meat if it is stored on their territory and the cutting and/or repacking is carried out there.

(ii) where meat is cut and/or repacked, the bags, cartons and other packaging material containing it shall bear particulars enabling it to be identified, including the net weight and the type and number of the cuts and may not be mixed with the other meat sold and must be in a frozen state when it is cut and repacked."

The regulation opening the sale may specify products not eligible for refunds.
It is clear from these regulations that in the circumstances outlined therein the cutting and re-packing of beef purchased from Intervention is lawful when authorised by the Intervention Authority.

Re-labelling is not only authorised but necessary because the regulations required that when meat is cut or repacked, the bags, cartons and other packaging material containing it shall bear particulars enabling it to be identified, including the net weight and the type and number of cuts.

It clearly emerged from the cross-examination of the various witnesses by Counsel for the Goodman Group of companies that it was their contention that such re-boxing, relabelling and removal of stamps or marks was authorised and legal and supervised.

During the course of its hearings dealing with Export Credit Insurance, it learned of allegations that intervention beef was used in the fulfilment of contracts for the export of beef to Iraq in respect of which Export Credit Insurance was granted.

To determine the truth or otherwise of this allegation and because of other matters in relation to the 'national interest', the Tribunal on the 22nd day of May, 1992 wrote to the Solicitor acting for the Department of Agriculture and Food seeking the following information:

"(i) The amount of stock sold into Intervention by each meat trader and the date when it was sold into Intervention

(ii) The amount of stock bought by each trader out of Intervention and the date it was bought out

(iii) A complete breakdown from the documentation of the origin of all meat exported by meat processors and the place to where it was exported during the year 1987 and 1988.

Following an exchange of correspondence between the Tribunal and the Department of Agriculture , information with regard to exports to Iraq during the years 1987 and 1988 was obtained on the 14th day of August 1992.

This information disclosed that in respect of exports to Iraq during this period by AIBP upon which export refund subsidies were paid viz 28,996.67012 tonnes (84%) was meat purchased from Intervention by AIBP, 1,877.38960 tonnes (6.5%) had been stored under APS Schemes and 2,727.44630 tonnes from other sources.

At this stage, the Tribunal had heard evidence of the findings of the Fisher Report, which is dealt with in the chapters on Export Credit Insurance, which established that of the tonnage exported to Iraq by AIBP during the years 1987 and 1988 declared for and subject to Export Credit Insurance 38% was sourced outside the jurisdiction of the Irish Republic in Northern Ireland and the United Kingdom.

Mr James Fairbairn was the Senior Manager in the International Division of the Goodman Group of companies which division had responsibility for exports and gave evidence:—

"(i) that the Meat Division was responsible for the purchase and processing of beef required for export and internal consumption.

(ii) that if the product was intended for export outside Europe, that the International Division was responsible for any reboxing or re-labelling required and deboning on rare occasions.

(iii) the International Division has attached to it a team or group of operatives who go, as required, to various plants for this purpose but are based in Ravensdale.

(iv) the International Division, when it purchased beef from Intervention would notify the Department of Agriculture of its intention to rebox and/or relabel at a particular time and place.

(v) the International Division has a standard box which is used in the reboxing of boneless beef for export to non-EEC countries.

(vi) the box, which contains intervention beef, contained the necessary particulars viz the sale reference number, net weight, type of cut and number of cuts.

(vii) the marking on this box would indicate that it came from AIBP.

(viii) that reboxing of meat purchased from the Irish Intervention Agency is reboxed within the State.

(ix) that reboxing of meat purchased from the NI Intervention Agency or the English Intervention Agency is required to be done there in accordance with the relevant Regulations.

(x) there is a difference between the boxes used by the company when exporting beef and the boxes used when it is being stored for intervention purposes.

(xi) there never has been any occasion where beef purchased from intervention for export to non-EEC countries has not been re-boxed.

(xii) the re-boxing renders the product much more presentable to a customer.

(xiii) the meat purchased from intervention would be two to three years old depending in what lot was allocated to the intervention contract."

This evidence, together with the evidence from the Department of Agriculture and Food with regard to the amount of intervention beef included in the exports of beef to Iraq in 1987 and 1988 provides the explanation for the amount of reboxing and relabelling given during the course of the Tribunal and for the fact that much of this reboxing and re-labelling was done by outside operatives who were not normally employed in the plants where such work was being done.

Though they would not welcome the designation it would appear reasonable to assume that the group referred to at (iii) were the group described by Mr McGuinness as the "A" team. Such a designation is merely descriptive and has no sinister undertones.

Having regard to the tonnage exported and the weight of each carton; it meant well over a million boxes required reboxing and re-labelling for the purpose of fulfilling these contracts alone and frequently such re-boxing and re-labelling had to be done expeditiously and within time constraints imposed by the requirement to fulfil various contracts.

The Intervention Regulation set out herein shows what must be set forth on the cartons in which the beef was placed as a result of reboxing.

In addition the contracts entered into between AIBP and the State Company for Foodstuff Trading and Iraqi Company for Agricultural Products Marketing for the export of beef to Iraq included the requirement that

"(i) the period between time of slaughter in Country of origin and delivery to buyers unloading place in Baghdad will not exceed 100 days,

(ii) the slaughtering must be by Islamic Rites with full bleeding,

(iii) the cuts are to be wrapped in polythene without holes and packed in cartons of uniform size not exceeding 33kgs nett weight. Each carton will contain only one type of cut and will bear on it in English the following information:—

Type of Beef/Cut
Country of Origin
Production date
Gross Weight and Nett Weight
Date of Expiry of beef for human consumption
Trade marks/Names of Buyers and Seller
L/C number"

Some of the contracts required that the words "slaughtered according to Islamic Rites" be included in the marking.

It is clear that 84% of the beef delivered by AIBP to Iraq during the years 1987 and 1988 did not comply with the written requirements of these contracts, that it was not slaughtered within 100 days of delivery but was purchased from intervention and was probably 2/3 years old and that it was impossible to determine whether it was slaughtered as required by the contracts by Islamic Rites with full bleeding.

It was contended by Counsel for AIBP that the terms of their contract with the Iraqi Trading companies was a matter of private law and that the Tribunal was not entitled to inquire into the terms thereof or the compliance or otherwise therewith and if the Tribunal were to concern itself with such, it would be acting ultra vires.

This Tribunal is only entitled, and the Oireachtas is only entitled by virtue of the terms of the Act under which this Tribunal is established, to inquire into definite matters of public importance and if this were a matter of purely private contract, the Tribunal would not be concerned nor would it concern itself therewith. However there is a public dimension to these contracts which necessitated inquiry.

The public dimension involved arises because of the use of beef purchased from the Intervention Agency to fulfil 84% of the requirement of the contract; the necessity of the Intervention Agency to ensure compliance with the terms of EEC Regulation 2824/85; the requirement in the contracts to produce certificates from the Department of Agriculture; the entitlement of the vendor/seller to subsidies by way of Export Refunds : the grants of Export Credit Insurance in respect of some of the contracts and the national interest in protecting the reputation of the quality of Irish beef, which is fundamental to the Agricultural Economy of this country.

Though the contracts were stated to require production of certificates from the Department of Agriculture it appeared from the evidence of Mr Aidan Connor, who from September 1987 onwards, was Deputy Chief Executive of the International Division of the Goodman Group, reporting directly to Mr Goodman, the Department of Agriculture was never requested by AIBP to issue such certificates and no such certificates were issued by them.

During the course of its inquiry into the operation of the export refund system the Tribunal became aware that the certifying authority in respect of exports of beef to Iraq was an organisation with Headquarters in Paris called Bureau Veritas together with its subsidiary Le Controle Technique of 5, Rue Chante Coq 92801 Puteaux - France (hereinafter called LTC).

The Tribunal entered into correspondence with Bureau Veritas, LTC and its Irish representative, Victor Broderick and received information from them, relative to their procedures with regard to certification, the contents of such certificates and copies thereof. The Tribunal compared the particulars contained in such records with the information contained in the Schedule dealing with exports of Intervention Beef to Iraq during the period September 1987 to December 1988.

Mons. Christian Peyron the Director of the Meat Department of LCT and Mr Victor Broderick, a Meat Inspector employed in Ireland by LCT, gave evidence before the Tribunal.

LCT was a subsidiary of Bureau Veritas. Mons. Peyron explained to the Tribunal that the Certificates of Inspection issued by LCT to both suppliers and Purchasers was a private certificate whereas the certificate issued by Bureau Veritas is a public document issued by them and recognised by the EEC as providing the proof of import required for payment of Export Refunds.

LCT had entered into two separate contracts with the two Purchasing Companies in Iraq viz The State Company for Foodstuff Trading the Iraqi Company for Agricultural Products Marketing to carry out the various inspection works which these two organisations required them to carry out on their behalf and on completion thereof and based thereon to issue an Inspection Certificate which was not to be issued unless the inspected commodities fully complied with the Supply Contract and the Letter of Credit terms.

It appears from the evidence that such Inspection Certificates are prepared in Paris based on information supplied by him or one of his fellow inspectors.

In order to appreciate the detail contained in such certificates, a copy of one of them is produced hereunder:

LE CONTROLE TECHNIQUE
INTERNATIONAL SURVEYORS

INSPECTION CERTIFICATE NO 01061

F.88.03.295 LCT 36886
ORIGINAL NO 1 PARIS LA-DEFENSE, LE September 25th 1987

LETTER OF CREDIT NO. 19546/87/109 OF RAFIDAIN BANK SAADOON BRANCH
CUSTOMER: STATE COMPANY FOR FOOD
STUFF TRADING BAGHDAD L/C NO: 19546/87/109 OF RAFIDAIN BANK SAADOON BRANCH

CONTRACT NO: BL32/87

SUPPLIER: ANGLO IRISH BEEF PACKERS GROUP LTD
14 CASTLE STREET ARDEE CO LOUTH IRELAND

INSPECTION DATES:	09.09.87 - 22.09.87	LOCATION:	IRELAND
COMMODITY:	HIND QUARTER MEAT OF STEER	QUANTITY:	2055,7416 NET M/TS
TRANSPORT:	M/V PACIFIC LADY	REF:	FLAG DUTCH
DEPARTURE:	25.09.87	FROM:	GREENORE-IRELAND
BILL OF LADING:	DATED 25.09.87	TO:	MERSIN - TURKEY

We hereby certify that we have inspected the animals before slaughtering that we attended slaughtering and deboning operations, that we supervised all steps until meat was delivered and loaded into the holds of refrigerated vessels M/V PACIFIC LADY.
We found the animals and carcasses according to specifications. The cuts were checked and found of good quality according to specifications of supply contract. All the technical specifications of frozen hind quarter meat of steer have been fulfilled by ANGLO IRISH BEEF PACKERS GROUP LTD.

CARTONS :87 826
NET METRIC TONNES :2055,7416
GROSS METRIC TONNES :2149,3738

" F.88.03.295 **INSPECTION CERTIFICATE PAGE 2** LCT 36886

ORIGINAL NO 1

LETTER OF CREDIT NO. 19546/87/109 OF THE RAFIDAIN BANK SAADOON BRANCH

1. - <u>INSPECTION IN SLAUGHTERHOUSES AND DEBONING HALLS</u>

Inspection took place in the following slaughterhouses and deboning halls:

SLAUGHTER HOUSES	EEC	SLAUGHTER HOUSES	EEC
AIBP (CAHIR)	300	KEPAK LTD	317
AIBP (NENAGH)	290	HORGAN MEATS LTD	330
AIBP (DUNDALK)	289	LIFFEY MEATS (CAVAN) LTD	325
AIBP (BAGENALSTOWN)	303	WESTERN MEAT PRODUCERS LTD	342
AIBP (WATERFORD)	344	SHANNON MEAT LTD	274
AIBP (DUBLIN)	333	AIBP NEWRY	NIS 9
KILDARE CHILLING CO.	268	FOYLE MEATS DERRY	NIS10
MASTER MEAT PACKERS		AIBP FERMANAGH	NIS19
(CLONMEL) LTD	336	ARDS MEAT AND	
SLANEY MEATS LTD	296	LIVESTOCK CO	NIS36
TUNNEY MEAT			
PACKERS LTD	295		
JAMES DOHERTY LTD	292	MASTER MEAT PACKERS	
HALAL MEAT PACKERS		(OMAGH LTD)	NIS4
(BALLYHAUNIS)	284	ABBEY MEAT PACKERS LTD	NIS14
		LAGAN MEATS BELFAST	NIS32

DEBONING HALLS	EEC
. RANGELAND MEATS LTD	717
. AIBP MEAT PROCESSORS (NI) NIC	65

II - **<u>QUALITY AND SPECIFICATIONS</u>**

Commodity : HIND QUARTER MEAT OF STEER.

SPECIFICATIONS:-

A. - BONELESS STEER MALE YOUNG BULL MEAT FROM HIND QUARTERS.
B. - ANIMAL AGE UP TO THREE YEARS MAX.
C. - MEAT IS FROZEN AND ROUND CUT ORIGINATING FROM THE HIND QUARTERS CUTS NAME (OUT SIDE/KNUCKLES/SIRLOIN RUMP/INSIDE AND STRIPLOIN).

LE CONTROLE TECHNIQUE.
International Surveyors

F.88.03.295 **INSPECTION CERTIFICATE PAGE 3** **LCT 36886**
ORIGINAL NO.1

LETTER OF CREDIT NO 19546/87/109 OF RAFIDAIN BANK SAADOON BRANCH

We found the meat according to specifications as follows:

1. The hind quarter meat of steer was derived from healthy cattle in good sound conditions, and free from infections and contagious diseases.

2. The slaughtering of the animals from which the hind quarter meat of steer has been derived was carried out with a sharp knife according to Islamic Rites with full bleeding.

3. The hind quarter meat of steer was well and quickly deep frozen according to modern technical methods at a maximum temperature of minus 35 C for a period not less than 36 hours.
   The hind quarter meat of steer was not exposed to thawing and refreezing operations at any stage of preparation or storage or transport.

4. The surface of the hind quarter meat of steer is free from any sticky substance, fungus and bacteria or any sign of putrefaction rancidity or abnormal or offensive odour.

5. Cuts are shipped without skins, heads, legs, entrails, interior fat, kidney fat and tails.

6. Cuts are clean, free from blood and refuses.

7. The animals were examined antemortem and postmortem.

8. The relevant hind quarter meat of steer was found fit for human consumption and is consumed locally by people of the producing country. No chemicals were added to the meat, and the meat is free of hormones, antibiotics, preservatives.
   The meat does not exceed radiation levels accepted by Iraq and international authorities namely 370 Bqs per Kilo.

9. Price : including interest of per net metric ton.

10. The time period from slaughtering process at the country of origin up to arrival at Buyers Stores will not exceed 100 days.

11. Dates of slaughter and inspection : from 09.09.87 - 22.09.87

LE CONTROLE TECHNIQUE.
International Surveyors

F.88.03.295 **INSPECTION CERTIFICATE PAGE 4** LCT 36886

ORIGINAL NO 1

LETTER OF CREDIT NO. 19546/87/109 OF RAFIDAIN BANK SAADOON BRANCH

III - **PACKING AND MARKING**

A) Each cut of meat is tightly wrapped in a transparent polyeytelene without holes.

B) The weight of each carton does not exceed thirty three (33) kilograms and cartons are suitable for transportation and exportation.

COMMERCIAL MARKING:

The cartons have been marked as specified in Letter of Credit NO. 19546/87/109 of RAFIDAIN BANK SAADOON BRANCH.

IV. - **LOADING SURVEY**

Loading took place in Greenore - Ireland.

Loading started on	23.09.87	at	08 HRS 00
Loading completed on	25.09.87	at	18 HRS 45

V. - **CONCLUSION**

Quality and specifications of this shipment on M/V PACIFIC LADY are in compliance with contract terms.

LE CONTROLE TECHNIQUE
Meat Department.
C. Peyron

LE CONTROLE TECHNIQUE
INTERNATIONAL SURVEYORS

The Certificates, copies of which were made available to the Tribunal, covered the period September 1987 to July 1988.

Mr Broderick in the course of his evidence said that:—

"(i) his job was to ensure that the meat that was exported to Iraq complied with the specifications in the contract,

(ii) His inspection covered the slaughterhouse, the boning hall, the cold store, pre-loading, the ports and the ships,

(iii) there were three other meat inspectors employed in Ireland by LTC,

(iv) In the lairage, his responsibility was to ensure that the animals were male, in good heath and not more than three years old,

(vi) in the slaughterhouse, his responsibility was to ensure that the animals were slaughtered in accordance with Islamic Rites,

(v) in the cold store, he was obliged to see that the temperature was correct, that the animals were male and that the meat was fresh,

(vi) in the boning hall, he was obliged to ensure that it was boned in accordance with specifications and when packaged in cartons it was subjected to blast freezing for a minimum of 36 hours before being placed in storage on pallets,

(vii) when the beef was about to be exported LCT would be notified of the point of export and the cold stores from which the beef would be transported,

(viii) they would go to the cold stores again to check the meat, take random samples and record the numbers of the cartons, to ensure that the meat being exported was the meat they had supervised,

(xi) they would be present at the point of export to check the unloading of the meat and its re-loading on to ships : again to ensure that the meat being exported corresponded with the meat the production of which they had supervised."

If these procedures were adopted, then LCT would be in a position to issue the Inspection Certificates and to give the information therein contained.

LCT was the only company in Ireland with a contract with the said Iraqi Trading Companies to inspect and certify meat for export to Iraq.
In his evidence Mr Broderick agreed that if the system of inspection which is outlined above had been implemented there was no way that intervention beef could have gone to Iraq during this period and that "under no circumstances" would he have certified Intervention Beef as being suitable for export to Iraq in compliance with the contract.

Mons. Peyron when dealing with the use of intervention beef for export to Iraq stated:

> "We didn't know. We haven't seen or haven't heard. Had we heard or seen, we would have told Baghdad and stopped issuing certificates"

However the uncontroverted facts remain, that between September 1987 and 31st of December 1988 AIBP exported 24,391.83422 tonnes and Dantean (Hibernian Meats)

8,999.2874 to Iraq. No reliance whatsoever can be placed on the Certificates issued by L.C.T. or the facts contained therein.

Both Mr Larry Goodman and Mr Aidan Connor of the Goodman Group stated that irrespective of the terms of the written contracts the Iraqi's were fully aware of the fact that intervention beef was being supplied to them in pursuance of the contracts.

Their evidence in this regard is confirmed by the evidence of Mr Oliver Murphy of Hibernia Meats and by the evidence of Mr Naser Taher of Taher Meats who stated:—

> "that so far as the Iraqi's were concerned that they knew they were purchasing intervention beef".

It was further confirmed by a letter 12/7/1993 received by the Tribunal from the French Company CED Viandes which was a major supplier of beef to the Iraqi market and which was the majority shareholder in Hibernia Meats Ltd which, inter alia, stated;

> "During this period there were only two customers for beef in Iraq namely The Iraqi Company for Products Marketing and the State Company for Foodstuffs Trading — both Iraqi state companies. Given that between (September 1980 to July 1988) Iraq was at war with Iran, the primary concern of these companies was to secure large quantities of beef from reliable sources at the right price, realising that it is not easy to supply product into a war zone."
>
> Very often their requirements were dictated by events in the war which gave rise to sudden surges in demand. While the normal process was for the Iraqis to source beef by way of public tender, on occasion, when necessity dictated, they would approach certain suppliers (of which we were one) in a form private tender seeking to secure beef on an urgent basis.
>
> During this period I was the primary person in CED Viandes responsible for concluding such contracts and I had the closest connection with Iraq.
>
> Although we were supplying beef in different and often extreme circumstances, and although the needs and requirements of the purchasers changed, the two state companies were obliged to use standard form contracts incorporating regulations and terms which they could not be seen to alter. In this regard the contracts concluded were often a matter of form rather than substance. For example, the term in such contracts which required that beef be not more than 90 days old when it arrived at the buyer's store was a term adapted from national Iraqi regulations concerning supply of beef. Iraq had traditionally experienced difficulties in properly freezing beef on the domestic market. This, of course, did not apply to beef which had been frozen in Ireland using more sophisticated blast freezing facilities. Such a term was not appropriate in any respect for contracts to be filled from Ireland yet it had to be included.
>
> When negotiating such contracts we would say to the Iraqis that we could not guarantee that we would comply with this type of contract stipulation as contained in the standard form contracts. Once the representatives of the two state companies were satisfied that the correct amount of beef could be supplied at the right price within the time frame stipulated they were happy. Their response was invariably to the effect of "you supply the beef we will worry about the domestic regulations".

The Iraqis were requesting (and obtaining) deliveries of full shiploads of beef immediately after opening Letters of Credit and consequently, must have realised that the beef was being supplied from existing stock. In this regard I am satisfied that the Iraqis knew that Irish beef to be supplied under the contracts would be sourced from intervention stocks. This meant not only were they getting the beef at the right price but also that it would be of a uniformly high quality and specification. I am aware that during this period representatives from Iraq travelled to Europe on a number of occasions to observe the supply process and were aware of the provenance of the beef. They were also keenly aware of the prices of intervention beef and the various market support schemes operating in the EC. When negotiating price and delivery arrangements in particular the Iraqis knew that the beef would be sourced in the main from stocks.

Thus, when contracts were concluded with the Iraqi state companies the important terms of the contracts were quantity, price and security of supply (often within a short time period). While the state companies were prepared to, and in practice did, waive strict compliance by us with certain terms of the standard form contracts, this was not something which they could publicly acknowledge for obvious reasons."

The Tribunal is satisfied having considered the evidence that all the requirements of Regulation 2824/85 were complied with, that authority for the repacking and relabelling of the beef purchased from intervention was given at the time of purchase thereof, that notification of the time and place of such repackaging or relabelling was given by or on behalf of the International Division to the Department of Agriculture , that the repacking and relabelling occurred in such places and at such times as were notified and that the repacking and the relabelling was supervised by the Agricultural Officers responsible by checking the movements into the plant and movements out and recording such movements and issuing the necessary certificates of movements in regard thereto. The actual re-boxing was not supervised on a permanent basis but by spot checks. And evidence has been given that re-boxing was carried out when no Agricultural Officer was present.

While the Tribunal is satisfied that from time to time reboxing did occur in the absence of an Agricultural Officer the Tribunal is further satisfied that such re-boxing or re-packaging was not carried out deliberately in the absence of an AO but because of the unavailability of the Agricultural Officer for one reason or another. In all plants the number of Agricultural Officers was inadequate.

As it was the intention of AIBP to claim subsidies by way of export refunds, it could not have been in the interests of AIBP not to comply with the relevant regulations and fail to obtain the necessary documentation showing compliance therewith which were necessary to substantiate claims to be made for the export refund subsidies.

To qualify for payment of export refunds a clear trail of the meat from its point of origin to its ultimate destination must be established.

Once the meat is purchased from intervention, the only obligation on the purchaser/exporter is to preserve its identity at all times and to provide evidence that it was fit for human consumption. It can as pointed out be re-cut, have any marking thereon removed and re-packaged to suit customer requirements.

However once that is done, Regulation 2730/79 requires as a condition for payment of refund, that the product has been imported in the unaltered state into a non-member countries for which a refund is eligible.

The Department of Agriculture was at all times aware of the fact that the beef purchased by the Goodman Group during 1987 and 1988 was intended for export to Iraq.

The CBF was also aware of the fact that beef purchased from Intervention stocks was being supplied to Iraq.

On the 30/9/1988, the CBF (the Irish Livestock and Meat Board) prepared a briefing note for the Department of Agriculture in anticipation of the Irish-Iraqi Joint Commission Talks due to be held in Baghdad in November 1988.
That briefing note included the paragraph:—

> "In recent years the product supplied to Iraq has largely been from Intervention Stocks with some APS. The market is mainly for frozen hindquarter boneless cuts. As the stocks of Intervention product decline the market is likely to move towards APS and possibly forequarter cuts as prices rise. The type of beef should not be mentioned to the Iraqis. At present, Islamic slaughter is a requirement of the market."

This information was for inclusion in the briefing documents for the delegation to the said Commission, including Mr Seamus Brennan TD, Minister for Trade, who was leading the delegation.

When this briefing note was considered by the Department of Agriculture , this paragraph was removed and the following substituted:—

> "The market is mainly for frozen hindquarters, boneless cuts. In some cases the exporters have availed of the EEC Aids to Storage Scheme prior to export. In view of rising price trends there may be some move towards some forequarter cuts."

It is significant that all reference to the use of Intervention Beef and Islamic slaughter was excised by the Department of Agriculture from the briefing note.

The explanation given for such excision by Mr Joseph Shorthall the Principal Officer in the Department of Agriculture who had made the alteration in the document was that he was aware from his experience at that time that there was going to be a significant move away from intervention, that there had at that time been a dramatic decrease in the quantities in Intervention, that the APS Scheme had been introduced and extended and that it was his belief that in respect of future exports, that they would be coming from a combination of the general commercial market and beef placed in-storage under the 1988 APS Scheme and that his purpose in amending the document was to provide briefing material which would indicate what he believed "the future was going to be" rather than indicate what had happened in the past.

Irrespective of Mr Shorthall's amendment of the document to be included in the briefing material for the Minister for Trade, Mr Seamus Brennan TD, it still purported to be a document emanating from the CBF and such amendment deprived the Minister of Trade of the information that:—

> "In recent years the product supplied to Iraq has *largely* been from Intervention stock, with some APS".

Instead of this important information the Minister and delegates were told:

> "In some cases, the Exporters have availed of the EEC aids to private storage scheme prior to export".

The important reference to the fact that the product supplied to Iraq had largely been from Intervention Stock was omitted.

This omission was of major significance because one of the major issues to be discussed at the meeting of the Irish-Iraqi Joint Commission was the issue of increasing of beef exports to Iraq, extending this period of credit for payment in respect thereof and the provision of Export Credit Insurance in respect thereof.

It appeared from the evidence given at this Tribunal that in reaching his decision to re-introduce Export Credit Insurance in respect of beef exports to Iraq in 1987 and grant the applications for Export Credit Insurance to Goodman International and Hibernia Meats in 1987 and 1988, the then Minister for Industry and Commerce and now Taoiseach Albert Reynolds TD and the officials of his Department believed that the beef in respect of which Export Credit Insurance was granted for export to Iraq was commercial beef, the purchase and processing of which would confer substantial benefits on the Irish Economy. This fact influenced Mr Reynolds to re-introduce in the national interest Export Credit Insurance in respect of beef to Iraq. If he had been or made aware of the fact that the beef being exported was largely beef purchased from Intervention stock with little if any benefit to the Irish Agricultural Economy then his decisions may have been different.

The fact that the beef exported to Iraq in 1987 and 1988 by AIBP and Hibernia Meats Ltd consisted of beef purchased from Intervention stocks was a fact extremely relevant to any negotiations with the Iraqi Government who were in 1988 pressing to have the level of Export Credit Insurance available for Irish exports to Iraq increased.

The substantial benefits which could accrue to the Irish economy if such an increase were granted were dependant on the sale of commercial beef and not on the sale of beef purchased from intervention stock and the situation in regard to the amount of intervention beef included in the contracts should have been disclosed to the Minister in order to enable him to evaluate the benefits to the Irish economy of such exports and their entitlement to or qualification for Export Credit Insurance.

It is clear from the evidence of Mr Laurence Goodman and Mr Aidan Connor of Goodman International, Mr Oliver Murphy of Hibernia Meats Ltd., and confirmed in the letter dated the 12th day of July 1993 from its parent company CED Viandes and perhaps more significantly the evidence of Mr Naser Taher of Taher Meats, that the Iraqi customers were aware of the fact that intervention beef was being used to fulfil a substantial portion of the contracts and had waived compliance with the terms of the written contracts.

The EEC., and the Beef Management Division thereof and the Department of Agriculture were fully aware of the fact that intervention beef was being exported to Iraq and there was no abuse of the Export Refund Subsidy System in regard to these exports.

The Tribunal has heard evidence from many witnesses with regard to the appearance of a small percentage of the meat being reboxed for export and the grottiness of the cartons in which it was contained.

Cartons which had been stored in Intervention Cold stores for an extended period would of necessity be damaged and present a grotty appearance and meat which had been frozen for such a period would not be as attractive in appearance as fresh meat though its quality would not be affected thereby.

The allegations made by Deputies MacGiolla and Rabbitte hereinbefore referred to that;

(i) the Goodman Group of Compnaies were abusing the system under which subsidies were paid

(*a*) by having the labels on meat changed in different parts of the country by a team moving about to do this job on behalf Goodman Companies,

(*b*) by the maintenance of an entire production line in Nenagh designed for taking stamps from frozen carcases and re-stamping and re-labelling them, and

(*c*) by carrying out repackaging and re-stamping operations in Goodman plants in operations heavily subsidised by the Irish Taxpayer, and therby putting Ireland's reputation for quality at risk.

were based on a lack of understanding or appreciation of the Export Refund Subsidy System and Regulations, the EC Regulations with regard to the sale of beef out of Intervention Stocks, the fact that the beef being re-packaged and relabelled was such intervention beef and the re-packaging and relabelling of same was duly authorised.

Having regard to the quantities involved, the volume of complaints was minimal and to a considerable extent is explained by the facts set forth above.

The meat being exported was certified to be fit for human consumption.

There was no breach by the Goodman companies of the Regulations governing the Export Sales Refund system and no abuse of the Scheme.

(*f*) *Abuse of Intervention Systems by use of bogus stamps to alter the classification of animals.*

On the ITV programme broadcast on the 13/5/1991 the male presenter referred to the IB4 forms, which had been produced in the Programme and went on to say that:—

> "World in Action has obtained these IB4 forms, they relate to Intervention contracts at one Goodman factory in 1987. Some of these forms have been duplicated to show an increase of up to 14 kilos for every animal. Payments are also based on the quality of each animal. This is assessed by a veterinary official who marks the carcass with

> an indelible grading stamp. The stamps are the property of the officials, who keep them securely under lock and key in each factory"

He then went on to say that:—

> "Goodmans own promotional videos made great play of this official grading system."

and quoted from such video as follows:—

> "Rigorous inspection and grading by ministry officials followed by careful selection and assessment by the companies own personnel means Anglo Irish customers get the beef that they specify."

The male presenter then went on to say:—

> "But in Goodman factories they used their own bogus stamps to change the grades".

On the programme Patrick McGuinness stated:—

> "It was very easy to change the grades with a knife you cut off the grade that is marked on the animal and you can then put any other grade you like on it. You would have your own stamps at the factory".

He further stated that:—

> "all grading stamps were supposed to be tightly controlled by the Department of Agriculture".

In the course of his evidence before this Tribunal, Mr McGuinness stated that:—

> "(i) Grading stamps are held by the Classification Officer of the Department of Agriculture;
>
> (ii) The Classification Officers grade the animal and apply a Grading Stamp to both the hind quarter and the forequarter;
>
> (iii) The price per kilo of meat being sold to the Intervention Agency depends on the classification;
>
> (iv) There were five intervention grades and the official stamp is a composite one and has the five grades on it and the appropriate one is applied."

He produced and identified in evidence the stamps which had been referred to and shown on the ITV programme and which had been given the producers thereof by him.

He stated that:—

> "(i) These stamps had been in the possession of John Connolly, the Plant Manager of AIBP Waterford;
>
> (ii) They would be given for use to George Williams, the loading bay supervisor;
>
> (iii) These stamps had been obtained by having them reproduced by cutting off pieces of meat from carcases upon which the official stamps had been properly placed and duplicates made;

(iv) That these pieces of meat had been cut from carcasses by Patsy O'Halloran, the Production Manager;

(v) Patsy O'Halloran had these pieces of meat in the office, together with a duplication of the EC 344 stamp and a duplication of the Hal-Al stamp which had been stamped on a certificate in the office;

(vi) These samples were sent to Gene Lamb's stationery for reproduction;

(vii) He was informed by Mr Patsy McGuinness that the stamps had been reproduced there."

He further stated that:—

"(viii) These were the first set of stamps prepared for Waterford when he was there;

(ix) That he observed the process of cutting the original stamp off the carcase, putting on the new stamp, the removal of the label containing the killing docket details applied to the forequarter and the hindquarter which show the kill number plus the grade and a new label put on;

(x) The procedure was followed for the purpose of placing in intervention carcasses that were not eligible for intervention."

On the 16th day of October 1987, Mr William O'Connor an Agricultural Officer on duty at AIBP Waterford loading bay observed carcases being taken out of the chill room and being cut in readiness for export and noticed 15 sides with similar classification marks on them.

The said stamps appeared to be wet and fresh and smaller than the official grading stamps.

He contacted Mr Padraig Feeney, the Classification Officer and showed the carcasses to him. He confirmed that the stamps were wet, were smaller than the official stamps and that no carcasses had been re-graded that morning.

He sought an explanation of the occurrence from Mr Williams, the foreman in the loading bay and Mr O'Halloran, the Production Manager but none was forthcoming.

He then reported the incident to Mr John Comerford who in company with another official went to the loading bay and observed that by this time the bits of meat with the stamps on them had been removed from the carcasses and were on the ground.

The carcases were still there but the stamps had been removed. On the instruction of Mr Michael Staff SAO, detention labels were placed on the carcases.

Mr Staff in company with Mr Andrew McCarthy went to the loading bay and observed the 15 hindquarters hanging on a rail: on one of them was a stamp CR3 and there were on the ground under them pieces of meat with imprints of classification stamps on them.

On comparing the genuine classification stamps with the ones on the meat, the ones on the meat were smaller.

The matter was dealt with then by Mr McCarthy.

This matter had been reported to Mr McCarthy by Mr Feeney, a Classification Officer, who had been contacted by Mr Willie O'Connor.

He had asked George Williams to hold the carcases while he went to contact Mr McCarthy.

When he returned he found that some of the carcases had been quartered and loaded on to a van, the stamps cut off and thrown on the ground.
Mr Andrew McCarthy was the Regional Supervisor and responsible for the classification operation in a number of plants in the South East including AIBP Waterford.

When he arrived on the scene he observed that:

(i) the stamps were not the classification stamps used by the Department.

(ii) they were smaller and had a different surrounding pad.

(iii) some of the carcasses were being put in a van and the stamps were being removed from them.

Mr McCarthy then requested Mr Williams to detain the carcases but he refused and continued loading the van.

Mr McCarthy then threatened to involve the Gardaí and the owner of the van said that he didn't want the carcases anyway.

Mr Williams then left the loading bay but before he left he removed the intervention stamps from the carcasses in the van from which the stamps had not previously been removed.

Mr McCarthy stated that the carcases were non-intervention type carcases, were poor quality and would not qualify for intervention.

The classification on the labels corresponded to the stamps on the carcases but Mr McCarthy stated that the labels were incorrect, he wouldn't expect any Classification Officer to put that type of classification on these carcases, that they were not border-line cases but considerably out of the intervention categories.

Mr McCarthy then discussed the incident with Mr Connolly, the Plant Manager and Mr McGuinness.

Mr Connolly refused to accept that the stamps were different whereas Mr McGuinness did.

Mr McCarthy suspended intervention classification at the plant and when contacted on the 'phone by Mr Gerry Thornton of the Meat Division indicated that he would not permit the resumption of such classification until the stamps were found and handed over.

Mr Thornton did not accept that there were any bogus stamps in the factory but undertook to investigate the matter.

Mr McCarthy contacted his superior Mr Dermot Ryan of the Department of Agriculutre and informed him of what had been discussed and the action taken in the suspension of intervention classification.

Mr Ryan agreed with the suspension but at about 4 pm. he contacted Mr McCarthy and informed him that he had been talking to Mr Thornton and his own superiors and directed that, though the bogus stamps had not been discovered, that as there was 200 cattle in the plant the kill and classification should proceed.

It appears from Mr Dermot Ryan's evidence on this issue that:—

"(i) when Mr McCarthy withdrew classification at 10 a.m. he reported the matter to him;

(ii) he at that time was in the AIBP plant at Ravensdale and reported the matter by telephone to the Beef Division at the Department of Agriculture but was unable to obtain any real guidance on this matter as the responsible people were in Brussels;

(iii) as he was in the AIBP plant at Ravensdale, he discussed the matter on a number of occasions during the day with Mr Peter Goodman and Mr Gerry Thornton who couldn't explain the stamp markings on the carcases but undertook to investigate the matter;

(iv) he was concerned about the number of cattle in the lairage at Waterford and suggested that they be moved to Cahir or Bagenalstown;

(v) when this proved impractical he authorised resumption of the kill and classification of carcases at 4 p.m"

Mr Thornton stated that he carried out an investigation but was unable to ascertain who was responsible.

The Tribunal had dealt with this matter in some detail because:—

(i) it establishes the use of bogus stamps at the AIBP plant in Waterford;

(ii) it establishes that the stamps were applied either in the chill room or as the carcases were being taken out of the chill room;

(iii) it established that in addition to the stamps being applied to the carcases the original labels containing the kill number and grades were replaced; and;

(iv) the effect of such changes in the stamping on the carcase and the labels attached thereto was to give the appearance that the carcases were eligible for intervention when, according to Mr McCarthy they obviously were not;

(v) the action taken by Mr McCarthy in stopping the kill and classification was indicative of the seriousness of the irregularity;

(vi) it confirms the evidence of Mr McGuinness, particularly at (ix) above;

Mr Connolly, the Plant Manager, in evidence stated that:—

"(i) he knew nothing about the stamps or how they made their way into the plant;

(ii) they were not kept in his office; and

(iii) he did not know who used them on that particular day;

(iv) investigations were carried out but they were unable to locate the stamps or identify anybody who may have put the stamps on the beef."

Mr Gerry Thornton stated in evidence with regard to this issue that:—

"(i) while in Longford on the morning of the 16th October 1987 he received a call from Mr Connolly on his car phone and as the reception was poor, he stated that he would ring the factory;

(ii) in doing so, he spoke to Mr McGuinness who informed him that:

(*a*) classification had been withdrawn because carcases had been found on the loading bank with stamps that seemed to be different from the normal stamp;

(*b*) 200 cattle could not be slaughtered because of the withdrawal of the classification process;

(iii) he immediately contacted Mr McCarthy the Regional Classification Officer in Waterford who informed him that bogus grading stamps had been used on carcases;

(iv) he informed Mr McCarthy that he did not believe that this could have happened and undertook to investigate the incident in addition to the investigations being carried out by Mr McCarthy;

(v) he subsequently spoke to Mr Dermot Ryan in Ravendsdale and reiterated his position;

(vi) he discussed the problem with regard to the 200 cattle which had been held back from slaughter and eventually it was agreed to have the kill resumed rather than having the cattle kept in pens over the weekend;

(vii) he carried out an investigation but was unable to locate the stamps or to ascertain who had applied them."

Both Mr Connolly's and Mr Thornton's evidence is at complete variance with and contradictory to the evidence of Mr McGuinness who stated that:—

"(i) after the incident occurred Mr Connolly came to his (Mr McGuinness) office and told him what had happened;

(ii) from his office Mr Connolly contacted Mr Gerry Thornton by phone and informed him of the problem and then returned to his (Connolly's) office;

(iii) some fifteen minutes later he (Mr McGuinness) was contacted on the phone by Gerry Thornton;

(iv) he informed Gerry Thornton of what had happened;

(v) Gerry Thornton was ringing in the car phone and as his signal faded he had to ring again, about four times in all;

(vi) by the end of the phone calls he had instructed Mr McGuinness to remove the stamps from the plant;

(vii) he knew that the stamps were kept in Mr Connolly's office in the right hand side desk drawer;

(viii) he went to John Connolly's office, informed him of his conversation with Gerry Thornton, collected the stamps from John Connolly and brought them to his home."

During the course of his cross-examination by Counsel for the Goodman Group of companies he stated that:—

"(i) he had handled the stamps before the 16th October 1987;

(ii) on a few occasions when George Williams would be using them he went with him and held the stamp in the chill room;

(iii) he did this to get a knowledge of what was going on in the line of production;

(iv) he brought the IB4s and the stamps home because of Gerry Thornton's instruction to "get rid of the stamps."

Counsel for the Goodman Group however suggested to Mr McGuinness that it was he, Mr McGuinness, who had acquired the stamps in the first place; that he had phoned Mrs Susan McGuinness, the wife of Patsy McGuinness (no relation), asking her to arrange to get some stamps made for him; that she agreed to do this: that he sent to her details of five stamps on a plain sheet of paper: that she arranged to have them made in Dundalk: that when they were ready, they arranged to meet in the car park of a hotel outside Dundalk: that the package containing the stamps were handed over to him and he paid £71.06 for them.

Mr McGuinness denied ordering the stamps or submitting any material in regard thereto to Mrs McGuinness but does admit to a vague recollection of collecting some package.

Mrs McGuinness gave evidence in support of the suggestions put to Mr McGuinness by Counsel and said that the details of the stamps were CO3, CO4, CR3, CR4 and CU3 and were obtained from Devaney's in Dundalk.

It is clear from the evidence adduced before this Tribunal that the five rubber stamps were ordered from Devanney Supplies Ltd of Dundalk and manufactured by August Engraving Company.

At different times this company manufactured for AIBP, Arabic stamps, and stamps indicating the numbers allocated to plants by the EEC.

In the course of his evidence before this Tribunal Mr Gerry Thornton stated that:—

> "From the evidence of Mrs Susan McGuinness it is now my belief that Mr McGuinness procured and used the bogus grading stamps himself".

This is an attempt by the Senior Management of the Goodman Group to disassociate itself from any irregularities or improper practices in plants under their control.

It is not supported by the facts.

John Meaney was employed by AIBP at Waterford from September 1987 to March 1991. He was eighteen years of age when he started work in the loading bay and worked there until September 1988 when, during the Iranian contracts season he worked at the killing scales.

During this period the slaughtering of cattle according to the Islamic Rite would be carried out by an Iranian slitting the throat of the animal. Depending on the grade of the animal, the animal would be stamped with an Iranian stamp. This stamp was kept by an Iranian Inspector.

While he was engaged on this work he was asked by Mr John Connolly to make a copy of the Iranian stamp on a piece of paper.

On one occasion when the Iranian left the scales to visit another part of the plant, he left the stamp at the scales.

Mr Meaney stated that he took the stamp, imprinted it on a piece of white paper and gave the white paper to Mr Connolly.

Subsequently a stamp became available and was used by factory operatives, under the instruction of either Mr Connolly, John Kelleher or Patsy O'Halloran, in the absence of the Iranian Inspector.

One or other of these Managers was present when the stamp was being used and they would indicate the carcases which were to be stamped and this was done either in the chillers in the early morning or in the loading bay.

Mr Meaney's evidence is accepted by the Tribunal, and establishes Mr Connolly's involvement in procuring of a bogus Iranian stamp and his, Mr Kelleher's and Mr O'Halloran's involvement in the use to which it was applied.

The plant at Waterford was not the only one in respect of which there was evidence of the use of bogus stamps for various purposes.

Mr Frank Whelan gave evidence before the Tribunal that:—

"(i) he had been employed as a factory worker in the AIBP Plant at Nenagh;

(ii) during this period he did various jobs at the Plant;

(iii) before he left the factory he was engaged in the boning and trimming of beef;

(iv) when engaged in the trimming of beef for the 'Arab trade', an Islamic Inspector was present and would stamp the carcases which he accepted;

(v) he and another worker were trimming beef, the production manager spoke to the other worker, who said to Mr Whelan that he had to get the stamp from the Inspector for a few minutes;

(vi) sometime later when the production line stopped, the worker took the stamp ostensibly for the purpose of washing the fat off it;

(vii) he did so at a hose situated in a corner of the slaughterhouse; then went into the offal room : when he returned from the offal room he again washed the stamp before returning it to the Iranian Inspector;

(viii) he told Mr Whelan that the production manager had made an impression of the stamp on a cardboard box;

(ix) subsequently an Islamic stamp became available to workers in the plant who applied it to carcases which had been rejected by the inspector;

(v) he, Mr Whelan, used the stamp on a number of occasions in the absence of the Islamic Inspector, usually in the chill room, early in the morning at the direction of the production manager."

Another example with regard to the use of bogus stamps was discovered in the AIBP plant in Cloghran/Ballymun on the 7th day of December 1987.

John Mitchell, an Agricultural Officer was engaged on lambing duties of the plant when he was approached by a factory employee who wanted to borrow an ink-pad. When asked the purpose for which it was required he stated that it was for stamping lambs heads, Mr Mitchell asked him did he require the stamp and was informed that he already had a stamp.

Mr Mitchell reported this incident to his superiors.

Mr Patrick Connolly, the Veterinary Inspector at the Plant stated in evidence that:—

"(i) Mr Mitchell reported that he had found an employee with a bogus health stamp;

(ii) health stamps are the standard health stamps used by the Department of Agriculture to stamp carcasses and labels;

(iii) these stamps contained the code number of the Plant (333 in this case), the letters EEC and IRELAND;

(iv) Mr Matthews went to the factory store and recovered a second stamp and an invoice from August Engraving Company in Burgh Quay, which showed 3 stamps had been ordered from them by AIBP Ballymun;

(v) he retained possession of the two bogus stamps for the best part of two years, when he destroyed them."

Mr Connolly reported the matter to his superior Mr Bennett SVI but no action was taken.

Mr Delaney, the manager of the plant, told them that he didn't know anything about it. It was the foreman in the sheep division who ordered the stamps.

Though it appears that this incident was not regarded by Mr Connolly as a serious matter, which the Tribunal finds difficult to understand, it clearly shows that bogus health stamps were ordered on behalf of AIBP Ballymun and used or sought to be used by their employees. There can be no innocent explanation for the deliberate ordering and use of bogus stamps.

The Tribunal in the section of its Report dealing with the "Eirfreeze Investigation" has referred to the use of a bogus CU2 stamp at the AIBP plant in Nenagh and at the Eirfreeze Cold Store.

Both Deputy Rabbitte and Deputy Spring had in the course of speeches in Dáil Eireann on the 28th August 1990 and the 15th day of May 1991 referred to the conviction of Mr N. Quinn, who was described as a close aide of Mr Goodman, in 1987. Mr Quinn on the 17th day of September 1987 pleaded guilty to the charge that:—

> "On or about the 28th day of October 1983 at the Department of Agriculture, at Agriculture House, Kildare Street, Dublin 2 in the County of the City of Dublin did utter to Gabriel Curley there, forged documents to wit seven Forwarders, Bills of Lading and eight European Economic Community Customs Entry Certificates known as Annex 11 Proof forms, knowing them to be forged and with intent to defraud contrary to Section 6(1)(2) of the Forgery Act 1913."

This matter has been dealt with by the appropriate Court and its relevance before this Tribunal is that the proceedings which led to the conviction arose out of the discovery by a Customs and Excise official of the Foreign Post Section, Cork on the 19th day of September 1983, of an undeclared package addressed to Coleman's Printer's, Clarkes Bridge Cork. Having received authority from this addressee, he, on the 28th September, 1983, opened the package and discovered that it contained two rubber hand stamps. Having made an impression he realised that they were East London (South Africa) Customs stamps. He then ascertained from the consignee that they had been ordered by Cahir Meat Packers, Limited, Cahir Co Tipperary. He was then authorised to and did release the stamps to Coleman Printers.

Mr Hickey of the Beef Export Refunds Section of the Department of Agriculture was informed by telephone of the importation of the two East London stamps by Coleman Printers Ltd to the order of Cahir Meat Packers Ltd and requested by Customs and Excise to inspect record of claims lodged for payment of Export Refunds to establish if any such claims had ben lodged by Cahir Meat Packers Ltd in respect of exports of beef to South Africa as East London is a port in that country and if so, to inspect all documents supporting such claims for the presence thereon of impressions of East London Customs stamps.

On the 28th day of October 1983, Mr Curley of the Department of Agriculture had received from a representative of Cahir Meat Packers Ltd documentation including Proof of Import (Annex 11) and transport documents (Bills of Lading) in respect of a claim for Export Refunds relating to the consignment of eight container loads of boneless beef exported by the company to South Africa, the amount involved being approximately £150,000.

A comparison of the impressions taken of the stamps discovered in the Cork Foreign Post Section and those appearing on the Annex 11s and Forwarders Bills of Lading presented by Cahir Meat Packers to the Department of Agriculture as proof of the arrival of the beef in South Africa disclosed that the impressions on the latter documents were made by one of the stamps which had been released by the Customs and Excise official in Cork.

Inquiries made disclosed that these stamps were ordered by the then Transport Manager of AIBP Cahir with the consent of Mr Quinn at the request of their consignee.

As this matter has been dealt with by the Circuit Criminal Court, the Tribunal does not intend to deal further with the facts but it is a further illustration of the ordering, procuring and use of bogus stamps for an illegal purpose.

The Tribunal has dealt with these five incidents for the purpose of illustrating that, at least, it was not unusual for different plants to order and use duplicate and/or bogus stamps and that their use was not confined to Waterford during the time Mr McGuinness was employed there.

The Tribunal is prepared to accept that the bogus stamps provided by Mr McGuinness on the ITV programme were ordered from his office, either by himself or Mr Patsy O'Halloran, the Production Manager but if ordered by Mr McGuinness, it was at the request of Mr O'Halloran, who had brought the materials necessary for the preparation of the "art work", upon which the duplicate stamps were prepared, to the office as described by Mr McGuinness.

It may well be that when the bogus stamps were ready for collection, Mrs McGuinness contacted Mr McGuinness's office in Waterford and that he arranged to collect them from her though it is difficult to understand the necessity for the unusual arrangements for their collection though the Tribunal is not convinced of this.

It is not of fundamental importance whether or not Mr McGuinness himself ordered and collected the stamps. If he did so he was acting on behalf of AIBP, not on his own behalf.

He was employed by AIBP in Waterford as the financial accountant and in charge of the office staff with administrative duties.

While he was interested in, he had no role to play in the production activities of the plant and could not interfere. Mr Connolly was the Plant Manager, Mr O'Halloran was the Production Manager and Mr George Williams was the foreman of the loading bay and it is inconceivable that Mr McGuinness could have interfered with the activities of the Plant by producing and using bogus stamps without their knowledge and approval.

Having regard to all the circumstances, the Tribunal is satisfied that:

(i) the Plant Manager Mr Connolly was fully aware of the existence of the stamps;

(ii) he kept custody of them and released them for use as required;

(iii) the stamps were used with his approval;

(iv) the purpose of the use of the bogus grading stamps was to upgrade the classification of carcases to render them eligible for intervention;

(v) the stamps were removed from Mr Connolly's office by Mr McGuinness on the instructions of Mr Gerry Thornton; and;

(vi) there is no basis for the suggestion made by Mr Thornton in his evidence that the bogus grading stamps were used by Mr McGuinness himself and by no other person.

The Tribunal has not received any other evidence in respect of the use of bogus official stamps but is satisfied that they were used in Waterford until their use was discovered on the 16th October 1987, in Nenagh and at the Eirfreeze Cold Store as described in this Report and bogus East London Customs stamps were procured by AIBP Cahir in September 1983.

These incidents, serious though they are, are not sufficient to justify a finding by the Tribunal that the use of bogus grading stamps was institutionalised throughout all the AIBP plants.

(*g*) Alleged removal of Classification Officer

It has already been pointed out during the course of this Report there are five grades of animals eligible for intervention and the grading thereof is the responsibility of the Department of Agriculture.

Grading is important because the amount of payment to the farmer and to the processor depends on the grade allocated to each animal.

Classification is to some extent a subjective exercise based as it is on visual inspection of the carcase and disputes arose from time to time between Classification Officers and those aggrieved by the Classification, the farmer and the processor.

It is of importance that the independence and the integrity of the Classification Officers be maintained and supported in the performance of their duties and that they should not be subjected to intimidation by factory personnel or management.

In the course of the statement submitted by Mr McGuinness to the Tribunal he stated that:—

> "I believe at least one grading official — Patrick Feeney — was transferred from the Waterford plant because he had become too obstreperous".
>
> In the course of his evidence Mr McGuinness stated, with regard to grading stamps abuses, that:—
>
> "The only situation where I ever became aware that the Department of Agriculture officials were aware to some abuses going on, was in Waterford. That came about as a result of a series of incidents. First of all there was a general suspicion within the A.O.'s of the location, by a Classification Officer on several occasions refusing to grade the animals and some incidents with grading the animals with a huge amount of stamps".

Mr McGuinness had forgotten the name of the Classification Officer concerned: he had been the individual who had stopped stamping or he had been over-stamping and according to Mr McGuinness' evidence:—

> "there was considerable determination within the plant between John Connolly and Gerry Thornton to try and get him removed".

In reply to a question by the Tribunal he stated that the particular Classification Officer had been removed.

It transpired however, that no grading or Classification Officer had been removed from Waterford and this illustrates the difficulty of dealing with Mr McGuinness' evidence some of which is based on hearsay and some on what he himself actually observed.

It appeared from the cross-examination of Mr McGuinness by Counsel for the State Authorities that the Agricultural Officer who had multiple stamped the carcases to ensure that the stamp could not be removed was one Martin Long and not Padraig Feeney as Mr McGuinness had believed and that no complaints had been made about him and he had not been removed from Waterford.

Mr McGuinness accepted that it may have been Martin Long who had multiple stamped the carcases as T.B. reactors, he stated:—

> "I was under the impression it was Padraig Feeney who did this"

He had told the Tribunal that John Connolly, the Plant Manager and Gerry Thornton, the head of the Meat Division:—

> "had tried to get rid of this official and made complaints about him".

While Mr McGuinness may have been vague about the circumstances he was undoubtedly right in this statement.

In the course of a meeting with Mr O'Mahony, Secretary to the Department of Agriculture and Food on the 30th day of July 1987 in connection with the Waterford/Ballymun investigation, Mr Larry Goodman availed of the occasion to complain to the Secretary regarding the standard of classification at AIBP and complained specifically about Mr Feeney.

On the 5th August, 1987, the Secretary wrote to Mr Larry Goodman as follows:—

> "5th August 1987
> Mr Larry Goodman
> Chairman
> Anglo Irish Beef Ltd
> Ravensdale
> Dundalk
>
> "Dear Larry
>
> "When you called to see me on 30 July about another matter, you expressed some dissatisfaction with the classification of cattle by Department staff at your Group's Waterford factory. I have since checked the position with the supervisory staff here at headquarters.

"Classification is to some extent a subjective exercise, based as it is on visual inspections alone. The possibility of human error, or of 'drift' from the norm, is therefore a real one. We try to provide against this by means of regular visits by supervisory staff — including headquarters staff — to all factories to ensure that each officer's work is satisfactory and consistent within narrow tolerance limits.

"We recognise that there will always be some variation between one factory and another in the percentage of cattle falling into particular cells or 'boxes' of the grid. The very idea of a national average implies that some factories will be below the norm and some above it, though obviously these relative positions are all liable to vary over time.

"We also recognise that the results at your Wateford plant may appear disappointing by comparison with those at other units within your Group. However, our people who have looked into the matter are quite satisfied that the classification at Waterford has been well up to standard. If it has departed at all from the norm, it has been on the side of leniency rather than of over-strictness. The explanation for the ineligibility of some cattlefor intervention may, therefore, lie in the quality of those cattle rather than in the quality of the classification.

"Yours sincerely

J.O'Mahony
Secretary"

In spite of this complaints continued to be made by Messrs Connolly, Thornton and Mr Peter Goodman throughout 1988.

AIBP Waterford stopped slaughtering on 5/5/1988 and did not recommence until 15/9/'88. During this period extensive renovations were carried out at the plant.

Immediately after the closure of the plant on the 5/5/1988 Mr Peter Goodman wrote to the Minister for Agriculture and Food as follows:—

"AIBP Meat Division
Ravensdale
Dundalk
Co. Louth
9th May 1988

PG:AM

Mr Michael O'Kennedy TD,
Minister for Agriculture
Office of the Minister for Agriculture
Kildare Street
Dublin 2

"Dear Minister

AIBP Waterford is closed for annual holidays until 30th May 1988. Regretfully, we will not be re-opening the plant and I feel it is important that you are made aware of the reason.

Farmers and suppliers in Waterford's catchment area have lost confidence in the Department's classification at the plant, so much so, in fact, that we are compelled to buy cattle on a flat basis, i.e, guaranteeing the price before slaughter.

The attitude of the graders is if the factory is paying flat why worry how they are graded. We are not prepared to stand this unnecessary and punitive cost any more.

In the past, as one would expect, problems arose in other AIBP plants but they were normally sorted out quickly by the people in Dublin, i.e. Mr Dermot Ryan and colleagues. However, in this instance, the problem has not been sorted out and the reason is the local supervisor, Mr Andrew McCarthy, seems to have a personal interest, for reasons unknown to us, that AIBP Waterford is harshly and unfairly treated. The two graders normally grading in Waterford have said that they have to follow Mr McCarthy's instructions.

Over the past eighteen months, I have had a number of my own experienced graders go to Waterford and they all agree that the grading is tough. Statistically the classification people in Dublin will say that Waterford is not much worse than the rest of the country, but statistics can hide a multitude.

"On Thursday, 5th May, at 1.00 p.m. I received a further complaint from Waterford. At 5.30 p.m. I arrived into Waterford unannounced to see the situation at first hand for myself. What I saw convinced me that there is no point in re-opening Waterford until something is done to sort out the grading problem. Somebody from Dublin arrived on Friday and, in the company of Mr McCarthy, looked at the carcases. They regraded a number of cattle but I cannot and will not accept that all carcases that deserved to be regarded were. We cannot run a business successfully where success or failure depends on the attitude of the local Classification Officers.

Yours sincerely

Peter Goodman
Deputy Chairman

C.C. Mr Donal Creedon"

Again, this letter contains the threat that if they do not get their own way with regard to classification (and now it is Mr Feeney's supervisor Mr Andrew McCarthy who is accused of harsh and unfair treatment) they will not re-open the plant.

A copy of this letter was sent to Mr Creedon, who had succeeded Mr O'Mahony as Secretary and who replied to Mr Peter Goodman as follows:—

"22 June 1988

"Mr Peter Goodman
Deputy Chairman
AIBP Meat Division
Ravensdale
Co. Louth

Dear Mr Goodman

You sent me a copy of your letter of 9 May addressed to the Minister about the Department's carcase classification work at your Waterford factory. Your Chairman had correspondence with my predecessor on the same subject last year.

We have looked into your complaint, just as we look into all complaints. I do not doubt that it was made in good faith, or that your local management may sometimes feel hard done by in the classification of carcases. It would be a miracle were it otherwise.

Classification is an inexact science (or "to some extent a subjective exercise", as Mr O'Mahony put it). Just about every decision a grader has to take is a marginal decision: is the carcase eligible or ineligible for intervention? is it an R3 at such-an-such price or an R4 at so-and-so price? With tens of thousands of cattle passing along the line every year there is simply no possibility of a one hundred per cent meeting of minds between grader, producer and factory management. That's why we rotate staff, to the extent that resources permit. That's why we employ supervisors. That's why we have a national standards panel to keep the performance of our graders under continuous review. That's why we have to investigate complaints like the present one.

You may take it that every effort will continue to be made to be fair to your Waterford factory — and to every other factory in the country. It is simply not constructive, however, to single out the work and attitudes of individual officers for special criticism. They are all members of the same team. Our controls, we are satisfied, are adequate to ensure high standards and their impartial application at every factory.

There are a couple of other points I would like to make clear. Firstly, as far as were are concerned classification determines what we can and cannot buy into intervention. The Department is not a party to the contract between farmer and factory. Secondly, if the EC Commission's recent statements of intention are anything to go by, intervention may in the future play a much less crucial role in the management of the beef market than it has done for the past fifteen years. The Classification problem — if there is a problem — will to that extent solve itself.

Yours sincerely
D. Creedon
Secretary"

Mr Peter Goodman replied as follows on the 27th June 1988:—

AIBP Meat Division
Ravensdale
Dundalk
Co. Louth
27th June 1988

PG:AM

Mr Donal Creedon
Secretary
Department of Agriculture & Food
Dublin 2

Dear Mr Creedon

Thank you for your letter of 22nd June and I would like to respond to some of the points you have made. I have been around livestock and carcases all my life and feel I am competent to judge carcases under any conditions and form an objective opinion as to the accuracy or otherwise of the classification.

I personally went to Waterford unannounced to see with my own eyes if the complaints and problems which I had been hearing about were justified. What I saw vindicated the complaints made to me on the grading that day and I have no reason to doubt the other complaints which I have received throughout the year.

I have in the past suggested the rotation of Classification Officers between the various plants and I also suggest the rotation of Area Supervisors. Your letter would indicate that this is happening but it is the exception rather than the rule to have any Classification Officers rotated and, on no occasion, have Area Supervisors been rotated.

I accept the Department is not a party to the contract between the farmer and factory but, over the last couple of years, we have tended to buy cattle on a graded basis and this grading is done by the Classification Officers. Where a number of suppliers, as has happened in Waterford have loaded cattle at random, some for AIBP Waterford and some for another local plant, one would expect that the grades would be reasonably in line, however, I have correspondence that would indicate that the grades are more severe in AIBP than in the other plant in the area covered by the same Supervisor. You will be aware that news like this spreads like wild fire and the net result that we are compelled to buy cattle on a flat basis or else guarantee the price pre-slaughter despite the grade.

I very much regret having to respond in the above vein but, because of the vast amount of money involved, I feel justified in the action I have taken.

Yours sincerely

Peter Goodman
Deputy Chairman "

No officer was transferred from Waterford and the plant re-commenced slaughtering on the 15th September 1988.

It appears, however, that because of the level of complaints from AIBP and due to the proximity of AIBP and Dawn Meats the rotation of officers between these plants during 1987 and 1988 was increased and in 1988, classification officers from other plants were on duty for approx. 40% of kill days at AIBP.

While the representations made by AIBP did not result in the actual removal of any Classification Officer, it did result in an increase in the level of the rotation of such officers as between different plants.

(h)Abuses of the Irish Tax System

In relation to the payment of tax by the Goodman Group, Mr McGuinness had stated.

"The company had a wide scheme of under the counter payments. Cheques were made out against bogus invoices, endorsed by Goodman Employees and cashed at local branches of the Allied Irish Bank. These cheques were payable quarterly in March, June, September and December of each year. They were paid to everyone in the company from the floor up and amounted approximately to 3 million pounds per year".

The Tribunal has dealt in detail with this allegation in the Chapter of this Report dealing with Tax Evasion and Tax Avoidance from which it is clear that his allegation in this regard has been substantiated with regard to payments within the jurisdiction of the State.

The Tribunal is not concerned with and does not intend to deal with or report on tax evasion which may or may not have been practised outside this jurisdiction.

(i) Mr McGuinness had alleged that the abuses which he had outlined were institutionalised within all the factories and that Larry Goodman "set the tone."

The use of the phrase 'set the tone' would seem to imply that Larry Goodman was aware of and authorised the practises referred to in evidence by Mr McGuinness and dealt with during the course of this Report.

The only evidence given by Mr McGuinness relating to the personal involvement of Mr Larry Goodman in any of the matters of which he gave evidence was in relation to a meeting of the management of the Goodman Group of companies held on the 28th day of March 1986 at Ardee in the County of Louth.

Meetings at Ardee, Co. Louth

It was the practice of the management of the Goodman Group of companies to have an annual review of the performance of each company in the Group during the preceding year.

These meetings were usually held in February/March of each year and Patrick McGuinness, as the Plant Accountant, attended two of these meetings during the course of his employment by the Goodman Group, one in respect of the performance of the Plant at Newry and the other in respect of the Plant at Waterford.

The first of such meetings was held at the company headquarters in Ardee on the 28th day of March 1986. When the accounts in respect of the year ended the 31st day of December 1985 of the Newry plant and its performance during that year were reviewed. Present at this meeting were Mr Larry Goodman, Mr Peter Goodman, Mr Brian Britton, Financial Controller of the Group, Mr Nobby Quinn, Manager of the Newry plant, Mr Patrick McGuinness and one other person.

According to Mr McGuinness' evidence, there was

> "A general discussion about the performance of the company at Newry for that particular year. The discussion which was opened up by Mr Goodman, involved a particular topic was that the abattoir was showing a very sizeable gross margin for the year based on a percentage turnover and the Boning Hall, which is where all the intervention boning had been undertaken, was not showing a large or a reasonable profit."

During the course of such discussion, Mr McGuinness stated that he attributed the profits made by the abattoir to the weights added on in the boning hall because:

> "Essentially 100% of the benefit was being passed to the abattoir because the invoices were regarded as a sale out of the abattoir even though the recording of the weight was done in the deboning hall stage."

He stated that it was resolved that a charge be instituted in the abattoir which would, in effect, be a transfer of profit from the abattoir to the deboning hall operation in order to enable the boning hall to obtain some benefit from the adding on of weights.

He stated that during the course of the discussion, the upping of the weights was discussed and it was accepted by those present that the practice had been originated in Newry.

The fact of such discussion was vehemently denied by those present at the meeting, other than Mr McGuinness and it was established during the course of the cross-examination of Mr McGuinness by Counsel for Mr Larry Goodman, that the premise upon which it was based was incorrect, that in fact the profits from the deboning hall considerably exceeded those of the abattoir, but that the target profit set for the abattoir was exceeded by £79,021 and the target figure for the profit for the boning hall was exceeded by over £250,000.

In addition, it was established that no boning charge was made to the account of the abattoir.

In view of the denials made by the other persons present at the meeting, the Tribunal is not satisfied to accept Mr McGuinness' evidence with regard to the details of the discussions or that there was any particular reference to the upping of weights on the IB4s in the Newry plant and in particular accepts the evidence of Mr Larry Goodman that the question of the "upping the weights" was not discussed with him or in his presence by Mr McGuinness on this occasion or any time.

Mr McGuinness also gave evidence with regard to the annual review in respect of the Waterford plant for the year ended 31st December 1986 which again was held in the company's headquarters at Ardee in or about the month of February 1987.

Present at this meeting were Mr Peter Goodman, Mr Gerry Thornton, Mr John Connolly, Manager of the Waterford plant, Mr David Murphy, Accountant, and Mr Aidan Connor of the International Division.
Again Mr McGuinness stated that in the course of the discussion he referred to the weights being added on in the boning hall. When he did so, Mr Peter Goodman is alleged to have said:

> "Don't get caught — perhaps you should take out more meat."

By this Mr McGuinness stated that he meant that more meat should be transferred to the company's own stock from the intervention cuts of meat.

Again, those present, Gerry Thornton, John Connolly, David Murphy and Aidan Connor denied that Mr McGuinness had informed them of the weights being added on or that Mr Peter Goodman had made the statement attributed to him by Mr McGuinness.

In regard to this meeting the Tribunal accepts the evidence of Mr McGuinness because it is most probable that any review of the performance of the company in the Waterford Plant for the year ended the 31st December 1986 would have involved a review or discussion of the difficulties created for the Group by the Customs and Excise investigation which was then in progress into the irregularities, involving the over-declaration of weights, in respect of the APS contracts even though these over-declarations had been made by the sub-contractors, Daltina Traders Ltd.

The discovery by the Customs and Excise authorities of the over-declaration of weights, and the part subsequently played by the Department of Agriculture officials in the investigation thereof, led to the risk of greater attention being given by Department of Agriculture officials to the weights being recorded on the IB4s and to a greater risk of detection.

In these circumstances the reaction of Mr Peter Goodman as described by Mr McGuinness consisting of a warning to avoid detection and a suggestion that more meat be taken out and transferred to own stock was not an unexpected one. In view of the policy of the group to transfer into Intervention storage beef representing a yield of 68% or slightly in excess thereof and to regard any meat in excess of such yield as "trimmings" which they claimed they were entitled to retain and transfer to its own stock.

This practice and the purported justification therefor has already been dealt with in detail in this Report.

While this practice was not followed at the Waterford Plant, while Mr McGuinness was there the policy of transferring excess yields to the Company's own stock was in accordance with the information given to him by Mr Nobby Quinn while he was in Newry and referred to herein.

The allegations made on the ITV programme by Mr McGuinness related to abuses of the system under which subsidies are paid by the European Economic Community consisting of the Aids to Private Storage Scheme, the Intervention System and the Export Refund Subsidy and he alleged that the abuses were institutionalised within all the factories.

The Tribunal had dealt in detail with the investigation carried out by the Customs and Excise authorities and the Department of Agriculture officials into the operation of the 1986 APS Scheme not only in the Waterford and Cloghran plants owned by AIBP but in all other plants operated by AIBP and the only abuses or irregularities discovered were the abuses and irregularities in Waterford and Ballymun/Cloghran.

The other plants were, having regard to the discoveries in Waterford and Clogram/Ballymun, subjected to a careful and thorough investigation and all their operations were found to be in order.

The Tribunal has already stated that the AIBP management personnel were not aware of the over declaration of weights in the boning hall production sheets and the APS yield sheets or of the presence of trimmings in the cartons of plate and flank in Waterford until the matter was drawn to their attention by the officers of Customs and Excise carrying out the investigation and that such abuses and irregularities were carried out by employees of Daltina Traders Ltd to whom the de-boning of beef had been sub contracted.

The Tribunal has dealt in detail with the joint investigation carried out by the Customs and Excise authorities into the operation of the 1988 APS scheme in pursuance of which AIBP had contracted to place in Private Storage 42,383 tonnes of beef representing 31.9% of the beef placed in storage in pursuance of the Scheme.

Despite being subjected to the careful and thorough examination and investigation described in the evidence and referred to in this Report, no irregularities were discovered.

As stated by Miss Harvey of the Department of Agriculture

(*a*) all cuts of meat were physically present;

(*b*) everything was in accordance with hygiene and quality requirements;

(*c*) all weights were correct;

(*d*) there was no inclusion of extraneous matter;

(*e*) every other regulation was complied with, save for a dispute with regard to the use of a continuous sheet of paper to accomplish individual wrapping, which was regarded by the EC Commission as a breach of regulation and in respect of which the fine of £90,228.78 was imposed.

While AIBP were undoubtedly liable for the abuses and irregularities committed by Daltina Traders Ltd in respect of which penalties the region of £1,084,866 were imposed there is no evidence to suggest any systematic abuse of the APS scheme, institutionalised or otherwise by AIBP and the allegations of such abuse are unfounded.

The ITV programme, Mr McGuinness and Deputies Rabbitte and McGiolla had alleged abuses of the Export Refund Subsidy system as outlined in this Report.

The Tribunal has dealt in detail with these allegations in the course of the Report and is satisfied that there was no abuse of the Export Refund Subsidy Regulations in respect of the export of intervention beef to Iraq and no breach of the Intervention Regulations with regard to the re-packaging and re-labelling of cartons of this product.

As stated in the Report, the EEC, the Beef Management Division thereof and the Department of Agriculture were fully aware of the fact that intervention beef was being exported: the Export Refund System Regulations provided for the payment of Export Refund Subsidies in respect of the export of intervention beef to the Third World; including Iraq and the Intervention Regulations permitted the re-packaging, re-boxing and re-labelling of cartons of beef purchased from intervention when permission therefor was obtained.

Consequently the Tribunal is satisfied that there was no abuse by AIBPI of the Export Refund Subsidy System or Regulations and allegations in respect thereof are unfounded.

This is the only public element relevant to the issues raised with regard to the export of beef to Iraq and the allegations made in respect thereof and which entitled the Tribunal to make inquiries in regard thereto.

The terms of the contracts made between AIBPI and the purchasing authorities in Iraq are undoubtedly a matter of private concern and normally would not have been the subject of inquiry by the Tribunal.

Because of the conflict between the terms of the contract which stipulated the nature of the beef to be supplied in pursuance thereof as set forth in this Report and the use of intervention beef in the supplies delivered in pursuance thereof, the Tribunal did inquire lest the reputation which Irish beef justifiably enjoys would be damaged by the inclusion of intervention beef in lieu of beef slaughtered within 90-100 days of delivery but is satisfied from the evidence adduced before it, that the requirements of the contracts in this regard were waived for the reasons set forth in this Report and that the beef supplied was and was certified to be fit for human consumption.

With regard to alleged abuses of the Intervention system, the Tribunal has set forth in detail the evidence with regard to the alleged abuses thereof consisting of;

(i) the adding of weights to the IB4 forms in Waterford and the falsification of such forms by the re-writing thereof.

(ii) the policy of all Goodman plants engaged in the deboning of sides of beef for Intervention purposes to deliver to the Intervention Agency only 68%, or slightly in excess thereof, and to retain as Company stock any meat achieved as a result of such deboning in excess of such percentage.

(iii) the limited use of bogus stamps to alter the classification of animal, and

(iv) the switching of carcases and the substitution of inferior grades of animals for animals with the appropriate grade for intervention purposes.

With the exception of (ii) above, the above abuses were limited. The evidence with regard thereto is set forth in this Report and is not such as to establish that the said practices were widespread throughout all the factories or were practiced at all times and were known to or authorised by the management of the Group as distinct from the Plant Managers of the plants concerned.

It is only right that it should be emphasised by the Tribunal that for the reasons outlined in this Report the finding by the Tribunal that the Goodman Companies are obliged to place in storage all meat achieved as a result of deboning for intervention other than fat, bone and 'certain small trimmings is strongly contested by the Goodman Companies who maintained that by virtue of the terms of the deboning contract and the practice in the trade that they are entitled to retain any yield obtained by them in excess of 68% and to transfer such additional yield to its own stock.

The abuses and malpractice's which occurred in Rathkeale and which have been outlined in this Report constitute serious offences and an abuse of the Intervention system but the Tribunal has held that there is no evidence to suggest that the AIBP management were aware of the fraudulent activities being carried out by management of the Plant at Rathkeale and the records furnished weekly to AIBP management at Ravensdale did not disclose such offences. The contents of the weekly returns submitted to management in accordance with established procedures did not contain all the material shown on the daily Costing Sheets which would have given all necessary information.

While the evidence before this Tribunal has established many irregularities and malpractices as outlined in this Report, it has not been established that they were carried on in all plants or with the knowledge of Mr Laurence Goodman and the management of the Group but they must accept responsibility therefor for failing to exercise effective control and supervision of the personnel employed by them and ensuring compliance with the requirements of all relevant regulations applicable where public funds are concerned. The Tribunal has already dealt with the allegation with regard to tax evasion in the course of this Report and is satisfied that it was practiced in all plants with the knowledge of the management of the Group and the allegation in respect of such practices have been fully substantiated.

There has not been established any basis for the allegation made in the ITV programme that Mr Larry Goodman and his companies had 'the right connections at the right places that could basically control any investigation that would be put in place'. There is no evidence to suggest that any investigation carried out by any of the relevant authorities including the Department of Agriculture, the Revenue Commissioners, the Customs and Excise authorities and the Garda Siochana were at any time or in anyway controlled or sought to be controlled by any "connections", political or otherwise. Indeed, all the evidence is to the contrary.

Index to Chapter 24

Pages

CHAPTER TWENTY-FOUR

Other Companies

The allegations made on the ITV programme and made in Dáil Eireann, (other than one allegation made by Deputy Desmond) related to the Goodman Group of Companies. The Tribunal considered it necessary to inquire into the activities of the other companies involved in the beef processing industry including the registered cold stores to ascertain whether or not there existed any illegal activities fraud or malpractice in or in connection with the beef processing industry in these companies.

The Tribunal with the assistance of the Department of Agriculture and Food prepared a list of those companies believed to be engaged in the beef processing industry. The Tribunal wrote to those companies on the 27th of June 1991 in the following terms:-

> "Dear Sirs,
>
> The Government of Ireland by Resolution passed by Dáil Eireann on the 24th day of May, 1991 and by Seanad Eireann on the 29th day of May, 1991, established a Tribunal of Inquiry, which Tribunal of Inquiry was appointed by Warrant of the Minister for Agriculture and Food dated the 31st day of May, 1991.
>
> The Terms of Reference of the Tribunal are as follows:-
>
> 1. To inquire into the following definite matters of urgent public importance:
>
>   1. Allegations regarding illegal activities, fraud and malpractice in and in connection with the beef processing industry made or referred to (a) in Dáil Eireann and (b) in a television programme transmitted by ITV on May 13th, 1991.
>
>   2. Any matters connected with or relevant to the matters aforesaid which the Tribunal considers it necessary to investigate in

connection with its inquiries into the matters mentioned at 1. above.

2. To make such recommendations (if any) as the Tribunal having regard to its findings thinks proper.

The Tribunal now requests that you immediately send to the Tribunal, at the above address, all material documentary or otherwise in your possession relevant to the matters referred to in the Terms of Reference. Furthermore, the Tribunal requests that you furnish the names and addresses of all persons who are able to assist the Inquiry in relation to the matters referred to above.

We would appreciate an early reply.

Yours faithfully

____

Mr Justice Liam Hamilton
President of the High Court
Sole Member of the Tribunal of Inquiry"

At the same time the Tribunal sought to obtain a list of employees employed in those companies and in its material terms requested as follows:

"The Tribunal would appreciate if you would furnish it with a list of names and addresses of all your staff to include the full-time, part-time staff as well as sub-contractors.

The Tribunal would appreciate an early reply."

The above correspondence was sent to the following 53 companies.

1	Ashbourne Meats Processors Ltd	27.	Honeyclover Limited
2.	Arax (Jamestown) Ltd	28.	IMP Limited
3.	Agra Trading Ltd	29.	Heritage Foods Limited
4.	Blanchvac Ltd	30.	Irish Casings Limited
5.	Ballywalter Ltd	31.	Heyer and Sinnat Ltd
6.	Baltinglass Meats Ltd	32.	Ox-Fleischandelgesellschafts Ltd
7.	Barford Meats Ltd	33.	Slaney Meats
8.	C.H. Foods Ltd	34.	Western Meat Producers Ltd
9.	Colso Cold Stores	35.	N.W.L.
10.	Continental Beef Packers Ltd	36.	Nordic Cold Store Limited
11.	Dawn Meats	37.	Q.K. Cold Stores Ltd
12.	D.J.S. / Doherty Meats Carrigans Ltd	38.	Norish PLC
13.	Dehymeats Limited	39.	Michael Purcell Meats Ltd
14.	Eurowest Limited	40.	Redways Ltd
15.	Freezomatic Ltd	41.	Purcell Foods Ltd
16.	Avrich T/a Freshland Foods Ltd	42.	Master Meats / Classic Meats Ltd
17.	Goldstar Meats Limited	43.	Cloon Foods Ltd
18.	Goudhurst Ltd and Hampton Meats	44.	UMP / Halal Meats Ltd
19.	Kildare Chilling Limited	45.	Tara Meats Ltd
20.	Kepak Limited	46.	Rangeland Meats Ltd
21.	Liffey Meats Limited	47.	Meadow Meats Ltd
22.	KMP Co-op. (Midleton) Ltd	48.	Tunney Meats Ltd
23.	Kerry Co-op Cold store	50.	Taher Meats Ltd
24.	Lixsteed Ltd	51.	Autozero / Tallaght Cold Store
26.	Hibernia Meats Limited	52.	Horgan Meats Ltd
		53	Transfreeze Ltd

In view of the evidence which had been adduced before the Tribunal in relation to the allegations made involving the Goodman Group of companies and the matter referred to therein, the Tribunal caused the following letter to be sent on the 8th day of April 1993 to all the other companies engaged in the beef processing industry.

"8 April 1993

Dear Sirs

<u>Re: Tribunal of Inquiry - Beef Processing Industry</u>

The Government of Ireland by Resolution passed by Dail Eireann on the 24th day of May, 1991 and by Seanad Eireann on the 29th day of May, 1991, established a Tribunal of Inquiry, which Tribunal of Inquiry was appointed by Warrant of the Minister for Agriculture and Food dated the 31st day of May, 1991.

The Terms of Reference of the Tribunal are as follows:-

1. To inquire into the following definite matters of urgent public importance:

    1. Allegations regarding illegal activities, fraud and malpractice in and in connection with the beef processing industry made or referred to (a) in Dail Eireann and (b) in a television programme transmitted by ITV on May 13th, 1991.

    2. Any matters connected with or relevant to the matters aforesaid which the Tribunal considers it necessary to investigate in connection with its inquiries into the matters mentioned at 1. above.

2. To make such recommendations (if any) as the Tribunal having regard to its findings thinks proper.

The Tribunal, pursuant to its inquiries into its Terms of Reference and since its appointment has been concentrating on the main beef processor in the industry. The Tribunal is now directing its inquiries to other processors who are important in the industry but may not have as much of a share of the market.

The Tribunal requests the following information from you, concerning the company and request that you note that such information may well be required to be given in evidence to this Tribunal.

1. The nature of the business operated by your company:

    (a) is it solely in the cattle business
    (b) as such, is it involved in:
    (i) commercial
    (ii) intervention
    (iii) exports

2. In respect of premises does it have?:-

    (a) its own slaughter house;
    (b) its own deboning hall
    (c) its own cold store

3. If it has none of the above in general:

    (a) where does it slaughter its beef?
    (b) where does it debone it? and
    (c) whose cold store does it use?

4. In respect of employees:-

   (a) how many employees have you?
   (b) do you engage sub-contractors?
   (c) if so, for what purpose?
   (d) how do you pay your employees?
   (f) do you pay all PAYE, PRSI?

5. Are there any other bonuses or payments made to employees which are not subject to PAYE or PRSI?

6. For how long has your company been in business?

7. Have you taken over any other business' connected with the beef processing industry?.

8. In respect of Intervention beef, on the assumption that your firm debones it, please indicate:-

   (a) what records are available in respect of deboning operations?
   (b) make available to the Tribunal all daily job costing documentation;
   (c) make available all weekly job costing documentation;
   (e) show all records kept by the company of beef above the 68% kept and processed by the company for its own purposes in respect of the years 1987 to-date.

9. In respect of intervention deboning indicate all returns made by the company to the Department of Agriculture & Food on the 1st of January 1987 to-date.

10. If the company sells to Third Countries please indicate:-

    (a) all sales to Third Countries from the 1st of January 1986 to-date.
    (b) in respect of such export all refunds claimed and paid;
    (c) in respect of such exports, whether and how much, of such exports was intervention;
    (f) in respect of such exports to each country how much was beef slaughtered and processed within the 26 counties
    (g) beef slaughtered and processed within the six counties;
    (h) beef slaughtered and processed outside of either of the above:
    (i) indicate whether such beef not slaughtered within Ireland was English, European or non-European.

11. In respect of beef exported by the company, when does the company purchase beef from intervention for export?

12. In respect of any beef exported in what boxes does the beef be exported?.

13. What markings are put on the boxes by the company?.

14. What markings are requested by the customer?.

15. What facilities does the company have for re-boxing?.

16. What proportion of re-boxing takes place without supervision?.

17. What proportion of re-labelling takes place without supervision?.

18. In respect of stamps, apart from intervention grading stamps:-

    (a) what other stamps are used by the company?

    (b) what stamps are provided by customers for use by the company?

    (c) what customers use their own stamps for beef?

The Tribunal appreciates that there is a large amount of information sought in relation to the above but requests such information be made available immediately.

The Tribunal intends resuming its public hearings on the 11th of May, next and will be writing, after the Easter break, to indicate the order and probable time when your company will be required to give evidence to the Tribunal.

The company should note that the Tribunal may request a visit to your companies premises.

The company should note that the Tribunal is also requesting files from the Department of Agriculture and other State Authorities concerning any irregularities known to them concerning the company and when and if furnished with such files will communicate further with the company concerning these matters.

The Tribunal would appreciate an early response and thanks you for your co-operation in anticipation.

Yours faithfully

____

Mr Justice Liam Hamilton
President of the High Court
Sole Member of the Tribunal of Inquiry

The Tribunal obtained the various lists of employees of the companies to whom it had written requesting such information and on the 30th of April 1993 the Tribunal wrote to each of those employees in the following terms:

"30th April 1993

Re: Tribunal of Inquiry - Beef Processing Industry

Dear Sir

The Government of Ireland by Resolution passed by Dail Eireann on the 24th day of May, 1991 and by Seanad Eireann on the 29th day of May, 1991, established a Tribunal of Inquiry, which Tribunal of Inquiry was appointed by Warrant of the Minister for Agriculture and Food dated the 31st day of May, 1991.

The Terms of Reference of the Tribunal are as follows:-

1. To inquire into the following definite matters of urgent public importance:

   1. Allegations regarding illegal activities, fraud and malpractice in and in connection with the beef processing industry made or referred to (a) in Dail Eireann and (b) in a television programme transmitted by ITV on May 13th, 1991.

   2. Any matters connected with or relevant to the matters aforesaid which the Tribunal considers it necessary to investigate in connection with its inquiries into the matters mentioned at 1. above.

2. To make such recommendations (if any) as the Tribunal having regard to its findings thinks proper.

Without being exhaustive and in general terms the following matters are matters which are forming the basis of the inquiries being made by the Tribunal of Inquiry.

(a) Irregularities into the meat processing business.

(b) Method of payment of employees.

(c) Non disclosure of payment of employees.

(d) Whether contract of services exist between any meat company and the employee.

The Tribunal has become aware that you are/were an employee of a meat processing firm and that accordingly you may/may not be in a position to give evidence to the Tribunal having regard to the Terms of Reference.

The Tribunal would appreciate if you would make available to it any documentary or other material or any evidence by way of statements in relation to the matter referred to above and/or the Terms of Reference of the Tribunal.

The Tribunal wishes to inform you that any statement that you wish to make may be made either by yourself or with the assistance of a solicitor or in such other way as you might wish to make it.

The Tribunal would appreciate an early reply at this time.

Yours faithfully,

____

Mr Justice Liam Hamilton,
President of the High Court,
Sole Member of the Tribunal of Inquiry.

The contents of these letters indicate the nature of the inquiries made by the Tribunal.

As part of its inquiries, the Tribunal also wrote to the Veterinary Inspectors and Agricultural staff employed in each of the plants operated by the companies identified as carrying on business in the beef processing industry. In essence the Tribunal sought from these and received, from these persons, statements setting out any irregularities, fraud or malpractice which they knew or were aware had been carried on in the company or companies to which they were attached. The reply from the Veterinary and Agricultural staff formed the basis of evidence subsequently given to the inquiry of irregularities in the beef processing industry.

The Tribunal received, from the Department of Agriculture and Food, files in respect of 47 companies containing particulars of irregularities reported to it and upon which action had been taken by them. These files related to the following companies:-

Agra Meat Packers Ltd.
Anglo Irish Beef Processors
Arax Jamestown
Ashbourne Meats Processors Ltd.
Autozero Ltd
Avrich Ltd
Ballywalter Meats Ltd
Baltinglass Meats Ltd
Blanchvac Ltd
Cahir Meat Packers
Clover Meats
Continental Beef Packers
Dawn Meats Ltd
DJS Meats
Doherty, Carrigans
Eurowest Foods
Freezomatic Cold Store
Gatehill Traders
Goudhurst Ltd
Heritage Foods
Heyer Meats
Hibernia Meats
International Ltd
Horgan Meats
IMP
Kepak Ltd
Kildare Chilling Co. Ltd.
KMP
Liffey Meats
Lyons & Co.
Master Meats Packers Ltd.
Meade Lonsdale
Meadow Meats
Nenagh Chilled Meats Ltd
NWL
OxFleischandels GMBH
Purcell Exports Ltd
Rangeland Meats
Sallyview Estates Ltd
Shannon Meats
Sinnat Ltd
Slaney Meats
Taher Meats Ltd
Tara Meats Ltd
Transfreee Ltd
Tunney Meats
UMP/Halal
Western Dromod

In addition the Tribunal wrote to the various State Authorities and in particular the Department of Agriculture, the Department of Industry and Commerce, the Revenue Commissioners and the Central Statistics Office for information relating to the beef processing industry in the following terms:-

"11th August 1992

Secretary
Office of the Revenue Commissioners
Dublin Castle
Dublin 2

RE: Tribunal of Inquiry - Beef Processing Industry

Dear Secretary

The Tribunal is trying to obtain certain information which is basically statistical in connection with the beef processing industry and seeks your assistance insofar as your Department may be able to assist in supplying the information listed hereunder.

1. **National Herd**

The Tribunal would appreciate if you would make available to it in respect of each of the above years:-

(a) The size of the national herd.
(b) Differentiate between the different types of cattle making up the national herd.
(c) In particular identify the numbers of live animals (prime steers) available for export.

2. **Slaughter**

(a) Identify the number of animals that were slaughtered in each year referred to above.
(b) Identify the different types of animals that were slaughtered in each year referred to above, particularly identifying the number of prime steers slaughtered.
(c) Advise whether monthly figures are available.
(d) Identify the number of licensed slaughter houses together with the ownership in the twenty six counties.
(e) Where possible identify the number of animals slaughtered at each licensed slaughter house.

3. **Cattle prices**

(a) Supply monthly statistics on cattle prices for the years referred to above.
(b) If possible supply similar prices for the UK, France, Germany and Brazil.

4. **Intervention**

(a) In respect of each beef processing trader indicate on a yearly/monthly basis the amount of beef put into intervention by each trader/slaughterer.

(b) Indicate the price paid to the trader/slaughterer for the beef put into intervention.

(c) Indicate the amount of beef put into APS by each trader.

(d) Indicate the amount paid to each trader for the beef put into APS by him.

(e) Identify the traders that obtained a payment by reason of their facilities/storage being available for APS from either the Department or the EEC.

(f) Indicate the amounts paid to the traders for their facilities/storage being available for intervention from either the Department/EEC.

(g) In respect of each trader identify the amount purchased by him from intervention together with the price paid by him for such amounts.

(h) If possible identify which portion of beef sold out of intervention would be subject to export refund, if claimed.

(i) Indicate the amount of export refund paid to each trader for each year.

(j) In respect of each year indicate the number of tenderings in respect of intervention sales would be held.

(I) Indicate the total amount sold into intervention by each trader in the years 1984 to 1991.

(m) Indicate the amount bought out of intervention by each trader in the years 1984 to 1991.

5. **Re Iraq**

(a) In respect of Iraq identify the total market available to world beef processors in the years 1984 to 1991.

(b) Where possible identify which countries supplied the beef to Iraq during the various years.

(c) In respect of Ireland identify the traders that supplied Iraq with beef in the various years.

(d) In respect of Irish traders indicate, where known, the price obtained by each Irish trader.

(e) In respect of Irish traders indicate the amount of export refunds paid to each of them.

(f) In respect of Irish traders indicate the amount of export credit insurance granted to each of them.

(g) In respect of irish traders indicate the amount of guarantees claimed or paid to each Irish trader.

(h) In respect of irish traders indicate the amount of export sales relief claimed by each of them and in respect of each countries.

6. In respect of the following countries:-

Morocco
Iran
Libya
Syria
Egypt
Russia

Supply similar information.

The Tribunal appreciates that the above information being sought may not be in the hands of any one Department and is accordingly sending this letter to the Department of Agriculture and Food, Department of Industry and Commerce, the Revenue Commissioners and the Central Statistics Office.

The Tribunal would appreciate in respect of each Department, if they would answer all information where possible, or such information as is within their power or procurement. It may well be that other questions will follow from the information supplied and the Tribunal would therefore appreciate an early response to enable it to consider and process the replies.

The Tribunal looks forward to hearing from you and thanks you for your co-operation in anticipation.

Yours faithfully,

____

Christina Loughlin
Solicitor to the Tribunal of Inquiry"

Arising from the above letter and also from other correspondence between the Tribunal and the Department of Agriculture and Food the Tribunal obtained inter alia:-

1) Particulars of yields achieved by the companies engaged in Intervention Deboning;

2) Results of defatting analysis, carried out by Department Officials;

3) Particulars of payments by way of Export Refunds to each of the companies between 1984 to 1990 when available;

4) Particulars of the export of beef by each of the companies and the destination to which the beef was exported together with the status of the beef for the years 1984 - 1990 where applicable;

(5) Particulars of purchases from intervention by each of the companies from 1984 to 1991 where applicable;

6) Particulars of the APS Scheme for 1984 to 1989, giving particulars of the companies involved and the amount, the tonnage, contracted and the aid paid to each individual company.

(7) Particulars of sales from Intervention to each individual company from 1987 to 1989 inclusive where applicable.

As a result of receiving this information the Tribunal prepared a book of documents, which was served on each individual company which involved the preparation and service of approximately 53 Books of Documents.

The Tribunal heard oral evidence from the following 46 companies:-

Agra Trading Ltd
Arax (Jamestown)
Ashbourne Meat Processors Ltd
Autozero Ltd / Tallaght Cold Stores Ltd
Avrich t/a Freshland Foods Ltd
Ballywalter Ltd
Baltinglass Meats Ltd
Barford Meats Ltd
Blanchvac Ltd
CH Foods Ltd
Cloon Foods Ltd
Continental Beef Packers Ltd
DJS Meats Ltd
Doherty's (Carrigans) Ltd
Eurowest Ltd
Freezomatic Ltd
Goudhurst Ltd / Hampton Meats Ltd.
Hibernia Meats Ltd
Honey Clover Ltd
Horgan Meats Ltd
Irish Meat Packers
Kepak Ltd
Kildare Chilling Ltd
KMP Co-op. (Midleton) Ltd
Liffey Meats Ltd
Master Meats Ltd / Classic Meats Ltd
Meadow Meats Ltd
Nordic Cold Stores Ltd
Norish Plc
NWL
OxFleischandels GMBH
Purcell Foods Ltd
Michael Purcell Meats Ltd
Q.K. Cold Store Ltd
Rangeland Meats Ltd
Slaney Meats Ltd
Taher Meats Ltd
Tara Meats Ltd
Transfreeze Ltd
Tunney Meats Ltd
United Meat Packers Ltd / Halal Meats Ltd.
Western Meat Producers Ltd.
Heyer / Sinnat Ltd
Lixsteed Ltd
Dawn Meats Ltd

While the Tribunal received and considered the files in respect of the 47 companies already referred to and heard oral evidence in respect of the companies listed above, it is relevant to this Report to point out that the evidence disclosed that in 1990 the market share of companies engaged in the beef processing industry was as follows:

COMPANY	MARKET SHARE 1990
AIBP	28.9
United Meat Packers	12.8
Kepak	7.3
Classic	7.1
Meadow Meats	6.5
Liffey Meats	6.4
Kildare Chilling	5.2
Agra	4.8
Hibernia	4.4
Dawn	4.3
IMP	NO MARKET SHARE FOR 1990
DJS (Tallaght)	
Horgan	

Much of the evidence in the case of the companies from whom evidence was heard, related to (minor) irregularities and minor infringements of regulations discovered by the Department of Agriculture and dealt with by them whether by rejection of meat, suspension of boning operations, by fine or by warnings.

Ms Bríd Cannon an Assistant Principal Officer in the Department of Agriculture and Food produced a table showing the number of forfeited recoveries and financial penalties imposed by the Department during the years 1981-1991 which were generally in respect of breaches of regulations which caused no harm but required financial correction.

This table differentiates between Beef Refunds and Beef APS and is shown overleaf.

TABLE 1

Forfeitures/Recoveries 1981 - 1990

	BEEF REFUNDS		BEEF APS	
	No. of cases	Amount (£)	No. of cases	Amount (£)
1981	7	276,732	-	
1982	9	101,013	n.a.	16,958
1983	n.a.	n.a.	n.a.	867
1984	10	439,682	1	1,425
1985	14	288,526	34	16,068
1986	47	184,908	68	41,870
1987	65	963,476	115	643,513
1988	87	832,342	80	352,541
1989	68	1,743,039	135	486,359
1990	62	2,794,816	158	165,961

In addition Ms Cannon produced a Table with regard to irregularities reported to the EEC in accordance with the provisions of Regulation 283/72 and gave evidence with regard thereto.

This table shows the name of the Company, the nature of the irregularity, the period of the irregularity, the estimated amount involved, the present position with regard thereto and is shown overleaf as follows:-

TABLE 11
IRREGULARITIES REPORTED TO EC UNDER REGULATION 283/72 ARTICLE 3 REPORTS IN PERIOD 1980 - 1991 BEEF SECTION

No. (year)	Company	Nature of Irregularitiy	Period of Irregularity	Est. Amount involved including regulatory penalty where appropriate IR£	Present Position
81	Prinde Ltd Dublin	Non payment of UK MCA on veal carcases exported to UK	November 1974 / October 1975	3,173.29	It proved impossible to recover amount. Amount met by the E. Commission
81	Co. Registered in Isle of Man	Non payment of MCA on beef exported from Ireland.	July 1975 / March 1976	75,857.87	It proved impossible to recover amount. Amount met by the E. Commission
81	Kildare Chilling Co. Ltd.,	Boxed beef and offals misdescribed as offals to avoid payment of UK MCA.	November / February 1981	92,697.49	£92,697.49 recovered from trader. **Case closed.**
82	Shannon Meats Ltd., Rathkeale Co. Limerick.	Understatement of number of beef hindquarters in APS contract.	November / December 1981	(No. APS paid)	Security of £16,958.25 forfeited by trader.
83	Dublin Meat Packers Ltd Cloghran Co. Dublin	Diversion of beef exported to Lebannon possible use of false customs stamps.	April 1982 / May 1983		Case submitted to Gardai for investigation. No prosecution resulted.
83	DJS Meats Ltd. Cookstown Industrial EState, Tallaght, Co. Dublin.	The plant deboning intervention beef failed to place in final store all yield produced from their deboning operations	April 1982 / March 1983	24,032.15	Amount in Full recovered from trader. **Case Closed**
84	Clover Meats Ltd., (no longer trading)	Failure to place in store full yield of beef produced from deboning of intervention beef	January 1983 / July 1993	40,906.80	Amount recovered on 7/2/'85. **Case closed.**
85	Cahir Meat Packers Ltd.,	Use of forged South African Stamps to validate export documentation in order to claim Export Refunds	May 1983 / October 1983	163,000.00 (Not paid to trader)	Case heard in Dublin Circuit Criminal Court in Sept. 1987. fine of £8,000 and two year suspended sentence on Norbert Quinn. Manager of the plant who was prosecuted in a personal capacity. **Case closed**

TABLE 11
IRREGULARITIES REPORTED TO EC UNDER REGULATION 283/72
ARTICLE 3 REPORTS IN PERIOD 1980 - 1991 BEEF SECTION

No. (year)	Company	Nature of Irregularitiy	Period of Irregularity	Est. Amount involved including regulatory penalty where appropriate IR£	Present Position
85	IMP Ltd., Leixlip and Midleton (no longer trading)	Beef originally imported into Egypt but not for home use - and subsequently exported to Trinidad. Possible use of forged Egyptian import document.	April 1985 - May 1985	943,405.60	Case referred to the Gardia for investigation. Criminal proceeding not pursued. £775,000 recovered from company; balance is secured and will be subject of force majeure application.
1/87	Dawn Meats Ltd. Cahir Meats Ltd., CH Food Ltd. Kildare Chilling Co. Ltd. Slaney Meats (Int) Ltd. IMP Ltd. (Leixlip and Midleton) (no longer trading). Rangeland Meats Ltd.	Export refunds claimed on consignments of beef rejected subsequently on entry to USA and Canada, a number of these consignments were re-exported to the Community falsely described.	1982 / 1985	72,234.48 121,594.75 150,133.06 497,035.11 367,785.99 261,428.75	Civil action against five of the companies involved is at an advanced stage. Amount owed by IMP recovered during the winding up of the company. Outstanding securities have also been forfeited in respect of two of the companies.Of the £1.47m. involved a total of £784,103 has been recovered to date.
2/87	AIBP Ltd (Waterford and Ballymun.)	Production records of beef deboned under certain APS contracts overstated. Export refund declaration overstated.	September 1986/ February 1987	1,100.00	Amount recovered. Case referred to Gardai for investigation.
88	AIBP Ltd. Waterford	Attempted use of false intervention. Stamps in order to place ineligible carcases into intervention.	October 1987	-----	Case referred to Gardai for investigation who directed that no further action be taken. **Case closed.**

TABLE 11
IRREGULARITIES REPORTED TO EC UNDER REGULATION 283/72
ARTICLE 3 REPORTS IN PERIOD 1980 - 1991 BEEF SECTION

No. (year)	Company	Nature of Irregularitiy	Period of Irregularity	Est. Amount involved including regulatory penalty where appropriate IR£	Present Position
88	Master Meat Packers Ltd. Clonmel (No. longer trading)	Replacement of carcase classification lables by labels bearing false information and slaughter line weights incorrectly recorded by factory operative.	September 1987	------	Case referred to Gardai for investigation who directed that no further action be taken. **Case closed**
3/88	AIBP Ltd., Ballymun.	Classification labels taken from Steer Carcases and transferred to bull carcases in an attempt to place ineligible carcases into intervention.	December 1987	______	Case referred to Gardai for investigation who directed that no further action be taken. **Case closed.**
4/88	Horgan Meats Ltd., (no longer trading)	Diversion to Zimbabwe of beef exported to South Africa	November 1985 / July 1986	462,00	Papers being finalised for issue of proceedings for recovery of amounts paid.
5/88	DJS Meats Ltd.	Diversion to Zimbabwe and Zaire of beef exported to South Africa.	November 1985 - July 1985	462,00	Papers being finalised for issue of proceedings for recorvery of amount paid.
5/88	Dawn Meats Ltd.	Diversion to Zimbabwe and Zaire of beef exported to South Africa.	1985 - 1986	1,090,000	Ditto
6/88	Dawn Meats Ltd.	Diversion to Zimbabwe of beef exported to South Africa	1986	328,000	ditto
7/88	Heyer Meats Ltd.	Diversion to Zimbabwe of beef exported to South Africa	1985 - 1986	113,000	ditto
8/88	Rangeland Meats Ltd.	Diversion to Zimbabwe, Zaire and Swaziland of beef exported to South Africa.	1985 - 1986	113,000	ditto

TABLE 11

IRREGULARITIES REPORTED TO EC UNDER REGULATION 283/72 ARTICLE 3 REPORTS IN PERIOD 1980 - 1991 BEEF SECTION

No. (year)	Company	Nature of Irregularitiy	Period of Irregularity	Est. Amount involved including regulatory penalty where appropriate IR£	Present Position
9/88	Gatehill Traders Ltd.	Diversion to Zaire of beef exported to South Africa	1986	_____	ditto
16/88	Transfreeze Ltd.	Unauthorised removal from Cold Store of boxes of intervention beef	October 1988 - December 1988	45,300	Case referred to Gardai for investigation. Amount recovered from trader.
5/89	United Meat Packers (Exports) Ltd.	Inclusion of ineligible pieces in beef bonded under APS and absence of individual wrapping for export refund entitlement.	September 1988 - December 19888	1,400,000	Demands issued on 17.5.91 for repayment of monies. Company has applied for injunction to restrain us from going to guarantors. PPS referred to CSSO for consideration of further proceedings.
6/89	Agra Trading Ltd.	Ditto	September 1988 - December 1988	529,000	Monies recovered. Papers referred to CSSO for consideration of futher proceedings.
7/89	Hibernia Meats (Int) Ltd.	Ditto	September 1988 - December 1988	15,000,000	As in case no. 5/89
8/89	Taher Meats Ltd. (No longer trading)	Ditto	September 1988 - December 1988	93,000	Demand issued on 17.5.91 for repayment of monies. Papers referred to CSSO as in case 5/89
11/8	Horgan Meats Ltd.	Inflation of weights misdescription of product etc to increase UK MCA payments	March 1986 - August 1988	168,829.90	Amount of irregularity of £32,9898.68 which remains to be collected from trader. Claims to cover this amount have been held.

TABLE 11
IRREGULARITIES REPORTED TO EC UNDER REGULATION 283/72
ARTICLE 3 REPORTS IN PERIOD 1980 - 1991 BEEF SECTION

No. (year)	Company	Nature of Irregularitiy	Period of Irregularity	Est. Amount involved including regulatory penalty where appropriate IR£	Present Position
15/88	Master Trade (Exports) Ltd Clonmel (C/0 Classic Meats)	Incorrect customs declaration of 197 cartons of fresh beef of which 28 cartons were frozen.	August 1988	6,628,82	Monies recovered. Administrative warning issued to trader by Customs authorities.
16/88	Master Trade (Exports) Ltd Clonmel (C/0 Classic Meats)	Incorrect customs declaration for quantity of beef produced one day after payment declaration was lodged with Customs.	October 1988	6,991.99	Monies recovered. Administrative warning issued to trader by Customs authorities.
17/89	Master Trade (Exports) Ltd. Clonmel (C/0) Classic Meats).	Quantity of beef produced was less than amount declared on payment declaration etc.	September 1988	8,505.77	Ditto.
1/90	Jenkinson Cold Store	Unauthorised substitution boxes containing offal in place of forequarter cuts of intervention beef.	Yet to be determined	Not yet determined.	
3/90	AIBP	Use of unusual wrapping method in respect of male hindquarter beef produced under Regulation 1964/82	September 1988 - December 1988	90,000	Demands issued on 17/5/91 for recovery of monies. Papers referred to CSSO as in case no 5/89
2/91	Liffey Meats Ltd., Ballyjamesduff, Co. Cavan	Possible Misdeclaration of beef being exported to UK (Monetary Compensatory Amounts)	1987 - 1988	To be determined	--------
3/91	Liffey Meats Ltd., Ballyjamesduff. Co. Cavan	Non individual wrapping of some product, found in routine control check by Customs.	June 1991	To be determined	-------

TABLE 11
IRREGULARITIES REPORTED TO EC UNDER REGULATION 283/72
ARTICLE 3 REPORTS IN PERIOD 1980 - 1991 BEEF SECTION

No. (year)	Company	Nature of Irregularitiy	Period of Irregularity	Est. Amount involved including regulatory penalty where appropriate IR£	Present Position
4/91	Anglo Irish Beef Processors Ltd., Ravensdale. Co. Louth	Non-individual wrapping of product, found in routine control check by Customs	June 1991	To be determined	------
5/91	J. Doherty Ltd, Carrigans Co. Donegal.	Possible alteration of intervention production records (Form IB4)	12 April 1991	To be determined	_____

Ms Cannon also produced a Table containing particulars of a number of other cases which though regarded as serious were not reported to the EEC Commission. This table is as follows:-

Other Serious Financial Penalties Imposed 1981 -1990
(Cases not reported as irregularities)

Dawn Meats	1983 - Intervention boning Spec. not complied with	Defective cuts returned to vendors; value recovered (£51,000)
NWL/Hibernia	1984 - Container stolen ex. intervention export	£36,000 in securities forfeit
Tunney	1984 - Intervention spec. not complied with.	Beef returned to company; £31,000 recovered.
AIBP (Barrow)	1984 - Ineligible quarters offered for intervention	Quarters rejected; value (£26,000) recovered.
AIBP Ltd.	1989 - Late lodgement of docts and discrepancies in same - Lebanon	£376,000 in refund securities forfeit.

Kildare Meats	1989 - Dept. not satisfied with docts. lodged re Lebanon	£366,000 in refund securities forfeit
AIBP	1990 - Beef rejected in N.I., returned to Ireland	£800 in MCAs recovered.

This list includes a number of references to cases involving AIBP companies which were specifically dealt with in the Chapters on Waterford/Ballymun, 1986 APS, the 1988 APS and Carousel.

The references to Cases No. 5-8/89 were also dealt with in the chapter on the 1988 APS.

Of particular relevance to the inquiry was the boning yields achieved and returned by the companies engaged in Intervention Boning and as appears from the table produced by Mr Mullen Assistant Principal Officer of the Department of Agriculture these are as follows:-

Average Boneless Yields (1983 - 1991)

FACTORY	1983	1984	1985	1986	1987	1988	1989	1990	1991 (Oct).
Tara Meats (Kilbeggan)	-	-	-	-	68.23	68.39	68.31	68.65	68.46
AIBP Carlow	66.45	67.94	68.10	68.34	68.65	68.38	68.8	68.53	68.47
Master Meats Bandon	-	-	-	-	68.47	68.5	68.48	68.40	68.84
KMP Co-op	-	-	-	69.2	68.79	68.99	68.77	68.79	68.53
Hibernia Athy	-	-	-	69.52	67.76	-	68.29	68.43	68.93
AIBP Waterford	-	-	-	68.25	68.84	68.76	68.56	68.43	68.34
Ashbourne	-	-	-	68.20	68.42	68.16	68.2	68.18	68.34
Hibernia Sallins	-	-	-	68.47	68.43	68.59	-	68.30	70.43
Meadow Meats Ferrybank	-	-	-	-	-	68.59	68.83	68.86	68.31
United	-	-	-	-	-	68.25	68.2	68.14	68.40
Taher	-	-	-	-	-	-	68.65	68.25	69.56
Doherty	66.39	67.35	68.15	68.21	68.19	-	68.82	69.06	68.33
Blanchvac	-	-	-	-	-	-	68.22	68.14	68.78
Baltinglass	-	-	-	-	-	-	69.01	68.32	69.17
Arax Leitrim	-	-	-	-	-	-	-	-	69.05
Kildare	-	-	-	-	-	-	69.39	69.91	68.13
AIBP Longford	-	-	-	-	-	68.16	68.24	68.13	68.49
AIBP Nenagh	-	-	69.24	68.48	68.24	68.24	68.31	68.61	68.66
Shannon Meats Rathkeale	66.31	-	67.72	68.16	68.16	68.27	68.25	68.61	68.31
Master Meats Clonmel	-	-	-	-	-	-	68.61	68.38	68.24
Master Meats Longford	-	-	-	-	-	-	68.5	68.12	68.76
Tara (Tallaght)	66.33	-	68.26	68.05	68.21	68.51	68.36	68.53	68.56
Tunney	-	-	-	-	-	-	68.17	68.30	68.47
Western (Dromod)	66.05	68.01	68.34	68.14	68.18	68.27	68.42	68.37	68.32
Meadow, Rathdowney	-	65.99	-	-	68.01	68.03	68.96	68.30	68.14
AIBP Cahir	-	-	-	-	-	-	68.57	68.10	68.33
Slaney	-	-	-	-	-	-	68.31	68.15	68.31
Halal Ballyhaunis (UMP)	66.07	67.57	68.25	67.98	67.98	68.09	67.88	68.12	68.20
Kepak Clonee	-	-	-	-	68.17	-	68.19	68.10	68.12
Dawn	66.16	66.54	68.11	68.17	68.06	68.24	68.13	68.12	68.10
Rangeland	66.28	67.24	67.87	68.06	68.21	68.14	68.09	68.09	68.32
Halal Sligo (UMP)	-	67.34	68.30	68.26	68.23	-	68.27	67.79	68.11
Liffey	-	-	68.17	68.19	-	68.47	68.21	68.37	68.22
Halal Ballaghadereen (UMP)	66.21	67.39	-	-	-	68.21	68.4	69.13	68.34
Agra	-	-	-	-	68.34	-	68.35	68.04	68.91
AIBP Dublin	-	-	66.31	68.03	-	68.73	-	-	68.08
Ox Fleisch	-	65.47	67.85	-	68.07	-	-	-	-
Kepak Longford	-	-	-	67.93	-	68.05	67.81	-	-
Eurowest	-	-	-	-	68.4	-	68.4	-	-
Kildare Store	66.12	66.07	68.42	68.30	68.62	68.5	68.77	-	-
Master (Clonmel) Store	-	-	68.25	68.67	68.21	68.42	68.42	-	-
Meadow (Rathdowney) Store	66.02	-	66.67	68.06	68.4	68.03	68.02	-	-
AIBP (Cahir) Store	66.47	67.92	68.16	68.12	68.01	68.48	68.31	-	-
Slaney Store	66.43	68.35	68.31	68.38	68.24	68.35	68.28	-	-
DJS Meats	66.02	67.62	69.97	67.95	68.94	68.28	68.58	-	-
Kepak Store (Clonee)	66.76	68.43	68.22	68.17	68.95	68.1	68.11	-	-
Agra Cold Store	-	-	-	69.20	68.23	68.88	68.27	-	-
KMP Co-op (Clones)	67.17	67.49	68.21	68.32	-	68.11	-	-	-
Horgan	-	67.86	68.34	68.46	68.14	68.21	-	-	-
Goudhurst	-	-	68.38	68.29	68.08	68.42	-	-	-
Ms Lyons	-	68.06	68.00	68.17	-	-	-	-	-
Roscrea	65.56	-	68.08	68.06	-	-	-	-	-
IMP (Middleton)	66.42	67.54	68.22	69.15	-	-	-	-	-
Purcell (Sallins)	66.31	66.55	68.09	68.28	-	-	-	-	-
IMP (Leixlip)	66.36	67.2	68.13	-	-	-	-	-	-
Clonmel Foods	66.51	67.50	-	-	-	-	-	-	-
Clover Meats	66.75	67.15	-	-	-	-	-	-	-
Dublin Meats	66.10	-	-	-	-	-	-	-	-
Premier Meats	66.49	-	-	-	-	-	-	-	-
Western Meats (Cork)	66.98	-	-	-	-	-	-	-	-
Liffey (Sligo)	66.43	-	-	-	-	-	-	-	-

This table is self explanatory and established that, with a number of rare exceptions, the yields returned since 1985 are in the region of 68% / 69%

Under the relevant specifications on the deboning contract, the amount of fat permitted to be retained on forequarter is 10% and on Plate and Flank is 30%.

In this regard the result of extensive tests carried out under Department of Agriculture supervision in 1992 showed the following average results in respect of the undermentioned companies:

FACTORY	FQ%	PF%
Tara Meats Tallaght	10.66	30.15
Baltinglass Meats	9.91	23.72
Eurowest Sallins	10.45	24.20
Hibernia Sallins	8.71	-
Eurowest Athy	9.75	24.47
Hibernia Athy	11.78	27.23
Liffey Meats	9.47	23.84
AIBP Nenagh	8.18	22.75
Western Meats Dromod	11.56	25.63
KMP Midleton	10.17	23.24
UMP Sligo	10.40	26.79
UMP Ballaghaderreen	9.93	28.96
Meadow Meats Waterford	10.77	25.94
Meadow Meats Clones	11.27	21.28
Tunney Meats Clones	12.79	26.72
Meadow Meats Rathdowney	10.13	25.80
Dawn Foods Carrolls Cross	7.21	19.66
Blanchvac Tallaght	12.30	17.28
Ashbourne Meats	11.57	27.12
Master Meats Clonmel	10.00	25.63
Kildare Chilling	10.63	21.38
Continental Meats	12.18	27.16
Oxfleisch	7.99	20.11
Freshland Meats	10.13	20.96
Agra Meats Cork	10.39	26.90
Rangeland Meats	11.68	28.48
Kepak Clonee	9.99	25.57
AIBP Cahir	7.01	24.64
Slaney	9.21	17.37
UMP Charleville	14.54	24.31
UMP Ballyhaunis	13.28	24.01
Tara Meats Kilbeggan	16.89	29.72
Kepak Ballymahon	13.08	33.69
Master Meats Ballymahon	16.84	32.18
Master Meats Bandon	13.97	25.46
AIBP Rathkeale	9.97	25.18
AIBP Dublin	6.30	23.13
AIBP Waterford	10.14	20.10
AIBP Longford	10.08	24.68

The table showing the results of extensive tests carried out under Department of Agriculture supervision in 1992 is of considerable importance particularly the column thereof which shows the % of fat left on the forequarters.

The permitted level under the relevant regulations is 10% and while amounts in excess of 10% and below 12% may be due to careless and negligent boning or the quality of the carcase, any percentage in excess of 12% is indicative of harvesting or misappropriation of beef which should have gone into intervention though not conclusive of that fact, provided that the number of cartons examined was sufficient to establish a fair average.

The companies which exceeded 12% fat were:-

Blanchvac Tallaght	12.30%
Continental Meats	12.18%
Tunney Meats	12.79%
UMP Charleville	14.54%
Tara Meats Kilbeggan	16.89%
Kepak Ballymahon	13.08%
Master Meats Ballymahon	16.84%
Master Meats Bandon	13.97%

The main evidence suggesting illegal activities, fraud or malpractice, available and adduced before the Tribunal was the evidence contained in the files submitted by the Department of Agriculture, Customs & Excise, the Veterinary and Agricultural staff, the Revenue Commissioners, together with the oral evidence of the officials from those respective Departments and the evidence of some employees and former employees connected with these companies.

All these irregularities were dealt with as considered appropriate by the Department of Agricultural and where necessary were reported to the EC Commission.

The records of each company, including intervention records, APS records, and Export Refund records, were examined regularly by the Audit Section of the Department of Agriculture which reported on and investigated discrepancies discovered during such examinations and made recommendations in regard thereto.

It is on the basis of the above that the Tribunal conducted its inquiries into those companies involved in the beef processing industry other than the Goodman International Group, and the Tribunal considers it appropriate that it should report on each of the companies from whom it heard evidence.

AIDS TO PRIVATE STORAGE SCHEME

The Tribunal, as one of the matters which it enquired into and obtained information from the Department of Agriculture concerned the details of the Aids to Private Storage Scheme between the years 1984 to 1989. The Tribunal here, sets out the details of those schemes, the companies involved, the tonnage contracted for and the aid paid as follows:

1984 APS SCHEME (2267/84)

CONTRACTOR	CONTRACTED TONNAGE	BONE-IN STORED	BONELESS STORED	TOTAL STORED	AID PAID
1. Anglo	22,691	5,909	11,706	17,615	£7,558.267.15
2. Purcells	15,166	9,731	2,186	11,917	£5,839,503.90
3. Hibernia	7,070	397	4,942	5,339	£3,291,857.36
4. Slaney	4,990	-	3,676	3,676	£1,855,208.88
5. Horgans	5,009	-	3,762	3,762	£1,980,287.13
6. Agra	5,001	618	3,625	4,243	£1,972,769.20
7. Kildare	4,516	-	3,248	3,248	£1,581,000.71
8. IMP Cork	2,342	-	1,624	1,624	£ 678,683.28
9. Halal	1,850	-	1,423	1,425	£ 709,988.40
10. IMP Leixlip	1,685	-	1,045	1,045	£ 475,793.98
11. Dawn	970	60	526	586	£ 235,705.12
12. D. Heyer	700	150	448	598	£ 235,705.12
13. DJS	88	-	69	69	£ 28,680.79
14. Shannon	60	-	47	47	£ 20,928.00
15. Tara	60	-	46	46	£ 15,761.98
16. Liffey	20	-	15	15	£ 8,114.12
17. C. Hurvitz	20	-	-	-	
TOTALS	72,288	16,865	38,388	55,253	£20,488.255.20

1985 APS SCHEME (952/85)

CONTRACTOR	CONTRACTED TONNAGE	BONE-IN STORED	BONELESS STORED	TOTAL STORED	AID PAID
1. PURCELL	4,940	966	3,015	3,981	£ 1,767,354.79
2. HIBERNIA	4,511	-	3,425	3,425	£1,707,504,26
3. ANGLO	4,040	1,514	1,905	3,419	£1,276,198.85
4. HORGANS	480	-	356	356	£ 140,901,37
5. SLANEY	282	-	210	210	£ 70,358,58
6. KILDARE	220	-	150	150	£ 44,617.12
7. IMP CORK	200	-	91	91	£ 25,471.66
8. RANGELAND	160	-	123	123	£ 40,134.87
9. AGRA	40	-	29	29	£ 10,850.76
10.CAHIR MEAT PACKERS	40	-	33	33	£ 8,741,32
TOTALS	**14,915**	**2,480**	**9,337**	**11,817**	**£3,092,133.58**

1985 APS SCHEME (2223/86)

CONTRACTOR	CONTRACTED TONNAGE	BONE-IN STORED	BONELESS STORED	TOTAL STORED	AID PAID
1. ANGLO	29,008	9,059	14,011	23,070	£8,967,115.98
2. HIBERNIA	14,045	-	10,263	10,263	£5,136,382.80
3. PURCELLS	13,577	7,382	4,200	11,582	£4,401,827,56
4. AGRA	8,074.5	253	5,920	6,173	£2,761,153.66
5. KILDARE	4,480	-	3,314	3,314	£1,465,634.81
6. HORGANS	3,750	-	2,787	2,787	£ 971,558.40
7. SINNAT	2,950	539	1,848	2,387	£1,096,531.18
8. HALAL	1,933	-	1,473	1,473	£ 615,223.26
9. SLANEY	1,640	-	1,189	1,189	£ 489,709.25
10. IMP CORK	610	-	409	409	£ 171,412.31
11. D HEYER	447	-	351	351	£ 205,011.93
12. WALDRON	210	-	164	164	£ 49,579.62
13. DJS	100	-	78	78	£ 24,512.35
14. TUNNEYS	70	70	-	70	£ 34,273.56
15. IMP LEIXLIP	20	-	15	15	£ 5,202.64
TOTALS	80,914.5	17,303	46,022	69,325	£26,395,129.31

1986 APS SCHEME (2651/86)

CONTRACTOR	CONTRACTED TONNAGE	BONE-IN STORED	BONELESS STORED	TOTAL STORED	AID PAID
1. ANGLO	16,397	3,365	8,538	11,903	£4,266.073.74
2. AGRA	10,327	1,798	5,683	7,481	£2,505,153.05
3. HIBERNIA	8,264	582	5,649	6,231	£2,438,185.79
4. KILDARE	7,348	-	5,336	5,336	£1,931,562.38
5. HALAL	6,200	-	4,172	4,172	£1,411,926.63
6. SLANEY	4,303	268	3,085	3,353	£1,057,163,07
7. MASTER	3,620	-	2,992	2,992	£1,076,077.64
8. HORGANS	1,120	-	2,553	2,553	£ 896,302.00
9. SHANNON	965	-	787	787	£ 277,719.47
10. LIFFEY	864	-	527	527	£ 71,273.20
11 DJS	800	-	508	508	£ 159,406.80
12. MEADE LONSDALE	600	-	547	547	£ 241,845.45
13. KEPAK	515	-	480	480	£ 129,631.26
14. D. HEYER	260	150	237	237	£ 64,026.38
15. DAWN	250	-	76	226	£ 79,317.06
16. J. DOHERTY	27	-	163	163	£ 55,297.63
17. PURCELLS			18	18	£ 7,943.42
TOTALS	66,827	6,163	41,351	47,514	£16,670,904.97

1987 APS SCHEME (2437/87)

CONTRACTOR	CONTRACTED TONNAGE	BONE-IN STORED	BONELESS STORED	TOTAL STORED	AID PAID
1. ANGLO	15,859	15,401	12	15,413	£4,285,423.91
2. HIBERNIA	15,622	10,053	2,188	12,241	£4,251,036.83
3. HALAL	10,880	-	7,704	7,704	£3,570,658.56
4. TAHER	10,589	-	7,393	7,393	£3,393,649.25
5. MASTERTRADE	7,295	5,745	872	6,617	£2,201,473.22
6. AGRA	6,367	-	4,342	4,342	£2,029,775.78
7. KILDARE	4,615	-	3,353	3,353	£1,488,593.83
8. K.M.P.	3,420	2,487	632	3,119	£ 962,529,29
9. KEPAK	1,265	-	854	854	£ 337,652,31
10. SLANEY	1,120	89	750	839	£ 346,149,78
11. MEADE LONSDALE	700	-	486	486	£ 273,120.45
12. HORGANS	550	-	309	309	£ 136,793.19
13. DAWN	500	489	-	489	£ 126,261.34
14. RANGELAND	490	-	366	366	£ 119,687,53
15. DJS	230	-	154	155	£ 60,930,37
16. J. DOHERTY'S	120	-	82	82	£ 32,830.11
17. N.W.L.	100	-	74	74	£ 39,302.78
TOTALS	79,722	34,264	29,571	63,835	£23,55.868.53

1988 APS SCHEME (2675/88)

CONTRACTOR	CONTRACTED TONNAGE	BONE-IN STORED	BONELESS STORED	TOTAL STORED	AID PAID
1. ANGLO	42,383	34,065	5,016	39,081	£13,620,244.46
2. AGRA	25,978.5	-	17,973	17,973	£10,605,237.06
3. U.M.P.	19,927	-	14,903	14,903	£8,434,643.19
4. HIBERNIA	18,335	4,658	9,609	14,267	£7,488,198.58
5. TAHER	9,598	-	6,900	6,900	£4,067,900.50
6. KILDARE	7,770	-	5,758	5,758	£3,508,178.93
7. LIFFEY	3,395	-	2,437	2,437	£1,224,700.78
8. SLANEY	2,234	-	1,689	1,689	£ 938,939.87
9. K.M.P.	1,000	-	672	672	£ 395,061.42
10. KEPAK	720	-	528	528	£ 295,061.01
11. DJS	700	-	531	531	£ 302,774.52
12. J. DOHERTY	700	-	543	543	£ 288,396.42
13. MEADOW	210	-	144	144	£ 65,090.33
14. RANGELAND	60	-	45	45	£ 24,620.06
TOTALS	133,010.5	38,723	66,748	105,471	£51,259,604.13

1989 APS SCHEME (2965/89)

CONTRACTOR	CONTRACTED TONNAGE	BONE-IN STORED	BONELESS STORED	TOTAL STORED	AID PAID
1. ANGLO	17,900	10,340	4,750	15,090	£5,576,739.54
2. U.M.P.	11,920	11,664	-	11,664	£5,048,342.60
3. HIBERNIA	9,820	4,940	3,366	8,306	£3,297,500.40
4. AGRA	5,931	-	4,213	4,213	£1,907,043,58
5. SLANEY	5,000	-	3,758	3,758	£1,837,021.16
6. KILDARE	4,600	-	3,384	3,384	£1,777,171.63
7. K.M.P.	4,300	4,193	-	4,193	£1,285,205.17
8. LIFFEY	3,560	-	2,567	2,567	£1,195,124.26
9. DAWN	3,550	2,092	1,045	3,137	£1,282,427.68
10. TUNNEY	3,420	-	2,506	2,566	£1,305,348.16
11. WEDDEL	3,000	-	2,174	2,174	£ 986,285.30
12. TAHER	2,840	883	1,314	2,197	£1,018,344.95
13. KEPAK	2,593	2,099	250	2,349	£ 754,020.11
14. MEADOWN	2,000	-	1,438	1,438	£ 724,166.24
15. RANGELAND	312	-	232	232	£ 110,937.46
16. ASHBOURNE	45	-	33	33	£ 15,503.04
TOTALS	**80,791**	**36,211**	**31,090**	**67,301**	**£27,121,182.10**

As pointed out in the chapter of this Report dealing with the 1988 APS scheme, fourteen different companies availed of the scheme; six of such companies between them supplying 2.5% of the contracted production, AIBP supplied 31.9% and the remaining 7 companies supplying between them 65.6% of the contracted production as follows:-

Agra	19.5%
UMP	15.0%
Hibernia	13.8%
Taher	07.2%
Kildare Meats	5.8%
Liffey Meats	02.6%
Slaney Meats	01.7%

These seven companies, together with AIBP and KMP were included in the Sampling Procedure carried out during the course of such investigation and in the case of:-

Slaney Meats and Liffey Meats - all samples were found to be in full compliance with the regulatory requirements.

KMP and Kildare Meats - a small number of boxes sampled were found to contain one or two pieces of meat not individually wrapped.

AIBP - all samples were found to be in full compliance with the regulatory requirements with the exception of a disputed failure to comply with a regulation requiring the individual wrapping of each individual piece of meat.

Serious breaches of regulations with regard to the inclusion of trimmings in the cartons stored and failure to individually wrap pieces of meat were discovered in the cases of:

Agra Trading Ltd
Hibernia Meats Ltd
United Meat Packers Ltd
Taher Meats Ltd

The penalties imposed on each of these companies were £529,817.45, £1,525,748.93, £1,418,148.55 and £96,613.00 respectively.

These four companies had included in the cartons of beef for the 1988 APS scheme trimmings of a similar nature as to those included by Daltina when deboning for the 1986 contract at the AIBP plant in Waterford.

Having regard to the disclosures which resulted from the discoveries made by the Customs and Excise officers in Waterford in 1986/1987 with regard to the inclusion of trimmings and the lessons learnt therefrom, it is surprising that the inclusion of trimmings by these four companies was not detected during the deboning and packaging of the beef prior to being placed in storage but had to wait until an extensive and expensive sampling procedure had to be carried out.

Agra Trading Limited

The evidence to the Tribunal in respect of this company was given by Mr Seamus Fogarty and Mr Maurice Mullen of the Department of Agriculture and Food, Mr Con Healy of the Revenue Commissioners, Mr John Meagher and Mr Owen McCarthy and Mr Denis Mahony of Customs & Excise, Mr John Murray, Veterinary Inspector and James Linnane, Veterinary Inspector, Mr Eugene Regan, Mr Paul Murphy, Directors of Agra Trading Ltd., Mr Michael Behan, ex-Managing Director of Agra Trading Ltd. and Mr Edward Gleeson, ex-General Manager of Agra Trading, Mr Noel Flood, Financial Controller of Agra Trading Limited.

Mr Eugene Regan advised the Tribunal that Agra Trading Ltd., was the trading arm of the Group involved in Third Country Trade or non-EEC Trade. Agra Meat Packers Ltd., deals with the inter-EC trade. Both companies are part of the same group of companies but have separate management and control. Agra Meat Packers Ltd., operates approved meat export premises under EC Council Directives 64/433 and 77/99 consisting of facilities for slaughtering, deboning together with a cold store, with the EEC No. 329. These premises are also approved for the purposes of EC Directive 88/657. The company Agra Trading Ltd., is exclusively involved in non-EEC trading and the company purchases its product from Intervention and from Agra Meat Packers Ltd., as well as from a wide range of EC approved plants in Ireland to fulfil specific contracts in Third Countries. Agra Meat Packers Ltd., is primarily involved in manufacturing slaughtering and processing of cattle and it also exports exclusively to destinations within the EC. This company would also produce beef for intervention. There are approximately 185 full-time employees working for both companies. The company employs sub-contractors mainly for deboning purposes when the company is extremely busy. The employees are paid subject to a deduction of PAYE and PRSI and the company pays all PAYE and PRSI to the Revenue Commissioners as appropriate. The company has been in existence since 1975. The Department of Agriculture and Food gave evidence to the Tribunal of the intervention deboning yields achieved by the company from 1986 to 1992 as follows:-

Intervention Yields							
	1986	**1987**	**1988**	**1989**	**1990**	**1991**	**1992**
Agra	_	_	_	68.35	68.37	68.34	69.57
Agra Cold Store	69.20	68.94	68.88	68.27	_	_	_

Mr Maurice Mullen of the Department of Agriculture and Food gave evidence of a defatting analysis carried out at the Group's premises at Watergrasshill, Co. Cork on various dates between the 10th April 1991 and the 12th of February 1992 with the following overall results.

OVERALL RESULTS	FQ %	PF %
No. of Boxes Defatted	25	10
No. Overfat	13	3
Average	10.39	26.90%
Range	6.10% - 15.34%	17.74% / 38.39%

The Department sought compensation of £586.38 from the company which was paid. The Department did further defatting analysis of the forequarter product in January of 1993 on two dates, the 10th/11/'92 and the 13th/11/'92. On both these days the fat level was well within specification being 6.76 percent on the 10th/11/'92 and 9.42 percent on the 13th/11/'92.

The Tribunal was furnished with details of the exports of beef to Third World Countries for the year 1984 to 1990 and they are as follows:-

Company	Destination	1984 STATUS			1985 STATUS		
		INT	APS	OTHER	INT	APS	OTHER

Agra Trading Limited	Zaire	_	-	259,652.4	-	-	148,480.20
	Iran	-	-	625,556.0	-	-	205,053.00
	South Africa	-	-	5,703.0	-	-	-
	Gabon	-	-	19,296.0	-	-	15,819.70
	Algeria	-	-	10,114.0	-	-	-
	Third Countries	38,059.0	-	-	-	-	-
	U.K. (Provisions)	18,003.0	-	-	-	-	-
	G.B.R.	-	-	85,066.7	-	-	15,452.00
	Egypt	-	584,155.1	3,790,888.6	-	2,824,887.80	90,217.60
	Cyprus	-	-	-	-	12,020.90	-
	West Africa	-	-	-	-	-	41,196.00
	Ivory Coast	-	-	-	-	29,779.50	41,316.90
	Victualling	-	-	-	50,976.0	-	-
Export Refunds Paid		**£14,456.187**			**£N/A**		

Company	**Destination**	**1986 STATUS**			**1987 STATUS**		
		INT	**APS**	**OTHER**	**INT**	**APS**	**OTHER**
Agra Trading Limited	Algeria	-	831,416.90	-	-	-	-
	Egypt	-	2,745,464.34	1,896.90	2,226,152.70	3,515,755.33	1,731,920.66
	Gabon	-	-	19,522.00	-	-	-
	Iran	-	488,391.30	-	-	1,894,743.86	-
	Israel	-	-	133,535.20	-	-	553,969.70
	Russia	-	-	46,039.80	-	-	-
	Zaire	16,025,664.10	-	134,341.87	-	-	-
	Cyprus	-	-	-	-	27,564.14	-
	Gibralter	-	-	-	13,006.0	-	-
	Malta	-	-	-	56,578.4	-	-
	Canaries	-	-	-	-	19,092.52	-
	South Africa	-	-	-	-	1,535,646.72	-
Export Refunds Paid		**£N/A**			**£21,690,383.4**		

Company	Destination	1988 STATUS			1989 STATUS		
		INT	APS	OTHER	INT	APS	OTHER
Agra Trading Limited	Canaries	151,737.9	54,059.59	6,119.92	-	366,528.26	-
	Cyprus	205,424.4	253,869.33	13,781.6	-	495,135.59	14,976.22
	Egypt	56,956.0	2,382,685.46	166,981.38	-	7,830,239.71	375,959.90
	Malta	119,574.0	-	13,971.92	-	10,035,.40	-
	New Caledonia	-	13,262.38	-	-	-	-
	Saudi Arabia	-	-	1,037.4	-	-	-
	South Africa	46,988.0	930,379.50	362,781.47	-	-	-
	Zaire	-	-	79,868.39	-	-	-
	French Army Victualling	4,228.0	-	-	-	-	-
	Sweden	19,985.0	-	-	-	20,084.12	-
	Finland	-	-	-	-	56,082.22	7,645.30
	Romania	-	-	-	-	187,383.00	574,746.42
	West Germany Victualling	-	-	-	-	-	955,523.10
	UK Victualling	-	-	-	-	-	169,832.70
	Iraq	-	-	-	-	3,029,492.16	-
	Nigeria	-	-	-	-	-	227,245.00
	Algeria	-	-	-	-	2,032.70	486,669.30
	Yugoslavia	-	-	-	-	-	100,265.00
	USSR	-	-	-	-	5,132,077.32	101,295.00
Export Refunds Paid		**£19,208,685.29**			**£25,667,034.22**		

Company	Destination	1990 STATUS		
		INT	APS	OTHER
Agra Trading Limited	Israel	-	-	633.10
	Tahiti	-	21,503.15	-
	French Polynesia	-	14,708.60	-
	New Caledonia	-	-	13,459.10
	Gambia	-	13,553.90	13,056.87
	Finland	-	708,937.92	19,720.60
	Iraq	-	45,835.90	-
	Romania	-	6,081.80	-
	Mauritius	-	26,829.50	600,564.90
	Algeria	-	1,876,473.70	-
	Zaire	-	324,310.89	81,798.64
	Saudi Arabia	-	23,650.08	455,293.49
	Cyprus	-	55,702.00	290,341.69
	Canaries	79,542.80	82,351.14	62,166.37
	Gabon	46,977.00	55,084.29	245,011.30
	The Ivory Coast	-	57,241.70	10,495.32
				44,681.61
EXPORT REFUNDS PAID		**£N/A**		

Agra Trading Ltd was one of the companies dealt with during the course of the investigation into the operation of the 1988 APS Scheme and who were obliged to refund the sum of £529,000 by way of penalty.

Mr Seamus Fogarty gave evidence to the Tribunal of irregularities as follows:

(1) On the 6th day of February 1986, an inspection of the deboning of intervention beef in the boning hall revealed certain defects in the trimming of the fillets with consequent loss in meat yield, knife marks were observed on the striploins and excessive fat was noted on the rump cuts.

(2) On the 18th of April 1986, a further inspection of intervention deboning revealed defects in the rumps and plate cuts. Other than that, the quality of the deboning was good and the defects pointed out were corrected immediately.

(3) On the 23rd day of May 1986, an inspection for deboning for intervention was carried out and numerous knife marks were noted on the briskets and the skin was not cut in accordance with intervention specifications.

Complaints were made to the foreman and a subsequent examination disclosed that the standard of deboning had improved.

(4) Between the 26th day of January and the 15th day of June 1988, it was discovered that a number of primal cuts were missing from intervention deboning production during that period. The number of cuts over this period of nearly 6 months which were missing was 19, and the explanation given, which was accepted by the Department, was that this was due to recording and/or packing error.

(5) In January 1992 the Classification staff reported that there were some variation in grades on production for the 17th December 1991 which had been loaded out from Freshford on the 17th December 1991 for deboning in Agra and it was noted that 5 carcase numbers had different grades on the IB1s and the IB4s.

Carcase No.	**IB1 IB4**
651	U4R4
6489	U3R3
6499	U4R4
6492	U3R3
6604	U4R4

The matter was fully investigated by the Departmental staff as they suspected that there had been a substitution of carcases but as a result of their investigations and the detailed explanation given by the company, the Department was satisfied that the problem arose because of labelling problems and not by a substitution of the carcases.

(6) An examination of the intervention stock in store, carried out by the audit team in July 1988 disclosed that 44 boxes of fillets were missing from the intervention store.

The explanation given by the company was that a former employee had been convicted of the theft of 4 boxes of fillets but they could not say if all the losses were attributable to theft.

Arrangements were immediately put in train to install a proper stock control system and the company co-operated with the audit team.

The company paid to the Department the sum of £22,635.73p in respect of the missing boxes and the Department did not store beef in that store again until Autumn of 1990 when they were satisfied that the new system of stock control had been put in place.

Mr Maurice Mullen of the Department of Agriculture and Food gave evidence of Audit Reports which were examinations into aspects of the business conducted by Agra Trading Group. The evidence was in respect of audits carried out in May of 1983, the 3rd and 4th of December 1985 and the 21st and 22nd of January 1986. The 25th of July of 1988 and the 9th of October 1990, the 11th of March, the 13th of March to the 15th of March, 1991 and while problems of varying kinds were disclosed during the course of these audits the evidence also established that all of these matters were remedied in discussions between the Department of Agriculture and the company.

The evidence of Mr Mullen with regard to details of the audit reports in respect of this company was detailed and showed the care and attention given to its investigations by the Audit Team of the Department of Agriculture.

BLANCHVAC LIMITED

The evidence in respect of this company was given to the Tribunal by Mr Kevin Galligan, Senior Agricultural Officer, with the Department of Agriculture and Food, Mr Eugene McGee, Higher Agricultural Officer with Department of Agriculture and Food, Mr Patrick Gregan, Veterinary Surgeon, Mr Maurice Mullen, Department of Agriculture and Food and the Chief Executive of Blanchvac Mr Laurence Montgomery.

The company has been established since 1981, and it owns its own premises at 55 Cookstown Industrial Estate in Tallaght. It is there that the company operates a deboning premises which is an approved meat export premises under EC Council Directive 64/433 and 77/99 and it has the EEC No. 526. The company is involved in the deboning of beef and sale of lamb and is also involved in commercial, intervention and exporting of beef. The company employs approximately 39 full-time employees and engages one firm of sub-contractors for the purpose of off-loading beef. The employees are paid by cheque and the company pays all PAYE and PRSI to the Revenue Commissioners as appropriate. If the sub-contractor does not have a C2 form then the company deducts in respect of tax, 35% from any sum due to the sub-contractor.

Mr Mullen, Department of Agriculture and Food gave evidence of the deboning yields achieved by the company in the years 1989 to 1991 (October) and they were as follows:

- 1989 - 68.22%
- 1990 - 68.14%
- 1991 (Oct) - 68.33%

The Department of Agriculture and Food carried out a defatting analysis on product produced by the company on various dates between the 21st March 1991 and the 20th February 1992 and the results were as follows:-

OVERALL RESULTS	FQ %	PF %
No. of Boxes Defatted	25	10
No. Overfat	18	-
Average	12.30	17.28%
Range	4.87% to 18.83%	19.18% / 27.99%

As a result of the high fat level on the forequarter compensation in the sum of £2,398.27 was sought from the company. The Tribunal has expressed the view that a fat content in excess of 12% maybe indicative though not conclusive of harvesting or misappropriation of beef the property of the Department of Agriculture. The Tribunal sought, from Mr Mullen, his view, as to a possible explanation for the range of 6.87% (which was very low) to 20.36% (which is extremely high) on the forequarter. Mr Mullen, told the Tribunal:-

> "There may be a number of explanations for it. Obviously, one is that just too much fat is left on, which is the bottom line. It may have been that the company were trying to, in defatting a substantial block of beef that would go into a number of boxes, they were trying to get a fat level of around 10 percent and they had mis-calculated their average. That may have been the answer. It might have been shoddiness, shoddy work, but in overall terms, still extremely high,"

The Tribunal is satisfied that there was no mis-appropriation of beef, the property of the Department of Agriculture and Food by the company.

Mr Mullen also gave evidence of exports by the company of beef to Third World Countries in the year 1990 as follows:-

Company	Destination	1990 STATUS		
		INT	APS	OTHER
Blanchvac	Malta Mauritius Ivory Coast West Africa	_ _ _ _	_ _ _ _	16,196.10 52,009.8 105,938.90 25,520.0
EXPORT REFUNDS PAID		**£N/A**		

The Tribunal received oral evidence from members of the agricultural staff of the Department of Agriculture that the amount of commercial beef physically in the Irish Cold Stores at Tallaght did not correspond with the Movement Certificate records in the Veterinary office in April of 1990. Evidence was given that a request by the company for an Export Certificate was refused by the Department and that Veterinary staff from the Department carried out a detailed investigation into the discrepancy. The investigation revealed that two separate loads, the first of 704 cartons delivered on the 16th of March of 1990 and on the second of 892 cartons delivered on the 27th of March of 1990 had entered into the cold store without a Movement Certificate. The cartons contained forequarter and plate and flank. The investigation revealed that the beef had been sent from Blanchvac Limited to an unapproved cold store in Co. Kildare and from there to the National Cold Store.

When the company was asked to give an explanation to the Department of Agriculture Officials the company explained that there had been a dispute at National Cold Store, which had prevented them shipping the product to that store in accordance with normal procedure. The company said that they had been warned that they would be black-listed if they shipped the beef to any other cold store in the immediate vicinity. They felt therefore, that they had no alternative other than to ship it outside the area for interim storage.

While the product had been stored in an unapproved cold store, it was not possible to obtain a movement certificate prior to its entry into the National Cold Store. However, the product entered the National Cold Store, when the permanently based Agricultural Officer at the store, was away working on a temporary relief basis at other stores in the vicinity. Neither the cold store or the company had informed the Agricultural Officer of the arrival of this product. The Blanchvac beef, the subject matter of the discrepancy, was eventually disposed of on the home market.

UNITED MEAT EXPORTERS LIMITED (HALAL)

The evidence in relation to this company was given by a number of witnesses including Mr Martin Blake, Mr Seamus Fogarty, Mr Patrick Joseph O'Connor, Mr Eamonn O'Donovan, Mr Patrick Garvey, Mr George Collins, Mr Martin Macken, all of the Department of Agriculture and Food. Mr Brian O'Beirn an Accountant with the company, Mr Gerard Butler, Mr Gerard Fogarty, both of the Department of Agriculture and Food, Mr George McLoughlin and Mr Sean Stapleton, both officers in Customs and Excise. Mr John Melville, Mr Aidan Nevin, of the Department of Agriculture and Food, Mr Sean O'Horan, Mr Maurice Mullen of the Department of Agriculture and

Food and Mr Sean Clarke, Chief Executive of the company. Prior to the company going into Receivership, the company operated inter alia a number of beef processing plants in Ballaghaderreen, Co. Roscommon, Ballyhaunis, Co. Mayo, Deepwater Quay, Co. Sligo and Charleville, Co. Mayo, Charleville, Co. Cork and Camolin in Co. Wexford.

The premises operated by the company, at these locations, provided the facilities of slaughtering, deboning and a cold store and were approved meat export premises pursuant to EC Council Directives 64/433 and 77/99.

Mr Maurice Mullen of the Department of Agriculture and Food gave evidence to the Tribunal of the deboning yields achieved by the company at their plants in Sligo, Ballaghaderreen, Ballyhaunis and Charleville as follows:-

Factory	1983	1984	1985	1986	1987	1988	1989	1990	1991
Halal Sligo (UMP)	-	67.34	68.30	63.26	63.21	-	63.27	68.12	63.32
Halal - (UMP) Ballaghaderreen	66.21	67.39	-	-	-	68.21	68.4	67.79	68.22
United Meat Packers (Charleville)	-	-	-	-	-	68.25	68.2	68.14	68.31
Halal (UMP) Ballyhaunis	66.97	67.57	68.25	68.07	67.98	68.09	67.88	68.10	68.31

The Department of Agriculture and Food gave the results of defatting analysis carried out on intervention product deboned by the company's plants on various dates between the 3rd June 1991 and the 6th March 1992 with the following overall results:-

PLANT	OVERALL RESULTS	FQ %	PF %
Sligo	No. of Boxes Defatted	25.16	10
	No. Overfat	10	2
	Average	40	26.79
	Range	5.99 to 16.26	15.81 to 37.12
Ballaghadereen	No. of Boxes Defatted	25	10
	No. Overfat	12	5
	Average	9.93	28.96
	Range	3.41 to 17.30	19.96 to 35.92
Charleville	No. of Boxes Defatted	25	10
	No. Overfat	23	1
	Average	14.54	24.31
	Range	9.16 to 20.32	18.18 to 37.74
Ballyhaunis	No. of Boxes Defatted	25	10
	No. Overfat	21	0
	Average	13.28	24.01
	Range	7.05 to 19.85	20.07 to 29.45

As a result of the overfat levels on the forequarter produced in Sligo, Charleville and Ballyhaunis, the Department sought compensation as follows:

Plant	FQ. - AMOUNT £.
Sligo	411.,65
Charleville	4,746.27
Ballyhaunis	5,992.47

The above sums were not paid. The Department is consulting with the Receiver of the company.

Mr Maurice Mullen gave evidence of the exports by the companies from 1984 to 1990 inclusive:-

Company	Destination	1984 STATUS			1986 STATUS		
		INT	APS	OTHER	INT	APS	OTHER
Halal Meat Packers / UMP Limited.	Cyprus	–	–	52,450.8	–	–	16,850.0
	Egypt	–	–	–	–	1,321,570.4	778,383.9
	Saudi Arabia	–	–	–	–	–	26,343.4
	U. Arab Emirates	–	–	–	–	9,923.0	74,408.1
	Bahrain	–	–	–	–	22,505.1	23,057.1
	South Africa	–	–	–	–	52,620.9	630,412.2
EXPORT REFUNDS PAID		£200,596.75			£9,816,777.85		

Company	Destination	1987 STATUS			1988 STATUS		
		INT	APS	OTHER	INT	APS	OTHER
Halal Meat Packers/ UMP Limited	South Africa	–	3,104,019.7	1,355,749.10	-	36,625.0	215,018.7
	Egypt	–	1,081,558.80	2,289,500.90	802,205.3	1,385,785.1	1,478,114.6
	Saudi Arabia	–	–	–	15,983.00	–	-
	Finland	–	–	–	–	41,005.60	-
	Sweden	–	–	–	–	-	42,359.10
	Algeria	–	–	–	–	3,999,976.50	28,285.7
EXPORT REFUNDS PAID		£14,180,020.20			£27,614,595.00		

Company	Destination	1989 STATUS			1990 STATUS		
		INT	APS	OTHER	INT	APS	OTHER
Halal Meat Packers/UMP Limited	Cyprus	–	867,198.90	78,638.90	–	–	285,452.1
	Canary Islands	–	388,444.80	–	–	–	–
	Norway	–	37,456.90	–	–	–	–
	Malta	–	282,063.90	60,686.10	–	3,589.0	42,751.7
	Hong Kong	–	32,589.10	–	–	–	–
	Finland	–	57,242.00	–	–	–	–
	Sweden	–	10,480.70	80,150.60	–	–	–
	Saudi Arabia	–	83,007.60	–	–	–	45,037.1
	Egypt	–	8,083,015.90	93,993.80	–	–	–
	Iraq	–	3,685,721.60	–	–	–	–
	Romania	–	-	42,713.10	–	178,490.4	–
	Bulgaria	–	99,891.90	47,876.80	–	–	–
	Iran	–	–	6,283,564.40	–	11,660,378.0	24,413.2
	South Africa	–	–	90,422.20	–	–	69,633.6
	Israel	–	–	-	–	–	200,045.4
	Ivory Coast	–	–	-	–	–	254,144.8
	Zaire	–	–	-	–	–	315,127.2
EXPORT REFUNDS PAID		**£33,410,001.25**			**£N/A**		

In the course of his speech in Dail Eireann on the 9th of March 1989 Deputy Barry Desmond referred to a major investigation into the Charleville Plant of the Halal - Associated United Meat Packers Limited in relation to Export Refunds.

The investigation referred to by Deputy Desmond originated from the examination by the Customs and Excise authorities in Waterford into APS product stored there, but which had been deboned at the Charleville plant of UMP Limited.

This investigation was subsequently incorporated into the overall investigation carried out by the Customs and Excise authorities and the Department of Agriculture into the operation of the 1988 APS Scheme.

The Tribunal was furnished by the Department of Agriculture and Food with a breakdown of the production for the 1988 Aids to Private Storage Scheme in which this company took part as follows:-

UNITED MEAT PACKERS (UMP) - APS

	A				B	
	1	2	3	4	5	6
Production Unit	Total quantity of P/F Produced (Tonnes)	No. of Boxes of P/F produced (Boxes)	No. of boxes sampled (Boxes)	Sample as % of boxes produced %	Total No. of contracts produced	No. of contracts sampled
Ballyhaunis	1,618	62,866	1,083	1.7	73	67
Charleville	1,258	56,216	1,961	3.5	76	58
Ballaghaderreen	330	14,536	262	1.8	25	22
Sligo	839	36,354	838	2.3	55	50
Tara	143	5,893	83	1.4	8	8

	C					
	7	8	9	10	11	12
Production Unit	No. of boxes with trim) (Boxes)	Quantity of trim involved (kgs)	Trims as % of sample (%)	Trims Box average by sample (kgs)	No. of Contracts with trims	Range of average quantity of trims by contract sampled (Kgs/box)
Ballyhaunis	7	18.5	0.07	-	-	-
Charleville	222	839.0	1.9	0.43	28	0.01-2.65
Ballaghaderreen	15	10.4	0.17	0.04	6	0.01-0.19
Sligo	2	2.1	0.01	-	-	-
Tara	5	9.9	0.49	0.12	2	0.29

	D			E	
	13	14	15	16	17
Production Unit	No. of boxes with NIW pieces (Boxes)	Total Quantity of NIW pieces (Kgs)	NIW pieces as % of sample (%)	Extra polated quantity of trim (Tonnes)	Extra polated quantity of pieces N/W (Tonnes)
Ballyhaunis	748	2,434	8.7	-	141.0
Charleville	236	1,009	2.3	23.9	28.9
Ballaghaderreen	184	577	9.7	0.6	32.0
Sligo	256	773	4.0	-	33.6
Tara	60	170	8.4	0.7	12.0

As a result of serious breaches of the relevant regulations by the inclusion of trimmings in the cartons placed for storage and the failure to individually wrap each piece of meat the overall penalty imposed on UMP was £1,418,140.55p of which £162,558.77p was attributable to breaches which occurred in Charleville.

This matter has been dealt with in the chapter of this Report dealing with the 1988 APS Scheme.

On the 8th of August 1989 the Veterinary Inspector Mr Blake discovered 8 boxes of forequarter and five boxes of plate and flank in a holding room at the UMP premises at Ballaghaderreen. These boxes were discovered subsequent to production and the weighing out of the beef. However, the IB6s and IB7s had not been checked and signed by the Department's Officials. As a result amended forms were lodged and accepted. The Department Officials informed management that the forms should be completed properly and product should be available for inspection and that if a similar situation arose again the product would be rejected. The company suggested that there had been an error by management which explanation was accepted by the Department of Agriculture though it was suspected that the company had been attempting to syphon off beef, the property of the Department.

On the 15th of August 1989, in the course of a transfer of intervention beef from UMP Ballyhaunis to permanent storage in the Autozero Cold Store a number of additional cartons (seven in all) were discovered in a holding room. The Department's staff having undertaken a check of the production records for that day and a check of the transfer records requested an explanation from management. The company management explained that the error may have been made in completing the IB7s or that the factory staff failed to send the seven boxes to Autozero Cold Store with the rest of the days consignment.

The Department staff prepared new IB7 and IB6 forms and these were forwarded unsigned to Headquarters. The Department were suspicious that the seven boxes were seen by the company as being in excess of the 68% yield requirement and there may have been a temptation on the part of the company to withhold some of the extra meat. The one thing that precluded the Department from concluding that this was a deliberate attempt to syphon off meat was the fact that one box contained shin and shank. This is a specific cut and can be checked.

In 1991 the Department considered suspending the Ballaghaderreen plant because the company did not have the IB7 forms available for inspection when these were requested. On one occasion this coincided with missing cartons of beef which formed the basis for the recommended suspension. However the plant ceased operations shortly afterwards.

Mr Fogarty explained, in evidence, that it was coming to the stage when there were too many plausible explanations and the Department was coming to the conclusion that the wrongful removal of beef was taking place. Mr Fogarty decided to suspend deboning operations at Ballaghaderreen, but a fire occurred at the premises which rendered the suspension unnecessary.

Mr P. J. O'Connor, the Veterinary Inspector in charge of UMP Sligo recalled the most serious incident in this plant, which took place on the 26th of November of 1990. Normally, the boxes of intervention beef would go into the blast freezer, which is adjacent to the boning hall but on this occasion twelve boxes were found in a cold store 50 metres, away where they should not have been. Furthermore these boxes had not been entered on the IB7. The factory blamed the occurrence on inexperienced operatives. A letter issued from Headquarters to the manager in the Sligo plant and a further letter of warning was issued to Mr Sean Clarke, Chief Executive of Halal at the Groups Headquarters.

Mr Shay Fogarty, of the Department of Agriculture, recalled a visit by the Department's Audit Team to UMP on the 24th to the 26th of March 1987 as a result of an anonymous 'phone call that had been received. This 'phone call alleged that cuts were being substituted. Despite a full examination by the Department of Agriculture no evidence of substitution was found.

Mr Blake, Veterinary Inspector in Charge of the UMP/Halal Ballyhaunis plant recalled that on the 4th of September 1989 the defatting analysis for the plate and flank showed up at 34.49% and 40.65% in respect of two boxes and a 16.4% in respect of a forequarter box.

On the 5th of September, 1989, Mr Blake wrote to the factory management, informing them that they were being suspended from intervention deboning for the time being. Subsequently in consultation with headquarters it was decided to reject the forequarter and plate and flank from that particular days production. Two trial production runs which Mr Blake allowed on the 6th of September, showed up more acceptable defatting allowed levels. Intervention deboning remained suspended at the plant for a period of two weeks in September 1989.

Mr Blake referred to the fact that on an occasional basis intervention sides or quarters without intervention stamps were found in the boning hall. This arose in the 1985/ 1986 period and again in 1990 / 1991. Mr Blake's policy in all these cases was to reject any beef without a stamp. Mr Blake wrote to the management on the 8th of October 1990 and on the 24th of July 1991 highlighting the problem and confirming that in all cases unstamped quarters would be rejected. This notification was in addition to many discussions between both parties on the subject. Mr Blake explained to the Tribunal that even though it may have been a procedural problem arising from the fact that the operator placed the stamp on one quarter only the onus was on the contractor to weigh-in for intervention only those quarters which had been stamped as being eligible.

Mr Fogarty recalled a visit by Mr Butler, Regional Supervisory Veterinary Inspector and a Mr Gavigan, Regional Supervisory Agricultural Officer on the 22nd of May 1990 to the Ballyhaunis plant. Mr Butler's report mentioned that the intervention specification was not being met. Shin and shanks had to be returned for skinning. Two boxes of forequarter were defatted and showed 13.7% and 14.6% fat and Mr Butler rejected all the forequarter production done for that day. Mr Butler queried these matters with the Management and advised them, because of his findings that day, and those found earlier by Mr Gavigan, that if there was a similar finding on a subsequent occasion he would have no alternative but to recommend suspension from intervention deboning.

On the 12th of June 1990 at a meeting between Mr Butler and Mr Gavigan and Mr Sean Clarke of Halal, the latter, complained that Mr Gavigan was too severe in his inspections and he alleged that Mr Gavigan selectively checked forequarters for defatting. Mr Gavigan refuted the allegations.

With regard to the operation of this company in Ballyhaunis, Ballaghaderreen and Sligo, the Tribunal heard evidence from the Department of Agriculture officials, including members of the veterinary staff employed in these companies dealing with irregularities which had been discovered by them and dealt with by them.

Explanations were given by the Company in regard to each individual incident and blame sought to be attached to carelessness on the part of operatives.

However the overall effect of the evidence suggests that there was in each of the plants misappropriation of beef which should have properly been placed in intervention storage, the substitution of carcases, the placing of ineligible carcases in intervention and incorrect recording on the IB7's.

Each individual incident was dealt with by the Department of Agriculture as appropriate but before more serious action could be taken by the Department of Agriculture the company was placed in receivership.

At the time of such receivership a considerable amount was due to the Revenue Commissioners by this company and it would appear that such sum is now irrecoverable.

TARA MEATS

The evidence to the Tribunal in respect of this company was given by inter alia Mr Bernard Kelly and Mr Daniel Brady, ex-Employee of Tara Meats, Mr Patrick Ennis and Mr Paul McLoughlin employees of Tara Meats. Mr Seamus Fogarty, Mr James Sheridan and Mr Maurice Mullen, all of the Department of Agriculture and Food. Mr Patrick Gregan and Mr John Matthews and Mr John Melville and Mr Patrick Sexton, all Veterinary Inspectors of the Department of Agriculture and Food, Mr Arthur Ormsby, ex-Factory Manager of Tara Meats and Mr Tony Dunne, Chief Executive of Tara Meats Ltd.

This company was incorporated in 1973 and operated in the beef processing business. Tara Meats Ltd., is the parent company in Tallaght and there is a separate company, Tara Meats Ltd., Kilbeggan. These companies operate approved meat export premises under EC Council Directive 64/433 and 77/99 in respect of Tara Meats Ltd., Tallaght a deboning plant with EEC No. 310 and in respect of Tara Meats Ltd., Kilbeggan a deboning plant with EEC No. 521. The company employs approximately 60 people in Tallaght and over 80 people in Kilbeggant and apart from one occasion in 1987 at Kilbeggan, the company does not make use of sub-contractors. The employees, who are mainly full-time are paid their weekly wages subject to all proper deductions and the company pays all PAYE and PRSI to the Revenue Commissioners as appropriate.

In 1980 the company set up an investment company called Metara Investments Ltd., for the purpose of making full use of Export Sales Relief which were intended to be paid as bonuses to executives through the medium of a tax free dividend. This would not have resulted in a loss to the Revenue as this money would have been paid in the ordinary way to the shareholders, who to a large extent were the executives with tax benefits in any case.

In 1984, the company sought ways in which the benefits of the Export Sales Relief Scheme could be extended to the employees of the company and the company was advised that overtime payment earned by full-time employees during busy periods which might last from 3 to 8 weeks, could be paid by Metara Investments Ltd without the deduction of income tax.

This Scheme was introduced by the company for this period and dividends were paid from 1980 to 1990 when the Scheme ended as exports sales relief, in this country, ceased in April 1990. The Scheme was only adopted in respect of Tara Meats Ltd., Tallaght.

Mr Maurice Mullen of the Department of Agriculture and Food gave details, to the Tribunal, of the intervention deboning yields returned by the company from 1985 to 1991 (October) and they were as follows:-

	1985	1986	1987	1988	1989	1990	1991 (Oct)
Tara (Tallaght)	68.26	68.05	68.21	68.51	68.36	68.53	68.76
Tara Meats Kilbeggan	-	-	68.23	68.89	68.31	68.65	68.79

The Department of Agriculture and Food told the Tribunal the results of the defatting analysis carried out by the Department in the middle of 1992 in respect of the two premises as follows:-

The Department did a defatting analysis from the 22nd of February 1991 to the 30th of April 1992 at Tara Tallaght with the following overall results.

OVERALL RESULTS	FQ %	PF %
No. of Boxes Defatted	25.0	10
No. Overfat	12.0	5
Average	10.66	30.15%
Range	3.56.% to 20.65	13.5% to 41.30%

The Department's results in respect of Tara Kilbeggan were, in respect of the forequarter, 16.89% and in respect of the plate and flank 29.72%.

On foot of these findings the Department requested compensation of £4,396.42 from Tara Meats, Kilbeggan and their deboning bond in the sum of £50,000 was declared forfeit.

A complaint by the Department of Agriculture with regard to the operation of this plant was the lack of boning weigh-in facilities at the point of entry to the boning hall.

It appeared that intervention quarters were weighed into the plant and recorded on the IB4 at the point of unloading from where they passed into the chill room and from there into the boning hall proper.

Up to 1992 this practice continued and as stated by Mr Ferris SSVI in his letter dated the 27th of September 1985 to the Intervention Agency.

> "Under ideal conditions where there are no further quarters in the plant and the meat is deboned immediately upon arrival, there is no need for further weighing of the quarters. However in practice, the situation is often very different. It frequently happens in both plants that the quarters arrived the previous day to being actually deboned. These quarters are stored overnight in the same chill with non-intervention beef having been entered on the IB4's on the day of

> arrival. These quarters are then deboned next day with no further check to ensure that there was no mixing with a non-intervention quarters in the same chill.
>
> There is obviously a great danger of substitution and with a reduction in our staff in both plants to one agricultural officer in each there is a great danger of irregularities occurring".

On the 8th of November 1985 Mr Maurice Mullen of the Beef Division wrote to Tara Meats stating that;

> "the procedures for deboning intervention beef require that beef must be weighed into the boning hall itself. I understand that in your plant the beef is weighed into the plant, onto the IB4 at the point of unloading from where it is brought into a chill and from there into the boning hall proper. This, I also understand, can involve storing the beef overnight and sometimes in the same chill with non-intervention beef.
>
> "Please arrange to provide weighing in facilities at the entrance to the boning hall proper as soon as possible. Beef should then be weighed onto the IB4 and pass directly into the boning hall. Any beef held overnight in chills should also be weighed on to the IB4 when it is actually being moved into the boning hall."

The company pointed out that this requirement would involve major alterations and eventually these alterations were carried out to the satisfaction of the Department of Agriculture.

Mr Bernard Kelly who had been employed by Tara Meats (Dublin) Limited for a period of 10 years gave evidence before the Tribunal and alleged that hindquarters rejected by the Supervisor or Quality Controller in the company for commercial purposes were boned out and placed in intervention boxes.

He stated that the general workers were told to take out cuts from the intervention hindquarters which were kept separate and then brought back to the factory at a later time and vacuum packed into Tara Meat boxes.

His evidence in this regard was challenged by a number of employees of the company and it would seem to the Tribunal that the confusion arose because of the particular circumstances of that company where both intervention and commercial beef were kept in the same chill after weighing in. This situation has now been rectified and there is no evidence of continued malpractices in respect of this company.

KEPAK

The evidence to the Tribunal, in respect of this company, was given by Mr Jon Roberts a former Kepak employee, Mr James Higgins a Veterinary Inspector, Mr Patrick Ledwith and Mr Liam Lynam, of the Department of Agriculture and Food, Mr Benny Bennett, former Supervisory Inspector, Mr Michael Durkan, Supervisor of the Department of Agriculture and Food, Mr Kilian Unger, Veterinary Inspector, Mr

Seamus Fogarty and Mr Maurice Mullen both of the Department of Agriculture and Food.

Evidence by the company was given by Ms. Joan Coughlan, Mr Brian Finnegan, Mr Edward Noonan, employees of Kepak. Ms Bernadette McCann, an ex-employee of Kepak, Mr Martin Finuncane, Factory Manager, Mr Brian Donohoe, Finance Controller, Mr John Horgan, Deputy Chief Executive and Mr Liam McGreal, Managing Director of Kepak Ltd.

The company's main business is beef processing and it operates approved meat export premises under EC Council Directive 64/433 and EC Council Directive 77/99 at Athleague Co. Roscommon where it has a slaughtering facility with EEC No. 313, at Clonee, Co. Meath, where it has a slaughtering, deboning and cold store with EEC No. 317, at Hacketstown, Co. Carlow, where it has a slaughtering and deboning facility with EEC No. 346, and at Ballymahon, Co. Longford where it has a deboning and cold store facility with EEC No. 533. The Longford premises are also approved under EEC Directive 88/657.

Mr Maurice Mullen of the Department of Agriculture gave details of the intervention deboning yields achieved by the group for the following years 1985 to 1991.

	1985	1986	1987	1988	1989	1990	1991 (Oct)
Kepak Store Clonee	68.22	68.17	68.24	68.1	68.19	68.15	68.20
Kepak Longford		67.93	68.07	68.05	67.81	-	-

Mr Mullen also gave evidence to the Tribunal of a defatting analysis carried out by the Department on various dates between the 11th of April of 1991 and the 25th of March 1992 with the following overall results.

OVERALL RESULTS	FQ %	PF %
No. of Boxes Defatted	25.0	10
No. Overfat	12.	2
Average	9.99	25.57%
Range	3.34% / 15.27%	11.63% / 32.21%

The Department carried out a further defatting analysis in respect of intervention production produced by Kepak, Clonee Ltd., on the 15th of December 1992 when the fat level for forequarter was 9.97%and on the 18th of December 1992 when the fat level on forequarter was 10.80%

The Department rejected the day's production for the 18th of December 1992 as the forequarter exceeded the permitted level.

The results for the defatting analysis in Kepak, Ballymahon, for 1992 on the forequarter was a yield of 13.808% and on the plate and flank a yield of 33.69%. The Department, as a result of these findings, sought compensation from the company of £714.43 and declared the deboning Bond in the sum of £25,000 forfeit. The company is disputing the forfeiture of the deboning Bond by the Department.

The 1993 examination at Ballymahon, by the Department for intervention production produced by the company showed a fat level of 9.51% on the 13th of November 1992 and 10.13% on the 16th of November 1992. Again by reason of the level achieved on the 16th of November, the Department rejected the forequarter production for that particular day. The Department carried out a further analysis on the 23rd of May 1993 when the relevant levels for forequarter was 7.42% and for plate and flank it was 23.82%. Both of these were within the permitted level and no further action was required.

Mr Mullen explained the forfeiting of the Bond as follows:

> "As part of its contract a company undertaking intervention boning must place, with the Department, a bond, a performance bond, a security of £50,000 and the Department can have recourse to that bond in the event of faults being observed and for which the company doesn't pay compensation on so its a performance bond in place. When we examined the production of this nature we may deem it to be serious and we would declare that a forfeiture of it, of that nature. We do similar things in the APS area as well. The performance bonds in respect of

those for non compliance, we can declare those from time to time depending on the gravity.

Mr Mullen, also gave evidence in relation to the exports of beef by Kepak Ltd., as follows:-

Company	Destination	1985 STATUS			1986 STATUS		
		INT	APS	OTHER	INT	APS	OTHER
Kepak Exports Limited	West Africa	–	–	17,532.7	–		-
	Saudi Arabia	–	–	–	382,850.0		110,384.6
	Zaire	–	–	–	–		235,257.9
	UAE	–	–	–	37,737.5		4,774.8
	Bahrain	–	–	–	121,732.3		63,037.5
	Oman	–	–	–	5,697.5		18,065.5
	South Africa	–	–	–	–		58,129.9
	Cyprus	–	–	–	71,255.7		6,042.5
	Dubai	–	–	–	8,146.8		-
	Malta	–	–	–	–		19,943.3
EXPORT REFUNDS PAID		£61,318.02			£1,830,742.		

Company	Destination	1987 STATUS			1988 STATUS		
		INT	APS	OTHER	INT	APS	OTHER
Kepak Exports Limited	Cyprus	504,920.93	–	102,941.6	357,207.0	–	19,793.5
	Bahrain	272,189.8	–	–	20,109.50	–	–
	Gibraltar	55,843.40	–	4,138.4	87,027.8	–	–
	Saudi Arabia	647,277.9	–	164,520.3	33,424.5	–	11,007.9
	South Africa	–	61,601.1	188,614.8	–	348,492.5	1,182,382.6
	West Africa	–	43,651.8	–	–	–	–
	Malta	144,445.4	–	28,937.6	4,833.0	–	–
	Egypt	–	–	37,253.5	–	–	–
	Sweden	43,792.4	–	–	60,746.5	–	–
	United Arab Emirates	11,860.0	–	–	38,433.	–	–
	Grand Canaria	141,404.8	–	–	147,913.0	–	–
	Las Palmas	14,010.0	–	–	–	–	–
	Malta	5,968.0	–	–	–	–	–
	Saudi Arabia	46,941.0	–	–	–	–	–
	French Polynesia	–	–	–	13,813.00	–	–
	Algeria	–	–	–	–	294,318.7	–
	Victualling	–	–	–	–	–	16,923.0
	Finland	–	–	–	19,774.5	–	–
EXPORT REFUNDS PAID		£3,279,208.03			£2,491,344.76		

Company	Destination	1989 STATUS			1990 STATUS		
		INT	APS	OTHER	INT	APS	OTHER
Kepak Exports Limited	Cyprus	7,021.0	6,339.9	249,785.8	12,888.0	29,826.6	–
	Romania	–	–	141,931.5	–	13,205.0	59,523.8
	South Africa	–	–	41,105.1	–	–	–
	Iran	–	–	282,482.0	–	–	1,816,928.4
	Malta	–	–	12,968.3	–	–	–
	Saudi Arabia	–	147,533.5	154,335.0	–	–	–
	Liberia	–	–	13,048.5	–	–	–
	United Arab Emirates	–	–	286,616.1	–	–	18,651.8
	Bahrain	–	–	88,164.9	40,003.0	–	21,446.7
	Zaire	–	–	22,000.0	–	–	32,000.0
	Ships Stores	3,999.5	–	–	–	–	–
	Ware House Victualling	4,955.0	–	–	–	–	–
	Ivory Coast	–	–	–	–	–	250,210.4
	Egypt	–	–	–	–	140,439.7	322,839.7
EXPORT REFUNDS PAID		**£8,519,373.20**			**£N/A**		

Jon Roberts had been employed by Kepak Ltd., as a Boning Hall Manager, in Ballymahon, Co. Longford from April of 1991 until January of 1992. Prior to that, he had worked with Classic Meats / Master Meats in the same plant. Kepak Ltd., took over the Master Meats plant in April of 1991.

The evidence of this witness had to be treated with caution because he committed perjury before the Tribunal in respect of evidence he gave on the alleged death of his mother. He gave evidence with regard to a number of facts alleging the misappropriation of intervention beef, the switching of intervention beef into commercial cartons, and he further alleged that the Agricultural Officers engaged in the plant were in collusion with the plant management in furthering these irregularities. He also gave evidence of payments by the company to him while he was a boner, and a boning hall manager, part of which payments were not subjected to the appropriate tax deduction. He also suggested that what he called "silence money" was paid to him while he was Boning Hall Manager, of amounts between £250.00 / £350.00 on four or five occasions. He suggested that these were paid by cheque made out in fictitious names. He stated that while the actual wage cheque was paid on a Friday at lunch-time, on a Friday evening the boners employed by the company got back handers or cash as well which cash payments were not subject to PAYE AND PRSI deductions.

Mr Robert's evidence was contested in its entirety by both the company and the Agricultural Officers, who worked in the plant, whilst Mr Roberts was there. These Agricultural Officers, together with employees of the company, named by Mr Roberts, and the plant management gave evidence refuting totally Mr Robert claims.

The essence of Mr Robert's allegation was that when beef, which had been slaughtered in the Athleague plant, was transferred for deboning purposes to the Ballymahon plant, he would receive instructions from the plant manager to retain between 12 and 20 forequarters for use in connection with the fulfilment of commercial contracts. He stated that this occurred, four out of five days in the week when he was present, when intervention deboning was being carried on in the Ballymahon plant. He told the Tribunal a Ms Bernie Dalton was the employee who held back the tags from the beef when they were being weighed in to the boning hall on arrival from the Athleague plant and that she also filled in the form in the scales area just outside the boning hall. He stated:-

> "The beef came in with tags, weights tags, and numbers on them and she would go round and say, if there was 20 forequarters to be held back, she would go and take the tags off them and she would put them in her pocket and they were kept aside. The stamps would be cut off them and she would carry on weighing in intervention as normal. Then the tags at a later date would be inserted into the forms."

The documentation that came with each load, Mr Roberts stated, that this was adjusted. He explained:-

> "The normal way it was done was when intervention came in, they had green tags and numbers on them, and the amount being held back, the tags were taken off and kept separate from the intervention going in to be deboned. The forms then, at the end of the evening or whenever, the ones that had been kept back, they were inserted into the forms then in the evening.

Mr Maurice Mullen of the Department of Agriculture and Food, at the request of the Tribunal, carried out an analysis of the documentation with regard to intervention for the purpose of ascertaining whether or not the carcases which had been recorded on the IB1 form, at Athleague, were recorded on the IB4 forms, which are completed on entry into the deboning hall. The analysis carried out by the Department showed:-

1. all the quarters coming into Kepak, Longford, were all listed on IB1 forms in the plants of slaughter were subsequently listed on the IB4 forms completed in Longford;

2. There was no evidence of forequarters missing;

3. It is clear that the quarters were all actually weighed into the boning hall.

Ms Bernadette McCann neé Dalton told the Tribunal that she had commenced employment in the Ballymahon plant in 1986., She had worked both as a Quality Controller and on the weighing scales. Her responsibility was to weigh the quarters of beef and record their weights in the intervention documentation on the 1B4s and always did so correctly. Mrs McCann denied categorically that intervention beef was stolen by the company or that a certain number of quarters were diverted from intervention to commercial production. She told the Tribunal that it was a matter for the company when quarters arrived in the factory in Ballymahon in containers whether the beef was to be used for commercial or intervention. Ms McCann's evidence was confirmed by Mr Finuncane and Ms Joan Coughlan.

Mr Brian Donohoe, Financial Controller from the 23rd of September 1991 outlined to the Tribunal the method of payment to employees at that time. He did so by way of an example:

> "If I just explain it by taking an example of a boner. The basic rate of payment was 80 pence per quarter boned for the boners. To determine the wages per

Saturday, it was decided to ensure attendance at work on Saturday and ensure full production - the scheme was put in that it was going to be one and a half times the gross salary with no tax deduction. The tax deduction, or the tax would be fully borne by the company and not by the employee. So, if you just take an example, and I just want to take two people that are named by Mr Roberts in his statement,Denis O'Meara and Billy Byrne. I will just take an example of Saturday the 5th of October 1991. Mr O'Meara boned 61 quarters and Mr Byrne boned 48 quarters. The basic rate was 87/80 pence per quarter. The overtime, take home rate, which would be, in that case, £1.20p. So, therefore, we guaranteed Mr O'Meara a take home pay for that day of £73.20 and Mr Byrne, a take-home pay for that day of £57.50. So, therefore, on the following week we had to recalculate the gross so that both of these people would take home those amounts. Of course, it depended on their tax position. If they were paying tax at the rate of 29%, as it was at that time, or the highest rate of 48%. So as we'll see later on, the cost of the company of Mr Bryne's wages for that day was substantially higher than Mr O'Meara's. Basically for Mr O'Meara, he was paying tax at 29%, PRSI at 7.75% so therefore his total tax liability would be 36.75%. His total take-home pay would be 63.25%. Therefore, the £73.20, the wages for that day would represent a 3.25% So recalculating or re-grossing upwards, Mr O'Meara's gross wages for that day was £115.73 and the company bore the difference between that and his take-home pay of tax and PRSI of £42.53. Mr Byrne, on the other hand was paying tax at the rate of 48% so therefore, his net take-home as a percentage of his gross was only 44.25%. So re-grossing Mr Byrne's wages it meant that even though he had boned 13 quarters less than Mr O'Meara, it cost the company £130.17 or £15 more to pay Mr Byrne. So the cost per quarter, boned for Mr O'Meara, was £190 and for Mr Byrne was £271."

Mr Donohoe, produced the originals of the gross calculation of the wages per quarter boned for each day, showing the calculation of the gross, the transfer of the gross to pay-slip which would be accompanying the pay-slip to the employee.

In respect of the allegation made by Mr Roberts, relating to one cheque on a Friday morning and another in the afternoon, Mr Donohoe said:-

"I don't think that would be hardly credible now considering that we re-grossed his wages and he paid £300 tax. We were hardly, if we were going to pay, split the payment and pay tax. We would hardly split the tax at 48%. It doesn't seem a credible suggestion."

With regard to the loyalty or "silence money," payments:

"When I joined the company in September 1991, the first payments, or the loyalty payments, had been made in June 1991 and the second one was just made, I think, the week that I came to the plant. So, I had no decision making input into that, into the payment of the loyalty payments, and they are actually, currently - well, the first point to make about them, is that there was words here used yesterday like "underhand" or "back hand payments". or whatever. All of these payments are fully identified in our check payments books, they are posted correctly in our monthly accounts, posted correctly in our yearly accounts and to make any reference to these payments as being underhand or whatever is totally, totally false. Now there is an example of our check journal, I think it is further on in the statement and it just shows the names of the people. Mr Roberts is actually named and £350 which was paid to him in June of 1991. So there is no question, at all, of underhand payments. Fully, open, clear, and in the books and they are currently under negotiation with the Revenue as to whether a tax liability attaches to them or not."

Mr Donohoe, also gave evidence in relation to restrictive covenants, which had been paid by the company to certain employees as follows:-

"When I came to the company in September of 1991, which was, I suppose at the start of the busy season and the basic wage structure as explained yesterday, by Mr Roberts was in place. There was a particular problem at that time because there was a lack of local skills in the area, people, a pool of talent of trained young people in the area to carry out the boning activity. In the back end of 1991 the company undertook a recruitment policy. So we recruited approximately 25 trainees, but their training could not start until January 1992. This training scheme was subsequently approved for a Training Support Scheme by Fás, for which we received a training grant. Therefore it left us in a position at the end of September 1991, where we had to maintain production in the plant. There was plenty of beef for boning. We wanted to keep the place boning, we had just taken it over and we needed additional boners. So, these people were to be - they were to be key people in the plant for those couple of months. They were to be on short term contract, but we did not want to change our pay structure to accommodate them. So, we used a legitimate method of payment, a payment by restrictive Covenant, a system which I have come across many times before. I have worked in the taxation area for a number of years and the nature of these payments was that they undertook, that the persons undertook to stay with the company for a fixed period of time, where upon we would pay them a certain amount of money to stay over that period. I was advised, at that time, that because they were coming in new to the company, it wasn't a good idea to pay the money up front to them because we didn't know who they were or the nature of the boners. They could move on to a different factory next week without any - it would be very difficult to get recourse to the payment. So, we spread the payments out over a busy season and each of them signed a restrictive covenant saying that they would stay with the company and the payments were made to them at the end of each week in accordance with the Agreement that Mr Finuncane had made with these boners. Now these restrictive Covenants, as I say, I have come across them many times before and where they are being used as a completely, - there was provision in the Finance Act up as far as 1992. The 1992 Finance Act abolished restrictive Covenants, but prior to that they were perfectly a legitimate method of payment to keep people in any business and I have come across it before. So, I used that method of payment in those months. That method of payment ceased at the end of 1991 because we took on our trainees in January of 1992. Production would have lessened anyway because it wasn't a busy season and by the time we came to the busy season, again in 1992, our trainees were very much on line and were able to take all of the available boning themselves. Now this issue, we have had correspondence with the Revenue. The Revenue are fully aware of these payments and no assessments have been made as yet. It is currently still in negotiation, with the Revenue and no payments have been made or no

> assessments have been made as yet. These may be the payments that Mr Roberts was referring to. I want to explain the facts of the case and not the backhanders or whatever. These are fully - they are available in the books, they were posted correctly to boning payroll, in the monthly accounts. Posted correctly in the yearly accounts, and the question of a tax liability is still under negotiation and I would be very confident, based on my past experience, that we will win that case. I have no doubt that we will fight the case very strongly."

Mr Benny Bennett, a Superintending Veterinary Inspector in the Eastern Meat region, told the Tribunal, that as a result of a telephone call he received on the 20th of March of 1990 in which he was advised, that there was a proposed unauthorised re-boxing of beef due to take place in Kepak, Clonee. Mr Bennett and a Mr Michael Durkan, went to the premises and after contacting the late Mr Keating they were allowed into the premises. They immediately proceeded to the rear of the premises where there was a container, a refrigerated container backed up to the load area where there was a light on. Mr Bennett saw, "a number of men, who were obviously engaged in re-boxing operation or a relidding or whatever you would call it." Mr Bennett fully investigated the matter speaking to Mr Collins, who appeared to be in charge, and later to the late Mr Noel Keating. As a result of his inquiries and discussions, Mr Bennett satisfied himself that a commercial re-boxing operation was going on and particularly that it was not Department of Agriculture intervention beef involved.

Mr Fogarty, an Assistant Principal Officer in the Department of Agriculture gave the Tribunal a list of minor irregularities between the years 1982 and 1990. Mr Fogarty accepted that on each occasion upon which defects were brought to the attention of the management they took steps to remedy them.

Mr Liam McGreal, a Managing Director of the Kepak Group gave evidence to the Tribunal that the company was committed to complying with the regulations. He denied all of Mr Robert's evidence.

The Tribunal is satisfied, that there was no basis for the allegations made by Mr Roberts, in relation to the operation of this company at their premises in Ballymahon in the County of Longford. The Tribunal accepts the evidence of the company in response to these allegations and particularly the evidence of Ms McCann and Mr Donohoe.

The Tribunal further accepts the statement by the Managing Director, that the company was committed to complying with the regulations.

MASTER MEATS/CLASSIC MEATS

Mr Pascal Phelan was Chairman of the Master Meat Group from approximately 1983 until he disposed of his interests therein on the 16th of September 1988.

The Group consisted of premises in Bandon, Clonmel, Freshford, two factories in Northern Ireland and a head office in Dublin.

Subsequent to the 16th September 1988 the Group traded as Classic Meats and Mr Norbert Quinn was Managing Director of the Group from that date until April 1991.

He was Managing Director of the Clonmel, Bandon, Freshford and Ballymahon Plants and the Omagh Plant in Northern Ireland.

The evidence of the irregularities in the operation of these plants both during the time when they were operated by Master Meats and subsequently by Classic Meats was provided by officials from the Department of Agriculture and discovered by them.

They were all dealt with as considered appropriate.

Many of them were of a serious nature and during the period pre September 1988 when Mr Pascal Phelan was Chairman of the Master Meat Group. Included at the Clonmel Plant were instances of careless boning, excess of fat in cartons, withholding of intervention meat for commercial purposes, missing cuts from various productions of intervention beef, stealing of EEC Health Certificates from the Veterinary office and improper use thereof, removal of stamps from condemned carcases, an attempt to include such carcases for processing duplication of carcase numbers on IB4's, recording of incorrect weights on IB4's, removal of classification labels and replacement thereof with labels showing a better grade and refusals to re-weigh carcases when requested.

After the change of ownership in September 1988, the level of irregularities diminished but again there were instances of pieces of intervention meat being packed into commercial boxes, of intervention forequarter being withheld, the removal of stamps from condemned carcases and attempts to use such carcases in processing, the unlawful use of Veterinary control labels and the use of a bogus stamp (Classification) in the Bandon Plant on the 14th day of May 1990 and the removal of classification labels from carcases and the substitution therefor of labels which had been removed from Intervention type animals.

In respect of the period August to November 1988, which period overlapped the change of ownership of the plant, the Customs and Excise Authorities carried out an inspection, both documentary and physical, of product placed under their control at the Plant in Clonmel and their report showed:

"(a) ....product bonded (i.e. Placed under customs warehouse control under Art. 5 of Regulation 565/80 on 3rd October 1988 was not actually produced until the following day;

(b) actual production for a customs bonding of 9th November 1989 was 105 cartons short of what was declared to Customs;

(c) elaborate attempts were made by the company to cover up the shortfall at (b) by applying veterinary labels missing from the Department's veterinary office at the plant to product already bonded on the 22nd September 1988;

(d) some 28 cartons of beef bonded as fresh or chilled beef on 17th of August came from 1987 production and could not have been fresh or chilled at the time of bonding. In addition the production records in respect of this bonding were altered by the use of Tippex and overwriting to give the impression that more beef was produced than was in fact the case;

(e) the dates of production entered by the company on two boning forms were incorrect;

(f) the company failed to produce for physical inspection 249.5 kgs of beef declared as being placed under control on 24th October 1988;

(g) there were minor discrepancies, i.e. shortfalls and overages, in respect of a number of bondings."

This Report indicated that in addition to over declarations of weight of the meat placed in storage there was the unlawful use of veterinary labels, which were missing from the Department of Agriculture's office at the Plant in an attempt to mislead the Customs and Excise Authorities.

The company had declared meat for loading before it was produced: declared product as fresh/chilled when it was frozen, had failed to comply with National Customs Regulations and had abused the Department of Agriculture Veterinary procedures.

Penalties were imposed by the Department of Agriculture in respect of these breaches and the company was warned that if there were any future transgressions the status of the warehouse as a Customs approved warehouse would be withdrawn.

Mr Maurice Mullen of the Department of Agriculture and Food gave evidence to the Tribunal of the Deboning Yields achieved by the company in respect of intervention deboning in respect of the company's premises in Clonmel, Longford, Bandon, Clonmel (Store) for the years 1985 to 1991 inclusive as follows:-

Plant	1985	1986	1987	1988	1989	1990	1991 (Oct.)
Master Meats Clonmel	–	–	–	–	68.61	68.38	68.31
Master Meats Longford	–	–	–	–	68.05	68.12	68.24
Master Meats Bandon	–	–	68.47	68.5	68.48	68.40	68.47
Master Meats Clonmel (Store)	68.25	68.67	68.62	68.42	68.42		

The Department of Agriculture and Food gave evidence to the Tribunal of a Defatting Analysis carried out by them on intervention product produced by the companies on various dates between the 3rd of September 1991 and the 18th of February 1992 in respect of forequarter and plate and flank with the following results:

PLANT	FQ %	PF %
Clonmel	10.0	25.63
Ballymahon	16.84	32.18
Bandon	13.97	25.46%

The Department, on foot of the overfat findings sought compensation and was paid £4,410.80 and £6,073,79 in respect of the Ballymahon and Bandon premises respectively.

Additionally, the Department considered the overfat levels to be very serious and therefore declared forfeit the Intervention Deboning Bonds in the sums of £50,000 in respect of the Bandon plant and £25,000 in respect of the Ballymahon plant.

Mr Maurice Mullen of the Department of Agriculture and Food told the Tribunal of the exports to various countries achieved by the Group for the years 1985 to 1990 inclusive, together with the Export Refunds paid in respect of some of those years.

The following are the particulars:-

Company	Destination	1985 STATUS			1986 STATUS		
		INT	APS	OTHER	INT	APS	OTHER
Master Trade Limited	West Germany Victualling	17,009.0	–	–	–	–	–
	Jordan	–	–	–	–	–	192,361.3
	Egypt	–	–	–	–	–	79,831.1
EXPORT REFUNDS PAID		**£11,929.31**			**£5,908,399.75**		

Company	Destination	1987 STATUS			1988 STATUS		
		INT	APS	OTHER	INT	APS	OTHER
Master Trade (Exports) Limited.	Egypt	–	524,957.6	3,269,877.1	–	–	–
	Iran	–	–	858,776.3	–	44,986,622.0	2,537,893.7
	Jordan	–	–	19,675.0	–	–	–
	Saudi Arabia	–	46,545.7	–	–	23,431.2	–
	South Africa	–	1,648,308.6	358,639.3	–	105,420.7	734,515.3
	Cyprus	–	–	–	–	16,309.2	34,704.7
	Switzerland	–	–	–	–	–	189.7
EXPORTS REFUND PAID		**£11,844,754.78**			**£2,051.746.48**		

Company	Destination	1989 STATUS			1990 STATUS		
		INT	APS	OTHER	INT	APS	OTHER
Master Trade (Exports) Limited	South Africa	–	–	1,549,544.4	–	–	35,733.9
	Ivory Coast	–	–	70,701.0	–	–	935,726.80
	Romania	–	–	964,802.2	–	–	175,867.30
	Egypt	–	–	42,039.6	–	–	43,217.30
	Iraq	–	–	246,355.6	–	–	–
	Malta	–	–	–	–	–	93,788.80
	Saudi Arabia	–	–	–	–	–	132,114.90
	French Polynesia	–	–	–	–	–	14,252.70
	Congo	–	–	–	–	–	335,631.00
	Cyprus	–	–	–	–	–	14,429.4
	West Africa	–	–	–	–	–	64,346.4
	Zaire	–	–	–	–	–	38,472.5
EXPORT REFUND PAID		**£3,651,600.56**			**£N/A**		

Company	Destination	1990 STATUS		
		INT	APS	OTHER
Master Meat Packers	West Africa Ivory Coast	_ _	_ _	49,264.70 309,508.70
EXPORT REFUNDS PAID		£N/A		

Mr Mullen of the Department of Agriculture and Food gave evidence of the involvement of the company in the 1988 Aids to Private Storage Scheme, whereby the company produced meat on a sub-contract basis for Agra Trading Ltd., and Hibernia Meats Ltd.

The Department carried out an examination of plate and flank produced by the company for both these contractors and it showed a serious level of infringement of the Aids to Private Storage Scheme and Export Refund regulations. Dealing with the production for Agra Trading Ltd., the position was as follows:-

1. In **Master Meats, Bandon** produced approximately 129 tonnes of plate and flank for Agra Trading Ltd., and of which 244 boxes were examined. There was 10.75 kilogrammes of trimmings representing .18% of the plate and flank product examined in 9 boxes. There was 67.775 kilogrammes of pieces of beef not individually wrapped representing 1.16% of plate and flank produce sampled in 31 boxes.

2. In **Master Meats Ballymahon**, produced approximately 182 tonnes of plate and flank for Agra Trading Ltd., of which 470 boxes were examined. The officials of the Department of Agriculture and Food discovered a substantial number of boxes containing a level of trimmings equal to or greater than 3 kilograms. Furthermore, in 229 boxes a substantial portion of the beef was not individually wrapped. Codfat was found in 8 boxes.

3. **Master Meats, Clonmel**, produced approximately 186 tonnes of plate and flank for Agra Trading Limited. The Department of Agriculture and Food found trimmings weighing 2.92 kilogrammes representing .03% of plate and flank sampled in two of the boxes of a sampled size of 402 boxes. Non-individually wrapped pieces were found in 35 boxes weighing in total 175.39 kilogrammes representing 1.99% of the plate and flank sampled.

Because the Department of Agriculture and Food considered the levels of infringements serious, particularly in Master Meats, Ballymahon, accordingly the Aids to Private Storage Securities were declared forfeit in addition to the recoupment sought in respect of the Aids to Private Storage and Export Refunds.

In respect of the production for Hibernia Meats Ltd., the position was as follows:

1. In **Master Meats Bandon**, produced approximately 256 tonnes of plate and flank for Hibernia. 127 boxes were examined. 2 boxes contained trimmings greater than 3 kilogrammes and 10 boxes also contained trimmings. Most of the trimmings were contained in boxes from a single contract. 10 boxes contained pieces not individually wrapped (representing 1.5% of the plate and flank sampled).

2. **Master Meats Ballymahon**, produced 387.5 tonnes of plate and flank for Hibernia. Some 535 cartons were examined. There were 117 boxes found with trimmings equal to or greater than three kilogrammes. Trimmings were also found in 61 other boxes. Non-individually wrapped pieces were found in 14 boxes weighing in total, 71.18 kilogrammes representing .5% of the plate and flank sampled.

3. **Master Meats Clonmel**, produced 250 tonnes of plate and flank for Hibernia. There were 483 boxes examined. Trimmings were found in 77 boxes and non-individually wrapped pieces were found in 88 boxes totalling 593.74 killogrammes and representing 4.7% of the plate and flank sampled.

The Department of Agriculture and Food in respect of Master Meats Ballymahon forfeited the security in addition to the recoupment sought in respect of Aids to Private Storage and Export Refunds.

TAHER MEATS LTD

The evidence in relation to this company was given by Ms Brid Cannon, Assistant Principal Officer of the Department of Agriculture and Food, Mr Michael Downey, Higher Officer, Customs and Excise, Mr Wilfred Woolett, Veterinary Inspector of the Department of Agriculture and Food, Mr Maurice Mullen, Assistant Principal Officer of the Department of Agriculture and Food, Mr Godfrey Higgins, Manager of the Interests of Taher Meats Ltd., and Mr Naser Taher, Principal of Taher Meats Ltd.

The company operated in the beef processing industry in the period 1987 to 1990 with a premises in Roscrea, Co. Tipperary, where the company had facilities which included a slaughter house, deboning hall and a cold store. These premises were approved meat export premises under EC Council Directive 64/433 and 77/99. The company employed, during this time, approximately 150 employees at the peak of the season. The company also engaged sub-contractors for cleaning, hygiene, deboning services. The employees were paid with all appropriate deductions for PAYE and PRSI.

The Department of Agriculture and Food gave details of the intervention deboning yields returned by the company for the years 1989 to 1991-(October) as follows:-

	1989	1990	1991 (Oct.)
Taher Meats Ltd.,	68.65	68.25	68.40

Mr Mullen also gave evidence of the exports by this company together with the Export Refunds paid to this company as follows:

EXPORT REFUNDS PAID FOR 1987 = £ 9,784,509.38

Company	Destination	1988 STATUS			1989 STATUS		
		INT	APS	OTHER	INT	APS	OTHER
Taher Meats Limited	Iraq	643,709.5	–	–	1,003,321.0	951,446.1	17,245.5
	Egypt	–	7,098,352.3	2,279,640.9	–	4,838,267.3	–
	South Africa	1,961,099.3	–	–	–	–	63,017.6
	Canary Islands	24,861.0	–	20,007.0	–	52,651.4	–
	Jordan	–	–	–	–	514,887.7	–
	Tahiti	–	–	–	–	112,924.6	–
	Bahrain	–	–	–	–	116,246.1	–
	Cyprus	–	–	–	–	91,702.1	–
	Hong Kong	–	–	–	–	25,746.7	–
	Finland	–	–	–	–	20,134.3	–
	Yugoslavia	–	–	–	–	38,507.7	–
	Norway	–	–	–	–	12,434.9	–
		£15,202,597.99			**£6,304,266.01**		

Company	Destination	1990 STATUS		
		INT	APS	OTHER
Taher Meats Limited	Bulgaria	–	829,923.4	–
	Canary Islands	–	14,203.6	–
	Sweden	–	8,219.1	–
	Jordan	–	95,057.8	–
	Iraq	–	341,015.6	–
	Egypt	–	508,012.70	–
EXPORT REFUNDS PAID		**£N/A**		

This was one of the companies who were found to have committed breaches of the regulations under the 1988 APS Scheme and who were penalised in the sum of £96,613 in respect thereof.

The irregularities consisted of the inclusion of ineligible trimmings in cartons of beef placed in storage under this scheme and the failure to individually wrap all pieces of meat in such cartons.

Evidence was adduced before this Tribunal with regard to this company concerning a shipment of 548.0823 tonnes of beef on the N.V. Yehya from Foynes on the 30th day of September 1989.

It appears from the evidence of Miss Bríd Cannon of the Department of Agriculture and Michael J Downey, Officer of Customs & Excise, that what happened in this case was:

(1) Taher Meats Ltd shipped a consignment of 548.0823 tonnes of beef on the M.V. Yehya from Foynes on the 30th day of September 1989.

(2) The beef was late 1988 production which had been removed from warehousing control for export on various dates from 2nd August 1989 onwards and its intended destination as evidenced by the declarations on the D & C forms and by the Bill of Lading was Egypt.

(3) It was certified by the Irish Customs Authorities as having left EC territory on the 30/9/89.

(4) The export refund value of the consignment was £1.055m and the advance payment securities lodged in respect thereof totalled £1.266m.

(5) The vessel was scheduled to call to the port of Imogen in Netherlands to take on board a consignment of frozen fish.

(6) During the course of the voyage, the engine developed trouble and the ship called to the port of Cobh to effect some repairs.

(7) It then proceeded to Imogen. On arrival at Imogen, a persistent smell of diesel oil was found to be emanating from the hold and further investigation showed that there had been a leak of fuel oil into the hold which appeared to have damaged part of the cargo.

(8) It was agreed that the cargo would have to be removed from the hold to assess the damage and to prevent further damage.

(9) Taher Meats made application for permission to re-enter the meat temporarily under Customs suspensory procedure in the Netherlands for the purpose of conducting this assessment.

(10) In anticipation of the grant of such permission, 517.348 tonnes of meat was placed in a cold store in Flushing, Netherlands, under Customs Suspensory Control.

(11) The balance of the meat was detained on board the ship for survey and after examination, it was deemed to be unfit for human consumption by the Dutch Health Authorities and was sent for destruction.

(12) A survey of the balance of the product was carried out by the Dutch Customs Authorities and certain infringements in the regulations were discovered. The infringements were largely similar to the infringements already discovered in respect of the beef in the Joint Department/Customs 1988 Beef Sampling Operation.

(13) Taher Meats sought and obtained permission to repackage the meat stored in Flushing and approval in respect of same was granted by the authorities, subject to certain specific requirements, including limitation on the type of repackaging and relabelling, the need for liaison with Dutch authorities and availability of an official account of operations conducted together with appropriate linkage of the product back to the product temporarily re-imported as required by the regulations.

(14) The final operation was supervised by Veterinary Inspectors from the Department of Agriculture who certified the meat fit for human consumption and the beef was eventually re-exported from Vlissingen in the Netherlands on the 23rd July 1990.

(15) The consignment, however, was refused entry to Egypt on health grounds.

(16) Following rejection of the meat in Egypt, Taher Meats sought an alternative market for the consignment. It was eventually imported into Jordan on the 23rd December 1990 and satisfactory proofs of import into Jordan were lodged on the 31st December 1990.

(17) Certification from the Jordanian Veterinary Authorities with regard to the fitness of the beef for human consumption was received in November 1992.

In order to adjudicate on export refund entitlement, the Department of Agriculture needed to consider the following issues:

a) Whether temporary re-import into the Community was justifiable.

b) Whether proper procedures were observed during the import, storage, repackaging and re-export operations.

c) The significance of the investigation by Dutch Customs.

d) The status of product allegedly sent for destruction on the instruction of the Dutch Health Authorities.

e) Whether an extension of the 12 month deadline for import into a Third Country was justified.

f) The fitness of the meat re-exported for human consumption.

The Department of Agriculture was satisfied that the temporary re-import of the beef was justified as it arose purely from an accident of fuel oil leaking into the hold of the vessel.

The Department of Agriculture was satisfied that all proper procedures were observed during the import, storage, repackaging and re-export operations.

The Dutch Customs report showed that non-individually wrapped meat had been included in the consignment and this was consistent with the findings of the Irish Joint Sampling operation. Some of the bondings examined in the Netherlands were covered by the Joint Sampling Operation already referred to.

The Dutch Customs report referred to the presence of forequarter, flank, brisket and shank meat in the consignment and expressed the view that a portion of this meat was not eligible for export refund. The Department of Agriculture did not agree with this interpretation of the regulations and was satisfied that the meat included in the consignment was eligible for export refunds.

With regard to the product destroyed, Taher Meats Ltd sought relief on the 20% advance payment premium for this beef on force majeure grounds and this claim is being considered by the Department of Agriculture.

The Department of Agriculture was satisfied that in all the circumstances the extension of the 12 month regulatory deadline was justified and approval for the late import of the product into a Third Country was granted.

Given that the rejection of the product by the Egyptian authorities occurred at a time when they had major BSE related difficulties with Irish beef, the Department of Agriculture did not place serious emphasis on the rejection of the beef by the Egyptian authorities.

It had been certified by the Department of Agriculture Veterinary staff as fit for human consumption before being re-exported from the Netherlands and subsequently by the Jordanian veterinary authorities.

The situation which led to this investigation was caused by the incident hereinbefore referred to, namely the engine trouble on the transporting ship and the seepage of oil into the hold, was dealt with as appropriate by the authorities and is not in any way indicative of any malpractice on the part of this company.

HIBERNIA MEATS LTD.

The evidence to the Tribunal in respect of Hibernia Meats Ltd., was given by Ms Brid Cannon, Assistant Principal Officer with the Department of Agriculture and Food, Declan Holmes, Supervisory Agricultural Officer with the Department of Agriculture and Food. Rory Godson, Business Editor of the Sunday Tribune, John Boothman, Veterinary Inspector, Peadar O'Duinn, Inspector of Taxes, Revenue Commissioners, Mr Maurice Mullen Assistant Principal Officer of the Department of Agriculture and Food and James Quinn, Chief Executive of Hibernia Meats Ltd.

Hibernia Meats Ltd, operated approved meat export premises under EC Council Directive 64/433 and 77/99 at Athy, Co Kildare, and Sallins, Co Kildare, where they had facilities for slaughtering, deboning and a cold store. The company's main business in the beef processing industry was in respect of commercial, intervention and the export of beef. Hibernia Meats Ltd, exported substantial quantities of beef already referred to in the Export Credit Insurance Chapter of this Report to Iraq on the benefit of contracts obtained by CED Viandes in France. This company, together with Mr Tom McAndrews subsequently purchased the Hibernia Meats Ltd., facility and set up the company Eurowest Foods Ltd.

Mr Maurice Mullen of the Department of Agriculture and Food gave evidence, to the Tribunal, of the Intervention Deboning Yields achieved by Hibernia Meats Ltd., in their premises in both Sallins and Athy in the years 1986 to 1991 inclusive and they were as follows:-

	1986	1987	1988	1989	1990	1991
Hibernia Meats	69.52	67.76	-	68.29	68.43	68.93
Hibernia (Sallins)	68.47	68.43	68.59	_	68.30	63.34

The Department of Agriculture and Food gave details of a Defatting Analysis carried out by the Department in respect of Intervention deboned product produced by the company from their Sallins plant for the 22nd of November of 1990 and the 30th of 1990 in respect of forequarter only. In this examination there were ten boxes defatted of which two showed overfat. The range of fat level was between 3.41% and 16.79% with the average of 8.71% which is within the permitted level.

The Department also gave details of a defatting analysis carried out on intervention deboned product by Hibernia, Athy, on various dates between the 1st of May of 1991 and the 29th of August 1991. This examination was carried out on both forequarter and plate and flank beef and had the following overall results.

OVERALL RESULTS	FQ %	PF %
No. of Boxes Defatted	40.0	12
No. Overfat	26.0	4
Average	11.78%	27.23%
Range	3.85% / 19.27%	18.68% / 39.99%

Since the fat level on the forequarter beef averaged in excess of the 10% the Department of Agriculture and Food sought compensation from the company in the sum of £4,177.06. At the time that Mr Mullen gave evidence to the Tribunal the company was disputing this amount.

Mr Mullen also gave evidence to the Tribunal of the beef exported by the company from 1984 to 1990 inclusive and these are set out hereunder:-

Company	Destination	1984 STATUS			1985 STATUS		
		INT	APS	OTHER	INT	APS	OTHER
Hibernia Meats Limited	USSR	391,390.9	–	–	–	–	–
	Ivory Coast	–	–	27,751.0	–	393,395.90	38,739.70
	Egypt	–	–	564,709.4	723,3337.90	1,792,525.80	52,422.90
	Zaire	–	–	78,880.5	–	99,831.60	–
	Saudi Arabia	–	–	325,281.9	–	294,655.60	94,363.90
	Cyprus	–	–	46,147.2	102,506.80	484,311,135.90	37,772.20
	Gabon	–	–	–	–	25,021.30	19,626.60
	Iraq	–	–	–	–	99,529.10	-
	Oman	–	–	–	–	–	13,264.80
	Togo	–	–	–	–	–	16,538.90
EXPORT REFUNDS PAID		**£7,844,396.56**			**£29,521,595.32**		

Company	Destination	1986 STATUS			1987 STATUS		
		INT	APS	OTHER	INT	APS	OTHER
Hibernia Meats Limited	Malta	83,878.9	20,589.7	24,938.8	–	–	–
	Cyprus	77,024.4	284,604.3	247,517.3	157,610.10	12,730.1	61,621.80
	Egypt	1,455,904.70	4,944,877.22	1,488,236.33	208,262.20	3,733,368.50	394,511.70
	Canary Islands	1,932.00	38,321.??	1,702.10	80,000.00	–	–
	Turkey	403,207.90	–	–	–	–	–
	Iraq	1,014,708.00	1,952,517.00	1,094,457.60	5,028,427.8	–	893,422
	Togo	–	20,928.00	19,583.50	–	–	–
	Algeria	–	998,797.30	55,823.90	–	–	229,885.40
	Sweden	–	12,806.30	27,431.50	71,307.30	–	–
	Norway	10,102.20	10,199.40	–	–	–	–
	Angola	–	2,770.10	8,617.40	–	499,100.80	33,796.90
	French Polynesia	–	–	19,659.60	–	–	–
	Tahiti	198,820.50	20,200.40	–	–	–	–
	Ivory Coast	–	413,200.30	16,451.40	–	–	68,577.00
	Zaire	802,439.60	164,656.30	17,187.70	–	–	–
	Saudi Arabia	39,783.10	180,562.70	93,599.60	98,410.40	29,198.10	479,594.18
	Victualling	–	–	–	–	–	–
	Iran	–	–	–	413,484.60	3,747,458.80	1,504,252.40
	USSR	–	–	–	–	–	–
	Kuwait	–	–	–	–	8,657.40	–
	Morocco	–	–	–	4,079,683.23	500,000.00	–
	Bahrain	–	–	–	17,643.60	–	24,358.20
	Yugoslavia	–	–	–	1,524,357.15	–	3,905.90
	Quatar	–	–	–	–	–	11,494.10
	Israel	–	–	–	8,691,884.04	–	1,399,672.60
					13,117.10	–	
					–	–	
EXPORT REFUNDS PAID		**£17,649,712.37**			**£36,470,083.90**		

Company	Destination	1988 STATUS			1989 STATUS		
		INT	APS	OTHER	INT	APS	OTHER
Hibernia Meats Limited	Iran	–	5,041,107.8	802,400.3	–	4,658,286.00	4,253,742.10
	Cyprus	25,434.9	–	1,943.9	28,576.80	202,482.70	51,116.20
	Malta	47,887.4	22,787.8	11,773.1	–	58,594.10	–
	Saudi Arabia	77,328.4	23,660.5	60,497.2	–	79,306.70	68,688.94
	Iraq	909,507.0	275,753.2	2,077,686.8	–	6,323,169.78	230,321.08
	Ivory Coast	117,302.4	26,550.3	–	55,564.90	–	5,652.90
	Egypt	–	305,969.7	22,891.6	1,048,158.60	2,731,060.36	798,791.40
	Qatar	9,175.2	4,076.8	–	–	–	–
	South Africa	5,834.5	249,653.3	63,997.1	37,855.50	–	23,538.70
	Morocco	1,484,647.5	–	–	–	–	–
	Canary Islands	–	–	–	55,188.60	28,614.20	–
	Victualling	–	–	–	31,544.90	–	–
	USSR	–	–	–	4,299,130.10	–	–
	Yugoslavia	–	–	–	–	–	23,355.10
	Nigeria	–	–	–	–	–	44,813.30
	Romania	–	–	–	–	–	11,144.88
	Gibraltar	–	–	–	–	8,019.10	–
	Bahrain	–	–	–	–	65,911.30	–
	Hong Kong	–	–	–	–	118,008.20	–
EXPORT REFUNDS PAID		**£36,552.686.99**			**£28,045,425.25**		

Company	Destination	1990 STATUS		
		INT	APS	OTHER
Hibernia Meats Limited	Yugoslavia	–	–	139,335.62
	Iraq	–	1,599,205.5	68,785.1
	Iran	–	4,745,884.60	420,211.6
	Bahrain	22,995.0	69,369.47	11,535.4
	Malta	832.3	54,904.5	13,752.06
	Oman	–	14,324.7	–
	Cyprus	48,141.0	92,934.83	45,270.2
	Canary Island	-	42,008.8	35,372.36
	Egypt	-	1,095,885.6	1,070,971.7
	Hong Kong	-	–	26,772.1
	Victualling	8,881.0	–	46,998.8
	USSR	16,939,617.59	–	–
	Sweden	7,976.0	11,932.4	–
	Saudi Arabia	–	–	98,182.0
	Turkey	–	–	21,494.3
	Ivory Coast	–	79,518.4	–
	Norway	–	20,198.4	–
	Finland	18,992.0	-	–
	Gabon	14,149.5	8,087.9	15,254.9
	£N/A	**£N/A**		

Mr O'Duinn, Inspector of Taxes at the Revenue Commissioners gave evidence concerning a Tax avoidance scheme put in place by the company in the years 1987 to 1990. Hibernia Meats International Ltd, had earnings on tax free profits from the export of beef and paid out tax free dividends in the four years to November of 1990 of £7,555.00. A company called Mettlehorn Ltd, a company incorporated on the 11th of May 1988 received approximately £683,000 of this dividend. This company paid out, between the 11th of May of 1988 to the 31st of December 1989 £483,324 and from the 1st of January 1990 to the 31st of December 1990 £191,725. These sums were paid to approximately 100 employees of the Hibernia Group who had subscribed for a variety of different classes of shares in the company Mettlehorn Ltd. There were approximately 12 different classes of non-voting shares in this company.

The view of the Revenue Commissioners was that this was a scheme whereby the dividends were a substitute for payment which would otherwise be treated as a remuneration subject to income tax. On the basis of the figures above, the tax would have been in the region of £250,000. It has always been a Revenue policy to challenge these schemes and it is viewed as tax avoidance as opposed to tax evasion.

Hibernia Meats Ltd., is one of the companies who were found to have committed serious breaches of regulations with regard to the 1988 APS scheme resulting in a penalty of £1,525,748.93p.

This matter has been dealt with in the Chapter of this Report dealing with the 1988 APS scheme and the right of the Department of Agriculture, as Intervention Authority, to impose this penalty, is the subject of proceedings in the High Court.

Mr Rory Godson, a journalist, gave evidence before the Tribunal that, having been informed that lorry loads of Irish beef had been rejected in Baghdad at the end of 1988 and the early part of 1989, he, on the 19th day of May 1989, travelled to Mersin in Turkey, which is the port to which the meat had been exported.

Having made inquiries there, he went to a warehouse, where he states there was a considerable amount of Irish beef being held.

He ascertained that the beef had been exported by Hibernia Meats.

He noted a roller assembly line with boxes of meat being pushed down and labels were being stripped off the packages of meat.

On his return, he sought information from the Department of Agriculture for the purpose of ascertaining how much intervention beef was being exported to Iraq but no information was forthcoming.

It appears from the evidence of Ms Cannon of the Export Refunds Section of the Department of Agriculture that:

(1) 477 tonnes of beef was despatched out of Ireland on the 30th day of November 1988 for export to Iraq.

(2) 43 tonnes were despatched on the 16th February 1989 and 500 tonnes despatched on the 22nd day of April 1989.

(3) The original 477 tonnes was rejected by the Iraqi authorities and returned to the warehouse in Mersin.

(4) The latter two consignments were intended originally for Iraq and were stored in a cold store in Mersin awaiting onward shipment when the rejection notice for the first consignment was received.

(5) Hibernia Meats were reluctant to take a chance on onward shipment to Iraq in the light of the rejection of the previous load.

(6) In June and July 1989, Hibernia made informal approaches to the Department concerning difficulties they were experiencing with a rejection by their Iraqi customer of a consignment of their beef and inquired about the procedures to be followed for bringing this consignment back to Rotterdam with a view to correcting the alleged defects.

(7) By letter dated the 18th day of August 1989 an application was made to the Department of Agriculture for permission to re-import the beef.

(8) They stated that it was their original intention to re-label the product in Mersin for despatch to another customer in the Mediterranean area but became dissatisfied with the progress on re-labelling and decided that the best course would be to move the product to Rotterdam to continue the operation of re-labelling.

(9) A total of 1,020 tonnes was imported under Customs Suspensory procedure into Rotterdam.

(10) The Investigations Branch of Customs monitored the case in association with Dutch Customs and with the Department of Agriculture.

(11) The Veterinary Services of the Department of Agriculture conducted an inspection of a portion of the product in September 1989 and while the meat was deemed to be fit for human consumption, Veterinary re-certification of the inspected meat was refused because there had been a break in the Veterinary control chain while the product was outside the Community.

(12) Hibernia applied for an extension of the 12 month period for import on 15th November 1989 and as the Department of Agriculture was not satisfied with regard to the reasons advanced by Hibernia for the re-import of meat, the Department of Agriculture sought the advice of the Commission concerning re-import in February 1991.

Certain theories were raised by the Department and replied to by Hibernia Meats but by the 11th November 1992, no advice had yet been received from the Commission.

This is obviously the case referred to by Mr Godson and again illustrates that the matter is being dealt with by the Department of Agriculture in accordance with the regulations and procedures.

The Tribunal, in the course of its inquiries in August of 1992, received documentation from Hibernia Meats Ltd., through their solicitors, Messrs Arthur Cox & Co. The documentation related to beef exports by Hibernia Meats Ltd., to Iraq. One of the documents supplied was a Veterinary Certificate for export of meat to Iraq and a number of these documents were supplied.

Mr Sean O'Connor, Deputy Director of Veterinary Services, responsible for Veterinary Control of Meat Production told the Tribunal that the Veterinary Certificate for export of meat to Iraq, was the standard basic meat inspection certificate. It is in the following terms:-

AN ROINN TALMHAIOCHTA AGUS BIA, ARAS TALMHAIOCHTA, BAILE ATHA CLIATH. 2
DEPARTMENT OF AGRICULTURE AND FOOD, AGRICULTURE HOUSE, DUBLIN 2

TEL, 789001
TELEX 83407
REF

VETERINARY CERTIFICATE FOR EXPORT OF MEAT TO IRAQ

1. **IDENTIFICATION OF THE MEAT**

Meat From:
Type:
Type of Packing:
Quantity:
Net Weight: Gross Weight:

Consignor: ______________________________

____

Consignee ______________________________

____

By Ship:

2. **ORIGIN OF THE MEAT**

Meat originates from abattoirs/boning plants inspected and approved under Co. Directive 64/433 EEC and 72/462/EEC

3. **DESTINATION OF THE MEAT**

Meat shipped from ireland to State Company for Foodstuff Trading Baghdad/Iraq Shipment by refrigerated container.

4. **HEALTH CERTIFICATE**

The meat described at 1. above has been produced at premises inspected and approved by the Irish Department of Agriculture and Food - a competent authority of the EEC - and after ante-mortem and post-mortem inspection has been found free from disease and fit for human consumption.

On the basis of routine sampling the meat is free from antibiotic, hormone preservative materials. The meat is in every respect, fit for human consumption in all countries including Ireland.

Signature: ____________________

Official Title: ____________________

Date: ____________________

Amongst the Veterinary Certificates, for export of meat to Iraq, furnished to the Tribunal, by Hibernia Meats Ltd., was a certificate dated the 25th February of 1988 prepared in respect of a shipment to be exported on the MV Ice Flower. This Veterinary Certificate differed from the basic Veterinary Certificate in that paragraph

4 entitled Health Certificate had a number of additions which made it read in the following terms:-

> "The meat described at 1 above, has been produced at premises inspected and approved by the Irish Department of Agriculture and Food - a competent authority of the EEC - and after ante-mortem and post-mortem inspection has been found free from disease and fit for human consumption. Animals were slaughtered within 90 days before arriving at the Buyer's stores.
>
> On the basis of routine sampling the meat is free from antibiotic, hormone preservative materials. The meat is in every respect, fit for human consumption in all countries including Ireland at the date of shipment."

This document bore the signature of Susan A. McKeever, Veterinary Inspector and it was dated the 25th of February 1988.

In evidence, Ms McKeever, told the Tribunal that in respect of the above shipment, at the request of the company she had signed two certificates. She said:-

> "the trade may request two certificates for any consignment if they wish. We only issue one certificate to cover any one carton of beef. We are not issuing two certificates for the same quantities that if the trade require two certificates for two separate customers, for two separate ports, that would not be a problem to us."

Accordingly, in respect of this shipment, two certificates were sought by Hibernia Meats Ltd., to cover the total showing of 4,968 cartons and the second for 72,777 cartons. Ms McKeever, told the Tribunal that she issued the two certificates in the form of the standard basic meat inspection certificate and that at the time that she signed the two certificates that the following words were not on them:-

> "Animals were slaughtered within 90 days before arriving at the buyers store."
>
> and;
>
> "at the date of shipment."

Ms McKeever, further told the Tribunal that when the Certificates issued, that is after they had been signed, dated and stamped, they would be given to a representative of the company exporting the beef, in this case, Hibernia Meats Ltd.,

Mr Oliver Murphy, formally with Hibernia Meats Ltd., gave evidence to the Tribunal in relation to the Certificate and particularly in relation to the addition which appeared on the certificate in respect of the MV Ice Flower, for the 25th of February of 1988 and told the Tribunal:-

> "That nobody in Hibernia Meats, or in the company which I ran, had anything to do with that particular Certification."

The Tribunal was informed that the person in the company, normally responsible for liaising with the Department was Mr Jim Quinn and that once the Department had given the documents to the company, they would be forwarded to their principal, CED Viandes in Paris. Mr Murphy told the Tribunal:-

> "two copies go with the shipment and the original goes to the bank in Paris and the company keeps a number of copies."

Mr James Quinn, Chief Executive of HMIL Ltd., formerly Hibernia Meats Ltd., gave evidence to the Tribunal of having examined the two certificates and accepted that the certificate for the MV Ice Flower, dated the 25th of February 1988, containing the phrases:

> "animals were slaughtered within 90 days before arriving at the buyers' stores."
>
> and;
>
> "at the time of shipment"

was not the certificate received by the company from the Department and was not the certificate that they had submitted to their associates CED Viandes in France.

As a result of the certificate being made available to the Tribunal, the Tribunal caused inquiries to be made by the Garda Siochana. As a result of their inquiries a further five certificates were produced similar to the certificate containing the additional phrases.

Mr Quinn had seen these additional five certificates. He accepts that these certificates were not the certificates which were signed by Susan McKeever at the time they were made available to his company.

> "Q. And clearly whoever interfered with them or put in the additional words is representing that the Department of Agriculture were certifying something which they weren't certifying.?
>
> "A. That would appear to be the case. Yes.
>
> "Q. And furthermore, doing it in the name of your company, Dantean International Ltd.,"?.
>
> "A. Well, or perhaps doing it in the name of the Department of Agriculture.
>
> "Q. Or both,?
>
> "A. Yes, Or both.
>
> "Q. Now is that a serious matter?
>
> "A. No doubt!."

Mr Quinn continued:

"Well I am satisfied, beyond any doubt whatsoever, that there was no abuse of the certs. which we received from the Department. We sent on to Paris, and as far as I am concerned, my function is, the territorial extent of my function is Ireland and what might or mightn't have happened off-shore, while it is of concern to me, is not a matter I am in a position to investigate.

The Tribunal, in the course of further inquiry, for the purpose of seeking to ascertain how these additions had come to be added to this certificate, wrote to CED Viandes on the 16th of June 1993, enclosing transcripts of the evidence relating to this Certificate and requesting CED Viandes for assistance in explaining the person(s) who added or interfered with Veterinary Certificates for the export of meat to Iraq enclosed with this letter. The company were invited to give evidence to the Tribunal but declined such invitation and replied by letter of the 12th of July 1993, and in respect of this issue, Mr Franz Klees, for and on behalf of CED Viandes, S.A. wrote:-

"I have no knowledge of any alleged alteration of or additions to the Veterinary Certificates enclosed with your letter of the 16th of June and I was unaware that there might have been any additions to such Veterinary Certificates until the matter became an issue in the Tribunal.

During the period in question, I was not responsible for or associated with the administrative side of the business which would include dealing with Veterinary Certificates. This was the responsibility of the then President of CED Viandes, Mr Rageszzi who died suddenly on the 9th of April 1988.

I can say that any amendments to such Veterinary Certificates were not carried out with my knowledge, consent, or approval. Furthermore, I wish to advise the Tribunal that immediately after having been made aware of it I raised the issue with such existing members of the staff of CED Viandes, who are employed by us in an administrative capacity during the period in question. None of the staff has any further knowledge concerning such alleged alterations. On page 2 of your letter, of 16th June last, you state that "the only place to which the certificates appear to have been sent were to the company CED Viandes" but this is not in fact the case. The documents were also presented to the Iraqi Authorities and in particular the Iraqi Embassy for Legalisation. While, as I have said, have no knowledge as to how, where or when, the Veterinary Certificates were allegedly altered it would be wrong to assume that the Iraqi purchasers of the beef had no interest in altering or amending such certificates. In circumstances where it was crucial for Iraq's war effort to have beef cleared from Mersin to Turkey to Iraq, the purchasers would have had an interest in overcoming what might have been difficulties for them in importing the beef within the required time frame."

The Tribunal, is accordingly unable to make finding as to who or where the Veterinary Certificates for the export of meat to Iraq, produced to the Tribunal were interfered with. However, the Tribunal is satisfied and accepts completely the evidence of Ms McKeever, that she did not alter the Certificates and the evidence of Mr Oliver Murphy and Mr Jim Quinn, that they were not altered by the company or whilst in the company's possession.

The Tribunal is satisfied that the Certificates were altered when they left the jurisdiction of the Republic.

TUNNEY MEATS

This company was the proprietor of a plant at Clones in the County of Monaghan. On the 9th day of July 1991 Mr Philip Smyth of Sach's Hotel in the City of Dublin contacted Mr Aidan McNamara of the Department of Agriculture by telephone and made certain allegations with regard to the deboning of intervention beef by Tunney Meats at their factory in Clones particularly during the period September 1990 - May 1991 when he alleged that the only business being carried out by the plant at that time was the deboning of beef for Intervention and that in spite of this they had exported to the United Kingdom approximately 400,000 lbs worth of forequarter beef and alleged that this beef was properly the property of the Agriculture and had been "syphoned off" during the deboning of intervention beef.

He called to see Mr McNamara on the following day namely the 10th day of July 1991 and repeated the allegation and alleged that the IB7 forms which had been supplied to the Department as part of the relevant documentation had been altered to ensure and to show that the yield in boneless beef was no greater than 68.08%. Whereas in fact it had been greater and the balance had been syphoned off by Tunney Meats for commercial purposes.

He was in possession of certain documents including photostat copies of invoices, IB7's and IB4's and undertook to forward copies thereof to Mr McNamara which he did.

Mr McNamara contacted Mr Gerard Mulligan the Veterinary Inspector in charge of the plant in Clones and informed him of the nature of the allegations and included copies of Health Certificate's which had been signed by Mr Mulligan in respect of the export of the beef to the United Kingdom.

The Department of Agriculture sought his observations as to whether there was a possibility that this beef should have become the property of the Intervention Agency or whether it can definitely be shown to be genuine commercial product.

On the 22nd of July 1991 Mr Mulligan replied to the Beef Intervention Section and stated that;

> "All intervention beef produced and stored at the premises in Clones in the period September 1990 to May 1991 was directly supervised by my staff and all records of production and storage are available. All the intervention beef is stored at the premises."

This did not really deal with the query and on the 30th day of July 1991 as a result of a telephone conversation from the Department of Agriculture. He again wrote stating

> "I would like to point out that although Tunney's processed beef for intervention there, 4,115 animals were slaughtered for commercial purposes between January 1990 and May 1991. There would also have been a certain amount of intervention rejects boned in this period as well as beef in storage from previous

years. To the best of my knowledge the beef in question exported to the UK came from non-intervention grade bovines."

On the 23rd of August 1991 Ms Flahive of the Beef Intervention Section of the Department of Agriculture wrote to Mr Smyth as follows:

"I refer to your letter of 10th July 1991 regarding intervention operations at Tunney's Meat Plant, Clones, Co Monaghan.

"Inquiries have been made in the matter and the position is that all beef exported to the UK during the period mentioned would appear to have come from animals slaughtered for commercial purposes or from non-intervention grade bovines."

Mr Smyth had been in touch with the office of the Tribunal and consequent to this reply from the Department of Agriculture made a considerable amount of documentation or copies thereof available to the Tribunal.

This documentation which included a notebook alleged to have been kept by Mrs Margaret Potter, who was employed by Tunney Meats as a clerk and who is responsible for the preparation of APS and Intervention documentation, copy health certificates, copy IB4s and IB7s which was alleged to show that:

(1) During a particular period between the 26th day of September 1988 and the first day of October 1988 during the operation of the APS Scheme it was usual for Tunney Meats to increase the weights of the carcases being placed in intervention by 2 kilos per side, that is four kilos per carcase, and to also add in two sides to the day's production and

(2) During deboning for intervention purposes it was the practice to remove or harvest for their own commercial purposes any meat which was not required to show the minimum yield of 68% of meat from the quarters.

In order to prove this documentation and to establish the meaning thereof it was necessary to hear oral testimony from the said Mrs Margaret Potter.

The testimony given by her was denied and contradicted by the boning hall manager of Tunney Meats who was her brother, Mr Ronnie Flanagan, by Mr Michael Connolly the Manager of the Plant and by Mr John Copas, the Managing Director of Tunney Meats.

A problem which concerned the Tribunal was how these documents came to be in the possession of Mr Philip Smyth, who made them available to the Tribunal.

During the course of her evidence Mrs Potter denied having made them available to Mr Smyth and did not appear to be fully conversant with their contents and in a position to give evidence with regard to their meaning.

When Mrs Potter had denied making these documents available to Mr Smyth, the Tribunal was concerned to ascertain how they came into his possession and having been sworn, Mr Smyth stated that he had obtained them from Mrs Potter and that he had made three separate payments to her in respect of documentation supplied by her to him, £500 on the first occasion, then when some documents were supplied a payment of £2,000 and when further documents were supplied a payment of £3,000.

Subsequent to this evidence and having received independent legal advice Mrs Potter admitted receiving the said payments other than the initial payment of £500 and that she had supplied the documents to Mr Smyth.

In these circumstances the evidence of Mrs Potter had to be approached with caution particularly in view of the denials by those alleged to have been involved.

The relevant documents however did show the addition of the kilo per quarter namely four kilos per carcase and the addition of the two sides on the days in question and her evidence in this regard is accepted by the Tribunal.

The Tribunal also accepts her evidence that during 1990 and 1991 that it was the practice when deboning for intervention purposes, particularly when there was a large kill, to appropriate for commercial purposes the red meat yield in excess of the 68% necessary to be obtained and that this was done in the manner in which she described namely that at lunchtime the IB7s would be made available to her and she would be in a position to calculate the number of boxes that would be produced, the number of boxes which were necessary to provide the yield of 68% and to calculate the number of boxes of meat that could be taken without interfering with that yield.

She stated when there was a kill involving 500 to 600 cartons of meat that it would be possible to use between 20 and 40 cartons for commercial purposes.

The Tribunal deprecates the activities of Mr Philip Smyth in regard to this matter and deprecates his approaches to Mrs Potter and the payments made to her by him and his attempts to purchase the testimony of other witnesses as part of his campaign against Mr Hugh Tunney with whom he was in conflict.

CLOON FOODS LIMITED

The evidence, in relation to this company, was given by Mr Seamus Fogarty, Department of Agriculture and Food and Mr Michael Behan, Managing Director of Cloon Foods Ltd. The company was established in April 1991 and it purchased the assets of Master Meat Packers (Clonmel) Limited with premises at Upper Irish Town, Clonmel, Co. Tipperary. The company operated a slaughtering, deboning and cold store facility with approved meat export premises under EC Council Directives 64/433 and 77/99 with EEC No. 336. The company is involved in the commercial intervention and export of beef. The company employs approximately 130 full-time employees who are paid by cheque and the company pays appropriate PAYE and PRSI to the Revenue Commissioners as appropriate.

The Veterinary office had recorded the issue of 4 IB4s on Friday the 24th of April 1992. The forms were signed on the top left-hand corner by the issuing officer and he also initialled the Register in which they are recorded. The following Tuesday the IB4s for that day's production were about to be issued when Agricultural Officers noticed that two additional IB4s had already been issued. It transpired that the two IB4s had been taken from the Veterinary office and substituted for two that had been issued and noted the previous Friday by the Department officials. The staff on the ground were adamant that they had not initialled the later IB4s at the time of issue and pointed out they had not signed those later IB4s either. They brought this matter to the attention of the Management, who could offer no realistic explanation. They reported the matter to headquarters and it was decided to refer the matter to the Gardai for investigation.

A letter issued from the Department to Cloon Foods, informing the company that it was being suspended from deboning operations on the 19th of May of 1992.

At an initial meeting between the company and headquarters staff in the Department the company repeated that they could not explain what had happened. They did offer to make further enquiries. At a follow-up meeting on the 5th of June the company explained that they had again re-interviewed staff about the matter and that two members of staff had admitted interfering with intervention documents. The reason was that a mistake had been made in supplying meat for tender. An excess had been supplied and the staff took it upon themselves to recover the over-supply of meat. The way they went about that was to gain access to the Veterinary office somehow and take two more IB4s and replace the ones that had been issued.

Two operatives, in the factory, made statements to the Gardai. These were supplied to the Department. The statements indicated that the operatives had taken this action on their own initiative and the company insisted that they had no part in the matter. Representatives of the company confirmed that they had improved their own supervisory system at the plant. The Department's Veterinary Inspector confirmed that new arrangements had been put in place by the company and following that report, in September of 1992, the company were re-instated for deboning purposes.

DJS. MEATS / DOHERTY'S CARRIGANS

The evidence in relation to these companies is given by Mr Seamus Fogarty, the Department of Agriculture and Food, Mr Diarmuid P. O'Ceallaigh, Mr Maurice Mullen, Department of Agriculture and Food and Mr Seamus Hand, former Managing Director of DJS Meats Ltd.

Mr Seamus Hand gave evidence to the Tribunal that he had been a shareholder in DJS Meats Ltd., which commenced operations in a meat premises at Tallaght in 1978 at which time the company was mainly involved in commercial contract deboning. In 1985, DJS Meats Ltd., purchased James Doherty Carrigans, which had an approved meat export premises under EEC Council Directive 64/433 and 77/99. DJS Meats carried on business there from October of 1985 to approximately June of 1989. Mr Hand told the Tribunal that in 1989 DJS Meats Ltd., sold the Tallaght plant, and in June of 1989 the Carrigans premises were also sold. At all times the main business carried on by the company was commercial intervention and export business.

The premises at Tallaght was solely involved in the beef processing business as a deboning premises. There was no cold store and the company had the facilities of the national cold store in Tallaght. The company purchased beef from various slaughter houses and deboned it in their own premises and then subsequently stored it in the National Cold Store. The company employed approximately 85 employees at the Tallaght premises and the company paid all PAYE and PRSI to the Revenue Commissioners as appropriate having deducted it from the wages of the employees. The company did not employ sub-contractors in the Tallaght premises.

The meat export premises at Carrigans in Co. Donegal, consisted of a slaughtering facility, deboning facility and cold store facility with a registered EEC No. of 292. The company was wholly involved in the beef business being concerned with commercial, intervention and the export of beef. The company employed approximately 75 who were paid weekly and the company paid all PAYE and PRSI to the Revenue Commissioners as appropriate. The company did not employ sub-contractors.

Mr Maurice Mullen of the Department of Agriculture and Food, gave evidence to the Tribunal of the Intervention Yields achieved by the companies in both the Tallaght premises and the Carrigans premises for the year 1983 to 1989 as follows:-

	1983	1984	1985	1986	1987	1988	1989
Tallaght	66.02%	67.62%	67.97%	67.95%	68.01%	68.28%	68.58%
Carrigans	____	____	____	68.21%	68.19%	-	68.82%

Mr Mullen gave further details to the Tribunal about the companies' exports of beef from 1984 to 1989 together with the export refunds paid to the company:-

Company	Destination	1984 STATUS			1985 STATUS		
		INT	APS	OTHER	INT	APS	OTHER
DJS Meats Limited	Benin	_	_	14,130.4	_	_	_
	Saudi Arabia	_	_	125,408.8	_	_	40,537.9
	Togo	_	_	15,540.2	_	_	_
	Zaire	_	_	64,651.2	_	_	102,861.1
	Ivory Coast	_	_	14,986.8	_	_	44,215.67
	Malta	_	_	_	_	_	71,994.8
	Cyprus	_	_	_	_	_	31,081.4
	Gabon	_	_	_	_	_	16,373.8
	Mauritius	_	_	_	_	_	14,686.5
	Zimbabee	_	_	_	_	_	45,035.5
EXPORT REFUNDS PAID		**£416,823.37**			**£530,405.79**		

Company	Destination	1986 STATUS			1988 STATUS		
		INT	APS	OTHER	INT	APS	OTHER
DJS Meats Limited	South Africa	_	_	637,608.5	_	150,507.1	40,117.4
	Malta	_	19,959.60	-	_	_	_
	Zaire	_	-	38,401.30	_	_	_
	Saudi Arabia	_	-	82,315.50	_	_	_
	Gabon	_	-	43,881.70	_	_	_
	Cyprus	_	-	-	_	_	46,659.2
EXPORT REFUNDS PAID		**£1,411,000.00**			**£744,944.08**		

EXPORT REFUNDS PAID FOR 1987 - £429,337.78	**Quantities. - 312,351.2 Kgs.**

Company	Destination	1989 STATUS		
		INT	APS	OTHER
DJS Meats Limited	Egypt Iraq Cyprus Malta	– – – –	333,480.30 178,633.90 3,480.10 4,675.60	– – 4,275.80 -
EXPORT REFUNDS PAID		£575,127.06		

Mr Mullen gave details of the Export Refunds paid to the company in respect of the Carrigans plant in 1987, 1988 and 1989 as follows.

The company exported 57.4 tonnes of commercial beef to Egypt for which the company was paid export refunds of £187,176.41. In 1988 the company exported 80 tonnes of APS beef to South Africa for which the company was paid £718,433.22. In 1989 the company exported 553.1 tonnes of which 351.8 tonnes was APS, 201.3 tonnes was commercial and the company was paid £489,848.85.

During 1980 the Department received two anonymous letters alleging fraud, that is stealing intervention prime cuts. The European Commission apparently received a similar letter from the same source in August of 1981. On foot of these allegations the Internal Audit Unit of the Department carried out a full scale inquiry. The investigation concentrated on production from the period August 1979 to July 1981. No irregularities were confirmed from the investigation.

The Gardai were also asked to conduct an investigation into the matter. The Garda Investigation also failed to establish that there was any truth in the allegations contained in the anonymous letters. The Commission were informed of the results of these investigations.

In October of 1982, a letter was received from a former employee, alleging that during the period he was employed at DJS Meats, there was a systematic stealing of intervention stocks. On foot of these allegations the Internal Audit Unit of the Department carried out an examination of product from different periods in 1982 and 1983. The investigation established that cuts were missing. The company claimed that it was pilferage. The value of the missing cuts was put at approximately £45,000. This sum and another of £9,000 approximately, to cover the costs of the investigation were recovered from the company. The deboning licence, was withdrawn from the 18th of March until the 13th of June of 1983. On advice received from the Chief State Solicitors' office it was decided not to attempt prosecution. The matter was reported as an irregularity to the European Commission.

Evidence was given by Mr Maurice Mullen of the Department of Agriculture, concerning the Customs' examination in the Spring of 1987 of product bonded by DJS Meats, Tallaght, in late 1986. The Customs' examination had established that trimmings had been included in boxes of beef on which advance payment refunds had been made.

The relevant D+C forms were amended by Customs and returned to the Department of Agriculture. Securities of approximately £12,700 are still held and forfeitures are to be applied.

The Tribunal was given evidence in relation to Doherty's Carrigans prior to it being taken over by AIBP in June of 1989. The evidence, which was given to the Tribunal, concerned the company while it was under the ownership of DJS Meats.

On the 22nd of March 1989 Veterinary staff in Doherty's Carrigans, discovered intervention product in the cold store that was not recorded on the IB7s. It was decided to debar the company from intervention deboning immediately. At this time, full carcase intervention was in operation due to the imminent changeover to the new intervention tendering system which became effective from the 1st of April 1989. The company was given permission to debone for intervention again on the 31st of March of 1989. However, they were effectively excluded from Intervention until July of 1989 as Ireland was triggered out of intervention until that time.

HORGAN MEATS

The evidence to the Tribunal in respect of this company was given by Mr James Clarke, Surveyor of Customs and Excise, Ms Mary Harvey, Principal Officer with the Department of Agriculture and Food, Mr Maurice Mullen, Assistant Principal Officer, Department of Agriculture and Food, together with Mr Peter Horgan, Director of Horgan Meats Ltd. The company operated its own slaughter house, deboning hall, and cold store, which were approved meat exports premises under EC Council Directive 68/433 and 77/99. The company operated in the years 1984 until it ceased trading in 1989. The company employed in peak season, approximately 220 to 230 employees including contract boners. The employees were represented by Trade Unions, were paid weekly and the wages were subject to PAYE and PRSI deductions which were paid by the company to the Revenue Commissioners as appropriate. Sub-contractors were paid cash without any deductions of income tax.

The Department of Agriculture and Food gave details to the Tribunal of the exports by the company for the years 1984 to 1987 inclusive and they are as follows:-

Company	**Destination**	**1984 STATUS**			**1985 STATUS**		
		INT	**APS**	**OTHER**	**INT**	**APS**	**OTHER**
Horgan Meats Limited	Egypt	–	–	1,077,956.61	–	1,675,334.12	1,619,494.98
	Algeria	–	–	3,158,341.8	–	–	642,316.90
	Saudi Arabia	–	–	–	–	789,740.28	268,523.43
	Dubai	–	–	–	–	4,200.59	14,486.60
	Zaire	–	–	–	–	47,997.92	31,701.74
	Cyprus	–	–	–	–	3,697.10	33,125.60
EXPORT REFUNDS PAID		**£15,49,530.85**			**£17,012,867.46**		

Company	Destination	1986 STATUS			1987 STATUS		
		INT	APS	OTHER	INT	APS	OTHER
Horgan Meats Limited	Zaire	–	–	171,475.0	–	83,050.0	–
	Malta	–	–	20,465.0	–	–	–
	Mauritius	–	–	110,187.3	–	–	13,398.8
	Togo	–	–	72,610.2	–	21,000.0	–
	Ivory Coast	–	–	35,007.4	–	–	335,944.3
	Egypt	–	619,037.08	1,321,754.74	–	602,836.00	1,137,472.0
	Algeria	–	992,494.80	11,199.5	–	–	–
	Saudi Arabia	–	39,279.60	128,640.1	–	–	51,711.0
	Dubai	–	25,135.40	13,127.4	–	–	–
	South Africa	–	–	245,307.7	–	–	–
	Cyprus	–	–	15,567.8	–	–	163,857.6
	Israel	–	–	–	–	9,762.1	–
EXPORT REFUNDS PAID						**£12,798,458.57**	

EXPORT REFUNDS PAID FOR 1988 = 55,670.34 - SAUDI ARABIA

Ms Mary Harvey, Principal Officer, Department of Agriculture and Food, testified in relation to irregularities in connection with MCA payments to Horgan Meats Ltd. The evidence, given by her, related to claims for the period March to August of 1986. At that time a very thorough Customs investigation revealed that the company were:-

1. over-declaring weight;
2. mis-declaring product some of which was eligible for MCA claim purposes.

The Department of Agriculture and Food established that a sum of £22,263,02 had been incorrectly claimed by the company by reason of the over-declaration of weight. It was further established by the Department that a sum of £20,300.55 was wrongly claimed by the company by reason of the mis-declaration of ineligible product. The Department sought and obtained a total of £42,563.57 from the liquidator of the company. This was recovered by deducting this sum from other monies due by the Department to the company. This irregularity was fully reported to the European Commission in 1989.

It was accepted by Mr Peter Horgan, in evidence, that there had been mistakes during that period by the company.

DAWN MEATS

The evidence to the Tribunal, in respect of this company was given by Mr Seamus Fogarty, Assistant Principal Officer, Department of Agriculture and Food, Mr David Tantrum, and Mr Brennock, Veterinary Inspectors with the Department of Agriculture and Food, Mr Maurice Mullen, Assistant Principal Officer, Department of Agriculture and Food and Mr Dan Browne, Managing Director of Dawn Meats Ltd.

The company has been in existence since approximately 1981 and involved in the beef processing industry, operating approved meat export premises under EC Council Directives 64/433 and 77/99. The company operates as Dawn Meats (Exports) Ltd., at Grannagh in Co. Waterford with an EEC No. 350 and as Dawn Meats Ltd., Carroll's Cross, Co. Waterford with an EEC No. 318 where it has a deboning and cold store facility. It operates a slaughtering and deboning facility in Waterford.

The company employs approximately 300 employees full-time who are paid by cheque and the company pays the appropriate PAYE and PRSI to the Revenue Commissioners as appropriate. There are no payments to employees which are not subjected to appropriate deductions. The company does not engage sub-contractors for deboning or similar processes, mainly only for haulage, maintenance and building.

Mr Maurice Mullen of the Department of Agriculture and Food gave the Tribunal details of the intervention deboning yields achieved and returned by the company Dawn Meats Ltd., from 1985 to 1991 (October) and they were as follows:-

	1985	1986	1987	1988	1989	1990	1991 (Oct)
Dawn Meats Ltd.,	68.11%	68.16%	68.17%	68.24%	68.13%	68.12%	68.12

The Department of Agriculture and Food gave details of a defatting analysis carried out at Dawn Meats Ltd., Carroll's Cross, on various dates between the 6th of March of 1991 and the 22nd of February of 1992 with the following overall results.

OVERALL RESULTS	FQ	PF
No. of boxes defatted	25	10
No. Overfat	2	-
Average	7.21%	19.66%
Range	2.98% - 11.99%	8.65% - 28.94%

As these did not disclose levels of overfat no penalty was incurred.

The Department of Agriculture and Food gave details, in evidence, of the exports together with the amount of Export Refunds paid to the company for the years 1984 to 1990 inclusive as follows:-

Company	Destination	1984 STATUS			1985 STATUS		
		INT	APS	OTHER	INT	APS	OTHER
Dawn Meats Limited	Zaire	–	–	484,471.75	–	332,706.30	949,805.60
	Saudi Arabia	–	–	473,946.00	–	-	866,617.90
	Ivory Coast	–	–	173,279.55	–	-	368,497.10
	Africa	–	–	12,525.20	–	-	-
	Egypt	–	–	65,303.60	–	-	-
	Gabon	–	–	-	–	34,789.00	58,507.20
	Canada	–	–	-	–	70,282.40	807,847.40
	Zimbabee	–	–	-	–	-	50,000.00
	Oman	–	–	-	2,510.00	-	2,592.80
EXPORT REFUNDS PAID							

Company	Destination	1986 STATUS			1987 STATUS		
		INT	APS	OTHER	INT	APS	OTHER
Dawn Meats Limited	Zaire	–	–	1,113,382.40	–	-	110,349.2
	South Africa	–	–	203,937.40	–	38,440.1	191,752.1
	Saudi Arabia	–	–	353,333.40	–	–	-
	Ivory Coast	–	–	110,923.70	–	–	-
	Egypt	–	–	170,546.70	–	–	64,515.0
	The Republic of Guinea	–	–	37,587.50	–	–	350,512.9
	Canada	–	–	22,249.60	–	–	-
	Israel	–	–	–	–	–	1,259,449.11
	The Republic of Angola	–	–	–	–	–	87,067.8
EXPORT REFUNDS PAID					**£2,746,761.45**		

Company	Destination	1988 STATUS			1989 STATUS		
		INT	APS	OTHER	INT	APS	OTHER
Dawn Meats Limited	Iran	–	–	493,492.10	–	–	–
	Republic of Guinea	–	–	11,467.00	–	–	–
	Sierra Le One	–	–	14,025.00	–	–	–
	Ghana	–	–	–	–	–	42,822.50
	South Africa	–	–	–	–	–	79,825.00
	Egypt	–	–	–	–	–	122,102.60
	Romania	–	–	–	–	–	198,000.00
EXPORT REFUNDS PAID		**£189,525.58**			**£4,125,994.65**		

Company	Destination	1990 STATUS		
		INT	APS	OTHER
Dawn Meats Limited	Malta	_	39,979.40	14,633.90
	Iraq	_	260,175.00	14,000.00
	Iran	_	1,850,452.60	286,259.00
	Zaire	_	-	66,502.00
	Saudi Arabia	_	72,075.00	27,925.00
	Cyprus	_	18,223.20	8,024.10
	Gibraltar	_	-	1,758.40
	Ivory Coast	_	-	750,600.00

On the 21st of October 1983 Mr Kearns, an Agricultural Officer in Dawn Meats, discovered two pallets of intervention forequarter and plates day coded the 20th of October 1983 in the Dawn Meats assembly area.

One of the pallets had a tag attached marked "surplus intervention". After checking the intervention form with Mr Carroll, another Agricultural Officer, it was confirmed that the entire production for that day had gone into intervention. Mr Kearns invited the plant manager to inspect the pallets of beef in question. The pallets were gone when they arrived to carry out the inspection but were later found in one of the Dawn Meats freezers with another pallet containing intervention boxes dated the 20th of October of 1983. The manager offered no explanation. Mr Kearns signed a Movement Permit for the boxes (102 in all) to Q.K. Cold Store. That evening, Mr Kearns and Mr Carroll saw in excess of 100 boxes of intervention beef dispersed through pallets of commercial beef in the Dawn Meats Cold Store. Mr Kearns and Mr Carroll were satisfied that the beef had not been brought back from Intervention. When they returned on the next working day, the 24th of October, the intervention boxes had been removed and the store had been rearranged.

Mr Kearns asked Mr Staff, a Supervisory Agricultural Officer, headquartered at Clover Meats to investigate the situation. Mr Staff examined the 102 boxes in Q.K. Cold Store and found them in their frozen state to be of intervention standard. He saw approximately 100 empty intervention boxes in the premises of Dawn Meats turned inside out. These had a variety of day codes with broken seals and were stacked ready for use. He also saw some of these boxes filled with commercial fillets and striploins. Mr Brown, Managing Director of the plant, explained that the 102 boxes were used as a float to make a yield of 66% when actual yields fell below this figure.

On the 1st of November 1983, Mr Staff and two other Agricultural Officers opened two of the 102 and six from the regular intervention stock. They found excessive fat and unacceptable trimmings therein. These boxes were further examined by Mr Ferris, Senior Superintending Veterinary Inspector, and Mr Deevy, Senior Veterinary Inspector with the same results. A further 25 boxes were examined and eleven of these had excessive trimmings. Of ten boxes defatted four contained excessive fat and two were very close to the limit.

On the 9th of November 1983 Mr Kearns and Mr O'Carroll, Agricultural Officers at the plant, found approximately 30 boxes of intervention beef in Dawn Meats Cold Store dispersed through three pallets of commercial beef. All the boxes were examined and some were found to have been opened with some of their probable contents missing. The production dates on these boxes vary. Mr Browne explained that these might be intervention beef that had been mislaid.

On the 23rd of November 1983 Mr Kearns and Mr O'Carroll, found five boxes of flank that were not recorded in the production records. Mr Browne suggested that these also were also mislaid intervention.

A list of these incidents were sent to Mr Browne asking for an explanation. Two reminders were issued. Mr Brown replied. Although his observations were considered in detail and the facts checked his observations did not fully explain the various incidents and much were left unaccounted for.

A number of Agricultural officers were satisfied beyond doubt that yields at Dawn Meats were being tailored. Mr Ferris also agreed with this.

All these incidents were subjected to intensive investigation and the Department sought legal advice as to how it should proceed. The advice given was that it would be hard to bring a criminal case as proof would be difficult. It was further complicated by the fact that Dawn Meats had acted as sub-contractors to other meat companies. After a lot of investigation the company was suspended from deboning on the 20th of August 1984. A settlement was finally agreed with the company which allowed the suspension to be lifted on the 23rd of May of 1985 provided that they undertook to:-

1) buy back 707 boxes of suspect beef ;
2) to pay the cost of the Department's investigation; and;
3) to indemnify the Department from any complaints arising from the sale of beef produced by Dawn Meats.

MEADOW MEATS

The evidence in relation to this Company was given to the Tribunal by Mr John Comerford, Mr Martin Long, Agricultural Officers, employed by the Department of Agriculture and Food at the company's plant. Mr Dermot Kiersey, Veterinary Surgeon attached to the Department of Agriculture and Food, Mr Dermot Ryan, Senior Agricultural Officer with the Department of Agriculture and Food, Mr Seamus Fogarty, Mr Aidan McNamara, Mr Maurice Mullen, all attached to the Department of Agriculture and Food. Mr Bill Deevy, Senior Veterinary Inspector with the Department of Agriculture and Food. Mr John Phelan, Veterinary Inspector, Department of Agriculture and Food and Mr Thomas Nolan, former Managing Director of Meadow Meats Ltd.

Mr Nolan was Managing Director of Meadow Meats Ltd, from 1980 to 1991. The company had operated approved meat export premises under EEC Council Directive

64/433 and 77/99 at Rathdowney, Co. Laois, where they had a slaughtering, deboning and cold store facility. The premises was also an approved premises under EEC Directive 88/486. In 1988 the company leased a boning hall in Waterford which they subsequently purchased which said premises were an approved meat export premises under EEC Council Directive 64/433 and 77/99. They operated only a deboning plant in Waterford with a registered EEC No. of 525. The company employed approximately 370 employees between the two premises and they were paid subject to the appropriate PAYE and PRSI which was subsequently paid to the Revenue Commissioners as appropriate. The company employed approximately six to ten contract boners who were paid on a rate per quarter and on foot of an invoice submitted by the contractor but without any deductions of PAYE or PRSI.

Mr Maurice Mullen gave evidence to the Tribunal of the Intervention Deboning Yields achieved by Meadow Meats Ltd., in both its premises, Rathdowney and Waterford as follows:-

	1985	1986	1987	1988	1989	1990	1991
Meadow Meats Waterford	-	-	-	68.59%	68.83%	69.86%	70.43%

	1985	1986	1987	1988	1989	1990	1991
Meadow Meats (Rathdowney)	66.57%	68.06%	-	68.03%	68.02%	-	-

The Department of Agriculture and Food gave evidence of a defatting analysis carried out on Intervention product in respect of this company between the 30th of April 1991 and the 12th of February 1992 with the following overall results.

OVERALL RESULTS	FQ %	PF %
No. of Boxes Defatted	25.0	10
No. Overfat	11.0	1
Average	10.13	25.80%
Range	3.41% / 17.88%	15.78% / 30.18%

The amounts of compensation is a matter of discussion between the Department and the company.

Mr Maurice Mullen, Department of Agriculture and Food also gave evidence in relation to the export of beef by Meadow Meats Ltd., for the years 1986 to 1990 inclusive :

Company	Destination	1986 STATUS			1987 STATUS		
		INT	APS	OTHER	INT	APS	OTHER
Meadow Meats Ltd.,	Zaire	–	–	79,640.0	–	–	–
	Swaziland	–	–	12,292.5	–	–	–
	West Africa	–	–	–	–	–	41,607.5
	South Africa	–	–	–	–	–	425,700.0
EXPORT REFUNDS PAID		£177,132.87			£69,401.35		

EXPORT REFUNDS PAID FOR 1989 = £2,846,007.87

Company	Destination	1989 STATUS			1990 STATUS		
		INT	APS	OTHER	INT	APS	OTHER
Meadow Meats Ltd.,	Egypt	–	–	100,800.9	–	323,440.7	75,379.9
	Ivory Coast	–	5,907.1	126,995.0	–	23,361.7	2,250.0
	Liberia	–	–	13,337.5	–	–	–
	United Arab Emirates	–	–	21,587.5	–	26,155.6	55,843.5
	Romania	–	–	254,127.5	–	–	101,217.5
	Saudi Arabia	–	19,858.2	–	–	–	226,674.8
	Cyprus	–	38,110.8	–	–	22,759.6	13,602.8
	Iraq	–	23,476.2	–	–	447,146.0	427,405.7
	Andorra	–	–	–	–	13,050.3	-
	Bahrain	–	–	–	–	–	44,585.3
	Gibraltar	–	–	–	–	–	1,598.2
	Hong Kong	–	–	–	–	11,185.7	-
	Malta	–	–	–	–	107,876.6	14,437.9
	Zaire	–	–	–	–	-	632,872.6
EXPORT REFUNDS PAID		£2,846,007.87			£N/A		

Mr John Comerford an Agricultural Officer in Meadow Meats, Waterford told the Tribunal that on the 13th of March 1990 he observed the application of commercial lids to intervention boxes. He became suspicious immediately, stopped production and informed his superiors. He also sought an explanation from the factory management but as none was forthcoming he left matters in the hands of his superiors. The cuts being put into the boxes were forequarter cuts LMCs and fillets. As the forequarter yield had been down on previous days he was somewhat concerned. One possibility for the low yields would be the quality of the cattle, others being of cuts from intervention or the actual boning process itself.

Mr Martin Long an Agricultural Officer, told the Tribunal of a similar incident that occurred between the 6th of March and the 9th of March 1990. During intervention deboning of forequarter product he observed Jewish fillets and LMCs being taken out. When the meat destined for intervention is being deboned these cuts are normally left attached to the forequarter, whereas in commercial boning they are taken out separately. As he was not sure if intervention could be boned that way, he brought it to the attention of Mr Dermot Kiersey his superior, and also discussed the matter with his colleagues, Mr John Comerford and Mr James Meade.

At the time, as part of a further check Mr Dermot Ryan, Supervisory Agricultural Officer, inspected the chills and opened a number of cartons. While similar cuts were found there it was not possible to say if they were intervention or commercial. These matters were brought to the attention of headquarters. Mr Aidan McNamara told the Tribunal that it was decided to suspend the deboning licence until the end of the year. This action was endorsed by the Minister. In May of 1990, the Secretary of the Department received a letter from the company complaining about the severity of the penalty. In it, they stated that the value of meat in question was less than £200. Since the incident, they had lost intervention production deboning revenue in excess of £150,000. During cross-examination Mr Shay Fogarty agreed that this figure could be possible as intervention was very high at the time. He also stated that as there was another month involved there would have been a further £50,000 bringing the total penalty to £200,000.

Subsequently a meeting took place between the company and officials of the Department of Agriculture. Because of the BSE scare, the demand for commercial beef had fallen and this had led to a sharp drop in cattle prices. Consequently, the intervention safety net mechanism was put in place.

Mr Nolan, when asked in the Tribunal, for an explanation concerning the above matter stated:

> "I should really give the background to that. In Rathdowney, we concentrated on a supermarket base which was heavily hindquarter business, so one of the rationales setting up Waterford, was we could bone-out the forequarter in Waterford and develop the same sort of business with commercial people and we managed to develop a very profitable business with companies in the U.K. by boning out the forequarter in a detailed and specific way. Doing it for different manufacturers who would require different forequarter muscles and intervention was always a very small part of our business. The general manager, as far as I was concerned, made the case that it was easier to keep the boners doing the very accurate breakdown job. That they were used to rather than how we did the odd bit of intervention to change them over and do something different. So that although that may look to be somewhat an unusual system for our point of view, intervention was a very small part of our business and the logic was to keep saying "its difficulty to change boners from doing one specification in the morning to a different specification in the afternoon", so the simplest way of keeping throughput going, was to maintain the same sort of system."

It was accepted by Mr Nolan that it must have looked fairly deliberately done. But he reiterated and stated that it was not company policy to the Department of Agriculture & Food.

In the two tenders since the 12th of June of 1990, 13,400 tonnes of beef were purchased by the Intervention Agency and indications were that this volume of purchases would continue as there was no sign of an improvement in market prices. Nothing had been purchased in the period April to September 1989.

When the decision to impose the penalty was decided in April it was envisaged that the effects would be felt in the period September/December (effectively a four month suspension) as there is very little intervention activity in the summer. But in 1990 the situation was completely different. Due to the difficult market position that year, an equivalent penalty had been imposed in the period March/July. It was therefore decided to re-instate the plant from the 1st of August 1990.

RANGELAND MEATS LIMITED

The evidence, to the Tribunal, in respect of Rangeland Meats Ltd., was given by John Matthews, Veterinary Inspector with the Department of Agriculture and Food, Mr Seamus Fogarty, Assistant Principal Officer and Mr Maurice Mullen, both of the Department of Agriculture and Food, Mr Ben McArdle an ex-employee of Rangeland Meats Ltd., and Dr. Roger McCarrick, Managing Director of Rangeland Meats Ltd.

Rangeland Meats Ltd., has been involved in the beef processing industry since 1982 and is involved and has been involved in all aspects of it, that is, commercial, intervention and the export of beef. The company operates a deboning and cold store facility at Tullynamarla, Castleblayney, Co. Monaghan, which facilities are approved meat export premise under EEC Council Directive 68/433 and 77/99. The company has an EEC No. 717.

Mr Maurice Mullen of the Department of Agriculture and Food gave details of the intervention deboning yields achieved by the company from 1983 to 1991 (October).

	1983	1984	1985	1986	1987	1988	1989	1990	1991
Rangeland Meats Ltd.	66.28	67.24	67.87	68.11	68.06	68.14	68.09	68.10	68.10

The Department of Agriculture and Food carried out a defatting analysis on intervention deboned product produced by the company for various dates between the 7th of March 1991 and the 9th of March 1992 with the following overall results:-

OVERALL RESULTS	FQ	PF
No. of boxes defatted	25	10
No. Overfat	16	* 5
Average	11.68%	28.48%
Range	2.47% - 18.39%	11.27% - 37.30%

* **One box at 30.03% included in this figure.**

When questioned on the yields achieved by the company, over the last few years, Dr. McCarrick, in evidence, accepted that as far as the Minister was concerned, that he, the Minister, was entitled to all the red meat whether it is 68%, 69% or 70%. He continued:-

> "If you look at the intervention contract from a commercial operator's point of view, we have a minimum 68% to produce for the Government, otherwise we get penalised. But if, in producing that, wasn't at 68% or even at 68% we have too much fat in, we get penalised again. The commercial rate is something like 12p or 13p a lb. If you are subjected to any of those fines, you are out of business. So what we have to do is to make sure the Department determines fat levels on an on-going basis based on your average so if I get in a lot of R4H's which is 4s, that is the fat grade and the "H" is the higher level. If I get in a load and I get 68% on it I can guarantee, I won't be caught for the fat level. And when you were discussing with some of the previous witnesses, our yields of fat, there was one day, the 18th of March, which was one of the tests that Maurice Mullen did, when we came out with the yield of 15.5% for forequarter and 37% for plate and flank. The reason is that on that day, we had one load of beef and they were all 4Hs and the Department took a sample of that and that went in as an average. An a sample average was taken of that with the day's production that might have 10 times as much beef but no credit was given for the volume. So we have to do as operators do, when we have lean fat beef we are taking a risk on the fat level by getting 68%. When we get lean beef we have to trim our lean beef down to a level which is maybe 4% or 5% fat, so what you average the whole lot together on an average sample, it will work out at 10% on forequarter and 30% on plate and flank and that would be our instruction to the people on the floor and it is based on that that you stay in business".

Beef that is processed in Rangeland Meats comes from EC approved abattoirs throughout Ireland and mainly from the Republic of Ireland. The company employ, depending on the time of year, 180 to 200 employees, who at certain times of the year may be on a three day week. The company also employs subcontractors for cleaning, certain maintenance jobs, the installation of machinery, packing burgers and certain contract boners. The boners are on a full-time basis from contracting companies who provide Rangeland Meats Ltd., with the personnel. The full-time employees are paid weekly by cheque and the company pays all PAYE and PRSI to the Revenue Commissioners as appropriate. The sub-contractors are expected to have a C2 form, in which case the company pays all money due to them and the Revenue Commissioners collect from the sub-contractors. Where a C2 form is not available the company deducts 35% from their gross earnings and that is dealt with in the ordinary Revenue way.

The Tribunal heard evidence from Mr Bernard McArdle an ex-employee of the company who suggested that during a period 1985 to 1986 he was paid for work on a Saturday, in cash without any deduction of PAYE or PRSI.

Dr. McCarrick informed the Tribunal that so far as the company could check that Mr McArdle may have worked four Saturdays only in 1985 and while they were able to establish what its gross earnings were and from that calculate what net earnings were, the record did not clearly show whether his Saturday work had been subject in the normal way to PAYE and PRSI deductions.

Mr Maurice Mullen gave evidence of the exports of beef by the company and the export refunds paid as follows:-

Company	Destination	1984 STATUS			1985 STATUS		
		INT	APS	OTHER	INT	APS	OTHER
Rangeland Meats Ltd.	Lebanon	–	–	11,948.0	–	–	–
	Zaire	–	–	335,277.35	–	–	355,128.88
	West Africa	–	–	21,901.2	–	–	41,630.60
	Bahrain	–	–	3,181,432.4	–	–	846,1798.59
	Saudi Arabia	–	–	–	–	–	10,614.7
	Tahiti	–	–	–	–	–	66,105.
	South Africa	–	–	–	–	–	98,967.54
	Cyprus	–	–	–	–	–	56,110.40
	Unknown Country	–	–	–	–	–	8,203.30
	Gabon	–	–	–	–	–	16,257.60
	Ivory Coast	–	–	–	–	–	5,339.39
							14,954.07
EXPORT REFUNDS PAID		**£2,707.247.64**			**£1,283.304.02**		

Company	Destination	1986 STATUS			1987 STATUS		
		INT	APS	OTHER	INT	APS	OTHER
Rangeland Meats Ltd.,	Mauritius	–	–	14,921.6	–	–	–
	Zaire	–	–	63,188.6	–	–	8,201.30
	Egypt	–	–	442,466.74	–	–	62,518.00
	Saudi Arabia	–	–	113,563.4	–	–	10,009.10
	Cyprus	–	–	18,245.5	–	–	7,600.60
	West Africa	–	–	39,557.8	–	–	–
	Republic of Guinea	–	–	–	–	–	14,398.20
	North Africa	–	–	–	–	–	14,975.00
	South Africa	–	–	–	159,769.10	220,109.90	–
	East Africa	–	–	–	–	–	20,249.10
	Bahrain	–	–	–	–	–	2,722.00
	Iraq	–	–	–	–	–	149,566.4
EXPORT REFUNDS PAID		**£1,553,063.13**			**£1,586,115.16**		

Company	Destination	1988 STATUS			1989 STATUS		
		INT	APS	OTHER	INT	APS	OTHER
Rangeland Meats Limited	Durban	–	–	5,935.90	–	–	–
	Algeria	–	–	5,655.01	–	–	–
	Guinea	–	–	28,305.00	–	–	–
	Egypt	–	–	361,028.70	–	–	366,335.6
	Quatar	–	–	169.10	3,756.5	–	–
	Iraq	–	–	–	–	–	–
	Bahrain	800,067.1	7,016.80	204,227.9	901,469.9	–	448,831.5
	South Africa	–	–	1,054.0	–	–	–
	Grand Canaria	–	–	26,102.2	–	9,650.0	108,266.6
				-	107,048.3	–	–
				-	404,857.0	–	–
EXPORT REFUNDS PAID		£1,775,442.23			£3,530,327.98		

Company	Destination	1990 STATUS		
		INT	APS	OTHER
Rangeland Meats	Iraq	–	–	190,959.2
	Romania	–	–	182,890.50
	Tahiti (French Polynesia)	–	–	38,637.00
	West Africa	–	–	80,050.0
	South Africa	–	–	92,395.20
	Yugoslavia	–	–	32,652.3
	Gibraltar	–	–	569.60
EXPORT REFUNDS PAID		£N/A		

Mr Matthews, a Veterinary Inspector in Charge, described in evidence the layout of the boning hall in relation to the marshalling area. On the morning of the 25th of April 1990 on entering the marshalling area, he noticed cartons of forequarter meat with the name of a commercial client of the plant, marked on them, coming from one conveyor and intervention cuts coming from the other. At the time he arrived in Rangeland the Agricultural Officer was having his tea-break. As production had been ongoing since 8.o'clock that morning it would have been obvious if no intervention forequarter had been produced. During cross-examination he stated that it appeared to him that the incident could only have occurred while the Agricultural Officer was having his tea.

He rejected the production from intervention and suspended intervention deboning operations. Mr Matthews also testified that the product on the conveyors from the boning hall should be either commercial or intervention beef.

Dr Roger McCarrick, a Director of Rangeland, told the Tribunal that in all 72 sides were rejected. He stated that Mr Matthews had given him a choice, either accept Mr Matthews' decision to reject or contact the Intervention Section in Dublin. Dr McCarrick decided to accept the rejection as he knew he was in breach of the regulations in respect of having non-intervention beef in the boning hall during intervention processing.

LIFFEY MEATS LIMITED

The evidence to the Tribunal, in respect of this company, was given by Mr Peter Smyth, Mr Brendan Smyth, Mr Killian Unger, Veterinary Inspectors with the Department of Agriculture and Food, Mr Charles Corr, Superintending Veterinary Inspector with the Department of Agriculture and Food, Mr John Cassells, Veterinary Surgeon, Mr John Ferris, Senior Supervisory Inspector, Mr Sean O'Connor, Deputy Director Veterinary Inspector, Mr Gerard Dromey, Mr Aidan McNamara, Ms Mary Harvey, Mr Maurice Mullen, Mr Leo McTiernan, Mr Gerry McPhillips, Mr Frank Walls, all officials of the Department of Agriculture and Food. Mr Felix Loughran and Mr Peter McGovern, Mr Nollaig O'Broin, all officers with the Customs and Excise. Mr Christy Kett and Mr John Murphy and Mr Sean McNamara, all managers with Liffey Meats Ltd., and Mr Frank Mallon, Managing Director of Liffey Meats Ltd.

Mr Frank Mallon told the Tribunal that the company was initiated in 1974/1975 when it commenced operation beside the Liffey which is why the company was named "Liffey Meats". The company transferred operations to Ballyjamesduff, Co. Cavan in 1983, when it purchased the assets of Ballyjamesduff Chilling Ltd. They have been operating there ever since an approved meat export premises under EEC Council Directives 68/433 and 77/99.

The main business of the company is slaughtering, deboning and the export of beef to Third Countries. The company employs approximately 180 people and in a busy part of the year this would increase to 240/250. The company does not employ sub-contractors in the factory but they might be used for haulage. The company pays its employees either by cheque or cash whichever the particular employee requires but at all times the company make the appropriate deductions of PAYE and PRSI and makes all necessary payments to the Revenue Commissioners as appropriate.

Mr Maurice Mullen of the Department of Agriculture and Food gave evidence to the Tribunal of the Intervention Deboning Yields achieved by the Group for the period 1986 to 1991 (October) as follows:-

COMPANY	1985	1986	1987	1988	1989	1990	1991 (Oct)
Liffey Meats Limited	68.17	68.19	68.23	68.47	68.21	68.09	68.13

The Department of Agriculture and Food gave details of a defatting analysis carried out by them on intervention product produced by the company on various dates between the 22nd of February of 1991 and the 26th of March of 1992 with the following overall results:-

OVERALL RESULTS	FQ %	PF %
No. of Boxes Defatted	25.0	10
No. Overfat	12.0	1
Average	9.47	23.84%
Range	0.87% / 16.07%	18.39% / 32.87%

There was no penalty imposed by the Department since the results were below specification.

Mr Maurice Mullen gave details of the exports to Third World Countries conducted by the company for the years 1986 to 1990 inclusive together with the Export Refunds paid to the company.

Company	Destination	1986 STATUS			1987 STATUS		
		INT	APS	OTHER	INT	APS	OTHER
Liffey Meats	Israel	_	80,767.4	64,097.8	_	460,216.90	826,667.40
EXPORT REFUNDS PAID		£1.35m.			£712,466.70		

Company	Destination	1988 STATUS			1989 STATUS		
		INT	APS	OTHER	INT	APS	OTHER
Liffey Meats	French Polynesia	_	205,152.3	_	_	40,723.4	3,034.1
	Tahiti	_	14,235.1	_	_	_	_
	Cyprus	_	83,078.60	23,123.10	_	268,468.8	48,974.0
	Comores	_	99,964.3	_	_	_	_
	Malta	_	_	_	_	35,286.6	21,194.1
	Quatar	_	_	_	_	18,579.2	1,361.6
	Bahrain	_	_	_	_	94,480.7	5,704.8
	Egypt	_	_	_	_	418,969.60	1,543,647.52
	Saudi Arabia	_	_	_	_	_	960.8
	Iraq	_	_	_	_	265,921.7	114,384.4
	South Africa	_	_	_	_	149,207.0	_
	Gabon	_	_	_	_	37,972.5	_
	Finland					4,200.7	_
EXPORT REFUNDS PAID		£2,829,294.98			£4,027.086.84		

Company	Destination	1990 STATUS		
		INT	APS	OTHER
Liffey Meats	Egypt	_	2,134,897.1	720,808.1
	Zaire	_	_	107,431.4
	Bahrain	_	29,371.5	_
	Cyprus	_	67,594.3	16,613.1
	Malta	_	106,404.91	32,059.4
	Finland	_	35,291.7	1,620.3
	Oman	_	_	6,908.7
	Tahiti	_	13,542.0	-
	UAE	_	_	20,565.0
	Saudi Arabia	_	_	10,874.7
EXPORT REFUNDS PAID		**£N/A**		

Mr McNamara of the Department, told the Tribunal of a Control Inquiry Team visit to Liffey Meats on the 20th of January, 1993 where 61 cartons of forequarters were found to be missing and 30 cartons of plate and flank in excess of those recorded for the 18th of January. As a result the company was suspended from deboning from the 29th of January 1993 for two months.

Mr Corr, Area Supervisory Veterinary Inspector, Liffey Meats, elaborated on an incident concerning the possible substitution / theft of intervention beef in June of 1988. Mr Corr related that the incident, which took place on the 29th of June that year, concerned the finding of a number of cartons marked "Irish Intervention Forequarter Beef" in a container with the serial number PMT 5 on the loading bay mixed with boxes of commercial beef. The container was destined for the U.K. Mr Corr, examined and recorded the date on two intervention cartons. One had a date code of the 22nd of June and the other the 24th of June 1988. Management were asked to unload the container but since the unloading bay staff had gone it was decided to leave the unloading until the following morning. Mr Corr testified that before he left, the container was sealed. The off-loading was supervised the following morning, the 30th of June 1988 by Department personnel, fifteen cartons marked "Irish Intervention Beef" with the date codes erased were found during unloading. Ten other cartons marked "Chuck and Blades" with tampered veterinary control labels were also found. These boxes were detained and later examined by Mr Corr.

The Department officials were concerned for the following reasons:-

(a) This container had been loaded without supervision;
(b) It appeared that intervention had been mixed with commercial beef;
(c) The boxes detained appeared to have been tampered with while in a sealed container.

Two of the boxes identified by Mr Corr, the previous evening, were not recovered. The incident was reported to Mr Ferris, then S.S.V.I., on the 4th of July 1988 with a recommendation that the plant management be called up to headquarters and asked for an explanation. Mr Mallon, the Plant Manager, was written to and was then called to Agriculture House for a meeting which Mr Corr attended.

The plant management's explanation was that surplus boxes had been marked "Forequarter Beef" and were used inadvertently by the staff. Mr Corr added that he and his staff were suspicious that the doors of the container could have been taken off the hinges in the course of the night. To improve the Department's controls of intervention beef, in this plant, special veterinary control labels were introduced. These special labels are still in operation there.

As well as introducing the labelling system, Mr Corr, told the Tribunal that the Department increased its supervision of intervention beef after the incident. Mr Ferris told the Tribunal that the annual cost of the labelling system is £10,000 approximately in Liffey. The cost of extending this system to all plants in the country would be £750,000 approximately. As well as introducing the system of labels Mr Ferris explained that the possibility of taking cartons of beef from Liffey and some other plants as well and sending them for DNA identification was considered to see if it could be ascertained conclusively whether it was male or female beef.

It transpired after consultation with the State Laboratory that there was no conclusive way of proving whether the beef was male or female. Mr Ferris confirmed that it was not possible to ascertain definitely that substitution was taking place. Mr Patrick Leo McTiernan, Higher Agricultural Officer, who was temporarily assigned to the Liffey Meats plant found trimmings in boxes of plate and flank on the 10th of October 1986. Mr McTiernan discussed the matter with Mr Unger, Veterinary Inspector in Liffey Meats, and it was decided to hold back two pallets (80 boxes) of intervention plate and flank for re-inspection when the day's deboning was completed. The examination of the 80 boxes resulted in a finding of 330 kilos of trimmings and 18,019 kilos of plate and flank. Mr Brendan Brady the Boning Hall Manager, told Mr Unger that the mistake was due to the inexperience of the packers.

Mr O'Connor, Deputy Director of Veterinary Services, informed the Tribunal about a complaint he received in late 1983 from Mr Morris, a Regional Veterinary Officer in the Ministry of Agriculture, Fisheries and Food in the U.K. concerning poor quality beef produced for sale by Liffey Meats. Mr O'Connor investigated the matter and found that there was no record of the Department's staff having issued the Certificate accompanying the beef or having sealed the container. When a satisfactory explanation was not forthcoming from the factory the matter was referred to the Gardai.

The Gardai confirmed that the Certificate had been stolen from the Veterinary Office in Liffey Meats and an employee of Liffey Meats was subsequently convicted of the offence.

Ms Harvey told the Tribunal of an agreed settlement with the company on foot of a Customs' investigation into mis-declaration of product for MCA purposes on export to the U.K. The company had actually under claimed on some of their entitlements due to their mis-declaration. They had also over-paid charges in other cases. The company subsequently repaid £15,000 to Customs of which £12,577 was reimbursed to the

Department which was more than adequate to repay the amounts which were wrongfully claimed and paid to the company.

KMP CO-OP (MIDLETON) LIMITED

The evidence to the Tribunal, in respect of this company, was given by Mr John Murray and Mr John Matthews, Veterinary Inspectors with the Department of Agriculture and Food. Mr John Ferris, Senior Supervisory Inspector with the Department of Agriculture and Food, Mr Charles O'Connell, Senior Agricultural Officer with the Department of Agriculture and Food. Mr Seamus Fogarty and Mr Maurice Mullen of the Department of Agriculture and Food and Mr Dan O'Halloran, Business Development Manager of the Beef Division with the Kerry Group plc who was prior to June of 1992, General Manager at KMP Co-op Midleton, Co. Cork.

The company operates an approved meat export premises having a slaughtering and deboning under EC Council Directive 68/433 and 77/99. The company's product is stored in the Nordic Cold Stores in Midleton. The company employs approximately 200 people, mainly full-time and engages sub-contractors for maintenance, cleaning, catering, haulage, cattle procurement and deboning. The employees are paid by a bank transfer, monthly, or weekly by cheque with all PAYE and PRSI deducted. The company returns all PAYE and PRSI to the Revenue Commissioners as appropriate.

The sub-contractors are paid on a production of invoice basis without any deductions of tax. KMP Co-op (Midleton) Ltd. is a subsidiary of the Kerry Group Plc. The parent company took over Meadow Meats in mid-1991 and purchased the assets of Tunney Meats Ltd., in October of 1991. Mr Maurice Mullen of the Department of Agriculture and Food gave details of the Intervention Deboning Yields achieved by the company from 1986 to October 1991 as follows:-

	1986	1987	1988	1989	1990	1992. (Oct)
KMP Co-op Midleton	69.22	68.79	68.99	68.77	68.79	68.12

The Department of Agriculture and Food gave details of the defatting analysis performed by them on the companies intervention product for various dates between the 27th of February 1991 and the 27th of March 1992 with the following overall results.

OVERALL RESULTS	FQ %	PF %
No. of Boxes Defatted	25.0	10
No. Overfat	10	1
Average	10.17	23.24%
Range	5.89% / 17.08%	17.23% / 32.50%

As a result of this slightly overfat forequarter, the Department sought compensation and received from the company £86.37. The Department did further defatting analysis for the companies intervention production in respect of two different days, the 20th and the 25th of November of 1992 when a defatting analysis on forequarter beef showed fat levels of 7.72% and 6.28% respectively: both well within specification.

Mr Maurice Mullen of the Department of Agriculture and Food gave details of the export by KMP Co-op. (Midleton) Limited for the years 1987, 1988 and 1989 together with details of the Export Refunds paid to the company. They were as follows:-

Company	Destination	1987 STATUS			1988 STATUS		
		INT	APS	OTHER	INT	APS	OTHER
KMP Co-op. Midleton	Iran	–	1,262,189.48	86,432.19	–	–	64,393.35
	South Africa	–	–	–	–	–	181,217.00
EXPORT REFUNDS PAID		£3,843,358.47			£796,257.50		

Company	Destination	1989 STATUS		
		INT	APS	OTHER
KMP Co-op Midleton	Iran	–	–	4,392,343.97
	Saudi Arabia	–	71,983.89	19,839.50
	Bahrain	–	40,948.06	–
	Cyprus	–	73,013.88	–
	Egypt	–	467,985.79	–
	Romania	–	–	90,276.83
EXPORT REFUNDS PAID		£12,504,137.68		

Mr Ferris, Senior Supervisory Veterinary Inspector, gave evidence to the Tribunal relating to a surprise inspection carried out on the 29th of March 1989. He noted that some chains of fillets and the striploins were not being packed with the plate and flank but were being packed in cartons of trimmings. Also rump tails would appear to have been removed when they should have been packed with the plate and flank. Subsequently, on foot of Mr Ferris' report a circular issued to all meat plants, pointing out their responsibilities and that all meat resulting from the deboning of intervention is the property of the Department. Mr Murray, the Veterinary Inspector at the plant, was adamant that there was no systematic syphoning off of meat in this plant due to the viligance of himself and his staff.

Mr Shay Fogarty, gave evidence on the finding of 68 substandard cube rolls being packed into intervention boxes on the 18th of June 1991. The Supervisory Agricultural Officer, Mr C. O'Connell, suspected that an attempt at substitution might be taking place. The matter was reported to headquarters and the company was subsequently asked for an explanation. The company responded stating that they had problems with their vacuum packing machine the previous day with the result that all of the commercial production from that day had not been packed. When intervention operations commenced the following day inexperienced operatives had allowed this commercial beef into the packing area in error. The company were of the opinion that the problem would have come to light later in the day when the final reconciliations were done.

The Department were not totally convinced by this explanation. Therefore, a surprise inspection was carried out on the 3rd of September 1991 which concluded that the quality of the meat was good. A further surprise inspection on the 23rd of March 1992 again confirmed that everything was in order at the plant.

The Tribunal accepts the evidence of Mr Murray, the Veterinary Inspector in charge of KMP Co-op (Midleton) Ltd., when he said that he was satisfied that there was no systematic syphoning off of meat in this premises due to the viligance of himself and his staff.

FREEZOMATIC LIMITED

The evidence to the Tribunal in respect of this company was given by Mr Cecil Rothwell a Director of Freezomatic Ltd and Mr Maurice Mullen of the Department of Agriculture and Food. This company operates two cold stores at Cahir and Tipperary and Tycor, Co. Waterford. The cold stores operated by the company are approved for the purposes of EEC Directive 64/433 and in respect of the Cahir premises has the Vet. Control No. 12 and in respect of the Waterford premises has the Vet. Control No. 41.

The company stores both intervention and private beef and other goods. It does not operate either a slaughter house or a deboning hall. It employs approximately 12 full-time employees who are paid by cheque and the company pays PRSI and deducts PAYE remitting same to the Revenue Commissioners. The company engages sub-contractors purely for maintenance and building works. The company has operated since 1979.

The Department of Agriculture and Food carried out audits to the company on the 5th of November 1984 and the 13th and 14th of August of 1987 and these disclosed no evidence of any irregularity.

HEYER MEAT EXPORTS LTD / SINNAT LTD.

The evidence before the Tribunal was given by a Mr D. Heyer, Managing Director of the company and by Mr Maurice Mullen of the Department of Agriculture and Food. The company is exclusively a meat trader operating out of Brendan Road, Donnybrook, Co. Dublin. The company owns a minority interest in Sinnat Limited of which Mr Heyer is also Managing Director. The company does not operate a slaughter house, deboning hall or cold store and employs three (3) full-time employees who are paid by cheque with the company paying PAYE and PRSI to the Revenue Commissioners.

There was no evidence of any irregularity fraud or malpractice in or in connection with the beef processing industry.

The Tribunal was furnished by the Department of Agriculture and Food with details of the beef export by D. Heyer Meat Exports Ltd., / Sinnat Ltd., from 1984 to 1990 and these are set out below.

Company	Destination	1984 STATUS			1985 STATUS		
		INT	APS	OTHER	INT	APS	OTHER
D Heyer Meats Ltd. / Sinnat Ltd	Algeria	-	-	117,055.7	-	-	-
	Ivory Coast	_	-	59,175.9	-	-	-
	Spain	18,019.0	-	-	-	-	-
	UK Ships Victualling	-	-	-	19,006.0	-	-
	Zaire	-	-	-	-	-	-
	Iraq	-	-	-	-	19,698.4	-
	Lebanon	-	-	-	-	-	20,751.2
	Portugal	-	-	-	-	-	1,374.4
	Gabon	-	-	-	-	-	-
	Israel	-	-	-	-	-	-
		£598,625.32			£1,025,119		

Company	Destination	1986 STATUS			1987 STATUS		
		INT	APS	OTHER	INT	APS	OTHER
D Heyer Meats Ltd. / Sinnat Ltd	Libya	18,011.0	-	-	-	-	-
	Cyprus	-	16,596.2	7,032.0	-	-	-
	Egypt	-	157,765.0	-	-	-	-
	Togo	-	-	2,883.1	-	-	-
	Canary Islands	-	19,306.2	-	56.5	-	-
	United Arab Emeriates	-	8,251.2	-	-	-	-
	Zaire	-	37,769.8	106,809.4	-	-	-
	Ivory Coast	-	14,800.5	7,347.1	-	-	-
	French Polynesia	-	26,570.3	-	-	-	-
	Iraq	-	27,524.0	-	518.	-	-
	Saudi Arabia	-	-	16,046.8	-	-	-
	Israel	-	-	234,632.0	-	-	-
	Sweden	-	-	-	4.5	-	-
	United Arab Emirates	-	-	-	5.0	-	-
	South Africa	-	-	-	99.4	-	-
	St. Lucia	-	-	-	166.0	-	-
Sinnat Ltd	Algeria	-	638,879.40	1,358,803.90			
	Togo	-	-	1,813.20			
	Cyprus	-	16,016.00	-			
EXPORT REFUNDS PAID		£1,431,240.61			£796,7036.98		

Company	Destination	1988 STATUS			1989 STATUS		
		INT	APS	OTHER	INT	APS	OTHER
D Heyer Meats Ltd. / Sinnat Ltd	Sweden	20,799.80	-	-	-	-	-
	Gibraltar	5,008.00	-	-	-	-	-
	Cueta	6,996.00	-	-	-	-	-
	Canary Islands	20,904.00	-	-	-	-	-
EXPORT REFUNDS PAID		£12,184.00					

Neither company received any Export Credit Insurance from the Department of Industry and Commerce in respect of any of the exports referred to above.

Mr Heyer accepted that his company may have made an enquiry in respect of the availability of Export Credit Insurance in 1987 but the company never applied for Export Credit Insurance in respect of exports for which they were responsible.

OX-FLEISCH(H)ANDELSGESELLSCHAFTMBH

The evidence, to the Tribunal, on this company was given by Mr Colm O'Hagan, Financial Director of the company and by Mr Maurice Mullen of the Department of Agriculture and Food. The company is the Irish branch of a German registered company set up solely for the purpose of building a boning hall in Ireland.

The factory was opened in September of 1984 and has an approved meat export premises under Council Directives 64/433 and 77/99. The company has its premises in Carrickmacross in the County of Monaghan and has an EEC No. 508.

The company was opened to supply vacuum packed primal cuts to the German supermarket trade and the German "quality butcher" market. The company operates 52 weeks a year and employs up to 35 people. The employees are paid by cheque and the company pays PAYE and PRSI as appropriate to the Revenue Commissioners. The company undertook some intervention deboning since 1990 and the average intervention boning yields for this company were:-

- 1990 - 68.04%
- 1991 - 68.05%
- 1992 - 68.34%

Between the 9th of May 1991 and the 10th of March, 1992, the Department of Agriculture and Food carried out defatting of forequarter and plate and flank deboned by this company and the results were:-

OVERALL RESULTS	FQ %	PF %
No. of Boxes Defatted	25.0	10
No. Overfat	6.0	-
Average	7.99	20.11%
Range	3.12% / 16.36%	16.28% / 25.88%

The company purchases its carcases mainly from Kildare Chilling and Liffey Meats. The company does not keep Daily Job Costing or Weekly Job Costing records in respect of its intervention deboning and the only records are the forms from the Department of Agriculture and Food. i.e., the IB4s and 1B7s. etc. The trimmings from the deboning are sold as trimmings and have no value as cuts.

WESTERN MEATS PRODUCERS LIMITED

The evidence on this company was given to the Tribunal by Mr Denis Lyons, Managing Director of Western Meats Producers Limited. Mr Maurice Mullen of the Department of Agriculture and Food, Mr Peter Smyth, Mr Gerard Fogarty, Mr Dermot Ryan and Mr Seamus Fogarty, all of the Department of Agriculture and Food.

Western Meats Producers Limited commenced trading in 1985 from a premises in Dromod in County Leitrim, previously operated by a company M.J. Lyons Group which ceased trading. The company operates meat export premises approved under EC Council Directives 64 / 433 and 77/99 being a slaughtering and deboning premises with EEC number 342. The company operates in both the beef and pork business. The company employs approximately 90 full-time employees and engages sub-contractors for maintenance and transport services only. The company does not engage sub-contractor for slaughtering or deboning work within the factory. All of the employees are paid by cheque and the company pays all PAYE and PRSI to the Revenue Commissioners as appropriate.

The deboning yields returned by the company from 1985 to 1991 were as follows:-

- 1985 - 68.34%
- 1986 - 68.14%
- 1987 - 68.13%
- 1988 - 68.27%
- 1989 - 68.42%
- 1990 - 68.37%
- 1991 - 68.47%

Defatting analysis of the deboning in the factory was carried out between the 4th of February 1991 and the 30th of March 1992 and the overall results were as follows:-

OVERALL RESULTS	FQ %	PF %
No. of Boxes Defatted	25.0	10
No. Overfat	16.0	03
Average	11.56	25.63%
Range	7.49% / 20.44%	15.35% / 32.44%

Mr Mullen of the Department of Agriculture, told the Tribunal that by reason of the forequarter being over-fat that the Department of Agriculture and Food sought compensation from the company of £1,286.88.

The company did not export meat to any Third World Countries and received no Export Refund.

The Tribunal was given evidence of complaints investigated by the Department of Agriculture and Food in May of 1986, October of 1987, and March of 1989 and these were fully dealt with at the time by the Department.

TRANSFREEZE COLD STORES LIMITED

The evidence, about this company, was given to the Tribunal by Aidan McNamara, a Principal Officer in the Intervention Operations Division of the Department of Agriculture and Food and Mr Liam Fleming, Manager of the Company. The company operates a cold store at Santry Hall, Santry and is approved for the purposes of EC Directive 64/433 and has the Vet. Control Number of 36. The company is not engaged in the beef processing industry but is a purely warehousing company.

In October of 1988, the Management of Transfreeze Cold Store Limited, reported to the Department of Agriculture and Food the alleged theft of 197 boxes of intervention fillets. The matter was fully investigated by the Department of Agriculture and Food and the company fully co-operated with a view to determining where the problem lay. The company reimbursed the Department the sum of £48,359.17 as it was believed to have been an internal problem. Subsequent investigation by the Internal Audit Unit of the Department of Agriculture and Food ensured that there were no further losses. The matter was fully reported to Brussels as an irregularity.

BALTINGLASS MEATS LIMITED

The evidence to the Tribunal on this company was given by Mr James Walsh, Managing Director of Baltinglass Meats Limited and Mr Maurice Mullen of the Department of Agriculture and Food.

The company, Baltinglass Meats Limited, commenced operations in 1989. It operates a beef deboning plant and is an approved meat export premises under EC Council Directive 64/433 and 77/99. Its EEC No. is 523. The company does not operate a slaughtering facility and uses the Q.K. Cold Store in Naas and Bralco in Newbridge, for storage purposes. The company is involved in commercial, intervention and the export business. The company employs approximately 30 full-time employees who are paid by cheque after deductions of PAYE and PRSI which are paid to the Revenue as appropriate.

The company employed one sub-contractor for deboning purposes when the company was particularly busy. The company exported beef in 1989 as follows:-

Company	Destination	1989 STATUS		
		INT	APS	OTHER
Baltinglass Meats Limited	Saudi Arabia Qatar Bahrain	50,421.2 - 31,905.8	- - -	- 36,095.4 -
EXPORT REFUNDS PAID		172,830.33		

The company, apart from that year, did not export beef to Third Countries but concentrated on commercial beef to European Countries for sale in supermarkets.

Mr Mullen gave evidence of the defatting results carried out by the Department between the 13th of September 1991 and the 30th of March 1992 and the results were as follows:-

OVERALL RESULTS	FQ %	PF %
No. of Boxes Defatted	25.0	10
No. Overfat	09.0	01
Average	9.91	23.72%
Range	1.96% / 21.96%	18.54% / 32.43%

The Department of Agriculture and Food compared a randomly selected sample of IB6 forms submitted by the company with the original IB6 forms in the possession of the Department of Agriculture and Food and both sets were found to correspond. The corresponding "Boning Hall Intervention Sheets" furnished by the company also matched the IB6 forms.

The deboning yields achieved by the company in the deboning of intervention beef since 1989 were as follows:

- 1989 - 69.01%
- 1990 - 68.32%
- 1991 (Oct) - 68.78%

N.W.L. (IRELAND) LIMITED

The evidence to the Tribunal in respect of this company was given by Mr Maurice Mullen of the Department of Agriculture and Food and Mr Anthony McNicholl, Managing Director of the company N.W.L. (Ireland) Limited. The company is a meat trading company which purchases substantial amounts of beef from other parties including intervention which it subsequently exports. The details of exports by this company from 1986 to 1990 were as follows:

Company	Destination	1986 STATUS			1987 STATUS		
		INT	APS	OTHER	INT	APS	OTHER
N.W.L. (IRELAND) Limited	Ship Victualling Warehouse	39,964.0	-	-	-	-	-
	Gibraltar	54,060.3	-	-	12,012.0	-	-
	Canary Island	418,080.3	-	-	1,028.543.9	-	20,058.0
	Sweden	222,260.0	-	-	367,520.0	-	-
	Noway	52,945.0	-	-	297,362.7	-	20,079.5
	Maderia	59,022.0	-	-	-	-	-
	Angola	11,507.0	-	-	23,998.8	-	-
	Malta	199,858.0	-	-	632,184.7	-	-
	Hong Kong	16,681.0	-	-	-	-	-
	Congo	-	-	-	1,994.0	-	-
	Gibraltar	-	-	-	12,012.0	-	-
	Israel	-	-	-	19,927.0	-	-
	French Polynesia	-	-	-	20,979.0	-	-
	Cyprus	-	-	-	41,984.8	-	-
	Ship Stores	-	-	-	42,023.0	-	-
	Saudi Arabia	-	-	-	6,987.0	-	-
EXPORT REFUNDS PAID		£			1,886,789.42		

Company	Destination	1988 STATUS			1989 STATUS		
		INT	APS	OTHER	INT	APS	OTHER
N.W.L. (Ireland) Limited.	Malta	183,870.9	-	14,217.5	-	-	-
	Qatar	-	1,692.6	-	-	-	-
	Canary Islands	518,249.4	-	5,729.0	-	-	-
	French Polynesia	70,345.5	-	14,250.0	-	-	89,948.20
	Gozo	1,404.0	-	-	-	-	-
	Sweden	164,898.0	-	-	60,910,0	-	-
	Egypt	-	4,929.2	18,733.6	-	-	-
	Norway	90,671.3	-	-	-	-	-
	South Africa	3,028,388.4	-	526,914.7	-	-	617,326.72
	Finalnd	19,991.0	-	-	-	-	-
	Cyprus	55,074.5	-	-	-	-	-
	Iraq	-	29,380.0	-	-	-	-
	Tahiti	-	-	-	18,013.00	-	-
EXPORT REFUNDS PAID		£2,451,116.47			£2,802,606.94		

Company	Destination	1990 STATUS		
		INT	APS	OTHER
N.W.L. (Ireland) Limited	French Polynesia	-	_	203.911.50
	Sweden	20,940.00	_	-
	Zaire	-	_	228,814.02
EXPORT REFUNDS PAID				

The company employs 8 (eight) full-time employees who are paid on a monthly or fortnightly basis and the company pays PAYE and PRSI as appropriate to the

Revenue Commissioners. The company purchases a substantial amount of beef from intervention. This is mainly purchased from Irish intervention but approximately 10% or 15% would come from the United Kingdom mainly Northern Ireland. A small tonnage of intervention meat may have been purchased from other European Countries on an occasion. The company would notify the Department of Agriculture on any occasion upon which it was carrying out a re-boxing procedure and it would only be done under supervision.

AUTOZERO / TALLAGHT COLD STORE

The evidence to the Tribunal in respect of Autozero / Tallaght Cold Store was given by Mr Michael Phelan, Chief Executive of the company and Mr Eamonn O'Donovan a Higher Agricultural Officer in the Department of Agriculture and Food. The company operates cold store which are approved for the purpose EEC Directive 64/433. They are a cold store at Cabra, Dublin with Veterinary Control No. 1. and National Cold Store at Cooks Town Industrial Estate, Tallaght, with Veterinary Control No. 4 and Waterford Cold Stores at Christendom, Co. Waterford, with Veterinary Control No. H. The company employs approximately 63 people full-time employees and the company pays PAYE and PRSI to the Revenue Commissioners as appropriate on behalf of such employees.

HONEY CLOVER LIMITED

The evidence to the Tribunal in respect of this company was given by Mr Martin Blake, a Director of the company, Honey Clover Limited, and Honey Clover (Freshford) Limited and Mr Maurice Mullen of the Department of Agriculture and Food. The company, Honey Clover Limited, operates out of the IDA Industrial Estate in Navan, Co. Meath, where it operates an approved meat export premises under Council Directive 64/433 and 77/99 being a slaughtering and deboning operation with EEC No. 363. This premises was newly built in 1992. The company originally operated out of a premises at Grand Canal Street in Dublin where it first commenced business in 1986. A new company, Honey Clover (Freshford) Limited, was set up in 1991 when it purchased the assets of Master Meat Packers (Kilkenny) Limited, Freshford, Co. Kilkenny and the company operates an approved meat export premises under EEC Council Directive 64/433 and EC Council Directive 77/99 and in particular a slaughtering facility with EC No. 34A. The company employs, between Navan and Freshford, approximately 115 full-time employees who are paid by cheque and the company pays all PAYE and PRSI to the Revenue Commissioners as appropriate.

BARFORD MEATS LIMITED

The evidence to the Tribunal in respect of this company was given by Mr Seamus McQuirk, Managing Director of Barford Meats Limited and an admitted statement from Mr Liam Yore, Agricultural Officer with the Department of Agriculture and Food. The company was established in 1984 when it purchased the assets of Alpha Foods Limited which operated a deboning premises at Carrickmacross in the County of Monaghan. It operated an approved meat export premises under Council Directive 64/433 and Council Directive 77/99 and it was also approved under Council Directive 88/657. The company is a specialist meat processor. The company employs approximately 57 full-time employees who are paid by cheque and the company pays PAYE and PRSI to the Revenue Commissioners as appropriate.

C.H. FOODS LIMITED

The evidence to the Tribunal, in respect of this company, was given by Mr Carton, Managing Director of C.H. Foods Limited and Mr Maurice Mullen of the Department of Agriculture and Food. The company is mainly a meat trader and does not operate a slaughter house, a deboning premises or a cold store. The company employs four (4) full-time employees who are paid by cheque and the company pays PAYE and PRSI to the Revenue Commissioners as appropriate. The company has exported beef as follows:-

Company	Destination	1986 STATUS			1987 STATUS		
		INT	APS	OTHER	INT	APS	OTHER
C.H. FOODS LIMITED	Gibraltar	16,314	-	-	-	-	-
	Ships Victualling	-	-	-	34,363.8	-	-
EXPORT REFUNDS PAID		£160,702.68			£24,431.91		

Company	Destination	1988 STATUS			1990 STATUS		
		INT	APS	OTHER	INT	APS	OTHER
C.H. FOODS LIMITED	Ship Stores	5,011.0	-	-	-	-	-
	Cyprus	-	-	-	6,942.1	-	-
EXPORT REFUNDS PAID		£3911.05			£N/A		

CONTINENTAL BEEF PACKERS LIMITED

The evidence to the Tribunal in respect of this company was given by Mr Joseph Gordan, Managing Director of Continental Beef Packers Limited and Mr Maurice Mullen of the Department of Agriculture and Food. This company commenced business in approximately June of 1991 when it purchased the asset of Michael Purcell Foods Limited. The company is involved in a deboning and cold storage operation and runs an approved meat export premises under EC Council Directives 64/433 and Council Directives 77/99 at Wrensboro, Thurles, Co. Tipperary with the EEC No. 531. The cold storage operation is carried out at Dublin Road, Thurles, Co. Tipperary. The company employs approximately 81 full-time staff who are paid by cheque and the company pays PAYE and PRSI to the Revenue Commissioners as appropriate. The company also engages sub-contractors to carry out the deboning work and they are paid on the C2 System. If a sub-contractor has a C2 the company retains the full PAYE or PRSI. If he has no C2 then the company deducts 30%. The company's slaughtering facility is conducted by a separate company, Clonmel Chilling Limited. The deboning yields achieved by the company are:-

- 1991 - 68.77%
- 1992 - 68.67%

The figures for 1991 relate only to November/December of that year.

The defatting analysis carried out by the Department in 1991/1992 showed that the average fat levels for forequarter was 12.18% (per cent) and for plate and flank was 21.16%. The Department of Agriculture, in view of the fact that the forequarter was found to be over the fat level, sought compensation from the company of £829 and this sum was paid by the company.

A further defatting analysis was carried out by the Department of Agriculture in January and February of 1993. The company's product, from the following days was tested with the following results:-

DATE	AVERAGE %
11/11/'92	15.48
19/11/'92	18.27
03/11/'92	12.16
04/11/'92	13.71
06/11/'92	15.32
16/11/'92	13.05
24/11/'92	20.59

As a result of these tests the company's deboning licence was suspended on the 20th of February of 1993, reviewable in June of 1993.

The Department of Agriculture carried out further defatting analysis of intervention, boneless beef produced by the company and stored at QK Cold Store in Naas in respect of the following dates with the following results:

DATE	AVERAGE %
12/11/'92	22.16
20/11/'92	15.83
05/11/'92	17.23
17/11/'92	16.54

The view of the Department of Agriculture and Food on these results was expressed by Mr Maurice Mullen as follows:-

> "I am sure there is always an element of inexperience but I think in this case that the Department would be of the view that there is possibly two reasonable explanations for his:-
>
> (1) That they just didn't know the specification, didn't adhere to it; or;
>
> (2) they purposely left it over-fat.
>
> I am not making a comment which one it is."

Mr Gordon, the Managing Director, argued in evidence and by letter of the 20th of January, 1993,

> "I refer to your letter of 20th January 1993 re suspension from the deboning of Intervention Beef, and also to the letter to you from Mr John Smith, Chief Executive, I.M.P.A. of 25th January 1993.
>
> As a member of the I.M.P.A. this company fully supports the points Mr Smith is making and the conclusions reached in his letter of 25th January.
>
> However given our particular situation here in Thurles I wish to set out herein our view of the reasons by the action of the Department of Agriculture in our case is both unfair and unwarranted.
>
> As you are aware Continental Beef Packers Ltd is a deboning plant only and all Intervention deboned at the factory is slaughtered elsewhere. Indeed the company has not to date tendered to supply Intervention beef to the Department.
>
> As a consequence of the above it is clear that any yields achieved on intervention deboning will be affected by the quality and fat score of the incoming intervention beef. In the period October to December 1992 incoming loads of beef for intervention deboning were considerably biased towards the fatter grades of animal. Indeed nine days productions averaged 90% 4L and 4H grade cattle. Some loads could be expected from fat cattle as from their leaner counterparts. Obviously the abattoirs retained the leaner cattle to allow them to meet the yield targets more easily in their own deboning.

Not only were the cattle excessively fat but when deboned the flanks and middle ribs were extremely fat, and trimming of these cuts caused them to fall apart into smaller pieces of meat that were described by your inspectors as trimmings. This tendency in the beef received by our company to be overfat was described in detail to Mr John Matthews V.I. on the occasion of his visit at our plant (28.10.92). I remember in particular showing Mr Matthews a flank that was 90% visual fat and explaining to him that if it was trimmed up to specification all that would remain would be small pieces of meat. I also recall him agreeing that the beef we were receiving was abnormally fat (because of the level of fat on the incoming beef we actually failed to achieve the 68% target yield on the day of his visit.)

I also asked Mr Matthews to make a note of the fat beef we were receiving and he confirmed that there was a problem in the grading of cattle in several plants which the Department was tackling on an ongoing basis.

At the time of the defatting exercises also I pointed out to Mr David Lynch V.I. the middle ribs that would fall into small pieces of meat if the seam of fat were to be removed. He too agreed that the beef was abnormally overfat (in respect of the internal fat).

As a consequence of the first defatting exercise we wrote to all our suppliers of Intervention cattle for deboning pointing out that the percentage of fat 4 cattle being received by our plant had grossly exceeded the national average and that in future we would only accept loads for deboning that were equivalent to the national average for fat score.

I have spent in excess of three hours explaining the abnormal position our company has been placed in by the extremely fat cattle being delivered here, to Mr John Matthews and Mr David Lynch. However no account has been taken of our explanations in the Departments actions.

Since we have circulated our suppliers we have been receiving loads of Intervention beef in line with the national average, thus allowing more than satisfactory yields to be achieved in our deboning.

Due to the factors outlined above we believe that the actions of the Department in our case are unfair, and we request a meeting to discuss this issue as soon as possible.

Yours faithfully.

____

Joe Gordon Managing Director.

As was pointed out by Mr Mullen, and accepted by the Tribunal, that Mr Gordon's case that the grades of cattle that were being sent to debone were of a fatty nature such that it was pretty well impossible to achieve the 68% yield completely misses the Department of Agriculture and Food's point, which was that

> "the company in respect of the grade of cattle or carcases presented must meet the specification in the first instance and this may indeed make it harder to achieve a yield if there are such a high percentage of RH4s but it does not exonerate the company from achieving the 68% yield."

The company's deboning licence was suspended in January of 1993 until June of 1993 and in fact the company bought back the beef at the same price that the Department had paid the company for it. The company paid £236,053.13 for the deboned intervention cattle rejected by the Department.

This company does not export to Third World Countries directly but sells to Irish trading companies who may then export the product. However, the company has never purchased intervention beef.

NORDIC COLD STORAGE LIMITED

The evidence to the Tribunal, in respect of this company, was given by Mr Patrick Santry a Veterinary Inspector of the Department of Agriculture and Food together with a statement from Mr Thomas Butler, Managing Director of the company who was not required to give oral evidence.

Nordic Cold Storage Limited, operates a public cold storage facility at Midleton, Co. Cork, and is an approved cold store for the purposes of EEC Directive 64/433 having the Veterinary Control Number 51. The store commenced operations in June of 1968 and has provided blast freezing and cold storage services to a large number of food processing companies in every sector of the food industry. Intervention beef is one of the products stored in the Midleton store and while, for some years it has formed a significant part of the stock holding in that company it is not and never has been the primary product stored there.

As a public cold storage operator, Nordic Cold Storage Limited holds product on behalf of its customers. The company does not own or trade in any of the products stored. It neither imports nor exports any of the products. The Department of Agriculture and Food is provided with an office on the Midleton site, the office is manned by an Agricultural Officer of that Department. The company provides permanent inspection facilities for the Department of Agriculture and Food.

The company has stored product on behalf of several customers under the Aids to Private Storage Scheme. The company's employees are all subject to PAYE and PRSI deductions and same are fully discharged to the Revenue Commissioners as appropriate.

Q.K. COLD STORES LIMITED

The evidence in relation to this company was given by Mr Patrick Santry of the Department of Agriculture and Food and Mr Joe Walsh, Assistant Manager of the company's premises at Grannagh, Co. Waterford. The company operates a number of cold stores at Carroll's Cross, Co. Waterford, which is an approved cold store for the purposes of EEC Directive 64/433 with Veterinary Control Number 10 and Dublin Road, Naas, Co. Kildare, which is also an approved cold store for the purpose of EC Directive, 64/433 with a Veterinary Control Number 19 and a further premises at Grannagh, Co. Waterford, which is an approved premises for the purposes of EEC Directive 64/433 and has a Veterinary Control Number 32.

GOUDHURST LIMITED (in Receivership) and HAMPTON MEATS LTD.

The evidence in relation to these two companies was given by Mr Maurice Mullen, Department of Agriculture and Food, Mr James Clarke, Surveyor of Customs and Excise to the Revenue Commissioners, Mr Michael McGill, Higher Officer attached to the Investigation Branch of the Customs and Excise, Mr Gabriel Davey of the Department of Agriculture and Food, Mr Brian Donovan, Manager of Prime Meats Ltd., Mr Christy O'Brien and Mr John Mair, Receiver of Goudhurst Limited.

The company Goudhurst Limited, a meat plant, at Grand Canal Street in Dublin. The company let various parts of the premises to a variety of companies including Hampton Meats Limited. The company, Goudhurst Limited, went into receivership in October of 1988 and the assets were disposed of in May of 1989 when the premises were sold and subsequently the plant and equipment. Mr Christy O'Brien gave evidence to the Tribunal of a dispute between him and another individual concerning the ownership of shares of Goudhurst Ltd. Mr O'Brien told the Tribunal that Hampton Meats Limited had been set up for the purpose of retaining a shareholding of Goudhurst Limited.

Mr Maurice Mullen, Department of Agriculture, gave evidence to the Tribunal, of deboning having been carried out in the years 1985, 1986 and 1988 when yields of 68.3% for 1985, 68.29.% for 1986 and 68.42% for 1988, were achieved. Mr Mullen gave evidence of irregularities which he considered more attributable to day to day problems than to any particular course of conduct on the part of Goudhurst Ltd. Mr Mullen considered that Goudhurst performed their functions reasonably correctly and while.

> "we did have concerns but I think the limiting of the deboning allowed us to be satisfied that what came out was reasonable."

NORISH PLC.

The evidence in relation to this company was given to the Tribunal by Mr Maurice Mullen, of the Department of Agriculture and Food and Mr Paul Short, Secretary of Norish Plc. Norish Plc. is a group company which operates cold stores in Dublin, Cork, Kilkenny and the United Kingdom. The group runs cold stores which have been approved for the purpose of EEC Directive 64/433 at Ballyragget, Co. Kilkenny with Vet. Control Number 36 and at East Wall Road, Dublin, Vet. Control Number 65 and Bond Road, Dublin, Vet. Control No. 66. The business of Norish is purely that of a public cold store for the storage of beef and other products for the food industry. Its total storage capacity is 11.54 m-cu/ft of which 4 million cu/ft is in the United Kingdom. The group companies within the Republic of Ireland stores intervention beef and approximately 40% total capacity has been utilised for same. The Department of Agriculture have permanent staff and facilities at each of the groups cold stores. Customs and Excise is assisted by the Group whenever requested. When the company receives notification of a consignment to a cold store the Agricultural Officers are immediately notified. It is the Department of Agriculture officials who break the seal on any consignments of meat products delivered to the cold store for storage and they further oversee the off-loading inspection grading and temperature testing of the product following which they authorise the taking in of the product for cold storage or as the case may be the rejection. No seal is broken by any other person other than the Agricultural official.

The Customs and Excises officials make regular visits to the cold stores to discharge their various duties in connection with the CAP. There are assisted by company's employees but they would be under the supervision of Customs and Excise officials.

Within, the group, employs 59 full-time staff approximately within the Republic of Ireland and these employees are paid by cheque. The company deducts all PAYE and PRSI and remit same to the Revenue Commissioners as appropriate.

The company employs sub-contractors for the loading and off loading of containers and these sub-contractors are paid by cheque and the company complies fully with the requirements of Section 17 of the Finance Act, 1972 in respect of withholding tax.

LIXSTEED LIMITED

The evidence, to the Tribunal, in respect of this company was given by Mr Richard McCann, Managing Director of the company. The company has been in existence for approximately 8 years. It was set up for the purposes of supplying labour services generally and over the years it supplied a considerable amount of labour services to the beef industry in particular. It initially supplied loading and unloading services but this extended to include deboning and trimming. The company sub-contracted services to the processors and supplied the labour. It supplied these services to inter alia Eurowest, Kildare Meats Limited, DJS Meats Limited, Tara Meats and Q.K. Cold Stores Limited.

Generally speaking, when Lixsteed agreed to supply labour services it entered into a contract with the company which required its facility. Part of the agreement indicated that Lixsteed Limited and or its employees would be responsible for the tax affairs of the employees.

Mr McCann told the Tribunal that:

> "the company operated initially on the basis that the people that we were employing were sub-contractor to Lixsteed Ltd., and we felt, and had the view, the they were so but the Department of Social Welfare and indeed, the Revenue Commissioners did not agree with that and did not agree with what we were doing".

As a result of the dispute, between the two Departments and the company, the company agreed to pay a sum of money to the Revenue to regularise the tax affairs and agreed that they would be responsible for further PAYE and PRSI payments as they arose. The agreement with the Revenue Commissioners was reached in October of 1991.

EUROWEST LIMITED

The evidence to the Tribunal, in respect of this company, was given by Mr Maurice Mullen of the Department of Agriculture and Food. Eurowest Limited was formed in 1991 and was formed by C.E.D. Viandes, the French concern and a Mr Thomas McAndrew. The company leased the former Hibernia premises being the Abattoir at Sallins and the boning hall in Athy. Since they commenced operations in 1991 the average deboning yield are in respect of Athy:-

- 1991 - 68.53%
- 1992 - 68.77%

and in respect of Sallins:-

- 1991 - 68.90%
- 1992 - 68.91%

The Department of Agriculture and Food carried out a defatting analysis in mid-1992 and the average levels found on forequarter and plate and flank at Sallins were 10.45% for forequarter and 24.2% for plate and flank and in Athy 19.75% for forequarter and 24.47% for plate and flank. The Department sought compensation of £861.94 for the over-fat on the forequarters.

Further defatting analysis were carried out on the 4th and 11th of November 1992 and the 4th and 10th of December 1992 at Athy and the respective levels were 7.74% and 7.10% for forequarter and 7.13% and 8.75% for forequarter in Athy.

SLANEY MEATS INTERNATIONAL LIMITED / BALLYWALTER MEATS LIMITED

The evidence, in respect of the above companies, was given by Mr Gerard Fogarty of the Department of Agriculture and Food, Mr John Melville, Veterinary Inspector, Mr Maurice Mullen of the Department of Agriculture and Food and Mr Brendan Dunne, General Manager of Slaney Meats International Ltd. The company, Ballywalter Meats Ltd., changed its name to Slaney Cooked Meats Ltd., in 1989.

Slaney Meats International Ltd., operates approved meat export premises under EEC Council Directive 64/433 and 77/99 at Ryland, Enniscorthy, Co. Wexford, where it has slaughtering, deboning and cold store facilities with EEC No. 296. Slaney Cooked Meats Ltd., has an approved meat export premises under EC Council Directive 64/433 and 77/99 at Ryland, Enniscorthy, Co. Wexford where it processed meat products under EEC p328.

Since its inception in 1968 Slaney Meats International Ltd., has been involved in the cattle sheep and bye-product business, involving itself in the commercial intervention and export of beef and sheep. In 1992 it employed approximately 300 and 23 full-time employees who were paid by weekly cheque and the company paid all PAYE and PRSI to the Revenue Commissioners as appropriate. The company also engaged sub-contractor solely for plant maintenance and repair. The deboning yields returned by the company were:-

- 1989 - 68.31%
- 1990 - 68.30%
- 1991 to Oct. - 68.33%

The Department of Agriculture and Food carried out defatting analysis between the 8th of March 1991 and the 15th of April 1992 and the overall results were as follows:-

OVERALL RESULTS	FQ %	PF %
No. of Boxes Defatted	25.0	10
No. Overfat	07.0	-
Average	9.21	17.37%
Range	4.87% / 21.23%	11.34% / 23.41%

The Department of Agriculture and Food carried out further examinations of boxes of deboned meat which had been placed in the cold store at Q.K. Grannagh, which meat had been deboned by Slaney Meats International Limited. The examination into the boxes was carried out between the 23rd of November 1991 and the 30th of March 1992. There were 119 boxes examined of all cuts. That is fillet, striploin, insides, outsides, knuckles, rump, cube rolls, brisket, shin and shank. There were four boxes with minor defects.

The company exported beef as follows:-

Company	Destination	1984 STATUS			1986 STATUS		
		INT	APS	OTHER	INT	APS	OTHER
Slaney Meats Limited	Canada	_	39,052.0	3,194,995.0	-	17,175.0	131,688.0
	USA	-	-	451,060.5	-	-	702,187.0
	Cyprus	-	-	91,197.8	-	-	14,942.0
	Gabon	-	-	158,290.0	-	22,082.0	14,856.0
	Togo	-	-	25,966.0	-	80,811.0	383,160.0
	Zaire	-	-	296,318.0	-	147,418.0	321,230.0
	Algeria	-	-	1,231.000.0	-	-	-
	Egypt	-	-	163,451.0	-	793,727.0	2,254,415.0
	Tunisia	-	-	20,554.0	-	-	-
	Saudi Arabia	-	-	103,218.0	-	17,288.0	51,174.0
	Cameroon	-	-	-	-	2,148.0	24,794.0
	West Africa	-	-	-	-	-	9,996.0
	Congo	-	-	-	-	20,473.0	8,427.0
	Mauritius	-	-	-	-	61,498.0	88,271.0
	Israel	-	-	-	-	-	49,420.4
	Canada	-	-	-	-	17,175.0	131,688.0
	Ivory Coast	-	-	-	-	-	70,665.0
	Camores	-	-	-	-	-	39,963.0
		EXPORT REFUNDS PAID					

Company	Destination	1987 STATUS			1988 STATUS		
		INT	APS	OTHER	INT	APS	OTHER
Slaney Meats Limited	USA	-	-	756,649.0	-	-	178,030.00
	Saudi Arabia	-	18,140.0	20,951.0	-	30,534.00	8,688.00
	Algeria	-	12,838.0	508,839.0	-	530,525.00	343,184.00
	Ivory Coast	-	220,294.0	117,835.0	-	-	57,889.00
	Egypt	-	1,170.650.0	2,756,408.0	-	11,205.00	9,587.00
	Togo	-	20,850.0	-	-	-	-
	Tahiti	-	4,825.0	47,425.0	-	-	23,288.00
	Mauritius	-	26,696.0	106,484.0	-	-	-
	Djibouti	-	-	4,046.0	-	-	-
	West Africa	-	20,036.0	19,943.0	-	10,777.00	10,805.00
	Malta	-	20,363.0	-	-	-	-
	Israel	-	-	1,066,798.6	-	-	-
	Zaire	-	-	19,779.0	-	-	-
	Comores	-	-	_	_	-	21,143.00
	Cyprus	-	-				800.00
EXPORT CREDIT REFUNDS PAID		£6,550,752.16			£3,438,070.09		

Company	Destination	1989 STATUS			1990 STATUS		
		INT	APS	OTHER	INT	APS	OTHER
Slaney Meats Limited	South Africa	-	-	95,275.0	-	-	15,675.0
	French Polynesia	-	12,892.0	174,662.0	-	28,579.0	141,068.0
	Yugoslavia	-	20,537.0	-	-	-	-
	Egypt	-	1,029,097.0	427,590.0	-	303,997.0	353,596.0
	Malta	-	14,699.0	-	-	132,257.0	59,209.0
	Gabon	-	-	24,005.0	-	-	71,301.0
	Iraq	-	586,539.0	233,786.0	-	2,395,715.0	375,605.0
	Ivory Coast	-	-	134,259.0	-	-	312,800.0
	Cyprus	-	2,338.0	3,283.0	-	-	53,807.0
	Mauritius	-	-	63,850.0	-	-	81,332.0
	Zaire	-	-	80,000.0	-	-	148,000.0
	Liberia	-	-	24,000.0	-	-	-
	Congo	-	-	24,000.0	-	-	-
	Ghana	-	-	21,000.0	-	-	-
	Greenland	-	-	-	-	-	6,616.0
	Victualling	-	-	-	-	20,450.0	30,286.0
	New Caledonia	-	-	-	-	-	107,007.0
	Israel	-	-	-	-	-	161,387.0
	Guinea	-	-	-	-	-	31,000.0
	Algeria	-	-	-	-	185,955.0	775,332.0
	Saudi Arabia	-	-	-	-	8,727.0	92,258.0
	Jordan	-	-	-	-	-	23,175.0
	Bahrain	-	-	-	-	-	24,583.0
	Dubai	-	-	-	-	-	45,460.0
	U.A.E.	-	-	-	-	-	25,488.0
	Tahiti	-	-	-	-	11,937.0	2,646.0
	West Africa	-	-	-	-	-	11,124.0
	Andora	-	-	-	-	10,054.0	-
	Senegal	-	-	-	-	-	15,275.0
EXPORTS REFUNDS PAID		**£9,471,551.00**			**£N/A**		

The Department of Agriculture and Food gave evidence to the Tribunal of audits carried out on the company's premises on various dates including the 27th of May 1985, the 1st to the 3rd of December 1986 and the 14th to the 16th of November of 1988. These audits did not reveal evidence of any illegal activity, fraud or malpractice in or in connection with the beef processing industry in respect of this company.

MICHAEL PURCELL FOODS LIMITED

The evidence in respect of this company was given by Mr Maurice Mullen, Department of Agriculture and Food and Mr John Purcell, Former Managing Director of Michael Purcell Foods Ltd. The company carried on business in Thurles, Co. Tipperary offering the facility of contract deboners for a number of companies from 1984 to 1989 when the company ceased trading.

At that time the company employed approximately 75 full-time employees and the company paid all PAYE and PRSI to the Revenue Commissioners as appropriate.

The main involvement, so far as the Tribunal of Inquiry is concerned, was to the company's involvement in the 1988 Aids to Private Storage when Michael Purcell Foods Ltd., was involved as a sub-contractor on behalf of Taher Meats Ltd. The product processed by this company, on behalf of Taher Meats Ltd., was examined in the early part of 1989 as part of the Aids to Private Storage Joint Sampling Programme. The examination showed that the company produced in the region of 366 tonnes of plate and flank for Taher Meats. It did so under 20 contracts in all and product was sampled for 15 of these 20 contracts. The Department sampled a total of 300 boxes and trimmings were found in a number of boxes and it was also discovered that the boxes contained pieces of non-individually wrapped beef. Mr Mullen told the Tribunal that in six boxes the trimmings found were greater than or equal to three kilos. There were also six boxes with trimmings less than three kilos. Six boxes had between them 10.715 kilos of trimmings. There were boxes found with pieces that were not individually wrapped, in quantity approximately 161.76 kilos in 75 boxes. There were 17 boxes which had both trims and pieces which were not individually wrapped and the quantity of trims in those boxes was 44.68 kilos and the quantity of meat not individually wrapped was 69.58 kilos.

Mr Mullen described it as "a finding of concern".

Mr John Purcell, told the Tribunal that a dispute existed between his company and Taher Meats which is presently pending in the High Court. Mr Purcell indicated to the Tribunal that the work done by his company had been done on the basis of what they had been asked to do by Taher Meats and also having regard to the copy of the Aids to Private Storage Rules and Regulations which were faxed to them by Taher Meats approximately 2 weeks into production.

Mr Purcell indicated that Taher Meats had also provided supervisors on a daily basis to inspect and supervise the works being done by his company.

Since the matter is presently pending before the High Court, this Tribunal does not intend to make any finding or fact in relation to the issue between Taher Meats and Michael Purcell Foods Ltd.

PURCELL MEATS LIMITED

The evidence in relation to this company was given by Mr Maurice Mullen of the Department of Agriculture and Food and Mr John O'Meara, former Managing Director of the company.

The company was engaged in the beef processing industry until 1986 when it sold its premises.

Mr Maurice Mullen gave evidence to the Tribunal of the exports by this company in the years 1984 to 1986 together with the Export Refunds paid to this company for 1987 to 1989 inclusive.

Company	Destination	1984 STATUS		
		INT	APS	OTHER
Purcell Exports Ltd.	Libya	-	-	1,740,966.6
	Gran Canaria	-	-	154,273.3
	Tahiti	-	16,387.1	217,199.0
	Ivory Coast	-	-	20,069.4
	South Africa	-	-	26,460.8
	Zaire	-	32,894.5	-
	Togo	-	19,500.4	-
	Egypt	-	213,227.9	5,304,994.9
	Iran	-	861,417.7	1,1185,463.6

Company	Destination	1985 STATUS		
		INT	APS	OTHER
Purcell Exports Limited.	Algeria	-	3,165,132.0	5,425,482.2
	Libya	-	_	2,371,655.0
	Gran Canaria	-	143,873.0	99,861.0
	Egypt	-	3,403,201.9	3,588,292.4
	Cyprus	-	39,813.4	79,043.5
	Malta	-	259,746.6	82,819.2
	Iran	-	6,023,239.8	2,584,991.86
	Mauritius	-	37,348.4	65,547.6
	Tahiti	-	46,117.2	3,023.6
	Ivory Coast	-	17,969.4	24,829.6
	Gabon	-	-	5,586.3
	Gibraltar	-	9,901.9	-

Company	Destination	1986 STATUS		
		INT	APS	OTHER
Purcell Meats Limited	Libya	-	930,189.8	842,364.5
	Malta	-	-	41,979.5
	Cyprus	-	28,381.2	22,432.6
	Gran Canaria	-	26,806.3	27,517.6
	Egypt	-	101,593.1	57,464.9
	Iran	-	2,646,867.8	1,067,237.9
	Saudi Arabia	-	-	19,912.6
	Algeria	-	325,189.5	23,430.5

The Export Refunds received by the company for:

- 1987; £10,583,888.35.
- 1988; £ 5,933,867.93
- 1989; £14,882,027.72

IRISH MEAT PRODUCERS LIMITED

The evidence in relation to this company was given by Mr Donal Russell, Counsellor (Agricultural Affairs) at the Irish Permanent representation to the European Communities in Brussels, Mary Harvey of the Department of Agriculture and Food, John F. McArdle, Member of An Garda Siochana, Gabriel Curley, Executive Officer assigned to the Beef Export Refund Payment Section, Mr Maurice Mullen of the Department of Agriculture and Food and Mr Sean Barton.

Mr Maurice Mullen of the Department of Agriculture and Food gave evidence to the Tribunal of the intervention yield returned by the company for the years 1983 to 1986 in respect of two plants IMP (Midleton) and IMP (Leixlip). They were as follows:-

	1983	1984	1985	1986
IMP (Midleton)	66.42%	67.54%	68.22%	69.15%
IMP (Leixlip)	66.36%	67.02%	68.13%	-

Mr Mullen also gave details of the beef exports by this company to various companies for the years 1984 to 1986 inclusive and they were as follows:-

Company	Destination	1984 STATUS			1985 STATUS		
		INT	APS	OTHER	INT	APS	OTHER
Irish Meat Producers Ltd.,	Israel	–	–	1,039,699.5	–	–	–
	USA	–	–	375,525.3	–	–	125,649.31
	Zaire	–	–	271,423.8	–	146,352.7	625,568.1
	South Africa	–	–	278,555.5	–	35,950.8	27,209.0
	Canada	–	–	2,997,212.5	–	303,397.5	238,319.8
	Saudi Arabia	–	–	530,922.7	–	162,527.6	352,907.1
	Cyrus	–	–	24,232.5	–	–	9,048.6
	West Africa	–	–	13,808.6	–	–	–
	Victualling West Germany	–	–	40,603.7	–	–	–
	Ivory Coast	–	–	43,946.9	–	27,042.7	13,396.5
	Quatar	–	–	12,166.9	–	–	–
	Ghana	–	–	–	–	–	11,996.0
	Egypt	–	–	–	–	69,819.8	564,603.5
	Cameroon	–	–	–	–	–	18,184.2
	Malta	–	–	–	–	37,665.0	24,984.9
EXPORT REFUNDS PAID		**£7,345,869.18**			**£4,363,754.83**		

Company	Destination	1990 STATUS		
		INT	APS	OTHER
Irish Meat Packers Ltd.,	Saudi Arabia	28,445.1	332,905.3	100,565.6
	Zaire	–	–	21,342.9
	Malta	–	–	20,050.4

The only issue which the Tribunal concerned itself with in connection with this company was the shipment of 560 tonnes of boneless beef by Irish Meat Producers Limited to Egypt which meat subsequently went to Trinidad and on which Export Refunds of some £786,000 was paid. Mr Russell, told the Tribunal, that in 1984/1985, Irish Meat Producers Limited from the Midleton plant shipped 560 tonnes of beef to Egypt. They claimed advance payment in the refund on that shipment in the normal way. They lodged a guarantee for that advance payment and they were paid a sum of £786,000 on that shipment. Subsequently, all requirements of the refund regulations, were met and particularly the most important one which was the furnishing of proof of import into Egypt for home use. When that proof was provided to the Department, the Department was then in a position to release the guarantee back to Irish Meat Producers Limited, the documents showing proof of entry into the country of Egypt would have been included with the Bills of Lading for the shipment from Ireland to Egypt.

Subsequently, on the 15th of January, 1986, the Trinidad Meat Cottage Ice and Cold Storage Limited wrote to the Minister for Agriculture and Food, complaining about the quality of meat which had been exported to their country by a company called SM. International Traders. The letter indicated that the meat the subject matter of the complaint had come via Port Said in Egypt and further investigation established that the meat exported by Irish Meat Producers Limited to Egypt was the same meat which was subsequently exported to Trinidad to the order of the Trinidad Meat Cottage Ice and Cold Storage Limited. As a result it was decided by the Department of Agriculture to make a complaint to the Garda Fraud Squad.

It was clear, as a result of meeting between the Department of Agriculture and Irish Meat Producer Company Limited, that the company was not involved in any way in an alleged fraudulent transaction involving the Export Refunds. Export Refunds would have, and indeed were paid, on the export of this meat to Egypt, once there was proof of entry into Egypt for home use. At that time, there were no Export Refunds available for the export of meat to Trinidad for home use.

Garda John McArdle of the Garda Siochana (Fraud Squad) together with other members of the Gardai carried out a very full and thorough investigation which included visiting Egypt and Trinidad for the purposes of interviewing persons who were in a position to be of assistance. At the conclusion of the investigation by the Fraud Squad the papers were sent to the Director of Public Prosecution in May of 1988. On the 20th of December, 1991, the Director of Public Prosecutions notified Inspector Murphy of the Fraud Squad:-

> "It has been concluded that unless some significant development in this matter has taken place, since your submission of this file, a prosecution of any person,......, would not be a viable proposition and that should not be initiated".

ARAX LIMITED

The evidence to the Tribunal, in respect of Arax Limited, was given by Mr Maurice Mullen, Department of Agriculture and Food and Mr Shay Fogarty, Department of Agriculture and Food, Mr Michael Sheehan, Veterinary Inspector and Mr Terry Hanlon, General Manager of the company.

A new premises being a factory and cold store with a capacity of 2,000 to 2,500 tonnes was built in late 1989 early 1990 by a company Pechenga Limited. The company changed its name to Arax Limited in August of 1990, it commenced accepting beef for storage in or about February/March of 1991 and commenced deboning operation in or about July of 1991. The premises being a deboning and cold store premises are approved Meat Export Premises under EC. Council Directive 64/433 and 77/99 with EEC No. 529. The company employs approximately 40 to 50 employees who are paid at an hourly rate on a weekly basis. The company pays all PAYE and PRSI to the Revenue Commissioners as appropriate. The Department carried out a defatting analysis in December of 1992 on two days and the fat levels reached were 9.42% on forequarter and 8.5% again for forequarter on the 9th of December. The maximum is 10%. The company returned average deboning yields for 1991/ January 1992 of 69.17%.

Mr Fogarty of the Department of Agriculture and Food gave evidence of minor problems in September of 1991, which he was satisfied was a result of inexperience on the part of a new employee as the company was in the process of commencing operation. Further examinations were carried out in October 1991 and November of 1991 and again minor problems were encountered and sorted out by the Department with the company. The Tribunal accepts Mr Fogarty's evidence in this regard.

ASHBOURNE MEAT PROCESSOR LIMITED.

The evidence in relation to this company was given by Mr John Matthews, a Veterinary Inspector, Mr Maurice Mullen of the Department of Agriculture and Food, and Mr Danny Houlihan, Joint Managing Director of the company.

The company employs 65 full-time employees who are paid by cheque. The company pays all PAYE and PRSI to the Revenue Commissioners as appropriate. The company employs sub-contractors only in connection with the facilities of transport, shipping, couriers, maintenance and occasional re-packing.

The company operates a deboning premises at Naas Industrial Estate, Co. Kildare which is an approved meat export premises under EC Council Directive 64/433 and EC Council Directive 77/99 with the EEC No. 512. The company has been in operation since about June of 1985.

The deboning yields achieved by the company in the years 1986 to 1991 (October) when:-

- 1986 - 68.20%
- 1987 - 68.02%
- 1988 - 68.16%
- 1989 - 68.2%
- 1990 - 68.18%
- 1991 - 68.8%.

The Department of Agriculture and Food gave evidence of a defatting analysis carried out by them between the 7th of March of 1991 and the 26th of March 1992 with the following results.

OVERALL RESULTS	FQ %	PF %
No. of Boxes Defatted	25.00	10
No. Overfat	25.19	1
Average	11.57	27.12%
Range	6.75% / 18.25%	22.54% / 35.41%

The Department levied a charge in respect of overfat on the forequarter for the sum of £1,013.48. The company exported beef in the year 1986 to 1989 as follows:-

Company	Destination	1985 STATUS			1986 STATUS		
		INT	APS	OTHER	INT	APS	OTHER
Ashbourne Meats	Saudi Arabia	-	-	116,255.37	-	-	561,930.90
	Kuwait	-	-	-	-	-	13,695.35
	Cyprus	-	-	-	-	-	6,687.20
£N/A					£N/A		

Company	Destination	1987 STATUS			1988 STATUS		
		INT	APS	OTHER	INT	APS	OTHER
Ashbourne Meats	Dubai	14,759.0	-	13,969.1	-	-	82,807.3
	Kuwait	-	-	15,092.4	-	-	-
	Saudi Arabia	-	-	445,686.25	37,112.1	-	973,302.0
	Bahrain	-	-	-	-	-	40,026.9
	UAE	-	-	-	-	-	26,115.2
EXPORT REFUNDS PAID		£955,134.57					

Company	Destination	1989 STATUS			1990 STATUS		
		INT	APS	OTHER	INT	APS	OTHER
Ashbourne Meats	Saudi Arabia	-	-	2,108,233.30	-	26,367.9	4,236,144.4
	Bahrain	-	-	84,627.50	-	-	162,492.0
	Phillipines	-	-	66,918.40	-	-	-
	Dubai	-	-	-	-	-	156,537.6
	Jordan	-	-	-	-	-	23,996.7
	Ghana	-	-	-	-	-	23,930.2
	Oman	-	-	-	-	-	15,451.3
	Kuwait	-	-	-	-	-	3,815.1
	U.A.E.	-	-	-	-	-	3,188.4
EXPORT REFUNDS PAID		£5,285,160.25			£		

The company took part in the 1988 Aids to Private Storage Programme and deboned beef under that programme for Agra Trading Limited. Mr Matthews, Veterinary Inspector, gave evidence to the Tribunal that this plant produced 117 tonnes of plate and flank. One contract was examined, that is about 40 boxes, of which 23 boxes were found to contain 67.6 kilograms of non-individually wrapped product. There were no trimmings found in any of the boxes examined. The contract was penalised to the sum of £20,675.34 and this sum was paid by Ashbourne Meat Processors Ltd.

Mr Danny Houlihan, Joint Managing Director of Ashbourne Meat Processors Ltd., told the Tribunal, that the 1988 Aids to Private Storage Scheme involved a contract between that company and Agra Trading Ltd., to debone and pack beef on behalf of Agra Trading Ltd., The contract was carried out on foot of a specification provided by Agra to Ashbourne in the Autumn of 1988 between the 15th of September and the 25th of November in their premises at Ballyjamesduff, Co. Cavan. Ashbourne Meats Processors Ltd., did not become aware of the problem concerning the non-individually wrapped pieces of meat until September of 1989 when they were contacted by Agra Trading Ltd., as a result the company carried out a full investigation into how the problem arose and Mr Houlihan explained to the Tribunal:

> "that particular contract, that particular time of the year is a very, very, very, chaotic time of the year in meat business when one is trying probably more work done than he should and as a result we do have more temporary and probably inexperienced staff doing the job and it is a question of human error.
>
> There was no financial advantage to Ashbourne Meats in not individually wrapping the cuts. We had no gain out of it, it was human error."

FRESHLAND FOODS LTD.,

The evidence to the Tribunal in respect of this company was given by Mr Wilfred Woollett, a Veterinary Inspector, Mr Maurice Mullen of the Department of Agriculture and Food and a Mr Seamus Hand, Managing Director of Freshland Foods Ltd. The company purchased the assets and stocks of Avrich Limited in April of 1991 and commenced operation as Freshland Foods Ltd. In October of 1991 a new 5,000 ton cold store was built and the company operates approved meat premises under EC. Council Directive 64/433 and 77/99 at Roscrea, Co. Tipperary, where it has a slaughtering and deboning facility and cold store facilities with and EEC No. of 359.

The company employs approximately 180 employees at peak season but 70 employees off-peak. The full-time employees are paid with all deductions of PAYE and PRSI being made by the company and the company pays all PAYE and PRSI to the Revenue Commissioners as appropriate. The sub-contractors are paid by invoice and on foot of the C2 form in compliance with the Revenue Commissioners' directions.

Since the company commenced operations in 1991 deboning yields returned for the company are:

- 1991 - 68.55%
- 1992 - 69.02%

The Department of Agriculture and Food, between the 21st of October of 1991 and the 5th of June of 1992 carried out defatting analyses in respect of the companies production with the following results.

OVERALL RESULTS	FQ %	PF %
No. of Boxes Defatted	25.00	10
No. Overfat	14.00	-
Average	10.13	20.96%
Range	3.27% / 15.48%	11.70% / 29.52

The company was required to pay compensation of £352.21 in respect of the forequarter being above the 10%.

KILDARE CHILLING COMPANY LIMITED

The evidence in relation to this company was given by Mr Seamus Fogarty of the Department of Agriculture and Food, Mr John Boothman, Veterinary Inspector, Mr Peter Burke, Farmer, Mr Gerard Fogarty of the ERAD Section of the Department of Agriculture and Food, Mr Dermot Ryan, Department of Agriculture and Food, Mr Maurice Mullen, Department of Agriculture and Food, Mr Victor Whelan, Veterinary Inspector and Mr Tom McParland, Managing Director of the company.

The company has been in existence for approximately 25 years, during that period it has been operated by different owners and the present owners took over in approximately 1989 when Mr McParland became Managing Director. The company operates from Kildare where it has a slaughtering, deboning and cold store facilities, which premises are approved meat export premises under EC Council Directive 64/433 and 77/99. The company's EEC No. is 268. The company operates in both the beef and lamb business and its beef is sold commercially as well as to intervention and export markets.

The company employs 150 to 160 full-time employees who are paid weekly by cheque. The company deducts all PAYE and PRSI and makes the payments to the Revenue Commissioners as appropriate. The company benefits from the use of sub-contractors for loading boxes from the cold store and for plant maintenance. If the sub-contractor does not produce a C2 form to the company pursuant to Revenue regulations, the company deducts 35% before payment.

The Department of Agriculture and Food carried out a defatting analysis between the 1st of the 3rd 1991 and the 20th of the February 1992 with the following overall results.

OVERALL RESULTS	FQ %	PF %
No. of Boxes Defatted	25.00	10
No. Overfat	13.00	-
Average	10.63	21.38%
Range	5.52% / 16.14%	15.19% / 29.16

The Department of Agriculture and Food sought compensation of £1,222.18 from the company by reason of the over-fat on the forequarter. The Department carried out further defatting analysis on the 11th of December 1980 to 1992 and the 16th of December 1992 on the forequarter only and the production fat levels were 9.93 and 8.97 respectively. These were both within specification.

Mr Maurice Mullen gave evidence to the Tribunal that the returns of the deboning yields from this company between 1983 and 1991.

	1983	1984	1985	1986	1987	1988	1989	1990	1991
Kildare Store	66.12	66.07	68.42	68.30	68.4	68.5	68.77	-	-
Kildare	-	-	-	-	-	-	69.39	69.91	69.05

The Tribunal was given evidence of the export of meat by the company from 1984 to 1990 inclusive and they were as follows:_

Company	Destination	1984 STATUS			1985 STATUS		
		INT	APS	OTHER	INT	APS	OTHER
Kildare Chilling Meats Limited	South Africa	-	-	39,177.9	-	-	-
	USA	-	-	664,671.0	-	-	625,369.0
	Canada	-	-	772,122.78	-	387,628.9	362,469.1
	Algeria	-	863,001.5	4,016,560.9	-	1,368,236.7	249,406.6
	Gabon	-	-	-	-	25,956.6	46,644.1
	Zaire	-	-	-	-	191,934.2	28,763.3
	Ivory Coast	-	-	-	-	9,702.0	2,929.0
	Togo	-	-	-	-	6,033.5	3,098.4
	Egypt	-	-	-	-	-	552,904.6
		£13,472,000.00			£8,095,579.00		

Company	Destination	1986 STATUS		
		INT	APS	OTHER
Kildare Meats	UAE	_	-	8,705.4
	Saudi Arabia	_	-	270,7893.4
	South Africa	_	456,194.5	300,089.3
	USA	_	69,977.6	30,224.1
	Gabon	_	81,847.6	-
	Zaire	_	-	14,326.9
	Egypt	_	1,749,139.1	2,722,243.51
	Cyprus	_	-	10,332.5
	Israel	_	-	119,378.2
	Ivory Coast	_	304,934.2	13,503.7
	Iran	_	-	192,161.6
	Canada	_	-	13,608.0
		_	-	
EXPORT REFUNDS PAID		£13,725,740.6		

Company	Destination	1987 STATUS			1988 STATUS		
		INT	APS	OTHER	INT	APS	OTHER
Kildare Meats Limited	Israel	-	-	1,009,092.7	-	-	-
	South Africa	624,380.0	759,425.1	701,870.7	-	4,416,249.70	2,450,956.6
	Egypt	-	1,174,569.7	1,186,539.3	-	-	5
	Saudi Arabia	-	74,370.6	14,192.9	-	44,177.4	25,772.82
	Algeria	-	918,913.1	1,822,428.0	-	474,417.00	52,241.5
							222,348.40
EXPORT REFUNDS PAID		£9,871,525.66			£10,816,998.65		

Company	Destination	1989 STATUS			1990 STATUS		
		INT	APS	OTHER	INT	APS	OTHER
Kildare Meats Limited	South Africa	-	324,195.7	74,175.3	-	-	158,449.90
	Cyprus	-	176,472.2	52,130.6	-	654,030.0	6,059.80
	Malta	-	29,284.1	-	-	91,219.8	5,005.30
	Egypt	-	2,504.52	50,093.5	-	587.5	308,073.90
	Iran	-	9.1	30,956.1	-	2,079,360.43	945,798.90
	Saudi Arabia	-	2,484,667.5	37,081.3	-	7,527.5	33,199.30
	Sweden	-	131,260.6	15.0	-	-	-
	Romania	-	30,888.8	19,757.0	-	-	-
	Tahiti	-	-	-	-	-	17,851.50
	Yoguslavia	-	-	-	-	-	46,323.90
	West Africa	-	-	-	-	8,044.1	1,914.50
	Ivory Coast	-	-	-	-	-	6,739.70
	Algeria	-	-	-	-	17,203.5	671,593.90
	Canaries	-	-	-	-	-	828.10
	Israel	-	-	-	-	30,678.9	164,794.40
	Iraq	-	-	-	-	-	-
EXPORT REFUNDS PAID		**£8,519,273.20**			**£N/A**		

Mr Shay Fogarty, Assistant Principal Officer, Intervention Operations Division, Department of Agriculture and Food gave evidence, to the Tribunal, in relation to a number of occasions upon which the company's deboning licence was suspended. There were as follows:-

1. On 17/11/1982 the company's deboning licence was suspended for two weeks for unsatisfactory deboning over the previous six months;

2. On 26/01/1983 the company's deboning licence was suspended from the 31st of January 1983 to the 12th of May of 1983 due to unsatisfactory deboning;

3. On 17/02/1984 a warning letter was issued to the company as a result of an inspection carried out by Mr Ferris on the 13th of February 1984, when he observed minor cutting faults on pieces of meat/beef.

4. On the 21/02/1990 the company's total days deboning was rejected by reason of unsatisfactory deboning.

5. On the 07-19th/10/1984, 490 cartons of beef were rejected from intervention. However, as this matter is the subject of High Court proceedings the Tribunal declined to inquire into the reasons therefor.

All these incidents occurred prior to the appointment of Mr McParland as Managing Director in 1989.

Mr Boothman, Veterinary Inspector attached to the company gave evidence that in October of 1988 Mr Padraic Naughton reported that he had discovered 13 TB Reactor had been deboned under an APS contract. This should not have been done and it was fully investigated by the Department of Agriculture and Food and a penalty was imposed on the company of £2,500. The company, at all times, maintained that what had happened was a mistake on their part.

The Tribunal had considered it necessary to inquire into the activities of these companies and did so as outlined in this Report.

It has reported on the irregularities and malpractices in respect of which it has received evidence under the heading of the Companies involved.

In the case of many of the Companies there was no evidence of any irregularities or malpractices as appears from lack of reference to any such irregularities or malpractices in this Report relating to such companies.

CHAPTER TWENTY-FIVE

Agricultural Recommendations

The Tribunal has during the course of this Report referred to and set forth the numerous EC Regulations dealing with Aids to Private Storage, Intervention, Export Refunds and Monetary Compensation Amounts and referred to many irregularities which were disclosed.

However the number of irregularities disclosed and the failure to prevent or discover same due to the inadequacy of the enforcement of the control system laid down by EC Regulation, and the Department of Agriculture, due mainly to inadequacy in the number of staff employed by the Department of Agriculture in the various meat processing plants throughout the country, mainly at peak periods, must be viewed in the context of the huge volume of meat passing through the boning plants to be stored, either under the Intervention system or the Aids to Private Storage Scheme (APS) and the seasonality of much of the activity in this regard.

The problems created and the steps taken by the Department of Agriculture to alleviate such problems are set forth in a statement submitted to the Tribunal by the Secretary to the Department of Agriculture, Food and Forestry and confirmed in evidence by him on the 12th July 1993.

The Tribunal considers that this statement should be printed in full because it illustrates the extent of the responsibilities of the Department of Agriculture with regard to the operation of the system whereby supports are given under the CAP, the manner in which the Department of Agriculture has fulfilled its responsibilities and the steps taken by them to deal with problems arising in connection with the operation of the system. The tables referred to in the statement have already been set forth in this Report.

Statement by Mr Michael Dowling, Secretary, Department of Agriculture, Food and Forestry

On Ireland's accession to the European Community, the Minister for Agriculture was appointed on 29 January 1973 as the Intervention Agent for the European Communities under the European Communities Act, 1972. As the Intervention Agency, the Minister had responsibility for the operation of both direct (e.g. purchase of beef off the Market) and indirect (e.g. export refunds and Aid to Private Storage) intervention measures in Ireland.

In the autumn of 1973 the market situation was such that for the first time beef was offered to the Minister for purchase at the intervention price. Between 1973 and 1992 2,116,508 tonnes of beef were purchased by the Minister for Agriculture in his role as Intervention Agent. This was equivalent to over 12 million sides of beef or 6.3 million cattle. The Department made payments totalling £4.097 billion in respect of the purchase of this beef. Average annual payments for the purchase of beef amounted to about £200m. In 1991 payments of £591m were made in respect of 262,000 tonnes of beef, accounting for the disposal of almost three quarters of a million cattle it was the highest ever annual intake into intervention. The quantities purchased and payments made each year are given in Exhibit I.

In the same period intervention beef, either in bone-in, boneless or canned form was sold from intervention stocks by the Department, at a total sales value of £1.657 billion. On average, beef to the value of £82,000,000 was sold each year. In 1992 sales of over 170,000 tonnes were effected at a sales value of over £200m.

In addition the Department paid aid towards the private storage of 569,152 tonnes of beef. The total aid payments amounted to £195,588,140.

Payments of export refunds and monetary compensatory amounts in the period 1973 to 1992 totalled £2.464 billion.

In addition variable premium payments amounted to £92,671,534 for the years 1975 to 1990.

The total financial responsibility of the Department in respect of the above measures from 1973 to 1992 amounted to £8.506 billion.

Separate from the above, there were other payments including ancillary payments viz. storage, handling and the financing or interest costs of intervention. Total beef production at meat export premises in the period 1973 to 1992 amount to 7.3m. tonnes equivalent to 20.5m. cattle and all of this was subject to Department supervision. By any measurement, the scale of the operation of the intervention measures in the beef sector in Ireland was enormous. It involved the Department in dealing with hundreds of companies in the beef industry including up to 50 meat factories and over 20 cold stores. It ranged from the organisation of cold storage ships, to the movement of large quantities of beef for storage in Continental stores, to the organisation of large-scale sales of intervention beef to third countries. It ranged from such minute items as control of trimmings of less than 100 grammes, to the financial accounting of individual sales as high as £46m. All of the expenditure had to be accounted for by the Department to the EC Commission on behalf of the Minister as the Commission's agent.

For this purpose the Department presented annual accounts to the EC Commission, which were then subject to clearance-of-account audit by the EAGGF auditors. This process involved in-depth analysis of the procedures and transactions of each year. In addition to the EAGGF annual audit, the Department's system was subject to audit by the Comptroller & Auditor General, the European Court of Auditors, Commission Special Investigations, and its own Internal Auditors. In the period 1973 to 1992 disallowances by the EC amounted to £3.07m or only 0.04% of total expenditure on beef intervention measures.

In implementing the various direct and indirect intervention measures in Ireland the Department had to observe a myriad of EC Regulations, which were subject to almost constant amendments. The complexity of these Regulations has been clearly demonstrated in the course of this Tribunal. In addition the lack of clarity in many of them, which gave rise to numerous questions of interpretations has also been evident. The Commission itself admitted as much in the statement of Mons Jacquot, Director of the EAGGF to the Tribunal. [Vol. 16, Book 4]. He said that "the Commission hardly spends a day without its services being questioned on the interpretations of one or another provision" of its Regulations [Page 91 Vol. 16, Book 4]. In addition he pointed out that "given the sheer volume of market-related legislation concerning the operations of relevant Regulations, questions of interpretation inevitably arise" [Page 70 Vol. 16, Book 4]. The EC Commission, in recognition of the difficulties encountered by Member States, established a group of experts known as the Lachaux group to examine the possibilities of simplifying the mechanisms of the common organisation of the markets and the Regulations governing them.

The Regulations and controls had to be exercised primarily in meat factories and cold stores which by their physical nature could hardly be described as being conducive to strict regulation in the detail required.

In a typical boning hall, for example, an average of 100 boners operate on a rapid line of production of boneless meat.

In peak periods, up to 350 sides of beef were deboned daily (between 8 am and 6 pm) resulting in 1,500 boxes of product. These practical difficulties had to be recognised.

The Department implemented the Regulations and controls at administrative and technical level. It deployed on average, 80 administrative staff at headquarters, processing the various payments under the intervention, APS, export refunds and MCA schemes. There were, on average, 60 Veterinary Officers and 234 Agricultural Officers located around the country carrying out the veterinary and intervention controls at plant level. The staff deployed in meat factories were part of a hierarchical structure subject to regional and headquarters supervision. Exhibit V details the veterinary responsibilities of Department staff at meat plants.

At each meat export-approved premises the Department's veterinary and technical staff attended on a permanent basis during production, with responsibility to supervise beef production activities, both commercial and intervention.

Additional staff are drafted into meat plants to cater for the normal busy period from September to December. These officers are transferred on a rotational basis, thus most officers have worked in meat plants during the last number of years. Approximately 80-100 temporary staff are transferred each year at the peak period. Staff transferred into meat plants are usually deployed on lairage, stamping and loading duties which they can become familiar with, within a relatively short time, and this releases the more experienced officers for the more complex duties in the boning halls. In the quiet period of the year even though slaughtering would be for 2-3 days only per week, the regular staff would continue to be involved in boning halls and loading bays but the need for additional assistance from outside does not usually arise.

Staffing policy will continue in this vein with emphasis on drawing from a pool of trained Agricultural Officers in the period of high slaughterings. I will deal with the issue of training in more detail later.

The control system operated by the Department in respect of the intervention measures extended from documentary/administrative checks by its various headquarters sections to the spot checks on all beef production in meat plants. The Department's system of control was not static but subject to review and strengthening since its introduction. The first such major review took place in 1977 with the introduction of new purchasing and control forms (the IB series with which the Tribunal is now familiar). In 1984 a new deboning contract was introduced. 1988 saw a change of emphasis to post factum spot checks on product in store under the Aids to Private Storage Scheme. This change of emphasis arose directly from the experience gained from the operation of earlier APS schemes. This approach was subsequently adopted by the EC Commission in the form of an amendment to the relevant regulation.

The operation of the intervention system was not an end in itself but it formed part of the EC Common Agricultural Policy, the objective of which was to protect farmers income by relieving the market of excess supplies.

Ireland, for a variety of reasons such as its peripheral location, small domestic market and seasonality of production, has relied heavily on these measures in the beef sector. The difficulties in implementing the complex intervention measures were exacerbated by the fact that most of the cattle slaughterings in Ireland were concentrated in a short number of weeks in the Autumn period. The Department, nevertheless, succeeded in ensuring that the system has remained operative and in this respect prevented on a number of occasions the complete collapse of the cattle trade in Ireland, particularly in the fall of 1990, with obvious adverse consequences for farm income and the national economy.

In recent years and especially from 1990 a combination of difficult market conditions, notably the BSE problem and the Gulf crisis, resulted in a dramatic increase in the quantities of beef offered to intervention. In 1991 a record 262,000 tonnes had to be purchased by the Intervention Agency. In the same year sales of almost 112,000 tonnes were overseen by the Department. As a result the workload imposed on the Department increased dramatically. It was at this stage and, recognising the need to strengthen controls, that a complete re-organisation of the administrative arrangements for the operation of the system became necessary within the Department.

In October 1990 the control functions which resided with the separate Commodity Divisions of the Department was changed and centralised in a specialised Intervention Unit. The setting up of this Unit brought together in a coherent and distinct administrative structure all of the Department's intervention activities. Control Enquiry Teams, operating a system of surprise ad hoc inspections, were an additional feature of the new unit. Further restructuring of the chain of command and control of beef intervention operations right down to plant level took place with the establishment within the Intervention Unit of a Beef Controls Division which was given line responsibility, under the Intervention Unit Assistant Secretary for all intervention control. This involved control of and support for the permanent presence, unannounced visits and post factum checks. Since then over 100 unannounced inspections have taken place. [Exhibit VI details the inspections to date]. This type of inspection is now an integral part of the control system alongside the permanent supervision of production on the factory floor. In fact, Mons Jacquot, Director of the EAGGF acknowledged that in establishing a control system involving spot checks and permanent supervision Ireland was "providing a indication of the right road". Post factum checking has been intensified with the examination of product right up to the time of sale out of intervention. Twenty five additional Agricultural Officer staff have been recruited and assigned to boning halls to increase the level of inspection of intervention intake, deboning and the supervision of weighing. New instructions have been issued to staff at meat plants which require them to exercise closer supervision of all weighing of intervention beef both bone-in and cartons of boneless beef. All IB forms have been serialised and are being recorded when issued in an official register by the Department staff, so that all forms, including cancelled forms are accounted for.

Other major improvements have also been effected. A new general intervention contract has been concluded with all the meat factories. In addition the deboning contract has been updated. The intervention contract places fairly and squarely on the trade the responsibility for compliance with EC Regulations on beef intervention. The revised deboning contract provides for agreed punitive penalties. Up to now, because there was no specific basis in law for extrapolation, this necessary control has been challenged by some in the industry. There is now a provision for extrapolation enshrined in the contracts. Exhibit VII are specimens of the new general contract and the new deboning contract together with details of improvements therein. In addition the beef purchasing agreements have been revised. In putting in place these improved control arrangements the Department was guided by two considerations in particular;

(*a*) that the primary responsibility for the proper operation of and conformity was the provisions of the various Regulations rests with the operators themselves and the new contract should underpin this principle,

(*b*) the command and control structures within which Irish controls operate needed to be more streamlined, more resources needed to be committed, specialisation in the control of intervention activities needed to be developed and the emphasis shifted from the permanent presence as the principal control instrument.

These changes were also in line with the views of the EC Commission as to the improvements required in the system.

An increase of £1m in the 1992 Exchequer allocation was made for the purpose of improving the beef intervention system.

In the meantime the policy of intervention had become the subject of detailed review at EC level. It was recognised that the policy had served the Community well for many years, but increasing Community expenditure, difficulties in stock disposals and the weakness of the measure as a support for farm incomes throughout the Community, together with a trend towards liberalisation of trade in agricultural produce made change inevitable.

In May 1992 the EC Council of Ministers agreed the reform of the Common Agricultural Policy. Essentially this resulted in a major re-orientation of the support mechanisms away from price support to direct income aids payable to farmers.

Accordingly the intervention system in the beef sector was modified to the extent that its use in future years will be substantially curtailed. In fact the Council of Ministers has decided as part of the CAP Reform measures, that the ceiling for normal intervention intake shall reduce from 750,000 metric tonnes this year, to 350,000 metric tonnes in 1997.

Taken together with the likely outcome of the GATT negotiations the clear implication is that the beef industry will have to rely to a far greater extent on the market for its returns in future years.

Having regard to the complexity of the intervention measures, the scale and diversity of the operations, the resources available, and the limited level of disallowances by the EC Commission, in general the administration of the various measures by the Department has been effective. That is not to say of course that there have not been weaknesses in the controls or no irregularities. It would be surprising if this were not the case in an undertaking of this magnitude.

It has to be pointed out, however, that the Department was not remiss over the years in penalising meat factories when breaches of the Regulations were discovered as is evidenced from the details provided to the Tribunal [Vol. 8A and 8B].

It is true of all policing systems that determined wrong-doers will always find a way to breach the best controls. What matters is the response of the "police" to the latest breach.

In this regard the Department's consistent approach has been to attempt to improve the system when weaknesses were uncovered.

The EC Commission has also followed this approach by way of amendments of Regulations to close off loop holes and tighten up on the administration of schemes. At EC level, the Department always supported such measures and in particular gives full support to the Commission's efforts in the fight against fraud. It will continue to be the Department's policy to ensure compliance with all the Regulations for which it has responsibility.

Tribunals, of their nature quite rightly concentrate their attentions on wrongdoing or suspected wrongdoing, however, a mistaken impression could be drawn of the Irish beef industry if reference was not made to the many positive aspects of the industry. Irregularities have been associated with a very small percentage of overall production.

Irish product has an excellent worldwide image. This derives from being produced in a clean and natural environment. Irish beef sells at a premium on the supermarket shelves of Europe.

Vacuum-packed sales of Irish beef have increased dramatically in recent years:—

1984 — 25,606 tonnes (carcase weight equivalent) Value £61m.
1992 — 113,300 tonnes (C.W.E.) Value £265m.

This more than four-fold increase has been achieved against a backgroup of falling EC consumption, increased pressure from competing meats, alternative protein sources and the growth in the convenience food sector. Such a striking commercial performance could not have taken place if the quality of the product were in doubt.

Complaints from importing countries concerning the quality of the commercial or intervention product have been minimal and the list of export destinations and new buyers continues to expand.

Prior to the establishment of the separate Intervention Unit in the Department, the question of transferring the intervention activities to an autonomous Intervention Agency was given consideration. This was referred to in Mr Mockler's evidence to the Tribunal. [Vol. 30B, Q.421, P.108].

This question was always rather finely balanced. The relative merits and demerits of each type of structure had to be weighed up carefully. The most obvious drawback to the control of intervention operations remaining within the Department is the possible conflict of interest for the Department as both regulator and promoter of the beef industry. However there is a counter argument that an autonomous agency tends to become isolated from policy changes effected at EC level. The need for clear and effective two way communication between the policy makers and the policy executioners is an essential requirement for the effective operation of the complex measures involved in intervening in markets. In this regard I can do no better than to refer to the evidence of the Director of EAGGF (M. Jacquot) [Vol. 137A, Q.17, P.17] who obviously has broad experience of the systems operating in all Member States of the Community. He referred to the fact that the same possible conflict of interest could be perceived as arising within the EC Commission itself as both controllers and promoters but went on to underline the necessity for policy makers and policy executioners to be working together because of the "complicated Regulations that are involved" in common organisation of the markets. Mons Jacquot acknowledged that "it is a good thing that the Intervention Agency in Ireland is dependant on the Department of Agriculture".

There were clearcut operational reasons why the Department decided not to set up a separate intervention agency on accession to the community. Given the universally accepted market outlook at the time — particularly for beef — it was expected that the principal market support activity would be payment of aids and subsidies rather than physical intervention. The existing divisional structure of the Department was adequately organised for this type of activity. In the event a whole variety of international and domestic factors ranging from oil crises to production changes to changes in consumer tastes and habits resulted in market support activity being very much different to what was anticipated.

The Department has responded to this over the years by adjusting its approach and its controls on a gradual basis. This culminated in the complete overhaul from the second half of 1990 of the whole system beginning with the establishment of the specific Intervention Unit and going on to changed control arrangements and practices, the provision of extra resources and the putting in place of the new contracts.

The new system, which was devised as part of the Department's policy of gradually upgrading controls in line with experience, has been accepted and welcomed in principle by the Commission. While the Commission has yet to examine it in detail on the ground, it clearly acknowledged in the course of the Tribunal that in establishing a control system involving spot checks and permanent supervision, Ireland was "providing an indication of the right road" [Vol. 138B, Q.73, P.51]. I take this as a clear reference to the fact that the present system exceeds that applied in other Member States and should be followed as an example.

The new Intervention Unit in practice has most of the characteristics of a separate Agency. It provides efficiencies and specialisation, effective control and a clear divorcing of the Department's intervention activities from its policy activities , while at the same time avoiding a number of practical problems particularly in terms of staff redeployment or recruitment which would result from the establishment of an agency completely outside the Department. A reluctance on the part of existing staff to transfer to an autonomous agency would lead to a loss of expertise built up over the past twenty years in operating these complex measures. Furthermore, the transfer of staff from other activities to intervention activities at peak periods would be more difficult where two separate agencies are concerned.

Having regard to all these factors and to the fact that intervention will play a diminishing role in the future, the case for making the major change of establishing a completely independent agency is less than convincing at this point.

A major priority in future years will be the operation of the direct income aids which is completely different from the present intervention system. The Director of EAGGF referred to the re-orientation in the Community support system, whereby "aid will be far more closely related to actual producers than it is to traders" [Vol. 138A, Q.2, P.5] and to the necessity of directing resources to this area. This is receiving the highest priority in the Department. A major computerisation project (which will be the largest in the State) in underway. The headquarters of the premia payments operation is being decentralised.

Notwithstanding the emphasis on these schemes the Department will continue to improve the operation of the intervention system and, as the need arises, strengthen the Intervention Unit and further streamline and make more effective its control systems and its staff training programmes. Without breaking its link with the Department its operational independence could be further strengthened through formally establishing it as an Executive Agency with the Department's Secretary remaining responsible to the Public Accounts Committee for its expenditure and the Minister responsible to the Dail for its overall policy.

In 1992 a Systems Investigations and Training Section was established within the Intervention Unit to ensure that structures, procedures and accountability of the intervention system was maintained at the highest level of efficiency and effectiveness in conformity with the relevant EC Regulations.

The system work of this new Section involves the continuous review of operating practice and procedures and involves evaluating the appropriateness, clarity and security of documents, the procedures for keeping records and control instruments such as stamps and the appropriateness of boxes, strappings, labels and markings. It also reviews the clarity and consistency of instructions to officers.

The investigations work of this Section involves in-depth examination of suspected irregular activities or investigations with a view to further improving control efficiency. This work differs from the immediacy of the on-the-spot inspections and shall be undertaken where appropriate in conjunction with the Gardai and Customs inspectorate.

On training the Section has been given specific responsibility to identify and respond to the training needs of those involved in the intervention system. It has established the following programme of activities:—

—a two week intensive course for new staff;

—seminars for all staff involved in control work introducing new deboning contracts, specifications, general contract, purchase agreements and control documents;

—a training video has been produced and is currently being used at training seminars — it covers all the control aspects of intervention beef and deboning operations;

—a series of refresher courses is being organised to update staff on new developments and procedures;

—all existing instructions to staff in meat plants are being reviewed with a view to finalising a consolidated instruction manual.

There is an ongoing computerisation programme in place in the Department and one of the areas receiving priority is the further computerisation of the intervention system.

The new system will result in the design and development of a system which will support the requirements and operation of the Intervention Unit in relation to beef.

- The system will;
- Record all purchases;
- Support storage;
- Support Sales, Contracts, Withdrawals and Invoicing;

- Record and account for securities;
- Record transfers, losses and damages;
- Provide Stock Control; and
- Support the Annual Account requirements in relation to Beef.

With regard to the disclosure of irregularities in the past, the view has been that there are a number of difficulties in publicising the results of actions taken in this regard. In the first instance criminal proceedings may well follow and these could be prejudiced by premature disclosures. There is also the difficulty of distinguishing between different types of irregularity, some breaches may be relatively minor in nature yet the fines involved can be very significant. In this regard, the Tribunal has heard ample evidence on the fines associated with non-individually wrapped product. The publication of such an irregularity could give a completely wrong impression to the general public and unjustifiably damage a company's reputation. Subject to these constraints we would, however, favour the disclosure of serious irregularities.

The preparation of statutory provisions for automatic access to a company's commercial records are at an advanced stage of preparation within the Department. In fact a similar type provision already exists in the new deboning contract already referred to. The new provision should come into effect later this year. Procedures relating to the certification and labelling of product are also being examined at the present time.

I might mention that the Commission, as a result of a review of intervention operations throughout the Community, last week put forward proposals to amend the basic Commission Regulation (No. 859/89) governing intervention.

The main changes proposed, which will come into effect later this year, are,

(i) Community-wide standardisation of the specification for intervention cuts,

(ii) a change in carcase dressing involving a reduction in fat levels in the carcase as presented for acceptance into intervention,

(iii) a requirement to debone all carcases at a plant other than the plant of the tenderer or a reduction in the amount of deboning which can be carried out at the plant of the tenderer to 1,000 tonnes per week nationally plus 50% of the remainder.

Finally, I would emphasise that on balance the Department's controls in a highly complex area have worked reasonably well over the years. We had adjusted them to meet weaknesses where these have appeared. For a variety of reasons the adjustments made over the past three to four years have been more fundamental and extensive than ever before. We believe that the controls operated though the Intervention Unit bear comparison with those anywhere else in the Community. In making all these changes we have worked from our own experience, the advice of the Commission and more recently, the evidence of the Tribunal. We will continue to work in this way and, in particular, to have regard to the final conclusion and recommendations of the Tribunal.

It is clear from this statement and from the evidence of Mr Dowling that the Department of Agriculture realised that "the command and control structures within which the Irish controls operate, needed to be more streamlined, more resources needed to be committed, specialisation in the control of intervention activities needed to be developed and the emphasis shifted from the permanent presence as the principal control instrument".

Prior to 1990, the Department of Agriculture had relied on what was referred to as 'the permanent presence' of Agricultural Officers in the deboning halls and detailed examinations of supporting documentation as the most effective way of exercising supervision and control of the operations in the boning halls attached to the various meat plants throughout the country.

This reliance on 'the permanent presence' and the failure to carry out any checks on the product after it had been accepted into either private or public storage had been the subject of some criticism by the EAGGF Division of the Commission of the European Communities.

In the course of his submission to the Tribunal Mons Jacquot, Director of EAGGF (FEOGA) had stated that:—

"—Control through a system of permanent presence could have a validity in exceptional circumstances though even then only for a limited period. Over any length of time in such a system one of the essential qualities of a controller (i.e. independence) is inevitably reduced.

—Within the two measures under review its predominance constitutes a misdirection of resources. It is true that the sequence of deboning, following the acceptance of beef into Private and Public Storage is a key phase in both measures. To concentrate all, or the predominant control effort into visual survey of the said phase with supporting documentation, while ignoring other possibilities (e.g. substitution, particularly in the case of Public Storage) within the audit field reflects a very limited conception of the control task involved. (Controls operated by the Customs Authorities does not come within the scope of my observations.)

—While control instructions from the Department make mention of back up physical checks such controls were never systematically integrated into the overall control programme and probably more importantly, the trade in general would appear to have developed confidence that it would not be subject to such further scrutiny. The "permanent presence" remained the only real barrier to impropriety."

Dealing with beef placed in Private Storage under an APS Scheme, Regulation (EEC) 2965/89 of the 29th September 1989 had provided 'inter alia' at Article 12 that

"5. On entry into storage, the intervention agencies shall conduct checks in particular to ensure that products stored are eligible for the aid and to prevent any possibility of substitution of products during storage under contract.

6. The national authorities responsible for controls shall undertake:

(*a*) for each contract, a check on the compliance with all the obligations in Article 3(2) of Regulation (EEC) No 1091/80:

(*b*) an unannounced check to see that the products are present in the store. The sample concerned must be representative and must correspond to at least 10% of the overall quantity under contract for a private storage aid measure. Such checks must include, in addition to an examination of the accounts referred to in paragraph 3, a physical check of the weight and type of product and their identification. Such physical checks must relate to at least 5% of the quantity subject to the unannounced check:

(*c*) a check to see that the products are present at the end of the storage period under contract."

The checks on the beef placed in private storage in pursuance of the 1986 and 1988 APS scheme, as described in this Report and carried out by the Customs and Excise authorities in conjunction with the Department of Agriculture pre-dated this regulation requiring such checks to be mandatory.

While this regulation relates to beef placed in private storage and does not relate to beef placed in public storage i.e. intervention, the new control system outlined by Mr Dowling provides for such checks to be made on beef placed in public storage and such investigation of and checks on beef placed in public storage will undoubtedly strengthen the control system.

The steps that have been taken and control systems introduced by the Department of Agriculture as outlined in Mr Dowling's statement to the Tribunal go very far towards dealing with the weaknesses in the system disclosed during the course of the Tribunal and have rendered unnecessary many of the recommendations which might otherwise have been made by the Tribunal.

A control system however, is only effective if adequate trained staff are available to enforce it. Mr Dowling, in his statement, said:—

> "At each meat export approved premises the Department's veterinary and technical staff attended on a permanent basis during production, with responsibility to supervise beef production activities both with commercial and intervention. Additional staff are drafted into meat plants to cater for the normal busy period from September to December".

While this is undoubtedly true, a recurring theme throughout the evidence and emphasised in many of the submissions received by the Tribunal and indeed accepted by many of the witnesses from the Department of Agriculture was the inadequacy of the number of staff provided by the Department of Agriculture for the proper exercise of the controls required to be exercised and this undoubtedly led to a laxity in the supervision and control of the system.

There was, from 1989 to 1990, a massive increase in the amount of beef placed in intervention and in the quantities deboned from 77,515 tonnes and 59,415 tonnes to 230,638 tonnes and 214,339 tonnes respectively.

As a result, the workload increased dramatically and the need to strengthen controls was realised.

The administrative arrangements for the operation of the system within the Department of Agriculture was completely re-organised.

A specialised Intervention Unit was established and all the control functions, which resided within the separate commodity divisions of the Department were changed and centralised in this special Intervention Unit.

Control Enquiry Teams were established and they were obliged to operate a system of unannounced inspections of the beef, in private and in public storage.

Within the Intervention Unit, a Beef Controls Division was established and this Division was given responsibility for all intervention control. This involved control of and support for the permanent presence, unannounced visits and post factum checks. This type of inspection is now an integral part of the control system alongside the permanent supervision of production on the factory floor and is in complete accordance with and indeed in advance of many of the requirements of the EEC Commission.

This checking of the beef in private and public storage has been intensified with the examination of product right up to the time of sale out of intervention.

Twenty five (25) additional Agricultural Officer staff have been recruited and assigned to boning halls to increase the level of inspection of intervention intake, deboning and the supervision of weighing.

New instructions have been issued to staff at meat plants which require them to exercise close supervision of all weighing of intervention beef, both bone-in and cartons of boneless beef.

All IB forms have been serialised and are being recorded when issued in an official register by the Department staff, so that all forms, including cancelled forms, are accounted for.

If this provision had been introduced earlier the system of re-writing of IB4 forms, described in detail in the Report, would not have been possible.

In addition a new general intervention contract has been concluded with all the meat factories and the deboning contract has been updated.

In his statement Mr Dowling said that;

> "The intervention contract places fairly and squarely on the trade the responsibility for compliance with EC Regulations on beef intervention".

The Department of Agriculture is clearly responsible however for ensuring that such compliance is effected and the control system should ensure compliance with the Regulations.

In addition, a new training system, as described by Mr Dowling, has been introduced and the operation of the intervention system is being computerised which should render the operation of the system more expeditious and more efficient.

In addition in 1992 a Systems Investigations and Training Section was established within the Intervention Unit and charged with the responsibility of ensuring that structures, procedures and accountability of the intervention system is maintained at the highest level of efficiency and effectiveness in conformity with the relevant EC Regulations.

This section's work involves the continuous review of operating practice and procedures and involves evaluating the appropriateness, clarity and security of documents, the procedures for keeping records and control instruments such as stamps and the appropriateness of boxes, strappings, labels and markings.

The investigation works of this Section is intended to involve in-depth examinations of suspected irregular activities or investigations with a view to further improving control efficiency.

The work of this Section is intended to differ from the immediacy of the on-the-spot inspections and intended to be undertaken where appropriate in conjunction with the Gardai and Customs inspectorate.

The establishment of this section by the Department of Agriculture, with the responsibilities outlined herein, is an important and positive step but all necessary steps should be taken to ensure that it is adequately staffed.

As stated many of the changes which have been effected by the Department of Agriculture since 1990 and in particular the establishment of the Systems Investigations and Training Section in 1992 would have been the subject of recommendations by the Tribunal.

It was submitted to the Tribunal that there was a conflict of interest between the Department of Agriculture and Food as the Department responsible for the development of agricultural policy in the State and as the Intervention Agency of the European Community and that as a consequence of such conflict, a separate Intervention Agency should be established.

The Tribunal gave serious consideration to this submission but having regard to the fact that the new Intervention Unit has as stated by Mr Dowling most of the characteristics of a separate agency and the evidence that intervention will play a diminishing role in the future, and is satisfied that no useful purpose would be served, at this stage, by the establishment of a separate Intervention Agency.

The Tribunal has received from the persons and organisations named in Appendix 6 submissions with regard to the recommendations which they considered should be made by this Tribunal.

The submissions were constructive and well-considered and were of assistance to the Tribunal.

Many of them related to matters which has already been dealt with by the Department of Agriculture and many of them related to matters which were outside the remit of this Tribunal.

The Tribunal wishes to acknowledge its appreciation of the efforts of those persons and organisations.

Having considered the steps taken by the Department of Agriculture as outlined in Mr Dowling's statement to streamline and improve the control systems against the background of the evidence adduced before the Tribunal, the Tribunal is satisfied that the Department of Agriculture has addressed the problems disclosed therein and has set up a system and established controls which, if fully implemented and sufficient trained staff employed, would greatly lessen the possibility of irregularities and malpractices occurring in the future and recommends that these systems and controls be rigorously enforced. Such systems should particularly ensure that all documents or certificates required to be completed by Departmental staff, either Veterinary Inspectors or Agricultural Officers, be completed in total by them and not signed until they have been so completed.

The Tribunal, however, recommends that the system introduced by the Department of Agriculture and Food should require:—

(*a*) that the quarters being weighed into a boning hall for either Public or Private Storage should be weighed in by an employee of the Department and all relevant documentation, in respect thereof, should be prepared and signed by him.

(*b*) That subsequent to deboning and cutting in accordance with specifications of the quarters of beef, and in addition to weighing the cartons containing the cuts being placed in public storage, the balance of meat remaining should be examined by an Officer of the Department to ensure that identifiable pieces of meat are not included in such balance and such meat should be weighed and its weights recorded.

The Tribunal has noted the proposed amendments to the basic Commission Regulation No. 859/89 governing intervention and considers that the proposed amendments are eminently desirable.

CHAPTER TWENTY-SIX

Costs

The Order appointing the Tribunal, which was made by the Minister for Agriculture and Food on the 31st day of May, 1991, provided that the Tribunal of Inquiry (Evidence) Act as adapted by or under subsequent enactments, and the Tribunals of Inquiry (Evidence)(Amendment) Act shall apply to the Tribunal.

Section 2 (*b*) of the Tribunals of Inquiry (Evidence) Act 1921 provided that;

"A Tribunal to which this Act is so applied as aforesaid:—

(*b*) shall have power to authorise the representation before them of any person appearing to them to be interested to be by counsel or solicitor or otherwise, or to refuse to allow such representation".

In pursuance of the said Act, the Tribunal authorised the representation before it of the persons set out in detail in Appendix 3 attached to this Report.

This Appendix is in a tabular form showing;

(i) the name of the person represented,

(ii) date upon which representation was granted,

(iii) Solicitors representing such parties,

(iv) Where applicable Counsel representing such parties,

(v) the persons who sought and were granted costs and those who did not seek costs or were refused costs.

Section 6 (1) of the Tribunals of Inquiry (Evidence)(Amendment) Act 1979 provides that:

"6. (1) Where a Tribunal, or, if the tribunal consists of more than one member, the chairman of the tribunal, is of the opinion that, having regard to the findings of the tribunal and all other relevant matters, there are sufficient reasons rendering it equitable to do so, the tribunal or the chairman, as the case may be, may by order direct that the whole or part of the costs of any person appearing before the tribunal by counsel or solicitor, as taxed by a Taxing

Master of the High Court, shall be paid to the person by any other person named in the order.

(2) Any sum payable pursuant to an order under this section shall be recoverable as a simple contract debt in any court of competent jurisdiction."

The Tribunal has in the course of its introductory chapter to this Report referred to the statement of Lord Justice Salmon made in the course of the Report of the Royal Commission on Tribunals of Inquiry (1966) that a person who is involved in an inquiry should normally have his legal expenses met out of public funds and the statement of the late Mr Justice McCarthy, concurred with by the Chief Justice, in the case of Goodman International and Laurence Goodman -v- The Tribunal, that "ordinarily, any party permitted to be represented at the inquiry should have their costs paid out of public funds."

The Tribunal is satisfied that in the exercise of its discretion to award the whole or part of the costs of any party appearing before the Tribunal, it cannot have regard to any of its findings on the matters being inquired into by it but is only entitled to consider the "conduct of or on behalf of that party at, during or in connection with the inquiry" and that unless such conduct so warrants, a party, permitted to be represented at the inquiry should have their costs paid out of public funds.

The Houses of the Oireachtas had, by resolution, considered it expedient to establish a Tribunal for inquiring into definite matters considered by them to be of urgent public importance viz allegations regarding illegal activities, fraud and malpractice in and in connection with the beef processing industry, made or referred to in Dáil Eireann and on the television programme transmitted on the 13th day of May, 1991 and any matters connected with or relevant to these allegations which the Tribunal considered it necessary to investigate.

As appears from this Report, the allegations were serious and covered many areas of public life and the food processing industry. The persons, who were afforded the right of representation before the Tribunal were entitled to be represented before the Tribunal and the Tribunal was so satisfied before authorising their representation in accordance with the provisions of Section 2(*b*) of the Tribunals of Inquiry (Evidence) Act, 1921. The nature and extent of the representation authorised by the Tribunal varied as between the interests of the persons granted representation in the specific matters being inquired into by the Tribunal.

Over the entire period of the public sittings of the Tribunal there was only one instance of any conduct by or on behalf of any party at during or in connection with the inquiry that would entitle the Tribunal to disallow any party their costs of appearing before the Tribunal.

In the course of this Report the Tribunal was obliged to and did express its disapproval of the activities of Mr Philip Smith in approaching Mrs Potter, an employee of Tunney Meats, in making substantial payments to her to obtain documents, the property of her employer, and to use same in the course of his submission to this Tribunal and his attempt to purchase the testimony of other witnesses as part of a personal dispute with Mr Hugh Tunney.

In these circumstances it would be inequitable to award Mr Smith his costs and the Tribunal has made no order as to his costs.

Having regard to the nature, extent and length of the inquiry it would be inequitable to require that persons, necessarily appearing at or before the Tribunal should be required to pay their own costs of such appearances and as the Houses of the Oireachtas had considered it expedient to establish the Tribunal, the Tribunal considers it equitable that the Minister for Finance should pay, out of monies provided by the Oireachtas the costs of the persons named in Appendix 3 in the manner appearing in the separate orders made by the Tribunal, under Section 6 of the Act of 1979 and which have been filed in the Central Office of the High Court, copies of which are annexed to this Chapter, which said orders provide, inter alia, that the cost awarded thereby shall, as required by Section 6(1) of the Tribunals of Inquiry (Evidence)(Amendment) Act, 1979 be taxed by a Taxing Master of the High Court; that the costs awarded shall be taxed on a party and party basis: that the statutory provisions and the Rules of the Superior Courts relating to the taxation of costs in an action in the High Court (including the provisions relating to review and appeal) shall, in so far as practicable, apply : that the Minister for Finance be at liberty to attend and be heard on the said taxation and any review or appeal in relation to it.

The Tribunal was faced with a difficult problem with regard to securing the attendance before the Tribunal of Patrick McGuinness who resided in London, Ontario, Canada and who was not amenable to the jurisdiction of the Tribunal and whose attendance at the Tribunal could not be enforced by service of a subpoena.

As the allegations made in the ITV programme were based mainly on information supplied by Mr McGuinness, who had been employed by the Goodman Group, the Tribunal considered that Mr McGuinness' evidence was vital to its inquiries and that it would be open to criticism if it did not take all reasonable steps to secure his attendance.

Mr McGuinness was unwilling to travel from London, Ontario without a guarantee that all his expenses, including travel to and from Dublin, and incidental living expenses and all legal costs, including travel and living disbursements would be borne by the Tribunal.

Granada Television Ltd had offered to extend the representation of its Counsel to Mr McGuinness but he remained unwilling to appear before the Tribunal unless he was represented by his own Canadian lawyers.

In an attempt to secure Mr McGuinness' attendance at the Tribunal, the Tribunal was obliged to authorise three separate visits to him in Canada by leading Counsel to the Tribunal in an attempt to ensure his attendance.

Eventually in view of the importance of Mr McGuinness' evidence, the Tribunal was obliged to accept the conditions imposed by Mr McGuinness and to treat his legal costs as expenses, necessarily incurred in connection with his attendance before the Tribunal.

The payments authorised by the Tribunal in respect of such attendance were:—

1. Amount paid to Mr McGuinness in respect of loss of earnings (23 days at 188 Canadian dollars per day) £ 2356.38
2. Travel Expenses of Mr McGuinness and his lawyers from Canada to Dublin £ 9290.00
3. Hotel Expenses of Mr McGuinness and his lawyers while in Dublin £ 5756.23
4. Legal costs £47,767.44

£65,170.05

Vouchers and receipts in respect of the foregoing items of expenditure were produced to and examined by the Office of the Comptroller and Auditor General.

Before authorising the payment of the legal costs incurred by Mr McGuinness as expenses necessarily incurred in connection with his appearance before the Tribunal and necessary preparatory work, including the preparation of submissions to the Tribunal, the Tribunal satisfied itself that such costs were fair and reasonable.

Mr McGuinness was represented by the firm of Lerner and Associates, Barristers and Solicitors, of London, Canada and in particular by Mr John Judson QC and Mr Ian Leach, Barrister-at-Law.

Their charges in respect of preparatory work and consultations were $250 per hour by Mr Judson QC and $95 per hour by Mr Leach BL and in respect of appearances before the Tribunal $2000 per day by Mr Judson QC and $1000 per day by Mr Leach.

In addition they claimed in respect of outlay such as postage charges, copying charges, telephone charges, binding charges and other incidental expenses.

Taking all relevant factors into consideration the Tribunal was satisfied that the legal expenses incurred by Mr McGuinness were fair and reasonable and in the opinion of the Tribunal were such as would be allowed by the Taxing Master of the High Court if such costs were taxed pursuant to an Order of this Tribunal.

Counsel on behalf of Patrick Boyhan, Con Howard, Patrick J Kenny, Michael MacElligott and Patrick Murphy (as representing themselves and all other members of the United Farmers Association) applied to the Tribunal for a full legal representation at the inquiry. These applications cited that the UFA were entitled to such full legal representation as they represented the public, the farmers, their organisation, their witnesses and that they needed to protect their interests and witnesses by being present throughout the proceedings before the Tribunal.

The UFA had approached the Tribunal and sought to adduce evidence before the Tribunal through witnesses.

The UFA was in the position of a willing witness who had approached the Tribunal and sought to give evidence.

No allegations whatsoever were made against the UFA.

The Tribunal accepted that they had an interest but a limited one and agreed to grant such limited representation on the basis that they would be entitled to be present at the Tribunal and participate in the workings thereof when witnesses produced by them were giving evidence before the Tribunal and that they would be entitled to cross-examination of such witnesses.

On the 26th of July 1991, Counsel on behalf of the UFA again sought full representation before the Tribunal. The Tribunal refused the application and stated in reply to submissions by Counsel that:—

> "The Tribunal has considered the submissions that are being made on behalf of his clients and if and when these submissions are being considered by the Inquiry or any witness referred to in his submissions are being called, it is only right and proper that limited representation be given to the UFA when and if these matters are being dealt with by the Tribunal and they will be notified in ample time if these matters which are referred to in the submissions are being considered by the Tribunal in order to enable them to be represented on a limited basis while these matters are being dealt with."

On the 30th day of September 1991, Counsel on behalf of the UFA made a further application in regard to representation which was refused and in the course of refusing the said application, the Tribunal stated to Counsel:—

> "Your interest and your representation before this Tribunal was granted on a limited basis, that was made quite clear. You will be heard when any witnesses whom you have made available to the Tribunal will be dealt with by the Tribunal and that is the basis of the representation and it is the only representation which you have got and it is the only representation you will get."

Counsel on behalf of the UFA attended the Tribunal on many occasions when witnesses other than witnesses introduced by the UFA were being examined and occasionally was permitted to cross-examine such witnesses. However, their attendance on these occasions was on a voluntary basis and not in accordance with the representation granted to them. It would be inequitable if the Tribunal were to award costs to the UFA on the basis of their entire attendance before the Tribunal. They are only entitled to costs in accordance with the representation granted to them and having regard to all relevant matters the Tribunal considers it equitable that these applicants be awarded costs in respect of attendance before this Tribunal on 25 days.

Index to Costs Orders

Pages

Pages

ANNEXE

IN THE MATTER OF the Tribunals of Inquiry (Evidence) Acts, 1921 and 1979; and IN THE MATTER OF a Tribunal of Inquiry established pursuant to Resolutions of Dail Eireann passed on the 24th May 1991, and by Seanad Eireann on the 29th May 1991, to inquire into allegations regarding illegal activities fraud and malpractice in and in connection with the beef processing industry by order of the Minister for Agriculture and Food made on the 31st day of May 1991.

AND WHEREAS it was provided by the said order that the Tribunals of Inquiry (Evidence) Acts 1921 and 1979 applied to the Tribunal

WHEREAS it is provided by Section 2 (*b*) of the Tribunals of Inquiry (Evidence) Act, 1921 that a Tribunal to which the Act is applied shall have power to authorise the representation before it by solicitor or counsel or otherwise of any person appearing to it to be interested; and

WHEREAS Goodman International and Subsidiary Companies (hereinafter referred to as "the applicants") applied to the said Tribunal to be represented before it by counsel instructed by solicitor and on the 21st June 1991 such authorisation under the said section was granted; and

WHEREAS it is provided by section 6 of the Tribunals of Inquiry (Evidence)(Amendment) Act, 1979, that if a Tribunal is of opinion that having regard to its findings and all other relevant matters there are sufficient reasons rendering it equitable to do so it may by Order direct that the whole or part of the costs of any person appearing before the Tribunal by counsel or solicitor as taxed by a Taxing Master of the High Court be paid by any other person named in the Order; and

WHEREAS by section 4 of the said Act of 1979 it is provided that a Tribunal may make such Order as it considers necessary for the purpose of its functions; and

WHEREAS the Tribunal referred to in the title hereof is satisfied that having regard to its findings and all other relevant matters there are sufficient reasons rendering it equitable that the Minister for Finance should pay the costs of the applicants as hereinafter appearing

IT IS HEREBY ORDERED that a Taxing Master of the High Court do tax in the manner hereinafter appearing the applicants' costs of appearing before the said Tribunal by three counsel instructed by solicitor.

In relation to the taxation of the applicants' costs the Tribunal DOTH ORDER as follows:

(*a*) That the statutory provisions and the Rules of the Superior Courts relating to the taxation of costs in an action in the High Courts (including the provisions relating to review and appeal) shall, in so far as is practicable, apply;

(*b*) That the costs of employing a solicitor and three counsel be taxed on a party and party basis.

(*c*) That the costs do include such witnesses' expenses as relate to the those witnesses who actually gave oral evidence before the Tribunal on the proposal of the applicants;

(*d*) That the amount of witnesses' expenses be (*a*) such fee for the preparation of any statement prepared by the witness which was received by the Tribunal as part of the witnesses' testimony as the Taxing Master (or on review or appeal, the Court) considers reasonable; (*b*) such fee (not already paid) for attending before the Tribunal on such days as the witness gave oral evidence as the Taxing Master (or, on review or appeal, the Court) considers reasonable and (*b*) the travelling and subsistence expenses (not already paid) reasonably incurred for the purpose of giving oral evidence before the Tribunal.

(*e*) That the Minister for Finance be at liberty to attend and be heard on the said taxation and any review or appeal in relation to it.

(*f*) That the Minister for Finance do pay the applicants' costs when taxed and ascertained.

Dated this 29th day of July 1994.

Signed Liam Hamilton

The Tribunal

ANNEXE

IN THE MATTER OF the Tribunals of Inquiry (Evidence) Acts, 1921 and 1979; and IN THE MATTER OF a Tribunal of Inquiry established pursuant to Resolutions of Dail Eireann passed on the 24th May 1991, and by Seanad Eireann on the 29th May 1991, to inquire into allegations regarding illegal activities fraud and malpractice in and in connection with the beef processing industry by order of the Minister for Agriculture and Food made on the 31st day of May 1991.

AND WHEREAS it was provided by the said order that the Tribunals of Inquiry (Evidence) Acts 1921 and 1979 applied to the Tribunal

WHEREAS it is provided by Section 2 (*b*) of the Tribunals of Inquiry (Evidence) Act, 1921 that a Tribunal to which the Act is applied shall have power to authorise the representation before it by solicitor or counsel or otherwise of any person appearing to it to be interested; and

WHEREAS Mr Laurence Goodman (hereinafter referred to as "the applicant") applied to the said Tribunal to be represented before it by counsel instructed by solicitor and on the 21st June 1991 such authorisation under the said section was granted; and

WHEREAS it is provided by section 6 of the Tribunals of Inquiry (Evidence)(Amendment) Act, 1979, that if a Tribunal is of opinion that having regard to its findings and all other relevant matters there are sufficient reasons rendering it equitable to do so it may by Order direct that the whole or part of the costs of any person appearing before the Tribunal by counsel or solicitor as taxed by a Taxing Master of the High Court be paid by any other person named in the Order; and

WHEREAS by section 4 of the said Act of 1979 it is provided that a Tribunal may make such Order as it considers necessary for the purpose of its functions; and

WHEREAS the Tribunal referred to in the title hereof is satisfied that having regard to its findings and all other relevant matters there are sufficient reasons rendering it equitable that the Minister for Finance should pay the costs of the applicant as hereinafter appearing

IT IS HEREBY ORDERED that a Taxing Master of the High Court do tax in the manner hereinafter appearing the applicant's costs of appearing before the said Tribunal by two counsel instructed by solicitor.

In relation to the taxation of the applicant's costs the Tribunal DOTH ORDER as follows:

(*a*) That the statutory provisions and the Rules of the Superior Courts relating to the taxation of costs in an action in the High Courts (including the provisions relating to review and appeal) shall, in so far as is practicable, apply;

(*b*) That the costs of employing a solicitor and two counsel be taxed on a party and party basis.

(*c*) That the costs do include such witnesses' expenses as relate to the those witnesses who actually gave oral evidence before the Tribunal on the proposal of the applicant;

(*d*) That the amount of witnesses' expenses be (*a*) such fee for the preparation of any statement prepared by the witness which was received by the Tribunal as part of the witnesses' testimony as the Taxing Master (or on review or appeal, the Court) considers reasonable; (*b*) such fee (not already paid) for attending before the Tribunal on such days as the witness gave oral evidence as the Taxing Master (or, on review or appeal, the Court) considers reasonable and (*b*) the travelling and subsistence expenses (not already paid) reasonably incurred for the purpose of giving oral evidence before the Tribunal.

(*e*) That the Minister for Finance be at liberty to attend and be heard on the said taxation and any review or appeal in relation to it.

(*f*) That the Minister for Finance do pay the applicant's costs when taxed and ascertained.

Dated this 29th day of July 1994.

Signed Liam Hamilton

The Tribunal

ANNEXE

IN THE MATTER OF the Tribunals of Inquiry (Evidence) Acts, 1921 and 1979; and IN THE MATTER OF a Tribunal of Inquiry established pursuant to Resolutions of Dail Eireann passed on the 24th May 1991, and by Seanad Eireann on the 29th May 1991, to inquire into allegations regarding illegal activities fraud and malpractice in and in connection with the beef processing industry by order of the Minister for Agriculture and Food made on the 31st day of May 1991.

AND WHEREAS it was provided by the said order that the Tribunals of Inquiry (Evidence) Acts 1921 and 1979 applied to the Tribunal

WHEREAS it is provided by Section 2 (*b*) of the Tribunals of Inquiry (Evidence) Act, 1921 that a Tribunal to which the Act is applied shall have power to authorise the representation before it by solicitor or counsel or otherwise of any person appearing to it to be interested; and

WHEREAS Mr Dick Spring TD and Mr Barry Desmond MEP (hereinafter referred to as "the applicants") applied to the said Tribunal to be represented before it by counsel instructed by solicitor and on the 30th September 1991 and the 9th March 1992 respectively such authorisation under the said section was granted; and

WHEREAS it is provided by section 6 of the Tribunals of Inquiry (Evidence)(Amendment) Act, 1979, that if a Tribunal is of opinion that having regard to its findings and all other relevant matters there are sufficient reasons rendering it equitable to do so it may by Order direct that the whole or part of the costs of any person appearing before the Tribunal by counsel or solicitor as taxed by a Taxing Master of the High Court be paid by any other person named in the Order; and

WHEREAS by section 4 of the said Act of 1979 it is provided that a Tribunal may make such Order as it considers necessary for the purpose of its functions; and

WHEREAS the Tribunal referred to in the title hereof is satisfied that having regard to its findings and all other relevant matters there are sufficient reasons rendering it equitable that the Minister for Finance should pay the costs of the applicants as hereinafter appearing

IT IS HEREBY ORDERED that a Taxing Master of the High Court do tax in the manner hereinafter appearing the applicants' costs of appearing before the said Tribunal by two counsel instructed by solicitor.

In relation to the taxation of the applicants' costs the Tribunal DOTH ORDER as follows:

(*a*) That the statutory provisions and the Rules of the Superior Courts relating to the taxation of costs in an action in the High Courts (including the provisions relating to review and appeal) shall, in so far as is practicable, apply;

(*b*) That the costs of employing a solicitor and two counsel be taxed on a party and party basis, as being necessary and proper for enforcing before the Tribunal the applicants' right to ensure that all the facts concerning the allegations made by them were disclosed so far as the applicants were concerned;

(*c*) That the costs do include such witnesses' expenses as relate to the those witnesses who actually gave oral evidence before the Tribunal on the proposal of the applicants;

(*d*) That the amount of witnesses' expenses be (*a*) such fee for the preparation of any statement prepared by the witness which was received by the Tribunal as part of the witnesses' testimony as the Taxing Master (or on review or appeal, the Court) considers reasonable; (*b*) such fee (not already paid) for attending before the Tribunal on such days as the witness gave oral evidence as the Taxing Master (or, on review or appeal, the Court) considers reasonable and (*b*) the travelling and subsistence expenses (not already paid) reasonably incurred for the purpose of giving oral evidence before the Tribunal.

(*e*) That the Minister for Finance be at liberty to attend and be heard on the said taxation and any review or appeal in relation to it.

(*f*) That the Minister for Finance do pay the applicants' costs when taxed and ascertained.

Dated this 29th day of July 1994.

Signed Liam Hamilton

The Tribunal

ANNEXE

IN THE MATTER OF the Tribunals of Inquiry (Evidence) Acts, 1921 and 1979; and IN THE MATTER OF a Tribunal of Inquiry established pursuant to Resolutions of Dail Eireann passed on the 24th May 1991, and by Seanad Eireann on the 29th May 1991, to inquire into allegations regarding illegal activities fraud and malpractice in and in connection with the beef processing industry by order of the Minister for Agriculture and Food made on the 31st day of May 1991.

AND WHEREAS it was provided by the said order that the Tribunals of Inquiry (Evidence) Acts 1921 and 1979 applied to the Tribunal

WHEREAS it is provided by Section 2 (*b*) of the Tribunals of Inquiry (Evidence) Act, 1921 that a Tribunal to which the Act is applied shall have power to authorise the representation before it by solicitor or counsel or otherwise of any person appearing to it to be interested; and

WHEREAS Mr Zachariah Al Taher, Mr Augustine Fitzpatrick and Mr Naser Taher (hereinafter referred to as "the applicants") applied to the said Tribunal to be represented before it by counsel instructed by solicitor and on the 26th July 1991 (to Mr Z Al Taher) and the 2nd June 1992 (to Mr A Fitzpatrick and Mr N Taher) such authorisation under the said section was granted; and

WHEREAS it is provided by section 6 of the Tribunals of Inquiry (Evidence)(Amendment) Act, 1979, that if a Tribunal is of opinion that having regard to its findings and all other relevant matters there are sufficient reasons rendering it equitable to do so it may by Order direct that the whole or part of the costs of any person appearing before the Tribunal by counsel or solicitor as taxed by a Taxing Master of the High Court be paid by any other person named in the Order; and

WHEREAS by section 4 of the said Act of 1979 it is provided that a Tribunal may make such Order as it considers necessary for the purpose of its functions; and

WHEREAS the Tribunal referred to in the title hereof is satisfied that having regard to its findings and all other relevant matters there are sufficient reasons rendering it equitable that the Minister for Finance should pay the costs of the applicants as hereinafter appearing

IT IS HEREBY ORDERED that a Taxing Master of the High Court do tax in the manner hereinafter appearing the applicants' costs of appearing before the said Tribunal by one counsel instructed by solicitor.

In relation to the taxation of the applicants' costs the Tribunal DOTH ORDER as follows:

(*a*) That the statutory provisions and the Rules of the Superior Courts relating to the taxation of costs in an action in the High Courts (including the provisions relating to review and appeal) shall, in so far as is practicable, apply;

(*b*) That the costs of employing a solicitor and one counsel be taxed on a party and party basis.

(*c*) That the costs do include such witnesses' expenses as relate to the those witnesses who actually gave oral evidence before the Tribunal on the proposal of the applicants;

(*d*) That the amount of witnesses' expenses be (*a*) such fee for the preparation of any statement prepared by the witness which was received by the Tribunal as part of the witnesses' testimony as the Taxing Master (or on review or appeal, the Court) considers reasonable; (*b*) such fee (not already paid) for attending before the Tribunal on such days as the witness gave oral evidence as the Taxing Master (or, on review or appeal, the Court) considers reasonable and (*b*) the travelling and subsistence expenses (not already paid) reasonably incurred for the purpose of giving oral evidence before the Tribunal.

(*e*) That the Minister for Finance be at liberty to attend and be heard on the said taxation and any review or appeal in relation to it.

(*f*) That the Minister for Finance do pay the applicants' costs when taxed and ascertained.

Dated this 29th day of July 1994.

Signed Liam Hamilton

The Tribunal

ANNEXE

IN THE MATTER OF the Tribunals of Inquiry (Evidence) Acts, 1921 and 1979; and IN THE MATTER OF a Tribunal of Inquiry established pursuant to Resolutions of Dail Eireann passed on the 24th May 1991, and by Seanad Eireann on the 29th May 1991, to inquire into allegations regarding illegal activities fraud and malpractice in and in connection with the beef processing industry by order of the Minister for Agriculture and Food made on the 31st day of May 1991.

AND WHEREAS it was provided by the said order that the Tribunals of Inquiry (Evidence) Acts 1921 and 1979 applied to the Tribunal

WHEREAS it is provided by Section 2 (*b*) of the Tribunals of Inquiry (Evidence) Act, 1921 that a Tribunal to which the Act is applied shall have power to authorise the representation before it by solicitor or counsel or otherwise of any person appearing to it to be interested; and

WHEREAS Amalgamated Transport and General Workers' Union (hereinafter referred to as "the applicants") applied to the said Tribunal to be represented before it by counsel instructed by solicitor and on the 26th July 1991 such authorisation under the said section was granted; and

WHEREAS it is provided by section 6 of the Tribunals of Inquiry (Evidence)(Amendment) Act, 1979, that if a Tribunal is of opinion that having regard to its findings and all other relevant matters there are sufficient reasons rendering it equitable to do so it may by Order direct that the whole or part of the costs of any person appearing before the Tribunal by counsel or solicitor as taxed by a Taxing Master of the High Court be paid by any other person named in the Order; and

WHEREAS by section 4 of the said Act of 1979 it is provided that a Tribunal may make such Order as it considers necessary for the purpose of its functions; and

WHEREAS the Tribunal referred to in the title hereof is satisfied that having regard to its findings and all other relevant matters there are sufficient reasons rendering it equitable that the Minister for Finance should pay the costs of the applicants as hereinafter appearing

IT IS HEREBY ORDERED that a Taxing Master of the High Court do tax in the manner hereinafter appearing the applicants' costs of appearing before the said Tribunal by one counsel instructed by solicitor.

In relation to the taxation of the applicants' costs the Tribunal DOTH ORDER as follows:

(*a*) That the statutory provisions and the Rules of the Superior Courts relating to the taxation of costs in an action in the High Courts (including the provisions relating to review and appeal) shall, in so far as is practicable, apply;

(*b*) That the costs of employing a solicitor and one counsel be taxed on a party and party basis.

(*c*) That the costs do include such witnesses' expenses as relate to the those witnesses who actually gave oral evidence before the Tribunal on the proposal of the applicants;

(*d*) That the amount of witnesses' expenses be (*a*) such fee for the preparation of any statement prepared by the witness which was received by the Tribunal as part of the witnesses' testimony as the Taxing Master (or on review or appeal, the Court) considers reasonable; (*b*) such fee (not already paid) for attending before the Tribunal on such days as the witness gave oral evidence as the Taxing Master (or, on review or appeal, the Court) considers reasonable and (*b*) the travelling and subsistence expenses (not already paid) reasonably incurred for the purpose of giving oral evidence before the Tribunal.

(*e*) That the Minister for Finance be at liberty to attend and be heard on the said taxation and any review or appeal in relation to it.

(*f*) That the Minister for Finance do pay the applicants' costs when taxed and ascertained.

Dated this 29th day of July 1994.

Signed Liam Hamilton
The Tribunal

ANNEXE

IN THE MATTER OF the Tribunals of Inquiry (Evidence) Acts, 1921 and 1979; and IN THE MATTER OF a Tribunal of Inquiry established pursuant to Resolutions of Dail Eireann passed on the 24th May 1991, and by Seanad Eireann on the 29th May 1991, to inquire into allegations regarding illegal activities fraud and malpractice in and in connection with the beef processing industry by order of the Minister for Agriculture and Food made on the 31st day of May 1991.

AND WHEREAS it was provided by the said order that the Tribunals of Inquiry (Evidence) Acts 1921 and 1979 applied to the Tribunal

WHEREAS it is provided by Section 2 (*b*) of the Tribunals of Inquiry (Evidence) Act, 1921 that a Tribunal to which the Act is applied shall have power to authorise the representation before it by solicitor or counsel or otherwise of any person appearing to it to be interested; and

WHEREAS Services Industrial Professional Technical Union (hereinafter referred to as "the applicants") applied to the said Tribunal to be represented before it by counsel instructed by solicitor and on the 26th July 1991 such authorisation under the said section was granted; and

WHEREAS it is provided by section 6 of the Tribunals of Inquiry (Evidence)(Amendment) Act, 1979, that if a Tribunal is of opinion that having regard to its findings and all other relevant matters there are sufficient reasons rendering it equitable to do so it may by Order direct that the whole or part of the costs of any person appearing before the Tribunal by counsel or solicitor as taxed by a Taxing Master of the High Court be paid by any other person named in the Order; and

WHEREAS by section 4 of the said Act of 1979 it is provided that a Tribunal may make such Order as it considers necessary for the purpose of its functions; and

WHEREAS the Tribunal referred to in the title hereof is satisfied that having regard to its findings and all other relevant matters there are sufficient reasons rendering it equitable that the Minister for Finance should pay the costs of the applicants as hereinafter appearing

IT IS HEREBY ORDERED that a Taxing Master of the High Court do tax in the manner hereinafter appearing the applicants' costs of appearing before the said Tribunal by one counsel instructed by solicitor.

In relation to the taxation of the applicants' costs the Tribunal DOTH ORDER as follows:

(*a*) That the statutory provisions and the Rules of the Superior Courts relating to the taxation of costs in an action in the High Courts (including the provisions relating to review and appeal) shall, in so far as is practicable, apply;

(*b*) That the costs of employing a solicitor and one counsel be taxed on a party and party basis;

(*c*) That the costs do include such witnesses' expenses as relate to the those witnesses who actually gave oral evidence before the Tribunal on the proposal of the applicants;

(*d*) That the amount of witnesses' expenses be (*a*) such fee for the preparation of any statement prepared by the witness which was received by the Tribunal as part of the witnesses' testimony as the Taxing Master (or on review or appeal, the Court) considers reasonable; (*b*) such fee (not already paid) for attending before the Tribunal on such days as the witness gave oral evidence as the Taxing Master (or, on review or appeal, the Court) considers reasonable and (*b*) the travelling and subsistence expenses (not already paid) reasonably incurred for the purpose of giving oral evidence before the Tribunal.

(*e*) That the Minister for Finance be at liberty to attend and be heard on the said taxation and any review or appeal in relation to it.

(*f*) That the Minister for Finance do pay the applicants' costs when taxed and ascertained.

Dated this 29th day of July 1994.

Signed Liam Hamilton

The Tribunal

ANNEXE

IN THE MATTER OF the Tribunals of Inquiry (Evidence) Acts, 1921 and 1979; and IN THE MATTER OF a Tribunal of Inquiry established pursuant to Resolutions of Dail Eireann passed on the 24th May 1991, and by Seanad Eireann on the 29th May 1991, to inquire into allegations regarding illegal activities fraud and malpractice in and in connection with the beef processing industry by order of the Minister for Agriculture and Food made on the 31st day of May 1991.

AND WHEREAS it was provided by the said order that the Tribunals of Inquiry (Evidence) Acts 1921 and 1979 applied to the Tribunal

WHEREAS it is provided by Section 2 (*b*) of the Tribunals of Inquiry (Evidence) Act, 1921 that a Tribunal to which the Act is applied shall have power to authorise the representation before it by solicitor or counsel or otherwise of any person appearing to it to be interested; and

WHEREAS Mr Patrick Boyhan, Mr Con Howard, Mr Patrick J Kenny, Mr Michael McElligott and Mr Patrick Murphy (hereinafter referred to as "the applicants") applied to the said Tribunal to be represented before it by counsel instructed by solicitor and on the 26th July 1991 such authorisation under the said section was granted; and

WHEREAS it is provided by section 6 of the Tribunals of Inquiry (Evidence)(Amendment) Act, 1979, that if a Tribunal is of opinion that having regard to its findings and all other relevant matters there are sufficient reasons rendering it equitable to do so it may by Order direct that the whole or part of the costs of any person appearing before the Tribunal by counsel or solicitor as taxed by a Taxing Master of the High Court be paid by any other person named in the Order; and

WHEREAS by section 4 of the said Act of 1979 it is provided that a Tribunal may make such Order as it considers necessary for the purpose of its functions; and

WHEREAS the Tribunal referred to in the title hereof is satisfied that having regard to its findings and all other relevant matters there are sufficient reasons rendering it equitable that the Minister for Finance should pay the costs of the applicants as hereinafter appearing

IT IS HEREBY ORDERED that a Taxing Master of the High Court do tax in the manner hereinafter appearing the applicants' costs of appearing before the said Tribunal by two counsel instructed by solicitor on twenty five sitting days of the Tribunal

In relation to the taxation of the applicants' costs the Tribunal DOTH ORDER as follows:

(*a*) That the statutory provisions and the Rules of the Superior Courts relating to the taxation of costs in an action in the High Courts (including the provisions relating to review and appeal) shall, in so far as is practicable, apply;

(*b*) That the costs of employing a solicitor and two counsel be taxed on a party and party basis.

(*c*) That the costs do include such witnesses' expenses as relate to the those witnesses who actually gave oral evidence before the Tribunal on the proposal of the applicants;

(*d*) That the amount of witnesses' expenses be (*a*) such fee for the preparation of any statement prepared by the witness which was received by the Tribunal as part of the witnesses' testimony as the Taxing Master (or on review or appeal, the Court) considers reasonable; (*b*) such fee (not already paid) for attending before the Tribunal on such days as the witness gave oral evidence as the Taxing Master (or, on review or appeal, the Court) considers reasonable and (*b*) the travelling and subsistence expenses (not already paid) reasonably incurred for the purpose of giving oral evidence before the Tribunal.

(*e*) That the Minister for Finance be at liberty to attend and be heard on the said taxation and any review or appeal in relation to it.

(*f*) That the Minister for Finance do pay the applicants' costs when taxed and ascertained.

Dated this 29th day of July 1994.

Signed Liam Hamilton

The Tribunal

ANNEXE

IN THE MATTER OF the Tribunals of Inquiry (Evidence) Acts, 1921 and 1979; and IN THE MATTER OF a Tribunal of Inquiry established pursuant to Resolutions of Dail Eireann passed on the 24th May 1991, and by Seanad Eireann on the 29th May 1991, to inquire into allegations regarding illegal activities fraud and malpractice in and in connection with the beef processing industry by order of the Minister for Agriculture and Food made on the 31st day of May 1991.

AND WHEREAS it was provided by the said order that the Tribunals of Inquiry (Evidence) Acts 1921 and 1979 applied to the Tribunal

WHEREAS it is provided by Section 2 (*b*) of the Tribunals of Inquiry (Evidence) Act, 1921 that a Tribunal to which the Act is applied shall have power to authorise the representation before it by solicitor or counsel or otherwise of any person appearing to it to be interested; and

WHEREAS Mr Pat Rabbitte TD and Mr Tomás MacGiolla (hereinafter referred to as "the applicants") applied to the said Tribunal to be represented before it by counsel instructed by solicitor and on the 26th July 1991 such authorisation under the said section was granted; and

WHEREAS it is provided by section 6 of the Tribunals of Inquiry (Evidence)(Amendment) Act, 1979, that if a Tribunal is of opinion that having regard to its findings and all other relevant matters there are sufficient reasons rendering it equitable to do so it may by Order direct that the whole or part of the costs of any person appearing before the Tribunal by counsel or solicitor as taxed by a Taxing Master of the High Court be paid by any other person named in the Order; and

WHEREAS by section 4 of the said Act of 1979 it is provided that a Tribunal may make such Order as it considers necessary for the purpose of its functions; and

WHEREAS the Tribunal referred to in the title hereof is satisfied that having regard to its findings and all other relevant matters there are sufficient reasons rendering it equitable that the Minister for Finance should pay the costs of the applicants as hereinafter appearing

IT IS HEREBY ORDERED that a Taxing Master of the High Court do tax in the manner hereinafter appearing the applicants' costs of appearing before the said Tribunal by two counsel instructed by solicitor.

In relation to the taxation of the applicants' costs the Tribunal DOTH ORDER as follows:

(*a*) That the statutory provisions and the Rules of the Superior Courts relating to the taxation of costs in an action in the High Courts (including the provisions relating to review and appeal) shall, in so far as is practicable, apply;

(*b*) That the costs of employing a solicitor and two counsel be taxed on a party and party basis, as being necessary and proper for enforcing before the tribunal the applicants' right to ensure that all the facts concerning the allegations made by them were disclosed so far as the applicants were concerned;

(*c*) That the costs do include such witnesses' expenses as relate to those witnesses who actually gave oral evidence before the Tribunal on the proposal of the applicants;

(*d*) That the amount of witnesses' expenses be (*a*) such fee for the preparation of any statement prepared by the witness which was received by the Tribunal as part of the witnesses' testimony as the Taxing Master (or on review or appeal, the Court) considers reasonable; (*b*) such fee (not already paid) for attending before the Tribunal on such days as the witness gave oral evidence as the Taxing Master (or, on review or appeal, the Court) considers reasonable and (*b*) the travelling and subsistence expenses (not already paid) reasonably incurred for the purpose of giving oral evidence before the Tribunal.

(*e*) That the Minister for Finance be at liberty to attend and be heard on the said taxation and any review or appeal in relation to it.

(*f*) That the Minister for Finance do pay the applicants' costs when taxed and ascertained.

Dated this 29th day of July 1994.

Signed Liam Hamilton
The Tribunal

ANNEXE

IN THE MATTER OF the Tribunals of Inquiry (Evidence) Acts, 1921 and 1979; and IN THE MATTER OF a Tribunal of Inquiry established pursuant to Resolutions of Dail Eireann passed on the 24th May 1991, and by Seanad Eireann on the 29th May 1991, to inquire into allegations regarding illegal activities fraud and malpractice in and in connection with the beef processing industry by order of the Minister for Agriculture and Food made on the 31st day of May 1991.

AND WHEREAS it was provided by the said order that the Tribunals of Inquiry (Evidence) Acts 1921 and 1979 applied to the Tribunal

WHEREAS it is provided by Section 2 (*b*) of the Tribunals of Inquiry (Evidence) Act, 1921 that a Tribunal to which the Act is applied shall have power to authorise the representation before it by solicitor or counsel or otherwise of any person appearing to it to be interested; and

WHEREAS Mr Patrick Quearney and Mr Bernard Kelly (hereinafter referred to as "the applicants") applied to the said Tribunal to be represented before it by solicitor and on the 16th September 1992 and the 11th May 1993 respectively such authorisation under the said section was granted; and

WHEREAS it is provided by section 6 of the Tribunals of Inquiry (Evidence)(Amendment) Act, 1979, that if a Tribunal is of opinion that having regard to its findings and all other relevant matters there are sufficient reasons rendering it equitable to do so it may by Order direct that the whole or part of the costs of any person appearing before the Tribunal by counsel or solicitor as taxed by a Taxing Master of the High Court be paid by any other person named in the Order; and

WHEREAS by section 4 of the said Act of 1979 it is provided that a Tribunal may make such Order as it considers necessary for the purpose of its functions; and

WHEREAS the Tribunal referred to in the title hereof is satisfied that having regard to its findings and all other relevant matters there are sufficient reasons rendering it equitable that the Minister for Finance should pay the costs of the applicants as hereinafter appearing

IT IS HEREBY ORDERED that a Taxing Master of the High Court do tax in the manner hereinafter appearing the applicants' costs of appearing before the said Tribunal Tribunal by solicitor.

In relation to the taxation of the applicants' costs the Tribunal DOTH ORDER as follows:

(*a*) That the statutory provisions and the Rules of the Superior Courts relating to the taxation of costs in an action in the High Courts (including the provisions relating to review and appeal) shall, in so far as is practicable, apply;

(*b*) That the costs of employing a solicitor be taxed on a party and party basis.

(*c*) That the costs do include such witnesses' expenses as relate to those witnesses who actually gave oral evidence before the Tribunal on the proposal of the applicants;

(*d*) That the amount of witnesses' expenses be (*a*) such fee for the preparation of any statement prepared by the witness which was received by the Tribunal as part of the witnesses' testimony as the Taxing Master (or on review or appeal, the Court) considers reasonable; (*b*) such fee (not already paid) for attending before the Tribunal on such days as the witness gave oral evidence as the Taxing Master (or, on review or appeal, the Court) considers reasonable and (*b*) the travelling and subsistence expenses (not already paid) reasonably incurred for the purpose of giving oral evidence before the Tribunal.

(*e*) That the Minister for Finance be at liberty to attend and be heard on the said taxation and any review or appeal in relation to it.

(*f*) That the Minister for Finance do pay the applicants' costs when taxed and ascertained.

Dated this 29th day of July 1994.

Signed Liam Hamilton

The Tribunal

ANNEXE

IN THE MATTER OF the Tribunals of Inquiry (Evidence) Acts, 1921 and 1979; and IN THE MATTER OF a Tribunal of Inquiry established pursuant to Resolutions of Dail Eireann passed on the 24th May 1991, and by Seanad Eireann on the 29th May 1991, to inquire into allegations regarding illegal activities fraud and malpractice in and in connection with the beef processing industry by order of the Minister for Agriculture and Food made on the 31st day of May 1991.

AND WHEREAS it was provided by the said order that the Tribunals of Inquiry (Evidence) Acts 1921 and 1979 applied to the Tribunal

WHEREAS it is provided by Section 2 (*b*) of the Tribunals of Inquiry (Evidence) Act, 1921 that a Tribunal to which the Act is applied shall have power to authorise the representation before it by solicitor or counsel or otherwise of any person appearing to it to be interested; and

WHEREAS Mr Liam Marks (hereinafter referred to as "the applicant") applied to the said Tribunal to be represented before it by counsel instructed by solicitor and on the 26th July 1991 such authorisation under the said section was granted; and

WHEREAS it is provided by section 6 of the Tribunals of Inquiry (Evidence)(Amendment) Act, 1979, that if a Tribunal is of opinion that having regard to its findings and all other relevant matters there are sufficient reasons rendering it equitable to do so it may by Order direct that the whole or part of the costs of any person appearing before the Tribunal by counsel or solicitor as taxed by a Taxing Master of the High Court be paid by any other person named in the Order; and

WHEREAS by section 4 of the said Act of 1979 it is provided that a Tribunal may make such Order as it considers necessary for the purpose of its functions; and

WHEREAS the Tribunal referred to in the title hereof is satisfied that having regard to its findings and all other relevant matters there are sufficient reasons rendering it equitable that the Minister for Finance should pay the costs of the applicant as hereinafter appearing

IT IS HEREBY ORDERED that a Taxing Master of the High Court do tax in the manner hereinafter appearing the applicant's costs of appearing before the said Tribunal by one counsel instructed by solicitor.

In relation to the taxation of the applicant's costs the Tribunal DOTH ORDER as follows:

(*a*) That the statutory provisions and the Rules of the Superior Courts relating to the taxation of costs in an action in the High Courts (including the provisions relating to review and appeal) shall, in so far as is practicable, apply;

(*b*) That the costs of employing a solicitor and one counsel be taxed on a party and party basis.

(*c*) That the costs do include such witnesses' expenses as relate to those witnesses who actually gave oral evidence before the Tribunal on the proposal of the applicant;

(*d*) That the amount of witnesses' expenses be (*a*) such fee for the preparation of any statement prepared by the witness which was received by the Tribunal as part of the witnesses' testimony as the Taxing Master (or on review or appeal, the Court) considers reasonable; (*b*) such fee (not already paid) for attending before the Tribunal on such days as the witness gave oral evidence as the Taxing Master (or, on review or appeal, the Court) considers reasonable and (*b*) the travelling and subsistence expenses (not already paid) reasonably incurred for the purpose of giving oral evidence before the Tribunal.

(*e*) That the Minister for Finance be at liberty to attend and be heard on the said taxation and any review or appeal in relation to it.

(*f*) That the Minister for Finance do pay the applicant's costs when taxed and ascertained.

Dated this 29th day of July 1994.

Signed Liam Hamilton

The Tribunal

ANNEXE

IN THE MATTER OF the Tribunals of Inquiry (Evidence) Acts, 1921 and 1979; and IN THE MATTER OF a Tribunal of Inquiry established pursuant to Resolutions of Dail Eireann passed on the 24th May 1991, and by Seanad Eireann on the 29th May 1991, to inquire into allegations regarding illegal activities fraud and malpractice in and in connection with the beef processing industry by order of the Minister for Agriculture and Food made on the 31st day of May 1991.

AND WHEREAS it was provided by the said order that the Tribunals of Inquiry (Evidence) Acts 1921 and 1979 applied to the Tribunal

WHEREAS it is provided by Section 2 (*b*) of the Tribunals of Inquiry (Evidence) Act, 1921 that a Tribunal to which the Act is applied shall have power to authorise the representation before it by solicitor or counsel or otherwise of any person appearing to it to be interested; and

WHEREAS Mr James Fairbairn (hereinafter referred to as "the applicant") applied to the said Tribunal to be represented before it by counsel instructed by solicitor and on the 30th September 1991 such authorisation under the said section was granted; and

WHEREAS it is provided by section 6 of the Tribunals of Inquiry (Evidence)(Amendment) Act, 1979, that if a Tribunal is of opinion that having regard to its findings and all other relevant matters there are sufficient reasons rendering it equitable to do so it may by Order direct that the whole or part of the costs of any person appearing before the Tribunal by counsel or solicitor as taxed by a Taxing Master of the High Court be paid by any other person named in the Order; and

WHEREAS by section 4 of the said Act of 1979 it is provided that a Tribunal may make such Order as it considers necessary for the purpose of its functions; and

WHEREAS the Tribunal referred to in the title hereof is satisfied that having regard to its findings and all other relevant matters there are sufficient reasons rendering it equitable that the Minister for Finance should pay the costs of the applicant as hereinafter appearing

IT IS HEREBY ORDERED that a Taxing Master of the High Court do tax in the manner hereinafter appearing the applicant's costs of appearing before the said Tribunal by one counsel instructed by solicitor.

In relation to the taxation of the applicant's costs the Tribunal DOTH ORDER as follows:

(*a*) That the statutory provisions and the Rules of the Superior Courts relating to the taxation of costs in an action in the High Courts (including the provisions relating to review and appeal) shall, in so far as is practicable, apply;

(*b*) That the costs of employing a solicitor and one counsel be taxed on a party and party basis.

(*c*) That the costs do include such witnesses' expenses as relate to those witnesses who actually gave oral evidence before the Tribunal on the proposal of the applicant;

(*d*) That the amount of witnesses' expenses be (*a*) such fee for the preparation of any statement prepared by the witness which was received by the Tribunal as part of the witnesses' testimony as the Taxing Master (or on review or appeal, the Court) considers reasonable; (*b*) such fee (not already paid) for attending before the Tribunal on such days as the witness gave oral evidence as the Taxing Master (or, on review or appeal, the Court) considers reasonable and (*b*) the travelling and subsistence expenses (not already paid) reasonably incurred for the purpose of giving oral evidence before the Tribunal.

(*e*) That the Minister for Finance be at liberty to attend and be heard on the said taxation and any review or appeal in relation to it.

(*f*) That the Minister for Finance do pay the applicant's costs when taxed and ascertained.

Dated this 29th day of July 1994.

Signed Liam Hamilton

The Tribunal

ANNEXE

IN THE MATTER OF the Tribunals of Inquiry (Evidence) Acts, 1921 and 1979; and IN THE MATTER OF a Tribunal of Inquiry established pursuant to Resolutions of Dail Eireann passed on the 24th May 1991, and by Seanad Eireann on the 29th May 1991, to inquire into allegations regarding illegal activities fraud and malpractice in and in connection with the beef processing industry by order of the Minister for Agriculture and Food made on the 31st day of May 1991.

AND WHEREAS it was provided by the said order that the Tribunals of Inquiry (Evidence) Acts 1921 and 1979 applied to the Tribunal

WHEREAS it is provided by Section 2 (*b*) of the Tribunals of Inquiry (Evidence) Act, 1921 that a Tribunal to which the Act is applied shall have power to authorise the representation before it by solicitor or counsel or otherwise of any person appearing to it to be interested; and

WHEREAS Stokes Kennedy Crowley (hereinafter referred to as "the applicants") applied to the said Tribunal to be represented before it by counsel instructed by solicitor and on the 7th November 1991 such authorisation under the said section was granted; and

WHEREAS it is provided by section 6 of the Tribunals of Inquiry (Evidence)(Amendment) Act, 1979, that if a Tribunal is of opinion that having regard to its findings and all other relevant matters there are sufficient reasons rendering it equitable to do so it may by Order direct that the whole or part of the costs of any person appearing before the Tribunal by counsel or solicitor as taxed by a Taxing Master of the High Court be paid by any other person named in the Order; and

WHEREAS by section 4 of the said Act of 1979 it is provided that a Tribunal may make such Order as it considers necessary for the purpose of its functions; and

WHEREAS the Tribunal referred to in the title hereof is satisfied that having regard to its findings and all other relevant matters there are sufficient reasons rendering it equitable that the Minister for Finance should pay the costs of the applicants as hereinafter appearing

IT IS HEREBY ORDERED that a Taxing Master of the High Court do tax in the manner hereinafter appearing the applicants' costs of appearing before the said Tribunal by two counsel instructed by solicitor.

In relation to the taxation of the applicants' costs the Tribunal DOTH ORDER as follows:

(*a*) That the statutory provisions and the Rules of the Superior Courts relating to the taxation of costs in an action in the High Courts (including the provisions relating to review and appeal) shall, in so far as is practicable, apply;

(*b*) That the costs of employing a solicitor and two counsel be taxed on a party and party basis.

(*c*) That the costs do include such witnesses' expenses as relate to those witnesses who actually gave oral evidence before the Tribunal on the proposal of the applicants;

(*d*) That the amount of witnesses' expenses be (*a*) such fee for the preparation of any statement prepared by the witness which was received by the Tribunal as part of the witnesses' testimony as the Taxing Master (or on review or appeal, the Court) considers reasonable; (*b*) such fee (not already paid) for attending before the Tribunal on such days as the witness gave oral evidence as the Taxing Master (or, on review or appeal, the Court) considers reasonable and (*b*) the travelling and subsistence expenses (not already paid) reasonably incurred for the purpose of giving oral evidence before the Tribunal.

(*e*) That the Minister for Finance be at liberty to attend and be heard on the said taxation and any review or appeal in relation to it.

(*f*) That the Minister for Finance do pay the applicants' costs when taxed and ascertained.

Dated this 29th day of July 1994.

Signed Liam Hamilton

The Tribunal

ANNEXE

IN THE MATTER OF the Tribunals of Inquiry (Evidence) Acts, 1921 and 1979; and IN THE MATTER OF a Tribunal of Inquiry established pursuant to Resolutions of Dail Eireann passed on the 24th May 1991, and by Seanad Eireann on the 29th May 1991, to inquire into allegations regarding illegal activities fraud and malpractice in and in connection with the beef processing industry by order of the Minister for Agriculture and Food made on the 31st day of May 1991.

AND WHEREAS it was provided by the said order that the Tribunals of Inquiry (Evidence) Acts 1921 and 1979 applied to the Tribunal

WHEREAS it is provided by Section 2 (*b*) of the Tribunals of Inquiry (Evidence) Act, 1921 that a Tribunal to which the Act is applied shall have power to authorise the representation before it by solicitor or counsel or otherwise of any person appearing to it to be interested; and

WHEREAS Mr Sean Sheelan and Mr Kevin J McDonald (hereinafter referred to as "the applicants") applied to the said Tribunal to be represented before it by solicitor and on the 20th November 1991 such authorisation under the said section was granted; and

WHEREAS it is provided by section 6 of the Tribunals of Inquiry (Evidence)(Amendment) Act, 1979, that if a Tribunal is of opinion that having regard to its findings and all other relevant matters there are sufficient reasons rendering it equitable to do so it may by Order direct that the whole or part of the costs of any person appearing before the Tribunal by counsel or solicitor as taxed by a Taxing Master of the High Court be paid by any other person named in the Order; and

WHEREAS by section 4 of the said Act of 1979 it is provided that a Tribunal may make such Order as it considers necessary for the purpose of its functions; and

WHEREAS the Tribunal referred to in the title hereof is satisfied that having regard to its findings and all other relevant matters there are sufficient reasons rendering it equitable that the Minister for Finance should pay the costs of the applicants as hereinafter appearing

IT IS HEREBY ORDERED that a Taxing Master of the High Court do tax in the manner hereinafter appearing the applicants' costs of appearing before the said Tribunal by solicitor.

In relation to the taxation of the applicants' costs the Tribunal DOTH ORDER as follows:

(*a*) That the statutory provisions and the Rules of the Superior Courts relating to the taxation of costs in an action in the High Courts (including the provisions relating to review and appeal) shall, in so far as is practicable, apply;

(*b*) That the costs of employing a solicitor be taxed on a party and party basis.

(*c*) That the costs do include such witnesses' expenses as relate to those witnesses who actually gave oral evidence before the Tribunal on the proposal of the applicants;

(*d*) That the amount of witnesses' expenses be (*a*) such fee for the preparation of any statement prepared by the witness which was received by the Tribunal as part of the witnesses' testimony as the Taxing Master (or on review or appeal, the Court) considers reasonable; (*b*) such fee (not already paid) for attending before the Tribunal on such days as the witness gave oral evidence as the Taxing Master (or, on review or appeal, the Court) considers reasonable and (*b*) the travelling and subsistence expenses (not already paid) reasonably incurred for the purpose of giving oral evidence before the Tribunal.

(*e*) That the Minister for Finance be at liberty to attend and be heard on the said taxation and any review or appeal in relation to it.

(*f*) That the Minister for Finance do pay the applicants' costs when taxed and ascertained.

Dated this 29th day of July 1994.

Signed Liam Hamilton

The Tribunal

ANNEXE

IN THE MATTER OF the Tribunals of Inquiry (Evidence) Acts, 1921 and 1979; and IN THE MATTER OF a Tribunal of Inquiry established pursuant to Resolutions of Dail Eireann passed on the 24th May 1991, and by Seanad Eireann on the 29th May 1991, to inquire into allegations regarding illegal activities fraud and malpractice in and in connection with the beef processing industry by order of the Minister for Agriculture and Food made on the 31st day of May 1991.

AND WHEREAS it was provided by the said order that the Tribunals of Inquiry (Evidence) Acts 1921 and 1979 applied to the Tribunal

WHEREAS it is provided by Section 2 (*b*) of the Tribunals of Inquiry (Evidence) Act, 1921 that a Tribunal to which the Act is applied shall have power to authorise the representation before it by solicitor or counsel or otherwise of any person appearing to it to be interested; and

WHEREAS Mr Eamonn Mackle (hereinafter referred to as "the applicant") applied to the said Tribunal to be represented before it by counsel instructed by solicitor and on the 25th November 1991 such authorisation under the said section was granted; and

WHEREAS it is provided by section 6 of the Tribunals of Inquiry (Evidence)(Amendment) Act, 1979, that if a Tribunal is of opinion that having regard to its findings and all other relevant matters there are sufficient reasons rendering it equitable to do so it may by Order direct that the whole or part of the costs of any person appearing before the Tribunal by counsel or solicitor as taxed by a Taxing Master of the High Court be paid by any other person named in the Order; and

WHEREAS by section 4 of the said Act of 1979 it is provided that a Tribunal may make such Order as it considers necessary for the purpose of its functions; and

WHEREAS the Tribunal referred to in the title hereof is satisfied that having regard to its findings and all other relevant matters there are sufficient reasons rendering it equitable that the Minister for Finance should pay the costs of the applicant as hereinafter appearing

IT IS HEREBY ORDERED that a Taxing Master of the High Court do tax in the manner hereinafter appearing the applicant's costs of appearing before the said Tribunal by one counsel instructed by solicitor.

In relation to the taxation of the applicant's costs the Tribunal DOTH ORDER as follows:

(*a*) That the statutory provisions and the Rules of the Superior Courts relating to the taxation of costs in an action in the High Courts (including the provisions relating to review and appeal) shall, in so far as is practicable, apply;

(*b*) That the costs of employing a solicitor and one counsel be taxed on a party and party basis.

(*c*) That the costs do include such witnesses' expenses as relate to those witnesses who actually gave oral evidence before the Tribunal on the proposal of the applicant;

(*d*) That the amount of witnesses' expenses be (*a*) such fee for the preparation of any statement prepared by the witness which was received by the Tribunal as part of the witnesses' testimony as the Taxing Master (or on review or appeal, the Court) considers reasonable; (*b*) such fee (not already paid) for attending before the Tribunal on such days as the witness gave oral evidence as the Taxing Master (or, on review or appeal, the Court) considers reasonable and (*b*) the travelling and subsistence expenses (not already paid) reasonably incurred for the purpose of giving oral evidence before the Tribunal.

(*e*) That the Minister for Finance be at liberty to attend and be heard on the said taxation and any review or appeal in relation to it.

(*f*) That the Minister for Finance do pay the applicant's costs when taxed and ascertained.

Dated this 29th day of July 1994.

Signed Liam Hamilton
The Tribunal

ANNEXE

IN THE MATTER OF the Tribunals of Inquiry (Evidence) Acts, 1921 and 1979; and IN THE MATTER OF a Tribunal of Inquiry established pursuant to Resolutions of Dail Eireann passed on the 24th May 1991, and by Seanad Eireann on the 29th May 1991, to inquire into allegations regarding illegal activities fraud and malpractice in and in connection with the beef processing industry by order of the Minister for Agriculture and Food made on the 31st day of May 1991.

AND WHEREAS it was provided by the said order that the Tribunals of Inquiry (Evidence) Acts 1921 and 1979 applied to the Tribunal

WHEREAS it is provided by Section 2 (*b*) of the Tribunals of Inquiry (Evidence) Act, 1921 that a Tribunal to which the Act is applied shall have power to authorise the representation before it by solicitor or counsel or otherwise of any person appearing to it to be interested; and

WHEREAS Mr Norbert Quinn (hereinafter referred to as "the applicant") applied to the said Tribunal to be represented before it by counsel instructed by solicitor and on the 27th November 1991 such authorisation under the said section was granted; and

WHEREAS it is provided by section 6 of the Tribunals of Inquiry (Evidence)(Amendment) Act, 1979, that if a Tribunal is of opinion that having regard to its findings and all other relevant matters there are sufficient reasons rendering it equitable to do so it may by Order direct that the whole or part of the costs of any person appearing before the Tribunal by counsel or solicitor as taxed by a Taxing Master of the High Court be paid by any other person named in the Order; and

WHEREAS by section 4 of the said Act of 1979 it is provided that a Tribunal may make such Order as it considers necessary for the purpose of its functions; and

WHEREAS the Tribunal referred to in the title hereof is satisfied that having regard to its findings and all other relevant matters there are sufficient reasons rendering it equitable that the Minister for Finance should pay the costs of the applicant as hereinafter appearing

IT IS HEREBY ORDERED that a Taxing Master of the High Court do tax in the manner hereinafter appearing the applicant's costs of appearing before the said Tribunal by one counsel instructed by solicitor.

In relation to the taxation of the applicant's costs the Tribunal DOTH ORDER as follows:

(*a*) That the statutory provisions and the Rules of the Superior Courts relating to the taxation of costs in an action in the High Courts (including the provisions relating to review and appeal) shall, in so far as is practicable, apply;

(*b*) That the costs of employing a solicitor and one counsel be taxed on a party and party basis.

(*c*) That the costs do include such witnesses' expenses as relate to those witnesses who actually gave oral evidence before the Tribunal on the proposal of the applicant;

(*d*) That the amount of witnesses' expenses be (*a*) such fee for the preparation of any statement prepared by the witness which was received by the Tribunal as part of the witnesses' testimony as the Taxing Master (or on review or appeal, the Court) considers reasonable; (*b*) such fee (not already paid) for attending before the Tribunal on such days as the witness gave oral evidence as the Taxing Master (or, on review or appeal, the Court) considers reasonable and (*b*) the travelling and subsistence expenses (not already paid) reasonably incurred for the purpose of giving oral evidence before the Tribunal.

(*e*) That the Minister for Finance be at liberty to attend and be heard on the said taxation and any review or appeal in relation to it.

(*f*) That the Minister for Finance do pay the applicant's costs when taxed and ascertained.

Dated this 29th day of July 1994.

Signed Liam Hamilton

The Tribunal

ANNEXE

IN THE MATTER OF the Tribunals of Inquiry (Evidence) Acts, 1921 and 1979; and IN THE MATTER OF a Tribunal of Inquiry established pursuant to Resolutions of Dail Eireann passed on the 24th May 1991, and by Seanad Eireann on the 29th May 1991, to inquire into allegations regarding illegal activities fraud and malpractice in and in connection with the beef processing industry by order of the Minister for Agriculture and Food made on the 31st day of May 1991.

AND WHEREAS it was provided by the said order that the Tribunals of Inquiry (Evidence) Acts 1921 and 1979 applied to the Tribunal

WHEREAS it is provided by Section 2 (*b*) of the Tribunals of Inquiry (Evidence) Act, 1921 that a Tribunal to which the Act is applied shall have power to authorise the representation before it by solicitor or counsel or otherwise of any person appearing to it to be interested; and

WHEREAS Mr Jerry O'Callaghan (hereinafter referred to as "the applicant") applied to the said Tribunal to be represented before it by counsel instructed by solicitor and on the 9th December 1991 such authorisation under the said section was granted; and

WHEREAS it is provided by section 6 of the Tribunals of Inquiry (Evidence)(Amendment) Act, 1979, that if a Tribunal is of opinion that having regard to its findings and all other relevant matters there are sufficient reasons rendering it equitable to do so it may by Order direct that the whole or part of the costs of any person appearing before the Tribunal by counsel or solicitor as taxed by a Taxing Master of the High Court be paid by any other person named in the Order; and

WHEREAS by section 4 of the said Act of 1979 it is provided that a Tribunal may make such Order as it considers necessary for the purpose of its functions; and

WHEREAS the Tribunal referred to in the title hereof is satisfied that having regard to its findings and all other relevant matters there are sufficient reasons rendering it equitable that the Minister for Finance should pay the costs of the applicant as hereinafter appearing

IT IS HEREBY ORDERED that a Taxing Master of the High Court do tax in the manner hereinafter appearing the applicant's costs of appearing before the said Tribunal by one counsel instructed by solicitor.

In relation to the taxation of the applicant's costs the Tribunal DOTH ORDER as follows:

(*a*) That the statutory provisions and the Rules of the Superior Courts relating to the taxation of costs in an action in the High Courts (including the provisions relating to review and appeal) shall, in so far as is practicable, apply;

(*b*) That the costs of employing a solicitor and one counsel be taxed on a party and party basis.

(*c*) That the costs do include such witnesses' expenses as relate to those witnesses who actually gave oral evidence before the Tribunal on the proposal of the applicant;

(*d*) That the amount of witnesses' expenses be (*a*) such fee for the preparation of any statement prepared by the witness which was received by the Tribunal as part of the witnesses' testimony as the Taxing Master (or on review or appeal, the Court) considers reasonable; (*b*) such fee (not already paid) for attending before the Tribunal on such days as the witness gave oral evidence as the Taxing Master (or, on review or appeal, the Court) considers reasonable and (*b*) the travelling and subsistence expenses (not already paid) reasonably incurred for the purpose of giving oral evidence before the Tribunal.

(*e*) That the Minister for Finance be at liberty to attend and be heard on the said taxation and any review or appeal in relation to it.

(*f*) That the Minister for Finance do pay the applicant's costs when taxed and ascertained.

Dated this 29th day of July 1994.

Signed Liam Hamilton

The Tribunal

ANNEXE

IN THE MATTER OF the Tribunals of Inquiry (Evidence) Acts, 1921 and 1979; and IN THE MATTER OF a Tribunal of Inquiry established pursuant to Resolutions of Dail Eireann passed on the 24th May 1991, and by Seanad Eireann on the 29th May 1991, to inquire into allegations regarding illegal activities fraud and malpractice in and in connection with the beef processing industry by order of the Minister for Agriculture and Food made on the 31st day of May 1991.

AND WHEREAS it was provided by the said order that the Tribunals of Inquiry (Evidence) Acts 1921 and 1979 applied to the Tribunal

WHEREAS it is provided by Section 2 (*b*) of the Tribunals of Inquiry (Evidence) Act, 1921 that a Tribunal to which the Act is applied shall have power to authorise the representation before it by solicitor or counsel or otherwise of any person appearing to it to be interested; and

WHEREAS Autozero Limited (hereinafter referred to as "the applicants") applied to the said Tribunal to be represented before it by counsel instructed by solicitor and on the 11th December 1991 such authorisation under the said section was granted; and

WHEREAS it is provided by section 6 of the Tribunals of Inquiry (Evidence)(Amendment) Act, 1979, that if a Tribunal is of opinion that having regard to its findings and all other relevant matters there are sufficient reasons rendering it equitable to do so it may by Order direct that the whole or part of the costs of any person appearing before the Tribunal by counsel or solicitor as taxed by a Taxing Master of the High Court be paid by any other person named in the Order; and

WHEREAS by section 4 of the said Act of 1979 it is provided that a Tribunal may make such Order as it considers necessary for the purpose of its functions; and

WHEREAS the Tribunal referred to in the title hereof is satisfied that having regard to its findings and all other relevant matters there are sufficient reasons rendering it equitable that the Minister for Finance should pay the costs of the applicants as hereinafter appearing

IT IS HEREBY ORDERED that a Taxing Master of the High Court do tax in the manner hereinafter appearing the applicants' costs of appearing before the said Tribunal by one counsel instructed by solicitor.

In relation to the taxation of the applicants' costs the Tribunal DOTH ORDER as follows:

(*a*) That the statutory provisions and the Rules of the Superior Courts relating to the taxation of costs in an action in the High Courts (including the provisions relating to review and appeal) shall, in so far as is practicable, apply;

(*b*) That the costs of employing a solicitor and one counsel be taxed on a party and party basis.

(*c*) That the costs do include such witnesses' expenses as relate to those witnesses who actually gave oral evidence before the Tribunal on the proposal of the applicants;

(*d*) That the amount of witnesses' expenses be (*a*) such fee for the preparation of any statement prepared by the witness which was received by the Tribunal as part of the witnesses' testimony as the Taxing Master (or on review or appeal, the Court) considers reasonable; (*b*) such fee (not already paid) for attending before the Tribunal on such days as the witness gave oral evidence as the Taxing Master (or, on review or appeal, the Court) considers reasonable and (*b*) the travelling and subsistence expenses (not already paid) reasonably incurred for the purpose of giving oral evidence before the Tribunal.

(*e*) That the Minister for Finance be at liberty to attend and be heard on the said taxation and any review or appeal in relation to it.

(*f*) That the Minister for Finance do pay the applicants' costs when taxed and ascertained.

Dated this 29th day of July 1994.

Signed Liam Hamilton

The Tribunal

ANNEXE

IN THE MATTER OF the Tribunals of Inquiry (Evidence) Acts, 1921 and 1979; and IN THE MATTER OF a Tribunal of Inquiry established pursuant to Resolutions of Dail Eireann passed on the 24th May 1991, and by Seanad Eireann on the 29th May 1991, to inquire into allegations regarding illegal activities fraud and malpractice in and in connection with the beef processing industry by order of the Minister for Agriculture and Food made on the 31st day of May 1991.

AND WHEREAS it was provided by the said order that the Tribunals of Inquiry (Evidence) Acts 1921 and 1979 applied to the Tribunal

WHEREAS it is provided by Section 2 (*b*) of the Tribunals of Inquiry (Evidence) Act, 1921 that a Tribunal to which the Act is applied shall have power to authorise the representation before it by solicitor or counsel or otherwise of any person appearing to it to be interested; and

WHEREAS Ms Angela Magee and Ms Imelda Murray (hereinafter referred to as "the applicants") applied to the said Tribunal to be represented before it by counsel instructed by solicitor and on the 17th January 1992 such authorisation under the said section was granted; and

WHEREAS it is provided by section 6 of the Tribunals of Inquiry (Evidence)(Amendment) Act, 1979, that if a Tribunal is of opinion that having regard to its findings and all other relevant matters there are sufficient reasons rendering it equitable to do so it may by Order direct that the whole or part of the costs of any person appearing before the Tribunal by counsel or solicitor as taxed by a Taxing Master of the High Court be paid by any other person named in the Order; and

WHEREAS by section 4 of the said Act of 1979 it is provided that a Tribunal may make such Order as it considers necessary for the purpose of its functions; and

WHEREAS the Tribunal referred to in the title hereof is satisfied that having regard to its findings and all other relevant matters there are sufficient reasons rendering it equitable that the Minister for Finance should pay the costs of the applicants as hereinafter appearing

IT IS HEREBY ORDERED that a Taxing Master of the High Court do tax in the manner hereinafter appearing the applicants' costs of appearing before the said Tribunal by one counsel instructed by solicitor.

In relation to the taxation of the applicants' costs the Tribunal DOTH ORDER as follows:

(*a*) That the statutory provisions and the Rules of the Superior Courts relating to the taxation of costs in an action in the High Courts (including the provisions relating to review and appeal) shall, in so far as is practicable, apply;

(*b*) That the costs of employing a solicitor and one counsel be taxed on a party and party basis.

(*c*) That the costs do include such witnesses' expenses as relate to those witnesses who actually gave oral evidence before the Tribunal on the proposal of the applicants;

(*d*) That the amount of witnesses' expenses be (*a*) such fee for the preparation of any statement prepared by the witness which was received by the Tribunal as part of the witnesses' testimony as the Taxing Master (or on review or appeal, the Court) considers reasonable; (*b*) such fee (not already paid) for attending before the Tribunal on such days as the witness gave oral evidence as the Taxing Master (or, on review or appeal, the Court) considers reasonable and (*b*) the travelling and subsistence expenses (not already paid) reasonably incurred for the purpose of giving oral evidence before the Tribunal.

(*e*) That the Minister for Finance be at liberty to attend and be heard on the said taxation and any review or appeal in relation to it.

(*f*) That the Minister for Finance do pay the applicants' costs when taxed and ascertained.

Dated this 29th day of July 1994.

Signed Liam Hamilton

The Tribunal

ANNEXE

IN THE MATTER OF the Tribunals of Inquiry (Evidence) Acts, 1921 and 1979; and IN THE MATTER OF a Tribunal of Inquiry established pursuant to Resolutions of Dail Eireann passed on the 24th May 1991, and by Seanad Eireann on the 29th May 1991, to inquire into allegations regarding illegal activities fraud and malpractice in and in connection with the beef processing industry by order of the Minister for Agriculture and Food made on the 31st day of May 1991.

AND WHEREAS it was provided by the said order that the Tribunals of Inquiry (Evidence) Acts 1921 and 1979 applied to the Tribunal

WHEREAS it is provided by Section 2 (*b*) of the Tribunals of Inquiry (Evidence) Act, 1921 that a Tribunal to which the Act is applied shall have power to authorise the representation before it by solicitor or counsel or otherwise of any person appearing to it to be interested; and

WHEREAS Mr Sean Barrett TD, Mr John Bruton TD, Senator Tom Raftery and Mr Paul Connaughton TD (hereinafter referred to as "the applicants") applied to the said Tribunal to be represented before it by counsel instructed by solicitor and on the 9th March 1992 (to Mr S Barrett, Mr J Bruton and Senator T Raftery) and the 13th January 1993 (to Mr P Connaughton) such authorisation under the said section was granted; and

WHEREAS it is provided by section 6 of the Tribunals of Inquiry (Evidence)(Amendment) Act, 1979, that if a Tribunal is of opinion that having regard to its findings and all other relevant matters there are sufficient reasons rendering it equitable to do so it may by Order direct that the whole or part of the costs of any person appearing before the Tribunal by counsel or solicitor as taxed by a Taxing Master of the High Court be paid by any other person named in the Order; and

WHEREAS by section 4 of the said Act of 1979 it is provided that a Tribunal may make such Order as it considers necessary for the purpose of its functions; and

WHEREAS the Tribunal referred to in the title hereof is satisfied that having regard to its findings and all other relevant matters there are sufficient reasons rendering it equitable that the Minister for Finance should pay the costs of the applicants as hereinafter appearing

IT IS HEREBY ORDERED that a Taxing Master of the High Court do tax in the manner hereinafter appearing the applicants' costs of appearing before the said Tribunal by one counsel instructed by solicitor.

In relation to the taxation of the applicants' costs the Tribunal DOTH ORDER as follows:

(*a*) That the statutory provisions and the Rules of the Superior Courts relating to the taxation of costs in an action in the High Courts (including the provisions relating to review and appeal) shall, in so far as is practicable, apply;

(*b*) That the costs of employing a solicitor and one counsel be taxed on a party and party basis, as being necessary and proper for enforcing before the Tribunal and applicants' right to ensure that all the facts concerning the allegations made by them were disclosed so far as the applicants were concerned;

(*c*) That the costs do include such witnesses' expenses as relate to those witnesses who actually gave oral evidence before the Tribunal on the proposal of the applicants;

(*d*) That the amount of witnesses' expenses be (*a*) such fee for the preparation of any statement prepared by the witness which was received by the Tribunal as part of the witnesses' testimony as the Taxing Master (or on review or appeal, the Court) considers reasonable; (*b*) such fee (not already paid) for attending before the Tribunal on such days as the witness gave oral evidence as the Taxing Master (or, on review or appeal, the Court) considers reasonable and (*b*) the travelling and subsistence expenses (not already paid) reasonably incurred for the purpose of giving oral evidence before the Tribunal.

(*e*) That the Minister for Finance be at liberty to attend and be heard on the said taxation and any review or appeal in relation to it.

(*f*) That the Minister for Finance do pay the applicants' costs when taxed and ascertained.

Dated this 29th day of July 1994.

Signed Liam Hamilton

The Tribunal

ANNEXE

IN THE MATTER OF the Tribunals of Inquiry (Evidence) Acts, 1921 and 1979; and IN THE MATTER OF a Tribunal of Inquiry established pursuant to Resolutions of Dail Eireann passed on the 24th May 1991, and by Seanad Eireann on the 29th May 1991, to inquire into allegations regarding illegal activities fraud and malpractice in and in connection with the beef processing industry by order of the Minister for Agriculture and Food made on the 31st day of May 1991.

AND WHEREAS it was provided by the said order that the Tribunals of Inquiry (Evidence) Acts 1921 and 1979 applied to the Tribunal

WHEREAS it is provided by Section 2 (*b*) of the Tribunals of Inquiry (Evidence) Act, 1921 that a Tribunal to which the Act is applied shall have power to authorise the representation before it by solicitor or counsel or otherwise of any person appearing to it to be interested; and

WHEREAS Mr Desmond O'Malley TD (hereinafter referred to as "the applicant") applied to the said Tribunal to be represented before it by counsel instructed by solicitor and on the 13th March 1992 such authorisation under the said section was granted; and

WHEREAS it is provided by section 6 of the Tribunals of Inquiry (Evidence)(Amendment) Act, 1979, that if a Tribunal is of opinion that having regard to its findings and all other relevant matters there are sufficient reasons rendering it equitable to do so it may by Order direct that the whole or part of the costs of any person appearing before the Tribunal by counsel or solicitor as taxed by a Taxing Master of the High Court be paid by any other person named in the Order; and

WHEREAS by section 4 of the said Act of 1979 it is provided that a Tribunal may make such Order as it considers necessary for the purpose of its functions; and

WHEREAS the Tribunal referred to in the title hereof is satisfied that having regard to its findings and all other relevant matters there are sufficient reasons rendering it equitable that the Minister for Finance should pay the costs of the applicant as hereinafter appearing

IT IS HEREBY ORDERED that a Taxing Master of the High Court do tax in the manner hereinafter appearing the applicant's costs of appearing before the said Tribunal by two counsel instructed by solicitor.

In relation to the taxation of the applicant's costs the Tribunal DOTH ORDER as follows:

(*a*) That the statutory provisions and the Rules of the Superior Courts relating to the taxation of costs in an action in the High Courts (including the provisions relating to review and appeal) shall, in so far as is practicable, apply;

(*b*) That the costs of employing a solicitor and two counsel be taxed on a party and party basis, as being necessary and proper for enforcing before the Tribunal and applicants' right to ensure that all the facts concerning the allegations made by him were disclosed so far as the applicant were concerned;

(*c*) That the costs do include such witnesses' expenses as relate to those witnesses who actually gave oral evidence before the Tribunal on the proposal of the applicant;

(*d*) That the amount of witnesses' expenses be (*a*) such fee for the preparation of any statement prepared by the witness which was received by the Tribunal as part of the witnesses' testimony as the Taxing Master (or on review or appeal, the Court) considers reasonable; (*b*) such fee (not already paid) for attending before the Tribunal on such days as the witness gave oral evidence as the Taxing Master (or, on review or appeal, the Court) considers reasonable and (*b*) the travelling and subsistence expenses (not already paid) reasonably incurred for the purpose of giving oral evidence before the Tribunal.

(*e*) That the Minister for Finance be at liberty to attend and be heard on the said taxation and any review or appeal in relation to it.

(*f*) That the Minister for Finance do pay the applicant's costs when taxed and ascertained.

Dated this 29th day of July 1994.

Signed Liam Hamilton

The Tribunal

ANNEXE

IN THE MATTER OF the Tribunals of Inquiry (Evidence) Acts, 1921 and 1979; and IN THE MATTER OF a Tribunal of Inquiry established pursuant to Resolutions of Dail Eireann passed on the 24th May 1991, and by Seanad Eireann on the 29th May 1991, to inquire into allegations regarding illegal activities fraud and malpractice in and in connection with the beef processing industry by order of the Minister for Agriculture and Food made on the 31st day of May 1991.

AND WHEREAS it was provided by the said order that the Tribunals of Inquiry (Evidence) Acts 1921 and 1979 applied to the Tribunal

WHEREAS it is provided by Section 2 (*b*) of the Tribunals of Inquiry (Evidence) Act, 1921 that a Tribunal to which the Act is applied shall have power to authorise the representation before it by solicitor or counsel or otherwise of any person appearing to it to be interested; and

WHEREAS Mr Ted O'Reilly (hereinafter referred to as "the applicant") applied to the said Tribunal to be represented before it by counsel instructed by solicitor and on the 30th March 1992 such authorisation under the said section was granted; and

WHEREAS it is provided by section 6 of the Tribunals of Inquiry (Evidence)(Amendment) Act, 1979, that if a Tribunal is of opinion that having regard to its findings and all other relevant matters there are sufficient reasons rendering it equitable to do so it may by Order direct that the whole or part of the costs of any person appearing before the Tribunal by counsel or solicitor as taxed by a Taxing Master of the High Court be paid by any other person named in the Order; and

WHEREAS by section 4 of the said Act of 1979 it is provided that a Tribunal may make such Order as it considers necessary for the purpose of its functions; and

WHEREAS the Tribunal referred to in the title hereof is satisfied that having regard to its findings and all other relevant matters there are sufficient reasons rendering it equitable that the Minister for Finance should pay the costs of the applicant as hereinafter appearing

IT IS HEREBY ORDERED that a Taxing Master of the High Court do tax in the manner hereinafter appearing the applicant's costs of appearing before the said Tribunal by one counsel instructed by solicitor.

In relation to the taxation of the applicant's costs the Tribunal DOTH ORDER as follows:

(*a*) That the statutory provisions and the Rules of the Superior Courts relating to the taxation of costs in an action in the High Courts (including the provisions relating to review and appeal) shall, in so far as is practicable, apply;

(*b*) That the costs of employing a solicitor and one counsel be taxed on a party and party basis.

(*c*) That the costs do include such witnesses' expenses as relate to those witnesses who actually gave oral evidence before the Tribunal on the proposal of the applicant;

(*d*) That the amount of witnesses' expenses be (*a*) such fee for the preparation of any statement prepared by the witness which was received by the Tribunal as part of the witnesses' testimony as the Taxing Master (or on review or appeal, the Court) considers reasonable; (*b*) such fee (not already paid) for attending before the Tribunal on such days as the witness gave oral evidence as the Taxing Master (or, on review or appeal, the Court) considers reasonable and (*b*) the travelling and subsistence expenses (not already paid) reasonably incurred for the purpose of giving oral evidence before the Tribunal.

(*e*) That the Minister for Finance be at liberty to attend and be heard on the said taxation and any review or appeal in relation to it.

(*f*) That the Minister for Finance do pay the applicant's costs when taxed and ascertained.

Dated this 29th day of July 1994.

Signed Liam Hamilton

The Tribunal

ANNEXE

IN THE MATTER OF the Tribunals of Inquiry (Evidence) Acts, 1921 and 1979; and IN THE MATTER OF a Tribunal of Inquiry established pursuant to Resolutions of Dail Eireann passed on the 24th May 1991, and by Seanad Eireann on the 29th May 1991, to inquire into allegations regarding illegal activities fraud and malpractice in and in connection with the beef processing industry by order of the Minister for Agriculture and Food made on the 31st day of May 1991.

AND WHEREAS it was provided by the said order that the Tribunals of Inquiry (Evidence) Acts 1921 and 1979 applied to the Tribunal

WHEREAS it is provided by Section 2 (*b*) of the Tribunals of Inquiry (Evidence) Act, 1921 that a Tribunal to which the Act is applied shall have power to authorise the representation before it by solicitor or counsel or otherwise of any person appearing to it to be interested; and

WHEREAS Mr John Stanley (hereinafter referred to as "the applicant") applied to the said Tribunal to be represented before it by counsel instructed by solicitor and on the 21st May 1992 such authorisation under the said section was granted; and

WHEREAS it is provided by section 6 of the Tribunals of Inquiry (Evidence)(Amendment) Act, 1979, that if a Tribunal is of opinion that having regard to its findings and all other relevant matters there are sufficient reasons rendering it equitable to do so it may by Order direct that the whole or part of the costs of any person appearing before the Tribunal by counsel or solicitor as taxed by a Taxing Master of the High Court be paid by any other person named in the Order; and

WHEREAS by section 4 of the said Act of 1979 it is provided that a Tribunal may make such Order as it considers necessary for the purpose of its functions; and

WHEREAS the Tribunal referred to in the title hereof is satisfied that having regard to its findings and all other relevant matters there are sufficient reasons rendering it equitable that the Minister for Finance should pay the costs of the applicant as hereinafter appearing

IT IS HEREBY ORDERED that a Taxing Master of the High Court do tax in the manner hereinafter appearing the applicant's costs of appearing before the said Tribunal by one counsel instructed by solicitor.

In relation to the taxation of the applicant's costs the Tribunal DOTH ORDER as follows:

(*a*) That the statutory provisions and the Rules of the Superior Courts relating to the taxation of costs in an action in the High Courts (including the provisions relating to review and appeal) shall, in so far as is practicable, apply;

(*b*) That the costs of employing a solicitor and one counsel be taxed on a party and party basis.

(*c*) That the costs do include such witnesses' expenses as relate to those witnesses who actually gave oral evidence before the Tribunal on the proposal of the applicant;

(*d*) That the amount of witnesses' expenses be (*a*) such fee for the preparation of any statement prepared by the witness which was received by the Tribunal as part of the witnesses' testimony as the Taxing Master (or on review or appeal, the Court) considers reasonable; (*b*) such fee (not already paid) for attending before the Tribunal on such days as the witness gave oral evidence as the Taxing Master (or, on review or appeal, the Court) considers reasonable and (*b*) the travelling and subsistence expenses (not already paid) reasonably incurred for the purpose of giving oral evidence before the Tribunal.

(*e*) That the Minister for Finance be at liberty to attend and be heard on the said taxation and any review or appeal in relation to it.

(*f*) That the Minister for Finance do pay the applicant's costs when taxed and ascertained.

Dated this 29th day of July 1994.

Signed Liam Hamilton

The Tribunal

ANNEXE

IN THE MATTER OF the Tribunals of Inquiry (Evidence) Acts, 1921 and 1979; and IN THE MATTER OF a Tribunal of Inquiry established pursuant to Resolutions of Dail Eireann passed on the 24th May 1991, and by Seanad Eireann on the 29th May 1991, to inquire into allegations regarding illegal activities fraud and malpractice in and in connection with the beef processing industry by order of the Minister for Agriculture and Food made on the 31st day of May 1991.

AND WHEREAS it was provided by the said order that the Tribunals of Inquiry (Evidence) Acts 1921 and 1979 applied to the Tribunal

WHEREAS it is provided by Section 2 (*b*) of the Tribunals of Inquiry (Evidence) Act, 1921 that a Tribunal to which the Act is applied shall have power to authorise the representation before it by solicitor or counsel or otherwise of any person appearing to it to be interested; and

WHEREAS Mr Sean Clarke (United Meat Packers) and Mr Satar Mohammed Khalid (United Meat Packers) and Mr Sean Clarke (Halal) (hereinafter referred to as "the applicants") applied to the said Tribunal to be represented before it by counsel instructed by solicitor and on the 25th May 1992 [to Mr S Clarke (UMP)], the 24th June 1992 [to Mr S M Khalid] and the 10th June 1993 [to Mr S Clarke (Halal)] such authorisation under the said section was granted; and

WHEREAS it is provided by section 6 of the Tribunals of Inquiry (Evidence)(Amendment) Act, 1979, that if a Tribunal is of opinion that having regard to its findings and all other relevant matters there are sufficient reasons rendering it equitable to do so it may by Order direct that the whole or part of the costs of any person appearing before the Tribunal by counsel or solicitor as taxed by a Taxing Master of the High Court be paid by any other person named in the Order; and

WHEREAS by section 4 of the said Act of 1979 it is provided that a Tribunal may make such Order as it considers necessary for the purpose of its functions; and

WHEREAS the Tribunal referred to in the title hereof is satisfied that having regard to its findings and all other relevant matters there are sufficient reasons rendering it equitable that the Minister for Finance should pay the costs of the applicants as hereinafter appearing

IT IS HEREBY ORDERED that a Taxing Master of the High Court do tax in the manner hereinafter appearing the applicants' costs of appearing before the said Tribunal by one counsel instructed by solicitor.

In relation to the taxation of the applicants' costs the Tribunal DOTH ORDER as follows:

(*a*) That the statutory provisions and the Rules of the Superior Courts relating to the taxation of costs in an action in the High Courts (including the provisions relating to review and appeal) shall, in so far as is practicable, apply;

(*b*) That the costs of employing a solicitor and one counsel be taxed on a party and party basis.

(*c*) That the costs do include such witnesses' expenses as relate to those witnesses who actually gave oral evidence before the Tribunal on the proposal of the applicants;

(*d*) That the amount of witnesses' expenses be (*a*) such fee for the preparation of any statement prepared by the witness which was received by the Tribunal as part of the witnesses' testimony as the Taxing Master (or on review or appeal, the Court) considers reasonable; (*b*) such fee (not already paid) for attending before the Tribunal on such days as the witness gave oral evidence as the Taxing Master (or, on review or appeal, the Court) considers reasonable and (*b*) the travelling and subsistence expenses (not already paid) reasonably incurred for the purpose of giving oral evidence before the Tribunal.

(*e*) That the Minister for Finance be at liberty to attend and be heard on the said taxation and any review or appeal in relation to it.

(*f*) That the Minister for Finance do pay the applicants' costs when taxed and ascertained.

Dated this 29th day of July 1994.

Signed Liam Hamilton

The Tribunal

ANNEXE

IN THE MATTER OF the Tribunals of Inquiry (Evidence) Acts, 1921 and 1979; and IN THE MATTER OF a Tribunal of Inquiry established pursuant to Resolutions of Dail Eireann passed on the 24th May 1991, and by Seanad Eireann on the 29th May 1991, to inquire into allegations regarding illegal activities fraud and malpractice in and in connection with the beef processing industry by order of the Minister for Agriculture and Food made on the 31st day of May 1991.

AND WHEREAS it was provided by the said order that the Tribunals of Inquiry (Evidence) Acts 1921 and 1979 applied to the Tribunal

WHEREAS it is provided by Section 2 (*b*) of the Tribunals of Inquiry (Evidence) Act, 1921 that a Tribunal to which the Act is applied shall have power to authorise the representation before it by solicitor or counsel or otherwise of any person appearing to it to be interested; and

WHEREAS Mr Tim Boland (hereinafter referred to as "the applicant") applied to the said Tribunal to be represented before it by counsel instructed by solicitor and on the 26th May 1992 such authorisation under the said section was granted; and

WHEREAS it is provided by section 6 of the Tribunals of Inquiry (Evidence)(Amendment) Act, 1979, that if a Tribunal is of opinion that having regard to its findings and all other relevant matters there are sufficient reasons rendering it equitable to do so it may by Order direct that the whole or part of the costs of any person appearing before the Tribunal by counsel or solicitor as taxed by a Taxing Master of the High Court be paid by any other person named in the Order; and

WHEREAS by section 4 of the said Act of 1979 it is provided that a Tribunal may make such Order as it considers necessary for the purpose of its functions; and

WHEREAS the Tribunal referred to in the title hereof is satisfied that having regard to its findings and all other relevant matters there are sufficient reasons rendering it equitable that the Minister for Finance should pay the costs of the applicant as hereinafter appearing

IT IS HEREBY ORDERED that a Taxing Master of the High Court do tax in the manner hereinafter appearing the applicant's costs of appearing before the said Tribunal by one counsel instructed by solicitor.

In relation to the taxation of the applicant's costs the Tribunal DOTH ORDER as follows:

(*a*) That the statutory provisions and the Rules of the Superior Courts relating to the taxation of costs in an action in the High Courts (including the provisions relating to review and appeal) shall, in so far as is practicable, apply;

(*b*) That the costs of employing a solicitor and one counsel be taxed on a party and party basis.

(*c*) That the costs do include such witnesses' expenses as relate to those witnesses who actually gave oral evidence before the Tribunal on the proposal of the applicant;

(*d*) That the amount of witnesses' expenses be (*a*) such fee for the preparation of any statement prepared by the witness which was received by the Tribunal as part of the witnesses' testimony as the Taxing Master (or on review or appeal, the Court) considers reasonable; (*b*) such fee (not already paid) for attending before the Tribunal on such days as the witness gave oral evidence as the Taxing Master (or, on review or appeal, the Court) considers reasonable and (*b*) the travelling and subsistence expenses (not already paid) reasonably incurred for the purpose of giving oral evidence before the Tribunal.

(*e*) That the Minister for Finance be at liberty to attend and be heard on the said taxation and any review or appeal in relation to it.

(*f*) That the Minister for Finance do pay the applicant's costs when taxed and ascertained.

Dated this 29th day of July 1994.

Signed Liam Hamilton
The Tribunal

ANNEXE

IN THE MATTER OF the Tribunals of Inquiry (Evidence) Acts, 1921 and 1979; and IN THE MATTER OF a Tribunal of Inquiry established pursuant to Resolutions of Dail Eireann passed on the 24th May 1991, and by Seanad Eireann on the 29th May 1991, to inquire into allegations regarding illegal activities fraud and malpractice in and in connection with the beef processing industry by order of the Minister for Agriculture and Food made on the 31st day of May 1991.

AND WHEREAS it was provided by the said order that the Tribunals of Inquiry (Evidence) Acts 1921 and 1979 applied to the Tribunal

WHEREAS it is provided by Section 2 (*b*) of the Tribunals of Inquiry (Evidence) Act, 1921 that a Tribunal to which the Act is applied shall have power to authorise the representation before it by solicitor or counsel or otherwise of any person appearing to it to be interested; and

WHEREAS Agra Trading Limited (hereinafter referred to as "the applicants") applied to the said Tribunal to be represented before it by counsel instructed by solicitor and on the 27th May 1992 such authorisation under the said section was granted; and

WHEREAS it is provided by section 6 of the Tribunals of Inquiry (Evidence)(Amendment) Act, 1979, that if a Tribunal is of opinion that having regard to its findings and all other relevant matters there are sufficient reasons rendering it equitable to do so it may by Order direct that the whole or part of the costs of any person appearing before the Tribunal by counsel or solicitor as taxed by a Taxing Master of the High Court be paid by any other person named in the Order; and

WHEREAS by section 4 of the said Act of 1979 it is provided that a Tribunal may make such Order as it considers necessary for the purpose of its functions; and

WHEREAS the Tribunal referred to in the title hereof is satisfied that having regard to its findings and all other relevant matters there are sufficient reasons rendering it equitable that the Minister for Finance should pay the costs of the applicants as hereinafter appearing

IT IS HEREBY ORDERED that a Taxing Master of the High Court do tax in the manner hereinafter appearing the applicants' costs of appearing before the said Tribunal by one counsel instructed by solicitor.

In relation to the taxation of the applicants' costs the Tribunal DOTH ORDER as follows:

(*a*) That the statutory provisions and the Rules of the Superior Courts relating to the taxation of costs in an action in the High Courts (including the provisions relating to review and appeal) shall, in so far as is practicable, apply;

(*b*) That the costs of employing a solicitor and one counsel be taxed on a party and party basis.

(*c*) That the costs do include such witnesses' expenses as relate to those witnesses who actually gave oral evidence before the Tribunal on the proposal of the applicants;

(*d*) That the amount of witnesses' expenses be (*a*) such fee for the preparation of any statement prepared by the witness which was received by the Tribunal as part of the witnesses' testimony as the Taxing Master (or on review or appeal, the Court) considers reasonable; (*b*) such fee (not already paid) for attending before the Tribunal on such days as the witness gave oral evidence as the Taxing Master (or, on review or appeal, the Court) considers reasonable and (*b*) the travelling and subsistence expenses (not already paid) reasonably incurred for the purpose of giving oral evidence before the Tribunal.

(*e*) That the Minister for Finance be at liberty to attend and be heard on the said taxation and any review or appeal in relation to it.

(*f*) That the Minister for Finance do pay the applicants' costs when taxed and ascertained.

Dated this 29th day of July 1994.

Signed Liam Hamilton
The Tribunal

ANNEXE

IN THE MATTER OF the Tribunals of Inquiry (Evidence) Acts, 1921 and 1979; and IN THE MATTER OF a Tribunal of Inquiry established pursuant to Resolutions of Dail Eireann passed on the 24th May 1991, and by Seanad Eireann on the 29th May 1991, to inquire into allegations regarding illegal activities fraud and malpractice in and in connection with the beef processing industry by order of the Minister for Agriculture and Food made on the 31st day of May 1991.

AND WHEREAS it was provided by the said order that the Tribunals of Inquiry (Evidence) Acts 1921 and 1979 applied to the Tribunal

WHEREAS it is provided by Section 2 (*b*) of the Tribunals of Inquiry (Evidence) Act, 1921 that a Tribunal to which the Act is applied shall have power to authorise the representation before it by solicitor or counsel or otherwise of any person appearing to it to be interested; and

WHEREAS Mr Oliver Murphy (hereinafter referred to as "the applicant") applied to the said Tribunal to be represented before it by counsel instructed by solicitor and on the 27th May 1992 such authorisation under the said section was granted; and

WHEREAS it is provided by section 6 of the Tribunals of Inquiry (Evidence)(Amendment) Act, 1979, that if a Tribunal is of opinion that having regard to its findings and all other relevant matters there are sufficient reasons rendering it equitable to do so it may by Order direct that the whole or part of the costs of any person appearing before the Tribunal by counsel or solicitor as taxed by a Taxing Master of the High Court be paid by any other person named in the Order; and

WHEREAS by section 4 of the said Act of 1979 it is provided that a Tribunal may make such Order as it considers necessary for the purpose of its functions; and

WHEREAS the Tribunal referred to in the title hereof is satisfied that having regard to its findings and all other relevant matters there are sufficient reasons rendering it equitable that the Minister for Finance should pay the costs of the applicant as hereinafter appearing

IT IS HEREBY ORDERED that a Taxing Master of the High Court do tax in the manner hereinafter appearing the applicant's costs of appearing before the said Tribunal by one counsel instructed by solicitor.

In relation to the taxation of the applicant's costs the Tribunal DOTH ORDER as follows:

(*a*) That the statutory provisions and the Rules of the Superior Courts relating to the taxation of costs in an action in the High Courts (including the provisions relating to review and appeal) shall, in so far as is practicable, apply;

(*b*) That the costs of employing a solicitor and one counsel be taxed on a party and party basis.

(*c*) That the costs do include such witnesses' expenses as relate to those witnesses who actually gave oral evidence before the Tribunal on the proposal of the applicant;

(*d*) That the amount of witnesses' expenses be (*a*) such fee for the preparation of any statement prepared by the witness which was received by the Tribunal as part of the witnesses' testimony as the Taxing Master (or on review or appeal, the Court) considers reasonable; (*b*) such fee (not already paid) for attending before the Tribunal on such days as the witness gave oral evidence as the Taxing Master (or, on review or appeal, the Court) considers reasonable and (*b*) the travelling and subsistence expenses (not already paid) reasonably incurred for the purpose of giving oral evidence before the Tribunal.

(*e*) That the Minister for Finance be at liberty to attend and be heard on the said taxation and any review or appeal in relation to it.

(*f*) That the Minister for Finance do pay the applicant's costs when taxed and ascertained.

Dated this 29th day of July 1994.

Signed Liam Hamilton

The Tribunal

ANNEXE

IN THE MATTER OF the Tribunals of Inquiry (Evidence) Acts, 1921 and 1979; and IN THE MATTER OF a Tribunal of Inquiry established pursuant to Resolutions of Dail Eireann passed on the 24th May 1991, and by Seanad Eireann on the 29th May 1991, to inquire into allegations regarding illegal activities fraud and malpractice in and in connection with the beef processing industry by order of the Minister for Agriculture and Food made on the 31st day of May 1991.

AND WHEREAS it was provided by the said order that the Tribunals of Inquiry (Evidence) Acts 1921 and 1979 applied to the Tribunal

WHEREAS it is provided by Section 2 (*b*) of the Tribunals of Inquiry (Evidence) Act, 1921 that a Tribunal to which the Act is applied shall have power to authorise the representation before it by solicitor or counsel or otherwise of any person appearing to it to be interested; and

WHEREAS Hibernia Meats (hereinafter referred to as "the applicants") applied to the said Tribunal to be represented before it by counsel instructed by solicitor and on the 15th June 1993 such authorisation under the said section was granted; and

WHEREAS it is provided by section 6 of the Tribunals of Inquiry (Evidence)(Amendment) Act, 1979, that if a Tribunal is of opinion that having regard to its findings and all other relevant matters there are sufficient reasons rendering it equitable to do so it may by Order direct that the whole or part of the costs of any person appearing before the Tribunal by counsel or solicitor as taxed by a Taxing Master of the High Court be paid by any other person named in the Order; and

WHEREAS by section 4 of the said Act of 1979 it is provided that a Tribunal may make such Order as it considers necessary for the purpose of its functions; and

WHEREAS the Tribunal referred to in the title hereof is satisfied that having regard to its findings and all other relevant matters there are sufficient reasons rendering it equitable that the Minister for Finance should pay the costs of the applicants as hereinafter appearing

IT IS HEREBY ORDERED that a Taxing Master of the High Court do tax in the manner hereinafter appearing the applicants' costs of appearing before the said Tribunal by one counsel instructed by solicitor.

In relation to the taxation of the applicants' costs the Tribunal DOTH ORDER as follows:

(*a*) That the statutory provisions and the Rules of the Superior Courts relating to the taxation of costs in an action in the High Courts (including the provisions relating to review and appeal) shall, in so far as is practicable, apply;

(*b*) That the costs of employing a solicitor and one counsel be taxed on a party and party basis.

(*c*) That the costs do include such witnesses' expenses as relate to those witnesses who actually gave oral evidence before the Tribunal on the proposal of the applicants;

(*d*) That the amount of witnesses' expenses be (*a*) such fee for the preparation of any statement prepared by the witness which was received by the Tribunal as part of the witnesses' testimony as the Taxing Master (or on review or appeal, the Court) considers reasonable; (*b*) such fee (not already paid) for attending before the Tribunal on such days as the witness gave oral evidence as the Taxing Master (or, on review or appeal, the Court) considers reasonable and (*b*) the travelling and subsistence expenses (not already paid) reasonably incurred for the purpose of giving oral evidence before the Tribunal.

(*e*) That the Minister for Finance be at liberty to attend and be heard on the said taxation and any review or appeal in relation to it.

(*f*) That the Minister for Finance do pay the applicants' costs when taxed and ascertained.

Dated this 29th day of July 1994.

Signed Liam Hamilton

The Tribunal

ANNEXE

IN THE MATTER OF the Tribunals of Inquiry (Evidence) Acts, 1921 and 1979; and IN THE MATTER OF a Tribunal of Inquiry established pursuant to Resolutions of Dail Eireann passed on the 24th May 1991, and by Seanad Eireann on the 29th May 1991, to inquire into allegations regarding illegal activities fraud and malpractice in and in connection with the beef processing industry by order of the Minister for Agriculture and Food made on the 31st day of May 1991.

AND WHEREAS it was provided by the said order that the Tribunals of Inquiry (Evidence) Acts 1921 and 1979 applied to the Tribunal

WHEREAS it is provided by Section 2 (*b*) of the Tribunals of Inquiry (Evidence) Act, 1921 that a Tribunal to which the Act is applied shall have power to authorise the representation before it by solicitor or counsel or otherwise of any person appearing to it to be interested; and

WHEREAS Mr Pascal Phelan (hereinafter referred to as "the applicant") applied to the said Tribunal to be represented before it by counsel instructed by solicitor and on the 22nd June 1992 such authorisation under the said section was granted; and

WHEREAS it is provided by section 6 of the Tribunals of Inquiry (Evidence)(Amendment) Act, 1979, that if a Tribunal is of opinion that having regard to its findings and all other relevant matters there are sufficient reasons rendering it equitable to do so it may by Order direct that the whole or part of the costs of any person appearing before the Tribunal by counsel or solicitor as taxed by a Taxing Master of the High Court be paid by any other person named in the Order; and

WHEREAS by section 4 of the said Act of 1979 it is provided that a Tribunal may make such Order as it considers necessary for the purpose of its functions; and

WHEREAS the Tribunal referred to in the title hereof is satisfied that having regard to its findings and all other relevant matters there are sufficient reasons rendering it equitable that the Minister for Finance should pay the costs of the applicant as hereinafter appearing

IT IS HEREBY ORDERED that a Taxing Master of the High Court do tax in the manner hereinafter appearing the applicant's costs of appearing before the said Tribunal by one counsel instructed by solicitor.

In relation to the taxation of the applicant's costs the Tribunal DOTH ORDER as follows:

(*a*) That the statutory provisions and the Rules of the Superior Courts relating to the taxation of costs in an action in the High Courts (including the provisions relating to review and appeal) shall, in so far as is practicable, apply;

(*b*) That the costs of employing a solicitor and one counsel be taxed on a party and party basis.

(*c*) That the costs do include such witnesses' expenses as relate to those witnesses who actually gave oral evidence before the Tribunal on the proposal of the applicant;

(*d*) That the amount of witnesses' expenses be (*a*) such fee for the preparation of any statement prepared by the witness which was received by the Tribunal as part of the witnesses' testimony as the Taxing Master (or on review or appeal, the Court) considers reasonable; (*b*) such fee (not already paid) for attending before the Tribunal on such days as the witness gave oral evidence as the Taxing Master (or, on review or appeal, the Court) considers reasonable and (*b*) the travelling and subsistence expenses (not already paid) reasonably incurred for the purpose of giving oral evidence before the Tribunal.

(*e*) That the Minister for Finance be at liberty to attend and be heard on the said taxation and any review or appeal in relation to it.

(*f*) That the Minister for Finance do pay the applicant's costs when taxed and ascertained.

Dated this 29th day of July 1994.

Signed Liam Hamilton

The Tribunal

ANNEXE

IN THE MATTER OF the Tribunals of Inquiry (Evidence) Acts, 1921 and 1979; and IN THE MATTER OF a Tribunal of Inquiry established pursuant to Resolutions of Dail Eireann passed on the 24th May 1991, and by Seanad Eireann on the 29th May 1991, to inquire into allegations regarding illegal activities fraud and malpractice in and in connection with the beef processing industry by order of the Minister for Agriculture and Food made on the 31st day of May 1991.

AND WHEREAS it was provided by the said order that the Tribunals of Inquiry (Evidence) Acts 1921 and 1979 applied to the Tribunal

WHEREAS it is provided by Section 2 (*b*) of the Tribunals of Inquiry (Evidence) Act, 1921 that a Tribunal to which the Act is applied shall have power to authorise the representation before it by solicitor or counsel or otherwise of any person appearing to it to be interested; and

WHEREAS Master Meats (Pre 16/9/1988) (hereinafter referred to as "the applicants") applied to the said Tribunal to be represented before it by counsel instructed by solicitor and on the 17th June 1993 such authorisation under the said section was granted; and

WHEREAS it is provided by section 6 of the Tribunals of Inquiry (Evidence)(Amendment) Act, 1979, that if a Tribunal is of opinion that having regard to its findings and all other relevant matters there are sufficient reasons rendering it equitable to do so it may by Order direct that the whole or part of the costs of any person appearing before the Tribunal by counsel or solicitor as taxed by a Taxing Master of the High Court be paid by any other person named in the Order; and

WHEREAS by section 4 of the said Act of 1979 it is provided that a Tribunal may make such Order as it considers necessary for the purpose of its functions; and

WHEREAS the Tribunal referred to in the title hereof is satisfied that having regard to its findings and all other relevant matters there are sufficient reasons rendering it equitable that the Minister for Finance should pay the costs of the applicants as hereinafter appearing

IT IS HEREBY ORDERED that a Taxing Master of the High Court do tax in the manner hereinafter appearing the applicants' costs of appearing before the said Tribunal by one counsel instructed by solicitor.

In relation to the taxation of the applicants' costs the Tribunal DOTH ORDER as follows:

(*a*) That the statutory provisions and the Rules of the Superior Courts relating to the taxation of costs in an action in the High Courts (including the provisions relating to review and appeal) shall, in so far as is practicable, apply;

(*b*) That the costs of employing a solicitor and one counsel be taxed on a party and party basis.

(*c*) That the costs do include such witnesses' expenses as relate to those witnesses who actually gave oral evidence before the Tribunal on the proposal of the applicants;

(*d*) That the amount of witnesses' expenses be (*a*) such fee for the preparation of any statement prepared by the witness which was received by the Tribunal as part of the witnesses' testimony as the Taxing Master (or on review or appeal, the Court) considers reasonable; (*b*) such fee (not already paid) for attending before the Tribunal on such days as the witness gave oral evidence as the Taxing Master (or, on review or appeal, the Court) considers reasonable and (*b*) the travelling and subsistence expenses (not already paid) reasonably incurred for the purpose of giving oral evidence before the Tribunal.

(*e*) That the Minister for Finance be at liberty to attend and be heard on the said taxation and any review or appeal in relation to it.

(*f*) That the Minister for Finance do pay the applicants' costs when taxed and ascertained.

Dated this 29th day of July 1994.

Signed Liam Hamilton

The Tribunal

ANNEXE

IN THE MATTER OF the Tribunals of Inquiry (Evidence) Acts, 1921 and 1979; and IN THE MATTER OF a Tribunal of Inquiry established pursuant to Resolutions of Dail Eireann passed on the 24th May 1991, and by Seanad Eireann on the 29th May 1991, to inquire into allegations regarding illegal activities fraud and malpractice in and in connection with the beef processing industry by order of the Minister for Agriculture and Food made on the 31st day of May 1991.

AND WHEREAS it was provided by the said order that the Tribunals of Inquiry (Evidence) Acts 1921 and 1979 applied to the Tribunal

WHEREAS it is provided by Section 2 (*b*) of the Tribunals of Inquiry (Evidence) Act, 1921 that a Tribunal to which the Act is applied shall have power to authorise the representation before it by solicitor or counsel or otherwise of any person appearing to it to be interested; and

WHEREAS Mr Owen Patten (hereinafter referred to as "the applicant") applied to the said Tribunal to be represented before it by counsel instructed by solicitor and on the 23rd June 1992 such authorisation under the said section was granted; and

WHEREAS it is provided by section 6 of the Tribunals of Inquiry (Evidence)(Amendment) Act, 1979, that if a Tribunal is of opinion that having regard to its findings and all other relevant matters there are sufficient reasons rendering it equitable to do so it may by Order direct that the whole or part of the costs of any person appearing before the Tribunal by counsel or solicitor as taxed by a Taxing Master of the High Court be paid by any other person named in the Order; and

WHEREAS by section 4 of the said Act of 1979 it is provided that a Tribunal may make such Order as it considers necessary for the purpose of its functions; and

WHEREAS the Tribunal referred to in the title hereof is satisfied that having regard to its findings and all other relevant matters there are sufficient reasons rendering it equitable that the Minister for Finance should pay the costs of the applicant as hereinafter appearing

IT IS HEREBY ORDERED that a Taxing Master of the High Court do tax in the manner hereinafter appearing the applicant's costs of appearing before the said Tribunal by one counsel instructed by solicitor.

In relation to the taxation of the applicant's costs the Tribunal DOTH ORDER as follows:

(*a*) That the statutory provisions and the Rules of the Superior Courts relating to the taxation of costs in an action in the High Courts (including the provisions relating to review and appeal) shall, in so far as is practicable, apply;

(*b*) That the costs of employing a solicitor and one counsel be taxed on a party and party basis.

(*c*) That the costs do include such witnesses' expenses as relate to those witnesses who actually gave oral evidence before the Tribunal on the proposal of the applicant;

(*d*) That the amount of witnesses' expenses be (*a*) such fee for the preparation of any statement prepared by the witness which was received by the Tribunal as part of the witnesses' testimony as the Taxing Master (or on review or appeal, the Court) considers reasonable; (*b*) such fee (not already paid) for attending before the Tribunal on such days as the witness gave oral evidence as the Taxing Master (or, on review or appeal, the Court) considers reasonable and (*b*) the travelling and subsistence expenses (not already paid) reasonably incurred for the purpose of giving oral evidence before the Tribunal.

(*e*) That the Minister for Finance be at liberty to attend and be heard on the said taxation and any review or appeal in relation to it.

(*f*) That the Minister for Finance do pay the applicant's costs when taxed and ascertained.

Dated this 29th day of July 1994.

Signed Liam Hamilton

The Tribunal

ANNEXE

IN THE MATTER OF the Tribunals of Inquiry (Evidence) Acts, 1921 and 1979; and IN THE MATTER OF a Tribunal of Inquiry established pursuant to Resolutions of Dail Eireann passed on the 24th May 1991, and by Seanad Eireann on the 29th May 1991, to inquire into allegations regarding illegal activities fraud and malpractice in and in connection with the beef processing industry by order of the Minister for Agriculture and Food made on the 31st day of May 1991.

AND WHEREAS it was provided by the said order that the Tribunals of Inquiry (Evidence) Acts 1921 and 1979 applied to the Tribunal

WHEREAS it is provided by Section 2 (*b*) of the Tribunals of Inquiry (Evidence) Act, 1921 that a Tribunal to which the Act is applied shall have power to authorise the representation before it by solicitor or counsel or otherwise of any person appearing to it to be interested; and

WHEREAS Mr Michael O'Kennedy TD (hereinafter referred to as "the applicant") applied to the said Tribunal to be represented before it by counsel instructed by solicitor and on the 8th September 1992 such authorisation under the said section was granted; and

WHEREAS it is provided by section 6 of the Tribunals of Inquiry (Evidence)(Amendment) Act, 1979, that if a Tribunal is of opinion that having regard to its findings and all other relevant matters there are sufficient reasons rendering it equitable to do so it may by Order direct that the whole or part of the costs of any person appearing before the Tribunal by counsel or solicitor as taxed by a Taxing Master of the High Court be paid by any other person named in the Order; and

WHEREAS by section 4 of the said Act of 1979 it is provided that a Tribunal may make such Order as it considers necessary for the purpose of its functions; and

WHEREAS the Tribunal referred to in the title hereof is satisfied that having regard to its findings and all other relevant matters there are sufficient reasons rendering it equitable that the Minister for Finance should pay the costs of the applicant as hereinafter appearing

IT IS HEREBY ORDERED that a Taxing Master of the High Court do tax in the manner hereinafter appearing the applicant's costs of appearing before the said Tribunal by one counsel instructed by solicitor.

In relation to the taxation of the applicant's costs the Tribunal DOTH ORDER as follows:

(*a*) That the statutory provisions and the Rules of the Superior Courts relating to the taxation of costs in an action in the High Courts (including the provisions relating to review and appeal) shall, in so far as is practicable, apply;

(*b*) That the costs of employing a solicitor and one counsel be taxed on a party and party basis.

(*c*) That the costs do include such witnesses' expenses as relate to those witnesses who actually gave oral evidence before the Tribunal on the proposal of the applicant;

(*d*) That the amount of witnesses' expenses be (*a*) such fee for the preparation of any statement prepared by the witness which was received by the Tribunal as part of the witnesses' testimony as the Taxing Master (or on review or appeal, the Court) considers reasonable; (*b*) such fee (not already paid) for attending before the Tribunal on such days as the witness gave oral evidence as the Taxing Master (or, on review or appeal, the Court) considers reasonable and (*b*) the travelling and subsistence expenses (not already paid) reasonably incurred for the purpose of giving oral evidence before the Tribunal.

(*e*) That the Minister for Finance be at liberty to attend and be heard on the said taxation and any review or appeal in relation to it.

(*f*) That the Minister for Finance do pay the applicant's costs when taxed and ascertained.

Dated this 29th day of July 1994.

Signed Liam Hamilton

The Tribunal

ANNEXE

IN THE MATTER OF the Tribunals of Inquiry (Evidence) Acts, 1921 and 1979; and IN THE MATTER OF a Tribunal of Inquiry established pursuant to Resolutions of Dail Eireann passed on the 24th May 1991, and by Seanad Eireann on the 29th May 1991, to inquire into allegations regarding illegal activities fraud and malpractice in and in connection with the beef processing industry by order of the Minister for Agriculture and Food made on the 31st day of May 1991.

AND WHEREAS it was provided by the said order that the Tribunals of Inquiry (Evidence) Acts 1921 and 1979 applied to the Tribunal

WHEREAS it is provided by Section 2 (*b*) of the Tribunals of Inquiry (Evidence) Act, 1921 that a Tribunal to which the Act is applied shall have power to authorise the representation before it by solicitor or counsel or otherwise of any person appearing to it to be interested; and

WHEREAS Mr Brian Britton (hereinafter referred to as "the applicant") applied to the said Tribunal to be represented before it by counsel instructed by solicitor and on the 8th September 1992 such authorisation under the said section was granted; and

WHEREAS it is provided by section 6 of the Tribunals of Inquiry (Evidence)(Amendment) Act, 1979, that if a Tribunal is of opinion that having regard to its findings and all other relevant matters there are sufficient reasons rendering it equitable to do so it may by Order direct that the whole or part of the costs of any person appearing before the Tribunal by counsel or solicitor as taxed by a Taxing Master of the High Court be paid by any other person named in the Order; and

WHEREAS by section 4 of the said Act of 1979 it is provided that a Tribunal may make such Order as it considers necessary for the purpose of its functions; and

WHEREAS the Tribunal referred to in the title hereof is satisfied that having regard to its findings and all other relevant matters there are sufficient reasons rendering it equitable that the Minister for Finance should pay the costs of the applicant as hereinafter appearing

IT IS HEREBY ORDERED that a Taxing Master of the High Court do tax in the manner hereinafter appearing the applicant's costs of appearing before the said Tribunal by one counsel instructed by solicitor.

In relation to the taxation of the applicant's costs the Tribunal DOTH ORDER as follows:

(*a*) That the statutory provisions and the Rules of the Superior Courts relating to the taxation of costs in an action in the High Courts (including the provisions relating to review and appeal) shall, in so far as is practicable, apply;

(*b*) That the costs of employing a solicitor and one counsel be taxed on a party and party basis.

(*c*) That the costs do include such witnesses' expenses as relate to those witnesses who actually gave oral evidence before the Tribunal on the proposal of the applicant;

(*d*) That the amount of witnesses' expenses be (*a*) such fee for the preparation of any statement prepared by the witness which was received by the Tribunal as part of the witnesses' testimony as the Taxing Master (or on review or appeal, the Court) considers reasonable; (*b*) such fee (not already paid) for attending before the Tribunal on such days as the witness gave oral evidence as the Taxing Master (or, on review or appeal, the Court) considers reasonable and (*b*) the travelling and subsistence expenses (not already paid) reasonably incurred for the purpose of giving oral evidence before the Tribunal.

(*e*) That the Minister for Finance be at liberty to attend and be heard on the said taxation and any review or appeal in relation to it.

(*f*) That the Minister for Finance do pay the applicant's costs when taxed and ascertained.

Dated this 29th day of July 1994.

Signed Liam Hamilton

The Tribunal

ANNEXE

IN THE MATTER OF the Tribunals of Inquiry (Evidence) Acts, 1921 and 1979; and IN THE MATTER OF a Tribunal of Inquiry established pursuant to Resolutions of Dail Eireann passed on the 24th May 1991, and by Seanad Eireann on the 29th May 1991, to inquire into allegations regarding illegal activities fraud and malpractice in and in connection with the beef processing industry by order of the Minister for Agriculture and Food made on the 31st day of May 1991.

AND WHEREAS it was provided by the said order that the Tribunals of Inquiry (Evidence) Acts 1921 and 1979 applied to the Tribunal

WHEREAS it is provided by Section 2 (*b*) of the Tribunals of Inquiry (Evidence) Act, 1921 that a Tribunal to which the Act is applied shall have power to authorise the representation before it by solicitor or counsel or otherwise of any person appearing to it to be interested; and

WHEREAS Mr Patrick Farrell (hereinafter referred to as "the applicant") applied to the said Tribunal to be represented before it by counsel instructed by solicitor and on the 16th September 1992 such authorisation under the said section was granted; and

WHEREAS it is provided by section 6 of the Tribunals of Inquiry (Evidence)(Amendment) Act, 1979, that if a Tribunal is of opinion that having regard to its findings and all other relevant matters there are sufficient reasons rendering it equitable to do so it may by Order direct that the whole or part of the costs of any person appearing before the Tribunal by counsel or solicitor as taxed by a Taxing Master of the High Court be paid by any other person named in the Order; and

WHEREAS by section 4 of the said Act of 1979 it is provided that a Tribunal may make such Order as it considers necessary for the purpose of its functions; and

WHEREAS the Tribunal referred to in the title hereof is satisfied that having regard to its findings and all other relevant matters there are sufficient reasons rendering it equitable that the Minister for Finance should pay the costs of the applicant as hereinafter appearing

IT IS HEREBY ORDERED that a Taxing Master of the High Court do tax in the manner hereinafter appearing the applicant's costs of appearing before the said Tribunal by one counsel instructed by solicitor.

In relation to the taxation of the applicant's costs the Tribunal DOTH ORDER as follows:

(*a*) That the statutory provisions and the Rules of the Superior Courts relating to the taxation of costs in an action in the High Courts (including the provisions relating to review and appeal) shall, in so far as is practicable, apply;

(*b*) That the costs of employing a solicitor and one counsel be taxed on a party and party basis.

(*c*) That the costs do include such witnesses' expenses as relate to those witnesses who actually gave oral evidence before the Tribunal on the proposal of the applicant;

(*d*) That the amount of witnesses' expenses be (*a*) such fee for the preparation of any statement prepared by the witness which was received by the Tribunal as part of the witnesses' testimony as the Taxing Master (or on review or appeal, the Court) considers reasonable; (*b*) such fee (not already paid) for attending before the Tribunal on such days as the witness gave oral evidence as the Taxing Master (or, on review or appeal, the Court) considers reasonable and (*b*) the travelling and subsistence expenses (not already paid) reasonably incurred for the purpose of giving oral evidence before the Tribunal.

(*e*) That the Minister for Finance be at liberty to attend and be heard on the said taxation and any review or appeal in relation to it.

(*f*) That the Minister for Finance do pay the applicant's costs when taxed and ascertained.

Dated this 29th day of July 1994.

Signed Liam Hamilton
The Tribunal

ANNEXE

IN THE MATTER OF the Tribunals of Inquiry (Evidence) Acts, 1921 and 1979; and IN THE MATTER OF a Tribunal of Inquiry established pursuant to Resolutions of Dail Eireann passed on the 24th May 1991, and by Seanad Eireann on the 29th May 1991, to inquire into allegations regarding illegal activities fraud and malpractice in and in connection with the beef processing industry by order of the Minister for Agriculture and Food made on the 31st day of May 1991.

AND WHEREAS it was provided by the said order that the Tribunals of Inquiry (Evidence) Acts 1921 and 1979 applied to the Tribunal

WHEREAS it is provided by Section 2 (*b*) of the Tribunals of Inquiry (Evidence) Act, 1921 that a Tribunal to which the Act is applied shall have power to authorise the representation before it by solicitor or counsel or otherwise of any person appearing to it to be interested; and

WHEREAS Mr Gene Lambe (hereinafter referred to as "the applicant") applied to the said Tribunal to be represented before it by counsel instructed by solicitor and on the 12th January 1993 such authorisation under the said section was granted; and

WHEREAS it is provided by section 6 of the Tribunals of Inquiry (Evidence)(Amendment) Act, 1979, that if a Tribunal is of opinion that having regard to its findings and all other relevant matters there are sufficient reasons rendering it equitable to do so it may by Order direct that the whole or part of the costs of any person appearing before the Tribunal by counsel or solicitor as taxed by a Taxing Master of the High Court be paid by any other person named in the Order; and

WHEREAS by section 4 of the said Act of 1979 it is provided that a Tribunal may make such Order as it considers necessary for the purpose of its functions; and

WHEREAS the Tribunal referred to in the title hereof is satisfied that having regard to its findings and all other relevant matters there are sufficient reasons rendering it equitable that the Minister for Finance should pay the costs of the applicant as hereinafter appearing

IT IS HEREBY ORDERED that a Taxing Master of the High Court do tax in the manner hereinafter appearing the applicant's costs of appearing before the said Tribunal by one counsel instructed by solicitor.

In relation to the taxation of the applicant's costs the Tribunal DOTH ORDER as follows:

(*a*) That the statutory provisions and the Rules of the Superior Courts relating to the taxation of costs in an action in the High Courts (including the provisions relating to review and appeal) shall, in so far as is practicable, apply;

(*b*) That the costs of employing a solicitor and one counsel be taxed on a party and party basis.

(*c*) That the costs do include such witnesses' expenses as relate to those witnesses who actually gave oral evidence before the Tribunal on the proposal of the applicant;

(*d*) That the amount of witnesses' expenses be (*a*) such fee for the preparation of any statement prepared by the witness which was received by the Tribunal as part of the witnesses' testimony as the Taxing Master (or on review or appeal, the Court) considers reasonable; (*b*) such fee (not already paid) for attending before the Tribunal on such days as the witness gave oral evidence as the Taxing Master (or, on review or appeal, the Court) considers reasonable and (*b*) the travelling and subsistence expenses (not already paid) reasonably incurred for the purpose of giving oral evidence before the Tribunal.

(*e*) That the Minister for Finance be at liberty to attend and be heard on the said taxation and any review or appeal in relation to it.

(*f*) That the Minister for Finance do pay the applicant's costs when taxed and ascertained.

Dated this 29th day of July 1994.

Signed Liam Hamilton

The Tribunal

ANNEXE

IN THE MATTER OF the Tribunals of Inquiry (Evidence) Acts, 1921 and 1979; and IN THE MATTER OF a Tribunal of Inquiry established pursuant to Resolutions of Dail Eireann passed on the 24th May 1991, and by Seanad Eireann on the 29th May 1991, to inquire into allegations regarding illegal activities fraud and malpractice in and in connection with the beef processing industry by order of the Minister for Agriculture and Food made on the 31st day of May 1991.

AND WHEREAS it was provided by the said order that the Tribunals of Inquiry (Evidence) Acts 1921 and 1979 applied to the Tribunal

WHEREAS it is provided by Section 2 (*b*) of the Tribunals of Inquiry (Evidence) Act, 1921 that a Tribunal to which the Act is applied shall have power to authorise the representation before it by solicitor or counsel or otherwise of any person appearing to it to be interested; and

WHEREAS Brennan Governey and Company (hereinafter referred to as "the applicants") applied to the said Tribunal to be represented before it by counsel instructed by solicitor and on the 21st January 1993 such authorisation under the said section was granted; and

WHEREAS it is provided by section 6 of the Tribunals of Inquiry (Evidence)(Amendment) Act, 1979, that if a Tribunal is of opinion that having regard to its findings and all other relevant matters there are sufficient reasons rendering it equitable to do so it may by Order direct that the whole or part of the costs of any person appearing before the Tribunal by counsel or solicitor as taxed by a Taxing Master of the High Court be paid by any other person named in the Order; and

WHEREAS by section 4 of the said Act of 1979 it is provided that a Tribunal may make such Order as it considers necessary for the purpose of its functions; and

WHEREAS the Tribunal referred to in the title hereof is satisfied that having regard to its findings and all other relevant matters there are sufficient reasons rendering it equitable that the Minister for Finance should pay the costs of the applicants as hereinafter appearing

IT IS HEREBY ORDERED that a Taxing Master of the High Court do tax in the manner hereinafter appearing the applicants' costs of appearing before the said Tribunal by one counsel instructed by solicitor.

In relation to the taxation of the applicants' costs the Tribunal DOTH ORDER as follows:

(*a*) That the statutory provisions and the Rules of the Superior Courts relating to the taxation of costs in an action in the High Courts (including the provisions relating to review and appeal) shall, in so far as is practicable, apply;

(*b*) That the costs of employing a solicitor and one counsel be taxed on a party and party basis.

(*c*) That the costs do include such witnesses' expenses as relate to those witnesses who actually gave oral evidence before the Tribunal on the proposal of the applicants;

(*d*) That the amount of witnesses' expenses be (*a*) such fee for the preparation of any statement prepared by the witness which was received by the Tribunal as part of the witnesses' testimony as the Taxing Master (or on review or appeal, the Court) considers reasonable; (*b*) such fee (not already paid) for attending before the Tribunal on such days as the witness gave oral evidence as the Taxing Master (or, on review or appeal, the Court) considers reasonable and (*b*) the travelling and subsistence expenses (not already paid) reasonably incurred for the purpose of giving oral evidence before the Tribunal.

(*e*) That the Minister for Finance be at liberty to attend and be heard on the said taxation and any review or appeal in relation to it.

(*f*) That the Minister for Finance do pay the applicants' costs when taxed and ascertained.

Dated this 29th day of July 1994.

Signed Liam Hamilton

The Tribunal

ANNEXE

IN THE MATTER OF the Tribunals of Inquiry (Evidence) Acts, 1921 and 1979; and IN THE MATTER OF a Tribunal of Inquiry established pursuant to Resolutions of Dail Eireann passed on the 24th May 1991, and by Seanad Eireann on the 29th May 1991, to inquire into allegations regarding illegal activities fraud and malpractice in and in connection with the beef processing industry by order of the Minister for Agriculture and Food made on the 31st day of May 1991.

AND WHEREAS it was provided by the said order that the Tribunals of Inquiry (Evidence) Acts 1921 and 1979 applied to the Tribunal

WHEREAS it is provided by Section 2 (*b*) of the Tribunals of Inquiry (Evidence) Act, 1921 that a Tribunal to which the Act is applied shall have power to authorise the representation before it by solicitor or counsel or otherwise of any person appearing to it to be interested; and

WHEREAS Mr Sean Goodwin, Mr Larry Kelly, Mr Sean Hartnett, Mr Matthew O'Doherty, Mr Anthony Butler, Mr Denis Murphy and Mr Thomas Keating (hereinafter referred to as "the applicants") applied to the said Tribunal to be represented before it by counsel instructed by solicitor and on the 28th January 1993 such authorisation under the said section was granted; and

WHEREAS it is provided by section 6 of the Tribunals of Inquiry (Evidence)(Amendment) Act, 1979, that if a Tribunal is of opinion that having regard to its findings and all other relevant matters there are sufficient reasons rendering it equitable to do so it may by Order direct that the whole or part of the costs of any person appearing before the Tribunal by counsel or solicitor as taxed by a Taxing Master of the High Court be paid by any other person named in the Order; and

WHEREAS by section 4 of the said Act of 1979 it is provided that a Tribunal may make such Order as it considers necessary for the purpose of its functions; and

WHEREAS the Tribunal referred to in the title hereof is satisfied that having regard to its findings and all other relevant matters there are sufficient reasons rendering it equitable that the Minister for Finance should pay the costs of the applicants as hereinafter appearing

IT IS HEREBY ORDERED that a Taxing Master of the High Court do tax in the manner hereinafter appearing the applicants' costs of appearing before the said Tribunal by one counsel instructed by solicitor.

In relation to the taxation of the applicants' costs the Tribunal DOTH ORDER as follows:

(*a*) That the statutory provisions and the Rules of the Superior Courts relating to the taxation of costs in an action in the High Courts (including the provisions relating to review and appeal) shall, in so far as is practicable, apply;

(*b*) That the costs of employing a solicitor and one counsel be taxed on a party and party basis.

(*c*) That the costs do include such witnesses' expenses as relate to those witnesses who actually gave oral evidence before the Tribunal on the proposal of the applicants;

(*d*) That the amount of witnesses' expenses be (*a*) such fee for the preparation of any statement prepared by the witness which was received by the Tribunal as part of the witnesses' testimony as the Taxing Master (or on review or appeal, the Court) considers reasonable; (*b*) such fee (not already paid) for attending before the Tribunal on such days as the witness gave oral evidence as the Taxing Master (or, on review or appeal, the Court) considers reasonable and (*b*) the travelling and subsistence expenses (not already paid) reasonably incurred for the purpose of giving oral evidence before the Tribunal.

(*e*) That the Minister for Finance be at liberty to attend and be heard on the said taxation and any review or appeal in relation to it.

(*f*) That the Minister for Finance do pay the applicants' costs when taxed and ascertained.

Dated this 29th day of July 1994.

Signed Liam Hamilton

The Tribunal

ANNEXE

IN THE MATTER OF the Tribunals of Inquiry (Evidence) Acts, 1921 and 1979; and IN THE MATTER OF a Tribunal of Inquiry established pursuant to Resolutions of Dail Eireann passed on the 24th May 1991, and by Seanad Eireann on the 29th May 1991, to inquire into allegations regarding illegal activities fraud and malpractice in and in connection with the beef processing industry by order of the Minister for Agriculture and Food made on the 31st day of May 1991.

AND WHEREAS it was provided by the said order that the Tribunals of Inquiry (Evidence) Acts 1921 and 1979 applied to the Tribunal

WHEREAS it is provided by Section 2 (*b*) of the Tribunals of Inquiry (Evidence) Act, 1921 that a Tribunal to which the Act is applied shall have power to authorise the representation before it by solicitor or counsel or otherwise of any person appearing to it to be interested; and

WHEREAS Mr Victor Broderick and Mr Christian Peyron (hereinafter referred to as "the applicants") applied to the said Tribunal to be represented before it by counsel instructed by solicitor and on the 2nd February 1993 such authorisation under the said section was granted; and

WHEREAS it is provided by section 6 of the Tribunals of Inquiry (Evidence)(Amendment) Act, 1979, that if a Tribunal is of opinion that having regard to its findings and all other relevant matters there are sufficient reasons rendering it equitable to do so it may by Order direct that the whole or part of the costs of any person appearing before the Tribunal by counsel or solicitor as taxed by a Taxing Master of the High Court be paid by any other person named in the Order; and

WHEREAS by section 4 of the said Act of 1979 it is provided that a Tribunal may make such Order as it considers necessary for the purpose of its functions; and

WHEREAS the Tribunal referred to in the title hereof is satisfied that having regard to its findings and all other relevant matters there are sufficient reasons rendering it equitable that the Minister for Finance should pay the costs of the applicants as hereinafter appearing

IT IS HEREBY ORDERED that a Taxing Master of the High Court do tax in the manner hereinafter appearing the applicants' costs of appearing before the said Tribunal by one counsel instructed by solicitor.

In relation to the taxation of the applicants' costs the Tribunal DOTH ORDER as follows:

(*a*) That the statutory provisions and the Rules of the Superior Courts relating to the taxation of costs in an action in the High Courts (including the provisions relating to review and appeal) shall, in so far as is practicable, apply;

(*b*) That the costs of employing a solicitor and one counsel be taxed on a party and party basis.

(*c*) That the costs do include such witnesses' expenses as relate to those witnesses who actually gave oral evidence before the Tribunal on the proposal of the applicants;

(*d*) That the amount of witnesses' expenses be (*a*) such fee for the preparation of any statement prepared by the witness which was received by the Tribunal as part of the witnesses' testimony as the Taxing Master (or on review or appeal, the Court) considers reasonable; (*b*) such fee (not already paid) for attending before the Tribunal on such days as the witness gave oral evidence as the Taxing Master (or, on review or appeal, the Court) considers reasonable and (*b*) the travelling and subsistence expenses (not already paid) reasonably incurred for the purpose of giving oral evidence before the Tribunal.

(*e*) That the Minister for Finance be at liberty to attend and be heard on the said taxation and any review or appeal in relation to it.

(*f*) That the Minister for Finance do pay the applicants' costs when taxed and ascertained.

Dated this 29th day of July 1994.

Signed Liam Hamilton

The Tribunal

ANNEXE

IN THE MATTER OF the Tribunals of Inquiry (Evidence) Acts, 1921 and 1979; and IN THE MATTER OF a Tribunal of Inquiry established pursuant to Resolutions of Dail Eireann passed on the 24th May 1991, and by Seanad Eireann on the 29th May 1991, to inquire into allegations regarding illegal activities fraud and malpractice in and in connection with the beef processing industry by order of the Minister for Agriculture and Food made on the 31st day of May 1991.

AND WHEREAS it was provided by the said order that the Tribunals of Inquiry (Evidence) Acts 1921 and 1979 applied to the Tribunal

WHEREAS it is provided by Section 2 (*b*) of the Tribunals of Inquiry (Evidence) Act, 1921 that a Tribunal to which the Act is applied shall have power to authorise the representation before it by solicitor or counsel or otherwise of any person appearing to it to be interested; and

WHEREAS Mr Michael Jinks (hereinafter referred to as "the applicant") applied to the said Tribunal to be represented before it by counsel instructed by solicitor and on the 16th September 1993 such authorisation under the said section was granted; and

WHEREAS it is provided by section 6 of the Tribunals of Inquiry (Evidence)(Amendment) Act, 1979, that if a Tribunal is of opinion that having regard to its findings and all other relevant matters there are sufficient reasons rendering it equitable to do so it may by Order direct that the whole or part of the costs of any person appearing before the Tribunal by counsel or solicitor as taxed by a Taxing Master of the High Court be paid by any other person named in the Order; and

WHEREAS by section 4 of the said Act of 1979 it is provided that a Tribunal may make such Order as it considers necessary for the purpose of its functions; and

WHEREAS the Tribunal referred to in the title hereof is satisfied that having regard to its findings and all other relevant matters there are sufficient reasons rendering it equitable that the Minister for Finance should pay the costs of the applicant as hereinafter appearing

IT IS HEREBY ORDERED that a Taxing Master of the High Court do tax in the manner hereinafter appearing the applicant's costs of appearing before the said Tribunal by one counsel instructed by solicitor.

In relation to the taxation of the applicant's costs the Tribunal DOTH ORDER as follows:

(*a*) That the statutory provisions and the Rules of the Superior Courts relating to the taxation of costs in an action in the High Courts (including the provisions relating to review and appeal) shall, in so far as is practicable, apply;

(*b*) That the costs of employing a solicitor and one counsel be taxed on a party and party basis.

(*c*) That the costs do include such witnesses' expenses as relate to those witnesses who actually gave oral evidence before the Tribunal on the proposal of the applicant;

(*d*) That the amount of witnesses' expenses be (*a*) such fee for the preparation of any statement prepared by the witness which was received by the Tribunal as part of the witnesses' testimony as the Taxing Master (or on review or appeal, the Court) considers reasonable; (*b*) such fee (not already paid) for attending before the Tribunal on such days as the witness gave oral evidence as the Taxing Master (or, on review or appeal, the Court) considers reasonable and (*b*) the travelling and subsistence expenses (not already paid) reasonably incurred for the purpose of giving oral evidence before the Tribunal.

(*e*) That the Minister for Finance be at liberty to attend and be heard on the said taxation and any review or appeal in relation to it.

(*f*) That the Minister for Finance do pay the applicant's costs when taxed and ascertained.

Dated this 29th day of July 1994.

Signed Liam Hamilton

The Tribunal

ANNEXE

IN THE MATTER OF the Tribunals of Inquiry (Evidence) Acts, 1921 and 1979; and IN THE MATTER OF a Tribunal of Inquiry established pursuant to Resolutions of Dail Eireann passed on the 24th May 1991, and by Seanad Eireann on the 29th May 1991, to inquire into allegations regarding illegal activities fraud and malpractice in and in connection with the beef processing industry by order of the Minister for Agriculture and Food made on the 31st day of May 1991.

AND WHEREAS it was provided by the said order that the Tribunals of Inquiry (Evidence) Acts 1921 and 1979 applied to the Tribunal

WHEREAS it is provided by Section 2 (*b*) of the Tribunals of Inquiry (Evidence) Act, 1921 that a Tribunal to which the Act is applied shall have power to authorise the representation before it by solicitor or counsel or otherwise of any person appearing to it to be interested; and

WHEREAS Mr Dermot Hanley (hereinafter referred to as "the applicant") applied to the said Tribunal to be represented before it by solicitor and on the 16th February 1993 such authorisation under the said section was granted; and

WHEREAS it is provided by section 6 of the Tribunals of Inquiry (Evidence)(Amendment) Act, 1979, that if a Tribunal is of opinion that having regard to its findings and all other relevant matters there are sufficient reasons rendering it equitable to do so it may by Order direct that the whole or part of the costs of any person appearing before the Tribunal by counsel or solicitor as taxed by a Taxing Master of the High Court be paid by any other person named in the Order; and

WHEREAS by section 4 of the said Act of 1979 it is provided that a Tribunal may make such Order as it considers necessary for the purpose of its functions; and

WHEREAS the Tribunal referred to in the title hereof is satisfied that having regard to its findings and all other relevant matters there are sufficient reasons rendering it equitable that the Minister for Finance should pay the costs of the applicant as hereinafter appearing

IT IS HEREBY ORDERED that a Taxing Master of the High Court do tax in the manner hereinafter appearing the applicant's costs of appearing before the said Tribunal by solicitor.

In relation to the taxation of the applicant's costs the Tribunal DOTH ORDER as follows:

(*a*) That the statutory provisions and the Rules of the Superior Courts relating to the taxation of costs in an action in the High Courts (including the provisions relating to review and appeal) shall, in so far as is practicable, apply;

(*b*) That the costs of employing a solicitor be taxed on a party and party basis.

(*c*) That the costs do include such witnesses' expenses as relate to those witnesses who actually gave oral evidence before the Tribunal on the proposal of the applicant;

(*d*) That the amount of witnesses' expenses be (*a*) such fee for the preparation of any statement prepared by the witness which was received by the Tribunal as part of the witnesses' testimony as the Taxing Master (or on review or appeal, the Court) considers reasonable; (*b*) such fee (not already paid) for attending before the Tribunal on such days as the witness gave oral evidence as the Taxing Master (or, on review or appeal, the Court) considers reasonable and (*b*) the travelling and subsistence expenses (not already paid) reasonably incurred for the purpose of giving oral evidence before the Tribunal.

(*e*) That the Minister for Finance be at liberty to attend and be heard on the said taxation and any review or appeal in relation to it.

(*f*) That the Minister for Finance do pay the applicant's costs when taxed and ascertained.

Dated this 29th day of July 1994.

Signed Liam Hamilton

The Tribunal

ANNEXE

IN THE MATTER OF the Tribunals of Inquiry (Evidence) Acts, 1921 and 1979; and IN THE MATTER OF a Tribunal of Inquiry established pursuant to Resolutions of Dail Eireann passed on the 24th May 1991, and by Seanad Eireann on the 29th May 1991, to inquire into allegations regarding illegal activities fraud and malpractice in and in connection with the beef processing industry by order of the Minister for Agriculture and Food made on the 31st day of May 1991.

AND WHEREAS it was provided by the said order that the Tribunals of Inquiry (Evidence) Acts 1921 and 1979 applied to the Tribunal

WHEREAS it is provided by Section 2 (*b*) of the Tribunals of Inquiry (Evidence) Act, 1921 that a Tribunal to which the Act is applied shall have power to authorise the representation before it by solicitor or counsel or otherwise of any person appearing to it to be interested; and

WHEREAS Arax Limited (hereinafter referred to as "the applicants") applied to the said Tribunal to be represented before it by counsel instructed by solicitor and on the 11th May 1993 such authorisation under the said section was granted; and

WHEREAS it is provided by section 6 of the Tribunals of Inquiry (Evidence)(Amendment) Act, 1979, that if a Tribunal is of opinion that having regard to its findings and all other relevant matters there are sufficient reasons rendering it equitable to do so it may by Order direct that the whole or part of the costs of any person appearing before the Tribunal by counsel or solicitor as taxed by a Taxing Master of the High Court be paid by any other person named in the Order; and

WHEREAS by section 4 of the said Act of 1979 it is provided that a Tribunal may make such Order as it considers necessary for the purpose of its functions; and

WHEREAS the Tribunal referred to in the title hereof is satisfied that having regard to its findings and all other relevant matters there are sufficient reasons rendering it equitable that the Minister for Finance should pay the costs of the applicants as hereinafter appearing

IT IS HEREBY ORDERED that a Taxing Master of the High Court do tax in the manner hereinafter appearing the applicants' costs of appearing before the said Tribunal by one counsl instructed by solicitor.

In relation to the taxation of the applicant's costs the Tribunal DOTH ORDER as follows:

(*a*) That the statutory provisions and the Rules of the Superior Courts relating to the taxation of costs in an action in the High Courts (including the provisions relating to review and appeal) shall, in so far as is practicable, apply;

(*b*) That the costs of employing a solicitor and one counsel be taxed on a party and party basis.

(*c*) That the costs do include such witnesses' expenses as relate to those witnesses who actually gave oral evidence before the Tribunal on the proposal of the applicants;

(*d*) That the amount of witnesses' expenses be (*a*) such fee for the preparation of any statement prepared by the witness which was received by the Tribunal as part of the witnesses' testimony as the Taxing Master (or on review or appeal, the Court) considers reasonable; (*b*) such fee (not already paid) for attending before the Tribunal on such days as the witness gave oral evidence as the Taxing Master (or, on review or appeal, the Court) considers reasonable and (*b*) the travelling and subsistence expenses (not already paid) reasonably incurred for the purpose of giving oral evidence before the Tribunal.

(*e*) That the Minister for Finance be at liberty to attend and be heard on the said taxation and any review or appeal in relation to it.

(*f*) That the Minister for Finance do pay the applicants' costs when taxed and ascertained.

Dated this 29th day of July 1994.

Signed Liam Hamilton

The Tribunal

ANNEXE

IN THE MATTER OF the Tribunals of Inquiry (Evidence) Acts, 1921 and 1979; and IN THE MATTER OF a Tribunal of Inquiry established pursuant to Resolutions of Dail Eireann passed on the 24th May 1991, and by Seanad Eireann on the 29th May 1991, to inquire into allegations regarding illegal activities fraud and malpractice in and in connection with the beef processing industry by order of the Minister for Agriculture and Food made on the 31st day of May 1991.

AND WHEREAS it was provided by the said order that the Tribunals of Inquiry (Evidence) Acts 1921 and 1979 applied to the Tribunal

WHEREAS it is provided by Section 2 (*b*) of the Tribunals of Inquiry (Evidence) Act, 1921 that a Tribunal to which the Act is applied shall have power to authorise the representation before it by solicitor or counsel or otherwise of any person appearing to it to be interested; and

WHEREAS Ashbourne Meats Limited (hereinafter referred to as "the applicants") applied to the said Tribunal to be represented before it by solicitor and on the 11th May 1993 such authorisation under the said section was granted; and

WHEREAS it is provided by section 6 of the Tribunals of Inquiry (Evidence)(Amendment) Act, 1979, that if a Tribunal is of opinion that having regard to its findings and all other relevant matters there are sufficient reasons rendering it equitable to do so it may by Order direct that the whole or part of the costs of any person appearing before the Tribunal by counsel or solicitor as taxed by a Taxing Master of the High Court be paid by any other person named in the Order; and

WHEREAS by section 4 of the said Act of 1979 it is provided that a Tribunal may make such Order as it considers necessary for the purpose of its functions; and

WHEREAS the Tribunal referred to in the title hereof is satisfied that having regard to its findings and all other relevant matters there are sufficient reasons rendering it equitable that the Minister for Finance should pay the costs of the applicants as hereinafter appearing

IT IS HEREBY ORDERED that a Taxing Master of the High Court do tax in the manner hereinafter appearing the applicants' costs of appearing before the said Tribunal by solicitor.

In relation to the taxation of the applicants' costs the Tribunal DOTH ORDER as follows:

(*a*) That the statutory provisions and the Rules of the Superior Courts relating to the taxation of costs in an action in the High Courts (including the provisions relating to review and appeal) shall, in so far as is practicable, apply;

(*b*) That the costs of employing a solicitor be taxed on a party and party basis.

(*c*) That the costs do include such witnesses' expenses as relate to those witnesses who actually gave oral evidence before the Tribunal on the proposal of the applicants;

(*d*) That the amount of witnesses' expenses be (*a*) such fee for the preparation of any statement prepared by the witness which was received by the Tribunal as part of the witnesses' testimony as the Taxing Master (or on review or appeal, the Court) considers reasonable; (*b*) such fee (not already paid) for attending before the Tribunal on such days as the witness gave oral evidence as the Taxing Master (or, on review or appeal, the Court) considers reasonable and (*b*) the travelling and subsistence expenses (not already paid) reasonably incurred for the purpose of giving oral evidence before the Tribunal.

(*e*) That the Minister for Finance be at liberty to attend and be heard on the said taxation and any review or appeal in relation to it.

(*f*) That the Minister for Finance do pay the applicants' costs when taxed and ascertained.

Dated this 29th day of July 1994.

Signed Liam Hamilton

The Tribunal

ANNEXE

IN THE MATTER OF the Tribunals of Inquiry (Evidence) Acts, 1921 and 1979; and IN THE MATTER OF a Tribunal of Inquiry established pursuant to Resolutions of Dail Eireann passed on the 24th May 1991, and by Seanad Eireann on the 29th May 1991, to inquire into allegations regarding illegal activities fraud and malpractice in and in connection with the beef processing industry by order of the Minister for Agriculture and Food made on the 31st day of May 1991.

AND WHEREAS it was provided by the said order that the Tribunals of Inquiry (Evidence) Acts 1921 and 1979 applied to the Tribunal

WHEREAS it is provided by Section 2 (*b*) of the Tribunals of Inquiry (Evidence) Act, 1921 that a Tribunal to which the Act is applied shall have power to authorise the representation before it by solicitor or counsel or otherwise of any person appearing to it to be interested; and

WHEREAS Tara Meats Limited (hereinafter referred to as "the applicants") applied to the said Tribunal to be represented before it by counsel instructed by solicitor and on the 11th May 1993 such authorisation under the said section was granted; and

WHEREAS it is provided by section 6 of the Tribunals of Inquiry (Evidence)(Amendment) Act, 1979, that if a Tribunal is of opinion that having regard to its findings and all other relevant matters there are sufficient reasons rendering it equitable to do so it may by Order direct that the whole or part of the costs of any person appearing before the Tribunal by counsel or solicitor as taxed by a Taxing Master of the High Court be paid by any other person named in the Order; and

WHEREAS by section 4 of the said Act of 1979 it is provided that a Tribunal may make such Order as it considers necessary for the purpose of its functions; and

WHEREAS the Tribunal referred to in the title hereof is satisfied that having regard to its findings and all other relevant matters there are sufficient reasons rendering it equitable that the Minister for Finance should pay the costs of the applicants as hereinafter appearing

IT IS HEREBY ORDERED that a Taxing Master of the High Court do tax in the manner hereinafter appearing the applicants' costs of appearing before the said Tribunal by one counsel instructed by solicitor.

In relation to the taxation of the applicants' costs the Tribunal DOTH ORDER as follows:

(*a*) That the statutory provisions and the Rules of the Superior Courts relating to the taxation of costs in an action in the High Courts (including the provisions relating to review and appeal) shall, in so far as is practicable, apply;

(*b*) That the costs of employing a solicitor and one counsel be taxed on a party and party basis.

(*c*) That the costs do include such witnesses' expenses as relate to those witnesses who actually gave oral evidence before the Tribunal on the proposal of the applicants;

(*d*) That the amount of witnesses' expenses be (*a*) such fee for the preparation of any statement prepared by the witness which was received by the Tribunal as part of the witnesses' testimony as the Taxing Master (or on review or appeal, the Court) considers reasonable; (*b*) such fee (not already paid) for attending before the Tribunal on such days as the witness gave oral evidence as the Taxing Master (or, on review or appeal, the Court) considers reasonable and (*b*) the travelling and subsistence expenses (not already paid) reasonably incurred for the purpose of giving oral evidence before the Tribunal.

(*e*) That the Minister for Finance be at liberty to attend and be heard on the said taxation and any review or appeal in relation to it.

(*f*) That the Minister for Finance do pay the applicants' costs when taxed and ascertained.

Dated this 29th day of July 1994.

Signed Liam Hamilton

The Tribunal

ANNEXE

IN THE MATTER OF the Tribunals of Inquiry (Evidence) Acts, 1921 and 1979; and IN THE MATTER OF a Tribunal of Inquiry established pursuant to Resolutions of Dail Eireann passed on the 24th May 1991, and by Seanad Eireann on the 29th May 1991, to inquire into allegations regarding illegal activities fraud and malpractice in and in connection with the beef processing industry by order of the Minister for Agriculture and Food made on the 31st day of May 1991.

AND WHEREAS it was provided by the said order that the Tribunals of Inquiry (Evidence) Acts 1921 and 1979 applied to the Tribunal

WHEREAS it is provided by Section 2 (*b*) of the Tribunals of Inquiry (Evidence) Act, 1921 that a Tribunal to which the Act is applied shall have power to authorise the representation before it by solicitor or counsel or otherwise of any person appearing to it to be interested; and

WHEREAS Rangeland Meats Limited (hereinafter referred to as "the applicants") applied to the said Tribunal to be represented before it by counsel instructed by solicitor and on the 12th May 1993 such authorisation under the said section was granted; and

WHEREAS it is provided by section 6 of the Tribunals of Inquiry (Evidence)(Amendment) Act, 1979, that if a Tribunal is of opinion that having regard to its findings and all other relevant matters there are sufficient reasons rendering it equitable to do so it may by Order direct that the whole or part of the costs of any person appearing before the Tribunal by counsel or solicitor as taxed by a Taxing Master of the High Court be paid by any other person named in the Order; and

WHEREAS by section 4 of the said Act of 1979 it is provided that a Tribunal may make such Order as it considers necessary for the purpose of its functions; and

WHEREAS the Tribunal referred to in the title hereof is satisfied that having regard to its findings and all other relevant matters there are sufficient reasons rendering it equitable that the Minister for Finance should pay the costs of the applicants as hereinafter appearing

IT IS HEREBY ORDERED that a Taxing Master of the High Court do tax in the manner hereinafter appearing the applicants' costs of appearing before the said Tribunal by one counsel instructed by solicitor.

In relation to the taxation of the applicants' costs the Tribunal DOTH ORDER as follows:

(*a*) That the statutory provisions and the Rules of the Superior Courts relating to the taxation of costs in an action in the High Courts (including the provisions relating to review and appeal) shall, in so far as is practicable, apply;

(*b*) That the costs of employing a solicitor and one counsel be taxed on a party and party basis.

(*c*) That the costs do include such witnesses' expenses as relate to those witnesses who actually gave oral evidence before the Tribunal on the proposal of the applicants;

(*d*) That the amount of witnesses' expenses be (*a*) such fee for the preparation of any statement prepared by the witness which was received by the Tribunal as part of the witnesses' testimony as the Taxing Master (or on review or appeal, the Court) considers reasonable; (*b*) such fee (not already paid) for attending before the Tribunal on such days as the witness gave oral evidence as the Taxing Master (or, on review or appeal, the Court) considers reasonable and (*b*) the travelling and subsistence expenses (not already paid) reasonably incurred for the purpose of giving oral evidence before the Tribunal.

(*e*) That the Minister for Finance be at liberty to attend and be heard on the said taxation and any review or appeal in relation to it.

(*f*) That the Minister for Finance do pay the applicants' costs when taxed and ascertained.

Dated this 29th day of July 1994.

Signed Liam Hamilton
The Tribunal

ANNEXE

IN THE MATTER OF the Tribunals of Inquiry (Evidence) Acts, 1921 and 1979; and IN THE MATTER OF a Tribunal of Inquiry established pursuant to Resolutions of Dail Eireann passed on the 24th May 1991, and by Seanad Eireann on the 29th May 1991, to inquire into allegations regarding illegal activities fraud and malpractice in and in connection with the beef processing industry by order of the Minister for Agriculture and Food made on the 31st day of May 1991.

AND WHEREAS it was provided by the said order that the Tribunals of Inquiry (Evidence) Acts 1921 and 1979 applied to the Tribunal

WHEREAS it is provided by Section 2 (*b*) of the Tribunals of Inquiry (Evidence) Act, 1921 that a Tribunal to which the Act is applied shall have power to authorise the representation before it by solicitor or counsel or otherwise of any person appearing to it to be interested; and

WHEREAS Slaney Meats Limited (hereinafter referred to as "the applicants") applied to the said Tribunal to be represented before it by solicitor and on the 12th May 1993 such authorisation under the said section was granted; and

WHEREAS it is provided by section 6 of the Tribunals of Inquiry (Evidence)(Amendment) Act, 1979, that if a Tribunal is of opinion that having regard to its findings and all other relevant matters there are sufficient reasons rendering it equitable to do so it may by Order direct that the whole or part of the costs of any person appearing before the Tribunal by counsel or solicitor as taxed by a Taxing Master of the High Court be paid by any other person named in the Order; and

WHEREAS by section 4 of the said Act of 1979 it is provided that a Tribunal may make such Order as it considers necessary for the purpose of its functions; and

WHEREAS the Tribunal referred to in the title hereof is satisfied that having regard to its findings and all other relevant matters there are sufficient reasons rendering it equitable that the Minister for Finance should pay the costs of the applicants as hereinafter appearing

IT IS HEREBY ORDERED that a Taxing Master of the High Court do tax in the manner hereinafter appearing the applicants' costs of appearing before the said Tribunal by solicitor.

In relation to the taxation of the applicants' costs the Tribunal DOTH ORDER as follows:

(*a*) That the statutory provisions and the Rules of the Superior Courts relating to the taxation of costs in an action in the High Courts (including the provisions relating to review and appeal) shall, in so far as is practicable, apply;

(*b*) That the costs of employing a solicitor be taxed on a party and party basis.

(*c*) That the costs do include such witnesses' expenses as relate to those witnesses who actually gave oral evidence before the Tribunal on the proposal of the applicants;

(*d*) That the amount of witnesses' expenses be (*a*) such fee for the preparation of any statement prepared by the witness which was received by the Tribunal as part of the witnesses' testimony as the Taxing Master (or on review or appeal, the Court) considers reasonable; (*b*) such fee (not already paid) for attending before the Tribunal on such days as the witness gave oral evidence as the Taxing Master (or, on review or appeal, the Court) considers reasonable and (*b*) the travelling and subsistence expenses (not already paid) reasonably incurred for the purpose of giving oral evidence before the Tribunal.

(*e*) That the Minister for Finance be at liberty to attend and be heard on the said taxation and any review or appeal in relation to it.

(*f*) That the Minister for Finance do pay the applicants' costs when taxed and ascertained.

Dated this 29th day of July 1994.

Signed Liam Hamilton
The Tribunal

ANNEXE

IN THE MATTER OF the Tribunals of Inquiry (Evidence) Acts, 1921 and 1979; and IN THE MATTER OF a Tribunal of Inquiry established pursuant to Resolutions of Dail Eireann passed on the 24th May 1991, and by Seanad Eireann on the 29th May 1991, to inquire into allegations regarding illegal activities fraud and malpractice in and in connection with the beef processing industry by order of the Minister for Agriculture and Food made on the 31st day of May 1991.

AND WHEREAS it was provided by the said order that the Tribunals of Inquiry (Evidence) Acts 1921 and 1979 applied to the Tribunal

WHEREAS it is provided by Section 2 (*b*) of the Tribunals of Inquiry (Evidence) Act, 1921 that a Tribunal to which the Act is applied shall have power to authorise the representation before it by solicitor or counsel or otherwise of any person appearing to it to be interested; and

WHEREAS Ballywalter Meats Limited (hereinafter referred to as "the applicants") applied to the said Tribunal to be represented before it by counsel instructed by solicitor and on the 18th May 1993 such authorisation under the said section was granted; and

WHEREAS it is provided by section 6 of the Tribunals of Inquiry (Evidence)(Amendment) Act, 1979, that if a Tribunal is of opinion that having regard to its findings and all other relevant matters there are sufficient reasons rendering it equitable to do so it may by Order direct that the whole or part of the costs of any person appearing before the Tribunal by counsel or solicitor as taxed by a Taxing Master of the High Court be paid by any other person named in the Order; and

WHEREAS by section 4 of the said Act of 1979 it is provided that a Tribunal may make such Order as it considers necessary for the purpose of its functions; and

WHEREAS the Tribunal referred to in the title hereof is satisfied that having regard to its findings and all other relevant matters there are sufficient reasons rendering it equitable that the Minister for Finance should pay the costs of the applicants as hereinafter appearing

IT IS HEREBY ORDERED that a Taxing Master of the High Court do tax in the manner hereinafter appearing the applicants' costs of appearing before the said Tribunal by one counsel instructed by solicitor.

In relation to the taxation of the applicants' costs the Tribunal DOTH ORDER as follows:

(*a*) That the statutory provisions and the Rules of the Superior Courts relating to the taxation of costs in an action in the High Courts (including the provisions relating to review and appeal) shall, in so far as is practicable, apply;

(*b*) That the costs of employing a solicitor and one counsel be taxed on a party and party basis.

(*c*) That the costs do include such witnesses' expenses as relate to those witnesses who actually gave oral evidence before the Tribunal on the proposal of the applicants;

(*d*) That the amount of witnesses' expenses be (*a*) such fee for the preparation of any statement prepared by the witness which was received by the Tribunal as part of the witnesses' testimony as the Taxing Master (or on review or appeal, the Court) considers reasonable; (*b*) such fee (not already paid) for attending before the Tribunal on such days as the witness gave oral evidence as the Taxing Master (or, on review or appeal, the Court) considers reasonable and (*b*) the travelling and subsistence expenses (not already paid) reasonably incurred for the purpose of giving oral evidence before the Tribunal.

(*e*) That the Minister for Finance be at liberty to attend and be heard on the said taxation and any review or appeal in relation to it.

(*f*) That the Minister for Finance do pay the applicants' costs when taxed and ascertained.

Dated this 29th day of July 1994.

Signed Liam Hamilton

The Tribunal

ANNEXE

IN THE MATTER OF the Tribunals of Inquiry (Evidence) Acts, 1921 and 1979; and IN THE MATTER OF a Tribunal of Inquiry established pursuant to Resolutions of Dail Eireann passed on the 24th May 1991, and by Seanad Eireann on the 29th May 1991, to inquire into allegations regarding illegal activities fraud and malpractice in and in connection with the beef processing industry by order of the Minister for Agriculture and Food made on the 31st day of May 1991.

AND WHEREAS it was provided by the said order that the Tribunals of Inquiry (Evidence) Acts 1921 and 1979 applied to the Tribunal

WHEREAS it is provided by Section 2 (*b*) of the Tribunals of Inquiry (Evidence) Act, 1921 that a Tribunal to which the Act is applied shall have power to authorise the representation before it by solicitor or counsel or otherwise of any person appearing to it to be interested; and

WHEREAS Dawn Meats Limited (hereinafter referred to as "the applicants") applied to the said Tribunal to be represented before it by counsel instructed by solicitor and on the 13th May 1993 such authorisation under the said section was granted; and

WHEREAS it is provided by section 6 of the Tribunals of Inquiry (Evidence)(Amendment) Act, 1979, that if a Tribunal is of opinion that having regard to its findings and all other relevant matters there are sufficient reasons rendering it equitable to do so it may by Order direct that the whole or part of the costs of any person appearing before the Tribunal by counsel or solicitor as taxed by a Taxing Master of the High Court be paid by any other person named in the Order; and

WHEREAS by section 4 of the said Act of 1979 it is provided that a Tribunal may make such Order as it considers necessary for the purpose of its functions; and

WHEREAS the Tribunal referred to in the title hereof is satisfied that having regard to its findings and all other relevant matters there are sufficient reasons rendering it equitable that the Minister for Finance should pay the costs of the applicants as hereinafter appearing

IT IS HEREBY ORDERED that a Taxing Master of the High Court do tax in the manner hereinafter appearing the applicants' costs of appearing before the said Tribunal by one counsel instructed by solicitor.

In relation to the taxation of the applicants' costs the Tribunal DOTH ORDER as follows:

(*a*) That the statutory provisions and the Rules of the Superior Courts relating to the taxation of costs in an action in the High Courts (including the provisions relating to review and appeal) shall, in so far as is practicable, apply;

(*b*) That the costs of employing a solicitor and one counsel be taxed on a party and party basis.

(*c*) That the costs do include such witnesses' expenses as relate to those witnesses who actually gave oral evidence before the Tribunal on the proposal of the applicants;

(*d*) That the amount of witnesses' expenses be (*a*) such fee for the preparation of any statement prepared by the witness which was received by the Tribunal as part of the witnesses' testimony as the Taxing Master (or on review or appeal, the Court) considers reasonable; (*b*) such fee (not already paid) for attending before the Tribunal on such days as the witness gave oral evidence as the Taxing Master (or, on review or appeal, the Court) considers reasonable and (*b*) the travelling and subsistence expenses (not already paid) reasonably incurred for the purpose of giving oral evidence before the Tribunal.

(*e*) That the Minister for Finance be at liberty to attend and be heard on the said taxation and any review or appeal in relation to it.

(*f*) That the Minister for Finance do pay the applicants' costs when taxed and ascertained.

Dated this 29th day of July 1994.

Signed Liam Hamilton
The Tribunal

ANNEXE

IN THE MATTER OF the Tribunals of Inquiry (Evidence) Acts, 1921 and 1979; and IN THE MATTER OF a Tribunal of Inquiry established pursuant to Resolutions of Dail Eireann passed on the 24th May 1991, and by Seanad Eireann on the 29th May 1991, to inquire into allegations regarding illegal activities fraud and malpractice in and in connection with the beef processing industry by order of the Minister for Agriculture and Food made on the 31st day of May 1991.

AND WHEREAS it was provided by the said order that the Tribunals of Inquiry (Evidence) Acts 1921 and 1979 applied to the Tribunal

WHEREAS it is provided by Section 2 (*b*) of the Tribunals of Inquiry (Evidence) Act, 1921 that a Tribunal to which the Act is applied shall have power to authorise the representation before it by solicitor or counsel or otherwise of any person appearing to it to be interested; and

WHEREAS Taher Meats (Naser Taher) (hereinafter referred to as "the applicants") applied to the said Tribunal to be represented before it by counsel instructed by solicitor and on the 14th May 1993 such authorisation under the said section was granted; and

WHEREAS it is provided by section 6 of the Tribunals of Inquiry (Evidence)(Amendment) Act, 1979, that if a Tribunal is of opinion that having regard to its findings and all other relevant matters there are sufficient reasons rendering it equitable to do so it may by Order direct that the whole or part of the costs of any person appearing before the Tribunal by counsel or solicitor as taxed by a Taxing Master of the High Court be paid by any other person named in the Order; and

WHEREAS by section 4 of the said Act of 1979 it is provided that a Tribunal may make such Order as it considers necessary for the purpose of its functions; and

WHEREAS the Tribunal referred to in the title hereof is satisfied that having regard to its findings and all other relevant matters there are sufficient reasons rendering it equitable that the Minister for Finance should pay the costs of the applicants as hereinafter appearing

IT IS HEREBY ORDERED that a Taxing Master of the High Court do tax in the manner hereinafter appearing the applicants' costs of appearing before the said Tribunal by one counsel instructed by solicitor.

In relation to the taxation of the applicants' costs the Tribunal DOTH ORDER as follows:

(*a*) That the statutory provisions and the Rules of the Superior Courts relating to the taxation of costs in an action in the High Courts (including the provisions relating to review and appeal) shall, in so far as is practicable, apply;

(*b*) That the costs of employing a solicitor and one counsel be taxed on a party and party basis.

(*c*) That the costs do include such witnesses' expenses as relate to those witnesses who actually gave oral evidence before the Tribunal on the proposal of the applicants;

(*d*) That the amount of witnesses' expenses be (*a*) such fee for the preparation of any statement prepared by the witness which was received by the Tribunal as part of the witnesses' testimony as the Taxing Master (or on review or appeal, the Court) considers reasonable; (*b*) such fee (not already paid) for attending before the Tribunal on such days as the witness gave oral evidence as the Taxing Master (or, on review or appeal, the Court) considers reasonable and (*b*) the travelling and subsistence expenses (not already paid) reasonably incurred for the purpose of giving oral evidence before the Tribunal.

(*e*) That the Minister for Finance be at liberty to attend and be heard on the said taxation and any review or appeal in relation to it.

(*f*) That the Minister for Finance do pay the applicants' costs when taxed and ascertained.

Dated this 29th day of July 1994.

Signed Liam Hamilton

The Tribunal

ANNEXE

IN THE MATTER OF the Tribunals of Inquiry (Evidence) Acts, 1921 and 1979; and IN THE MATTER OF a Tribunal of Inquiry established pursuant to Resolutions of Dail Eireann passed on the 24th May 1991, and by Seanad Eireann on the 29th May 1991, to inquire into allegations regarding illegal activities fraud and malpractice in and in connection with the beef processing industry by order of the Minister for Agriculture and Food made on the 31st day of May 1991.

AND WHEREAS it was provided by the said order that the Tribunals of Inquiry (Evidence) Acts 1921 and 1979 applied to the Tribunal

WHEREAS it is provided by Section 2 (*b*) of the Tribunals of Inquiry (Evidence) Act, 1921 that a Tribunal to which the Act is applied shall have power to authorise the representation before it by solicitor or counsel or otherwise of any person appearing to it to be interested; and

WHEREAS Avrich Limited T/A Freshlands Foods (hereinafter referred to as "the applicants") applied to the said Tribunal to be represented before it by counsel instructed by solicitor and on the 14th May 1993 such authorisation under the said section was granted; and

WHEREAS it is provided by section 6 of the Tribunals of Inquiry (Evidence)(Amendment) Act, 1979, that if a Tribunal is of opinion that having regard to its findings and all other relevant matters there are sufficient reasons rendering it equitable to do so it may by Order direct that the whole or part of the costs of any person appearing before the Tribunal by counsel or solicitor as taxed by a Taxing Master of the High Court be paid by any other person named in the Order; and

WHEREAS by section 4 of the said Act of 1979 it is provided that a Tribunal may make such Order as it considers necessary for the purpose of its functions; and

WHEREAS the Tribunal referred to in the title hereof is satisfied that having regard to its findings and all other relevant matters there are sufficient reasons rendering it equitable that the Minister for Finance should pay the costs of the applicants as hereinafter appearing

IT IS HEREBY ORDERED that a Taxing Master of the High Court do tax in the manner hereinafter appearing the applicants' costs of appearing before the said Tribunal by one counsel instructed by solicitor.

In relation to the taxation of the applicants' costs the Tribunal DOTH ORDER as follows:

(*a*) That the statutory provisions and the Rules of the Superior Courts relating to the taxation of costs in an action in the High Courts (including the provisions relating to review and appeal) shall, in so far as is practicable, apply;

(*b*) That the costs of employing a solicitor and one counsel be taxed on a party and party basis.

(*c*) That the costs do include such witnesses' expenses as relate to those witnesses who actually gave oral evidence before the Tribunal on the proposal of the applicants;

(*d*) That the amount of witnesses' expenses be (*a*) such fee for the preparation of any statement prepared by the witness which was received by the Tribunal as part of the witnesses' testimony as the Taxing Master (or on review or appeal, the Court) considers reasonable; (*b*) such fee (not already paid) for attending before the Tribunal on such days as the witness gave oral evidence as the Taxing Master (or, on review or appeal, the Court) considers reasonable and (*b*) the travelling and subsistence expenses (not already paid) reasonably incurred for the purpose of giving oral evidence before the Tribunal.

(*e*) That the Minister for Finance be at liberty to attend and be heard on the said taxation and any review or appeal in relation to it.

(*f*) That the Minister for Finance do pay the applicants' costs when taxed and ascertained.

Dated this 29th day of July 1994.

Signed Liam Hamilton
The Tribunal

ANNEXE

IN THE MATTER OF the Tribunals of Inquiry (Evidence) Acts, 1921 and 1979; and IN THE MATTER OF a Tribunal of Inquiry established pursuant to Resolutions of Dail Eireann passed on the 24th May 1991, and by Seanad Eireann on the 29th May 1991, to inquire into allegations regarding illegal activities fraud and malpractice in and in connection with the beef processing industry by order of the Minister for Agriculture and Food made on the 31st day of May 1991.

AND WHEREAS it was provided by the said order that the Tribunals of Inquiry (Evidence) Acts 1921 and 1979 applied to the Tribunal

WHEREAS it is provided by Section 2 (*b*) of the Tribunals of Inquiry (Evidence) Act, 1921 that a Tribunal to which the Act is applied shall have power to authorise the representation before it by solicitor or counsel or otherwise of any person appearing to it to be interested; and

WHEREAS Western Meat Producers Limited (hereinafter referred to as "the applicants") applied to the said Tribunal to be represented before it by counsel instructed by solicitor and on the 18th May 1993 such authorisation under the said section was granted; and

WHEREAS it is provided by section 6 of the Tribunals of Inquiry (Evidence)(Amendment) Act, 1979, that if a Tribunal is of opinion that having regard to its findings and all other relevant matters there are sufficient reasons rendering it equitable to do so it may by Order direct that the whole or part of the costs of any person appearing before the Tribunal by counsel or solicitor as taxed by a Taxing Master of the High Court be paid by any other person named in the Order; and

WHEREAS by section 4 of the said Act of 1979 it is provided that a Tribunal may make such Order as it considers necessary for the purpose of its functions; and

WHEREAS the Tribunal referred to in the title hereof is satisfied that having regard to its findings and all other relevant matters there are sufficient reasons rendering it equitable that the Minister for Finance should pay the costs of the applicants as hereinafter appearing

IT IS HEREBY ORDERED that a Taxing Master of the High Court do tax in the manner hereinafter appearing the applicants' costs of appearing before the said Tribunal by one counsel instructed by solicitor.

In relation to the taxation of the applicants' costs the Tribunal DOTH ORDER as follows:

(*a*) That the statutory provisions and the Rules of the Superior Courts relating to the taxation of costs in an action in the High Courts (including the provisions relating to review and appeal) shall, in so far as is practicable, apply;

(*b*) That the costs of employing a solicitor and one counsel be taxed on a party and party basis.

(*c*) That the costs do include such witnesses' expenses as relate to those witnesses who actually gave oral evidence before the Tribunal on the proposal of the applicants;

(*d*) That the amount of witnesses' expenses be (*a*) such fee for the preparation of any statement prepared by the witness which was received by the Tribunal as part of the witnesses' testimony as the Taxing Master (or on review or appeal, the Court) considers reasonable; (*b*) such fee (not already paid) for attending before the Tribunal on such days as the witness gave oral evidence as the Taxing Master (or, on review or appeal, the Court) considers reasonable and (*b*) the travelling and subsistence expenses (not already paid) reasonably incurred for the purpose of giving oral evidence before the Tribunal.

(*e*) That the Minister for Finance be at liberty to attend and be heard on the said taxation and any review or appeal in relation to it.

(*f*) That the Minister for Finance do pay the applicants' costs when taxed and ascertained.

Dated this 29th day of July 1994.

Signed Liam Hamilton

The Tribunal

ANNEXE

IN THE MATTER OF the Tribunals of Inquiry (Evidence) Acts, 1921 and 1979; and IN THE MATTER OF a Tribunal of Inquiry established pursuant to Resolutions of Dail Eireann passed on the 24th May 1991, and by Seanad Eireann on the 29th May 1991, to inquire into allegations regarding illegal activities fraud and malpractice in and in connection with the beef processing industry by order of the Minister for Agriculture and Food made on the 31st day of May 1991.

AND WHEREAS it was provided by the said order that the Tribunals of Inquiry (Evidence) Acts 1921 and 1979 applied to the Tribunal

WHEREAS it is provided by Section 2 (*b*) of the Tribunals of Inquiry (Evidence) Act, 1921 that a Tribunal to which the Act is applied shall have power to authorise the representation before it by solicitor or counsel or otherwise of any person appearing to it to be interested; and

WHEREAS Transfreeze Coldstores Limited (hereinafter referred to as "the applicants") applied to the said Tribunal to be represented before it by counsel instructed by solicitor and on the 18th May 1993 such authorisation under the said section was granted; and

WHEREAS it is provided by section 6 of the Tribunals of Inquiry (Evidence)(Amendment) Act, 1979, that if a Tribunal is of opinion that having regard to its findings and all other relevant matters there are sufficient reasons rendering it equitable to do so it may by Order direct that the whole or part of the costs of any person appearing before the Tribunal by counsel or solicitor as taxed by a Taxing Master of the High Court be paid by any other person named in the Order; and

WHEREAS by section 4 of the said Act of 1979 it is provided that a Tribunal may make such Order as it considers necessary for the purpose of its functions; and

WHEREAS the Tribunal referred to in the title hereof is satisfied that having regard to its findings and all other relevant matters there are sufficient reasons rendering it equitable that the Minister for Finance should pay the costs of the applicants as hereinafter appearing

IT IS HEREBY ORDERED that a Taxing Master of the High Court do tax in the manner hereinafter appearing the applicants' costs of appearing before the said Tribunal by one counsel instructed by solicitor.

In relation to the taxation of the applicants' costs the Tribunal DOTH ORDER as follows:

(*a*) That the statutory provisions and the Rules of the Superior Courts relating to the taxation of costs in an action in the High Courts (including the provisions relating to review and appeal) shall, in so far as is practicable, apply;

(*b*) That the costs of employing a solicitor and one counsel be taxed on a party and party basis.

(*c*) That the costs do include such witnesses' expenses as relate to those witnesses who actually gave oral evidence before the Tribunal on the proposal of the applicants;

(*d*) That the amount of witnesses' expenses be (*a*) such fee for the preparation of any statement prepared by the witness which was received by the Tribunal as part of the witnesses' testimony as the Taxing Master (or on review or appeal, the Court) considers reasonable; (*b*) such fee (not already paid) for attending before the Tribunal on such days as the witness gave oral evidence as the Taxing Master (or, on review or appeal, the Court) considers reasonable and (*b*) the travelling and subsistence expenses (not already paid) reasonably incurred for the purpose of giving oral evidence before the Tribunal.

(*e*) That the Minister for Finance be at liberty to attend and be heard on the said taxation and any review or appeal in relation to it.

(*f*) That the Minister for Finance do pay the applicants' costs when taxed and ascertained.

Dated this 29th day of July 1994.

Signed Liam Hamilton

The Tribunal

ANNEXE

IN THE MATTER OF the Tribunals of Inquiry (Evidence) Acts, 1921 and 1979; and IN THE MATTER OF a Tribunal of Inquiry established pursuant to Resolutions of Dail Eireann passed on the 24th May 1991, and by Seanad Eireann on the 29th May 1991, to inquire into allegations regarding illegal activities fraud and malpractice in and in connection with the beef processing industry by order of the Minister for Agriculture and Food made on the 31st day of May 1991.

AND WHEREAS it was provided by the said order that the Tribunals of Inquiry (Evidence) Acts 1921 and 1979 applied to the Tribunal

WHEREAS it is provided by Section 2 (*b*) of the Tribunals of Inquiry (Evidence) Act, 1921 that a Tribunal to which the Act is applied shall have power to authorise the representation before it by solicitor or counsel or otherwise of any person appearing to it to be interested; and

WHEREAS Blanchvac Limited (hereinafter referred to as "the applicants") applied to the said Tribunal to be represented before it by solicitor and on the 19th May 1993 such authorisation under the said section was granted; and

WHEREAS it is provided by section 6 of the Tribunals of Inquiry (Evidence)(Amendment) Act, 1979, that if a Tribunal is of opinion that having regard to its findings and all other relevant matters there are sufficient reasons rendering it equitable to do so it may by Order direct that the whole or part of the costs of any person appearing before the Tribunal by counsel or solicitor as taxed by a Taxing Master of the High Court be paid by any other person named in the Order; and

WHEREAS by section 4 of the said Act of 1979 it is provided that a Tribunal may make such Order as it considers necessary for the purpose of its functions; and

WHEREAS the Tribunal referred to in the title hereof is satisfied that having regard to its findings and all other relevant matters there are sufficient reasons rendering it equitable that the Minister for Finance should pay the costs of the applicants as hereinafter appearing

IT IS HEREBY ORDERED that a Taxing Master of the High Court do tax in the manner hereinafter appearing the applicants' costs of appearing before the said Tribunal by solicitor.

In relation to the taxation of the applicants' costs the Tribunal DOTH ORDER as follows:

(*a*) That the statutory provisions and the Rules of the Superior Courts relating to the taxation of costs in an action in the High Courts (including the provisions relating to review and appeal) shall, in so far as is practicable, apply;

(*b*) That the costs of employing a solicitor be taxed on a party and party basis.

(*c*) That the costs do include such witnesses' expenses as relate to those witnesses who actually gave oral evidence before the Tribunal on the proposal of the applicants;

(*d*) That the amount of witnesses' expenses be (*a*) such fee for the preparation of any statement prepared by the witness which was received by the Tribunal as part of the witnesses' testimony as the Taxing Master (or on review or appeal, the Court) considers reasonable; (*b*) such fee (not already paid) for attending before the Tribunal on such days as the witness gave oral evidence as the Taxing Master (or, on review or appeal, the Court) considers reasonable and (*b*) the travelling and subsistence expenses (not already paid) reasonably incurred for the purpose of giving oral evidence before the Tribunal.

(*e*) That the Minister for Finance be at liberty to attend and be heard on the said taxation and any review or appeal in relation to it.

(*f*) That the Minister for Finance do pay the applicants' costs when taxed and ascertained.

Dated this 29th day of July 1994.

Signed Liam Hamilton

The Tribunal

ANNEXE

IN THE MATTER OF the Tribunals of Inquiry (Evidence) Acts, 1921 and 1979; and IN THE MATTER OF a Tribunal of Inquiry established pursuant to Resolutions of Dail Eireann passed on the 24th May 1991, and by Seanad Eireann on the 29th May 1991, to inquire into allegations regarding illegal activities fraud and malpractice in and in connection with the beef processing industry by order of the Minister for Agriculture and Food made on the 31st day of May 1991.

AND WHEREAS it was provided by the said order that the Tribunals of Inquiry (Evidence) Acts 1921 and 1979 applied to the Tribunal

WHEREAS it is provided by Section 2 (*b*) of the Tribunals of Inquiry (Evidence) Act, 1921 that a Tribunal to which the Act is applied shall have power to authorise the representation before it by solicitor or counsel or otherwise of any person appearing to it to be interested; and

WHEREAS Baltinglass Meats Limited (hereinafter referred to as "the applicants") applied to the said Tribunal to be represented before it by solicitor and on the 19th May 1993 such authorisation under the said section was granted; and

WHEREAS it is provided by section 6 of the Tribunals of Inquiry (Evidence)(Amendment) Act, 1979, that if a Tribunal is of opinion that having regard to its findings and all other relevant matters there are sufficient reasons rendering it equitable to do so it may by Order direct that the whole or part of the costs of any person appearing before the Tribunal by counsel or solicitor as taxed by a Taxing Master of the High Court be paid by any other person named in the Order; and

WHEREAS by section 4 of the said Act of 1979 it is provided that a Tribunal may make such Order as it considers necessary for the purpose of its functions; and

WHEREAS the Tribunal referred to in the title hereof is satisfied that having regard to its findings and all other relevant matters there are sufficient reasons rendering it equitable that the Minister for Finance should pay the costs of the applicants as hereinafter appearing

IT IS HEREBY ORDERED that a Taxing Master of the High Court do tax in the manner hereinafter appearing the applicants' costs of appearing before the said Tribunal by solicitor.

In relation to the taxation of the applicants' costs the Tribunal DOTH ORDER as follows:

(*a*) That the statutory provisions and the Rules of the Superior Courts relating to the taxation of costs in an action in the High Courts (including the provisions relating to review and appeal) shall, in so far as is practicable, apply;

(*b*) That the costs of employing a solicitor be taxed on a party and party basis.

(*c*) That the costs do include such witnesses' expenses as relate to those witnesses who actually gave oral evidence before the Tribunal on the proposal of the applicants;

(*d*) That the amount of witnesses' expenses be (*a*) such fee for the preparation of any statement prepared by the witness which was received by the Tribunal as part of the witnesses' testimony as the Taxing Master (or on review or appeal, the Court) considers reasonable; (*b*) such fee (not already paid) for attending before the Tribunal on such days as the witness gave oral evidence as the Taxing Master (or, on review or appeal, the Court) considers reasonable and (*b*) the travelling and subsistence expenses (not already paid) reasonably incurred for the purpose of giving oral evidence before the Tribunal.

(*e*) That the Minister for Finance be at liberty to attend and be heard on the said taxation and any review or appeal in relation to it.

(*f*) That the Minister for Finance do pay the applicants' costs when taxed and ascertained.

Dated this 29th day of July 1994.

Signed Liam Hamilton

The Tribunal

ANNEXE

IN THE MATTER OF the Tribunals of Inquiry (Evidence) Acts, 1921 and 1979; and IN THE MATTER OF a Tribunal of Inquiry established pursuant to Resolutions of Dail Eireann passed on the 24th May 1991, and by Seanad Eireann on the 29th May 1991, to inquire into allegations regarding illegal activities fraud and malpractice in and in connection with the beef processing industry by order of the Minister for Agriculture and Food made on the 31st day of May 1991.

AND WHEREAS it was provided by the said order that the Tribunals of Inquiry (Evidence) Acts 1921 and 1979 applied to the Tribunal

WHEREAS it is provided by Section 2 (*b*) of the Tribunals of Inquiry (Evidence) Act, 1921 that a Tribunal to which the Act is applied shall have power to authorise the representation before it by solicitor or counsel or otherwise of any person appearing to it to be interested; and

WHEREAS Kildare Chilling Company Limited (hereinafter referred to as "the applicants") applied to the said Tribunal to be represented before it by counsel instructed by solicitor and on the 20th May 1993 such authorisation under the said section was granted; and

WHEREAS it is provided by section 6 of the Tribunals of Inquiry (Evidence)(Amendment) Act, 1979, that if a Tribunal is of opinion that having regard to its findings and all other relevant matters there are sufficient reasons rendering it equitable to do so it may by Order direct that the whole or part of the costs of any person appearing before the Tribunal by counsel or solicitor as taxed by a Taxing Master of the High Court be paid by any other person named in the Order; and

WHEREAS by section 4 of the said Act of 1979 it is provided that a Tribunal may make such Order as it considers necessary for the purpose of its functions; and

WHEREAS the Tribunal referred to in the title hereof is satisfied that having regard to its findings and all other relevant matters there are sufficient reasons rendering it equitable that the Minister for Finance should pay the costs of the applicants as hereinafter appearing

IT IS HEREBY ORDERED that a Taxing Master of the High Court do tax in the manner hereinafter appearing the applicants' costs of appearing before the said Tribunal by one counsel instructed by solicitor.

In relation to the taxation of the applicants' costs the Tribunal DOTH ORDER as follows:

(*a*) That the statutory provisions and the Rules of the Superior Courts relating to the taxation of costs in an action in the High Courts (including the provisions relating to review and appeal) shall, in so far as is practicable, apply;

(*b*) That the costs of employing a solicitor and one counsel be taxed on a party and party basis.

(*c*) That the costs do include such witnesses' expenses as relate to the those witnesses who actually gave oral evidence before the Tribunal on the proposal of the applicants;

(*d*) That the amount of witnesses' expenses be (*a*) such fee for the preparation of any statement prepared by the witness which was received by the Tribunal as part of the witnesses' testimony as the Taxing Master (or on review or appeal, the Court) considers reasonable; (*b*) such fee (not already paid) for attending before the Tribunal on such days as the witness gave oral evidence as the Taxing Master (or, on review or appeal, the Court) considers reasonable and (*b*) the travelling and subsistence expenses (not already paid) reasonably incurred for the purpose of giving oral evidence before the Tribunal.

(*e*) That the Minister for Finance be at liberty to attend and be heard on the said taxation and any review or appeal in relation to it.

(*f*) That the Minister for Finance do pay the applicants' costs when taxed and ascertained.

Dated this 29th day of July 1994.

Signed Liam Hamilton

The Tribunal

ANNEXE

IN THE MATTER OF the Tribunals of Inquiry (Evidence) Acts, 1921 and 1979; and IN THE MATTER OF a Tribunal of Inquiry established pursuant to Resolutions of Dail Eireann passed on the 24th May 1991, and by Seanad Eireann on the 29th May 1991, to inquire into allegations regarding illegal activities fraud and malpractice in and in connection with the beef processing industry by order of the Minister for Agriculture and Food made on the 31st day of May 1991.

AND WHEREAS it was provided by the said order that the Tribunals of Inquiry (Evidence) Acts 1921 and 1979 applied to the Tribunal

WHEREAS it is provided by Section 2 (*b*) of the Tribunals of Inquiry (Evidence) Act, 1921 that a Tribunal to which the Act is applied shall have power to authorise the representation before it by solicitor or counsel or otherwise of any person appearing to it to be interested; and

WHEREAS Barford Meats Limited (hereinafter referred to as "the applicants") applied to the said Tribunal to be represented before it by counsel instructed by solicitor and on the 21st May 1993 such authorisation under the said section was granted; and

WHEREAS it is provided by section 6 of the Tribunals of Inquiry (Evidence)(Amendment) Act, 1979, that if a Tribunal is of opinion that having regard to its findings and all other relevant matters there are sufficient reasons rendering it equitable to do so it may by Order direct that the whole or part of the costs of any person appearing before the Tribunal by counsel or solicitor as taxed by a Taxing Master of the High Court be paid by any other person named in the Order; and

WHEREAS by section 4 of the said Act of 1979 it is provided that a Tribunal may make such Order as it considers necessary for the purpose of its functions; and

WHEREAS the Tribunal referred to in the title hereof is satisfied that having regard to its findings and all other relevant matters there are sufficient reasons rendering it equitable that the Minister for Finance should pay the costs of the applicants as hereinafter appearing

IT IS HEREBY ORDERED that a Taxing Master of the High Court do tax in the manner hereinafter appearing the applicants' costs of appearing before the said Tribunal by one counsel instructed by solicitor.

In relation to the taxation of the applicants' costs the Tribunal DOTH ORDER as follows:

(*a*) That the statutory provisions and the Rules of the Superior Courts relating to the taxation of costs in an action in the High Courts (including the provisions relating to review and appeal) shall, in so far as is practicable, apply;

(*b*) That the costs of employing a solicitor and one counsel be taxed on a party and party basis.

(*c*) That the costs do include such witnesses' expenses as relate to the those witnesses who actually gave oral evidence before the Tribunal on the proposal of the applicants;

(*d*) That the amount of witnesses' expenses be (*a*) such fee for the preparation of any statement prepared by the witness which was received by the Tribunal as part of the witnesses' testimony as the Taxing Master (or on review or appeal, the Court) considers reasonable; (*b*) such fee (not already paid) for attending before the Tribunal on such days as the witness gave oral evidence as the Taxing Master (or, on review or appeal, the Court) considers reasonable and (*b*) the travelling and subsistence expenses (not already paid) reasonably incurred for the purpose of giving oral evidence before the Tribunal.

(*e*) That the Minister for Finance be at liberty to attend and be heard on the said taxation and any review or appeal in relation to it.

(*f*) That the Minister for Finance do pay the applicants' costs when taxed and ascertained.

Dated this 29th day of July 1994.

Signed Liam Hamilton

The Tribunal

ANNEXE

IN THE MATTER OF the Tribunals of Inquiry (Evidence) Acts, 1921 and 1979; and IN THE MATTER OF a Tribunal of Inquiry established pursuant to Resolutions of Dail Eireann passed on the 24th May 1991, and by Seanad Eireann on the 29th May 1991, to inquire into allegations regarding illegal activities fraud and malpractice in and in connection with the beef processing industry by order of the Minister for Agriculture and Food made on the 31st day of May 1991.

AND WHEREAS it was provided by the said order that the Tribunals of Inquiry (Evidence) Acts 1921 and 1979 applied to the Tribunal

WHEREAS it is provided by Section 2 (*b*) of the Tribunals of Inquiry (Evidence) Act, 1921 that a Tribunal to which the Act is applied shall have power to authorise the representation before it by solicitor or counsel or otherwise of any person appearing to it to be interested; and

WHEREAS C H Foods Limited (hereinafter referred to as "the applicants") applied to the said Tribunal to be represented before it by counsel instructed by solicitor and on the 21st May 1993 such authorisation under the said section was granted; and

WHEREAS it is provided by section 6 of the Tribunals of Inquiry (Evidence)(Amendment) Act, 1979, that if a Tribunal is of opinion that having regard to its findings and all other relevant matters there are sufficient reasons rendering it equitable to do so it may by Order direct that the whole or part of the costs of any person appearing before the Tribunal by counsel or solicitor as taxed by a Taxing Master of the High Court be paid by any other person named in the Order; and

WHEREAS by section 4 of the said Act of 1979 it is provided that a Tribunal may make such Order as it considers necessary for the purpose of its functions; and

WHEREAS the Tribunal referred to in the title hereof is satisfied that having regard to its findings and all other relevant matters there are sufficient reasons rendering it equitable that the Minister for Finance should pay the costs of the applicants as hereinafter appearing

IT IS HEREBY ORDERED that a Taxing Master of the High Court do tax in the manner hereinafter appearing the applicants' costs of appearing before the said Tribunal by one counsel instructed by solicitor.

In relation to the taxation of the applicants' costs the Tribunal DOTH ORDER as follows:

(*a*) That the statutory provisions and the Rules of the Superior Courts relating to the taxation of costs in an action in the High Courts (including the provisions relating to review and appeal) shall, in so far as is practicable, apply;

(*b*) That the costs of employing a solicitor and one counsel be taxed on a party and party basis.

(*c*) That the costs do include such witnesses' expenses as relate to the those witnesses who actually gave oral evidence before the Tribunal on the proposal of the applicants;

(*d*) That the amount of witnesses' expenses be (*a*) such fee for the preparation of any statement prepared by the witness which was received by the Tribunal as part of the witnesses' testimony as the Taxing Master (or on review or appeal, the Court) considers reasonable; (*b*) such fee (not already paid) for attending before the Tribunal on such days as the witness gave oral evidence as the Taxing Master (or, on review or appeal, the Court) considers reasonable and (*b*) the travelling and subsistence expenses (not already paid) reasonably incurred for the purpose of giving oral evidence before the Tribunal.

(*e*) That the Minister for Finance be at liberty to attend and be heard on the said taxation and any review or appeal in relation to it.

(*f*) That the Minister for Finance do pay the applicants' costs when taxed and ascertained.

Dated this 29th day of July 1994.

Signed Liam Hamilton

The Tribunal

ANNEXE

IN THE MATTER OF the Tribunals of Inquiry (Evidence) Acts, 1921 and 1979; and IN THE MATTER OF a Tribunal of Inquiry established pursuant to Resolutions of Dail Eireann passed on the 24th May 1991, and by Seanad Eireann on the 29th May 1991, to inquire into allegations regarding illegal activities fraud and malpractice in and in connection with the beef processing industry by order of the Minister for Agriculture and Food made on the 31st day of May 1991.

AND WHEREAS it was provided by the said order that the Tribunals of Inquiry (Evidence) Acts 1921 and 1979 applied to the Tribunal

WHEREAS it is provided by Section 2 (*b*) of the Tribunals of Inquiry (Evidence) Act, 1921 that a Tribunal to which the Act is applied shall have power to authorise the representation before it by solicitor or counsel or otherwise of any person appearing to it to be interested; and

WHEREAS Continental Beef Packers Limited (hereinafter referred to as "the applicants") applied to the said Tribunal to be represented before it by solicitor and on the 21st May 1993 such authorisation under the said section was granted; and

WHEREAS it is provided by section 6 of the Tribunals of Inquiry (Evidence)(Amendment) Act, 1979, that if a Tribunal is of opinion that having regard to its findings and all other relevant matters there are sufficient reasons rendering it equitable to do so it may by Order direct that the whole or part of the costs of any person appearing before the Tribunal by counsel or solicitor as taxed by a Taxing Master of the High Court be paid by any other person named in the Order; and

WHEREAS by section 4 of the said Act of 1979 it is provided that a Tribunal may make such Order as it considers necessary for the purpose of its functions; and

WHEREAS the Tribunal referred to in the title hereof is satisfied that having regard to its findings and all other relevant matters there are sufficient reasons rendering it equitable that the Minister for Finance should pay the costs of the applicants as hereinafter appearing

IT IS HEREBY ORDERED that a Taxing Master of the High Court do tax in the manner hereinafter appearing the applicants' costs of appearing before the said Tribunal by solicitor.

In relation to the taxation of the applicants' costs the Tribunal DOTH ORDER as follows:

(*a*) That the statutory provisions and the Rules of the Superior Courts relating to the taxation of costs in an action in the High Courts (including the provisions relating to review and appeal) shall, in so far as is practicable, apply;

(*b*) That the costs of employing a solicitor be taxed on a party and party basis.

(*c*) That the costs do include such witnesses' expenses as relate to the those witnesses who actually gave oral evidence before the Tribunal on the proposal of the applicants;

(*d*) That the amount of witnesses' expenses be (*a*) such fee for the preparation of any statement prepared by the witness which was received by the Tribunal as part of the witnesses' testimony as the Taxing Master (or on review or appeal, the Court) considers reasonable; (*b*) such fee (not already paid) for attending before the Tribunal on such days as the witness gave oral evidence as the Taxing Master (or, on review or appeal, the Court) considers reasonable and (*b*) the travelling and subsistence expenses (not already paid) reasonably incurred for the purpose of giving oral evidence before the Tribunal.

(*e*) That the Minister for Finance be at liberty to attend and be heard on the said taxation and any review or appeal in relation to it.

(*f*) That the Minister for Finance do pay the applicants' costs when taxed and ascertained.

Dated this 29th day of July 1994.

Signed Liam Hamilton

The Tribunal

ANNEXE

IN THE MATTER OF the Tribunals of Inquiry (Evidence) Acts, 1921 and 1979; and IN THE MATTER OF a Tribunal of Inquiry established pursuant to Resolutions of Dail Eireann passed on the 24th May 1991, and by Seanad Eireann on the 29th May 1991, to inquire into allegations regarding illegal activities fraud and malpractice in and in connection with the beef processing industry by order of the Minister for Agriculture and Food made on the 31st day of May 1991.

AND WHEREAS it was provided by the said order that the Tribunals of Inquiry (Evidence) Acts 1921 and 1979 applied to the Tribunal

WHEREAS it is provided by Section 2 (*b*) of the Tribunals of Inquiry (Evidence) Act, 1921 that a Tribunal to which the Act is applied shall have power to authorise the representation before it by solicitor or counsel or otherwise of any person appearing to it to be interested; and

WHEREAS Mr Christopher O'Brien and Mr Patrick Fox (hereinafter referred to as "the applicants") applied to the said Tribunal to be represented before it by counsel instructed by solicitor and on the 25th May 1993 such authorisation under the said section was granted; and

WHEREAS it is provided by section 6 of the Tribunals of Inquiry (Evidence)(Amendment) Act, 1979, that if a Tribunal is of opinion that having regard to its findings and all other relevant matters there are sufficient reasons rendering it equitable to do so it may by Order direct that the whole or part of the costs of any person appearing before the Tribunal by counsel or solicitor as taxed by a Taxing Master of the High Court be paid by any other person named in the Order; and

WHEREAS by section 4 of the said Act of 1979 it is provided that a Tribunal may make such Order as it considers necessary for the purpose of its functions; and

WHEREAS the Tribunal referred to in the title hereof is satisfied that having regard to its findings and all other relevant matters there are sufficient reasons rendering it equitable that the Minister for Finance should pay the costs of the applicants as hereinafter appearing

IT IS HEREBY ORDERED that a Taxing Master of the High Court do tax in the manner hereinafter appearing the applicants' costs of appearing before the said Tribunal by one counsel instructed by solicitor.

In relation to the taxation of the applicants' costs the Tribunal DOTH ORDER as follows:

(*a*) That the statutory provisions and the Rules of the Superior Courts relating to the taxation of costs in an action in the High Courts (including the provisions relating to review and appeal) shall, in so far as is practicable, apply;

(*b*) That the costs of employing a solicitor and one counsel be taxed on a party and party basis.

(*c*) That the costs do include such witnesses' expenses as relate to the those witnesses who actually gave oral evidence before the Tribunal on the proposal of the applicants;

(*d*) That the amount of witnesses' expenses be (*a*) such fee for the preparation of any statement prepared by the witness which was received by the Tribunal as part of the witnesses' testimony as the Taxing Master (or on review or appeal, the Court) considers reasonable; (*b*) such fee (not already paid) for attending before the Tribunal on such days as the witness gave oral evidence as the Taxing Master (or, on review or appeal, the Court) considers reasonable and (*b*) the travelling and subsistence expenses (not already paid) reasonably incurred for the purpose of giving oral evidence before the Tribunal.

(*e*) That the Minister for Finance be at liberty to attend and be heard on the said taxation and any review or appeal in relation to it.

(*f*) That the Minister for Finance do pay the applicants' costs when taxed and ascertained.

Dated this 29th day of July 1994.

Signed Liam Hamilton

The Tribunal

ANNEXE

IN THE MATTER OF the Tribunals of Inquiry (Evidence) Acts, 1921 and 1979; and IN THE MATTER OF a Tribunal of Inquiry established pursuant to Resolutions of Dail Eireann passed on the 24th May 1991, and by Seanad Eireann on the 29th May 1991, to inquire into allegations regarding illegal activities fraud and malpractice in and in connection with the beef processing industry by order of the Minister for Agriculture and Food made on the 31st day of May 1991.

AND WHEREAS it was provided by the said order that the Tribunals of Inquiry (Evidence) Acts 1921 and 1979 applied to the Tribunal

WHEREAS it is provided by Section 2 (*b*) of the Tribunals of Inquiry (Evidence) Act, 1921 that a Tribunal to which the Act is applied shall have power to authorise the representation before it by solicitor or counsel or otherwise of any person appearing to it to be interested; and

WHEREAS Norish Plc (hereinafter referred to as "the applicants") applied to the said Tribunal to be represented before it by solicitor and on the 25th May 1993 such authorisation under the said section was granted; and

WHEREAS it is provided by section 6 of the Tribunals of Inquiry (Evidence)(Amendment) Act, 1979, that if a Tribunal is of opinion that having regard to its findings and all other relevant matters there are sufficient reasons rendering it equitable to do so it may by Order direct that the whole or part of the costs of any person appearing before the Tribunal by counsel or solicitor as taxed by a Taxing Master of the High Court be paid by any other person named in the Order; and

WHEREAS by section 4 of the said Act of 1979 it is provided that a Tribunal may make such Order as it considers necessary for the purpose of its functions; and

WHEREAS the Tribunal referred to in the title hereof is satisfied that having regard to its findings and all other relevant matters there are sufficient reasons rendering it equitable that the Minister for Finance should pay the costs of the applicants as hereinafter appearing

IT IS HEREBY ORDERED that a Taxing Master of the High Court do tax in the manner hereinafter appearing the applicants' costs of appearing before the said Tribunal by solicitor.

In relation to the taxation of the applicants' costs the Tribunal DOTH ORDER as follows:

(*a*) That the statutory provisions and the Rules of the Superior Courts relating to the taxation of costs in an action in the High Courts (including the provisions relating to review and appeal) shall, in so far as is practicable, apply;

(*b*) That the costs of employing a solicitor be taxed on a party and party basis.

(*c*) That the costs do include such witnesses' expenses as relate to the those witnesses who actually gave oral evidence before the Tribunal on the proposal of the applicants;

(*d*) That the amount of witnesses' expenses be (*a*) such fee for the preparation of any statement prepared by the witness which was received by the Tribunal as part of the witnesses' testimony as the Taxing Master (or on review or appeal, the Court) considers reasonable; (*b*) such fee (not already paid) for attending before the Tribunal on such days as the witness gave oral evidence as the Taxing Master (or, on review or appeal, the Court) considers reasonable and (*b*) the travelling and subsistence expenses (not already paid) reasonably incurred for the purpose of giving oral evidence before the Tribunal.

(*e*) That the Minister for Finance be at liberty to attend and be heard on the said taxation and any review or appeal in relation to it.

(*f*) That the Minister for Finance do pay the applicants' costs when taxed and ascertained.

Dated this 29th day of July 1994.

Signed Liam Hamilton
The Tribunal

ANNEXE

IN THE MATTER OF the Tribunals of Inquiry (Evidence) Acts, 1921 and 1979; and IN THE MATTER OF a Tribunal of Inquiry established pursuant to Resolutions of Dail Eireann passed on the 24th May 1991, and by Seanad Eireann on the 29th May 1991, to inquire into allegations regarding illegal activities fraud and malpractice in and in connection with the beef processing industry by order of the Minister for Agriculture and Food made on the 31st day of May 1991.

AND WHEREAS it was provided by the said order that the Tribunals of Inquiry (Evidence) Acts 1921 and 1979 applied to the Tribunal

WHEREAS it is provided by Section 2 (*b*) of the Tribunals of Inquiry (Evidence) Act, 1921 that a Tribunal to which the Act is applied shall have power to authorise the representation before it by solicitor or counsel or otherwise of any person appearing to it to be interested; and

WHEREAS Horgan Meats Limited (in Liquidation) (hereinafter referred to as "the applicants") applied to the said Tribunal to be represented before it by counsel instructed by solicitor and on the 25th May 1993 such authorisation under the said section was granted; and

WHEREAS it is provided by section 6 of the Tribunals of Inquiry (Evidence)(Amendment) Act, 1979, that if a Tribunal is of opinion that having regard to its findings and all other relevant matters there are sufficient reasons rendering it equitable to do so it may by Order direct that the whole or part of the costs of any person appearing before the Tribunal by counsel or solicitor as taxed by a Taxing Master of the High Court be paid by any other person named in the Order; and

WHEREAS by section 4 of the said Act of 1979 it is provided that a Tribunal may make such Order as it considers necessary for the purpose of its functions; and

WHEREAS the Tribunal referred to in the title hereof is satisfied that having regard to its findings and all other relevant matters there are sufficient reasons rendering it equitable that the Minister for Finance should pay the costs of the applicants as hereinafter appearing

IT IS HEREBY ORDERED that a Taxing Master of the High Court do tax in the manner hereinafter appearing the applicants' costs of appearing before the said Tribunal by one counsel instructed by solicitor.

In relation to the taxation of the applicants' costs the Tribunal DOTH ORDER as follows:

(*a*) That the statutory provisions and the Rules of the Superior Courts relating to the taxation of costs in an action in the High Courts (including the provisions relating to review and appeal) shall, in so far as is practicable, apply;

(*b*) That the costs of employing a solicitor and one counsel be taxed on a party and party basis.

(*c*) That the costs do include such witnesses' expenses as relate to the those witnesses who actually gave oral evidence before the Tribunal on the proposal of the applicants;

(*d*) That the amount of witnesses' expenses be (*a*) such fee for the preparation of any statement prepared by the witness which was received by the Tribunal as part of the witnesses' testimony as the Taxing Master (or on review or appeal, the Court) considers reasonable; (*b*) such fee (not already paid) for attending before the Tribunal on such days as the witness gave oral evidence as the Taxing Master (or, on review or appeal, the Court) considers reasonable and (*b*) the travelling and subsistence expenses (not already paid) reasonably incurred for the purpose of giving oral evidence before the Tribunal.

(*e*) That the Minister for Finance be at liberty to attend and be heard on the said taxation and any review or appeal in relation to it.

(*f*) That the Minister for Finance do pay the applicants' costs when taxed and ascertained.

Dated this 29th day of July 1994.

Signed Liam Hamilton

The Tribunal

ANNEXE

IN THE MATTER OF the Tribunals of Inquiry (Evidence) Acts, 1921 and 1979; and IN THE MATTER OF a Tribunal of Inquiry established pursuant to Resolutions of Dail Eireann passed on the 24th May 1991, and by Seanad Eireann on the 29th May 1991, to inquire into allegations regarding illegal activities fraud and malpractice in and in connection with the beef processing industry by order of the Minister for Agriculture and Food made on the 31st day of May 1991.

AND WHEREAS it was provided by the said order that the Tribunals of Inquiry (Evidence) Acts 1921 and 1979 applied to the Tribunal

WHEREAS it is provided by Section 2 (*b*) of the Tribunals of Inquiry (Evidence) Act, 1921 that a Tribunal to which the Act is applied shall have power to authorise the representation before it by solicitor or counsel or otherwise of any person appearing to it to be interested; and

WHEREAS Freezomatic Coldstores Company (in Liquidation) (hereinafter referred to as "the applicants") applied to the said Tribunal to be represented before it by solicitor and on the 26th May 1993 such authorisation under the said section was granted; and

WHEREAS it is provided by section 6 of the Tribunals of Inquiry (Evidence)(Amendment) Act, 1979, that if a Tribunal is of opinion that having regard to its findings and all other relevant matters there are sufficient reasons rendering it equitable to do so it may by Order direct that the whole or part of the costs of any person appearing before the Tribunal by counsel or solicitor as taxed by a Taxing Master of the High Court be paid by any other person named in the Order; and

WHEREAS by section 4 of the said Act of 1979 it is provided that a Tribunal may make such Order as it considers necessary for the purpose of its functions; and

WHEREAS the Tribunal referred to in the title hereof is satisfied that having regard to its findings and all other relevant matters there are sufficient reasons rendering it equitable that the Minister for Finance should pay the costs of the applicants as hereinafter appearing

IT IS HEREBY ORDERED that a Taxing Master of the High Court do tax in the manner hereinafter appearing the applicants' costs of appearing before the said Tribunal by solicitor.

In relation to the taxation of the applicants' costs the Tribunal DOTH ORDER as follows:

(*a*) That the statutory provisions and the Rules of the Superior Courts relating to the taxation of costs in an action in the High Courts (including the provisions relating to review and appeal) shall, in so far as is practicable, apply;

(*b*) That the costs of employing a solicitor be taxed on a party and party basis.

(*c*) That the costs do include such witnesses' expenses as relate to the those witnesses who actually gave oral evidence before the Tribunal on the proposal of the applicants;

(*d*) That the amount of witnesses' expenses be (*a*) such fee for the preparation of any statement prepared by the witness which was received by the Tribunal as part of the witnesses' testimony as the Taxing Master (or on review or appeal, the Court) considers reasonable; (*b*) such fee (not already paid) for attending before the Tribunal on such days as the witness gave oral evidence as the Taxing Master (or, on review or appeal, the Court) considers reasonable and (*b*) the travelling and subsistence expenses (not already paid) reasonably incurred for the purpose of giving oral evidence before the Tribunal.

(*e*) That the Minister for Finance be at liberty to attend and be heard on the said taxation and any review or appeal in relation to it.

(*f*) That the Minister for Finance do pay the applicants' costs when taxed and ascertained.

Dated this 29th day of July 1994.

Signed Liam Hamilton

The Tribunal

ANNEXE

IN THE MATTER OF the Tribunals of Inquiry (Evidence) Acts, 1921 and 1979; and IN THE MATTER OF a Tribunal of Inquiry established pursuant to Resolutions of Dail Eireann passed on the 24th May 1991, and by Seanad Eireann on the 29th May 1991, to inquire into allegations regarding illegal activities fraud and malpractice in and in connection with the beef processing industry by order of the Minister for Agriculture and Food made on the 31st day of May 1991.

AND WHEREAS it was provided by the said order that the Tribunals of Inquiry (Evidence) Acts 1921 and 1979 applied to the Tribunal

WHEREAS it is provided by Section 2 (*b*) of the Tribunals of Inquiry (Evidence) Act, 1921 that a Tribunal to which the Act is applied shall have power to authorise the representation before it by solicitor or counsel or otherwise of any person appearing to it to be interested; and

WHEREAS Liffey Meats Limited (hereinafter referred to as "the applicants") applied to the said Tribunal to be represented before it by counsel instructed by solicitor and on the 26th May 1993 such authorisation under the said section was granted; and

WHEREAS it is provided by section 6 of the Tribunals of Inquiry (Evidence)(Amendment) Act, 1979, that if a Tribunal is of opinion that having regard to its findings and all other relevant matters there are sufficient reasons rendering it equitable to do so it may by Order direct that the whole or part of the costs of any person appearing before the Tribunal by counsel or solicitor as taxed by a Taxing Master of the High Court be paid by any other person named in the Order; and

WHEREAS by section 4 of the said Act of 1979 it is provided that a Tribunal may make such Order as it considers necessary for the purpose of its functions; and

WHEREAS the Tribunal referred to in the title hereof is satisfied that having regard to its findings and all other relevant matters there are sufficient reasons rendering it equitable that the Minister for Finance should pay the costs of the applicants as hereinafter appearing

IT IS HEREBY ORDERED that a Taxing Master of the High Court do tax in the manner hereinafter appearing the applicants' costs of appearing before the said Tribunal by one counsel instructed by solicitor.

In relation to the taxation of the applicants' costs the Tribunal DOTH ORDER as follows:

(*a*) That the statutory provisions and the Rules of the Superior Courts relating to the taxation of costs in an action in the High Courts (including the provisions relating to review and appeal) shall, in so far as is practicable, apply;

(*b*) That the costs of employing a solicitor and one counsel be taxed on a party and party basis.

(*c*) That the costs do include such witnesses' expenses as relate to the those witnesses who actually gave oral evidence before the Tribunal on the proposal of the applicants;

(*d*) That the amount of witnesses' expenses be (*a*) such fee for the preparation of any statement prepared by the witness which was received by the Tribunal as part of the witnesses' testimony as the Taxing Master (or on review or appeal, the Court) considers reasonable; (*b*) such fee (not already paid) for attending before the Tribunal on such days as the witness gave oral evidence as the Taxing Master (or, on review or appeal, the Court) considers reasonable and (*b*) the travelling and subsistence expenses (not already paid) reasonably incurred for the purpose of giving oral evidence before the Tribunal.

(*e*) That the Minister for Finance be at liberty to attend and be heard on the said taxation and any review or appeal in relation to it.

(*f*) That the Minister for Finance do pay the applicants' costs when taxed and ascertained.

Dated this 29th day of July 1994.

Signed Liam Hamilton
The Tribunal

ANNEXE

IN THE MATTER OF the Tribunals of Inquiry (Evidence) Acts, 1921 and 1979; and IN THE MATTER OF a Tribunal of Inquiry established pursuant to Resolutions of Dail Eireann passed on the 24th May 1991, and by Seanad Eireann on the 29th May 1991, to inquire into allegations regarding illegal activities fraud and malpractice in and in connection with the beef processing industry by order of the Minister for Agriculture and Food made on the 31st day of May 1991.

AND WHEREAS it was provided by the said order that the Tribunals of Inquiry (Evidence) Acts 1921 and 1979 applied to the Tribunal

WHEREAS it is provided by Section 2 (*b*) of the Tribunals of Inquiry (Evidence) Act, 1921 that a Tribunal to which the Act is applied shall have power to authorise the representation before it by solicitor or counsel or otherwise of any person appearing to it to be interested; and

WHEREAS Meadow Meats (post 25/5/1993) (hereinafter referred to as "the applicants") applied to the said Tribunal to be represented before it by counsel instructed by solicitor and on the 1st June 1993 such authorisation under the said section was granted; and

WHEREAS it is provided by section 6 of the Tribunals of Inquiry (Evidence)(Amendment) Act, 1979, that if a Tribunal is of opinion that having regard to its findings and all other relevant matters there are sufficient reasons rendering it equitable to do so it may by Order direct that the whole or part of the costs of any person appearing before the Tribunal by counsel or solicitor as taxed by a Taxing Master of the High Court be paid by any other person named in the Order; and

WHEREAS by section 4 of the said Act of 1979 it is provided that a Tribunal may make such Order as it considers necessary for the purpose of its functions; and

WHEREAS the Tribunal referred to in the title hereof is satisfied that having regard to its findings and all other relevant matters there are sufficient reasons rendering it equitable that the Minister for Finance should pay the costs of the applicants as hereinafter appearing

IT IS HEREBY ORDERED that a Taxing Master of the High Court do tax in the manner hereinafter appearing the applicants' costs of appearing before the said Tribunal by one counsel instructed by solicitor.

In relation to the taxation of the applicants' costs the Tribunal DOTH ORDER as follows:

(*a*) That the statutory provisions and the Rules of the Superior Courts relating to the taxation of costs in an action in the High Courts (including the provisions relating to review and appeal) shall, in so far as is practicable, apply;

(*b*) That the costs of employing a solicitor and one counsel be taxed on a party and party basis.

(*c*) That the costs do include such witnesses' expenses as relate to the those witnesses who actually gave oral evidence before the Tribunal on the proposal of the applicants;

(*d*) That the amount of witnesses' expenses be (*a*) such fee for the preparation of any statement prepared by the witness which was received by the Tribunal as part of the witnesses' testimony as the Taxing Master (or on review or appeal, the Court) considers reasonable; (*b*) such fee (not already paid) for attending before the Tribunal on such days as the witness gave oral evidence as the Taxing Master (or, on review or appeal, the Court) considers reasonable and (*b*) the travelling and subsistence expenses (not already paid) reasonably incurred for the purpose of giving oral evidence before the Tribunal.

(*e*) That the Minister for Finance be at liberty to attend and be heard on the said taxation and any review or appeal in relation to it.

(*f*) That the Minister for Finance do pay the applicants' costs when taxed and ascertained.

Dated this 29th day of July 1994.

Signed Liam Hamilton

The Tribunal

ANNEXE

IN THE MATTER OF the Tribunals of Inquiry (Evidence) Acts, 1921 and 1979; and IN THE MATTER OF a Tribunal of Inquiry established pursuant to Resolutions of Dail Eireann passed on the 24th May 1991, and by Seanad Eireann on the 29th May 1991, to inquire into allegations regarding illegal activities fraud and malpractice in and in connection with the beef processing industry by order of the Minister for Agriculture and Food made on the 31st day of May 1991.

AND WHEREAS it was provided by the said order that the Tribunals of Inquiry (Evidence) Acts 1921 and 1979 applied to the Tribunal

WHEREAS it is provided by Section 2 (*b*) of the Tribunals of Inquiry (Evidence) Act, 1921 that a Tribunal to which the Act is applied shall have power to authorise the representation before it by solicitor or counsel or otherwise of any person appearing to it to be interested; and

WHEREAS Kerry Meat Packers (hereinafter referred to as "the applicants") applied to the said Tribunal to be represented before it by counsel instructed by solicitor and on the 2nd June 1993 such authorisation under the said section was granted; and

WHEREAS it is provided by section 6 of the Tribunals of Inquiry (Evidence)(Amendment) Act, 1979, that if a Tribunal is of opinion that having regard to its findings and all other relevant matters there are sufficient reasons rendering it equitable to do so it may by Order direct that the whole or part of the costs of any person appearing before the Tribunal by counsel or solicitor as taxed by a Taxing Master of the High Court be paid by any other person named in the Order; and

WHEREAS by section 4 of the said Act of 1979 it is provided that a Tribunal may make such Order as it considers necessary for the purpose of its functions; and

WHEREAS the Tribunal referred to in the title hereof is satisfied that having regard to its findings and all other relevant matters there are sufficient reasons rendering it equitable that the Minister for Finance should pay the costs of the applicants as hereinafter appearing

IT IS HEREBY ORDERED that a Taxing Master of the High Court do tax in the manner hereinafter appearing the applicants' costs of appearing before the said Tribunal by one counsel instructed by solicitor.

In relation to the taxation of the applicants' costs the Tribunal DOTH ORDER as follows:

(*a*) That the statutory provisions and the Rules of the Superior Courts relating to the taxation of costs in an action in the High Courts (including the provisions relating to review and appeal) shall, in so far as is practicable, apply;

(*b*) That the costs of employing a solicitor and one counsel be taxed on a party and party basis.

(*c*) That the costs do include such witnesses' expenses as relate to the those witnesses who actually gave oral evidence before the Tribunal on the proposal of the applicants;

(*d*) That the amount of witnesses' expenses be (*a*) such fee for the preparation of any statement prepared by the witness which was received by the Tribunal as part of the witnesses' testimony as the Taxing Master (or on review or appeal, the Court) considers reasonable; (*b*) such fee (not already paid) for attending before the Tribunal on such days as the witness gave oral evidence as the Taxing Master (or, on review or appeal, the Court) considers reasonable and (*b*) the travelling and subsistence expenses (not already paid) reasonably incurred for the purpose of giving oral evidence before the Tribunal.

(*e*) That the Minister for Finance be at liberty to attend and be heard on the said taxation and any review or appeal in relation to it.

(*f*) That the Minister for Finance do pay the applicants' costs when taxed and ascertained.

Dated this 29th day of July 1994.

Signed Liam Hamilton
The Tribunal

ANNEXE

IN THE MATTER OF the Tribunals of Inquiry (Evidence) Acts, 1921 and 1979; and IN THE MATTER OF a Tribunal of Inquiry established pursuant to Resolutions of Dail Eireann passed on the 24th May 1991, and by Seanad Eireann on the 29th May 1991, to inquire into allegations regarding illegal activities fraud and malpractice in and in connection with the beef processing industry by order of the Minister for Agriculture and Food made on the 31st day of May 1991.

AND WHEREAS it was provided by the said order that the Tribunals of Inquiry (Evidence) Acts 1921 and 1979 applied to the Tribunal

WHEREAS it is provided by Section 2 (*b*) of the Tribunals of Inquiry (Evidence) Act, 1921 that a Tribunal to which the Act is applied shall have power to authorise the representation before it by solicitor or counsel or otherwise of any person appearing to it to be interested; and

WHEREAS Michael Purcell Foods Limited (hereinafter referred to as "the applicants") applied to the said Tribunal to be represented before it by counsel instructed by solicitor and on the 2nd June 1993 such authorisation under the said section was granted; and

WHEREAS it is provided by section 6 of the Tribunals of Inquiry (Evidence)(Amendment) Act, 1979, that if a Tribunal is of opinion that having regard to its findings and all other relevant matters there are sufficient reasons rendering it equitable to do so it may by Order direct that the whole or part of the costs of any person appearing before the Tribunal by counsel or solicitor as taxed by a Taxing Master of the High Court be paid by any other person named in the Order; and

WHEREAS by section 4 of the said Act of 1979 it is provided that a Tribunal may make such Order as it considers necessary for the purpose of its functions; and

WHEREAS the Tribunal referred to in the title hereof is satisfied that having regard to its findings and all other relevant matters there are sufficient reasons rendering it equitable that the Minister for Finance should pay the costs of the applicants as hereinafter appearing

IT IS HEREBY ORDERED that a Taxing Master of the High Court do tax in the manner hereinafter appearing the applicants' costs of appearing before the said Tribunal by one counsel instructed by solicitor.

In relation to the taxation of the applicants' costs the Tribunal DOTH ORDER as follows:

(*a*) That the statutory provisions and the Rules of the Superior Courts relating to the taxation of costs in an action in the High Courts (including the provisions relating to review and appeal) shall, in so far as is practicable, apply;

(*b*) That the costs of employing a solicitor and one counsel be taxed on a party and party basis.

(*c*) That the costs do include such witnesses' expenses as relate to the those witnesses who actually gave oral evidence before the Tribunal on the proposal of the applicants;

(*d*) That the amount of witnesses' expenses be (*a*) such fee for the preparation of any statement prepared by the witness which was received by the Tribunal as part of the witnesses' testimony as the Taxing Master (or on review or appeal, the Court) considers reasonable; (*b*) such fee (not already paid) for attending before the Tribunal on such days as the witness gave oral evidence as the Taxing Master (or, on review or appeal, the Court) considers reasonable and (*b*) the travelling and subsistence expenses (not already paid) reasonably incurred for the purpose of giving oral evidence before the Tribunal.

(*e*) That the Minister for Finance be at liberty to attend and be heard on the said taxation and any review or appeal in relation to it.

(*f*) That the Minister for Finance do pay the applicants' costs when taxed and ascertained.

Dated this 29th day of July 1994.

Signed Liam Hamilton

The Tribunal

ANNEXE

IN THE MATTER OF the Tribunals of Inquiry (Evidence) Acts, 1921 and 1979; and IN THE MATTER OF a Tribunal of Inquiry established pursuant to Resolutions of Dail Eireann passed on the 24th May 1991, and by Seanad Eireann on the 29th May 1991, to inquire into allegations regarding illegal activities fraud and malpractice in and in connection with the beef processing industry by order of the Minister for Agriculture and Food made on the 31st day of May 1991.

AND WHEREAS it was provided by the said order that the Tribunals of Inquiry (Evidence) Acts 1921 and 1979 applied to the Tribunal

WHEREAS it is provided by Section 2 (*b*) of the Tribunals of Inquiry (Evidence) Act, 1921 that a Tribunal to which the Act is applied shall have power to authorise the representation before it by solicitor or counsel or otherwise of any person appearing to it to be interested; and

WHEREAS Kepak Limited (hereinafter referred to as "the applicants") applied to the said Tribunal to be represented before it by counsel instructed by solicitor and on the 3rd June 1993 such authorisation under the said section was granted; and

WHEREAS it is provided by section 6 of the Tribunals of Inquiry (Evidence)(Amendment) Act, 1979, that if a Tribunal is of opinion that having regard to its findings and all other relevant matters there are sufficient reasons rendering it equitable to do so it may by Order direct that the whole or part of the costs of any person appearing before the Tribunal by counsel or solicitor as taxed by a Taxing Master of the High Court be paid by any other person named in the Order; and

WHEREAS by section 4 of the said Act of 1979 it is provided that a Tribunal may make such Order as it considers necessary for the purpose of its functions; and

WHEREAS the Tribunal referred to in the title hereof is satisfied that having regard to its findings and all other relevant matters there are sufficient reasons rendering it equitable that the Minister for Finance should pay the costs of the applicants as hereinafter appearing

IT IS HEREBY ORDERED that a Taxing Master of the High Court do tax in the manner hereinafter appearing the applicants' costs of appearing before the said Tribunal by one counsel instructed by solicitor.

In relation to the taxation of the applicants' costs the Tribunal DOTH ORDER as follows:

(*a*) That the statutory provisions and the Rules of the Superior Courts relating to the taxation of costs in an action in the High Courts (including the provisions relating to review and appeal) shall, in so far as is practicable, apply;

(*b*) That the costs of employing a solicitor and one counsel be taxed on a party and party basis.

(*c*) That the costs do include such witnesses' expenses as relate to the those witnesses who actually gave oral evidence before the Tribunal on the proposal of the applicants;

(*d*) That the amount of witnesses' expenses be (*a*) such fee for the preparation of any statement prepared by the witness which was received by the Tribunal as part of the witnesses' testimony as the Taxing Master (or on review or appeal, the Court) considers reasonable; (*b*) such fee (not already paid) for attending before the Tribunal on such days as the witness gave oral evidence as the Taxing Master (or, on review or appeal, the Court) considers reasonable and (*b*) the travelling and subsistence expenses (not already paid) reasonably incurred for the purpose of giving oral evidence before the Tribunal.

(*e*) That the Minister for Finance be at liberty to attend and be heard on the said taxation and any review or appeal in relation to it.

(*f*) That the Minister for Finance do pay the applicants' costs when taxed and ascertained.

Dated this 29th day of July 1994.

Signed Liam Hamilton

The Tribunal

ANNEXE

IN THE MATTER OF the Tribunals of Inquiry (Evidence) Acts, 1921 and 1979; and IN THE MATTER OF a Tribunal of Inquiry established pursuant to Resolutions of Dail Eireann passed on the 24th May 1991, and by Seanad Eireann on the 29th May 1991, to inquire into allegations regarding illegal activities fraud and malpractice in and in connection with the beef processing industry by order of the Minister for Agriculture and Food made on the 31st day of May 1991.

AND WHEREAS it was provided by the said order that the Tribunals of Inquiry (Evidence) Acts 1921 and 1979 applied to the Tribunal

WHEREAS it is provided by Section 2 (*b*) of the Tribunals of Inquiry (Evidence) Act, 1921 that a Tribunal to which the Act is applied shall have power to authorise the representation before it by solicitor or counsel or otherwise of any person appearing to it to be interested; and

WHEREAS Mr Jon Roberts (hereinafter referred to as "the applicant") applied to the said Tribunal to be represented before it by counsel instructed by solicitor and on the 3rd June 1993 such authorisation under the said section was granted; and

WHEREAS it is provided by section 6 of the Tribunals of Inquiry (Evidence)(Amendment) Act, 1979, that if a Tribunal is of opinion that having regard to its findings and all other relevant matters there are sufficient reasons rendering it equitable to do so it may by Order direct that the whole or part of the costs of any person appearing before the Tribunal by counsel or solicitor as taxed by a Taxing Master of the High Court be paid by any other person named in the Order; and

WHEREAS by section 4 of the said Act of 1979 it is provided that a Tribunal may make such Order as it considers necessary for the purpose of its functions; and

WHEREAS the Tribunal referred to in the title hereof is satisfied that having regard to its findings and all other relevant matters there are sufficient reasons rendering it equitable that the Minister for Finance should pay the costs of the applicant as hereinafter appearing

IT IS HEREBY ORDERED that a Taxing Master of the High Court do tax in the manner hereinafter appearing the applicant's costs of appearing before the said Tribunal by one counsel instructed by solicitor.

In relation to the taxation of the applicant's costs the Tribunal DOTH ORDER as follows:

(*a*) That the statutory provisions and the Rules of the Superior Courts relating to the taxation of costs in an action in the High Courts (including the provisions relating to review and appeal) shall, in so far as is practicable, apply;

(*b*) That the costs of employing a solicitor and one counsel be taxed on a party and party basis.

(*c*) That the costs do include such witnesses' expenses as relate to the those witnesses who actually gave oral evidence before the Tribunal on the proposal of the applicant;

(*d*) That the amount of witnesses' expenses be (*a*) such fee for the preparation of any statement prepared by the witness which was received by the Tribunal as part of the witnesses' testimony as the Taxing Master (or on review or appeal, the Court) considers reasonable; (*b*) such fee (not already paid) for attending before the Tribunal on such days as the witness gave oral evidence as the Taxing Master (or, on review or appeal, the Court) considers reasonable and (*b*) the travelling and subsistence expenses (not already paid) reasonably incurred for the purpose of giving oral evidence before the Tribunal.

(*e*) That the Minister for Finance be at liberty to attend and be heard on the said taxation and any review or appeal in relation to it.

(*f*) That the Minister for Finance do pay the applicant's costs when taxed and ascertained.

Dated this 29th day of July 1994.

Signed Liam Hamilton

The Tribunal

ANNEXE

IN THE MATTER OF the Tribunals of Inquiry (Evidence) Acts, 1921 and 1979; and IN THE MATTER OF a Tribunal of Inquiry established pursuant to Resolutions of Dail Eireann passed on the 24th May 1991, and by Seanad Eireann on the 29th May 1991, to inquire into allegations regarding illegal activities fraud and malpractice in and in connection with the beef processing industry by order of the Minister for Agriculture and Food made on the 31st day of May 1991.

AND WHEREAS it was provided by the said order that the Tribunals of Inquiry (Evidence) Acts 1921 and 1979 applied to the Tribunal

WHEREAS it is provided by Section 2 (*b*) of the Tribunals of Inquiry (Evidence) Act, 1921 that a Tribunal to which the Act is applied shall have power to authorise the representation before it by solicitor or counsel or otherwise of any person appearing to it to be interested; and

WHEREAS Mr Brian O'Beirn (hereinafter referred to as "the applicant") applied to the said Tribunal to be represented before it by counsel instructed by solicitor and on the 10th June 1993 such authorisation under the said section was granted; and

WHEREAS it is provided by section 6 of the Tribunals of Inquiry (Evidence)(Amendment) Act, 1979, that if a Tribunal is of opinion that having regard to its findings and all other relevant matters there are sufficient reasons rendering it equitable to do so it may by Order direct that the whole or part of the costs of any person appearing before the Tribunal by counsel or solicitor as taxed by a Taxing Master of the High Court be paid by any other person named in the Order; and

WHEREAS by section 4 of the said Act of 1979 it is provided that a Tribunal may make such Order as it considers necessary for the purpose of its functions; and

WHEREAS the Tribunal referred to in the title hereof is satisfied that having regard to its findings and all other relevant matters there are sufficient reasons rendering it equitable that the Minister for Finance should pay the costs of the applicant as hereinafter appearing

IT IS HEREBY ORDERED that a Taxing Master of the High Court do tax in the manner hereinafter appearing the applicant's costs of appearing before the said Tribunal by one counsel instructed by solicitor.

In relation to the taxation of the applicant's costs the Tribunal DOTH ORDER as follows:

(*a*) That the statutory provisions and the Rules of the Superior Courts relating to the taxation of costs in an action in the High Courts (including the provisions relating to review and appeal) shall, in so far as is practicable, apply;

(*b*) That the costs of employing a solicitor and one counsel be taxed on a party and party basis.

(*c*) That the costs do include such witnesses' expenses as relate to the those witnesses who actually gave oral evidence before the Tribunal on the proposal of the applicant;

(*d*) That the amount of witnesses' expenses be (*a*) such fee for the preparation of any statement prepared by the witness which was received by the Tribunal as part of the witnesses' testimony as the Taxing Master (or on review or appeal, the Court) considers reasonable; (*b*) such fee (not already paid) for attending before the Tribunal on such days as the witness gave oral evidence as the Taxing Master (or, on review or appeal, the Court) considers reasonable and (*b*) the travelling and subsistence expenses (not already paid) reasonably incurred for the purpose of giving oral evidence before the Tribunal.

(*e*) That the Minister for Finance be at liberty to attend and be heard on the said taxation and any review or appeal in relation to it.

(*f*) That the Minister for Finance do pay the applicant's costs when taxed and ascertained.

Dated this 29th day of July 1994.

Signed Liam Hamilton

The Tribunal

ANNEXE

IN THE MATTER OF the Tribunals of Inquiry (Evidence) Acts, 1921 and 1979; and IN THE MATTER OF a Tribunal of Inquiry established pursuant to Resolutions of Dail Eireann passed on the 24th May 1991, and by Seanad Eireann on the 29th May 1991, to inquire into allegations regarding illegal activities fraud and malpractice in and in connection with the beef processing industry by order of the Minister for Agriculture and Food made on the 31st day of May 1991.

AND WHEREAS it was provided by the said order that the Tribunals of Inquiry (Evidence) Acts 1921 and 1979 applied to the Tribunal

WHEREAS it is provided by Section 2 (*b*) of the Tribunals of Inquiry (Evidence) Act, 1921 that a Tribunal to which the Act is applied shall have power to authorise the representation before it by solicitor or counsel or otherwise of any person appearing to it to be interested; and

WHEREAS Mr Rory Godson (hereinafter referred to as "the applicant") applied to the said Tribunal to be represented before it by counsel instructed by solicitor and on the 15th June 1993 such authorisation under the said section was granted; and

WHEREAS it is provided by section 6 of the Tribunals of Inquiry (Evidence)(Amendment) Act, 1979, that if a Tribunal is of opinion that having regard to its findings and all other relevant matters there are sufficient reasons rendering it equitable to do so it may by Order direct that the whole or part of the costs of any person appearing before the Tribunal by counsel or solicitor as taxed by a Taxing Master of the High Court be paid by any other person named in the Order; and

WHEREAS by section 4 of the said Act of 1979 it is provided that a Tribunal may make such Order as it considers necessary for the purpose of its functions; and

WHEREAS the Tribunal referred to in the title hereof is satisfied that having regard to its findings and all other relevant matters there are sufficient reasons rendering it equitable that the Minister for Finance should pay the costs of the applicant as hereinafter appearing

IT IS HEREBY ORDERED that a Taxing Master of the High Court do tax in the manner hereinafter appearing the applicant's costs of appearing before the said Tribunal by one counsel instructed by solicitor.

In relation to the taxation of the applicant's costs the Tribunal DOTH ORDER as follows:

(*a*) That the statutory provisions and the Rules of the Superior Courts relating to the taxation of costs in an action in the High Courts (including the provisions relating to review and appeal) shall, in so far as is practicable, apply;

(*b*) That the costs of employing a solicitor and one counsel be taxed on a party and party basis.

(*c*) That the costs do include such witnesses' expenses as relate to the those witnesses who actually gave oral evidence before the Tribunal on the proposal of the applicant;

(*d*) That the amount of witnesses' expenses be (*a*) such fee for the preparation of any statement prepared by the witness which was received by the Tribunal as part of the witnesses' testimony as the Taxing Master (or on review or appeal, the Court) considers reasonable; (*b*) such fee (not already paid) for attending before the Tribunal on such days as the witness gave oral evidence as the Taxing Master (or, on review or appeal, the Court) considers reasonable and (*b*) the travelling and subsistence expenses (not already paid) reasonably incurred for the purpose of giving oral evidence before the Tribunal.

(*e*) That the Minister for Finance be at liberty to attend and be heard on the said taxation and any review or appeal in relation to it.

(*f*) That the Minister for Finance do pay the applicant's costs when taxed and ascertained.

Dated this 29th day of July 1994.

Signed Liam Hamilton

The Tribunal

ANNEXE

IN THE MATTER OF the Tribunals of Inquiry (Evidence) Acts, 1921 and 1979; and IN THE MATTER OF a Tribunal of Inquiry established pursuant to Resolutions of Dail Eireann passed on the 24th May 1991, and by Seanad Eireann on the 29th May 1991, to inquire into allegations regarding illegal activities fraud and malpractice in and in connection with the beef processing industry by order of the Minister for Agriculture and Food made on the 31st day of May 1991.

AND WHEREAS it was provided by the said order that the Tribunals of Inquiry (Evidence) Acts 1921 and 1979 applied to the Tribunal

WHEREAS it is provided by Section 2 (*b*) of the Tribunals of Inquiry (Evidence) Act, 1921 that a Tribunal to which the Act is applied shall have power to authorise the representation before it by solicitor or counsel or otherwise of any person appearing to it to be interested; and

WHEREAS Eurowest Foods Limited (hereinafter referred to as "the applicants") applied to the said Tribunal to be represented before it by counsel instructed by solicitor and on the 16th June 1993 such authorisation under the said section was granted; and

WHEREAS it is provided by section 6 of the Tribunals of Inquiry (Evidence)(Amendment) Act, 1979, that if a Tribunal is of opinion that having regard to its findings and all other relevant matters there are sufficient reasons rendering it equitable to do so it may by Order direct that the whole or part of the costs of any person appearing before the Tribunal by counsel or solicitor as taxed by a Taxing Master of the High Court be paid by any other person named in the Order; and

WHEREAS by section 4 of the said Act of 1979 it is provided that a Tribunal may make such Order as it considers necessary for the purpose of its functions; and

WHEREAS the Tribunal referred to in the title hereof is satisfied that having regard to its findings and all other relevant matters there are sufficient reasons rendering it equitable that the Minister for Finance should pay the costs of the applicants as hereinafter appearing

IT IS HEREBY ORDERED that a Taxing Master of the High Court do tax in the manner hereinafter appearing the applicants' costs of appearing before the said Tribunal by one counsel instructed by solicitor.

In relation to the taxation of the applicants' costs the Tribunal DOTH ORDER as follows:

(*a*) That the statutory provisions and the Rules of the Superior Courts relating to the taxation of costs in an action in the High Courts (including the provisions relating to review and appeal) shall, in so far as is practicable, apply;

(*b*) That the costs of employing a solicitor and one counsel be taxed on a party and party basis.

(*c*) That the costs do include such witnesses' expenses as relate to the those witnesses who actually gave oral evidence before the Tribunal on the proposal of the applicants;

(*d*) That the amount of witnesses' expenses be (*a*) such fee for the preparation of any statement prepared by the witness which was received by the Tribunal as part of the witnesses' testimony as the Taxing Master (or on review or appeal, the Court) considers reasonable; (*b*) such fee (not already paid) for attending before the Tribunal on such days as the witness gave oral evidence as the Taxing Master (or, on review or appeal, the Court) considers reasonable and (*b*) the travelling and subsistence expenses (not already paid) reasonably incurred for the purpose of giving oral evidence before the Tribunal.

(*e*) That the Minister for Finance be at liberty to attend and be heard on the said taxation and any review or appeal in relation to it.

(*f*) That the Minister for Finance do pay the applicants' costs when taxed and ascertained.

Dated this 29th day of July 1994.

Signed Liam Hamilton

The Tribunal

ANNEXE

IN THE MATTER OF the Tribunals of Inquiry (Evidence) Acts, 1921 and 1979; and IN THE MATTER OF a Tribunal of Inquiry established pursuant to Resolutions of Dail Eireann passed on the 24th May 1991, and by Seanad Eireann on the 29th May 1991, to inquire into allegations regarding illegal activities fraud and malpractice in and in connection with the beef processing industry by order of the Minister for Agriculture and Food made on the 31st day of May 1991.

AND WHEREAS it was provided by the said order that the Tribunals of Inquiry (Evidence) Acts 1921 and 1979 applied to the Tribunal

WHEREAS it is provided by Section 2 (*b*) of the Tribunals of Inquiry (Evidence) Act, 1921 that a Tribunal to which the Act is applied shall have power to authorise the representation before it by solicitor or counsel or otherwise of any person appearing to it to be interested; and

WHEREAS Lixsteed Limited (hereinafter referred to as "the applicants") applied to the said Tribunal to be represented before it by counsel instructed by solicitor and on the 16th June 1993 such authorisation under the said section was granted; and

WHEREAS it is provided by section 6 of the Tribunals of Inquiry (Evidence)(Amendment) Act, 1979, that if a Tribunal is of opinion that having regard to its findings and all other relevant matters there are sufficient reasons rendering it equitable to do so it may by Order direct that the whole or part of the costs of any person appearing before the Tribunal by counsel or solicitor as taxed by a Taxing Master of the High Court be paid by any other person named in the Order; and

WHEREAS by section 4 of the said Act of 1979 it is provided that a Tribunal may make such Order as it considers necessary for the purpose of its functions; and

WHEREAS the Tribunal referred to in the title hereof is satisfied that having regard to its findings and all other relevant matters there are sufficient reasons rendering it equitable that the Minister for Finance should pay the costs of the applicants as hereinafter appearing

IT IS HEREBY ORDERED that a Taxing Master of the High Court do tax in the manner hereinafter appearing the applicants' costs of appearing before the said Tribunal by one counsel instructed by solicitor.

In relation to the taxation of the applicants' costs the Tribunal DOTH ORDER as follows:

(*a*) That the statutory provisions and the Rules of the Superior Courts relating to the taxation of costs in an action in the High Courts (including the provisions relating to review and appeal) shall, in so far as is practicable, apply;

(*b*) That the costs of employing a solicitor and one counsel be taxed on a party and party basis.

(*c*) That the costs do include such witnesses' expenses as relate to the those witnesses who actually gave oral evidence before the Tribunal on the proposal of the applicants;

(*d*) That the amount of witnesses' expenses be (*a*) such fee for the preparation of any statement prepared by the witness which was received by the Tribunal as part of the witnesses' testimony as the Taxing Master (or on review or appeal, the Court) considers reasonable; (*b*) such fee (not already paid) for attending before the Tribunal on such days as the witness gave oral evidence as the Taxing Master (or, on review or appeal, the Court) considers reasonable and (*b*) the travelling and subsistence expenses (not already paid) reasonably incurred for the purpose of giving oral evidence before the Tribunal.

(*e*) That the Minister for Finance be at liberty to attend and be heard on the said taxation and any review or appeal in relation to it.

(*f*) That the Minister for Finance do pay the applicants' costs when taxed and ascertained.

Dated this 29th day of July 1994.

Signed Liam Hamilton
The Tribunal

ANNEXE

IN THE MATTER OF the Tribunals of Inquiry (Evidence) Acts, 1921 and 1979; and IN THE MATTER OF a Tribunal of Inquiry established pursuant to Resolutions of Dail Eireann passed on the 24th May 1991, and by Seanad Eireann on the 29th May 1991, to inquire into allegations regarding illegal activities fraud and malpractice in and in connection with the beef processing industry by order of the Minister for Agriculture and Food made on the 31st day of May 1991.

AND WHEREAS it was provided by the said order that the Tribunals of Inquiry (Evidence) Acts 1921 and 1979 applied to the Tribunal

WHEREAS it is provided by Section 2 (*b*) of the Tribunals of Inquiry (Evidence) Act, 1921 that a Tribunal to which the Act is applied shall have power to authorise the representation before it by solicitor or counsel or otherwise of any person appearing to it to be interested; and

WHEREAS Master Meats (post 16/9/1988) (hereinafter referred to as "the applicants") applied to the said Tribunal to be represented before it by counsel instructed by solicitor and on the 17th June 1993 such authorisation under the said section was granted; and

WHEREAS it is provided by section 6 of the Tribunals of Inquiry (Evidence)(Amendment) Act, 1979, that if a Tribunal is of opinion that having regard to its findings and all other relevant matters there are sufficient reasons rendering it equitable to do so it may by Order direct that the whole or part of the costs of any person appearing before the Tribunal by counsel or solicitor as taxed by a Taxing Master of the High Court be paid by any other person named in the Order; and

WHEREAS by section 4 of the said Act of 1979 it is provided that a Tribunal may make such Order as it considers necessary for the purpose of its functions; and

WHEREAS the Tribunal referred to in the title hereof is satisfied that having regard to its findings and all other relevant matters there are sufficient reasons rendering it equitable that the Minister for Finance should pay the costs of the applicants as hereinafter appearing

IT IS HEREBY ORDERED that a Taxing Master of the High Court do tax in the manner hereinafter appearing the applicants' costs of appearing before the said Tribunal by one counsel instructed by solicitor.

In relation to the taxation of the applicants' costs the Tribunal DOTH ORDER as follows:

(*a*) That the statutory provisions and the Rules of the Superior Courts relating to the taxation of costs in an action in the High Courts (including the provisions relating to review and appeal) shall, in so far as is practicable, apply;

(*b*) That the costs of employing a solicitor and one counsel be taxed on a party and party basis.

(*c*) That the costs do include such witnesses' expenses as relate to the those witnesses who actually gave oral evidence before the Tribunal on the proposal of the applicants;

(*d*) That the amount of witnesses' expenses be (*a*) such fee for the preparation of any statement prepared by the witness which was received by the Tribunal as part of the witnesses' testimony as the Taxing Master (or on review or appeal, the Court) considers reasonable; (*b*) such fee (not already paid) for attending before the Tribunal on such days as the witness gave oral evidence as the Taxing Master (or, on review or appeal, the Court) considers reasonable and (*b*) the travelling and subsistence expenses (not already paid) reasonably incurred for the purpose of giving oral evidence before the Tribunal.

(*e*) That the Minister for Finance be at liberty to attend and be heard on the said taxation and any review or appeal in relation to it.

(*f*) That the Minister for Finance do pay the applicants' costs when taxed and ascertained.

Dated this 29th day of July 1994.

Signed Liam Hamilton

The Tribunal

ANNEXE

IN THE MATTER OF the Tribunals of Inquiry (Evidence) Acts, 1921 and 1979; and IN THE MATTER OF a Tribunal of Inquiry established pursuant to Resolutions of Dail Eireann passed on the 24th May 1991, and by Seanad Eireann on the 29th May 1991, to inquire into allegations regarding illegal activities fraud and malpractice in and in connection with the beef processing industry by order of the Minister for Agriculture and Food made on the 31st day of May 1991.

AND WHEREAS it was provided by the said order that the Tribunals of Inquiry (Evidence) Acts 1921 and 1979 applied to the Tribunal

WHEREAS it is provided by Section 2 (*b*) of the Tribunals of Inquiry (Evidence) Act, 1921 that a Tribunal to which the Act is applied shall have power to authorise the representation before it by solicitor or counsel or otherwise of any person appearing to it to be interested; and

WHEREAS Irish Meat Producers (hereinafter referred to as "the applicants") applied to the said Tribunal to be represented before it by counsel instructed by solicitor and on the 22nd June 1993 such authorisation under the said section was granted; and

WHEREAS it is provided by section 6 of the Tribunals of Inquiry (Evidence)(Amendment) Act, 1979, that if a Tribunal is of opinion that having regard to its findings and all other relevant matters there are sufficient reasons rendering it equitable to do so it may by Order direct that the whole or part of the costs of any person appearing before the Tribunal by counsel or solicitor as taxed by a Taxing Master of the High Court be paid by any other person named in the Order; and

WHEREAS by section 4 of the said Act of 1979 it is provided that a Tribunal may make such Order as it considers necessary for the purpose of its functions; and

WHEREAS the Tribunal referred to in the title hereof is satisfied that having regard to its findings and all other relevant matters there are sufficient reasons rendering it equitable that the Minister for Finance should pay the costs of the applicants as hereinafter appearing

IT IS HEREBY ORDERED that a Taxing Master of the High Court do tax in the manner hereinafter appearing the applicants' costs of appearing before the said Tribunal by one counsel instructed by solicitor.

In relation to the taxation of the applicants' costs the Tribunal DOTH ORDER as follows:

(*a*) That the statutory provisions and the Rules of the Superior Courts relating to the taxation of costs in an action in the High Courts (including the provisions relating to review and appeal) shall, in so far as is practicable, apply;

(*b*) That the costs of employing a solicitor and one counsel be taxed on a party and party basis.

(*c*) That the costs do include such witnesses' expenses as relate to the those witnesses who actually gave oral evidence before the Tribunal on the proposal of the applicants;

(*d*) That the amount of witnesses' expenses be (*a*) such fee for the preparation of any statement prepared by the witness which was received by the Tribunal as part of the witnesses' testimony as the Taxing Master (or on review or appeal, the Court) considers reasonable; (*b*) such fee (not already paid) for attending before the Tribunal on such days as the witness gave oral evidence as the Taxing Master (or, on review or appeal, the Court) considers reasonable and (*b*) the travelling and subsistence expenses (not already paid) reasonably incurred for the purpose of giving oral evidence before the Tribunal.

(*e*) That the Minister for Finance be at liberty to attend and be heard on the said taxation and any review or appeal in relation to it.

(*f*) That the Minister for Finance do pay the applicants' costs when taxed and ascertained.

Dated this 29th day of July 1994.

Signed Liam Hamilton

The Tribunal

ANNEXE

IN THE MATTER OF the Tribunals of Inquiry (Evidence) Acts, 1921 and 1979; and IN THE MATTER OF a Tribunal of Inquiry established pursuant to Resolutions of Dail Eireann passed on the 24th May 1991, and by Seanad Eireann on the 29th May 1991, to inquire into allegations regarding illegal activities fraud and malpractice in and in connection with the beef processing industry by order of the Minister for Agriculture and Food made on the 31st day of May 1991.

AND WHEREAS it was provided by the said order that the Tribunals of Inquiry (Evidence) Acts 1921 and 1979 applied to the Tribunal

WHEREAS it is provided by Section 2 (*b*) of the Tribunals of Inquiry (Evidence) Act, 1921 that a Tribunal to which the Act is applied shall have power to authorise the representation before it by solicitor or counsel or otherwise of any person appearing to it to be interested; and

WHEREAS Tunney Meat Packers (in Voluntary Liquidation) Mr Hugh Tunney and Mr John Copas (hereinafter referred to as "the applicants") applied to the said Tribunal to be represented before it by counsel instructed by solicitor and on the 23rd June 1993 such authorisation under the said section was granted; and

WHEREAS it is provided by section 6 of the Tribunals of Inquiry (Evidence)(Amendment) Act, 1979, that if a Tribunal is of opinion that having regard to its findings and all other relevant matters there are sufficient reasons rendering it equitable to do so it may by Order direct that the whole or part of the costs of any person appearing before the Tribunal by counsel or solicitor as taxed by a Taxing Master of the High Court be paid by any other person named in the Order; and

WHEREAS by section 4 of the said Act of 1979 it is provided that a Tribunal may make such Order as it considers necessary for the purpose of its functions; and

WHEREAS the Tribunal referred to in the title hereof is satisfied that having regard to its findings and all other relevant matters there are sufficient reasons rendering it equitable that the Minister for Finance should pay the costs of the applicants as hereinafter appearing

IT IS HEREBY ORDERED that a Taxing Master of the High Court do tax in the manner hereinafter appearing the applicants' costs of appearing before the said Tribunal by one counsel instructed by solicitor.

In relation to the taxation of the applicants' costs the Tribunal DOTH ORDER as follows:

(*a*) That the statutory provisions and the Rules of the Superior Courts relating to the taxation of costs in an action in the High Courts (including the provisions relating to review and appeal) shall, in so far as is practicable, apply;

(*b*) That the costs of employing a solicitor and one counsel be taxed on a party and party basis.

(*c*) That the costs do include such witnesses' expenses as relate to the those witnesses who actually gave oral evidence before the Tribunal on the proposal of the applicants;

(*d*) That the amount of witnesses' expenses be (*a*) such fee for the preparation of any statement prepared by the witness which was received by the Tribunal as part of the witnesses' testimony as the Taxing Master (or on review or appeal, the Court) considers reasonable; (*b*) such fee (not already paid) for attending before the Tribunal on such days as the witness gave oral evidence as the Taxing Master (or, on review or appeal, the Court) considers reasonable and (*b*) the travelling and subsistence expenses (not already paid) reasonably incurred for the purpose of giving oral evidence before the Tribunal.

(*e*) That the Minister for Finance be at liberty to attend and be heard on the said taxation and any review or appeal in relation to it.

(*f*) That the Minister for Finance do pay the applicants' costs when taxed and ascertained.

Dated this 29th day of July 1994.

Signed Liam Hamilton

The Tribunal

ANNEXE

IN THE MATTER OF the Tribunals of Inquiry (Evidence) Acts, 1921 and 1979; and IN THE MATTER OF a Tribunal of Inquiry established pursuant to Resolutions of Dail Eireann passed on the 24th May 1991, and by Seanad Eireann on the 29th May 1991, to inquire into allegations regarding illegal activities fraud and malpractice in and in connection with the beef processing industry by order of the Minister for Agriculture and Food made on the 31st day of May 1991.

AND WHEREAS it was provided by the said order that the Tribunals of Inquiry (Evidence) Acts 1921 and 1979 applied to the Tribunal

WHEREAS it is provided by Section 2 (*b*) of the Tribunals of Inquiry (Evidence) Act, 1921 that a Tribunal to which the Act is applied shall have power to authorise the representation before it by solicitor or counsel or otherwise of any person appearing to it to be interested; and

WHEREAS Margaret Potter (hereinafter referred to as "the applicant") applied to the said Tribunal to be represented before it by counsel instructed by solicitor and on the 25th June 1993 such authorisation under the said section was granted; and

WHEREAS it is provided by section 6 of the Tribunals of Inquiry (Evidence)(Amendment) Act, 1979, that if a Tribunal is of opinion that having regard to its findings and all other relevant matters there are sufficient reasons rendering it equitable to do so it may by Order direct that the whole or part of the costs of any person appearing before the Tribunal by counsel or solicitor as taxed by a Taxing Master of the High Court be paid by any other person named in the Order; and

WHEREAS by section 4 of the said Act of 1979 it is provided that a Tribunal may make such Order as it considers necessary for the purpose of its functions; and

WHEREAS the Tribunal referred to in the title hereof is satisfied that having regard to its findings and all other relevant matters there are sufficient reasons rendering it equitable that the Minister for Finance should pay the costs of the applicant as hereinafter appearing

IT IS HEREBY ORDERED that a Taxing Master of the High Court do tax in the manner hereinafter appearing the applicant's costs of appearing before the said Tribunal by one counsel instructed by solicitor.

In relation to the taxation of the applicant's costs the Tribunal DOTH ORDER as follows:

(*a*) That the statutory provisions and the Rules of the Superior Courts relating to the taxation of costs in an action in the High Courts (including the provisions relating to review and appeal) shall, in so far as is practicable, apply;

(*b*) That the costs of employing a solicitor and one counsel be taxed on a party and party basis.

(*c*) That the costs do include such witnesses' expenses as relate to the those witnesses who actually gave oral evidence before the Tribunal on the proposal of the applicants;

(*d*) That the amount of witnesses' expenses be (*a*) such fee for the preparation of any statement prepared by the witness which was received by the Tribunal as part of the witnesses' testimony as the Taxing Master (or on review or appeal, the Court) considers reasonable; (*b*) such fee (not already paid) for attending before the Tribunal on such days as the witness gave oral evidence as the Taxing Master (or, on review or appeal, the Court) considers reasonable and (*b*) the travelling and subsistence expenses (not already paid) reasonably incurred for the purpose of giving oral evidence before the Tribunal.

(*e*) That the Minister for Finance be at liberty to attend and be heard on the said taxation and any review or appeal in relation to it.

(*f*) That the Minister for Finance do pay the applicant's costs when taxed and ascertained.

Dated this 29th day of July 1994.

Signed Liam Hamilton

The Tribunal

ANNEXE

IN THE MATTER OF the Tribunals of Inquiry (Evidence) Acts, 1921 and 1979; and IN THE MATTER OF a Tribunal of Inquiry established pursuant to Resolutions of Dail Eireann passed on the 24th May 1991, and by Seanad Eireann on the 29th May 1991, to inquire into allegations regarding illegal activities fraud and malpractice in and in connection with the beef processing industry by order of the Minister for Agriculture and Food made on the 31st day of May 1991.

AND WHEREAS it was provided by the said order that the Tribunals of Inquiry (Evidence) Acts 1921 and 1979 applied to the Tribunal

WHEREAS it is provided by Section 2 (*b*) of the Tribunals of Inquiry (Evidence) Act, 1921 that a Tribunal to which the Act is applied shall have power to authorise the representation before it by solicitor or counsel or otherwise of any person appearing to it to be interested; and

WHEREAS Derek Montgomery and Seamus Hand (hereinafter referred to as "the applicants") applied to the said Tribunal to be represented before it by counsel instructed by solicitor and on the 25th June 1993 such authorisation under the said section was granted; and

WHEREAS it is provided by section 6 of the Tribunals of Inquiry (Evidence)(Amendment) Act, 1979, that if a Tribunal is of opinion that having regard to its findings and all other relevant matters there are sufficient reasons rendering it equitable to do so it may by Order direct that the whole or part of the costs of any person appearing before the Tribunal by counsel or solicitor as taxed by a Taxing Master of the High Court be paid by any other person named in the Order; and

WHEREAS by section 4 of the said Act of 1979 it is provided that a Tribunal may make such Order as it considers necessary for the purpose of its functions; and

WHEREAS the Tribunal referred to in the title hereof is satisfied that having regard to its findings and all other relevant matters there are sufficient reasons rendering it equitable that the Minister for Finance should pay the costs of the applicants as hereinafter appearing

IT IS HEREBY ORDERED that a Taxing Master of the High Court do tax in the manner hereinafter appearing the applicants' costs of appearing before the said Tribunal by one counsel instructed by solicitor.

In relation to the taxation of the applicants' costs the Tribunal DOTH ORDER as follows:

(*a*) That the statutory provisions and the Rules of the Superior Courts relating to the taxation of costs in an action in the High Courts (including the provisions relating to review and appeal) shall, in so far as is practicable, apply;

(*b*) That the costs of employing a solicitor and one counsel be taxed on a party and party basis.

(*c*) That the costs do include such witnesses' expenses as relate to the those witnesses who actually gave oral evidence before the Tribunal on the proposal of the applicants;

(*d*) That the amount of witnesses' expenses be (*a*) such fee for the preparation of any statement prepared by the witness which was received by the Tribunal as part of the witnesses' testimony as the Taxing Master (or on review or appeal, the Court) considers reasonable; (*b*) such fee (not already paid) for attending before the Tribunal on such days as the witness gave oral evidence as the Taxing Master (or, on review or appeal, the Court) considers reasonable and (*b*) the travelling and subsistence expenses (not already paid) reasonably incurred for the purpose of giving oral evidence before the Tribunal.

(*e*) That the Minister for Finance be at liberty to attend and be heard on the said taxation and any review or appeal in relation to it.

(*f*) That the Minister for Finance do pay the applicants' costs when taxed and ascertained.

Dated this 29th day of July 1994.

Signed Liam Hamilton

The Tribunal

ANNEXE

IN THE MATTER OF the Tribunals of Inquiry (Evidence) Acts, 1921 and 1979; and IN THE MATTER OF a Tribunal of Inquiry established pursuant to Resolutions of Dail Eireann passed on the 24th May 1991, and by Seanad Eireann on the 29th May 1991, to inquire into allegations regarding illegal activities fraud and malpractice in and in connection with the beef processing industry by order of the Minister for Agriculture and Food made on the 31st day of May 1991.

AND WHEREAS it was provided by the said order that the Tribunals of Inquiry (Evidence) Acts 1921 and 1979 applied to the Tribunal

WHEREAS it is provided by Section 2 (*b*) of the Tribunals of Inquiry (Evidence) Act, 1921 that a Tribunal to which the Act is applied shall have power to authorise the representation before it by solicitor or counsel or otherwise of any person appearing to it to be interested; and

WHEREAS Cloon Meats Limited (hereinafter referred to as "the applicants") applied to the said Tribunal to be represented before it by counsel instructed by solicitor and on the 25th June 1993 such authorisation under the said section was granted; and

WHEREAS it is provided by section 6 of the Tribunals of Inquiry (Evidence)(Amendment) Act, 1979, that if a Tribunal is of opinion that having regard to its findings and all other relevant matters there are sufficient reasons rendering it equitable to do so it may by Order direct that the whole or part of the costs of any person appearing before the Tribunal by counsel or solicitor as taxed by a Taxing Master of the High Court be paid by any other person named in the Order; and

WHEREAS by section 4 of the said Act of 1979 it is provided that a Tribunal may make such Order as it considers necessary for the purpose of its functions; and

WHEREAS the Tribunal referred to in the title hereof is satisfied that having regard to its findings and all other relevant matters there are sufficient reasons rendering it equitable that the Minister for Finance should pay the costs of the applicants as hereinafter appearing

IT IS HEREBY ORDERED that a Taxing Master of the High Court do tax in the manner hereinafter appearing the applicants' costs of appearing before the said Tribunal by one counsel instructed by solicitor.

In relation to the taxation of the applicants' costs the Tribunal DOTH ORDER as follows:

(*a*) That the statutory provisions and the Rules of the Superior Courts relating to the taxation of costs in an action in the High Courts (including the provisions relating to review and appeal) shall, in so far as is practicable, apply;

(*b*) That the costs of employing a solicitor and one counsel be taxed on a party and party basis.

(*c*) That the costs do include such witnesses' expenses as relate to the those witnesses who actually gave oral evidence before the Tribunal on the proposal of the applicants;

(*d*) That the amount of witnesses' expenses be (*a*) such fee for the preparation of any statement prepared by the witness which was received by the Tribunal as part of the witnesses' testimony as the Taxing Master (or on review or appeal, the Court) considers reasonable; (*b*) such fee (not already paid) for attending before the Tribunal on such days as the witness gave oral evidence as the Taxing Master (or, on review or appeal, the Court) considers reasonable and (*b*) the travelling and subsistence expenses (not already paid) reasonably incurred for the purpose of giving oral evidence before the Tribunal.

(*e*) That the Minister for Finance be at liberty to attend and be heard on the said taxation and any review or appeal in relation to it.

(*f*) That the Minister for Finance do pay the applicants' costs when taxed and ascertained.

Dated this 29th day of July 1994.

Signed Liam Hamilton
The Tribunal

ANNEXE

IN THE MATTER OF the Tribunals of Inquiry (Evidence) Acts, 1921 and 1979; and IN THE MATTER OF a Tribunal of Inquiry established pursuant to Resolutions of Dail Eireann passed on the 24th May 1991, and by Seanad Eireann on the 29th May 1991, to inquire into allegations regarding illegal activities fraud and malpractice in and in connection with the beef processing industry by order of the Minister for Agriculture and Food made on the 31st day of May 1991.

AND WHEREAS it was provided by the said order that the Tribunals of Inquiry (Evidence) Acts 1921 and 1979 applied to the Tribunal

WHEREAS it is provided by Section 2 (*b*) of the Tribunals of Inquiry (Evidence) Act, 1921 that a Tribunal to which the Act is applied shall have power to authorise the representation before it by solicitor or counsel or otherwise of any person appearing to it to be interested; and

WHEREAS Mr Michael Connolly (hereinafter referred to as "the applicant") applied to the said Tribunal to be represented before it by counsel instructed by solicitor and on the 9th July 1993 such authorisation under the said section was granted; and

WHEREAS it is provided by section 6 of the Tribunals of Inquiry (Evidence)(Amendment) Act, 1979, that if a Tribunal is of opinion that having regard to its findings and all other relevant matters there are sufficient reasons rendering it equitable to do so it may by Order direct that the whole or part of the costs of any person appearing before the Tribunal by counsel or solicitor as taxed by a Taxing Master of the High Court be paid by any other person named in the Order; and

WHEREAS by section 4 of the said Act of 1979 it is provided that a Tribunal may make such Order as it considers necessary for the purpose of its functions; and

WHEREAS the Tribunal referred to in the title hereof is satisfied that having regard to its findings and all other relevant matters there are sufficient reasons rendering it equitable that the Minister for Finance should pay the costs of the applicant as hereinafter appearing

IT IS HEREBY ORDERED that a Taxing Master of the High Court do tax in the manner hereinafter appearing the applicant's costs of appearing before the said Tribunal by one counsel instructed by solicitor.

In relation to the taxation of the applicant's costs the Tribunal DOTH ORDER as follows:

(*a*) That the statutory provisions and the Rules of the Superior Courts relating to the taxation of costs in an action in the High Courts (including the provisions relating to review and appeal) shall, in so far as is practicable, apply;

(*b*) That the costs of employing a solicitor and one counsel be taxed on a party and party basis.

(*c*) That the costs do include such witnesses' expenses as relate to the those witnesses who actually gave oral evidence before the Tribunal on the proposal of the applicants;

(*d*) That the amount of witnesses' expenses be (*a*) such fee for the preparation of any statement prepared by the witness which was received by the Tribunal as part of the witnesses' testimony as the Taxing Master (or on review or appeal, the Court) considers reasonable; (*b*) such fee (not already paid) for attending before the Tribunal on such days as the witness gave oral evidence as the Taxing Master (or, on review or appeal, the Court) considers reasonable and (*b*) the travelling and subsistence expenses (not already paid) reasonably incurred for the purpose of giving oral evidence before the Tribunal.

(*e*) That the Minister for Finance be at liberty to attend and be heard on the said taxation and any review or appeal in relation to it.

(*f*) That the Minister for Finance do pay the applicant's costs when taxed and ascertained.

Dated this 29th day of July 1994.

Signed Liam Hamilton

The Tribunal

CHAPTER TWENTY-SEVEN

Acknowledgements

The Tribunal was established on the 31st day of May, 1991 and has completed its report on this day the 29th day of July 1994.

It held its first Public Sitting on the 21st day of June 1991 and its last Public Sitting on the 15th day of July, 1993. Since that date, the Tribunal has been engaged in the compilation of this Report.

It is obvious that the Tribunal could not have adequately carried out its function without the assistance of and co-operation from many interested parties. It is right that this Public Record of the Tribunal's work should contain an acknowledgement of such assistance and co-operation so willingly given by all parties represented before it or in any way involved in or concerned with the inquiries being conducted by it, and an expression of the deep sense of gratitude that the Tribunal feels for that assistance and co-operation.

The assistance and the co-operation was provided from so many sources, including Departments of State, Semi State and other public bodies, representative organisations, individual public servants and private citizens, including those involved in the beef processing industry, that it would be a task of mammoth proportions for the Tribunal to acknowledge separately and individually the assistance given over such a long period of time by so many different bodies, organisations and individuals. It would be invidious on the part of the Tribunal to single out any one person for acknowledgement and personal thanks as the willingness to assist was common to all and the Tribunal is deeply grateful for such assistance.

It would, however, be remiss of the Tribunal if it failed to publicly acknowledge the assistance given to it by the persons assigned to it from the public service.

The solicitor assigned to the Tribunal by the Attorney General was Ms Christina Loughlin, of the Chief State Solicitors office and the magnitude of the task which confronted Ms Loughlin can hardly be overstated: It involved the careful study of numerous submissions and Statements and the examination of a vast amount of material. She had to prepare Books of Evidence for service on the necessary parties, to ensure the presence of witnesses whose evidence was considered essential, to conduct correspondence on behalf of the

Tribunal and discharge many other functions which inevitably arose during so lengthy an enquiry, including instructing Counsel not only in relation to the enquiry but in connection with the High Court and Supreme Court proceedings to which the Tribunal was a party.

In all this she was ably assisted by Mrs Ann Foskin, Legal Staff officer of the Chief State Solicitors Office. In the performance of their duties they displayed a high degree of professional skill, dedication and meticulous attention to detail which contributed immeasurably to the Tribunal's work. They worked as a team and between them made a contribution to the workings of this Tribunal which the Tribunal will never forget and is deeply indebted to them.

Mr Ted McCarthy, a Registrar of the High Court, acted as the Tribunal's Registrar. In that capacity the length and complexity of the Tribunal's hearings imposed many functions on him, all of which he fulfilled with the efficiency with which those associated with him in the courts will be familiar.

He assisted in the editorial work associated with the Tribunal's Report, he prepared the orders made by the Tribunal and fulfilled many other functions. In the discharge thereof he displayed a high degree of professionalism, attention to detail and was at all times available and of considerable assistance to the Tribunal.

The Tribunal could not have functioned without the help and assistance of a committed staff in the Tribunal Office, the personnel of which changed from time, due to promotions and transfer of staff; the one thing that never changed was the level of efficiency and commitment shown by Áine Stapleton, Mary Doyle, Mary McKenna, Anthony Tyrell, Bernadette Geoghegan, Paula Hughes and Yvonne Faughnan. In addition, Bernadette Geoghegan, Paula Hughes and Yvonne Faughnan undertook the task of typing the many drafts of this Report and displayed meticulous attention to detail therein. The Tribunal is deeply grateful for the skill and speed with which they carried out this work.

The Tribunal is deeply indebted to Counsel for the Tribunal who had the extremely responsible and exacting task of ensuring that the evidence was presented to the Tribunal, fully, clearly and where possible in logical sequence and who discharged their responsibilities with consistent efficiency and thoroughness.

The Tribunal is also deeply indebted to Counsel and Solicitors for the parties appearing before the Tribunal who greatly assisted the Tribunal in its task and helped to ensure that efficient conduct of the Tribunal's sittings.

The enormous task of recording all the oral testimony and submissions and the transcribing of same overnight into Transcripts of Evidence was undertaken by Mr. Padhraig O'Fearghail and his team of stenographers and this task was performed with their usual skill and efficiency, again contributing to the efficient conduct of the Tribunal's sittings.

The Tribunal is also deeply indebted to the staff of the Office of Public Works based at Dublin Castle under the direction of Mr David Byers and Mr Tom Doyle for all their assistance and co-operation in making available the room at Dublin Castle for the holding of public sittings and offices for the carrying out of the administrative and secretarial work of the Tribunal and all the facilities provided by them.

Appendices

APPENDIX ONE

Book of Allegations

TRIBUNAL OF INQUIRY

BEEF PROCESSING INDUSTRY

Book of Allegations

Index to Book of Allegations

PAGES

FRAUDULENT PRACTICES

1. A journalist saw the removal of and changing of stamps, dressings and labels on beef carcasses in a plant on the 12th-13th January, 1989, which allegation was notified to the Department of Agriculture and Food.

—12th April, 1989, Barry Desmond.

2. Labels were being changed on meat in different parts of the country by a team moving about to do this job on behalf of Goodman companies.

—9th March, 1989, Tomás MacGiolla.

3. That Goodman companies or employees had been prosecuted at least 3 times within the past three years.

   (*a*) In 1987 a close aide was convicted of attempting to defraud the Department of Agriculture, having been found in possession of South African Customs stamps and was fined £8,000 and received a two year suspended sentence.

   (*b*) On 30th July, 1990 in the Dublin District Court, the Eirfreeze Company, owned by Goodman, was convicted and fined on two charges related to illegal labelling of meat carcasses in the North Wall in March, 1989.

   (*c*) In April 1989, a Goodman factory in County Waterford was fined £1,000 for irregularities in cattle documentation.

—28th August, 1990, Pat Rabbitte.

4. Maintaining an entire production line in Nenagh designed for taking stamps from frozen carcases and re-stamping and re-packaging them.

—15th May, 1991, Pat Rabbitte.

5. Evidence has been uncovered of substantial fraud involving E.C. payments reported to be as high as £40 million.

—28th August 1990, Pat Rabbitte.

6. (*a*) No action was taken following the interception of a "carousel" operation by Customs officers of a lorry containing boneless beef on an unapproved road near Castleblayney on its way back into the Republic in 1988. The lorry had left a Goodman plant near Wexford with a container load of boneless beef, crossed by ferry to Britain, travelled up the mainland and crossed to Northern Ireland at Larne. The driver explained to Customs officers that he was on his way to a Goodman plant near Enniskillen but had got lost. A carousel operation involves exporting meat on which export refunds are paid, secretly reimporting it and then reexporting it again to claim yet more export refunds.

—15th May, 1991, Pat Rabbitte.

   (*b*) That Pat Rabbitte has in his possession the name of the driver of the lorry involved.

—15th May, 1991, Pat Rabbitte.

7. That it was because this "very Fianna Fail state company had the inside political track that our international reputation for quality was put at risk in grotty repackaging and restamping operations in Goodman plants, in operations heavily subsidised by the Irish tax payer".

—24th May, 1991, Pat Rabbitte.

8. There is no question but that a serious fraud occurred in the beef processing industry.

—2nd March, 1989, Albert Reynolds.

9. Earlier in the '80s counterfeit South African customs stamps on their way to the AIBP plant in Cahir for the purpose of falsely stamping documents in order to convey the impression that meat had been received in South Africa, thereby qualifying that meat for EC subsidies, were intercepted by Customs Authorities in Cork.

—15th May, 1991, Dick Spring.

10. In 1987 a close aide of Mr Goodman (Nobby Quinn) was convicted of attempting to defraud the Department of Agriculture and Food having been found in possession of forged South African Customs stamps and was fined £8,000 and received a two year suspended jail sentence.

—15th May, 1991, Dick Spring.

11. (*a*) The Department gave advance notice of inspections at meat plants and in particular at Foynes on the 15th and 16th April, 1989.

—26th April, 1989, Dick Spring.

(*b*) Almost all the samples taken in Foynes had trimmings in them or were otherwise suspicious.

—26th April, 1989, Dick Spring.

12. The regulatory authorities turned a blind eye on (Goodman's) dubious business practices — the false labelling and accounting, the commercial arrangements involved in the disposal of offal and so on.

—28th August, 1990, Dick Spring.

13. (*a*) That the abuse of the E.C. subsidy system was "practically institutionalised" in all of the Goodman factories.

—Patrick McGuinness/World in Action.

(*b*) That all factories in the Goodman Group abused E.C. subsidy schemes, and that Mr Goodman "set the tone because he controlled the company very tightly".

—Patrick McGuinness/World in Action.

14. Fraudulently obtaining payment of E.C. Beef subsidies by the following means:—

(*a*) Forging of documents, representing same to be originals, whilst inserting therein higher quantities and grades than appearing in the originals.

—Patrick McGuinness/World in Action.

(*b*) "World in Action" obtained a number of original and forged duplicate IB4 documents which originated from a Goodman factory; some of the duplicates show an increase of up to 14 kgs for every animal.

(*c*) Keeping and using bogus grading stamps for the purpose of fraudulently altering the grades on produce whereby Department of Agriculture grading marks were cut off these carcases and false marks inserted in their place.

—Patrick McGuinness/World in Action.

(*d*) "World in Action" obtained a large number of bogus stamps which it was claimed came from Goodman factories.

(*e*) Switching products, i.e. removing intervention animals and substituting therefor defective, sub-standard and low grade animals.

—Patrick McGuinness/World in Action.

(*f*) Keeping and using bogus Halal stamps and falsely representing on the relevant forms that livestock had been slaughtered according to the Halal ritual so as to maximise the export refund subsidy available on Halal slaughtered produce.

—Patrick McGuinness/World in Action.

(*g*) Goodman Management employed a group known as "The A-Team" for a period of 18 months for the purpose of re-boxing old (intervention) meat as new at Ulster Cold stores, Craigavon.

—Thomas Ruddy/World in Action.

THE WATERFORD AND BALLYMUN INVESTIGATIONS

1. That a fine of £1.084 million had been imposed on Anglo Irish Beef Processors (International) Limited on 16th January, 1989, in respect of fraudulent practices previously raised in the Dail.

—15th March, 1989, Barry Desmond.

2. That there was a major Garda Fraud Squad investigation taking place into Goodman International; that the investigation concerned both the Waterford and Ballymun plants; that boxes were alleged to have been packed with offal and trimmings and to have been mislabelled; and that the Fraud Squad was investigating documentation forwarded to the Department of Agriculture and Food and Revenue Commissioners.

—9th March, 1989, Barry Desmond.

3. Deputy Desmond was trying to sabotage the entire beef industry in this country.

—15th March, 1989, Taoiseach.

4. Mr Goodman's denial of responsibility for Waterford and his blaming a sub contractor, Mr Marks, a man who had a notorious track record in Viking Meats and in Benburb Meats, stands contradicted by the evidence of Mr McGuinness who said there was a high level plan to obstruct the investigation at Waterford. Mr Goodman made the same type of denial in the case of his former employee, Mr Nobby Quinn, in whose possession the Customs and Excise officials discovered forged stamps and in respect of which he received a two year suspended sentence for falsification of documents. Again, despite denials, there was established through the Fair Trade Commission Inquiry that Mr Goodman controls Classic Meats through Mr Nobby Quinn.

—15th May, 1991, Pat Rabbitte.

5. Some of the details in a detailed brief made available to Deputy McCartan in 1987 outlining similar charges to those highlighted in the World in Action programme, were made available to the Minister for Agriculture and Food, Deputy O'Kennedy, but no action resulted.

—15th May, 1991, Pat Rabbitte.

6. That some of the Goodman employees interviewed on the World in Action programme had also been interviewed for a Today Tonight programme and that, although an affidavit was sworn validating their case, someone decided that it should not go out.

—15th May, 1991, Pat Rabbitte.

7. The Department of Agriculture and Food did not diligently assist the Garda Fraud Squad in relation to the Waterford and Ballymun investigations. Notwithstanding a formal request from the Department to the Fraud Squad in June 1987 asking for an investigation, the Department failed to release their file to the Fraud Squad despite innumerable requests until December, 1988 which was no more than two to three

weeks before the Department wrote to AIBP informing them of the penalties that had been decided upon for Waterford.

—15th May, 1991, Dick Spring.

8. In June 1987 a memorandum from the Department of Agriculture and Food requested the Garda Fraud Squad to investigate serious irregularities in Goodman's Waterford and Ballymun plants. The essential matters to be inquired into were, in the case of the Waterford plant, the false altering of weights and cartons, both before the customs investigation and during it, and the inclusion of beef trimmings in the cartons to maximise the weight, and in the case of the Ballymun plant, the false altering of case weights and documentation.

—15th May, 1991, Dick Spring.

9. No investigation appears to have been carried out by the Fraud Squad in relation to the Waterford plant or the Ballymun allegations.

—15th May, 1991, Dick Spring.

10. That penalties were imposed by the Minister for Agriculture and Food on AIBP, in relation to irregularities at its Waterford and Ballymun plants, as referred to in a letter from the Department to the company dated 16th January, 1989.

—12th April, Dick Spring.

11. That, although Goodman had blamed the fraud at Waterford on a sub-contractor (Mr Marks — Daltina Ltd.) and it succeeded in proceedings against the company, the fact that the judgement was entered in default of appearance by the defendant in May, 1990, and that no effort had been made in the intervening year to secure a High Court hearing into an assessment of the quantum of damages could result in Goodman getting tax relief on an uncollected judgement of £1 million.

—15th May, 1991, Dick Spring.

12. (*a*) That, despite the Fraud Squad making innumerable requests for the file — which included every detail of the Department of Agriculture and Food's investigation — these requests were ignored until December 1988.

—15th May, 1991, Dick Spring.

(*b*) That, in fact, no Fraud Squad investigation was carried out in relation to the Waterford plant, although a file was submitted to the Office of the DPP in respect of the Ballymun allegations.

—15th May, 1991, Dick Spring.

13. Notwithstanding their knowledge of the irregularities at Waterford and Ballymun and the prosecution of Mr Nobby Quinn in relation to the bogus South African stamps, the Department (and the Minister) was prepared to release bank guarantees of up to £20 million (frozen because of the irregularities at Waterford) as part of the overall deal (in the Examinership) last Autumn.

—15th May, 1991, Dick Spring.

14. In 1986 the AIBP factory in Waterford came under Customs scrutiny; at the time this investigation was kept secret.

—May, 1991, World in Action.

15. Customs men found:—

(*a*) Weights had been falsified.

(*b*) Third quality meat had been added to 70,000 boxes of frozen meat bound for the Middle East.

(*c*) On at least two occasions Goodman's own Managers tried to obstruct the Customs investigation notwithstanding that Goodman had always maintained that a sub-contractor was responsible and that he himself, was innocent of any wrongdoing.

(*d*) A plan was agreed at Senior Management level, within the Goodman Group, with Customs people at their Head Office to contain the damage resulting from the investigation. The plan involved selecting a sample of good meat for investigation.

(*e*) World in Action had a document stated to be the master plan which showed the locations where the boxes of good meat were supposed to be opened by Customs officials. The plan failed because local (Waterford) Customs agents became suspicious and kicked up a fuss.

(*f*) World in Action obtained a Customs case summary which highlighted a second attempt by Goodman employees to undermine the investigation. The case summary stated that "Goodman employees attempted to disguise the extended fraud by altering case weights at the cold store".

(*g*) Customs officials recommended the instigation of criminal proceedings, yet the Fraud Squad were unable to get their hands on the Customs report until 18 months later.

16. That notwithstanding a Garda Investigation had been recommended by Customs Officers in respect of Waterford, the Fraud Squad were unable to get their hands on the Department of Agricultures file on the matter for 18 months.

THE EIRFREEZE INVESTIGATION

1. The Eirfreeze plant located in the North Wall was shut down at 6 p.m. or 7 p.m. on Saturday, 4th March, 1989 by Inspectors from the Department of Agriculture and Food because of very serious illegal activities by a team acting on behalf of one of the Goodman companies — changing labels and dates of slaughter on meat.

—9th March, 1989, repeated 15th May, 1991, Tomás MacGiolla.

2. Deputy MacGiolla had been given detailed information prior to the 4th March, 1989, about the manner in which labels are changed, the use of angle grinders to cut off the old label and a blast freezing process after the replacement of the new label.

—15th May, 1991, Tomás MacGiolla.

3. There was an article in the Sunday Press of the 5th March, 1989 about the Eirfreeze incident but the story suddenly died. Previous allegations of a serious nature against Goodman employees were quietened down and hushed up.

—9th March, 1989, Tomás MacGiolla.

4. On the 10th March, 1989, (the day after Deputy MacGiolla's statement to the Dail,) the Goodman P.R. Company accused Deputy MacGiolla of seriously damaging the reputation of Goodman International and the whole meat industry, denied the Eirfreeze Plant had been shut down and stated that the charges made by Deputy MacGiolla were utterly false.

—15th May, 1991, Tomás MacGiolla.

5. At the hearing of the prosecution against Eirfreeze in the District Court on 30th July, 1990, Defence Counsel on behalf of the company pleaded guilty to two charges relating to illegal labelling of meat carcasses. It was stated in court that the Eirfreeze plant was shut down on Saturday night (4th March, 1989) and Department Inspectors took away 63 carcasses in which they found false CU2 labels which indicated the meat was from steers of good confirmation and of low fat, in other words, a high quality product, which was at variance with the original grading by the Department official.

—15th May, 1991, Tomás MacGiolla.

6. The Department of Agriculture and Food and the prosecuting Counsel seemed very reluctant to pursue the charges with any vigour. The Department withdrew one charge against Eirfreeze and two charges against AIBP. The issues of fraud and forgery against the company were not pursued as the Judge seemed to expect. Where fraud is suspected the Garda are notified but on this occasion the Garda were not informed of the fraud and forgery. Was any audit of the commercial and financial records of Eirfreeze or of any of the companies associated with AIBP, Eirfreeze or any of those companies carried out?

—15th May, 1991, Tomás MacGiolla.

7. The Department seemed very reluctant to pursue the charges against Eirfreeze (and AIBP) with any vigour on the 30th July, 1990 and in particular the issue of fraud and forgery about which the Garda were not involved.

—15th May, 1991, Tomás MacGiolla.

8. On 30th July, 1990 in the Dublin District Court, the Eirfreeze Company, owned by Goodman, was convicted and fined on two charges related to illegal labelling of meat carcasses in the North Wall in March, 1989.

—28th August, 1990, Pat Rabbitte.

TAX EVASION

1. Writing off £4 million in taxes in respect of under-the-table payments to Goodman employees was a wrong judgement on the part of the Revenue Commissioners.

—24th May, 1991, John Bruton.

2. (*a*) Because of Goodman's political connections, the Revenue Commissioners turned a blind eye to the type of "remuneration packages" enjoyed by senior executives and a non return of P.A.Y.E. and P.R.S.I. to the Exchequer for many workers because of the operation of the contract system for a large proportion of the Goodman workforce.

—28th August, 1990, repeated 15th May, 1991, Pat Rabbitte.

(*b*) A great many Goodman workers were on the dole.

(*c*) And a great many of them were being paid under the counter.

Deputy Rabbitte has in his possession official AIBP notepaper showing:—

Under-the-counter payments to a half dozen workers totalling £8,280 with a corresponding invoice made out to a fictional haulier and a further amount of £3,278 paid to a large number of workers with again a corresponding invoice made out to a fictional haulier; and

the distinction drawn between monies put through the books and total monies paid.

—15th May, 1991, Pat Rabbitte.

3. In the Finance Act, the Government made a special arrangement to enable Mr Goodman to avail of high coupon finance (in respect of Section 84 loans) to fund speculative ventures abroad.

—15th May, 1991, Pat Rabbitte.

4. Mr Goodman got special concessions in regard to tax from the Government. He got a concession of £4 million from the Revenue Commissioners which was 50% of the tax bill he owed and which did not include interest.

—Senate, 29th May, 1991, Thomas Raftery.

5. In return for the Revenue Commissioners agreeing not to take proceedings against Mr Goodman or his company in respect of large scale tax evasion practices going back over many years Goodman International paid the Revenue Commissioners £4 million in respect of all outstanding liabilities and penalties, a settlement which was by far the largest of its kind in the history of the State.

—15th May, 1991, Dick Spring.

6. The Government support for Goodman included changes in the tax laws to enable a substantial amount of Mr Goodman's income from beef processing to be taxed at 10% manufacturing rate. He had been given access to such large amounts of (unsecured) cash borrowings by the banks because of the explicit support given to

him by Fianna Fail in Government. There was the decision of the Fianna Fail Government in 1987 against the best professional advice to reinstate Export Credit Insurance and then subsequently to ensure Mr Goodman got the lions share of it. Further tax changes included provisions which made Section 84 financing for Mr Goodman more advantageous. The announcement of June 1987 that effectively Goodman was to be entrusted with the task of developing the Irish beef industry essentially as the only agent of the Irish State in the matter.

—28th August, 1990, Dick Spring.

7. That the Company had a wide scheme of under-the-table payments. Cheques were made out against bogus invoices, endorsed by Goodman employees and cashed at local branches of the Allied Irish Bank.

—May, 1991, Patrick McGuinness/World in Action.

8. These cheques were payable quarterly in March, June and September and December of each year. They were paid to everyone in the Company, from the shop floor up and amounting to approximately £3 million a year.

—May, 1991, Patrick McGuinness/World in Action.

GOODMAN AND THE BANKS

1. It was a requirement of Section 84 Finance that loans be used for working capital and because some of Goodman borrowing was used outside the State to fund speculative ventures, it amounted to tax evasion warranting prosecution.

—15th May, 1991, Pat Rabbitte.

2. The four major banks involved (in providing Section 84 Finance) Amsterdam Rotterdam Bank, Algemene Bank, Commerz Bank and Bank of Indosuez had misrepresented to them the financial soundness of the group of companies dated before the 29th of August when the Examiner was appointed, and (when) the Group was knowingly trading while insolvent, which was illegal.

—15th May, 1991, Pat Rabbitte.

3. Goodman obtained unsecured loans of hundreds of millions of pounds from the Banks which he then gambled on the Stock Exchange.

—Senate, 29th May, 1991, Thomas Raftery.

4. Goodman owed a grand total of just under £700 million in cash and guarantees.

—28th August, 1990, Dick Spring.

5. Altogether well in excess of 30 banks were "taken for a ride". The personal commitments Goodman offered his bankers to cover the Iraq exposure had been spent in foolish and greedy investments in the United Kingdom which have gone disastrously wrong, it represents a catalogue of commercial adventurism that is unacceptable by any business standards.

—28th August, 1990, Dick Spring.

6. The banks which lent money to Goodman were deceived by his assurances and representations that the purpose for which the money was being borrowed was to cover the working capital requirements of Goodman International, instead they were used to finance Goodman's empire building in the United Kingdom and elsewhere. The unsecured loans were from banks which thought they were lending for the routine (working capital) purposes of Goodman International.

—28th August, 1990, Dick Spring.

7. Mr Goodman has alleged privately in his recent dealings with bankers that he has been a victim of a massive internal fraud in the Company, accounting for perhaps £40 million of the missing money.

—28th August, 1990, Dick Spring.

8. The Government support for Goodman included changes in the tax laws to enable a substantial amount of Mr Goodman's income from beef processing to be taxed at 10% manufacturing rate. He had been given access to such large amounts of (unsecured) cash borrowings by the banks because of the explicit support given to

him by Fianna Fail in Government. There was the decision of the Fianna Fail Government in 1987 against the best professional advice to reinstate Export Credit Insurance and then subsequently to ensure Mr Goodman got the lions share of it. Further tax changes included provisions which made Section 84 financing for Mr Goodman more advantageous. The announcement of June 1987 that effectively Goodman was to be entrusted with the task of developing the Irish beef industry essentially as the only agent of the Irish State in the matter.

—28th August, 1990, Dick Spring.

9. I have been told by a number of sources that Mr Goodman has been guaranteed immunity from civil prosecution as part of his settlement with the banks and from criminal prosecution. I should like to know — and perhaps the Minister would be so kind as to put it on record of the House — if that immunity has been given, why it was given, when it was given and on what basis it was given.

—15th May 1991 — Dick Spring 1247

SECTION 84

1. In the Finance Act, the Government made a special arrangement to enable Mr Goodman to avail of high coupon finance (in respect of Section 84 loans) to fund speculative ventures abroad.

—15th May, 1991, Pat Rabbitte.

GOODMAN AND THE IDA

1. That "enormous political pressure from the highest possible level" was brought to bear on the Goodman Group and the IDA to announce the expansion programme of 1987 before details had been worked out, solely as a P.R. exercise for the Taoiseach and his Government.

—28th August, 1990, Sean Barrett.

2. That the decision on the part of the Government to rely solely on Goodman to develop the beef industry was "downright irresponsible" and was made at considerable expense to the taxpayer.

—18th December, 1990, Sean Barrett.

3. That when Goodman applied for assistance for a Five Year Plan for the Beef Industry, the grant package was rushed through by the IDA under political pressure and was rushed through the Department of Finance under similar political pressure — with the Taoiseach's own personal (and improper) interference.

—24th May, 1991, John Bruton.

4. In June of 1987, the Government decided against the wishes of the IDA to give £25 million to Laurence Goodman.

—9th March, 1989, Barry Desmond.

5. The Goodman Organisation was chosen as the hub around which Fianna Fail had built its development policy for the food industry, including beef, dairying and sugar, Government funding commitments to him of between £200 and £250 million in 1987 had given him "tremendous credit" in raising finance wherever he wished to go, and he had also received IDA grants of up to £25 million.

—9th March, 1989, Tomás MacGiolla.

6. The Taoiseach himself directly intervened with the IDA (who advanced up to £25 million in grants to Goodman) to drop the performance clause in the case of grants to the Goodman company.

—9th March, 1989, Tomás MacGiolla.

7. The entire board of the IDA at one stage threatened to resign over this grant to expand an industry that already had a surplus processing capacity.

—Senate, 29th May, 1991, Thomas Raftery.

EXPORT CREDIT INSURANCE SCHEME

1. In respect of Goodman's Export Credit Insurance policy declarations were made that only beef with its origins in the Republic of Ireland would be covered, nevertheless very large quantities of non-Irish beef were included in shipments purporting to be covered by that policy.

—28th August, 1990, Des O'Malley.

2. The provision by the State of Export Credit Insurance cover on the sale of beef to Iraq in 1987 and 1988 of an amount in excess of the amount actually exported was in breach of the terms of the Export Credit Insurance Scheme and constituted a substantial abuse amounting to a fraud on the tax payer, the scale of the abuse and of the potential liability of the State was unprecedented.

—10th May, 1989, Des O'Malley.

3. The provision in 1987 and 1988 of between one fifth and one third of all Export Credit Insurance cover available to just two companies both exporting beef to Iraq, with over 80% going to Goodman, amounted to an abuse of the scheme. The provision of the bulk of the cover under this scheme to one company to the exclusion of competitors excluded fair competition from within the State which aggravated the scandal.

—10th May, 1989, Des O'Malley.

4. Allowing just two companies, of which by far the larger and more substantial was Goodman, cover under the Export Credit Insurance Scheme for beef exports to Iraq for such large sums, so considerably in excess of their actual exports to that country, was an act of "blatant political favouritism".

—10th May, 1989, Des O'Malley.

5. Decisions by Ministers and, in some cases by the Government, in relation to Export Credit Insurance had the effect of strengthening further the already strong position of Goodman (to whom members of the Government were extremely personally close) as the dominant group within the beef processing and allied trades, contrary to the interests of farmers and employees, and of exporters in other business sectors.

—10th May, 1989, Des O'Malley.

6. The Export Credit Insurance Scheme was restored in 1987 in the teeth of professional and administrative opposition.

—24th May, 1991, Pat Rabbitte.

7. Conscious decisions were taken to give one conglomerate (Goodman) more than 80% of the available cover in that market, disadvantaging rivals and exporters in other products.

—24th May, 1991, Pat Rabbitte.

8. (*a*) The granting of Export Credit Insurance was a political decision and depended on whether "you were a member of the club".

   (*b*) On hearing that Halal had been granted their share of the newly restored Export Credit Insurance in 1987, Mr Goodman intervened with the Taoiseach who caused the Minister for Industry and Commerce of the day, Deputy Albert Reynolds, to get in touch with the Chief Executive of Halal and tell him he was sorry the whole thing was an error. Deputy Rabbitte furnished the name of the official who made the telephone call.

   (*c*) The Minister for Finance had stated on a radio programme on Sunday, 19th May, 1991, that the reason Halal had been refused Export Credit Insurance was because it had no completed contract of sale.

   (*d*) Insurance cover had been offered to a person who had no such contracts and no immediate contemplation of doing business with Iraq.

   (*e*) Deputy Rabbitte can provide the name of the person in the Beef Industry who provided the above information.

—24th May, 1991, Pat Rabbitte.

9. The decision taken in 1987 by the Fianna Fail Government to reinstate Export Credit Insurance was taken against the best professional advice available to the Government.

—28th August, 1990, Repeated 15th May, 1991, Dick Spring.

10. This decision granted a virtual monopoly of Export Credit Insurance to Mr Goodman's company.

—15th May, 1991, Dick Spring.

11. This decision (along with the decision to invest £30 million in a grandiose scheme for the development of the beef industry being undertaken by Goodman International) proved to be disastrous in terms of the beef industry.

—15th May, 1991, Dick Spring.

12. The Government support for Goodman included changes in the tax laws to enable a substantial amount of Mr Goodman's income from beef processing to be taxed at 10% manufacturing rate. He had been given access to such large amounts of (unsecured) cash borrowings by the banks because of the explicit support given to him by Fianna Fail in Government. There was the decision of the Fianna Fail Government in 1987 against the best professional advice to reinstate Export Credit Insurance and then subsequently to ensure Goodman got the lions share of it. Further tax changes included provisions which made Section 84 financing for Goodman more advantageous. The announcement of June 1987 that effectively Goodman was to be entrusted with the task of developing the Irish beef industry essentially as the only agent of the Irish State in the matter.

—28th August, 1990, Dick Spring.

13. Government representatives were left in no doubt from negotiations between the Taoiseach, the Minister for Industry and Commerce and Goodman relating to the reintroduction of Export Credit Insurance, that some of the meat to be covered by the scheme would come from sources outside the State, notwithstanding which the Government decided to proceed to grant cover to the maximum level.

—18th December, 1990, Sean Barrett.

ANTI-COMPETITIVE PRACTICES

1. That there was a conspiracy between "all the players" (the meat plants) in the beef processing sector to depress prices to a stage where huge profits were being made by processors while farmers suffered losses of up to £50 per head.

—18th December, 1990, Paul Connaughton.

GENERAL ALLEGATIONS

1. (*a*) That the abuse of the E.C. subsidy system was "practically institutionalised" in all of the Goodman factories.

—Patrick McGuinness/World in Action.

(*b*) That all factories in the Goodman Group abused E.C. subsidy schemes, and that Mr Goodman "set the tone because he controlled the company very tightly".

—Patrick McGuinness/World in Action.

2. There was a feeling within the Goodman Organisation that we were invincible because they had right connections in the right places that could basically control any investigation that would be put in place.

—Patrick McGuinness/World in Action.

3. There was official indifference to the climate of fraudulent practices that characterised the Goodman Group. According to one public official, the whole ethos was "do not interfere, do not make trouble, this man is doing a great job". If you hoped to be promoted the last thing you wanted to do was start shouting foul at Larry Goodman.

—15th May, 1991, Pat Rabbitte.

4. The regulatory and control procedures for the Irish Beef Industry are not satisfactory and in particular the Government have failed in their responsibility of rooting out those people who have turned the beef industry into an object of scandal and disgrace.

—15th May, 1991, Dick Spring.

5. The Government covered up the illegal and improper activities in the beef industry since 1987. The refusal to reveal any detail of investigations, the failure to investigate thoroughly the complacency about control and the willingness to take action amounted to a cover up.

—15th May, 1991, Dick Spring.

6. At the time of the press conference to announce the State investment of £30 million for the development of the beef industry being undertaken by Goodman International, the Garda Fraud Squad was being asked by the Department of Agriculture and Food to investigate serious irregularities in Goodman's Waterford and Ballymun plants.

—15th May, 1991, Dick Spring.

7. The pursuit of Goodman monopolistic ambitions over the last couple of years would not have been possible without the support of the Fianna Fail Government.

—28th August, 1990, Dick Spring.

SPECIAL LEGISLATION

1. That the extraordinary recall of the Dail and Seanad in August, 1990 "had as much to do with the integral link between Fianna Fail and the Goodman organisation as it has with protecting a key Irish industry".

—28th August, 1990, Pat Rabbitte.

2. The role played by professional advisers such as accountants who signed off a balance sheet (just four weeks before the Dail Debate on the Examinership) that revealed nothing of Goodman's difficulties should be investigated.

—28th August, 1990, Dick Spring.

3. The E.C. Agricultural Commissioner Mr Mac Sharry, abused his position as Commissioner to stall a Dutch Bank from petitioning for the liquidation of Goodman while the Irish Government pushed through emergency legislation to prevent liquidation.

—John Tomlinson, M.E.P./World in Action.

GOODMAN AND THE MINISTER FOR AGRICULTURE

1. That the statement of the Minister for Agriculture and Food had, within 20 to 30 minutes, been established to be "inadequate" in three material respects:—

   1. that the statement had not confirmed that matters had been referred to the Garda authorities;
   2. that the statement had not confirmed that figures (relating to the penalties imposed) claimed by Barry Desmond were indicative of the figures actually imposed; and;
   3. that the Minister was unable to clarify the status of the Garda investigations (into the company).

   —12th April, 1989, Dick Spring.

2. At the time of the press conference to announce the State investment of £30 million for the development of the beef industry being undertaken by Goodman International, the Garda Fraud Squad was being asked by the Department of Agriculture and Food to investigate serious irregularities in Goodman's Waterford and Ballymun plants.

   —15th May, 1991, Dick Spring.

3. That the Department of Agriculture and Food's enthusiasm for Mr Goodman was all the harder to understand given the fact that the Department was aware of the prosecution and conviction of Mr Nobby Quinn — a Goodman associate — arising from the importation of counterfeit duplicates of South African customs stamps.

   —15th May, 1991, Dick Spring.

4. Notwithstanding their knowledge of the irregularities at Waterford and Ballymun and the prosecution of Mr Nobby Quinn in relation to the bogus South African stamps, the Department (and the Minister) was prepared to release bank guarantees of up to £20 million (frozen because of the irregularities at Waterford) as part of the overall deal (in the Examinership) last Autumn.

   —15th May, 1991, Dick Spring.

5. It is alleged that financial penalties have been imposed on the beef sector on fifteen companies on twenty two occasions in the period 1983 — 1990. Five instances occurred in 1983, three in 1984, ten in 1989 and four in 1990. The total financial penalties imposed in respect of these occurrences amounted to £557,000 in 1983, £92,000 in 1984, £3.9 million in 1989, and £23,000 in 1990.

   —Reply to P. Q. put down for 28th May, 1991 from Minister for Agriculture to Deputy G. Mitchell.

GOODMAN AND THE TAOISEACH

1. The Companies (Amendment) Bill, 1990 represented only Goodman's third choice proposal arising from meetings held with the Taoiseach, the first being a £300 million rescue package which Mr Goodman demanded the Government should underwrite, the second involving an approach by Mr Goodman's friends in Cabinet to the EEC Commissioner, Ray Mac Sharry in an attempt to persuade him to bring forward an EEC plan that would be of similar assistance to Mr Goodman but which would be cosmetically packaged as being in the interest of the "total industry".

—28th August, 1990, repeated 15th May, 1991, Pat Rabbitte.

2. That Goodman successfully intervened with the Taoiseach to cause the Government to reverse a decision to increase the budget to be given to C.B.F, the meat marketing board, in 1988, in order to shut out the prospect of markets being expanded for his competitors.

—25th October, Pat Rabbitte.

3. On hearing that Halal had been granted their share of the newly restored Export Credit Insurance in 1987, Mr Goodman intervened with the Taoiseach who caused the Minister for Industry and Commerce of the day, Deputy Albert Reynolds, to get in touch with the Chief Executive of Halal and tell him he was sorry the whole thing was an error. Deputy Rabbitte furnished the name of the official who made the telephone call.

—24th May, 1991, Pat Rabbitte.

4. For his part Charles Haughey publicly promoted Goodman. At the very same time the customs investigators were warning that Goodman's operations were strongly suspected of involvement in fraud, the Irish Prime Minister was endorsing Goodman for millions in Irish and European grants.

—15th May, 1991, Dick Spring.

5. The pursuit of Goodman's monopolistic ambitions over the last couple of years would not have been possible without the support of the Fianna Fail Government.

—28th August, 1990, Dick Spring.

6. At the very same time that Customs Investigators were warning that Goodman's operations were strongly suspected of involvement in fraud, the Irish Prime Minister was endorsing Goodman for millions in Irish and European Grants.

—May, 1991, World in Action.

7. That a proposed major European Investigation into Goodman's Organisation did not happen after assurances were received from the Irish Authorities that they themselves had a wide ranging investigation into Goodman in hand; but that there is no evidence of any such investigation.

—May, 1991, World in Action.

8. The links between the Irish Prime Minister, Charles Haughey, and Europes Mr Meat go back a long way. Goodman gave money to Haughey's Fianna Fail Party. For his part, Charles Haughey publicly promoted Goodman.

—May, 1991, World in Action.

9. At the very same time that Customs Investigations were warning that Goodman's operations were strongly suspected of other involvement in fraud, the Irish Prime Minister was endorsing Goodman for millions in Irish and European Grants.

—May, 1991, World in Action.

INTIMIDATION

1. People in the Irish media (who knew of the matters alleged in the World in Action programme) were intimidated from publishing it by writs slapped on them by the Goodman organisation.

—Senate, 29th May, 1991, John Dardis.

2. Threats were made against individuals and members of the House (Desmond and MacGiolla) by a business man (Goodman) outside the House.

—15th March, 1989, Proinsias de Rossa.

3. Professor Raftery was aware of one person about whom Mr Goodman threatened to reveal aspects of his private life if he did not drop a charge of dishonest dealing that he was taking against Mr Goodman.

—Senate, 29th May, 1991, Thomas Raftery.

4. The farmers were reluctant to take on Mr Goodman (in respect of his dominant position in the beef processing sector in Ireland) because he collected for them the levies which kept their organisation going.

—Senate, 29th May, 1991, Thomas Raftery.

5. Threats were made against individuals and members of the House (Desmond and MacGiolla) by a business man (Goodman) outside the House.

—12th April, 1989, Dick Spring.

6. That Barry Desmond received threatening phone calls in the aftermath of his disclosures in the Dail in 1989, that penalties of £1.084 million had been imposed on a Goodman company, to the effect that he should "lay off Larry Goodman".

—May, 1991, World in Action/Barry Desmond, M.E.P.

GOODMAN AND PROSECUTION

1. The fact that very few cases have been referred (by the Department) to the Garda for full investigation, resulting in only one prosecution, indicates a major problem with regard to policing considering the admitted extent of fraud in the area of E.C. grants.

—24th February, 1988, Pat McCartan

2. The Department seemed very reluctant to pursue the charges against Eirfreeze (and AIBP) with any vigour on the 30th July, 1990 and in particular the issue of fraud and forgery about which the Garda were not involved.

—15th May, 1991, Tomás MacGiolla.

3. The Minister for Agriculture had misled, and failed to fully inform the public, of the circumstances of the £1 million fine on AIBP.

—12th April, 1989, Desmond O'Malley.

4. That fines initially totalling over £20 million for irregularities uncovered in the operation of the APS scheme were reduced to £3.6 million last week by the Minister. Four of the five companies were fined for serious and deliberate breaches of Regulations. The fifth and smallest fine was £90,000 for a technical breach and this was Anglo Irish.

—24th May, 1991, Brendan McGahon.

5. That Goodman companies or employees had been prosecuted at least three times within the past three years.

    (*a*) In 1987 a close aide was convicted of attempting to defraud the Department of Agriculture, having been found in possession of South African Customs stamps and was fined £8,000 and received a two year suspended sentence.

    (*b*) On 30th July, 1990 in the Dublin District Court, the Eirfreeze Company, owned by Goodman, was convicted and fined on two charges related to illegal labelling of meat carcases in the North Wall in March, 1989.

    (*c*) In April 1989, a Goodman factory in County Waterford was fined £1,000 for irregularities in cattle documentation.

—28th August, 1990, Pat Rabbitte.

GOODMAN AND CUSTOMS

1. The Department failed to make proper arrangements to give Customs officials sufficient notice of export consignments to allow them to carry out detailed examinations.

—21st. June, 1990, Eamon Gilmore.

2. Political interference in the work of Agricultural Officers and Customs men in attempting to investigate suspected breaches of EC Regulations.

—24th February, 1988, Pat McCartan.

3. Political interference in the work of Agricultural Officers and Customs men in attempting to investigate suspected breaches of EEC Regulations.

—9th March, 1989, Tomás MacGiolla.

4. That it has been suggested that Goodman was subjected to a lesser degree of Customs inspection than other commercial operations (especially in regard to container loads going North), and that he was able to virtually close off the port of Greenore to other people when he was exporting meat.

—15th May, 1991, Pat Rabbitte.

THE CYPRUS LOAN

1. That it was reported as a possibility that the £25 million lent by Goodman, the subject of a court case in Cyprus, was to be used for the procurement of arms.

—18th December, 1990, Sean Barrett.

2. That the origins of the sum of approximately £20 million now on deposit in an account in the Bank of Cyprus and the purposes for which this money was raised were a matter in respect of which allegations of irregular practice involving the Goodman Group of companies had been made and ought properly to be made the subject of a judicial inquiry.

—15th May, 1991, Tomás MacGiolla.

3. Goodman raised a loan of £20 million and put it on deposit, as a tax dodge, with Mercantile Credit in order to avoid D.I.R.T. tax. As a result of the problems in Mercantile Credit and other problems, the £20 million is now missing.

—28th August, 1990, Pat Rabbitte.

4. That Goodman had alleged to his bankers that he was the victim of a massive internal fraud, with up to £40 million missing.

—28th August, 1990, Dick Spring.

5. That the security and intelligence authorities of several countries were investigating the provenance and purpose of the Cyprus loan.

—18th December, 1990, Dick Spring.

GOODMAN AND CLASSIC MEATS

1. That the Government had been misled as to the ownership of Classic Meats by Goodman.

   —18th December, 1990, Sean Barrett.

2. That the Minister for Industry and Commerce failed to complete and publish the results of an Inquiry, for the purposes of the Mergers and Monopolies Acts, into the beneficial ownership of Classic Meats.

   —15th March, 1989, Michael McDowell.

3. Goodman tried to take over Master Meats (Classic Meats) in high secrecy, yet he continued to insist he was not the beneficial owner until meeting his bankers last week.

   —28th August, 1990, Pat Rabbitte.

4. Mr Nobby Quinn, a former Goodman employee, convicted on a charge of falsification of documents, subsequently re-emerged "in the guise" of Classic Meats,

   —15th May, 1991, Pat Rabbitte.

5. The Dail found out through the Fair Trade Commission Inquiry that Mr Goodman controls Classic Meats despite the denials.

   —15th May, 1991, Pat Rabbitte.

6. That, at his meeting with his creditors, Goodman had included as among his assets Classic Meats, a company in which he has previously denied having any interest.

   —28th August, 1990, Dick Spring.

GOODMAN AND THE EEC

1. That, for ten months, the EC Court of Auditors had not been furnished with an explanation into the irregularities raised by Barry Desmond

   —12th April, 1989, Barry Desmond.

2. In a special report 2-90, the E.C. Court of Auditors criticised the Department for failing to allow direct access to reports and documentation concerning audit work carried out under Community rules.

   —21st. June, 1990, Eamon Gilmore.

3. That the EEC Court of Auditors had collected abundant evidence to demonstrate that the EC subsidy schemes are very open to fraud and that the subsidy system has been wide open to abuse.

   —May, 1991, Joe Carey, Member European Court of Auditors/World in Action.

GOODMAN AND THE COMPANY LAW

1. The Goodman companies have broken the law in not making returns to the Company Registration Office contrary to Section 127 of the Companies Act, 1963.

—9th March, 1989, Tomás MacGiolla.

2. The non enforcement by the Government of the terms of the Mergers, Takeovers and Monopolies (Control) Act, 1978 in respect of the Master Meats Group (exemplified) the closeness and helpfulness exhibited by the Government to the Goodman Group.

—10th May, 1989, Desmond O'Malley.

GOODMAN AND CHRISTMAS PRESENTS

1. That Deputy Rabbitte has in possession a list of public servants who were in receipt of Christmas presents from Goodman, that the list included Customs personnel, civil servants at the Department of Agriculture, vets and certain civil servants at the Department of Industry and Commerce, that the list also included public servants in Northern Ireland, and that the acceptance of such gifts was not normal practice in the Civil Service and was, possibly, in breach of Regulations.

—15th May, 1991, Pat Rabbitte.

2. In Brussels presents were brought unashamedly to officials and in some cases were returned on the spot. Certain officials from Brussels had the use of Mr Goodman's helicopter when going on business trips. A former secretary of the Department of Agriculture is now on Mr Goodman's payroll.

—Senate, 29th May, 1991, Thomas Raftery.

GOODMAN AND GREENORE

1. That it has been suggested that Goodman was subjected to a lesser degree of Customs inspection than other commercial operations (especially in regard to container loads going North), and that he was able to virtually close off the port of Greenore to other people when he was exporting meat.

—15th May, 1991, Pat Rabbitte.

GOODMAN AND THE ICC

1. That Goodman had been allowed to "cherry pick" the best of the ICC property portfolio because he was on the "inside political track", before any other party became aware of the availability of those properties.

—28th August, 1990, Pat Rabbitte.

GOODMAN AND THE CBF

1. That Goodman successfully intervened with the Taoiseach to cause the Government to reverse a decision to increase the budget to be given to C.B.F, the meat marketing board, in 1988, in order to shut out the prospect of markets being expanded for his competitors.

—25th October, 1990, Pat Rabbitte.

GOODMAN AND FINE GAEL

1. Fine Gael's attitude to Goodman was uncommonly acquiescent, a consideration affecting their attitude being the receipt of a donation of £60,000 from Goodman in 1988.

—25th October, 1990, Pat Rabbitte.

GOODMAN AND FIANNA FAIL

1. Knowing the inside political track had enabled him to get access to exceptional lines of credit and to benefit from risky but profitable Middle East contracts, confident in the knowledge that he was guaranteed by the Government so long as Fianna Fail remained in power.

—28th August, 1990, Pat Rabbitte.

HALAL AND UNITED MEAT EXPORTERS

1. There has been a major investigation into the Charleville plant of the Halal-associated United Meat Packers Exports Company in relation to export refunds.

—9th March, Barry Desmond.

2. That four foreign owned Irish based companies like Halal, Hibernia Meats, Agra and Taher have not been mentioned in this House when Larry Goodman and his companies were vilified in a manner that no other public person has been vilified in this House in the 70 years of this State.

—24th May, 1991, Brendan McGahon.

3. There has been a major investigation into Charleville plant of the Halal-associated United Meat Packers Exports Company in relation to export refunds.

—9th March, 1989, Pat Rabbitte.

4. (*a*) The granting of Export Credit Insurance was a political decision and depended on whether "you were a member of the club".

   (*b*) On hearing that Halal had been granted their share of the newly restored Export Credit Insurance in 1987, Mr Goodman intervened with the Taoiseach who caused the Minister for Industry and Commerce of the day, Deputy Albert Reynolds, to get in touch with the Chief Executive of Halal and tell him he was sorry the whole thing was an error. Deputy Rabbitte furnished the name of the official who made the telephone call.

   (*c*) The Minister for Finance had stated on a radio programme on Sunday, 19th May, 1991 that the reason Halal had been refused Export Credit Insurance was because it had no completed contract of sale.

   (*d*) Insurance cover had been offered to a person who had no such contracts and no immediate contemplation of doing business with Iraq.

   (*e*) Deputy Rabbitte can provide the name of the person in the Beef Industry who provided the above information.

—24th May, 1991, Pat Rabbitte.

APPENDIX TWO

List of Witnesses

Chronological list of all witnesses heard by the Tribunal with transcript references.

WITNESS NO.	NAME	DESCRIPTION	TRANSCRIPT DAY NO.
W1	Derek Mockler	Deputy Secretary D/Agriculture	D5 D30 D216
W2	John Ferris	Senior Superintending Veterinary Inspector D/Agriculture	D5 D6 D44 D118 D204 D207
W3	Dermot Ryan	Agricultural Inspector D/Agriculture	D6 D10 D35 D198 D206 D216
W4	Maurice Mullen	Assistant Principal D/Agriculture	D7 D8 D22 D28 to D30 D36 D37 D121 D124 D125 D131 D132 D164 to D166 D189 D194 to D207 D209 D210 D212 to D214 D217 to D220.
W5	Brid Cannon	Assistant Principal D/Agriculture	D8 D158 D167 D197 D213
W6	Ken Brennan	Former Employee AIBP	D8 D9
W7	John Lynch	Detective Sergeant Garda Siochana	D9 D21
W8	Fionnuala Fenton	Employee AIBP	D9
W9	James Meade	Agricultural Officer D/Agriculture	D10
W10	Martin Long	Agricultural Officer D/Agriculture	D10 D21 D206
W11	William O'Connor	Agricultural Officer D/Agriculture	D10 D21
W12	Ger Kelly	Production Manager AIBP	D10
W13	Eric Kinlan	Deputy Superintendent Harbour Police Dublin Port	D12
W14	Michael Curtin	Sergeant Harbour Police Dublin Port	D12
W15	Liam Curtin	Sergeant Garda Siochana	D12
W16	Gabriel McIntyre	Sergeant Garda Siochana	D12 D14
W17	Patrick Gregan	Veterinary Inspector D/Agriculture	D12 D14 D195 D199
W18	Gabriel Mellett	Superintending Agricultural Officer D/Agriculture	D12
W19	Daniel Joseph Smith	Veterinary Inspector D/Agriculture	D12
W20	Nicholas Finnerty	Assistant Agricultural Inspector D/Agriculture	D12 D13

Chronological list of all witnesses cont./...

WITNESS NO.	NAME	DESCRIPTION	TRANSCRIPT DAY NO.
W21	Gabriel Davey	Higher Agricultural Officer D/Agriculture	D13 D202
W22	Daniel Gavigan	Regional Supervisory Officer D/Agriculture	D13
W23	Paul O'Keeffe	Principal Officer D/Agriculture	D13
W24	Joseph Matthews	Barrister	D13
W25	Brendan Murphy	Officer of Customs and Excise	D13 D34
W26	Diarmuid O'Ceallaigh	Officer of Customs and Excise	D13 D27 D220
W27	Tomás MacCraith	Higher Officer of Customs and Excise	D13 D27
W28	Benny Bennett	Superintending Veterinary Inspector D/Agriculture	D13 D209
W29	James O'Neill	Inspector of Customs and Excise	D13
W30	Pól S.O Cinneide	Higher Officer Customs and Excise	D14
W31	Denis Murphy	Foreman Eirfreeze	D14
W32	Pat O'Neill	Manager Eirfreeze	D14
W33	Patrick McGuinness	Former Accountant AIBP	D15 to D20
W34	Vera Asanin	Real Estate Agent	D20
W35	Ron Schmehr	Vice President United Proteins	D20
W36	Liam Healy	Superintending Agricultural Officer D/Agriculture	D21
W37	Richard Molloy	Agricultural Officer D/Agriculture	D21
W38	Leonard Buckley	Higher Agricultural Officer D/Agriculture	D21 D43 D167
W39	Michael Staff	Superintending Agricultural Officer D/Agriculture	D21
W40	John Comerford	Agricultural Officer D/Agriculture	D21 D206
W41	Padraig Feeney	Superintending Agricultural Officer D/Agriculture	D21
W42	Andrew McCarthy	Regional Beef Classification Supervisor D/Agriculture	D21 D216
W43	Joseph Shortall	Personnel Officer D/Agriculture	D21 D121 to D123 D216
W44	Seamus Fogarty	Assistant Principal D/Agriculture	D21 D122 D163 D194 to D196 D198 D200 D204 D206 D207 D209 to D112 D215 D216 D220
W45	Mark Murphy	Former Employee AIBP	D21 D22
W46	Michael J. Cashman	Chief Superintendent Garda Siochana	D22
W47	Seamus O Neachtain	Officer of Customs and Excise	D22
W48	Gerard Dromey	Principal Officer D/Agriculture	D22 D54 D102 D204
W49	John Fowler	Former Employee AIBP	D22
W50	John Wallace	Former Employee AIBP	D22
W51	John Meaney	Former Employee AIBP	D22
W52	Maurice Windle	Beef Classification Officer D/Agriculture	D23
W53	Eamonn Phelan	Beef Classification Officer D/Agriculture	D23
W54	James Gallagher	Beef Classification Officer D/Agriculture	D23 D48
W55	William Sean O'Connor	Deputy Director of Veterinary Services D/Agriculutre	D23 D136 D204

Chronological list of all witnesses cont./...

WITNESS NO.	NAME	DESCRIPTION	TRANSCRIPT DAY NO.
W56	Donal Creedon	Former Secretary D/Agriculture	D23 D30
W57	Michael Cronin	Assistant Principal D/Agriculture	D23
W58	John Hayes	Detective Garda Siochana	D23 D34
W59	Thomas Rogan	Employee AIBP	D23
W60	Philip De Vere Hunt	Farmer	D23
W61	Susan McGuinness	Principal of Rockstead Securities	D23
W62	Jerry O'Callaghan	Journalist	D24
W63	Brendan Solan	Former Employee AIBP	D24
W64	Edward Kelly	Agricultural Officer D/Agriculture	D24 D25
W65	Frank Brislane	Superintending Agricultural Officer D/Agriculture	D25 D47
W66	Brian Maguire	Officer of Customs and Excise	D25
W67	Eoin Prunty	Officer of Customs and Excise	D25
W68	Sean Brosnan	Higher Officer of Customs and Excise	D25
W69	Nollaig S. O Broin	Higher Officer of Customs and Excise	D25 D132 D204
W70	John McBennett	Employee Creggan Transport	D25
W71	Timothy Ralph	Former Employee AIBP	D26
W72	Richard Hanrahan	Officer of Customs and Excise	D26
W73	Sean R. O Briain	Officer of Customs and Excise	D26
W74	John Horan	Higher Officer of Customs and Excise	D26 D132 D212
W75	Micheal S. O Braonain	Surveyor of Customs and Excise	D26 D132 D215
W76	Micheal O Briain	Higher Officer of Customs and Excise	D26
W77	Norman Kehily	Collector of Customs and Excise	D26
W78	Augustine O'Connell	Officer of Customs and Excise	D26
W79	Michael Mullowney	Operations Manager Waterford Cold Stores	D27
W80	David Murphy	Investigation Officer of Customs and Excise	D27 D219
W81	Michael Igoe	Inspector of Customs and Excise	D27
W82	Michael McGill	Higher Officer of Customs and Excise	D27 D31 D159 D202
W83	Seamus Paircеir	Former Chairman Revenue Commissioners	D28
W84	Con Healy	Assistant Principal Revenue Commissioners	D28 D210 D215
W85	John Hallissey	Principal Officer Revenue Commissioners	D28
W86	Donal Russell	Principal Officer D/Agriculture	D30 D217
W87	James P. McCabe	Assistant Secretary D/Agriculture	D31
W88	Dermot Ryan	Superintending Agricultural Officer D/Agriculture	D31 D200 D206
W89	Mary Keane	Higher Officer of Customs and Excise	D31
W90	Declan O Dálaigh	Higher officer of Customs and Excise	D32
W91	Angela Magee	Former employee AIBP	D32
W92	Imelda Murray	Former employee AIBP	D32

Chronological list of all witnesses cont./...

WITNESS NO.	NAME	DESCRIPTION	TRANSCRIPT DAY NO.
W93	Raymond Watson	Former employee Daltina Traders Ltd	D32
W94	James O'Mahony	Former Secretary D/Agriculture	D33
W95	Laurence Oakes	Former Employee AIBP	D33
W96	William Meagher	Former Garda Siochana	D33
W97	Sean O Donnghaile	Officer of Customs and Excise	D34
W98	Hugh Sweeney	Agricultural Officer D/Agriculture	D34
W99	John O'Callaghan	Superintending Agricultural Officer D/Agriculture	D34
W100	Patrick Timmons	Superintending Agricultural Officer D/Agriculture	D34
W101	Patrick Connolly	Veterinary Inspector D/Agriculture	D34 D35
W102	Declan Comer	Veterinary Inspector D/Agriculture	D35
W103	Pat Kieran (Read)	Higher Agricultural Officer D/Agriculture	D35
W104	Patrick McGlinchey (Read)	Higher Agricultural Officer D/Agriculture	D35
W105	Kieran McGowan (Read)	Higher Agricultural Officer D/Agriculture	D35
W106	T. Mullany (Read)	Veterinary Inspector D/Agriculture	D35
W107	John Mitchell	Agricultural Officer D/Agriculture	D35 D45
W108	Eamonn Mackle	Manager Daltina Traders Ltd	D37 D38
W109	Liam Marks	Manager Daltina Traders Ltd	D38
W110	Pat Rice	Superintending Agricultural Officer D/Agriculture	D39
W111	John Hughes	Veterinary Inspector D/Agriculture	D39
W112	Conal O'Brien	Veterinary Inspector D/Agriculture	D39
W113	Gerard O'Brien	Former Employee AIBP	D39
W114	Bertie Wall	Former Chairman UFA	D40
W115	Paddy Boyhan	Member UFA	D40 D48
W116	Patrick Frisby	Member UFA	D40
W117	John Conroy	Member UFA	D40
W118	Paul Courtney	Member UFA	D40
W119	Sean Dalton	Veterinary Inspector D/Agriculture	D40 D41
W120	John Lynch	Former Manager AIBP	D41 D42
W121	Patrick Dunne	Former Employee AIBP	D42
W122	Denis O'Connor	Former Employee AIBP	D42
W123	John Horgan	Former Employee AIBP	D42 D162
W124	William Chawke	Former Employee AIBP	D43
W125	John Vaughan	Agricultural Officer D/Agriculture	D43 D168
W126	Richard Healy	Assistant Principal D/Agriculture	D44
W127	Brendan McInerney	Principal Officer D/Agriculture	D44
W128	David Lynch	Veterinary Inspector D/Agriculture	D45 D170
W129	Christy Jones	Former Employee AIBP	D45

Chronological list of all witnesses cont./...

WITNESS NO.	NAME	DESCRIPTION	TRANSCRIPT DAY NO.
W130	Shane Hourigan	Former Employee AIBP	D45
W131	Pat Liston	Former Employee AIBP	D45
W132	Tom Lyons	Agricultural Officer D/Agriculture	D45
W133	Denis Carroll	Superintending Agricultural Officer D/Agriculture	D45 D46 D167
W134	Joe Mangan	Veterinary Inspector D/Agriculture	D46 D167
W135	Pat O'Neill	Superintending Veterinary Inspector D/Agriculture	D46 D168
W136	Oliver Forde	Former Employee AIBP	D46
W137	Denis Galvin	Former Employee AIBP	D46
W138	Frank Whelan	Former Employee AIBP	D46
W139	John Ray	Former Employee AIBP	D46
W140	Patrick Sweeney	Former Employee AIBP	D47
W141	Andrew Conway	Veterinary Inspector D/Agriculture	D47 D167
W142	Patrick Santry	Veterinary Inspector D/Agriculture	D47 D201
W143	William O'Rourke	Agricultural Officer D/Agriculture	D47
W144	Con Howard	Farmer	D47
W145	Michael Tevlin	Farmer	D47
W146	Gerard Fogarty	Assistant Principal D/Agriculture	D47 D195 D198 D260 D212
W147	John Grier	Vice President UFA	D48
W148	Martin Corcoran	Temporary Veterinary Inspector D/Agriculture	D48
W149	James Shea (Read)	Temporary Veterinary Inspector D/Agriculture	D48
W150	John J. Walsh (Read)	Veterinary Inspector D/Agriculture	D48
W151	Peter Gallagher (Read)	Agricultural Officer D/Agriculture	D48
W152	J. S. Gubbins (Read)	Temporary Veterinary Inspector D/Agriculture	D48
W153	V. J. Ireton (Read)	Temporary Veterinary Inspector D/Agriculture	D48
W154	Michael O'Farrell	Veterinary Inspector D/Agriculture	D48
W155	Sean Donnelly	Executive Director IDA	D49 D50 D55 D155
W156	Martin Lowery	Chief Executive IDA	D50 D51 D123
W157	Pádraic White	Former Managing Director IDA.	D51 D52
W158	Joe McCabe	Non-Executive Chairman IDA	D52
W159	John Loughrey	Secretary D/Energy	D53 D54 D123
W160	Brendan Dowling	Economist	D54
W161	Phelim Molloy	Assistant Secretary D/Finance	D54 D55
W162	Padraig O-hUigínn	Secretary D/Taoiseach	D55 D123
W163	Fionnán Coleman	Former Assistant Principal D/Finance	D55
W164	Joe Walsh	Minister for Agriculture & Food	D56
W165	Brian Carey	Journalist Sunday Business Post	D57
W166	Damien Kiberd	Editor Sunday Business Post	D57

Rep
[illegible] 13/3/92

Chronological list of all witnesses cont./...

WITNESS NO.	NAME	DESCRIPTION	TRANSCRIPT DAY NO.
W167	Frank Mee	Company Secretary ICI	D58 to D61 D77
W168	John Donlon	Former Secretary D/Trade, Commerce & Tourism	D61 to D67
W169	Ted O'Reilly	Former Assistant Secretary D/Industry & Commerce	D68 to D71
W170	Joe Timbs	Principal Officer D/Industry & Commerce	D72 to D75
W171	Gerry Donnelly	Assistant Principal D/Industry & Commerce	D76 D77
W172	Michael Fahy	Principal Officer D/Industry & Commerce	D78 D79
W173	Liam Kilroy	Principal Officer D/Industry & Commerce	D79 D80
W174	Peter C. Fisher	Head of Consultancy Unit, D/Industry & Commerce	D80 D81
W175	Sean Barton	Employee ICI	D81 D217
W176	Patrick Howard	Principal Officer, D/Finance	D82
W177	John Swift	Ambassador UN	D82 to D85
W178	Patrick McCabe	Ambassador to Russia	D86
W179	Brendan Nevin	Assistant Principal D/Agriculture	D87 D88
W180	John Hanney	Former Executive Officer D/Agriculture	D88
W181	Finbarr Kelly	Former Private Secretary to Minister for Industry & Commerce	D88
W182	Ciarán O Cuinneagáin	Former Private Secretary to Minister of State for D/Industry and Commerce	D89
W183	John Stanley	Former Financial Controller Halal Meat Packers	D89
W184	Pat O'Malley	Politician	D90
W185	Sean Clarke	Former Chief Executive UMP	D91 D92 D212
W186	Tim Boland	Farmer / Member IFA	D92
W187	Dominic McBride	Higher Executive Officer D/Industry & Commerce	D92
W188	John Egan	Financial Director Agra Trading Group	D93
W189	Sean Barrett	TD	D94
W190	John Fanning	Higher Executive Officer D/Industry & Commerce.	D94
W191	Michael Noonan	Former Minister for Industry & Commerce	D95
W192	Augustine Fitzpatrick	Former Managing Director Taher Meats Ltd.	D96
W193	Naser Taher	Principal of Taher Group	D97 D98 D197
W194	Oliver Murphy	Chief Executive Hibernia Meats International	D98 D99 D173
W195	John Bruton	T.D.	D100
W196	Eugene Regan	Executive Director of Agra	D100 D101 D210

Chronological list of all witnesses cont./...

WITNESS NO.	NAME	DESCRIPTION	TRANSCRIPT DAY NO.
W197	Paddy Moore	Chief Executive CBF	D10 D102
W198	Sean McCarthy	Senator	D102
W199	Satar Mohammed Khalid	Former Marketing Director UMP	D102
W200	Pascal Phelan	Former Chairman Master Meat Packers Group	D103, D217
W201	Owen Patten	Public Relations Consultant	D103
W202	Desmond O'Malley	Minister for Industry & Commerce	D104 to D109
W203	Ray Burke	Former Minister for Industry & Commerce	D110
W204	Seamus Brennan	Minister for Education	D111
W205	Peter Fitzpatrick	Chartered Accountant	D112 D156
W206	Padraig O'Donghaile	Investigation Branch Revenue Commissioners	D112 D113 D157
W207	Peadar O'Duinn	Investigation Branch Revenue Commissioners	D113 D213
W208	Dermot Nally	Secretary to the Government	D114
W209	Sean Moriarty	Assistant Secretary Revenue Commissioners	D115
W210	Sean Mooney	Tax Partner Stokes Kennedy Crowley	D115 D156 D157
W211	John King	Manager Stokes Kennedy Crowley	D116
W212	Niall O'Carroll	Audit Partner Stokes Kennedy Crowley	D116 D157
W213	Sean Murray	Trustee Fine Gael	D116
W214	John O'Donnell	Former Branch Secretary ATGWU Newry	D116 D117 D145
W215	Ray Kavanagh	General Secretary Labour Party	D117
W216	Patrick Querney	General Secretary Workers Party	D117
W217	Paul McKay	Trustee / Joint Treasurer Progressive Democrats	D117
W218	Pat Farrell	General Secretary Fianna Fail	D117
W219	Michael O'Reilly	Branch Secretary ATGWU Dundalk	D117
W220	Brendan Hodgers	Branch Secretary ATGWU Dundalk	D117
W221	Sean Kelly	District Officer ATGWU Clonmel	D117
W222	Pat Collins	Detective Sergeant Garda Siochana	D117
W223	Hugh McCluskey	Superintending Agricultural Officer D/Agriculture	D118
W224	Dermot Butler	Veterinary Inspector D/Agriculture	D118
W225	Michael O'Kennedy	Former Minister for Agriculture	D119 D120 D189
W226	Michael Dowling	Secretary, D/Agriculture	D120 D121 D223
W227	Susan McKeever	Veterinary Inspector D/Agriculture	D122
W228	Charles J. Haughey	Former Taoiseach	D126 TO D129
W229	Ray Mac Sharry	European Commissioner	D130
W230	David Barry	Surveyor of Customs & Excise	D131 D132
W231	Thomas K. Sutton	Higher Officer of Customs & Excise	D132
W232	Albert Reynolds	Taoiseach	D133 TO D135
W233	Seamus O'Conchuir	Higher Officer of Customs & Excise	D136
W234	Michael F. MacDarach	Higher Officer of Customs & Excise	D136

Chronological list of all witnesses cont./...

WITNESS NO.	NAME	DESCRIPTION	TRANSCRIPT DAY NO.
W235	Daniel Kenny	Higher Officer of Customs & Excise	D136
W236	Denis Leahy	Higher Officer of Customs & Excise	D136
W237	David O'Callaghan	Personnel Officer Revenue Commissioners	D136
W238	Aodh O'Gallchoir	Personnel Division D/Agriculture	D136
W239	Patrick Cannon	Veterinary Inspector D/Agriculture	D136
W240	Aodoghán O Rahilly	Chief Executive - Greenore Ferry Services Ltd.	D136
W241	Michel Jacquot	Director of FEOGA	D137 D138
W242	Susan O'Keeffe	Journalist	D139 TO D141 D145
W243	John F. McArdle	Detective Garda Garda Siochana	D146 D150 D217
W244	Tomas MacGiolla	Former TD	D147 D150
W245	Dick Spring	TD	D149
W246	Pat McCartan	Former TD	D150
W247	Tony Byrne	Managing Director August Engraving Co.	D150
W248	Bernard Devanney	Director Kamsor Ltd	D150
W249	Gene Lambe	Principal of Genelambe Ltd	D150
W250	Dermot Kiersey	Veterinary Inspector D/Agriculture	D150 D206
W251	Thomas Raftery	Senator	D151
W252	Paul Connaughton	TD.	D151
W253	Pat Rabbitte	TD	D152 D153 D155
W254	Joe Meade	Senior Auditor/Comptroller & Auditor General	D154
W255	Mary Harvey	Principal Officer D/Agricultue	D154 D202 D204 D215 D217 D219
W256	John Purcell	Director of Audit Comptroller & Auditor General	D156
W257	Liam Murphy	Principal Officer D/Finance	D156
W258	Michael Sargent	Chartered Accountant Brennan Governey & Co.	D157
W259	Leo Roche	Manager ICC Bank	D157
W260	Frank Cassells	Revenue Commissioner	D158
W261	Michael Byrne	Former Contract Cleaner AIBP	D158
W262	James Clarke	Surveyor / Customs & Excise	D159 D200 D202
W263	Victor Broderick	Former Employee Le Controle Technique	D161
W264	Christian Peyron	Director of Meat Department Le Controle Technique	D161
W265	Gerard McGrath	Former Employee AIBP	D162
W266	Anthony Barry	Former Employee AIBP	D162
W267	William Harrington	Former Employee AIBP	D162
W268	Brian Kennedy	Former Employee AIBP	D162
W269	Denis Hogan	Former Employee AIBP	D162
W270	Raymond Gilbourne	Former Employee AIBP	D162
W271	Patrick Benson	Former Employee AIBP	D162
W272	Patrick Lynch	Former Employee AIBP	D162

Chronological list of all witnesses cont./...

WITNESS NO.	NAME	DESCRIPTION	TRANSCRIPT DAY NO.
W273	Thomas Murphy	Former Employee AIBP	D162
W274	Sean O'Donoghue	Former Employee AIBP	D162
W275	Kevin McKnight	Former Employee AIBP	D162
W276	Thomas Dunne	Former Employee AIBP	D162
W277	John O'Sullivan	Former Employee AIBP	D162
W278	Michael Jinks	Former Employee AIBP	D165
W279	Paul Cahill	Former Employee AIBP	D165
W280	James Higgins	Former Employee AIBP	D165
W281	John Collins	Former Employee AIBP	D165
W282	Michael Lawlor	Former Employee AIBP	D165
W283	James Leahy	Former Employee AIBP	D165
W284	Pat Lynch	Former Employee AIBP	D165
W285	Anthony Murphy	Former Employee AIBP	D165
W286	Peter O'Connor	Former Employee AIBP	D165
W287	Gerard O'Grady	Former Employee AIBP	D165
W288	Gerard Shinny	Former Employee AIBP	D165
W289	Barry Toomey	Former Employee AIBP	D165
W290	Patrick Lane	Former Employee AIBP	D165
W291	Liam Quirke	Former Employee AIBP	D165
W292	Colm Quirke	Former Employee AIBP	D165
W293	Liam O'Shaughnessy	Former Employee AIBP	D166
W294	Edward Vercker	Former Employee AIBP	D166
W295	Trevor Hayes	Former Employee AIBP	D166
W296	Treasa Ní Dhuinn	Assistant Principal Officer D/Agriculture.	D166
W297	Bart Brady	Assistant Secretary D/Agriculture	D167
W298	Fank Kenny	Superintending Veterinary Inspector D/Agriculture	D167
W299	Vincent Young	Microbiologist Eastern Health Board	D167
W300	Patrick Casey	Veterinary Inspector D/Agriculture	D167 D170 D215
W301	Patrick O'Mahony	Higher Officer of Customs & Excise	D168
W302	David Geary	Former Accountant AIBP	D168
W303	Barry Shanahan	Employee AIBP	D169
W304	Michael Dunne	Employee AIBP	D169
W305	Dermot Hanley	Former Accountant AIBP	D169
W306	Sean O'Shea	Employee AIBP	D169
W307	Michael Meehan	Employee AIBP	D169
W308	Ashling Melvin	Employee AIBP	D169
W309	Sean Geoghegan	Employee AIBP	D169
W310	Martin Finucane	Employee AIBP	D169

Chronological list of all witnesses cont./...

WITNESS NO.	NAME	DESCRIPTION	TRANSCRIPT DAY NO.
W311	Tom Keating	Employee AIBP	D171
W312	Sean Goodwin	Employee AIBP	D171 D174
W313	Cento Veljanovski	Economist	D173
W314	Joseph Gracey	Veterinary Surgeon	D174
W315	Larry Kelly	Gentleman	D174
W316	Sean Harnett	Employee AIBP	D174
W317	Matthew O'Doherty	Gentleman	D174
W318	Tony Butler	Accountant	D174
W319	Denis Murphy	Employee AIBP	D174
W320	Aidan Connor	Former Deputy Chief Executive of Goodman International	D175 D176
W321	Brian Britton	Former Financial Group Accountant Goodman International	D176 to D179
W322	Laurence Goodman	Chief Executive Goodman International	D179 TO D182
W323	Joe Harte	Agricultural Scientist	D183
W324	Gerry Maynes	Regional Sales Manager AIBP	D183 D184
W325	James Fairbairn	Senior Manager International Division AIBP	D184
W326	Patrick Birdy	Transport Manager AIBP	D184 D188
W327	Declan McDonnell	Group Projects Engineer AIBP	D184 D185
W328	Colin Duffy	General Manager AIBP	D185
W329	John Delaney	Manager AIBP	D185
W330	Gerry Thornton	Deputy Chief Executive Meat Division AIBP	D186
W331	David Murphy	Accountant AIBP	D186
W332	Patrick Lynch	Group Projects Manager AIBP	D187
W333	James Monaghan	Factory Manager AIBP	D187
W334	Peter Maguire	Plant Manager AIBP	D187
W335	George Mullan	Plant Manager AIBP	D187
W336	Kenneth Robinson	Manager AIBP	D187
W337	John Kelleher	General Manager AIBP	D187
W338	Colm O'Loughlin	Projects Manager AIBP	D188
W339	John Connolly	Manager AIBP	D188
W340	James Walsh	Former Manager AIBP	D188 D189
W341	Eoin Lambe	Administrator AIBP	D188
W342	Finbar McDonnell	Senior Production Manager AIBP	D188
W343	Barry Desmond	M.E.P.	D190
W344	Joe McCartin	M.E.P.	D190
W345	Declan MacPartlin	Public Relations Person	D190
W346	Michael Sheehan	Veterinary Inspector D/Agriculture	D194
W347	Terry Hanlon	General Manager Arax Ltd	D194
W348	John Matthews	Veterinary Inspector D/Agriculture	D194 to D196 D207

Chronological list of all witnesses cont./...

WITNESS NO.	NAME	DESCRIPTION	TRANSCRIPT DAY NO.
W349	Danny Houlihan	Managing Director Ashbourne Meats	D194
W350	Bernard Kelly	Former Employee Tara Meats	D194
W351	Patrick Ennis	Employee Tara Meats	D194
W352	Paul McLoughlin	Employee Tara Meats	D194
W353	Daniel Brady	Former Employee Tara Meats	D195
W354	Jim Sheridan	Higher Agricultural Officer D/Agriculture	D195
W355	Arthur Ormsby	Former Factory Manager Tara Meats Ltd	D195
W356	Tony Dunne	One of the owners of Tara Meats Ltd	D195
W357	Patrick Sexton	Veterinary Inspector D/Agriculture	D195
W358	John Melville	Veterinary Inspector D/Agriculture	D195 D212
W359	Brendan Dunne	General Manager Slaney Meats Limited	D195 D198
W360	Ben McArdle	Former Employee Rangeland Meats	D195
W361	Roger McCarrick	Managing Director Rangeland Meats Ltd	D196
W362	David Tantrum	Veterinary Inspector D/Agriculture	D196
W363	Maurice Brennock	Veterinary Inspector D/Agriculture	D196
W364	Dan Brown	Managing Director Dawn Meats	D196
W365	Michael J. Downey	Higher Officer Customs & Excise	D197
W366	Wilfred Woolett	Veterinary Inspector D/Agriculture	D197
W367	Godfrey Higgins	Manager of Taher Interests in Ireland	D197
W368	Seamus Hand	Managing Director Freshlands Ltd	D197 D220
W369	Peter Smith	Veterinary Inspector D/Agriculture	D198 D203 D204
W370	Denis Lyons	Managing Director Western Meats Producers Ltd	D198
W371	Aidan McNamara	Principal Officer D/Agriculture	D198 D204 D206 D218
W372	Liam Fleming	Manager Transfreeze Cold Stores Ltd	D198
W373	Patrick Canton	Head Of International Audit Unit D/Agriculture	D198
W374	Kevin Galligan	Superintending Agricultural Officer D/Agriculture	D199
W375	Eugene Magee	Higher Agricultural Officer D/Agriculture	D199
W376	Derek Montgomery	Chief Executive Blanchvac	D199
W377	Tony McNicholl	Managing Director N.W.L Ltd	D199
W378	Michael Phelan	Managing Director Autozero Ltd	D199
W379	Eamonn O'Donovan	Higher Agricultural Officer D/Agriculture	D199 D212
W380	John Boothman	Veterinary Inspector D/Agriculture	D200 D213
W381	Peter Bourke	Farmer	D200
W382	Victor Whelan	Veterinary Inspector D/Agriculture	D200
W383	Tom McParland	Managing Director Kildare Chilling	D200
W384	Martin Blake	Director of Honey Clover Limited	D201
W385	Seamus McQuinny	Managing Director Barford Meats Ltd	D201
W386	Patrick Carton	Managing Director CH Foods Ltd	D201
W387	Joseph Gordon	Managing Director Continental Beef Limited	D201
W388	Joe Walsh	Managing Director Q.K. Cold Stores. Ltd	D201

Chronological list of all witnesses cont./...

WITNESS NO.	NAME	DESCRIPTION	TRANSCRIPT DAY NO.
W389	Brian Donovan	Manager Prime Meats	D202
W390	Christy O'Brien	Director Goudhurst	D202
W391	John Mair	Receiver Goudhurst	D202
W392	Paul Shortt	Secretary Norish Plc Group	D202
W393	Peter Horgan	Director Horgan Meats	D202
W394	Cecil Rothwell	Director, Freezomatic	D203
W395	Charles J. Corr	Superintending Veterinary Inspector D/Agriculture	D204
W396	Brendan Smith	Veterinary Inspector D/Agriculture	D204
W397	John Cassells	Veterinary Surgeon	D204
W398	Felix Loughran	Officer of Customs & Excise	D204
W399	Peter McGovern	Officer of Customs & Excise	D204
W400	Detlef Hayer	Managing Director Sinnat Ltd	D205
W401	Leo McTiernan	Higher Agricultural Officer D/Agriculture	D205 D219
W402	Kilian Unger	Veterinary Inspector D/Agriculture	D205 D209
W403	Gerry McPhillips	Agricultural Officer D/Agriculture	D205
W404	Frank Walls	Higher Agricultural Officer D/Agriculture	D205
W405	John Murphy	Manager Liffey Meats	D205
W406	Frank Mallon	Managing Director Liffey Meats	D205
W407	Sean McNamara	Manager Liffey Meats	D205
W408	Christy Kett	Factory Manager Liffey Meats	D205
W409	Colm O'Hagan	Financial Director Oxfleisch	D206
W410	Bill Deevy	Superintending Veterinary Inspector D/Agriculture	D206
W411	John A. Phelan	Veterinary Inspector D/Agriculture	D206
W412	Thomas Nolan	Former Managing Director / Meadow Meats	D206
W413	John Purcell	Managing Director Michael Purcell Foods Ltd	D207
W414	John Murray	Veterinary Inspector D/Agriculture	D207 D210
W415	Charles O'Connell	Superintending Agricultural Officer D/Agriculture	D207
W416	Dan O'Halloran	Manager Kerry Group	D207
W417	Jon Roberts	Former Employee Kepak	D208
W418	James Higgins	Veterinary Inspector D/Agriculture	D208
W419	Patrick J. F. Ledwith	Higher Agricultural Officer D/Agriculture	D209
W420	Liam Lynam	Higher Agricultural Officer D/Agriculture	D209
W421	Michael Durkin	Regional Superintending Agricultural Officer D/Agriculture	D209
W422	Brian Donohoe	Financial Director Kepak	D209
W423	Martin Finucane	Factory Manager Kepak	D209

Chronological list of all witnesses cont./...

WITNESS NO.	NAME	DESCRIPTION	TRANSCRIPT DAY NO.
W424	Bernadette McCann	Former Employee Kepak	D209
W425	Edward Noonan	Employee Kepak	D209
W426	Brian Finnegan	Manager Kepak	D209
W427	John Horgan	Deputy Chief Executive Kepak	D209
W428	Joan Coughlan	Employee Kepak	D209
W429	Liam McGreal	Managing Director Kepak	D209
W430	John Meagher	Officer of Customs and Excise	D210
W431	James Linnane	Veterinary Inspector D/Agriculture	D210
W432	Owen McCarthy	Higher Officer of Customs and Excise	D210
W433	Denis Mahony	Officer of Customs and Excise	D210
W434	Paul Murphy	Executive Agra Trading Ltd	D211
W435	Michael Behan	Managing Director Agra Trading Ltd	D211 D220
W436	Edward Gleeson	General Manager Agra Meat Packers Ltd	D211
W437	Noel Flood	Financial Controller Agra Trading Ltd	D211
W438	Martin Blake	Veterinary Inspector D/Agriculture	D211
W439	Patrick Joseph O'Connor	Veterinary Inspector D/Agriculture	D211
W440	Brian O'Beirn	Accountant United Meat Packers	D211
W441	Gerald Michael Butler	Former Superintending Veterinary Inspector D/Agriculture	D212
W442	George McLoughlin	Officer of Customs and Excise	D212
W443	Sean Stapleton	Officer of Customs and Excise	D212
W444	Patrick J. Garvey	Veterinary Inspector D/Agriculture	D212
W445	George Collins	Veterinary Inspector D/Agriculture	D212 D216
W446	Martin Macken	Agricultural Officer D/Agriculture	D212
W447	Aidan Nevin	Superintending Agricultural Officer D/Agriculture	D212
W448	Declan J. Hayes	Officer of Customs and Excise	D212
W449	John Roycroft	Former Administrative Officer D/Finance	D212
W450	Declan Holmes	Superintending Agricultural Officer D/Agriculture	D213
W451	Rory Godson	Business Editor, Sunday Tribune	D213
W452	James Quinn	Chief Executive Hibernia Meats	D213 D214
W453	Richard McCann	Managing Director Lixsteed Ltd	D214
W454	John O'Meara	Liaison Officer Purcell Meats Ltd	D214
W455	Padraig O Gairbhin	Officer of Customs and Excise	D215
W456	Patrick J. Moroney	Officer of Customs and Excise	D215
W457	Timothy O'Donnell	Superintending Agricultural Officer D/Agriculture	D215
W458	Gerard Barry	Veterinary Inspector D/Agriculture	D215

Chronological list of all witnesses cont./...

WITNESS NO.	NAME	DESCRIPTION	TRANSCRIPT DAY NO.
W459	Mary J. Byrne	Assistant Principal D/Agriculture	D215
W460	Liam King	Assistant Principal D/Agriculture	D215
W461	Donal Murphy	Superintending Agricultural Officer D/Agriculture	D216
W462	William Maher	Gentleman	D216
W463	Norbert Quinn	Managing Director Master Meats Group (post September 1988)	D217
W464	Gabriel Curley	Executive Officer D/Agriculture	D217
W465	Gerald Mulligan	Veterinary Inspector D/Agriculture	D218
W466	Raymond McGovern	Former Employee Tunney Meats Ltd	D218
W467	Peter McAleer	Former Employee Tunney Meats Ltd	D219
W468	Margaret Potter	Former Employee Tunney Meats Ltd	D219 D220
W469	Philip Smyth	Business Executive	D219
W470	Ronnie Flanagan	Former Manager Tunney Meats Ltd	D220
W471	John Copas	Managing Director Tunney Meats Ltd	D220
W472	Vincent Reilly	Former Employee Tunney Meats Ltd	D221
W473	Jim Flanagan	Former Employee Tunney Meats Ltd	D221
W474	Michael Connolly	Former Manager Tunney Meats Ltd	D222
W475	Hugh Thomas Tunney	Former Principal Tunney Meats Ltd.	D222

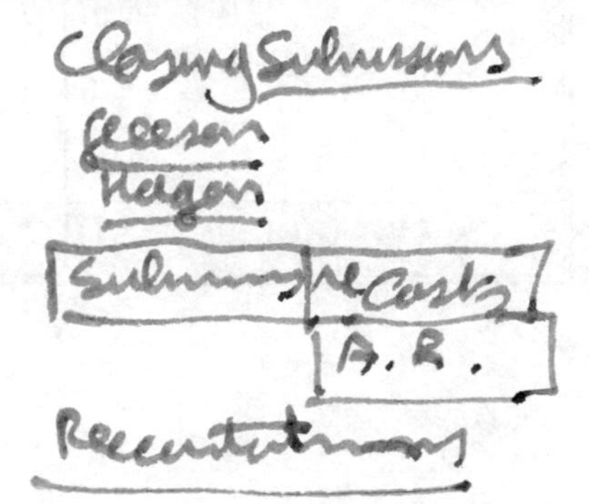

APPENDIX THREE

Legal Representation

Legal Representation before the Tribunal

Applicants/ Clients	Date granted	Solicitors	Counsel	Costs
The Tribunal	—	Ms C Loughlin (Chief State Solicitor's Office)	Mr E McGonigal SC Mr D Byrne SC Mr R Fullam BL Ms A O'Reilly BL	--
The Attorney General and all State Authorities	--	Chief State Solicitor (Mr J. Corcoran)	Mr H Whelehan SC (replaced by Mr C Maguire SC) Mr H Hickey SC Mr G Danaher BL Mr C Ó hOisín BL	--
Goodman International and Subsidiary Companies	21/6/1991	(A) A & L Goodbody (B) Rory O'Donnell & Co (Export Credit)	Mr D Gleeson SC Mr P Kelly SC Mr M Collins BL Mr I Finlay BL	Granted
Mr L Goodman	21/6/1991	(A) A & L Goodbody (B) Rory O'Donnell & Co (Export Credit)	Mr S McKenna SC Mr D O'Donnell BL	Granted
Ms S O'Keeffe (Granada Television)	Limited Representation 21/6/1991	McCann Fitzgerald	Mr N Fennelly SC	No Order (Not Claimed)
Mr B Solan (Granada Television)	Limited Representation 9/12/1991	Mr McCann Fitzgerald	Mr N Fennelly	No Order (Not Claimed)
Mr Dick Spring TD	Limited Representation 30/9/1991	Spring Murray & Co	Mr B McCracken SC Mr G Durcan BL Mr F O'Malley BL	Granted
Mr Barry Desmond MEP	Limited Representation 9/3/1992	Spring Murray & Co	Mr B McCracken SC Mr G Durcan BL Mr F O'Malley BL	
Mr R Kavanagh	Limited Representation 16/9/2992	Spring Murray & Co	Mr G Durcan BL	No Order (No application made)
Mr Z Al Taher	Limited Representation 26/7/1991	Murray Sweeney (Replaced by Bowler Geraghty & Co)	Mr F Clarke SC Mr W Shipsey BL	Granted
Mr A Fitzpatrick and Mr N Taher	Limited Representation 02/6/1992	Bowler Geraghty & Co	Mr F Clarke SC Mr F McEnroy BL	
ATGWU	Limited Representation 26/7/1991	Donal J Hamilton & Co	Mr D Hamilton SC Mr B Hickson BL	Granted
SIPTU	Limited Representation 26/7/1991	Bowler Geraghty & Co	Mr E Stewart SC Mr R Kean BL	Granted

Legal Representation before the Tribunal

<table>
<tr><th>Applicants/ Clients</th><th>Date granted</th><th>Solicitor</th><th>Counsel</th><th>Costs</th></tr>
<tr><td>Patrick Boyhan
Con Howard
Patrick J Kenny
Michael McElligott
Patrick Murphy</td><td>Limited Representation 26/7/1991</td><td>Henry P Kelly & Co</td><td>Mr P Callan SC
Mr D V Buckley SC
Mr M Gray BL</td><td>Granted (25 sittings days)</td></tr>
<tr><td>Mr Pat Rabbitte TD and Mr Tomás MacGiolla</td><td>Limited Representation 26/7/1991</td><td>Michael D White & Co</td><td>Mr A Hardiman SC
Mr T Clarke BL
Mr S Woulfe BL</td><td>Granted</td></tr>
<tr><td>Mr P Quearney</td><td>Limited Representation 16/9/1992</td><td>Michael D White & Co</td><td>—</td><td rowspan="2">Granted</td></tr>
<tr><td>Mr B Kelly</td><td>Limited Representation 11/5/1993</td><td>Michael D white & Co</td><td>—</td></tr>
<tr><td>Mr Eamon Gilmore TD</td><td>Limited Representation 4/11/1992</td><td>Michael D White & Co</td><td>—</td><td>No Order (Not Claimed)</td></tr>
<tr><td>Mr L Marks</td><td>Limited Representation 26/7/1991</td><td>Johnston Lavery & McGahon</td><td>Mr B White SC
Mr E Coffey BL</td><td>Granted</td></tr>
<tr><td>Mr J Fairbairn</td><td>Limited Representation 30/9/1991</td><td>Moran & Ryan</td><td>Mr R Weir QC
Mr K Haugh SC
Mr P Fogarty BL
Ms G Connolly BL</td><td>Granted</td></tr>
<tr><td>Irish Press Ltd</td><td>Contempt matter</td><td>William Fry</td><td>Mr P Shanley SC</td><td>No Order (Not Claimed)</td></tr>
<tr><td>Independent News- papers Ltd</td><td>Contempt matter</td><td>McCann Fitzgerald</td><td>Mr K Feeney SC</td><td>No Order (Not Claimed)</td></tr>
<tr><td>Irish Times Ltd</td><td>Contempt matter</td><td>Hayes & Sons</td><td>Mr R Nesbitt SC</td><td>No Order (Not Claimed)</td></tr>
<tr><td>Radio Telefís Eireann</td><td>Contempt matter</td><td>Eugene F Collins (Mr E Murphy)</td><td>Mr G Cooney SC</td><td>No Order (No application made)</td></tr>
<tr><td>Mr P Mc Guinness</td><td>Limited Representation 14/10/ 1991</td><td>Lerner & Associates (Canada)</td><td>Mr J Judson
Mr I Leach
(Canada)</td><td>—</td></tr>
<tr><td>Stokes Kennedy Crowley</td><td>Limited Representation 7/11/1991</td><td>Beauchamps</td><td>Mr C Condon SC
Mr P Shanley SC
Mr M Cush BL</td><td>Granted</td></tr>
<tr><td>Mr S Sheelan & Mr K J McDonald</td><td>Limited Representation 20/11/1991</td><td>MacGuill & Co</td><td>—</td><td>Granted</td></tr>
</table>

Legal Representation before the Tribunal

<table>
<tr><th>Applicants/ Clients</th><th>Date granted</th><th>Solicitor</th><th>Counsel</th><th>Costs</th></tr>
<tr><td>Mr E Mackle</td><td>Limited Representation 25/11/1991</td><td>C Mac An Ailí (replaced by Seamus Mallon & Co)</td><td>Mr S Moylan BL
Mr R Lyons BL</td><td>Granted</td></tr>
<tr><td>Mr N Quinn</td><td>Limited Representation 27/11/1991</td><td>Rory O'Donnell & Co</td><td>Mr G Cooney SC
Mr J Connolly BL</td><td>Granted</td></tr>
<tr><td>Mr M Murphy</td><td>Limited Representation 4/12/1991</td><td>Rory O'Donnell & Co</td><td>—</td><td>No Order (No application made)</td></tr>
<tr><td>Mr J O'Callaghan</td><td>Limited Representation 9/12/1991</td><td>Eugene F. Collins</td><td>Mr N Fennelly SC</td><td>Granted</td></tr>
<tr><td>Autozero Ltd
(Mr M Mullowney)</td><td>Limited Representation 11/12/1991</td><td>T. P. Robinson & Co (Mr T McGrath)</td><td>Mr K Feeney SC</td><td>Granted</td></tr>
<tr><td>Ms A Magee</td><td>Limited Representation 17/1/1992</td><td>Devaney & Ryan</td><td>Mr T O'Sullivan BL</td><td rowspan="2">Granted</td></tr>
<tr><td>Ms I Murray</td><td>Limited Representation 17/1/1992</td><td>Devaney & Ryan</td><td>Mr T O'Sullivan BL</td></tr>
<tr><td>Mr Sean Barrett TD
Mr John Burton TD
Senator Tom Raftery</td><td>Limited Representation 9/3/1992</td><td>J G O'Connor & Co</td><td>Mr J Nugent SC</td><td rowspan="2">Granted</td></tr>
<tr><td>Mr Paul Connaughton TD</td><td>Limited Representation 13/1/1993</td><td>J G O'Connor & Co</td><td>Mr J Nugent SC</td></tr>
<tr><td>Mr Michael Noonan TD</td><td>Limited Representation 29/5/1992</td><td>J G O'Connor & Co</td><td>Mr J Nugent SC</td><td>No Order (Not Claimed)</td></tr>
<tr><td>Mr S Murray</td><td>Limited Representation 15/9/1992</td><td>J G O'Connor & Co</td><td>Mr J Nugent SC</td><td>No Order (Not Claimed)</td></tr>
<tr><td>Mr Joe McCartin MEP</td><td>Limited Representation 2/4/1993</td><td>J G O'Connor & Co</td><td>Mr J Nugent SC</td><td>No Order (Not Claimed)</td></tr>
<tr><td>Mr Desmond O'Malley TD</td><td>Limited Representation 13/3/1992</td><td>Bell Brannigan O'Donnell & O'Brien</td><td>Mr A Hardiman SC
Mr G Hogan BL
Mr D McGuinness BL</td><td>Granted</td></tr>
<tr><td>Mr P McKay</td><td>Limited Representation 16/9/1992</td><td>Bell Brannigan O'Donnell & O'Brien</td><td>Mr G Hogan BL</td><td>No Order (No application made)</td></tr>
<tr><td>Mr D Kiberd
Mr B Carey</td><td>Contempt matter</td><td>Eugene F Collins</td><td>Mr P Finlay BL</td><td>No Order (No application made)</td></tr>
</table>

Legal Representation before the Tribunal

Applicants/ Clients	Date granted	Solicitors	Counsel	Costs
Mr Ted O'Reilly	Limited Representation 30/3/1992	Roger Greene & Sons	Mr G Humphreys BL	Granted
Mr J Stanley	Limited Representation 21/5/1992	Eugene F Collins	Mr D McCann BL	Granted
Mr S Clarke (United Meat Packers)	Limited Representation 25/5/1992	Crean O'Cleirigh & O'Dwyer	Mr L McKechnie SC Mr R Hastings BL	
Mr S M Khalid (United Meat Packers)	Limited Representation 24/6/1992	Crean O'Cleirigh & O'Dwyer	Mr L McKechnie SC Mr R Hastings BL	Granted
Mr S Clarke (Halal)	Limited Representation 10/6/1993	Crean O'Cleirigh & O'Dwyer	Mr L McKechnie SC Mr R Hastings BL	
Mr T Boland	Limited Representation 26/5/1992	John J O'Hare & Co	Mr P Geraghty SC Mr D Flanagan BL	Granted
Agra Trading Limited	Limited Representation 27/5/1992	William Fry	Mr R Nesbitt BL	Granted
Mr O Murphy	Limited Representation 27/5/1992	Arthur Cox	Mr C Allen BL	Granted
Hibernia Meats	Limited Representation 15/6/1993	Arthur Cox	Mr R Brady BL	Granted
Mr P Phelan	Limited Representation 22/6/1992	Robert Walsh & Co	Mr D O'Keeffee SC	Granted
Master Meats (Pre 16/9/1988)	Limited Representation 17/6/1993	Robert Walsh & Co	Mr H Hayden BL	Granted
Mr O Patten	Limited Representation 23/6/1992	John S O'Connor SC	Mr D McCullough SC Mr P Burns BL	Granted
Mr Michael O'Kennedy TD	Limited Representation 8/9/1992	Bruce St John Blake & Co	Mr P Geraghty SC Mr M Hanna BL	Granted
Mr B Britton	Limited Representation 8/9/1992	Gore & Grimes	Mr G Cooney SC Mr G Birmingham BL	Granted
Mr P Farrell	Limited Representation 16/9/1992	John S O'Connor & Co	Mr L Reidy SC Mr P Burns BL	Granted

Legal Representation before the Tribunal

Applicants/ Clients	Date granted	Solicitors	Counsel	Costs
European Commission	Limited Representation 9/10/1992	Matheson Ormsby Prentice	Mr M Daly SC Mr J O'Reilly BL	No Order (Not Claimed)
Mr Albert Reynolds TD	Limited Representation 4/11/1992	Patrick J Groarke & Son	Mr G Cooney SC Mr P Burns BL	No Order (Not Claimed)
Mr G Lambe	Limited Representation 12/1/1993	Donal O'Hagan & Co	Mr E Coffey BL	Granted
Brennan Governey & Co	Limited Representation 21/1/1993	Ivor Fitzpatrick & Co	Mr P Hunt BL	Granted
Mr S Goodwin Mr L Kelly Mr S Harnett Mr M O'Doherty Mr A Butler Mr D Murphy Mr T Keating	Limited Representation 28/1/1993	Holmes O'Malley & Sexton	Mr P MacEntee SC Mr P Gageby BL	Granted
Mr V Broderick and Mr C Peyron	Limited Representation 2/2/1993	Collins Crowley & Co	An tUas. S Ó Tuathail BL	Granted
Mr M Jinks	Limited Representation 16/9/1993	McGovern & Walsh	Mr B Grehan BL	Granted
Mr D Hanley	Limited Representation 16/2/1993	Richard R O'Hanrahan & Co	—	Granted
Arax Ltd	Limited Representation 11/5/1993	Lehane & Hogan	An tUas. S Ó Tuathail BL	Granted
Ashbourne Meats Ltd	Limited Representation 11/5/1993	Garrett Sheehan & Co	—	Granted
Tara Meats Ltd	Limited Representation 11/5/1993	Whitney Moore & Keller	Mr M Cush BL	Granted
Rangeland Meats Ltd	Limited Representation 12/5/1993	Donal J Hamilton & Co	Mr D Hamilton SC Mr B Hickson BL	Granted
Slaney Meats Ltd	Limited Representation 12/5/1993	John A Mernagh & Co	—	Granted
Ballywalter Meats Ltd	Limited Representation 18/5/1993	John A Mernagh & Co	Mr M Feehan SC	Granted
Dawn Meats Ltd	Limited Representation 13/5/1993	Nolan Farrell & Goff	Mr P Gageby BL	Granted

Legal Representation before the Tribunal

Applicants/ Clients	Date granted	Solicitors	Counsel	Costs
Taher Meats (N Taher)	Limited Representation 14/5/1993	Bryan F Fox & Co	Mr F McEnroy BL	Granted
Avrich Ltd T/A Freshlands Foods	Limited Representation 14/5/1993	Robert Walsh & Co	Mr P Fogarty BL	Granted
Western Meat Producers Ltd	Limited Representation 18/5/1993	Connellan	Mr E Walsh BL	Granted
Transfreeze Coldstores Ltd	Limited Representation 18/5/1993	Lennon, Heather & Co	Mr H Mohan BL	Granted
Blanchvac Ltd	Limited Representation 19/5/1993	Arthur Cox	—	Granted
Baltinglass Meats Ltd	Limited Representation 19/5/1993	Dermot Fullam	—	Granted
Kildare Chilling Co. Ltd	Limited Representation 20/5/1993	Donal McAuliffe & Co	Mr P Gageby BL	Granted
Barford Meats Ltd	Limited Representation 21/5/1993	Seamus Mallon & Co	Mr R Lyons BL	Granted
C H Foods Ltd	Limited Representation 21/5/1993	Barry Bowman & Co	Mr A Redmond BL	Granted
Continental Beef Packers Ltd	Limited Representation 21/5/1993	Hussey & O'Higgins	—	Granted
Mr C O'Brien and Mr P Fox	Limited Representation 25/5/2993	Martin E Marren & Co	Mr J Dillon BL	Granted
Norish Plc	Limited Representation 25/5/1993	Mason Hayes & Curran	—	Granted
Horgan Meats Ltd (in Liquidation)	Limited Representation 25/5/1993	P J O'Driscoll & Sons	Mr P McCarthy BL	Granted
Freezomatic Coldstores Co	Limited Representation 26/5/1993	Nolan, Farrell & Goff	—	Granted
Liffey Meats Ltd	Limited Representation 26/5/1993	Seamus Mallon & Co	Mr S Moylan BL	Granted

Legal Representation before the Tribunal

Applicants/ Clients	Date granted	Solicitors	Counsel	Costs
Meadow Meats (post 25/5/1993	Limited Representation 1/6/1993	Kerry Group Legal Department (Mr M O'Donoghue)	Mr D McDonald BL	Granted
Kerry Meat Packers	Limited Representation 2/6/1993	Kerry Group Legal Department (Mr M O'Donoghue)	Mr D McDonald BL	Granted
Michael Purcell Foods Ltd	Limited Representation 2/6/1993	Kieran T Flynn & Co	Mr D Kennedy BL	Granted
Kepak Ltd	Limited Representation 3/6/1993	Wm Smyth, O'Brien and Hegarty	Mr G Cooney SC Mr T Mallon BL	Granted
Mr J Roberts	Limited Representation 3/6/1993	John J Quinn & Co	Mr D O'Donovan BL	Granted
Mr B O'Beirn	Limited Representation 10/6/1993	C P Crowley & Co	Mr N McCarthy BL	Granted
Mr R Godson	Limited Representation 15/6/1993	Mohan & Co	Mr H Mohan BL	Granted
Eurowest Foods Ltd	Limited Representation 16/6/1993	Smithwick & Co	Mr T Mallon BL	Granted
Lixsteed Ltd	Limited Representation 16/6/1993	Paul N Beausang & Co	Mr P Finlay BL	Granted
Master Meats (Post 16/9/1988)	Limited Representation 17/6/1993	Gerrard Scallan & O'Brien	Mr R Brady BL	Granted
Irish Meats Producers	Limited Representation 22/6/1993	Arthur Cox	Mr J McBratney BL	Granted
Mr J McMahon	Limited Representation 22/6/1993	Garrett Sheehan & Co	Mr P MacEntee SC Mr F McEnroy BL	No Order (Not Claimed)
Tunney Meats Packers (in voluntary Liquidation)	Limited Representation 23/6/1993	McEntee & O'Doherty	Mr J Connolly SC Mr P Hanratty BL	Granted
Mr H Tunney	Limited Representation 23/6/1993	McEntee & O'Doherty	Mr J Connolly SC Mr P Hanratty BL	
Mr J Copas	Limited Representation 23/6/1993	McEntee & O'Doherty	Mr J Connolly SC Mr P Hanratty BL	

APPENDIX FOUR

List of Persons / Bodies

NO.	CORRESPONDENCE
1.	A.I.B.
2.	A.T.G.W.U.
3.	Abacus Capital Fund plc
4.	ABN-AMRO Bank
5.	Accountant
6.	Achille Bertrand, Avocat a la Cour
7.	Agra Meat Packers Ltd.
8.	Agricultural Officers (Various)
9.	Agricultural Credit Corporation plc
10.	Aktiengesellschaft
11.	Anderson Michael
12.	Arax Ltd
13.	Ashbourne Meat Processors
14.	Association
15.	Autozero
16.	Avonmore Foods
17.	Avrich Ltd
18.	Ballywalter Ltd.)
19.	Baltinglass Meats Ltd
20.	Bank of America
21.	Bank of Ireland
22.	Barford Meats Ltd.
23.	Barrett Seán T.D.
24.	Barry Gerald, The Sunday Tribune
25.	Behan Michael J
26.	Bell, Branigan, O'Donnell & O'Brien, Berrills Brian & Co Solicitors
27.	Blanchfield Paul
28.	Blatherwick D., British Ambassador
29.	BNP
30.	BOPA Ireland Ltd
31.	Bowler Geraghty & Co., Solicitors
32.	Brennan Governey & Co
33.	Brennan Ken
34.	Brennan Seamus
35.	Britton Brian
36.	Brown & McCann Solicitors
37.	Bruce St John Blake, Solicitors
38.	Bruton, John T.D
39.	Burke Ray T.D.
40.	Byrne Michael
41.	Byrne T.

NO.	CORRESPONDENCE
42.	Callan Bacon Co.
43.	Carlos John J & Co Solicitors
44.	CBF
45.	CED Viandes
46.	Central Statistics Office
47.	Chambers Philip
48.	Chief State Solicitor
49.	Ciarán Mac an Ailí, Solicitor
50.	Circuit Court Office
51.	Clancy Paddy, Sunday Press
52.	Classic Meats
53.	Cocking Mary
54.	Colemans Printers Ltd
55.	Collier Michael, Solicitor
56.	Collins, Crowley & Co., Solicitors
57.	Collins Eugene F. Solicitors
58.	Colso Enterprises
59.	Commissioned Officers
60.	Committee of Public Accounts
61.	Companies Registration Office
62.	Comptroller and Auditor General
63.	Conlan, Albert, ABN (Irl.) Ltd.
64.	Connaught Meat Producers ltd
65.	Connaughton Paul T.D.
66.	Connellan Solicitors
67.	Connolly Edward
68.	Connolly Frank, Sunday Tribune
69.	Continental Beef Packers Ltd.
70.	Cooke Nigel
71.	Coopers & Lybrand
72.	Cox Arthur Solicitors
73.	Craig Gardner
74.	Crean O'Cleirigh & O'Dwyer
75.	Crean Dermot
76.	Croskerrys Solicitors
77.	Curran Noel Business & Finance
78.	Curtin Bryan F. & Co Solrs
79.	Curtin Michael, Harbour Police
80.	Customs & Excise
81.	D. Heyer Meat Exports Ltd.
82.	Dairygold

NO.	CORRESPONDENCE
83.	Dardis John
84.	Dawn Meats Ltd. Kilmacthomas
85.	Dawn Meats Ltd. Grannagh
86.	Dawn Fresh Foods Ltd
87.	Dawnfresh Foods Ltd
88.	De Feu Sharpe, Solicitors
89.	Dehymeats Ltd
90.	Department of Finance
91.	Department of Industry & Commerce
92.	Department of Tourism & Trade
93.	Department of Agriculture & Food
94.	Department of Social Welfare
95.	Department of the Taoiseach
96.	Dept. of Agriculture, Nth.Ireland
97.	Desmond Barry M.E.P.
98.	Devaney & Ryan Solicitors
99.	Director of Public Prosecutions
100.	Doherty James
101.	Donlon John
102.	Dowdell Derek
103.	Doyle Avril
104.	Dukes Alan
105.	Dunne Brendan, Slaney Meats
106.	Dunphy Tony
107.	EC Court of Auditors
108.	EC Commission
109.	Emerald Meats Ltd
110.	Emerald Meats London ltd.
111.	Employees of Meat Plants (Various)
112.	Erin Foods
113.	Ernst & Young
114.	European Food Processing Service
115.	Eurowest Sallins
116.	Fianna Fáil
117.	Commerzbank GMBH
118.	Fine Gael
119.	Fitzgerald Garrett
120.	Fitzpatrick A.C.
121.	Fitzpatrick Peter, Legal Costs
122.	Flynn & Co Kieran T Solicitors
123.	Fox & Co. Bryan F. Solicitors

NO.	CORRESPONDENCE
124.	Fox John
125.	Freaney Oliver & Co.
126.	French E.J.
127.	Freshland Foods Ltd
128.	Gannon, Michael Bank of Ireland
129.	Garda Síochána
130.	Gerrard, Scallan & O'Brien, Solrs
131.	Gilmore Eamon, T.D.
132.	Gleeson John & Co
133.	Glentara Foods Ltd
134.	Godson Rory, Sunday Tribune
135.	Goodbody A & L, Solicitors
136.	Gore & Grimes, Solicitors
137.	Gorman Richard
138.	Granada Television
139.	Green Isle Food Group Ltd
140.	Greenore Ferry Services Ltd
141.	Groarke Patrick J & Son, Solicitor
142.	Guerin Veronica
143.	Halpin Joseph
144.	Hamilton Donal J & Co. Solicitors
145.	Hand Seamus
146.	Hanley Dermot
147.	Hanley Patrick
148.	Hanratty Joseph
149.	Harty Thomas
150.	Haughey Charles J. T.D.
151.	Hayes & Sons, Solicitors
152.	Heneghan Pat
153.	Henry P. Kelly & Co. Solicitors
154.	Hibernia Meats
155.	Hoey Vincent B & Co solicitors
156.	Holmes, O'Malley, Sexton Solrs
157.	Honey Clover (Freshford) Ltd
158.	Horan, Maurice, Gandon Corporate
159.	Horgan Meats Ltd
160.	Hourigan Shane
161.	Huggard A.
162.	Hussey & O'Higgins, Solicitors
163.	I.C.M.S.A.
164.	IDA Ireland

NO.	CORRESPONDENCE
165.	Impact
166.	Industrial Credit Corporation plc
167.	Inland Revenue, Northern Ireland
168.	Insurance Corporation of Ireland
169.	International Hide and Skin Co.
170.	Intervention Board, Reading
171.	Intervention Board for Agricultural
172.	Irish Ropes Ltd
173.	Irish Country Meats
174.	Irish Intercontinental Bank
175.	Irish Cold Stores Ltd
176.	Irish Casing Co. Ltd
177.	Irish Veterinary Union
178.	Irish Veterinary Association
179.	Islamic Foundation of Ireland
180.	Jacquot M EC Commission
181.	Johnston, Lavery & McGahon, Solrs
182.	Joint Arab-Irish Chamber of Commerce
183.	Jones Christy
184.	Kavanagh Paul A.
185.	Keane Michael, Sunday Press
186.	Kennedy Larry
187.	Kepak Longford
188.	Kepak Ltd Clonee
189.	Kepak Ltd
190.	Kerry Convenience Foods
191.	Kerry Group plc
192.	Kiberd Damien, Sunday Business Post
193.	Kiely Niall The Irish Times
194.	Kildare Chilling Co.
195.	Kilroys Solicitors
196.	KPMG Stokes, Kennedy Crowley
197.	Labour Party
198.	Lavery Kirby & Co Solicitors
199.	Lawlor Liam
200.	Le Controle Technique
201.	Lehane & Hogan Solicitors
202.	Lennon, Tonery & O'Neill Solicitors
203.	Lennon, Heather & Co Solicitors
204.	Lerner & Associates, Solicitors
205.	Limerick Cold Store Ltd

NO.	CORRESPONDENCE
206.	Lixsteed Ltd
207.	Loughrey John
208.	Lowery Martin, Coillte Teo
209.	Lowery Martin D
210.	Lynch Charlie
211.	M.J. Horgan & Sons, Solicitors
212.	MacCoille Cathal, Sunday Tribune
213.	MacGiolla Tomás
214.	MacGuill & Co Solicitors
215.	MacPartlin Declan
216.	MacSharry Ray EC Commissioner
217.	Mactavish Kenneth, Bank of America
218.	Maher, McAlinden, Gallagher Mallow Foods Ltd
219.	Maloney P.J. Marius, Theodore E.B., Amro
220.	Marren Martin E & Co Solicitors
221.	Marrow Meats Ltd
222.	Mason, Hayes & Curran Solicitors
223.	Master Meat Packers Group t/a Master Meat Packers (Clonmel) Ltd
224.	Master Pork
225.	Matheson, Ormsby, Prentice Solrs.
226.	McAleer Peter
227.	McAuliffe Donal T & Co Solicitors
228.	McCann Fitzgerald, Solicitors
229.	McCarren and Co. Ltd.
230.	McCartan Pat
231.	McCartin Joe
232.	McCourt James & Son Solicitors
233.	McDevitt Michael P.
234.	McDowell Michael
235.	McEntee & O'Doherty, Solicitors
236.	McGrath & Co Solicitors
237.	McInerney Michael
238.	McNally J. J. Ulster Bank Ltd.
239.	McNamara Matthew & Son Solicitors
240.	MCS Computers Ltd
241.	Millea Carol
242.	Millotte Mike, Sunday Tribune
243.	Moers Rose, Irish Press Newspapers
244.	Mohan & Company, Solicitors
245.	Molloy & Sherry Transport Ltd
246.	Monaghan Jim, AIBP Nenagh

NO.	CORRESPONDENCE
247.	Moran & Ryan Solicitors
248.	Morgan Daniel Solicitor
249.	Morrissey James, Sunday Business Muldowney & Co. Brendan T.
250.	Murphy Finbar Bank of Ireland
251.	Murphy Anthony Solicitor
252.	Murphy Thomas J
253.	Murphy John
254.	Murphy John
255.	Murray Sweeney, Solicitor
256.	Murray Imelda
257.	Nolan Farrell & Goff, Solicitors
258.	Nolan Thomas
259.	Noonan Michael (FG)
260.	Nordic Cold Storage Ltd
261.	Norish (Eirfreeze) Ltd
262.	Norish Food Care
263.	North Kerry Milk Products Ltd
264.	North Eastern Health Board
265.	North West Radio
266.	Northern Ireland Meat Exporters'
267.	NWL (Ireland) Ltd
268.	Ó Broin Niall
269.	O'Brien Daly Solicitors
270.	O'Callaghan Jerry
271.	O'Ceallaigh Padraig
272.	O'Connor & Co. J G Solicitors
273.	O'Donnell Rory Solicitors
274.	O'Driscoll & Sons P.J., Solrs
275.	O'Hagan & Co. Donal Solicitors
276.	O'Halpin Dr Eunan
277.	O'Hare & Co. John J. Solicitors
278.	O'Keeffe, Ned T.D.
279.	O'Mahony James Solicitor
280.	O'Malley Pat
281.	O'Malley Desmond T.D.
282.	O'Reilly Vincent
283.	O'Shea Fiona
284.	O'Toole Fintan, Irish Times
285.	Oliver, Michael Commerzbank Ox-Fleischhandelsgesellschaft mbh
286.	Paircéir Seamus
287.	Patten Owen K.

NO.	CORRESPONDENCE
288.	Poldys
289.	Post
290.	Power Meats Ltd.
291.	Premier By-Products
292.	Prendiville Paddy, The Phoenix
293.	Produce
294.	Progressive Democrats
295.	Purcell Ml. Foods Ltd
296.	Purcell Seamus
297.	Q.K. Cold Stores (Grannagh) Ltd
298.	Q.K. Cold Stores (Carroll's Cross)
299.	Quinn & Co. John J. Solicitors
300.	Quirke & Co., Solicitors
301.	Rabbitte Pat, T.D.
302.	Raftery Tom
303.	Raheenmore/Cloneyheighue Rangeland Meats Ltd
304.	Redi Chef Ltd
305.	Redways Ltd
306.	Reidy Stafford, Solicitors
307.	Representative Association of Residents' Associations
308.	Revenue Solicitor
309.	Revenue Commissioners
310.	Reynolds, Albert T.D.
311.	Rhatigan Brian
312.	Robinson Seán
313.	Rochford James MPC
314.	Roger Greene & Sons Solicitors
315.	Ronan, Daly, Jermyn, Solicitors
316.	Rory O'Donnell Solicitors
317.	Roscrea Cold Storage Ltd
318.	Royal Ulster Constabulary
319.	RTE
320.	Ruddy Thomas
321.	Ryan, Ritchie, EC Court of Auditors
322.	Ryan Tim, Irish Press Newspapers
323.	Ryan P.P. & Co.
324.	Rye Valley Foods Ltd
325.	S&D Meat Packers
326.	Sallyview Estates Ltd.
327.	Seamus Mallon & Co., Solicitors
328.	Shanahan, Ella, The Irish Times

NO.	CORRESPONDENCE
329.	Sheehan Garrett & Co. Solicitors
330.	Sheppard PC AIB Newry
331.	Sheridan Bryan, AIB
332.	Sheridan Ian
333.	Silver Hill Foods Ltd
334.	Sinnat Ltd
335.	Sinnott Hughes & Co
336.	SIPTU
337.	Slaney Cooked Meats Slaney Meats
338.	Smurfit Paribas Bank Ltd
339.	Smyth Peter VI
340.	Smyth O'Brien & Hegarty Solicitors
341.	Solan Brendan Solicitors
342.	Spring Murray Solicitors
343.	Spring Dick, T.D.
344.	Swissco Ltd
345.	T.L.S.
346.	Tallaght Cold Store Ltd
347.	Tara Meats (Kilbeggan) Ltd
348.	Tara Meats (Dublin) Ltd
349.	Tevlin Michael
350.	The Employment Appeals Tribunal
351.	Tiernan's Solrs - Creggan Transport
352.	Today Tonight, RTE
353.	Tomlinson John M.E.P.
354.	Toolan E.D.
355.	Transfreeze Cold Stores Ltd.
356.	Ulster Bank Ltd
357.	Ulster Bank Group
358.	Veterinary Inspectors (Various)
359.	Veterinary Officers' Association
360.	Veterinary Council
361.	Walsh Robert & Co. Solicitors
362.	Walsh Joe T.D.
363.	West David, Ulster Bank Ltd.
364.	Whelan Ken, Irish Press Newspapers
365.	White Michael D & Co., Solrs
366.	Whitney Moore & Keller Solrs
367.	Workers Party
368.	Wyse, Paul O. Freaney & Co

NAME	ADDRESS
Agra Meat Packers Ltd.,	Condonstown, Watergrasshill, Co. Cork.
Anglo Irish Beef Processors Ltd.,	t/a AIBP Waterford, Christendom, Co. Waterford.
Anglo Irish Pork Pocessors,	t/a AIPP Monaghan, Coolshannagh, Co. Monaghan.
Anglo Irish Beef Processors Ltd.,	t/a AIBP Rathkeale, Rathkeale, Co. Limerick.
Anglo Irish Pork and Bacon,	Coolshannagh, Co. Monaghan.
Anglo Irish Beef Processors Ltd.,	t/a AIBP Carlow, Royal Oak Rd., Bagenalstown, Co. Carlow.
Anglo-Irish Beef Processors Ltd.,	t/a AIBP Cahir, Cahir, Co. Tipperary.
Anglo-Irish Beef Processors Ltd.,	t/a AIBP Dundalk, Ravensdale, Dundalk,
Anglo-Irish Beef Processors Ltd.,	t/a AIBP Dublin, Cloghran, Swords, Co. Dublin.

NAME	ADDRESS
Anglo-Irish Beef Processors Ltd.,	t/a AIBP Nenagh, Nenagh, Co. Tipperary.
Anglo-Irish Beef Processors Ltd.,	t/a AIBP Longford, Bridge Street, Longford.
Anglo-Irish Beef Processors Ltd.,	t/a AIBP Dublin, Cloghran, Swords, Co. Dublin.
Anglo-Irish Beef Processors Ltd.,	t/a AIBP Nenagh, Nenagh, Co. Tipperary.
Anglo-Irish Beef Processors Ltd.,	t/a AIBP Longford, Bridge Street, Longford.
Ashbourne Meat Processors Ltd.,	Naas Industrial Estate, Naas, Co. Kildare.
Autozero Ltd.,	Cabra, Dublin 7.
Avonmore Foods PLC.,	t/a Irish Country Meats, Carrig, Roscrea, Co. Tipperary.
Baltinglass Meats Ltd.,	Baltinglass, Co. Wicklow.

NAME	ADDRESS
Barford Ltd.,	t/a Alpha Foods, Cloughvalley, Carrickmacross, Co. Monaghan.
Blanchvac Ltd.,	55 Cookstown Industrial Estate, Tallaght, Dublin 24.
Bopa Ireland Ltd.,	Clara, Co. Offaly.
Callan Processed Foods,	Callan, Co. Kilkenny.
Callan Bacon Company Ltd.,	Westcourt, Callan, Co. Kilkenny
Carroll Meats Manufacturing Ltd.,	Tullamore, Co. Offaly.
Connaught Meat Producers Ltd.,	t/a Kepak (Athleague), Athleague, Co. Roscommon.
Dawn Meats Ltd.,	Carrolls Cross, Co. Waterford.
Dawn Fresh Foods Ltd.,	Fethard, Co. Tipperary.
Dawn Farm Foods Ltd.,	Unit 60, Cherry Orchard Industrial Estate, Dublin 10.

NAME	ADDRESS
Dawn Meats Ltd.,	Grannagh, Co. Waterford.
Dehymeats Ltd.,	Ballast Quay, Sligo.
Dew Valley Meats,	Thurles, Co. Tipperary.
Doughlass Meats Ltd.,	Baltinglass, Co. Wicklow.
Duffy Meats Ltd.,	Hacketstown, Co. Carlow.
Eatwell Ltd.,	Cutlery Road, Newbridge, Co. Kildare
Eirefreeze Ltd.,	Little Island Co. Cork.
Emerald Meat Ltd.,	Emerald House, 8 Herbert Street, Dublin 2.
Erin Foods Ltd.,	Thurles, Co. Tipperary.
Feldhues Ltd.,	Rosslea, Monaghan Road, Clones, Co. Monaghan.

NAME	ADDRESS
Food Industries p.l.c.,	East Wall, Dublin 3.
Freezomatic Ltd.,	Tycor, Waterford.
Freezomatic Ltd.,	Cahir, Co. Tipperary.
Gentara Foods Ltd.,	Ashbury Industrial Estate, Roscrea, Co. Tipperary.
Goldstar Meats Ltd.,	Unit 27A, Dublin Industrial Estate, Glasnevin, Dublin 11.
Goudhurst Ltd.,	Beef Packers, 2 Upper Grand Canal Street, Dublin 2.
Grove Turkeys,	Smithboro, Co. Monaghan.
Halal Meat Packers (Ballyhaunis) Ltd.,	Clare Road, Ballyhaunis, Co. Mayo.
Halal Meat Packers (Ballaghaderreen) Ltd.,	Ballaghaderreen, Co. Roscommon.
Halal Meat Packers (Sligo) Ltd.,	Deepwater Quay, Sligo.

NAME	ADDRESS
Halal Meat Packers (Charleville) Ltd.,	Ardnageehy, Charleville, Co. Cork
Henry Denny & Sons (Irl) Ltd.,	Tralee, Co. Kerry.
Hibernia Coldstore Ltd.,	Sallins, Co. Kildare.
Hibernia Meats International Ltd.,	Woodstock Industrial Estate, Athy, Co. Kildare.
Hibernia Meats,	Cold Stores Ltd., Sallins, Co. Kildare.
Irish Country Bacon (Rooskey) Ltd.,	Rooskey, Carrick.on.Shannon, Co. Roscommon.
Irish Country Bacon (Cooked Meats) Ltd.,	Rooskey, Carrick.on.Shannon, Co. Roscommon.
Irish Cold Stores,	Ballymount Road, Tallaght, Dublin 24
Irish Ropes Ltd. Coldstore,	Newbridge, Co. Kildare.
Irish Horse Abattoir Investment Co. Ltd.,	Turnings Lower, Straffan, Co. Kildare.

NAME	ADDRESS
James Doherty (Carrigans) Ltd.,	Carrigans, Co. Donegal.
John Brennan & Sons Ltd.,	Dublin Street, Carlow.
John Murphy & Sons Ltd.,	Little Island, Co. Cork.
Kepak Ltd.,	Clonee, Co. Meath.
Kepak (Export) Ltd.,	Hacketstown, Carlow.
Kildare Chilling Company Ltd.,	Curragh Road, Kildare.
Kinsale Foods Ltd.,	Unit 5, Ballycurreen Industrial Estate, Kinsale Road, Cork.
KMP Co-op Society Ltd.,	Knockgriffin, Midleton, Co. Cork.
KMP Co-op Society Ltd.,	t/a Convenience Foods Ltd., Cookstown Industrial Estate, Tallaght, Co. Dublin.
Liffey Meats (Cavan) Ltd.,	Ballyjamesduff, Co. Cavan.

NAME	ADDRESS
Limerick Bacon Co. Ltd.,	Food Centre, Raheen, Limerick.
Lyonara Coldstore,	Clonminan Industrial Estate, Portlaoise, Co. Laois.
M & M Walsh,	Crossgallen Enterprise Centre, Ballysimon Road, Limerick.
Mallow Foods Ltd.,	Newberry, Mallow, Co. Cork.
Marrow Meats Ltd.,	Dromkeen, Co. Limerick.
Master Meat Packers (Clonmel) Ltd.,	Upper Irishtown, Clonmel, Co. Tipperary.
Master Meat Packers (Bandon) Ltd.,	Kilbrogan, Bandon, Co. Cork.
Master Pork Packers Ltd.,	Edenderry, Co. Offaly.
Master Meat Packers (Longford) Ltd.,	Rathmore, Ballymahon, Co. Longford.
Master Meat Packers (Longford) Ltd.,	Rathmore, Ballymahon, Co. Longford.

NAME	ADDRESS
Master Meat Packers (Bandon) Ltd.,	Upper Irishtown, Clonmel, Co. Tipperary.
Master Meat Packers (Kilkenny) Ltd.,	Freshford, Co. Kilkenny.
McCarren & Co. Ltd.,	Cavan.
Meadow Meats Ltd.,	Rathdowney, Co. Laois.
Meadow Meats (Waterford) Ltd.,	Christendom, Co. Waterford.
Mitchelstown Co-op Agrl Society Ltd.,	(t/a Galtee Food Products), Mitchelstown, Co. Cork.
Mitchelstown Co-op,	Agricultural Society Ltd., Dungarvan, Co. Waterford.
National Cold Store,	Cookstown Industrial Estate, Belgard Road, Tallaght, Dublin 24.
Norish Food City Ltd.,	Lough Egish, Co. Monaghan.

NAME	ADDRESS
Norish (Kilkenny) Ltd.,	Ballyconra North, Ballyragget, Co. Kilkenny.
North Kerry Co-op Cold Store,	Listowel, Co. Kerry.
O'C Coldstore Ltd.,	Unit 10, Clondalkin Industrial Estate, Crag Avenue, Clondalkin, Dublin 22.
Oakpark Food Ltd.,	Clogheen Road, Cahir, Co. Tipperary.
Ox Fleisch Handels GMBH,	Carrickmacross, Co. Monaghan.
Poldy's Fresh Foods Ltd.,	Naas Industrial Estate, Naas, Co. Kildare.
Poldys Fresh Foods (Portumna) Ltd.,	Portumna, Co. Galway.
Power Meats,	Sandyford Road, Dundrum, Dublin.

NAME	ADDRESS
Q.K. Cold Stores Ltd.,	Carrolls Cross, Co. Waterford.
Q.K. Coldstores (Grannagh) Ltd.,	Grannagh, Waterford.
Queally Pig Slaughtering Ltd.,	t/a Dawn Pork and Bacon, Grannagh, Co. Kilkenny.
Rangeland Meats Ltd.,	Tullynamalra, Lough Egish, Co. Monaghan.
Rangeland Meats Ltd.,	Tullynamarla, Castleblaney, Co. Monaghan.
Redi-Chef Ltd.,	Dublin Road, Thurles, Co. Tipperary
Redways Ltd.,	Sinnotshill, Castlebridge, Co. Wexford.
Roscrea Fresh Foods,	Parkmore Industrial Estate, Roscrea, Co. Tipperary.
Roscrea Coldstore Ltd.,	Roscrea, Co. Tipperary.

NAME	ADDRESS
Rye Valley Foods,	Carrickmacross, Co. Monaghan.
Silver Crest Foods,	Greencastle Parade, Coolock Industrial Estate, Dublin 17.
Slaney Cooked Meats,	Ryland, Enniscorthy, Co. Wexford.
Slaney Meats International,	Ryland, Enniscorthy, Co. Wexford.
Stanlow Trading Ltd.,	t/a Ellis Meats, Rockhill, Ballintra, Co. Donegal.
Swissco Ltd.,	Little Island, Co. Cork.
Taher Meats (Roscrea) Ltd.,	Roscrea, Co. Tipperary.
Tara Meats Ltd.,	Cookstown Industrial Estate, Tallaght, Dublin 24.
Transfreeze Ltd.,	Santry Hall, Santry, Dublin 9.
Tunney Meat Packers Ltd.,	Teehill, Clones, Co. Monaghan.

NAME	ADDRESS
United Meat Packers (Charleville) Ltd.,	Ardnageehy, Charleville, Co. Cork.
United Meat Packers (Wexford) Ltd.,	Baylands, Camolin, Co. Wexford.
Waterford Cold Stores,	Christendom, Co. Waterford.
Western Meat Producers Ltd.,	t/a Kepak (Athleague), Athleague, Co. Roscommon.
Western Meats Producers Ltd.,	Dromod, Co. Leitrim.
X.L. Meats Ltd.,	t/a Tara Meats, Kilbeggan, Co. Westmeath.

APPENDIX FIVE

List of Exhibits

LIST OF EXHIBITS

EXHIBITS		NAME	
Ex. 1	D5	Production Line, Schematic Diagram	Tribunal
Ex. 2	D6.	Health Stamp "Ireland, 324, EEC"	Tribunal
Ex. 3	D6	Classification Schedule I to V	A&L
Ex. 4	D7	IB4 form	Tribunal
Ex. 5	D7	Sample Documents of Purchase of Intervention Beef	Tribunal
Ex. 6	D7	Market Share Intervention & APS Utilisation	Tribunal
Ex. 7	D7	Letter Department of Agriculture to AIBP, Dundalk, dated 17/5/91	A.& L.
Ex. 8	D7	(Copy) P.Q. from Jim Mitchell T.D.	A.& L.
Ex. 9	D8	Forfeiture/Recoveries '81-'90 and Irregularities reported to EEC	Tribunal
Ex. 10	D9	Garda Book of Photos of IB4 - initialled	Tribunal
Ex. 11	D10	(Copy) Back of IB4	Tribunal
Ex. 12	D10	(Copy) Yellow Form	Tribunal
Ex. 13	D10	Excerpt from Instructions to A. O.'s	Tribunal
Ex. 14	D10	Classification Stamps - Long (A) CR3 C03 CR4	Tribunal
		Face (B) CU3	Tribunal
		Face (C) CU4	Tribunal
		Hand Stamp (D) CU3	Tribunal
Ex. 15A	D10	**Bogus** Classification Stamps - C03, CR4	Tribunal
		CR3,	Tribunal
		CU3,CU4	Tribunal
Ex. 15B	D10	**Bogus** Health Stamps "Ireland, 344 EEC"	Tribunal
EX. 15C	D10	**Bogus** "Islamic" stamp and "R4", "R3"	Tribunal
Ex. 16	D12	Certified Copy Sums (12)	Tribunal
Ex. 17	D12	(Copy) Conviction Orders (2)	Tribunal
Ex. 18	D12	Table of Deboning Yield 3/87 - 3/88	A.G.
Ex. 19	D13	Per Cent Weight Loss on Net Weight	A.& L.
Ex. 20	D13	Average Boneless Yields '83-'91	A.G.
Ex. 21	D14	6 - Export Documents	Tribunal
Ex. 22A	D14	12- Movement Documents (into Eirfreeze)	Tribunal
Ex. 22B	D14	6 - Movement Documents (Eirfreeze to Docks)	Tribunal
Ex. 23	D14	7 - Meat Inspection Certificates (originals)	Tribunal
Ex. 24	D14	6 - Meat Inspection Certificates	A. G.
Ex. 25	D15	4 - Original Documents (P & L Report and 3 weekly reports - Wtfd)	
Ex. 26	D15	Granada and Department IB4's (Originals)	Tribunal
Ex. 27	D15	Attempt Weight Reconciliation to Cold Meat (9 Original Pages)	Tribunal
Ex. 28	D15	(Copy) "Problems Identified by DOA Official"	Tribunal
Ex. 29	D15	(Copy) APS Contract Documents	Tribunal
Ex. 30	D16	4 Exact Copy Documents. Re Pallets	Tribunal
Ex. 31	D16	Invoices and Photocopies of Cheques	Tribunal
Ex. 32	D16	3 Documents - Days of week with hours for 3 individuals	Tribunal
Ex. 33	D16	6 Documents - Complimentary Slip and Expenses and Claims Sheets etc.	Tribunal
Ex. 34	D16	Livestock Remittance Dockets (origns.)	Tribunal
Ex. 35	D16	AIBP List of Names and Amts.	Tribunal
Ex. 36	D16	Cash Payment Schedule for Waterford	Tribunal
Ex. 37	D16	George McMillan Document.	Tribunal
Ex. 38	D16	1985 Amts. and Bankdraft (6 Documents)	Tribunal
Ex. 39	D16	Fax Sheets from Ravensdale to Waterford (re Management Salaries)	Tribunal
Ex. 40	D16	Compliment Slip, List of Names and expenses Claim Sheets etc.	Tribunal
Ex. 41	D16	Chart Marked (B) with carcases numbers, weights and price.	Tribunal
Ex. 42	D16	Folder of similar documents as went before	Tribunal
Ex. 43	D18	Alleged Move of Classification Officer (Copy document)	AG
Ex. 44	D18	Patrick McGuinness Letter (copy) to Mr McAndrews dated 19/1/84	A.& L.
Ex. 45	D18	Susan O'Keeffe's article in Independent	A.& L.
Ex. 46	D18	IB4 (copy) "Jan 87"	A.& L.
Ex. 47	D18	(Copy) letter IDB Northern Ireland Dated 25/11/91	A.& L.
Ex. 48	D19	2 Photographs of Waterford Office	A.& L.
Ex. 49	D19	(Copy) Weekly Records No's 50 and 51	A.& L.

List of Exhibits cont/.....

EXHIBITS		NAME	
Ex. 50	D19	(Copy) Carcase Weight Loss Chart (fig.1)	A. & L.
Ex. 51	D19	(Copy) Letter from Biochemist in Teagasc	A. & L.
Ex. 52	D19	(Copy) Carcase Weight Loss Chart (fig.2)	A. & L.
Ex. 53	D19	(Copy) Carcase Weight Loss Chart (fig.3)	A. & L.
Ex. 54	D19	(Copy) Single Sheet "Table A" October/November etc.	A. & L.
Ex. 55	D19	(Copy) Single Sheet "Table B"	A. & L.
Ex. 56	D19	(Copy) Carcase Weight Loss (fig.4)	A. & L.
Ex. 57	D19	(Copy) Weight Gain/Loss (fig.5)	A. & L.
Ex. 58	D19	(Copy) Expenses Sheet "(1)" "(C)"	A. & L.
Ex. 59	D19	(Copy) Expenses Sheet "(2)" "(D)"	A. & L.
Ex. 60	D19	(Copy) Expenses Sheet "(F)"	A. & L.
Ex. 61	D19	(Copy) Expenses Sheet "(G)"	A. & L.
Ex. 62	D19	(Copy) Expenses Sheet "(H)"	A. & L.
Ex. 63	D19	(Copy) Expenses Sheet "(I)"	A. & L.
Ex. 64	D19	(Copy) Expenses Sheet "(K)"	A. & L.
Ex. 65	D19	(Copy) AIBP Paper Expenses Sheet (M) and (L) (2 pages)	A. & L.
Ex. 66	D19	(Copy) Photograph of Coldstore	A. & L.
Ex. 67	D20	(Copy) Newry Factory Profits for 1985	A. & L.
Ex. 68	D20	Box (Intervention in N.I.)	A. & L.
Ex. 69	D20	Copy letter from S.K.C. Dated 7/6/88	A. & L.
Ex. 70	D21	IB4's and Yellow form for 19/1/87	Tribunal
Ex. 71	D21	IB4's (from witness 40) dated 19/2/88	Tribunal
Ex. 72	D21	Letter of 26/2/88 from AIBP to DOA	Tribunal
Ex. 73	D21	2 Pictures of Classified Carcases	A. & L.
Ex. 74	D21	(Copy) Monthly Reports from Brussels	A. & L.
Ex. 75	D22	(2 Copy Sheets) Re Container Movements	Tribunal
Ex. 76	D22	6 Meat Inspection Certs 1/3/89 (Originals)	Tribunal
Ex. 77	D22	Export Declaration Forms (Originals)	Tribunal
Ex. 78	D22	Letter DOA 28/11/91 / Export Forms	Tribunal
Ex. 79	D22	Payslips for John Wallace	Tribunal
Ex. 80	D23	Garda Book. of Photo of Stamps	Tribunal
Ex. 81	D23	Docket for Bogus Stamps (Devenney Ltd)	A. & L.
Ex. 82	D24	(Copy) Letter of B. Dolan to B. Desmond	A. & L.
Ex. 83	D24	(Copy) Reboxing Regulations	A. & L.
Ex. 84	D24	(Copy) "Contracts Reboxed in AIBP, Cahir"	A. & L.
Ex. 85	D25	(Original) Container records/DOA (17/2/89 onwards)	Tribunal
Ex. 86	D25	(Copy) 3 Movement Certs for above	Tribunal
Ex. 87	D25	Drawings/Sketch of Nenagh Factory	A. & L.
Ex. 88	D25	Sketch Map Border area	White Solr.
Ex. 89	D25	2 N.I. Seals broken (Carousel Opn.)	Tribunal
Ex. 90	D25	Export Documents - Re-No.89	Tribunal
Ex. 91	D25	Letter of 14/4/91 from AIBP to C & E	Tribunal
Ex. 92	D25	Copy receipt Gypsum Industries	A. & L.
Ex. 93	D26	Removal order Intervention Stock	A. & L.
Ex. 94	D26	Schedule of Weights (C. &. E. docs.)	Tribunal
Ex. 95	D27	Copies of Coldstore Plant (Mullowney)	Robinson Solr.
Ex. 96	D28	(Copy) Group Structure Goodman International Ltd.	Tribunal
Ex. 97	D28	(Copy) Letter from Goodman to Revenue Commissioners (18/3/87)	Tribunal
Ex. 98	D28	Memo and letters of Revenue Comms. (17/4/87)	Tribunal
Ex. 99	D28	Letter 19/1/87 of Revenue Comms. to DOA.	Tribunal
Ex. 100	D30	Notes of Meeting with Minister	Tribunal
Ex. 101	D31	A.P.S. Yield Sheets etc. (14/11/86)	Tribunal
Ex. 102	D31	A.P.S. Yield Sheets etc. (27/11/86)	Tribunal
Ex. 103	D31	A.P.S. Yield Sheets etc. (28/11/86)	Tribunal
Ex. 104	D31	A.P.S. Yield Sheets etc. (04/12/86)	Tribunal
Ex. 105	D31	A.P.S. Yield Sheets etc. (05/12/86)	Tribunal
Ex. 106	D31	A.P.S. Yield Sheets etc. (08/12/86)	Tribunal
Ex. 107	D31	A.P.S. Yield Sheets etc. (02/10/86)	Tribunal

List of Exhibits cont/.....

EXHIBITS		NAME	
Ex. 108	D31	A.P.S. Yield Sheets etc. (28/10/86)	Tribunal
Ex. 109	D31	A.P.S. Yield Sheets etc. (03/11/86)	Tribunal
Ex. 110	D31	A.P.S. Yield Sheets etc. (10/11/86)	Tribunal
Ex. 111	D31	A.P.S. Yield Sheets etc. (30/09/86)	Tribunal
Ex. 112	D31	A.P.S. Yield Sheets etc. (03/10/86)	Tribunal
Ex. 113	D31	A.P.S. Yield Sheets etc. (06/10/86) (Job 20)	Tribunal
Ex. 114	D31	A.P.S. Yield Sheets etc. (06/10/86) (Job 21)	Tribunal
Ex. 115	D31	A.P.S. Yield Sheet etc. (08/10/86)	Tribunal
Ex. 116	D31	A.P.S. Yield Sheet etc. (13/02/87)	Tribunal
Ex. 117	D31	A.P.S. Yield Sheet etc. (12/11/86)	Tribunal
Ex. 118	D31	A.P.S. Yield Sheet etc. (11/11/86)	Tribunal
Ex. 119	D31	A.P.S. Yield Sheet etc. (07/11/86)	Tribunal
Ex. 120	D31	A.P.S. Yield Sheet etc. (06/11/86) (Job 88)	Tribunal
Ex. 121	D31	A.P.S. Yield Sheet etc. (06/11/86) (Job 89)	Tribunal
Ex. 122	D31	A.P.S. Yield Sheet etc (17/12/86)	Tribunal
Ex. 123	D31	A.P.S. Yield Sheet etc (16/12/86)	Tribunal
Ex. 124	D31	A.P.S. Yield Sheet etc. (20/11/86)	Tribunal
Ex. 125	D31	A.P.S. Yield Sheet etc. (14/11/86)	Tribunal
Ex. 126.	D31	A.P.S. Yield Sheet etc. (4,7/10/86)	Tribunal
Ex. 127.	D31	A.P.S. Yield Sheet etc. (12/11/86) (Job 109)	Tribunal
Ex. 128	D31	A.P.S. Yield Sheet etc. (08/11/86)	Tribunal
Ex. 129	D31	A.P.S. Yield Sheet etc. (07/11/86) (Job 94)	Tribunal
Ex. 130	D31	A.P.S. Yield Sheet etc. (07/11/86) (Job 93)	Tribunal
Ex. 131	D31	A.P.S. Yield Sheet etc. (07/11/86) (Job 92)	Tribunal
Ex. 132	D31	A.P.S. Yield Sheet etc. (04/11/86) (Job 83)	Tribunal
Ex. 133	D31	A.P.S. Yield Sheet etc. (04/11/86) (Job 81)	Tribunal
Ex. 134	D31	A.P.S. Yield Sheet etc. (01/11/86) (Job 75)	Tribunal
Ex. 135	D31	A.P.S. Yield Sheet etc. (31/10/86)	Tribunal
Ex. 136	D31	A.P.S. Yield Sheet etc. (29/10/86) (Job 69)	Tribunal
Ex. 137	D31	A.P.S. Yield Sheet etc. (19/02/87) (Job 12)	Tribunal
Ex. 138	D31	A.P.S. Yield Sheet etc. (29/12/86) (Job 5)	Tribunal
Ex. 139	D31	A.P.S. Yield Sheet etc. (12/12/86) (Job 3)	Tribunal
Ex. 140	D31	A.P.S. Yield Sheet etc. (01/12/86) (Job 2)	Tribunal
Ex. 141	D31	A.P.S. Yield Sheet etc. (21/11/86) (Job 4, 5)	Tribunal
Ex. 142	D31	A.P.S. Yield Sheet etc. (11/12/86) (Job 20)	Tribunal
Ex. 143	D31	A.P.S. Yield Sheet etc. (12/12/86) (Job 21)	Tribunal
Ex. 144	D31	A.P.S. Yield Sheet etc. (13/12/86) (Job 22)	Tribunal
Ex. 145	D31	A.P.S. Yield Sheet etc. (15/12/86) (Job 23)	Tribunal
Ex. 146	D31	A.P.S. Yield Sheet etc. (01/10/86) (Job 16)	Tribunal
Ex. 147	D31	A.P.S. Yield Sheet etc. (07/10/86) (Job 23)	Tribunal
Ex. 148	D31	A.P.S. Yield Sheet etc. (08/10/86) (Job 27)	Tribunal
Ex. 149	D31	A.P.S. Yield Sheet etc. (09/10/86) (Job 29)	Tribunal
Ex. 150	D31	A.P.S. Yield Sheet etc. (10/10/86) (Job 30)	Tribunal
Ex. 151	D31	A.P.S. Yield Sheet etc. (14/10/86) (Job 38)	Tribunal
Ex. 152	D31	A.P.S. Yield Sheet etc. (17/10/86) (Job 43)	Tribunal
Ex. 153	D31	A.P.S. Yield Sheet etc. (17/10/86) (Job 44)	Tribunal
Ex. 154	D31	A.P.S. Yield Sheet etc. (18/10/86) (Job 46)	Tribunal
Ex. 155	D31	A.P.S. Yield Sheet etc. (18/10/86) (Job 47)	Tribunal
Ex. 156	D31	A.P.S. Yield Sheet etc. (20/10/86) (Job 49)	Tribunal
Ex. 157	D31	A.P.S. Yield Sheet etc. (20/10/86) (Job 50)	Tribunal
Ex. 158	D31	A.P.S. Yield Sheet etc. (20/10/86) (Job 51)	Tribunal
Ex. 159	D32	A.P.S. Yield Sheet etc. (19/11/86) (Job 2)	Tribunal
Ex. 160	D32	A.P.S. Yield Sheet etc. (24/11/86) (Job 5)	Tribunal
Ex. 161	D32	A.P.S. Yield Sheet etc. (25/11/86) (Job 6)	Tribunal
Ex. 162	D32	A.P.S. Yield Sheet etc. (21/12/86) (Job 12)	Tribunal
Ex. 163	D32	A.P.S. Yield Sheet etc. (03/12/86) (Job 13)	Tribunal
Ex. 164	D32	A.P.S. Yield Sheet etc. (09/12/86) (Job 18)	Tribunal
Ex. 165	D32	A.P.S. Yield Sheet etc. (19/12/86) (Job 27)	Tribunal
Ex. 166	D32	A.P.S. Yield Sheet etc. (20/12/86) (Job 28)	Tribunal

List of Exhibits cont/.....

EXHIBITS		NAME	
Ex. 167	D32	A.P.S. Yield Sheet etc. (22/12/86) (Job 29)	Tribunal
Ex. 168	D32	A.P.S. Yield Sheet etc. (29/12/86) (Job 31)	Tribunal
Ex. 169	D32	A.P.S. Yield Sheet etc. (30/12/86) (Job 32)	Tribunal
Ex. 170	D32	A.P.S. Yield Sheet etc. (06/01/87) (Job 34)	Tribunal
Ex. 171	D32	A.P.S. Yield Sheet etc. (15/01/87) (Job 36)	Tribunal
Ex. 172	D32	A.P.S. Yield Sheet etc. (16/01/87) (Job 37)	Tribunal
Ex. 173	D32	A.P.S. Yield Sheet etc. (20/01/87) (Job 39)	Tribunal
Ex. 174	D32	A.P.S. Yield Sheet etc. (27/01/87) (Job 40)	Tribunal
Ex. 175	D32	A.P.S. Yield Sheet etc. (03/02/87) (Job 41)	Tribunal
Ex. 176	D32	A.P.S. Yield Sheet etc. (11/02/87) (Job 43)	Tribunal
Ex. 177	D32	A.P.S. Yield Sheet etc. (19/02/87) (Job 46)	Tribunal
Ex. 178	D32	A.P.S. Yield Sheet etc. (20/02/87) (Job 47)	Tribunal
Ex. 179	D32	A.P.S. Yield Sheet etc. (24/02/87) (Job 49)	Tribunal
Ex. 180	D32	A.P.S. Yield Sheet etc. (26/02/87) (Job 50)	Tribunal
Ex. 181	D32	(Copy) Official Yield Sheet .... (86/5 Form)	A. & L.
Ex. 182	D32	(Copy) Production Sheets (10/12/90)	A. & L.
Ex. 183	D32	(Copy) Yield Sheets ...(11/08/87 - 02/12/91)	A. & L.
Ex. 184	D32	Note to Minister (copy) of 18/3/87	Tribunal
Ex. 185	D33	Table % Forfeiture/Recoveries '81/90	A. & L.
Ex. 186	D33	(Copy) Notice re-weights - 15/10/90	A. & L.
Ex. 187	D33	(Copy) Letter from DPP to C.S.S. 16/5/91	Tribunal
Ex. 188	D34	C & E. forms 977 (2 origs. + 12 copies)	Tribunal
Ex. 189	D34	C & E. Tabulation of Weights (Transfreeze)	Tribunal
Ex. 190	D34	Form 86/4M .... (2,722. kilos)	Tribunal
Ex. 191	D34	Form 86/4M .... (1,045.5 kilos)	Tribunal
Ex. 192A	D34	Form 86/4M................ (8,900.5 kilos)	Tribunal
Ex. 192B	D34	Form 86/4M................ (9,284.5 kilos)	Tribunal
Ex. 192C	D34	Form 86/4M................ (9,920.0 kilos)	Tribunal
Ex. 192D	D34	Form 86/4M................ (11,355.5 kilos)	Tribunal
Ex. 193A	D34	Form 86/5 .... (32,924.9 kilos)	Tribunal
Ex. 193B	D34	Form 86/6 .... (69 boxes)	Tribunal
Ex. 193C	D34	Form 86/6 .... (502 boxes)	Tribunal
Ex. 193D	D34	Form 86/6 .... (524 boxes)	Tribunal
Ex. 193E	D34	Form 86/6 .... (193 boxes)	Tribunal
Ex. 194	D34	Cert. No. 21045	Tribunal
Ex. 195	D34	Cert. No. 22517	Tribunal
Ex. 196A	D34	Form 86/6................. (43 boxes)	Tribunal
Ex. 196B	D34	Form 86/6................. (100 boxes)	Tribunal
Ex. 196C	D34	Form 86/6................. (370 boxes)	Tribunal
Ex. 197A	D34	Cert. No. 21046	Tribunal
Ex. 197B	D34	Cert. No. 22518	Tribunal
Ex. 198	D34	Form 86/6 (189 boxes)	Tribunal
Ex. 199A	D34	Cert. No. 22520	Tribunal
Ex. 199B	D34	Cert. No. 22520	Tribunal
Ex. 200	D34	Form 86/4M ....(7746.9 kilos)	Tribunal
Ex. 201	D34	Form 86/5...............(5754.5 Kilos)	Tribunal
Ex. 202A	D34	Cert. No................22522	Tribunal
Ex. 202B	D34	Cert. No................22522	Tribunal
Ex. 203A	D34	Form 86/5...............1444.6 kilos)	Tribunal
Ex. 203B	D34	Form 86/6...............23 Boxes	Tribunal
Ex. 203C	D34	Cert. No................21050	Tribunal
Ex. 203D	D34	Cert. No................22523	Tribunal
Ex. 204A	D34	Form 86/4M..............(2521 kilos)	Tribunal
Ex. 204B	D34	Form 86/5...............(1917.4 kilos)	Tribunal
Ex. 204C	D34	Form 86/6...............(71 boxes)	Tribunal
Ex. 205A	D34	Cert. No................27805	Tribunal
Ex. 205B	D34	Cert. No................22528	Tribunal
Ex. 205C	D34	Cert. No................22528	Tribunal
Ex. 206	D34	(Copy) Invoice 10/11/87 (August Co.)	Tribunal
Ex. 207	D34	(Copy) Note 20/5/91 D.O.A. Vets.	Tribunal

List of Exhibits cont/.....

EXHIBITS		NAME	
Ex. 208	D35	(Copy) Production Records (22/9/86 - 14/11/86)	A.& L.
Ex. 209	D36	IB1 forms etc. Re Path of carcases thro-system	Tribunal
Ex. 210	D36	(Copy) Appendix 2 with manuscript % fig.	A.& L.
Ex. 211	D36	(Copy) Diagram straight cut pistola	A.& L.
Ex. 212	D36	(Copy) Diagram 13 Rib pistola	A.& L.
Ex. 213	D36	(Copy) Diagram straight cut hind.	A.& L.
Ex. 214	D37	(Copy) Circular 1/82 re-deboning Yields	A.& L.
Ex. 215	D37	(Copy) Intervention Boning Yields	A.& L.
Ex. 216	D39	Pay Slips for W115	Tribunal
Ex. 217	D40	Weights and Measures Certs. '85-88	Tribunal
Ex. 218	D41	Meat Inspection Certs.........2/84 (3)	Tribunal
Ex. 219	D41	Boning Hall Job Costing Sheets	Tribunal
Ex. 220	D42	(Copy) Letter of 3/7/86 from D.O.A.	A.& L.
Ex. 221	D42	(Copy) Daily Classification Sheet, Rathkeale	A.& L.
Ex. 222	D42	(Copy) Yield Sheet 5/2/86, Rathkeale	A.& L.
Ex. 223	D42	(Copy) Yield Sheet 12/9/85, Rathkeale	A.& L.
Ex. 224	D42	IB6 form etc. for 3/12/90	Tribunal
Ex. 225	D43	(Copy) Shannon Meat Invoice/AIBP equiv.	A & L
Ex. 226	D43	(Copies) "Deboning of Intervention Beef" week ending 4/8/90 to 2/9/91	Tribunal
Ex. 227A	D43	(Copies) Tests of defatting	Tribunal
Ex. 227B	D43	(Copies) Circular of 1982 (2/4/82) etc.	Tribunal
Ex. 228	D44	(Copies) Annex V Table of Weightings	A & L
Ex. 229	D44	(Copies) Schedule of Break-down of 68% yield	A.& L.
Ex. 230	D44	(Copies) 16 Photos of Plate and Flank	A.& L.
Ex. 231	D45	(Copy) Council Regulation 1208/81	A.& L.
Ex. 232	D45	(Copy) Commission Regulation 563/82	A.& L.
Ex. 233	D45	Docs for 4/9/90	Wit 135
Ex. 234	D45	Docs for 26/10/90	Tribunal
Ex. 235	D46	Docs for 4/9/90 (further to ex.233)	Tribunal
Ex. 236	D47	(Copy) Statement of W142 re Irregularities.	Tribunal
Ex. 237	D47	4 Actual Seals	A. G.
Ex. 238	D47	(Copy) Classification Sheets 22/3/91 etc.	A & L.
Ex. 239	D48	(Copy) Chart of "Cattle Prices paid by Cahir, Steers, 1987 etc."	A & L
Ex. 240	D49	Detailed Proposal sub. to I.D.A. **(12/6/87)**	Tribunal
Ex. 241	D49	Letter of 11/3/88 DOA. to I.D.A. (Performance Clause).	Tribunal
Ex. 242	D49	Letter of 21/5/90 Mr Goodman to IDA regarding Plan (Copy)	Tribunal
Ex. 243	D50	Presentation to Taoiseach etc. 27/4/87 by I.D.A.	W.155
Ex. 244	D50	"The Food Industry" - a Report	A.& L
Ex. 245	D50	Chart/I.D.A. Grants to Beef Processing Industry	A.& L.
Ex. 246	D50	Chart / Feoga Grants approved	A.& L.
Ex. 247	D50	Chart / Grants - V - Market Share	A.& L.
Ex. 248	D50	Letter 17/8/87 I.D.A. to Revenue Commissioners	Tribunal
Ex. 249	D50	Letter 4/3/88 IDA to Goodman International	Tribunal
Ex. 250	D50	I.D.A. memo 8/3/88 re: telephone calls	Tribunal
Ex. 251	D50	Minutes. of I.D.A Board 10/6/87	Tribunal
Ex. 252	D51	(Copy) I.D.A. Minutes of 1/3/88	Tribunal
Ex. 253	D51	I.D.A. Minutes of 15/3/88	Tribunal
Ex. 254	D51	I.D.A. Minutes of 12/6/87	Tribunal
Ex. 255A	D51	Letter 8/8/88 Deputy Bruton to IDA (Mr. McCabe)	Tribunal
Ex. 255B	D51	I.D.A. acknowledgement of 10/8/88 plus copy letter to Taoiseach	Tribunal
Ex. 256A	D51	Brian Britton's Letter of 12/8/88 to P. White I.D.A.	Tribunal
Ex. 256B	D51	Copy letter 9/8/88 from Goodman International to Deputy Bruton	Tribunal
Ex. 257	D51	I.D.A. Letter 18/8/88 to Deputy Bruton	Tribunal
Ex. 258	D51	I.D.A. Letter 18/8/88 to Taoiseach	Tribunal
Ex. 259	D51	Letter 25/8/88 Deputy Bruton to Mr. McCabe I.D.A.	Tribunal
Ex. 260	D51	Memo (I.D.A.) of 13/9/88	Tribunal
Ex. 261	D51	Letter 20/9/88 I.D.A. to Deputy Bruton	Tribunal
Ex. 262	D51	Memo (I.D.A.) of 14/3/88 (Legal opinion)	Tribunal
Ex. 263	D51	Letter 14/5/87 S.K.C. to I.D.A.	A.& L.
Ex. 264	D51	Letter 14/5/87 Goodman International to I.D.A.	A.& L.

List of Exhibits cont/.....

EXHIBITS		NAME	
Ex. 265A	D51	Internal P. White (I.D.A.) Memo 10/3/88 + Government details	A.G.
Ex. 265B & C	D51	Letters/Goodman International to I.D.A. 2/3/88 and 7/3/88	A.G.
Ex. 266	D51	Irish Meat Exporters Association to P. White 17/6/87	A.G.
Ex. 267	D51	I.D.A. reply 18/6/87 to ex. 266	A.G.
Ex. 268	D51	I.D.A. Annex to Government	Tribunal
Ex. 269	D51	I.D.A. letter 6/8/87 to Larry Goodman	Tribunal
Ex. 270	D52	I.D.A. Memo 15/6/87 to Department of Agriculture.	Tribunal
Ex. 271	D52	Memo (Minister Walsh) to Larry Goodman 12/6/87.	Tribunal
Ex. 272	D52	Minutes 19/5/87 of I.D.A. (Handwritten)	I.D.A
Ex. 273	D52	Minutes 26/5/87 of I.D.A. (Typed)	1.D.A.
Ex. 274	D52	Memo of telephone calls 14/3/88	1.D.A.
Ex. 275	D52	Letter I.D.A. to Brian Britton 1/9/88	1.D.A.
Ex. 276	D53	Secret document 26/4/87 Department of An Taoiseach to Dept. of Agriculture	Tribunal
Ex. 277	D53	Document of 16/6/87 (D.O.A.) to Government.	Tribunal
Ex. 278	D53	D.O.A. Report of Goodman Meeting (24/2/88)	Tribunal
Ex. 279	D53	Notes to Secretary, Department of Agriculture 2/3/88	Tribunal
Ex. 280	D53	D. Nally's letter of 8/3/88 to Department of Agriculture.	Tribunal
Ex. 281	D53	Mr Molloy's minute to Mr Curran regarding Mr Loughrey	Tribunal
Ex. 282	D53	Aide Memoire Department of Agriculture 26/4/87 to Government	Tribunal
Ex. 283	D54	Note re Government meeting 27/10/89	Tribunal
Ex. 284	D54	Minutes of Meeting Department of Agriculture 24/4/87	Tribunal
Ex. 285	D54	Form A (Copy) **"Submission to Government".**	Tribunal
Ex. 286	D54	Memo Mr O'Connell to B. Brady 12/6/87	Tribunal
Ex. 287	D55	P. O'hUiginns memo to Taoiseach 8/3/88	Tribunal
Ex. 288	D55	Goodman International Grant Agreement	Tribunal
Ex. 289	D55	Handwritten copy of B. Brady's comments	Tribunal
Ex. 290	D56	Letter 28/7/87 from Minister for Food to Larry Goodman	A. G.
Ex. 291	D56	Letter July/87 Minister for Food to Taoiseach + Draft letter to Larry Goodman	A. G.
Ex. 292	D56	Letter 16/11/87 Brian Britton to Minister for Food.	A. G.
Ex. 293	D56	Letter 19/9/89 form Minister for Food to Larry Goodman.	A. G.
Ex. 294	D58	2/Agency (ICI) Agreements (Originals)	ICI
Ex. 295	D58	Minutes of meeting 8/4/86 ICI & Dept. of Ind. & Commerce and Appendix.	ICI
Ex. 296	D58	Telex 27/4/83 regarding Iraq Mr Leamy to Michael Walsh.	ICI
Ex. 297	D58	Telex 24/6/83 to ICI from Mr. Walsh	ICI
Ex. 298	D58	Telex to Michael Walsh of 7/6 refers from ICI	ICI
Ex. 299	D58	Telex of 26/6/85 marked "IRAQ"/ Dept. of Industry & Commerce to Mr Leamy	ICI ICI
Ex. 300	D58	Telex from Departments of Industry & Commerce to ICI of 20/2/86	ICI
Ex. 301	D58	Telex from Department of Industry and Commerce to ICI of 26/2/86	ICI
Ex. 302	D58	Telex of 23/5/86 to stop cover (Specific)	ICI
Ex. 303	D58	Telex of 23/5/86 to stop all cover	ICI
Ex. 304	D58	Telex from Department of Industry and Commerce to ICI of 10/6/87.	ICI

List of Exhibits cont/.....

EXHIBITS		NAME	
Ex. 305	D58	Report of 17/9/87 (of meeting)	ICI
Ex. 306	D58	Letter Department of Industry & Commerce to ICI of 29/9/87	ICI
Ex. 307	D58	Letter ICI to Department of Industry & Commerce 30/12/87	ICI
Ex. 308	D58	ICI Credit Assessment of Iraq and letter 28/1/88	ICI
Ex. 309	D58	ICI Report of 27/9/88	ICI
Ex. 310	D58	ICI Detailed check list	ICI
Ex. 311	D58	3 Policies of Export Credit Insurance	ICI
Ex. 312	D58	Fax of fax of 19/8/88 Department of Industry & Commerce to ICI	ICI
Ex. 313	D58	Memo 28/10/88 of ICI	ICI
Ex. 314	D58	ICI letter of 22/12/88 to Department of Industry & Commerce	ICI
Ex. 315	D58	Department of Industry & Commerce 17/2/89 to ICI	ICI
Ex. 316	D58	Bunch of Documents (Telex 27/20/87 to ICI letter 18/9)	ICI
Ex. 317	D59	Letter/Halal to ICI 26/2/88 (Copy)	Tribunal
Ex. 318	D59	Letter/Taher to ICI 13/11/87 (Copy)	Tribunal
Ex. 319	D61	Letter/J Timbs to Mr O'Hanlon ICI 7/3/88 (Copy)	A.G.
Ex. 320	D61	Credit Limit Approval form (Copy)	A & L
Ex. 321	D62	Halal to Albert Reynolds 3/12/87 - etc.	D.I.C.
Ex. 322	D62	Delegation order S.I. 166/1987	D. Spring
Ex. 323	D64	Letter 30/11/87 and draft Minister Brennan to M/Hamza (Copy)	D.I.C.
Ex. 324	D64	Original handwritten note of Albert Reynolds	Finance
Ex. 325	D64	Excerpt (2) from Dail Debates 6/88 and 3/89	D O'Malley
Ex. 326	D64	Unsigned Memo regarding Export Credit Insurance for Hibernia.	D.I.C.
Ex. 327	D65	Book/Decisions Export Credit ag. advice (Copy)	A.& L.
Ex. 328	D65	Book/National Interest considerations regarding Iraq.	A.& L.
Ex. 329	D65	Book CBF/Agriculture and Department of Industry & Commerce regarding use of Export Credit in Iraq.	A.& L.
Ex. 330	D65	Book AIBP Claim to Export Credit Allocation	A.& L.
Ex. 331	D66	Book Background documents on Departmental position regarding Iraqi cover.	A.& L.
Ex. 332	D66	Copy single page regarding Market Conditions.	A.& L.
Ex. 333	D66	List of words used regarding sourcing	A.& L.
Ex. 334	D66	Book "Index 2" regarding meetings of 11/9/87 etc.	A.& L.
Ex. 335	D66	Decisions as to whether Northern Ireland beef be treated as Irish.	A.& L.
Ex. 336	D67	Copy letter 19/2/86 Hibernia Meats to Alan Dukes, Minister for Justice,	A. G.
Ex. 337	D67	Copy overall Analysis of Cattle Price Index.	A. G.
Ex. 338	D68	Memo of Private Secretary (F. Kelly) to Secretary 2/6/88 etc.	D.I.C.
Ex. 339	D68	(Copies) bunch of/memo T. O'Reilly 15/12/88 up to letter of 14/12/88 of Halal to Taoiseach.	Tribunal
Ex. 340	D68	(Original) Memo Mr Timbs to Mr. O'Reilly 15/12/88 etc.	D.I.C.
Ex. 341	D68	(Original) Memo (Mr Kelly) to Mr. O'Reilly received on 14/12/88	D.I.C.
Ex. 342	D69	(Original) Memo Mr Walsh (Dept. of Industry & Commerce) to Mr McNally 10/11/86.	D.1.C.
Ex. 343	D69	(Original) Memo Mr Kilroy to Mr O'Reilly etc. 8/5/89.	D.I.C.
Ex. 344A	D69	Aide Memoire underlined in pen 2/9/87	D.I.C.
Ex. 344B	D69	Aide Memoire underlined by typewriter 2/9/87	D.I.C.
Ex. 345	D69	(Original) Memo of 2/9/87 to Secretary and Memo 31/8/87 of Mr McBride to Mr O'Reilly etc.	D.I.C.
Ex. 346	D69	Letter (Original) from A. G. to Department of Industry and Commerce regarding Export Credit Insurance.	D.I.C.
Ex. 347	D70	(Original) Letter AIBP to Department of Industry and Commerce 27/11/87	D.I.C.
Ex. 348	D71	Book of Documents (excerpts from) Pink cover.	A. G.
Ex. 349	D71	Correspondence between T. O'Reilly & Department of Industry & Commerce 21/1/92 etc.	A. G.

List of Exhibits cont/.....

EXHIBITS		NAME	
Ex. 350	D72	(Copy) letter 27/11/87 AIBP to Department of Industry & Commerce.	D.I.C
Ex. 351	D73	1st Original letter 27/11/87 to Department of Industry & Commerce.	A & L
Ex. 352	D73	Original memo of J. Donlon of 2/2/89	D.I.C
Ex. 353	D75	(Copy)) letter 18/7/85 A. Deasy to Minister in Iran.	A.G.
Ex. 354	D75	Original bunch of Hamza letters 11/87.	D.I.C
Ex. 355	D75	(Copy) Minister Brennan's Statement	D.I.C.
Ex. 356	D77	(Copy) Excerpts from Sunday Independent (26/4/92)	Tribunal
Ex. 357 A	D77	(Copies) Letter 28/4/92 P.S.S. to Dept. of Industry & Commerce	A.G.
Ex. 357 B	D77	Letter 1/5/92 Dept. of Industry and Commerce to John Corcoran	A.G.
Ex. 358	D77	Letter 3/88 Department of Industry and Commerce to Mr McCann, ICI.	ICI
Ex. 359	D79	Letter Mahon & McPhillips 23/2/84 etc.	A.G.
Ex. 360	D79	(Copy) letter 16/4/86 Mr Noonan to Mr Bruton.	A.G.
Ex. 361	D79	(Copy) Letter 2/5/83 ICI to Department of Industry & Commerce.	ICI
Ex. 362	D79	Original Note 8/4/88 of D. O'Mahony (ICI)	ICI
Ex. 363	D79	Original Note to Minister on Export Credit Insurance for Iraq	D.I.C.
Ex. 364	D80	(Copy) Report of meeting 5/9 Consultancy Unit/AIBP	D.I.C.
Ex. 365	D80	(Copy) Inspection Certificate 9/11/88 of Rafidain Bank	A & L
Ex. 366A	D81	(Copy) Table 1 AIBP Shipments to Iraq 87/89	A & L
Ex. 366B	D81	(Copy) Table 2 AIBP Shipments to Iraq 87/89 plus	A & L
Ex. 366C	D81	(Copy) Table 3 AIBP Shipments to Iraq 87/89 plus plus.	A & L
Ex. 366D	D81	(Copy) Bunch of documents that No. 2 A/C decided by Dept. of Industry and Commerce.	A.G.
Ex. 367	D83	(Copy) Telex 11/12/87 Dept. of Industry and Commerce to Embassy Baghdad	A. G.
Ex. 368	D83	(Copy) Bunch of documents .... 15/9/87 to 20/11/89 from Dept. of Foreign Affairs.	A. G.
Ex. 369	D84	Original Telex from Minister Brennan to Hamza 18/4/88	D.I.C
Ex. 370	D85	(Copy) Telex 10/9/85 Mr Goodman to Austin Deasy (Minister for Agriculture).	A.& L
Ex. 371	D85	(Copy) Bunch of documents regarding Russian matter 1986	A. G.
Ex. 372	D85	(Copy) Documents showing Goodman V Embassy same throughout.	A. G.
Ex. 373	D86	(Copy) Statement of P. McCabe (Ambassador).	Tribunal
Ex. 374	D88	(Copy) excerpts from McKinsey Study.	A.& L.
Ex. 375	D88	(Copy) excerpt from Doyle V. An Taoiseach - Supreme Court decision.	A.& L.
Ex. 376	D88	(Copy) Entries for Minister's diaries.	A. G.
Ex. 377	D89	(Copy) Transcript of "Morning Ireland" programme 7/4/92	D O'Malley
Ex. 378	D89	(Copy) Claim form of driver from Goodman's	A & L
Ex. 379	D90	(Copy) Excerpts from Dail Debates 18/12/90 Column 131	A & L
Ex. 380	D92	(Copy) Banque Nationale De Paris to Crean O'Cleirigh Solicitors - letters of 2/12/91	Tribunal
Ex. 381	D95	(Copy) Telex Baghdad Embassy to Foreign Affairs. 4/8/87	A.G.
Ex. 382	D97	(Copy) Contract in Arabic 22/12/88 Taher - V - Iraqis	Tribunal
Ex. 383	D97	2 Fax sheets of Tenders 22/3/88	N Taher
Ex. 384	D98	Anglo Irish, Taher etc. prices 25/9/89 of Tender.	N Taher
Ex. 385	D98	(Copy) receipt from F. F. 6/11/87	O. Murphy
Ex. 386	D98	(Copy) cheque to P.D's £10,000 26/7/89	D O'Malley
Ex. 387	D98	(Copy) Letter of P.D.'s 27/7/89 returning £10,000.	D O'Malley
Ex. 388	D98	(Copy) Schedule of Shipments to Iraq.	O. Murphy
Ex. 389	D99	(Copy) Veterinary Cert. of 10/11/88.	O. Murphy
Ex. 390	D100	(Copy) Set of documents beginning minutes of meeting 5/6/1985	A. G.
Ex. 391	D101	(Copy) CBF letter 12/11/'82 to Irish meat Exporters.	A. & L.
Ex. 392	D102	(Copy) CBF Grant-in-aid 1987-1992	A.G.
Ex. 393	D103	(Copy) Letter 9/3/'88 Master Meats to Albert Reynolds	Tribunal
Ex. 394	D104	(Copy) Excerpt Dail Reports 10/5/'89	Tribunal

List of Exhibits cont/.....

EXHIBITS		NAME	
Ex. 395	D105	(Copy) Excerpt Transcript D70, page 101 (T.O'Reilly)	A & L
Ex. 396	D105	(Copy) Excerpt Dail Report Excerpt 1/11/'89 Col. 1012	A & L
Ex. 397	D105	(Copy) Excerpt Transcript D75 (Joe Timbs evid)	A & L
Ex. 398	D105	(Copy) Excerpt Transcript D65 (J. Donlon's)	A & L
Ex. 399	D105	(Copy) Excerpt Transcript D66 (J. Donlon's)	A & L
Ex. 400	D105	(Copy) Excerpt Transcript D61 (J. Donlon's)	A & L
Ex. 401	D105	(Copy) Book regarding policy of selectivity	A & L
Ex. 402	D106	(Copy) Table regarding Grants -V- Market Share	A & L
Ex. 403	D106	(Copy) Book entitled "SCALE"	A & L
Ex. 404	D106	(Copy) Excerpt Dail Reports 18/12/90. Col. 150.	A & L
Ex. 405	D106	(Copy) Mr O'Cuinneagain (Private Secretary) to Mr Dully Dept of Industry & Commerce 18/12/'89.	A & L
Ex. 406	D106	(Copy) Excerpt Dail Reports 23/4/85 Col. 1419.	A & L
Ex. 407	D106	(Copy) excerpt Transcript D66 (Donlon)	A & L
Ex. 408	D106	(Copy) Excerpt Transcript D68 (O'Reilly)	A & L
Ex. 409	D106	(Copy) Excerpt Transcript D71 (O'Reilly)	A & L
Ex. 410	D106	(Copy) Excerpt Transcript D71 (O'Reilly)	A & L
Ex. 411	D106	(Copy) Excerpt House of Commons Committee 1988/1989	A & L
Ex. 412	D106	(Copy) Excerpt ECGD Report by Kemp	A & L
Ex. 413	D106	(Copy) EC Council Directive 1/2/71	A & L
Ex. 414	D106	(Copy) EC Council Decision 10/12/82	A & L
Ex. 415	D106	(Copy) EC proposal for Council Regulations 24/7/87	A & L
Ex. 416	D106	(Copy) Proposal for Council regulations (no date)	A & L
Ex. 417	D107	(Copy) Letters between P.D's and Mr Goodman 10/3/86 etc.	A & L
Ex. 418	D107	(Copy) Letter 24/10/88 P.D's to Mr Goodman	A & L
Ex. 419	D107	(Copy) Informal Government decision regarding Export Credit Insurance.	A. G.
Ex. 420	D107	(Copy) P806 from textbook regarding E.C. policy on export credit	A. G.
Ex. 421	D107	(Copy) Overall Cattle price index with source documents.	A. G.
Ex. 422	D107	(Copy) Excerpt from ESRI Report '91 page 2 & 3.	A. G.
Ex. 423	D107	(Copy) Excerpt Gray Murray Report 1991	A. G.
Ex. 424	D107	(Copy) Fahy Document regarding Export Credit Insurance 7/4/86.	A. G.
Ex. 425	D107	(Copy) Excerpt transcript D95 (Noonan T.D.,) page 24, 25.	A. G.
Ex. 426	D108	(Copy) Analysis on Iraqi Imports from EC.	A. G.
Ex. 427	D108	(Copy) Nally's letter 6/2/86 regarding Government approval on Export Credit.	A. G.
Ex. 428	D108	(Copy) Excerpt Gray Murray Report - May 1991.	A. G.
Ex. 429	D109	(Copy) Analysis of Potential Liabilities of State.	A. G.
Ex. 430	D109	(Copy) Breakdown Iraqi payments 9/88 - 1/89	A. G.
Ex. 431	D109	(Copy) Excerpt transcript D73B, p102 and definition.	A. G.
Ex. 432	D109	(Copy) Excerpt Gray Murray Report	D O'Malley
Ex. 433	D109	(Copy) Excerpt Annual Report Department of Agriculture & Food 1990.	D O'Malley
Ex. 434	D109	(Copy) Excerpt Report of Examiner 1990	D O'Malley
Ex. 435	D109	(Copy) Excerpts on Industrial Policy.	D O'Malley
Ex. 436	D109	(Copy) Excerpts Directory of Community Legislation in force.	A.& L.
Ex. 437	D118	(Copy) Invoice etc. Doherty Carrigans 25/10/90 for £2,375.00.	A.& L.
Ex. 438	D118	(Copy) Message 17/12/90 re stopping of "Dead Mondays"	A.& L.
Ex. 439	D118	(Copy) Invoices Dohertys Carrigans 11/4/91 etc.	A.& L.
Ex. 440	D118	(Copy) Kill Sheet & Sales W/E 7/4/91 Carrigans	A.& L.
Ex. 441	D118	(Copy) IB4 of 12/4/91 (ten)	A.& L.
Ex. 442	D123	(Copy) Commission Regulation EEC No. 3964/87 (Export Refunds)	A.& L
Ex. 443	D131	(Copy) True Story of 1988 APS	A.& L.
Ex. 444	D131	Photos 1 to 5 of box of plate and flank meat	A.& L.
Ex. 445	D132	Box for Meat (Empty)	C. & E.
Ex. 446	D132	Copy EEC Regulations 283/72 etc.	C. & E.
Ex. 447	D133	(Copy) letter McNamara and Sons, Solicitors, to Tribunal 25/9/92	Tribunal
Ex. 448	D139	(Copy) letter of 29/4/91 Granada to Mr Heneghan and reply of 30/4/91.	Tribunal
Ex. 449	D140	(Copy Dáil Debates Excerpts 15-16th May 1991.	A. & L.

List of Exhibits cont/.....

EXHIBITS		NAME	
Ex. 450	D140	(Copy) Visitors Pass for Oireachtas for Susan O'Keeffe	A. & L.
EX. 451	D140	(Copy) Excerpts Irish Independent 18/5/91.	A. & L.
Ex. 452	D140	(Copy) Excerpts Transcript RTE Radio programme 19/5/91.	A. & L.
Ex. 453	D140	(Copy) Excerpts Transcript RTE Radio programme 15/5/'91.	A. & L.
Ex. 454	D149	(Copy) Dáil Debates excerpts 12/4/89, Col. 1071	A. & L.
Ex. 455	D149	(Copy) Dáil Debates excerpts 28/8/'90 Col. 2080	A. & L.
Ex. 456	D149	(Copy) Excerpts Evening Herald, Evening Echo 20/6/'91	A. & L.
Ex. 457	D149	(Copy) Transcript "Today at 5" RTE Radio 21/5/91	A. & L.
Ex. 458	D149	(Copy) Excerpts Dáil Debates 15 - 16th May 1991	A. & L.
Ex. 459	D149	(Copy) Book of Excerpts from Newspapers 16-21/5/'91	A. & L.
Ex. 460	D149	(Copy) Excerpt Dáil Debates 28/5/'91	A. & L.
Ex. 461	D149	(Copy) Excerpt Transcript 1 30/9/91	A. & L.
Ex. 462	D149	(Copy) Transcript 30, Page 11	A. & L.
Ex. 463	D149	(Copy) Transcript 34 (Transcript 23) Page 30, 33.	A. & L.
Ex. 464	D149	(Copy) Dáil Debates 28/8/'90 Col. 2088	A. & L.
Ex. 465	D149	(Copy) 1 page excerpt Transcript 53	A. & L.
Ex. 466	D149	(Copy) Transcript 124 PP 15, 16	A. & L.
Ex. 467	D149	(Copy) Dáil Debates 14/11/'91 (Col 1604).	A. & L.
Ex. 468	D151	(Copy) Excerpt Senate Debates 29/5/91	A. & L.
Ex. 469	D151	(Copy) Tribunal letter, 17/7/91 and reply of Raftery 17/7/91	A. & L.
Ex. 470	D152	(Copy) Excerpt Transcript D136, page 42-44	A. & L.
Ex. 471	D153	(Copy) Excerpt Dáil Debate 15/5/'91	A. & L.
Ex. 472	D153	(Copy) Excerpt Irish Press 16/6/'91	A. & L.
Ex. 473	D153	(Copy) Excerpt Irish Times 17/5/'91	A. & L.
Ex. 474	D153	(Copy) Excerpt Irish Press 16/5/'91	A. & L.
Ex. 475	D153	(Copy) Excerpt Transcript D.5 page 161	A. & L.
Ex. 476	D153	(Copy) Excerpt Dáil Debates 15/5/91	A. & L.
Ex. 477	D153	(Copy) Excerpt Dáil Debates 15/5/91 (Col 1265).	A. & L.
Ex. 478	D153	(Copy) Excerpt Dáil Debates 24/5/91 (Col. 2467)	A. & L.
Ex. 479	D153	(Copy) Excerpt RTE Radio 18/5/91	A. & L.
Ex. 480	D153	(Copy) Excerpt Dáil Debates 28/8/'90	A. & L.
Ex. 481	D153	(Copy) Excerpt Transcript RTE Radio 26/10/90	A. & L.
Ex. 482	D153	(Copy) Excerpt Dáil Debates 28/8/90 (Col. 2101)	A. & L.
Ex. 483	D153	(Copy) Excerpt RTE RAdio 16/5/91	A. & L.
Ex. 484	D153	(Copy) Excerpt Today Tonight 16/5/'91	A. & L.
Ex. 485	D153	(Copy) Excerpt "Marketplace" 20/6/'91	A. & L.
Ex. 486	D153	(Copy) Excerpt Transcript D.128.	A. & L.
Ex. 487	D157	(Copy) Removal Order for Intervention Beef.	S.K.C.
Ex. 488	D157	(Copy) AIBP Northern Ltd., minutes of 23/12/87	S.K.C.
Ex. 489	D162	(Copy) Excerpt Council of Europe Decision	A. & L.
Ex. 490	D163	(Copy) Boneless Yields 1983-1991.	A. & L.
Ex. 491	D163	(Copy) Summary of Results of Trials - April 1984	A. & L.
Ex. 492	D163	(Copy) Summary Analysis of Primals (Cuts)	A. & L.
Ex. 493	D163	(Copy) Alternative analysis of figures.	A. & L.
Ex. 494	D164	(Copy) a) Appointment card b) Statutory Instrument} c) Search Warrant }	Tribunal
Ex. 495	D165	(Copy) Defatting Tables sorted by box	A. & L.
Ex. 496	D165	(Copy) Defatting Tables for whole country	A. & L.
Ex. 497	D165	(Copy) Defatting Tables sorted by % fat.	A. & L.
Ex. 498	D166	(Copy) Table of % defects from defatting	A. & L.
Ex. 499	D167	(Copy) Plan of Rathkeale Boning Hall	A. G.
Ex. 500	D169	(Copy) Weekly Job Cost Summary and Analysis	A. & L.
Ex. 501	D169	(Copy) Comparison of Deboning charges	A. & L.
Ex. 502	D169	(Copy) Blank Boning Sheet from another jurisdiction.	A. & L.
Ex. 503	D169	(Copy) Authorization of Brid Cannon	A. & L.
Ex. 504	D170	(Copy) Rathkeale production as per Casey.	A. & L.
Ex. 505	D170	(Copy) Table of weights (Casey wk.3)	A. & L.
Ex. 506	D170	(Copy) Table regarding Defatting	A. & L.
Ex. 507	D171	(Copy) Shannon Meats Quality Control Tables.	Holmes Solr

List of Exhibits cont/.....

EXHIBITS		NAME	
Ex. 508	D171	Table of cooking times of meat	Holmes Solr
Ex. 509	D173	(Copy) Circular 11/5/89 DoA.	A. G.
Ex. 510	D178	Document regarding Section 84 (High Coupon)	B Britton
Ex. 511	D178	(Copy) Goodman Statement 10/3/89	White Solr
Ex. 512	D181	(Copy) Excerpt Dail Debates 9/3/89	D Spring
Ex. 513	D182	(Copy) Analysis of Political Contributions	A. & L.
Ex. 514	D182	(Copy) Comparison Sheet Beef cost. Intervention v Fresh.	A. & L.
Ex. 515	D182	(Copy) Cattle Price Movements Sheet '86-90.	A. & L.
Ex. 516	D182	(Copy) Book of Statistics	A. & L.
Ex. 517	D182	(Copy) Sheet marked "Culliton"	A. & L.
Ex. 518	D182	(Copy) Correspondence Trib. and A.& L. re Mr Fairbairne.	Tribunal
Ex. 519	D184	(Copy) D.O.A. Circ. regarding reboxing 24/1/86 plus Cert.	A. G.
Ex. 520	D185	(Copy) Regulation EEC 2651/86 of 22/8/86 re. APS.	A. G.
Ex. 521	D186	(Copy) DoA. Circular 11/5/89 re. misappropriated beef.	A. G.
Ex. 522	D186	(Copy) Boning Hall Job Costing Sheets (3) 28/10/92	A. & L.
Ex. 523	D187	(Copy) Nenagh Job Costing Sheet 15/3/93	A. & L.
Ex. 524	D187	(Copy) Telex DoA. to AIBP 17/12/90	A. G.
Ex. 525	D187	(Copy) Boning Hall Sheet	A. G.
Ex. 526	D188	(Copy) Boning Hall Sheet Bagenalstown 28/2/93	A. & L.
Ex. 527	D188	(Copy) Results of Primal Examination (defatting)	A. & L.
Ex. 528	D190	(Copy) Book of correspondence for employees.	Tribunal
Ex. 529	D191	(Copy) Statement of Maurice Mullen re. boxing weights.	A. G.
Ex. 530	D194	2 Original Documents Metara Invests re. Shareholders.	White Solr
Ex. 531	D194	4 Docs. marked " Offloading Services"	White Solr
Ex. 532	D195	2 copy docs. re: shareholders Metara(D. Brady)	Tribunal
Ex. 533	D195	1 (copy) Slaney Meats reply to Political contributions.	Tribunal
Ex. 534	D201	(copy) 1 sheet "intervention Production Nov. 1992"	Hussey Solr
Ex. 535	D208	Cheques "forged" (W415)	Smith Solr
Ex. 546	D208	List of 3 "informants" (W415)	Quinn Solr
Ex. 537	D209	DoA. books from Kepak factory.	A. G.
Ex. 538	D209	Book of photos of factory.	Hegarty Solr
Ex. 539	D209	Letter regarding Roberts wages 26/8/91	Quinn Solr
Ex. 540	D209	Original cheques signed by Mr Finucane	Hegarty Solr
Ex. 541	D209	(Copy) Letters re. political contributions - Kepak.	Hegarty Solr
Ex. 542	D209	(Copy) Meat specifications 1988 A + B	Fry Solr
Ex. 543	D212	2 photocopies 1B7(5/9/91) + Blank IB8's	A. G.
Ex. 544	D213	2 Photos of meat in Mercin	R Godson
Ex. 545	D213	Customer's Arabic Label	C MacCoille
Ex. 546	D216	(Copy) Group of % rejection rate at Clonmel	Walsh Solr
Ex. 547	D128	Original small note book - 21/12/86 etc.	Mac An Aili Solr
Ex. 548	D218	Doc's infile arranged in certain order.	Mac An Aili Solr
Ex. 549	D219	3 original APS docs. 15/9/88	Mac An Aili Solr
Ex. 550	D220	2 orig. docs (re amounts and list of documents)	Tribunal
Ex. 551	D220	Original IB4's (2) + IB7(1) + IB7A (2) 24/4/92.	A. G.
Ex. 552	D221	Name of operative given by W470	W470

APPENDIX SIX

List of Submissions

List of people/organisations who made submissions in relation to Recommendations:

Dick Spring TD
Desmond O'Malley TD
Pat Rabbitte TD
Tomás MacGiolla TD
John Bruton TD
Pat O'Malley
ATGWU
SIPTU
CBF
Patrick Connolly VI Department of Agriculture
Richie Ryan - Court of Auditors
Faculty of Veterinary Medicine UCD
IMPACT
Andrew McCarthy - Regional Classification Supervisor
Irish Farmers Association
Irish Meat Processors Association
Irish Veterinary Association in association with the Irish Veterinary Union and the Local Authority Veterinary Officers Association
United Farmers Association
Cornelius Howard
Patrick J. O'Connor Registrar of Veterinary Council
Veterinary Officers Association
Veterinary Inspectors Department of Agriculture
Agricultural Officers Department of Agriculture
Michael Dowling Secretary of Department of Agriculture
Larry Goodman and Goodman International
Tom Harty
Revenue Commissioners - Customs & Excise Branch

Wt. P38856. 2,000. 8/94. Cahill. (M11763). G.Spl.